Chemistry in Your Life

Chemistry in Your Life

SECOND EDITION

Colin Baird
University of Western Ontario

W. H. Freeman and Company · New York

Senior Acquisitions Editor: Clancy Marshall
Developmental Editors: Rebecca Pearce/Hannah Thonet
Publisher: Susan Finnemore Brennan
Senior Marketing Manager: Krista Bettino
Media Editor: Victoria Anderson
Associate Editor: Amy Shaffer
Project Editor: Bradley Umbaugh
Design Manager: Diana Blume
Illustration Coordinator: Bill Page
Illustrations: Fine Line Studio
Photo Editor: Bianca Moscatelli
Photo Researcher: Brian Donnelly
Production Coordinator: Ellen Cash
Composition: Schawk, Inc.
Manufacturing: Quebecor

Library of Congress Control Number: 2005932533

Printed in the United States of America

First printing

Brief Contents

Contents

Sections in red are application-oriented sections. These sections enhance coverage of basic principles but do not introduce them. They can be omitted without losing the essential chemistry.

Contents

CHAPTER 5 **From Diamonds to Plastics:** Carbon's Elemental Forms, Addition Polymers, and Substituted Hydrocarbons **171**

Chemistry in Your Home: The synthetic products you use 171

Organic Molecules Containing Halogen Atoms 172

Addition Polymers 176

Properties of More Complex Polymers 183

The Recycling of Plastics 190

Elemental Carbon 194

Recycling of Tires 203

CHAPTER 6 **The Flavor of Our World:** The Oxygen-Containing Organic Compounds We Drink, Smell, and Taste **211**

Ethers 211

Alcohols 214

Contents

Contents

Contents

A Note from the Author

*C*hemistry in Your Life tries to help students "get" science. In a primarily qualitative way, the book illustrates a thorough coverage of chemical principles with their immediate application to students' lives. Building on the strength of the first edition's pedagogical resources, we've added content and created new tools to engage students in meaningful and interesting ways. Activities, experiments, exercises, fun facts, discussions, and research are all instrumental in illustrating how chemistry drives the world around them.

☀ New Features ☀

Group Activities

Students collaborate with their peers in a classroom setting and explore the physical applications of the chemical principles they've studied. Student groups are asked to share and discuss their findings with the class. Examples include:

- Brainstorming ways to conserve energy
- Determining the types of plastics in the world
- Proving Boyle's and Charles's gas laws using balloons
- Finding how much water's freezing point is depressed by salt
- Extracting strands of DNA from wheat germ
- Observing how temperature affects reaction rate

> **■ Group Activity:** The Speed of Chemical Reactions
> *Adapted from an experiment provided by Professor Bob Perkins, Kwantlen University College*
>
> Alka-Seltzer tablets contain three ingredients, two of which react vigorously with each other when they come into contact with water. The fizz that the reaction produces is caused by bubbles of carbon dioxide gas, which quickly rise and escape from the surface of the water. In this activity, you will observe how the temperature of the water affects the speed of the reaction. Assign one member of your group to be the timekeeper while the other members watch carefully to see when the reaction is over. Note: If you have a thermometer, take an exact reading of the temperature of the water you use in each experiment before the addition of the Alka-Seltzer tablet and after the reaction is complete.

Chemistry in Your Home

Learning doesn't stop when students leave the classroom. Students can explore these simple experimental activities using ordinary materials and expand their understanding of how chemistry—quite literally—affects them everywhere, every day.

> **■ Chemistry in Your Home:** Catalysts in potatoes
>
> Potatoes contain a substance that speeds the reaction that converts peroxides into water and oxygen. If you add a slice of raw potato to a little hydrogen peroxide (a bottle of which is available in a local drugstore), you can see oxygen bubbles forming. You can collect some of this oxygen by inverting a small drinking glass over the potato for a few minutes, then quickly removing the glass and covering it. Then light a match or candle's wick, blow it out, and quickly insert it into the glass. The oxygen makes the match or wick glow and may even cause it to relight.

New Content

Many students become interested in chemistry only when the material relates to their personal concerns. For that reason, such concerns have been used as focal points to teach the chemical principles that apply.

The Recycling of Plastics

The per capita annual use of plastics in North America is approximately 30 kilograms.

In the last quarter of the 20th century, plastics became the symbol of the "throwaway society," since many products—especially those used in packaging—were designed to be used once and then quickly discarded. Many environmentalists believed that waste plastic was a major culprit in the "garbage crisis." Indeed, plastics are the second-most-common constituent of municipal garbage, following paper and cardboard, though they follow by a large margin. Molded plastics take up more room in landfills—otherwise known as garbage dumps—than is indicated by their percentage by mass because they are not very dense. These plastics do, however, become compressed by the weight of the materials placed on top of them, as well as by compacting machinery before placement in the landfill (see Figure 5.10).

(a)

5.10 The recycling of plastic is a controversial issue

For a number of reasons, including the facts that landfills—especially throughout Europe—are reaching their capacity and that many citizens in developed countries are opposed to garbage incineration, many plastics now are collected from consumers and recycled. Some countries, such as Sweden and Germany, have made manufacturers

Figure 5.10 (a) Uncompressed plastics and (b) compressed plastics in a landfill. (Part a, Mark E. Gibson/ Visuals Unlimited; part b, Ken Lucas/ Visuals Unlimited)

(b)

A particular effort was made to cover organic chemistry relatively early in the text, thereby allowing popular topics such as fake fats, alcohol metabolism, vegetarian diets and the DNA basis of inherited characteristics to be discussed. New discussions include:

- Green chemistry
- The importance of sodium and potassium in the human diet
- How low-fat diets work; the glycemic index
- The dietary importance of folic acid and lycopene
- Hybrid electric cars
- Sulfur dioxide pollution control
- The controversy about banning methyl bromide
- Hair color and artificial hair coloring

Chemistry in Students' Lives

- The popular **Fact or Fiction?** marginal sidebars expose the error and confirm the truth behind a lot of conventional wisdom. Instructors have found these questions useful in capturing students' attention and in generating interest in new subjects. To capitalize on this, a multitude of new questions on such topics as dirty bombs, the ozone layer, trans fats, decaffeinated drinks, and the carat scales of diamonds and gold have been added.

???????????????? **?** ????????????
Fact or Fiction ?

Honey can kill germs.

Honey has been used historically to dress wounds. The presence of a small amount of hydrogen peroxide in honey is thought to be responsible in part for its antibiotic properties.

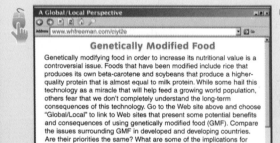

A Global/Local Perspective

Address www.whfreeman.com/ciyl2e

Genetically Modified Food

Genetically modifying food in order to increase its nutritional value is a controversial issue. Foods that have been modified include rice that produces its own beta-carotene and soybeans that produce a higher-quality protein that is almost equal to milk protein. While some hail this technology as a miracle that will help feed a growing world population, others fear that we don't completely understand the long-term consequences of this technology. Go to the Web site above and choose "Global/Local" to link to Web sites that present some potential benefits and consequences of using genetically modified food (GMF). Compare the issues surrounding GMF in developed and developing countries. Are their priorities the same? What are some of the implications for small-farm owners who grow their own food? Are GMFs grown or sold in your area? How are these foods or seeds labeled? Is nonmodified food affected when grown in the vicinity of GMF?

- **A Global/Local Perspective.** Students use what they have learned in the chapter along with information from Internet sources listed on this text's Web site, www.whfreeman.com/ciyl2e, to see how the principles they have just studied are relevant both at home and in the wider world.

- **Discussion Point.** Students are asked to develop arguments on both sides of a controversial issue involving chemistry in the real world.

▶ **Discussion Point:** Deciding whether to remain a gasoline-based society

Currently we are reliant upon gasoline derived from petroleum to fuel our vehicles. No matter how we reformulate this liquid, there are flaws with its use. In the coming decades, as petroleum runs out, society will face several choices regarding fuel for our vehicles, including (1) to remain gasoline-based by converting our ample supply of coal into synthetic gasoline, though, as we've seen, the use of coal involves environmental problems, or (2) to switch to fuels such as hydrogen, which could eventually be obtained using solar energy but which would be more expensive than gasoline for some time to come.

Develop arguments for and against each of these two options by using material in this chapter and available at some of the Web sites listed on the Web site for this book, as well as your own opinions regarding these issues.

A Note from the Author

- **Chemistry in Action.** This collection of video clips illustrates chemistry in students' lives.

- Many real-life examples illustrate the principles throughout every chapter.

Minimizing Mathematics

We have found that most topics can be taught—often to a fair level of sophistication—without the use of much mathematics. Therefore, in the main part of the text, we have minimized the use of mathematical equations and complex calculations. We have also avoided (except briefly when pH is considered) the mole concept and the calculations involving molar masses that usually accompany it.

- Most mathematical and advanced material is covered in **Taking It Further** sections at the end of the chapters or is available on the textbook's Web site, and can be incorporated or not as desired by the instructor. These sections expand on the concepts introduced in each chapter with more technical and quantitative detail. **New Taking It Further sections include material on calculating reaction heats from bond energies and problem solving with Boyle's and Charles's Laws.**

- **Taking It Further with Math** on the text's Web site, as well as Appendix A: Scientific Notation, provide the option of integrating more math into the course and include problems for student practice. Throughout the textbook there are icons that refer students to the Web site where they can examine the more quantitative aspects of chemistry and apply those concepts to solve chemical problems.

Developing the Chemistry

● **Tying Concepts Together:** Water

Water, the compound at the basis of life on Earth, is a good example to let us review and extend the various concepts about matter we have developed so far in this book. First consider that, individually, hydrogen and oxygen atoms are very tiny spheres of matter that have, at their centers, nuclei of charges $+1e$ and $+8e$, respectively. Around these nuclei swiftly travel electrons, tiny particles each having a charge of $-1e$ but very little mass. The electrons of oxygen atoms are divided between two shells, the inner shell having two electrons that play no role in bonding. The valence shell of each oxygen atom, before the oxygen forms a bond, contains six electrons. When it joins with hydrogen to form water molecules, each oxygen atom shares two of its six valence shell electrons and acquires a share in the electron associated with each of two hydrogen atoms. Because they share electrons in electron shells which must partially overlap each other, the two hydrogen atoms in every H_2O molecule are strongly and closely held to the oxygen atom b... together as... phase of the...

- Each chapter-opening photograph illustrates one or more of the concepts taught in the chapter and asks a question that invites students to think about the relation of the concepts to their lives.

- *Chapter Goals* listed on the first page of each chapter describe principles and applications students will explore in the chapter.

  - *Tying Concepts Together*, in many chapters, integrates concepts introduced in the chapter and extends them further.

  - *Worked Examples* and *Exercises* throughout the book provide many opportunities for students to solve problems related to the concepts just learned.

  - Three-tiered problems at the end of each chapter— *Review Questions*, *Understanding Concepts*, and *Synthesizing Ideas*—enable students to test their knowledge progressively.

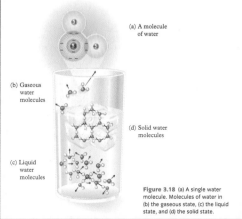

(a) A molecule of water

(b) Gaseous water molecules

(c) Liquid water molecules

(d) Solid water molecules

Figure 3.18 (a) A single water molecule. Molecules of water in (b) the gaseous state, (c) the liquid state, and (d) the solid state.

Figure 3.15 Destruction by fire of the hydrogen-filled airship *Hindenburg* in New Jersey in 1937, after its inaugural transatlantic flight. (Archive Photo Royalty/The Image Bank)

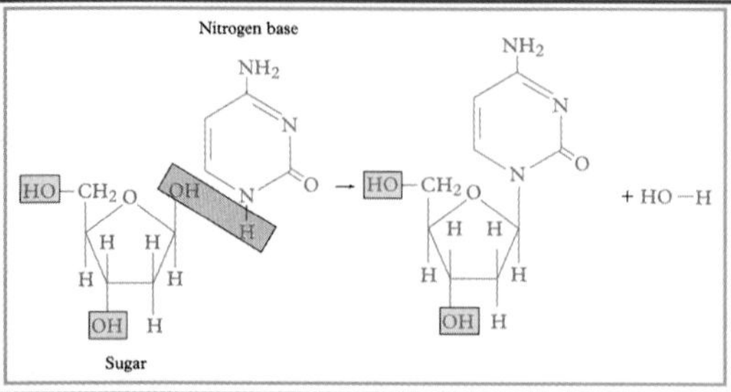

Visualizing Chemistry

The text and illustrations are closely coordinated so students learn visually as well as verbally. Students can see the molecular structure of familiar materials or follow "step-by-step" diagrams through complex processes.

Throughout the text students are referred to locations on the text's Web site where they can see visualizations of molecular structures and processes.

Media and Print Supplements

For the Student

NEW! ■ **Lab Manual** by Brian Arnerson (ISBN 0-7167-6956-5). This manual is designed to accompany the text and illustrate the chemical principles and applications discussed in each chapter.

■ A complete online learning center available at **www.whfreeman.com/ciyl2e** fully integrates interactivity, visualization, problem solving, and real-world examples with chemistry concepts in the textbook.

   ■ *Chapter Outlines.* Highlight key concepts and topics for every chapter.

   ■ *Visualizations.* QuickTime animations allow students to view motion, three dimensions, and atomic and molecular interactions, as well as to learn to visualize like chemists.

   ■ *A Global/Local Perspective.* Includes active hyperlinks that students can use to do the Global/Local exercises in the book.

   ■ *Web sites.* Hyperlinks for the Internet resources cited in the book.

   ■ *Chemistry in Action.* Commercial video clips courtesy of *Films for the Humanities and Sciences* that depict how chemistry is used in everyday life.

   ■ *Flash Cards.* Key words and definitions from the textbook set up in an interactive flash-card format.

   ■ *Online Quizzing.* 15 multiple-choice questions per chapter with **NEW!** appropriate feedback, including 20 percent new or revised questions.

   ■ *Tools.* Tools for chemical calculations, graphing, and exploration of periodic properties with an interactive periodic table.

   ■ *Taking It Further with Math.* Text and problems that provide information and practice of mathematical skills.

A Note from the Author

- **Students Solutions Manual** by Joseph Laurino, University of Tampa (ISBN 0-7167-7074-1). This booklet contains worked-out solutions to the odd-numbered end-of-chapter problems in the textbook as well as hints to help students with difficult concepts.

For the Instructor

- **Enhanced Instructor's Resource CD-ROM** (0-7167-0220-7). To help instructors create lecture presentations web sites, and other resources, this CD-ROM allows instructors to **search** and **export** all the resources contained below by key term or chapter:
  - All text images.
  - Visualizations, Flashcards and more
  - Instructor's Manual
  - Test Bank files
  - *In-Class Instructor Demonstrations.* Assists instructors who would like to add a visual element to lecture-only courses.

- **Instructor's Solutions Manual** by Joseph Laurino, University of Tampa (0-7167-6159-9). Contains worked-out solutions to every end-of-chapter problem in the textbook.

- **Printed Test Bank** by Charles Kotulski, University of Nevada, Las Vegas (0-7167-7075-X). Includes 50 to 70 multiple-choice questions per chapter, including 20 percent new or revised questions.

- **Computerized Test Bank** (0-7167-6181-5). With Windows and Mac versions on one disc, the CD-ROM allows for adding, editing, printing, and resequencing questions, and launching Diploma Online Testing, from the Brownstone Research Group.

- **Overhead Transparencies** (0-7167-6186-6). 100 full-color images from the textbook printed on acetates.

Flexibility

We have written the text to allow for maximum flexibility in its use. Applications sections, which do not introduce basic principles, are easily identified by red section numbers and headings, and can be omitted without losing any essential chemistry. Additionally, the various parts of the text are largely independent of one another. Once the basic concepts in Chapters 1–3 are covered, other topics can be covered essentially in any order, and to whatever extent, the instructor wishes. (Some material on water chemistry is best covered before air chemistry, however, and the material on covalent molecular structure in the first part of Chapter 4 is probably central to all courses.)

A Student-Friendly Text

Recognizing that students learn in many ways, we have created several features in this book to draw on many learning styles. We have developed these features carefully to promote learning and capture the imagination in the hope that students will enjoy chemistry as much as we do. Many of the topics, especially those in the biochemistry and energy chapters, may be deemed controversial, and we include them as a way to bring excitement and relevance to the discussion of chemical principles. It is our hope that this book will help students develop an appreciation for chemistry and how it affects every aspect of their lives. And, who knows, the world may have a few more students pursuing a degree in the sciences after they have read *Chemistry in Your Life*.

A Bit About the Author

Colin Baird (Ph.D., McGill University) has taught introductory chemistry at the University of Western Ontario many times in his career, both the general chemistry course for science majors and the chemistry course for students not majoring in science. Although his training and early research dealt with theoretical chemistry, in recent decades his interests have turned toward chemistry education and environmental chemistry. He is also the author of the best-selling W. H. Freeman text *Environmental Chemistry,* now in its third edition.

Acknowledgements

As I revised *Chemistry in Your Life,* I consistently relied upon the suggestions, advice, encouragement, and enthusiasm of my students and fellow colleagues. Any successes or accolades this text yields I attribute to the many people who helped bring this text to fruition.

Joseph Laurino of the University of Tampa wrote the *Instructor's* and *Student's Solutions Manuals,* as well as the In-Class *Instructor's Demonstrations.* Charles Kotulski of the University of Nevada Las Vegas created the *Test Bank* and *Online Quizzing* questions, and Robley Light of Florida State University checked the answers. Brian Arneson developed the *Lab Manual* to accompany this text and its philosophy. Myra Gordon and Duncan Hunter shared their extensive chemical knowledge and provided useful ideas and resources, and Hillary Austin diligently researched data in the literature.

Many thanks are due to the following professors for graciously donating their group activities for use in this text: Bob Perkins of Kwantlen University College; Rudy Luck of Michigan Technological University; Maria Dean of Coe College; and Kay Calvin of the University of Western Ontario.

I would also like to express my gratitude and appreciation to Wendy Gloffke for her talents in writing and shaping the first edition and to Susan Weisberg for her diligence, creativity, and perseverance in developing the first edition.

My colleagues have provided extensive critiques and immeasurable support of the second edition. I would to thank the following professors for their keen insight and efforts on both this text and the accompanying lab manual:

Georgia Arbuckle-Keil, *Rutgers University–Camden*
Joseph D. Augspurger, *Grove City College*
Susan E. Berkow, *University of Mary Washington*
David R. Bjorkman, *East Carolina University*
Susan L. Boyd, *Mount Saint Vincent University*
Paul Brandt, *North Central College*
Hindy E. Bronstein, *Fordham College at Lincoln Center*
Patrick E. Buick, *Florida Atlantic University*
Houston Byrd, *University of Montevallo*
Jerry W. Cannon, *Mississippi College*
Allan Childs, *Northwest College*
Thomas Corso, *Canisius College*
Cielito M. De Ramos-King, *Bridgewater State College*
Thomas D. Getman, *Northern Michigan University*

Marcia L. Gillette, *Indiana University-Kokomo*
Cliff Gottlieb, *Shasta College*
Tammy S. Gummersheimer, *Schenectady County Community College*
Barry H. Gump, *California State University, Fresno*
Midge Hall, *Clark State Community College*
Brendan Haynie, *Rowan University*
Al Hazari, *University of Tennessee, Knoxville*
Xiche Hu, *University of Toledo*
Raji Iyer, *Niagara University*
Laya Kesner, *University of Utah*
Angela G. King, *Wake Forest University*
Robley J. Light, *Florida State University*
Catherine MacGowan, *Armstrong Atlantic State University*
Edward Maslowsky, Jr., *Loras College*

Kenneth A. Marx, *University of Massachusetts, Lowell*
Audrey E. McGowin, *Wright State University*
Karen J. Nordell, *Lawrence University*
Linda Pallack, *Washington & Jefferson College*
Barbara Sweeting Pappas, *Ohio State University*
Darryl K. Reach, *University of Arkansas at Little Rock*
Joe M. Ross, *Central State University*
Kresimir Rupnik, *Louisiana State University*
Cindy Samet, *Dickinson College*
Ronald Sobczak, *George Washington University*

Dan M. Sullivan, *University of Nebraska*
Joseph C. Tausta, *State University of New York at Oneonta*
Christopher L. Truitt, *Texas Tech University*
George Uhlig, *College of Eastern Utah*
Kris Varazo, *Francis Marion University*
K. Vinodgopal, *Indiana University Northwest*
Doug Wendel, *Snow College*
Kim R. Woodrum, *University of Kentucky*
Martín G. Zysmilich, *George Washington University*

I remain indebted to the professors who offered their expertise and encouragement as I developed the first edition:

Adele Addington, *Virginia Military Institute*
John D. Anderson, *Midland College*
Cynthia Atterholt, *Western Carolina University*
David W. Ball, *Cleveland State University*
Richard C. Banks, *Boise State University*
Steven Bennett, *Bloomsburg University*
Kamala N. Bhat, *Alabama A&M University*
Joseph Bieron, *Canisius College*
Simon Bott, *University of Houston*
Bruce S. Burnham, *Rider University*
Francis Burns, *Grand Valley State University*
Grant F. Carruth, *Lipscomb University*
Kenneth Carter, *Truman State University*
Edward L. Case, *Clemson University*
Ann Chamberlain, *Monmouth University*
Feng Chen, *Rider University*
Kent Clinger, *Lipscomb University*
David J. Cohen, *J. Sargeant Reynolds Community College*
Cynthia H. Coleman, *State University of New York at Potsdam*
Jeanne M. Dorweiler, *Colorado State University*
Stephanie S. Flynn, *Francis Marion University*
Thomas A. Furtsch, *Tennessee Technological University*
Ana Gaillat, *Greenfield Community College*
Nancy Gardner, *California State University, Long Beach*
Paula Getzin, *Kean University*
Leslie L. Gillespie, *California State University, Fullerton*
Teresa D. Golden, *University of North Texas*
Theodore D. Goldfarb, *Stony Brook University*
Gabriel M. Grudis, *Sinclair Community College*
C. Alton Hassell, *Baylor University*

Chu-Ngi Ho, *East Tennessee State University*
Mark D. Jackson, *Florida Atlantic University*
Fred T. Johnson, *Brevard Community College*
James T. Johnson, *Sinclair Community College*
Paul Karr, *Wayne State College*
Cindy Kepler, *Bloomsburg University*
Joel Liebman, *University of Maryland-Baltimore County*
Joe Lutheran, *University of Akron*
Lawrence L. Mack, *Bloomsburg University*
Keith McCleary, *Adrian College*
Shelley D. Minteer, *Saint Louis University*
Villa Mitchell, *Lipscomb University*
Claire R. Olander, *Appalachian State University*
Jessica Orvis, *Georgia Southern University*
Deborah Otis, *Virginia Wesleyan College*
Mike Perona, *California State University-Stanislaus*
Michael J. Pikaart, *Hope College*
Greg Pippin, *Tennessee Technological University*
Albert C. Plaush, *Saginaw Valley State University*
Ann Randolph, *Rosemont College*
Jason Ribblett, *Ball State University*
Matthew E. Riehl, *Bethany Lutheran College*
Richard J. Rosso, *St. John's University*
Kathryn M. Rust, *Tennessee Technological University*
Maureen Scharberg, *San José State University*
James Schreck, *University of Northern Colorado*
Russell Selzer, *Western Connecticut State University*
Bryan L. Spangelo, *University of Nevada, Las Vegas*
Stacy Sparks, *University of Texas at Austin*
Gail Steehler, *Roanoke College*
Koni Stone, *California State University, Stanislaus*
Leon T. Venable, *Agnes Scott College*
Dawn Wiser, *Lake Forest College*

This text could not have been completed without the help and support of many talented people at W. H. Freeman and Company. I am particularly grateful to Clancy Marshall, my acquisitions editor, and to Rebecca Pearce and Hannah Thonet, my development editors, for their determination and encouragement. I would also like to acknowledge the efforts of my project editor, Bradley Umbaugh, and my production coordinator, Ellen Cash, in piecing this book together, and of Bianca Moscatelli and Brian Donnelly in searching for the perfect photographs. Bill Page worked with Fine Line Illustrations to create a new look for the illustrations, and Diana Blume's design enhanced the aesthetic quality of the text. Copy editor, David Roemer, and proofreader Anna Paganelli can be credited with the text's consistency and clarity, and Victoria Anderson managed the print and media supplements. Final thanks go to senior marketing manager Krista Bettino and publisher Craig Bleyer for their expertise and unstinting enthusiasm.

In this chapter you will learn:

- about familiar metals, such as mercury, gold, and nickel;

- how to distinguish among elements, compounds, and mixtures;

- the difference between homogenous and heterogeneous mixtures;

- about the different states of matter, including liquid crystals used for television and computer screens.

When you look at this campfire scene, can you see the chemistry behind it?

Every component of the picture can be related to chemistry. Sand, wood, the human body—even air—are all forms of matter. Chemical processes produce heat and light in the campfire, changing wood into ashes and allowing you to toast marshmallows. (Royalty-Free/Corbis)

The "Elemental" Foundation of Chemistry

Atoms, Molecules, Elements, Compounds, Mixtures, and States of Matter

Chemistry is concerned primarily with matter and its transformations. **Matter** is anything that has mass and occupies space—the pages of this book, the ground beneath your feet, and the air that you feel as wind on your cheeks.

Chemistry is vitally involved in almost every aspect of your life—the food you eat; the clothes, jewelry, and makeup you wear; the computers, cars, and DVDs you use; the drugs you might take, from aspirin to birth control pills; the quality of the air you breathe and of the water you drink. Chemistry involves the climate that you influence and the traits that you pass on to your children—indeed, the list goes on virtually indefinitely, as we shall see in this book. Chemistry is used in the development of computer chips, synthetic fibers, plastics, modern drugs, and perfumes. Progress in treating genetic diseases and in understanding how our actions create air pollution and modify the weather all depend on the continuing exploration by chemists and other scientists of the nature of matter.

Understanding the structure of matter has allowed chemistry to create the myriad of new and useful materials that are so vital to our modern society and to its future. Indeed, most of the world's population would not be living today without the creation of synthetic fertilizers to increase food production. Currently, insights into the chemical behavior of biological matter, such as genes, are providing hope for new medical advances and raising concerns about the impact of introducing new and modified organisms into the environment. Indeed, not all of chemistry's achievements have been used exclusively to improve the lot of humanity. Knowledge of how matter works has also been applied, for example, to the production of explosives and poisonous gases that have been used in warfare.

In this book, we shall explore the many ways in which chemistry affects *your* life. We shall see how the scientific method leads us to an understanding of the way matter is constructed and of how we can produce new substances having specific, desirable properties. We shall learn what occurs when living entities—including you—breathe, eat, exercise, and reproduce. We shall also consider some of the policy choices that will face you during your lifetime concerning such issues as the quality of your physical environment and the extent to which genetic engineering can modify living organisms. It is our aim to give you the scientific understanding that you will need to function as an informed citizen in the world.

Whenever you see this icon in this chapter, go to
www.whfreeman.com/ciyl2e

We begin our exploration of chemistry by discussing the general nature of matter at its smallest level—the elements and the atoms that form them.

Elements

1.1 Elements are the fundamental types of matter

Picture a bottle of oil-and-vinegar salad dressing that has been sitting on the shelf for a while, or a can of paint that hasn't been used for months. Under the influence of gravity, the mixtures that were uniformly smooth when you were using them have spontaneously separated into two layers. Why is it that a number of common types of matter separate like this into simpler, usually more uniform components? Throughout the ages chemists have been interested in separating all sorts of natural materials into their fundamental constituents. The techniques they have used to extract the individual components of the materials range from simply filtering liquid–solid mixtures or evaporating liquids to isolate solids dissolved in them, to harsher procedures including strong heating or exposure to intense light or to other chemicals.

About 100 substances have resisted *all* attempts to split them into two or more stable components. Fundamental types of matter that cannot be split into other stable entities are called **elements** (see Figure 1.1). Most of the matter you deal with in your everyday life is *not* in

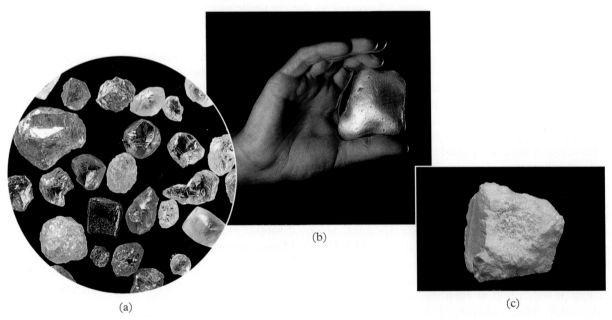

Figure 1.1 Samples of some naturally occurring elements. (a) Uncut diamonds from Zaire. (b) Gold nugget. (c) Sulfur. (None of these samples is 100% pure.) (Part a, Charles D. Winters; part b, Pascal Goetgheluck/Science Photo Library/Photo Researchers; part c, Mark A. Schneider/Photo Researchers)

Table 1.1	The top 10 elements in your body	
Element name	Percent of your body	
Oxygen	64.6%	
Carbon	18.0%	
Hydrogen	10.0%	
Nitrogen	3.1%	
Calcium	1.9%	
Phosphorus	1.1%	
Chlorine	0.40%	
Potassium	0.36%	
Sulfur	0.25%	
Sodium	0.11%	

Mirrors are pieces of glass coated on one side with a reflecting metal such as silver.

View the metallic properties, malleability and luster, at Chapter 1: Visualizations: Media Link 1.

The names of most metals end in *-um* or *-ium*. An exception is helium, which is not a metal.

Several other metals, such as cesium, melt just above room temperature.

its elemental form. However, three common examples of elements that you have probably seen are carbon in the form of diamonds, aluminum in the form of foil, and gold in the form of 24K jewelry. In addition to discovering that elements cannot be decomposed, chemists have also found that *all other substances are composed of combinations of two or more elements.* Thus *elements are the fundamental types of matter.* All known materials are either elements or, more commonly, are a combination of several elements. For example, table sugar is a combination of the three elements carbon, hydrogen, and oxygen, whereas water combines only hydrogen and oxygen. Complex materials such as soil and living cells contain dozens of elements. In human bodies, the most vital elements include hydrogen, carbon, oxygen, nitrogen, sulfur, and phosphorus (see Table 1.1).

1.2 Elements are classified as metals or nonmetals

What are the 100 or so materials that are called elements? Many of them are materials commonly known as **metals.** When asked to list the properties of metals, most people would use terms such as *shiny* and *hard*. Scientists have their own list, which runs as follows:

- *Shiny:* Metals possess a characteristic luster. Metals reflect light that is shined upon their surfaces.
- *Opaque:* Metals do not allow light to pass through them.
- *Malleable:* Metals can be hammered into shapes without fracturing.
- *Ductile:* A sample of the metal can be drawn into wire.
- *Conductive:* Metals conduct both heat and electricity.

You are familiar with many metals, such as gold, silver, iron, aluminum, and tin. You may be less familiar with other elements that are also metals, such as palladium, cadmium, vanadium, and cesium. We can categorize metals into several types having common characteristics or uses:

- Coinage metals such as gold, silver, and copper
- Structural metals such as iron, aluminum, and lead
- Shiny common metals, such as chromium, nickel, zinc, and tin
- Metals that are usually encountered only in combination with other elements. Examples include sodium, potassium, calcium, magnesium, barium, and radium

All the metallic elements are solid under normal, everyday conditions except mercury, which is a liquid. We will discuss metals in more detail later in the chapter.

Although most elements are metals, a sizable minority are not and are simply termed **nonmetals.** The best-known nonmetals are hydrogen, helium, carbon, nitrogen, oxygen, fluorine, neon, silicon, phosphorus, sulfur, chlorine, argon, radon, bromine, and iodine. These elements

View the structures of atomic, metallic, and solid copper at Chapter 1: Visualizations: Media Link 2.

generally do not have the characteristics of metals: they are not particularly shiny, malleable, or good conductors of heat and electricity.

Some elements change their nature under extreme conditions. For example, it is believed that hydrogen, a gas that most definitely is a nonmetal under normal conditions, becomes a metal under extremely high pressures. Only very recently have scientists been able to achieve pressures high enough to force this conversion to occur, and only for a fraction of a second. However, it is believed that in the interior of the planet Jupiter—which consists primarily of liquid and gaseous hydrogen—the pressure exerted from the matter that lies above is so great that hydrogen exists in a liquid metallic form.

1.3 Each element has its own symbol

For convenience, each element has been assigned a shorthand symbol, consisting of one or two letters, only the first of which is capitalized. Often the elemental symbol is based upon its English name. For example, the first letter of the English name is used for the elements Boron (B), Carbon (C), Fluorine (F), Hydrogen (H), Iodine (I), Nitrogen (N), Oxygen (O), Phosphorus (P), Sulfur (S), and Uranium (U). Common elements that have symbols corresponding to the first two letters of their name include Helium (He), Neon (Ne), Calcium (Ca), Nickel (Ni), Bromine (Br), and Argon (Ar). For many other elements with two-letter symbols, the second letter occurs somewhere in the English name—an example is Zn for Zinc. Still other elements have symbols based upon their Latin name—an example is Na, for sodium (from the Latin *natrium*). A complete list of elements and their symbols—and a number whose significance will become clear later—is given in the table on the page facing the back cover of this book. Frequently you will see the elements arranged in the periodic table (see inside front cover), which we will explain in Chapter 3.

Three-letter symbols are used temporarily for new synthetic elements before a final name has been agreed on.

The Atomic Nature of Matter

1.4 Matter is ultimately discrete, not continuous

Let's look a bit more closely at what it means to say that elements are the fundamental types of matter. Using a pair of scissors, you can cut a piece of aluminum foil into small pieces. In turn, you can cut each such piece into even smaller ones. With more specialized tools, it is possible to split even tiny bits of the aluminum into smaller ones, all of which still retain the properties that define the element aluminum. As far back as the time of the ancient Greeks, philosophers and scientists wondered whether there was any *limit* to the splitting of matter into smaller and smaller amounts that would still retain the properties of the larger units. At this lower limit of division, all the tiny particles would presumably be identical.

Through the power of modern technology, we have determined that there *is* indeed a limit to the divisibility of matter. Matter is *not* a continuous substance that can be divided indefinitely but consists of discrete, or separate, units. This confirms the view long held by scientists

Actually, in doing the "cutting" of the metal, you'd have to keep the foil in an oxygen-free atmosphere, as the surface of aluminum easily combines with oxygen.

By contrast, the width of a human hair is about 0.0001 meter.

and based on extensive amounts of indirect evidence. The limit to divisibility occurs when the dimensions of matter—the diameters of the particles—reach the incredibly small values of about 0.0000000001 meters, a distance far below the ability of any conventional "scissors" to cut it. At this limit, matter consists of spherical particles that cannot be split into two or more identical parts—that is, parts that are all of the same size and that behave identically. These particles are called **atoms,** a name derived from the Greek word for *indivisible*. Until recently, no one had ever actually observed an atom, though all indirect evidence pointed to their existence. In the latter part of the 20th century, scientists were able to "see" individual atoms through highly specialized types of microscopes. You can see a microphotograph of gold atoms in Figure 1.2b.

In any speck of an element just large enough for the human eye to see—about one-tenth of a millimeter (a hundredth of an inch) in size—there are about a million (1,000,000) atoms in any line along the speck. When dealing with such large numbers, scientists routinely express them in *scientific notation,* which makes it easy to convey very large and very small numbers. The number 1,000,000 is expressed in this system as 10^6, where the exponent 6 tells us the number of zeros that follow 1 when the number is written out in full. Similarly 100,000,000 is expressed as 10^8.

In this system, numbers less than 1 correspond to 10 raised to a negative power. The power is equal to the number of digits the decimal place needs to be moved to the right so there is only one nonzero digit before it; thus, 0.010 is equal to 10 raised to the power of negative 2. Similarly, our atomic diameter of 0.0000000001 meters is expressed as 10^{-10} meters, since the decimal place is moved 10 places to the right to convert 0.0000000001 to 1. (See Appendix A.1 for a further review of scientific notation if such symbolism is unfamiliar to you, and to see its extension to numbers in general.)

Atoms of different elements vary somewhat in size, but their diameters all fall in the range of a few tenths of a nanometer (nm), which is 10^{-9} meters. (This diameter is about 10,000 times smaller than the diameter of a human red blood cell.) As we shall see, atoms often attach themselves together to form structures whose dimensions are in the nanometer region. One of the exciting areas of modern science and engineering is *nanotechnology,* the development of incredibly tiny devices to manipulate matter one atom at a time and thereby assemble microscopic computers, machines for use in medicine, and other futuristic materials. Because of the importance of such structures to chemistry and to newly developing technology, we shall usually use the nanometer unit of length in this book when we discuss the sizes of atoms and of the groupings that they form.

A tenth of a nanometer is 10^{-10} meters.

View the atomic structures of hydrogen, carbon, nitrogen, oxygen, sulfur, copper, and silver at Chapter 1: Visualizations: Media Link 3.

Chemists consider atoms to be the fundamental unit of material composition and construction—as the letters of an alphabet are to words and as individual notes are to music. Atoms are essentially indestructible, both in number and in mass, unless we use very large amounts of energy to blast them apart. Elements are a particularly

100 nm

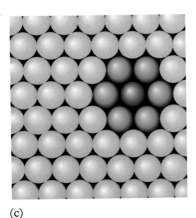

100 nm
(a)

5.0
nm

5.0
nm

(b)   0

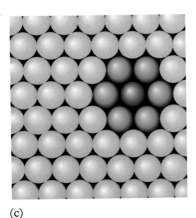

(c)

Figure 1.2 (a)–(b) A piece of gold foil at different levels of magnification. (c) A model for the atomic structure of gold. (Parts a, b, and c, Jacek Lipkowski/Canadian Chemical Society)

simple type of matter, since all their atoms are identical in size and in most aspects of their behavior.

Using the piece of modern technology called a *scanning tunneling microscope*, we can "see" atoms for ourselves. The photographs in Figure 1.2 are of the magnified surface of a piece of pure solid gold foil. If we zoom in to view a small portion of a single terrace of the segment (see Figure 1.2a), the topology of the gold surface still appears mainly continuous. However, zooming down to view a portion measuring only 5 nm (5×10^{-9} meters) long and wide reveals a simple pattern of regular rows of identical spherical particles—these are individual atoms of gold, each one measuring about 0.3 nanometers (3×10^{-10} meters) in diameter (see Figure 1.2b)!

1.5 Different atomic arrangements produce different types of matter

As we shall see many times in this book, the location of its atoms relative to each other plays an important role in determining the properties of a substance. The spatial arrangement of the atoms in all metallic elements is particularly simple and uniform, and in many cases it is identical to that of gold. The surface layer of gold is illustrated by the photograph in Figure 1.2b and the schematic diagram in Figure 1.2c. The atoms, represented by green circles in Figure 1.2c, that touch any given atom, represented by a blue circle, are known as its **nearest neighbors.**

In general, the structure of solid metals consists of a regular, repeating pattern of particles that extends indefinitely in at least two directions and usually in all three. We say that such substances—gold, for example—have an **extended network** of atoms. As we shall see later, extended networks are also present in some materials that are not

Figure 1.3 Different types of crystals. (Kristen Brochmann/Fundamental Photographs)

Chapter 1: The "Elemental" Foundation of Chemistry

simple elements. The arrangement of atoms in most solid *nonmetals* is more complicated than the arrangement of atoms in metals; some examples of these structures will be discussed later. Although all solid metals have simple atomic arrangements of their atoms in the solid, several different patterns of arrangement are known. Metals also differ in some of their other properties, including size—for example, atoms of copper are smaller and lighter than atoms of gold.

When very large numbers of particles are arranged in a regular pattern, with the same distances between nearest neighbors throughout, the solid they collectively form is a **crystal** (see Figure 1.3). In size, crystals range from those too small to see with the naked eye to those that measure a meter across! The solid form of most metals is crystalline. Many other substances, some with extended networks and others with alternative arrangements of particles, also form crystals. Some substances also form, under certain conditions, an alternative type of solid in which the arrangement of particles is much less ordered. The solids produced by such an arrangement are called **amorphous,** and they are not crystals because of their lack of order. (The word amorphous is derived from the Greek word *morph*, meaning shape, combined with the prefix *a*, which means without.) Amorphous solids usually lack the shiny, clear quality associated with materials in the crystalline solid state. The element silicon, for example, occurs in both a crystalline and an amorphous form in the solid state (see Figure 1.4).

View the structures of copper metal, crystalline sodium chloride, and the solid sulfur at Chapter 1: Visualizations: Media Link 4.

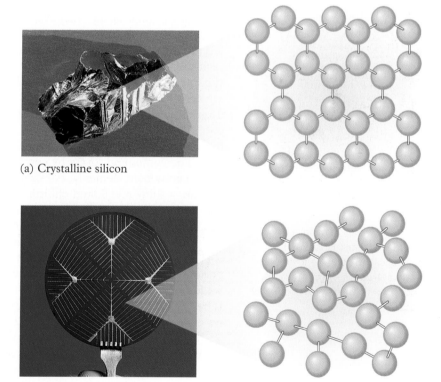

(a) Crystalline silicon

(b) Amorphous silicon

Figure 1.4 (a) Crystalline silicon. (b) Photovoltaic solar cell made of amorphous silicon. (Part a photo, Jeff J. Daly/Visuals Unlimited; part b photo, Rosenfield Images Ltd./Rainbow)

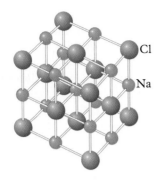

Figure 1.5 Atomic-level structure of sodium chloride.

Compounds and Mixtures

1.6 Compounds are composed of elements in a fixed ratio

When observed at the level of atoms, some substances are found to consist of a simple repeating pattern of the atoms of two elements, usually one of them a metal and the other a nonmetal. A common example is table salt, whose scientific name is sodium chloride. A schematic representation of the repeating pattern in salt is shown in Figure 1.5. The smaller spheres represent atoms of the element sodium and the larger spheres represent atoms of chlorine. As the illustration suggests, the repeating pattern extends almost indefinitely in three dimensions. This means salt has an extended network structure at the atomic level. Salt exists as crystals, because it consists of a *uniform* extended network structure.

Salt is but one of a huge number of substances that are composed of repeating units of several types of atoms in a simple ratio of small integers (whole numbers) such as 1:1, 2:1, and so on. Such substances maintain this fixed ratio of atoms throughout their structure, and, consequently, macroscopic samples of them also have the same total ratio of atoms. Thus, even a grain of salt that is large enough to see with the naked eye contains sodium and chlorine atoms in the 1:1 ratio.

Materials that consist of two or more types of atoms in a fixed ratio, with uniform composition throughout, and that *cannot* easily be separated into their pure component elements are called **compounds.** Millions of examples of compounds are known—including common substances such as table sugar, starch, water, and dry ice, as well as salt. Some compounds occur in nature, and many others have been produced synthetically in the laboratory. Most compounds are combinations of several nonmetals or of metals and nonmetals. Matter that is composed exclusively of one particular element or compound is known as a **pure substance.** It has a uniform composition down to the atomic level.

Usually, the characteristics of a compound are radically different from those of the individual elements of which the compound is composed. Formation of a compound is accompanied in many instances by the evolution of substantial amounts of energy in the form of heat or light. The compound forms by combining together only a specific ratio of the component elements. For example, when samples of the elements sodium, a shiny metal, and chlorine, a yellowish-green gas, are brought into contact, heat and light are emitted (see Figure 1.6). The sodium and chlorine become transformed into common salt, a material which has the physical characteristics of neither of the individual elements!

The effects on the human body also differ radically. Chlorine is a poisonous gas, and has been used in warfare since it can disable and kill soldiers by destroying their lungs. Sodium is a highly reactive metal that is best not to touch. But sodium chloride (*i.e.*, table salt) is a substance we eat every day, and a small amount of which is vital to good health.

The ratio of sodium to chlorine atoms is always 1:1 in any sample of sodium chloride. Any sodium or chlorine in excess of the amounts required to give the 1:1 atom ratio in salt would remain untransformed

(a)

(b)

(c)

Figure 1.6 (a) Sodium metal and (b) chlorine gas (c) combining to form table salt, sodium chloride. (a: Martyn F. Chillmaid/Photo Researchers, Inc.; b: Charles O. Winders/Photo Researchers, Inc.; c: Chip Clark)

and could be separated from the newly formed salt. The properties of most pure substances do not depend upon how much of the elements were mixed together in the first place, although in some cases several different substances can be formed by the same elements depending upon the proportions initially present in the mixture. Interestingly, we encounter no pure substances made of the elements sodium and chlorine in which the atomic ratio is *other* than 1:1.

Another example of a compound is arsenic oxide. This powdery substance is quite poisonous and was a common agent used for murder and suicide from Roman times to the Middle Ages. One of its component elements is oxygen, a gas that we depend upon for life. The other element, arsenic, is a grayish solid that is called a **metalloid** since it has some of the properties of a metal and some of a nonmetal. Most compounds of arsenic are toxic in some way to humans. For example, arsenic compounds dissolved in drinking water from wells in countries such as India, Chile, and Taiwan cause skin cancer. Although elemental arsenic itself is also toxic, when people speak of "arsenic" as being poisonous, they are usually referring to its compounds.

1.7 Mixtures contain substances in no fixed proportion

In contrast to the formation of a compound, it is often possible to mix together two or more pure substances in almost any proportion without any fundamental change occurring at the atomic level. No substantial gain or loss of energy, whether in the form of light or heat, accompanies the process. Such combinations of pure substances are called **mixtures.** In many cases, their properties are the average of those of the substances

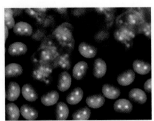

View sodium chloride and sugar dissolving in water at Chapter 1: Visualizations: Media Link 5.

that constitute the mixture. For example, air consists of a mixture predominantly of the elements nitrogen and oxygen in an atomic ratio of about 78:21, and it has properties that are similar to the weighted average of the properties of the two pure elements. It is also possible to prepare mixtures having proportions of 75:21, or 78:18, or any other values of nitrogen to oxygen, by simply mixing appropriate samples of the pure elements. Mixtures having ratios that differ slightly from 78:21 have properties only slightly different from that of natural air. Indeed, even the air you exhale has a higher ratio of nitrogen to oxygen than 78:21 since your lungs extract some of the oxygen from the air you inhale.

Similarly, you may know by experience that you can dissolve almost any proportion of sugar or salt in water. The sugar or salt can be recovered upon evaporation of the pure water, so it follows that there was no fundamental change in its composition when it dissolved. Indeed, most mixtures can be resolved into separate pure substances without subjecting the material to harsh conditions. In contrast, a compound usually cannot be resolved into its elements without supplying large quantities of energy—for example, by using high temperatures.

If a mixture has a uniform composition throughout its volume, it is called a **homogeneous mixture,** better known as a **solution.** For example, adding some salt to water and mixing thoroughly produces a solution. Similarly, adding pure alcohol to water and mixing produces a solution. In practice, the term *solution* is usually used for liquids, though some solid solutions exist. Homogeneous mixtures of gases, such as air, are also solutions but are rarely referred to by this name. An example of a solid homogeneous mixture is jewelry gold, in which some silver and copper are also present. When one substance readily dissolves in another, we say it is **soluble** in it. When a substance does not dissolve at all, it is said to be **insoluble.** If only a tiny amount will dissolve, leaving the bulk of the substance separate from the solution, we say it is "practically insoluble."

When you add sugar to black coffee and stir the mixture thoroughly, you obtain a solution, since it is uniform throughout (see Figure 1.7a). However, if you simply allow the sugar to remain at the bottom of the cup and slowly dissolve on its own, then, before spontaneous mixing has a chance to occur, the resulting mixture has different properties near the top of the liquid, where there is little dissolved sugar per gram of coffee, compared to the bottom, where it is mainly sugar. Mixtures of substances that do *not* have uniform properties throughout are called **heterogeneous mixtures.** Another example is oil-and-vinegar salad dressing, where small regions of oil droplets suspended in watery vinegar are obvious. Sometimes regions of one substance exist permanently beside regions of the other substances in such mixtures. In contrast to a homogeneous mixture, where the mixing extends right down to the atomic level, a heterogeneous mixture has zones where there is exclusively one type of substance or another (see Figure 1.7b). In many cases, you can see the heterogeneity of a mixture with your naked eye—think for example of wood with a grain, or beef in which the fat and muscle are clearly distinguishable, or rock in which separate components are

Figure 1.7 (a) Homogeneous and (b) heterogeneous mixtures. (Part a photo, George Semple for W. H. Freeman and Company; part b photo, Photodisc Blue/Getty Images)

 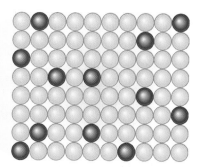

(a) A homogeneous mixture: coffee with sugar completely dissolved in it

 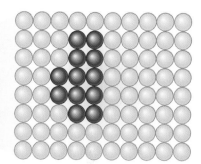

(b) A heterogeneous mixture: granite that contains visibly different types of rock

obvious. Separating such complex mixtures into components that are individually uniform is known as *purifying* the substances.

It is important to fully understand the fundamental difference between compounds on the one hand and simple mixtures of substances on the other. To summarize what we have described:

Compound	Mixture
Elements are present in only one unique ratio.	Elements are present in no fixed proportion.
Combination often occurs with emission of light and/or much heat.	Usually mixing produces no dramatic evidence of change.
Fundamental change occurs at the atomic level when the compound forms.	No fundamental change occurs at the atomic level.
Properties are quite distinct from those of component elements.	Properties often are an average of those of the components.
Cannot be readily resolved back to its components by simple techniques.	Can be readily resolved into components without drastic measures.

The conceptual relationships among the various forms of matter we have discussed so far are summarized in Figure 1.8.

Compounds and Mixtures

Figure 1.8 Classification scheme for the various forms of matter.

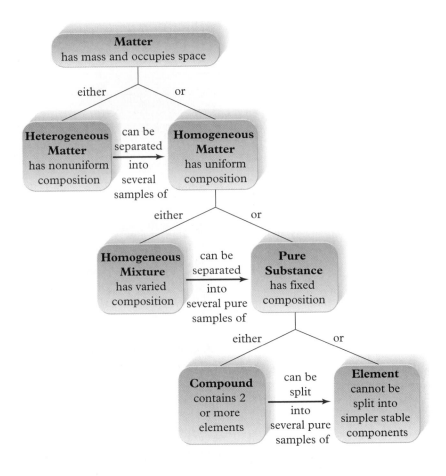

Exercise 1.1

Classify each of the following materials as either a homogeneous or a heterogeneous mixture, based upon what you observe about them, and explain your answers:

a) a U.S. quarter coin   **b)** freshly squeezed orange juice   **c)** clear tea

1.8 Some heterogeneous mixtures look like homogeneous ones

The classification of most materials, such as water, sugar, salt, and air, as liquids, solids, or gases seems straightforward to us. But what about jelly—is it a liquid or a solid? What about a cloud—is it a gas or a liquid? These situations, in which the substance seems to have some of the characteristic properties of two different states of matter, are found to invariably involve two (or more) different substances, present in a heterogeneous mixture. We are fooled into thinking that the mixture is a single, homogeneous substance because the particles involved are so small that we cannot distinguish them with our unaided eyes, even though they are much bigger than atoms.

Table 1.2 Types of colloids

A **solid** dispersed in a	is called a(n)	Common examples
gas	aerosol	smoke, dust in air
liquid	sol	mud, paint
solid	solid sol	some alloys (e.g., steel)
A liquid dispersed in a	is called a(n)	Common examples
gas	aerosol	fog, clouds
liquid	emulsion	milk, mayonnaise
solid	gel	jelly, shaving preparations
A gas dispersed in a	is called a	Common examples
liquid	foam	whipped cream, soap suds
solid	solid foam	ice cream, Styrofoam, popcorn, marshmallows

Many common substances that appear uniform to the naked eye are seen under magnification to actually be heterogeneous mixtures consisting of two or more types of materials, with one finely divided material dispersed in the other. The general name for such a substance is **colloid.** An example is homogenized milk, which under an ordinary microscope can be seen to consist of individual fat particles suspended in a watery substance. *Homogenized* milk is actually a *heterogeneous* mixture. The term *homogenized* comes from the fact that the butterfat is pre-mixed into the liquid, not separate as a layer of cream at the top.

In general, in a colloid, tiny particles of one substance, whether gas, liquid, or solid, are dispersed in the more prominent substance, whether gas, liquid, or solid. The dispersed particles in colloids have diameters in the range of 1 to 1000 nanometers (10^{-9} to 10^{-6} meters), whereas in solutions, the dissolved particles are of the size of atoms, that is, about 0.1 to 1 nm (10^{-10} to 10^{-9} meters). The substances in a colloid are not soluble in each other. They do not form a homogeneous solution and often eventually separate unless the mixture is kept agitated. The various types of colloids are listed in Table 1.2. Note that both clouds and jelly are colloids.

1.9 Emulsions are useful colloids

Colloids can be very useful and important in many areas of life. Particularly important commercially are **emulsions,** colloids in which both components are liquids. In many cases, one phase in the emulsion is water, and the other is an oily substance. Emulsions consisting of oil and water are commonly used as skin *moisturizers,* or *emollients* to give them their more scientific name. People use these products after washing—a process that removes much of the skin's own oils—to provide lubrication of the dry skin until the body gradually replenishes its natural oils. Using a pure oil, rather than an emulsion, would leave the skin feeling very greasy. The emulsion feels more natural to the skin, and it also supplies water, some of which is absorbed by the skin. A *lotion* is an emulsion of this type that feels more like a liquid, whereas a *cream* seems more like a solid.

Emulsions are also common in food products—ice cream and mayonnaise are two familiar examples. Ice cream is an emulsion of fat (cream) in a water solution containing sugar and flavorings. The two liquids in an emulsion are insoluble in each other and in many cases would spontaneously separate if a small amount of an **emulsifying agent,** a substance that is soluble in both of them and that therefore stabilizes the copresence of the two, were not present (see Figure 1.9). For example, egg yolk is used as the emulsifying agent in mayonnaise, because the egg yolk keeps water droplets evenly dispersed in the vegetable oil that is the dominant constituent.

Your body houses solutions and mixtures similar to the ones we have been describing. Blood, for instance, is a complex system that has characteristics of solutions, mixtures, and colloids. Water is the most prominent substance in blood and disperses compounds and biological substances such as enzymes, hormones, and cells. Your body also produces an emulsifying agent called bile that helps to keep fatty substances dispersed in watery body fluids.

■ Chemistry in Your Home: Emulsification in the kitchen

Egg yolks are not the only foods that act as emulsifying agents. Onions and mustard also assist in stabilizing oil-in-water emulsifications. Follow the recipe for the vinaigrette dressing below, but omit the mustard and onion. Put all the ingredients in a small glass bowl or jar and whisk or shake the jar to mix. Let the mixture stand for two minutes.

Emulsifying agent   Colloidal particle   Liquid

Figure 1.9 An emulsifying agent surrounds colloidal particles and helps keep them dispersed in a liquid. (Photo by Photodisc Green/Getty Images)

What happens to the oil and vinegar? Are they emulsified? Now add the onions and mustard and shake vigorously. Does the mixture look different? How? Let the dressing stand for two minutes. Do the oil and vinegar components separate? In this recipe there are 4 parts oil to 1 part vinegar. If you change this ratio, you can increase the time before the emulsification breaks down. Would you increase or decrease the amount of oil to increase the stability of the emulsification? Explain your choice. The recipe is:

1/4 cup olive oil (or any salad oil)

2 tablespoons red wine vinegar

1/2 teaspoon finely minced onion

1/8 teaspoon mustard

pinch salt (to taste)

pinch pepper (to taste)

View the molecular-level activity of gaseous, liquid, and solid water at Chapter 1: Visualizations: Media Link 6.

States of Matter

We know from common experience that there are three common states in which matter can exist. Your body contains matter in all three physical states: *solid, liquid, and gas* (see Figure 1.10). Bone, teeth, hair, skin, muscle, tendons, and cartilage are all solids. Blood and body fluids are liquids. Gases enter your lungs when you breathe in and leave when you exhale. You also find gases in the intestine, where they are produced as a by-product of bacterial activity; emission of these gases constitutes what we call *flatulence*.

There is in fact a fourth state of matter: plasma. It is discussed in Chapter 3. The unusual state of matter called the liquid crystal state is discussed in the *Taking It Further III* section at the end of this chapter.

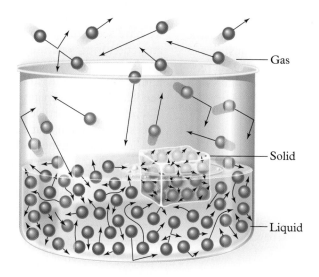

Figure 1.10 A schematic representation of the spacing and motion of the independent particles in gas, liquid, and solid states of matter.

States of Matter

1.10 The particles in solids and liquids are close together

Evidence provided by the microscope (discussed in section 1.4) has indicated that the atoms in solid metals lie very close together. As a consequence, it is not generally possible to force the atoms any closer to each other. Indeed, we know from experience—for example, with ice cubes—that it is very difficult to compress any solid, even if we apply considerable pressure to it. The same generalization about atoms lying close together must be true for liquids, since the volume occupied by a given mass of matter in the liquid state is almost identical with that for the solid state. Liquids, too, are difficult to compress. For example, we know from experience that applying pressure to try to squeeze liquid water into a smaller volume is a futile exercise!

The particles that make up a solid are quite fixed in location and execute only small motions about their average positions, rarely "getting anywhere" as a result. In contrast, particles in a liquid do travel. We know that motion of the particles past each other must occur, since we observe that when many liquids are combined—such as coffee and cream in a cup—they eventually mix together. If they happen to be insoluble in each other—such as oil and vinegar—they instead form two distinct layers. However, we can also conclude that the particles must remain close to other particles as they move, since, for a given mass of a substance, the liquid occupies a volume not much greater than the solid form would. Liquids and solids are called the **condensed states of matter** because they are much denser than gases; that is, their mass-to-volume ratio is much greater.

1.11 Most of a gas is empty space

The temperature of a substance is a measure of the average energy of the motion of its constituent particles.

At low temperatures, matter exists in the liquid or solid state, where, as we have seen, the particles are close together. Therefore, some force of attraction must operate between the particles to keep them from moving very far from each other. At some point, as the temperature of a liquid is raised, the substance becomes a gas, a state in which each particle lies on average at a relatively great distance from all others. Therefore, the particles must have enough energy to overcome the forces of attraction that occur between them. Although other factors are also involved, in general the greater the attractive force between its particles, the higher the temperature required for a substance to boil (convert to a gas).

The volume occupied by a given sample of matter is much greater when it exists as a gas rather than as a liquid or solid, because *most of the space in a gas is completely empty* at any instant (see Figure 1.10). The volume occupied by liquid water, for example, expands by a factor of about 1000 when it is converted, by boiling, into a gas at everyday pressure. The atoms themselves do *not* change in size when they become part of a gas rather than a liquid or solid; they remain the same size, but they are much farther apart from one another.

View gas particles hitting a surface at Visualizations: Chapter 1: Media Link 7.

1.12 The scientific model for a gas

Perhaps you have had the unpleasant experience of walking down the street and being accosted by the noxious smell of rotten eggs emanating from a nearby sewer. The odor, which is produced by the gaseous compound hydrogen sulfide, will reach your nose even if there is no wind. This experience confirms the existence of gases in air, and the theory that the particles in a gas are constantly in motion. If they were not, odors would not carry unless there was a wind current present. As the Roman poet Lucretius said 2000 years ago in his epic poem *The Nature of Things:*

> We can perceive the various scents of things
> Yet never see them coming to our nostrils

The scientific model for gases is that of independent, tiny particles traveling rapidly in straight-line motion through empty space, as a rocket ship travels through outer space. Owing to the rapid motion of its constituent particles, a gas quickly expands to fill completely whatever space is accessible to it. As a given gas particle travels through space, it occasionally collides with other gas particles or with the walls of its container if it is in one. These collisions result in a change in direction for the particles—much as a billiard ball changes direction when it hits another ball or hits the side of the pool table (see Figure 1.10).

One of the many pieces of evidence that led to the scientific model for gases is that gases are much easier to compress to a smaller volume than are liquids or solids. Compressing a gas corresponds only to reducing the amount of empty space that lies between the independent particles.

A piece of evidence that led to the notion that the particles in a gas are in constant motion is the fact that a gas exerts a force on the walls of whatever container it occupies. Technically, the **pressure** exerted by a gas is the amount of force that it exerts on a specified area of surface, say one square centimeter (see Figure 1.11a). For example, the helium gas atoms in a helium-filled balloon are in constant motion. As a

Figure 1.11 Gas particles in motion and exerting pressure on (a) a wall and (b) the inside surface of a balloon.

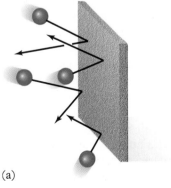

(a)

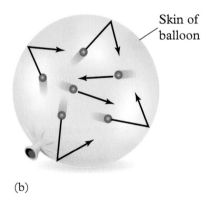

Skin of balloon

(b)

consequence of their movements, they often collide with the inside skin of the balloon (see Figure 1.11b). The pressure exerted on the balloon walls by this constant bombardment is sufficient to keep the balloon "blown up," even though the stretched elastic of the balloon's material is trying to contract and thereby collapse the interior. Indeed, the balloon collapses only when some of the helium leaks into the air outside the balloon.

1.13 The volume occupied by a gas depends on its pressure and temperature

As you know from experience, the more you squeeze a balloon and thereby reduce its volume, the more resistant it is to further reduction. This resistance occurs because the pressure exerted by the gas in the balloon increases as its volume decreases For the gas in the balloon—or in any other enclosed space—the pressure P exerted by the gas is inversely proportional to its volume $V;$ this is known in science as Boyle's Law.

As you are probably aware, it is important that the air pressure inside the rubber tires of vehicles be sufficiently high so that the tires remain well inflated. We commonly measure this pressure with a simple tire gauge, and force extra air into the tire if the pressure is found to be too low. Have you ever noticed that the tire pressure increases when the vehicle has been recently driven? This is because the tire—and the air inside it—have become hot. Experiments have shown that the pressure exerted by a gas increases in proportion to its temperature, provided that the volume is held constant. Alternatively, if the pressure is kept constant, the gas volume increases in proportion to its temperature: this is called Charles's Law. Both Boyle's and Charles's Laws are discussed in more detail in the *Taking It Further II* section at the end of this chapter.

1.14 Molecules are groups of tightly bound atoms

The particles that undergo independent motion in certain gases, such as neon and argon, are individual atoms. However, the nitrogen particles that travel in air are *not* individual atoms, but rather are *pairs* of nitrogen atoms that are strongly connected to each other. These pairs of atoms travel through space as intact units, even surviving collisions with other particles and with the container's walls. The same is true for oxygen, which also occurs as pairs of atoms (see Figure 1.12).

Indeed, a great many substances exist as **molecules,** collections of a relatively small number of atoms that are strongly bound to each other and that remain as intact units even when the material is melted or boiled. The elements hydrogen, fluorine, chlorine, bromine, and iodine, as well as nitrogen and oxygen, all consist of two-atom, or **diatomic,** molecules (see Figure 1.13a). Other substances are made up of diatomic molecules containing two different types of atom. For example, the substance carbon monoxide consists of diatomic molecules, each made up of one atom of carbon and one atom of oxygen.

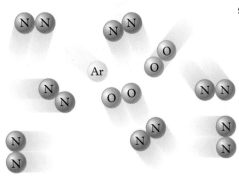

Figure 1.12 Some of the atomic (Ar) and molecular (N_2 and O_2) components of air. Notice that there is a lot of space between gas particles, but no space between atoms bound together as molecules.

Figure 1.13 Diatomic and polyatomic molecules.

Oxygen gas (O_2)

Nitrogen gas (N_2)

Carbon monoxide (CO)

(a) Examples of diatomic molecules

Phosphorus gas (P_4)

Water (H_2O)

Carbon dioxide (CO_2)

(b) Examples of polyatomic molecules

The alternative of listing the atom's symbol as a repetitive list—for example, NN—is cumbersome with large molecules and never used.

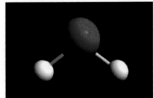

View representations of the molecular structures of water, oxygen, and sugar at Chapter 1: Visualizations: Media Link 8.

Taking It Further

For information and problems on close-packing molecular structures, go to Taking It Further at www.whfreeman.com/ciyl2e

Molecules that contain *more* than two atoms, whether the atoms are all the same or of more than one type, are called **polyatomic** molecules (see Figure 1.13b). Molecules of gaseous phosphorus consist of four phosphorus atoms. Carbon dioxide exists as polyatomic molecules having one carbon atom and two oxygen atoms each. Water consists of molecules containing two hydrogen atoms and one oxygen atom each. (Substances that exist as independent atomic units, such as helium, should presumably be called "molecules" as well; however, conventional chemical usage restricts the term to particles consisting of two or more atoms.)

Chemists indicate the elemental composition of molecules by a **molecular formula** that gives the symbol for each atom followed by a numerical subscript to indicate the number of that type of atom present in *one molecule*. Thus diatomic nitrogen, oxygen, and chlorine molecules are represented as N_2, O_2, and Cl_2, respectively. Where no subscript is shown, the implied value is one. The formula for water molecules, the well-known H_2O, means that each molecule consists of two hydrogen atoms and one oxygen atom. The molecular formula for carbon dioxide molecules is CO_2. Remember, the subscript applies only to the element that directly precedes it; thus the formula CO_2 for carbon dioxide specifies that there are two oxygen atoms, but only one carbon atom, in each molecule. Some compounds consist of molecules that have quite a large number of atoms. For example, a molecule of common table sugar (sucrose) is $C_{12}H_{22}O_{11}$, so there are 45 atoms per molecule! Notice that the atoms are written as a continuous string with no punctuation marks.

Exercise 1.2

Write out the formulas for molecules containing:
a) eight atoms of sulfur, whose symbol is S
b) three atoms of oxygen

Worked Example: The meaning of formulas

a) How many atoms of sulfur and how many of oxygen are contained in one molecule of sulfur dioxide, SO_2?
b) What would be the formula for a molecule that has three atoms of oxygen and one of sulfur?

Solution: a) The symbol for each element in the molecular formula is followed by a subscript, unless the number of atoms is one, in which case no subscript is shown. There is one atom of sulfur and two of oxygen in SO_2, since the subscript 2 refers only to the element whose symbol precedes it. **b)** If a molecule had three atoms of oxygen and one of sulfur, the formula would be SO_3 since, as in SO_2, the implied subscript to sulfur is one.

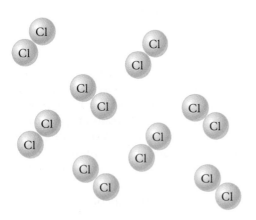

Figure 1.14 Position of the atoms in solid chlorine, Cl_2.

Exercise 1.3

a) How many atoms of nitrogen and how many of oxygen are contained in one molecule of the substance called nitrous oxide, the formula for which is N_2O? **b)** Each molecule of the gas called ammonia has one nitrogen atom and three hydrogen atoms. What is the formula for ammonia?

Molecules occur not only in the gas phase, but in liquids and solids as well. Liquid nitrogen consists of intact, individual N_2 molecules. The same is true for the other diatomic molecular gases O_2, H_2, F_2, Cl_2, Br_2, and I_2 when they are liquefied. Freezing these liquids produces solids that contain diatomic molecules that are clearly separated from each other. This is illustrated in Figure 1.14 for the case of solid chlorine.

Metals and Alloys

1.15 Metals have effects on the human body

Humans are amazingly sensitive to nickel, which can be carried into the body from the skin when even a small amount of it dissolves in sweat. Sensitivity to nickel is the main reason that some people develop rashes when they pierce some part of their body and attach stainless steel or gold-plated objects such as studs. These objects usually contain a tiny proportion of nickel, especially in the metal that lies below the surface. Once a human body has been sensitized to nickel, lifelong reaction to even tiny amounts can occur. For more information about nickel sensitivity and the use of hypoallergenic jewelry, visit the Web site for this book.

Gold, silver, and copper have all been used to treat diseases. Copper bracelets are sold to people suffering from arthritis. Silver present as a colloid in a liquid mixture is reputed to kill microorganisms such as bacteria (antibiotic effect) when ingested or applied to the skin. The established medical community does not agree that these treatments actually work, and the products containing them generally have not been subjected to tests by agencies such as the U.S. Food and Drug Administration (FDA).

(a)

(b)

Figure 1.15 (a) A human-made alloy. Frames for racing bicycles are built from lightweight, high-strength steel alloys that combine metals such as magnesium, molybdenum, and titanium with iron. (b) A natural alloy. This meteorite, composed primarily of iron and nickel, was found in Australia. (Part a, courtesy of Trek Bicycle Corporation; part b, Paul Bierman/Visuals Unlimited)

1.16 Alloys are mixtures of metals

Many metals, when melted, mix together easily in various proportions; such mixtures are called **alloys.** For example, stainless steel is an alloy consisting of 74% iron, 18% chromium, and 8% nickel. The properties of alloys are usually intermediate between those of their component metals, though this is not always true of characteristics such as hardness. Most alloys, including some of our most common industrial materials, are human-made (see Figure 1.15a). However, some do occur naturally, such as the meteorite in Figure 1.15b.

The presence of atoms of different size from the main metal in an alloy allows alloys to resist deforming if a force is applied to them. Thus, relatively soft elemental metals can be hardened by alloying them with another element. This is easy to understand for **interstitial alloys** such as steel, where small atoms such as those of carbon fit in the spaces *between* the rows of the main metal, iron, and prevent the iron atoms from gliding past each other. Iron has a structure in which there is room for small atoms of other elements to fit in the "holes" formed between the four adjacent atoms in any plane. A symbolic representation of the presence of such atoms in iron is illustrated in Figure 1.16a.

In many alloys, such as those formed between gold and silver or copper, atoms of the minority metal simply replace a few of those in the regular structure of the majority metal when the substance solidifies (Figure 1.16b). Thus in an alloy consisting of 80% gold atoms and 20% silver atoms, on

Figure 1.16 Atomic structure of two types of alloys. (a) In interstitial alloys, small atoms lie between larger ones. (b) In substitutional alloys, atoms of one element replace those of another since they are of similar size.

Interstitial atom

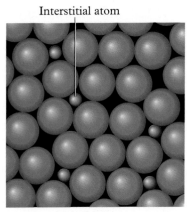
(a) Interstitial alloy

Substituted atom

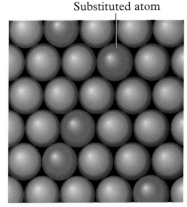
(b) Substitutional alloy

the average every fifth atom along any row in any direction in the structure is one of silver rather than gold. Such so-called **substitutional alloys** have a uniform composition throughout their structure and consequently they are solutions. However, unlike compounds, it is possible to prepare alloys with other proportions of the constituents—say 85% gold and 15% silver, or 95% gold and 5% silver. Examples of substitutional alloys include:

- *Brass:* an alloy of zinc (up to 40%) in copper
- *Bronze:* an alloy of metals such as tin and lead in copper
- *Coinage metals:* various alloys of gold, silver, copper, and nickel

1.17 Mercury is a controversial metal

Virtually all metals exist as solids at room temperature. Mercury is the only metallic element that is a liquid under normal conditions. If cooled to −39°C, it does freeze to a solid. Liquid mercury is shiny and metallic-looking.

You have probably heard a fair amount about the toxicity of mercury. As a liquid, it is not especially toxic when swallowed since most of it passes through the body unchanged. However, mercury vapor is highly toxic, as are all compounds of mercury that dissolve in water to form solutions. Once they enter the body, these forms of mercury can attack the brain and produce mental and physiological disturbances. An incident in Texarkana, on the Texas–Arkansas border, illustrated the hazards of handling mercury. Two teenagers stole 40 pounds of liquid mercury from a site where it had been used to make neon lights. They poured it over themselves and on floors in their homes, gave it out to friends, and even dipped cigarettes into the liquid and smoked them. Within days they began to exhibit the signs of mercury poisoning: coughing up blood, vomiting, breathing difficulties, and seizures. The end result was that eight contaminated homes were evacuated, a family dog was killed by the vapors, and more than 170 people in the town and surrounding areas received medical treatment for mercury exposure.

Mercury poisoning was much more common in the 19th century when workers who used mercury to cure felt hats developed twitches, spoke incoherently, and drooled as a result of long-term exposure to mercury vapors. These workers provided Lewis Carroll with a model for the Mad Hatter in *Alice in Wonderland*. These days, most of the mercury that enters the environment comes from the incineration of waste and sewage

> The word *vapor* is used to describe the gaseous form of a substance that is normally a liquid or solid.

> **Discussion Point:** Should we phase out the use of mercury?
>
> Do you think we should continue to use mercury-containing products? Use the resources on the Web site for this book to identify mercury-containing products in common use. What are the advantages and disadvantages of their use? Develop arguments both for and against the use of mercury.

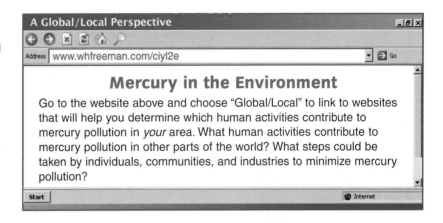

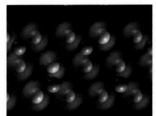

View the structure of a stainless steel alloy at Chapter 1: Visualizations: Media Link 9.

sludge, and the burning of coal. Recently, scrapping cars without removing the elemental mercury used in light switches and other components was identified as a significant source of the element to the environment.

Until quite recently, an alloy that most people had an intimate acquaintance with was the material used to fill cavities in decayed teeth. You may be surprised to learn that mercury was one of the metals used to fill teeth. Although mercury is a liquid at room and body temperatures, it forms many alloys, called *amalgams,* that are solid at normal temperatures. Those having melting points in the 60°C range are useful for fillings, since they can be placed in the decay cavity as a warm liquid metal without causing the patient pain. The liquid assumes the cavity shape as it cools and solidifies in place. Dental amalgam combines mercury with silver, which imparts resistance to tarnishing and mechanical strength, and about half as much tin, which readily amalgamates with mercury. When first placed in a tooth, and whenever the filling is involved in the chewing of food, a tiny amount of the mercury is vaporized. Some scientists believe that mercury exposure from this source causes long-term health problems in some individuals, but an expert panel of the U.S. National Institute of Health concluded that dental amalgams do not pose a health risk. A recent study of adults found that no measure of exposure to mercury—whether the level of the element in the urine or the number of dental fillings—correlated with any measure of mental functioning or fine motor control.

Recently, amalgams have been replaced by mercury-free materials such as composite resins, porcelain and ceramic overlays, and gold. The trend toward using alternative materials stems from concerns about the potential negative health impact of mercury fillings and from the more aesthetically pleasing look of tooth-colored composites. For more information and discussion about the issues surrounding mercury amalgams, visit the Web site for this book.

1.18 The ages of history

Historians refer to long blocks of time as ages, and they name each age to indicate the principal technological material used in that era. Much

Metals and Alloys

of prehistory consisted of the Stone Age. Near its end, people began to make some use of the gold, silver, and copper that they found in the elemental state here and there.

Copper sometimes was found in huge masses and was probably the first metal to be widely used. Although pure copper is relatively soft, it hardens when beaten and hammered. Knowledge of these characteristics allowed people in the late Stone Age to make objects including primitive blades, sickles, and daggers. Much greater amounts of copper became available when people learned, around 4000 B.C., to extract it from the naturally occurring compounds in which it is combined with oxygen or with sulfur. Rock mixtures of metal-containing compounds with other components are called *ores*. Eventually, presumably by accident, copper and tin ores were mixed during the extraction process, and consequently the copper–tin alloy called bronze was first produced (about 3000 B.C.). Trial and error eventually led to the optimum composition for the alloy, which is 11% tin. Because bronze is harder than copper but is easier to melt and cast into shapes, it became the preferred material (see Figure 1.17). The Bronze Age began about 2000 B.C.

The extensive use of a metal for the practical fabrication of large objects requires the use of its liquid form. Copper melts at 1083°C and brass at a lower temperature still; these temperatures were accessible using rudimentary expertise in building fires. However, iron, which is a superior structural metal, does not become liquid until 1535°C. Eventually the technology required to produce and use iron was discovered, and at about 1200 B.C., the Iron Age began. At some point, the discovery—again presumably by accident—was made that if some charcoal was added to the iron, a much superior alloy, steel, could be obtained.

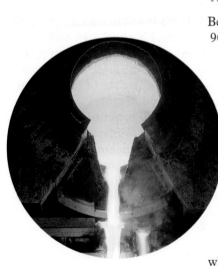

Figure 1.17 A Bronze Age artifact. (The Granger Collection)

1.19 Gold is a valuable and versatile element

Because it occurs in nature as the metal, in pieces that are as heavy as 90 kilograms, gold was one of the first elements to be discovered by humans. Gold jewelry manufactured about 3500 B.C. has been discovered in Mesopotamia (modern-day Turkey); not much later than this date, gold was used as currency. There are references to gold in the Old Testament. Alexander the Great invaded Persia (modern-day Iran) for gold, and later Cortés was willing to wipe out the Aztec civilization for it. Since gold does not tarnish and has always been valuable, it is recycled endlessly. Some of the gold in any piece of modern jewelry was probably first mined thousands of years ago!

The gold that is found in nature is not the pure element, but an alloy that contains some other metals. Some of the metallic impurities can be removed simply by melting the mixture. Other impurities require that air be blasted over the hot liquid, which converts the impurities into compounds that are not soluble in metal and that consequently can be skimmed off the pure, liquid gold (see Figure 1.18).

Figure 1.18 Liquid gold. (Georg Gerster/Photo Researchers)

Because the atoms in metals can readily be displaced relative to each other, solid metals can often be drawn into long wires and/or hammered into thin sheets. Very thin gold sheets, corresponding to only a few hundred layers of gold atoms, can be made (see Figure 1.19a). Such thin sheets of gold, called "gold leaf," are used for decoration on buildings, books, and other objects (see Figure 1.19b). Very thin gold coatings are used on the windows of some office buildings. In addition to being visually attractive, the coating reflects radiative heat, so it keeps the offices cool in the summer by reflecting incoming sunlight, and warm in the winter by retaining heat inside the building. You may be surprised to learn that edible gold foil is sometimes used to decorate elegant desserts. The ingestion of a very small amount of gold seems to have no negative effect on the body. In fact, gold is sometimes used as a therapeutic agent to treat arthritis.

Like other metals, gold is a good conductor of electricity. Because it does not tarnish, gold is often used in microelectronics to connect components electrically.

Commercial gold is an alloy of gold with silver and usually copper and other metals such as nickel. These alloys are much stronger than the pure metal, which is too soft to use alone for most items of jewelry. The relative proportions of silver and copper in the alloys must be carefully controlled if the color of the "gold" is to be the right shade of yellow. Too much silver gives gold a pale, whiteish look whereas too much of the reddish-brown metal copper makes it too reddish.

Sometimes people find that their skin turns green as a result of its contact with gold jewelry. This occurs because the copper in gold alloys can interact with moist, salty skin to transform the element into a compound that is green, just as copper roofs on buildings gradually turn green after exposure to air. Other people find that rubbing gold or silver jewelry on their skin produces black marks—these arise when the silver in the alloy similarly interacts with salty skin. In medieval times, such black marks were often connected with witchcraft.

Gold alloys are classified using the **carat** system, invented by the British about A.D. 1300, which gives the proportion of gold in the combination. The carat scale is based upon the number 24—pure gold is called 24-carat, whereas an alloy having 50% gold by mass is 12-carat. Although spelled "carat" rather than the European "karat," the number of carats is always indicated by the capital letter K following the number. The formula by which the number of carats for a sample of any gold alloy can be obtained is:

carats of gold = 24 × the fraction of mass of the sample that is gold

In other words,

carats of gold = 24 × (mass of gold in a sample / mass of the entire sample)

If a ring, for example, is 75% gold by mass, the fraction is 0.75 and the number of carats is 24 × 0.75 = 18. The ring is 18K gold. Most of the

The term *carat* also is used in a different sense to specify the mass, rather than purity, of diamonds.

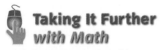

Taking It Further with Math

For information and problems on significant figures, go to Taking It Further with Math at www.whfreeman.com/ciyl2e.

(a)

(b)

Figure 1.19 (a) Gold can be hammered so thin that light passes through it, as illustrated here, where you can see the light of a candle through a thin sheet of gold. (b) Gold leaf can be used to coat cathedral domes, adding not only beauty but also protection from corrosion. (Part a, Chip Clark; part b, Thierry Borredon/Stone. Both from L. L. Jones and P. W. Atkins, Chemistry, 4th ed. © 2000 by L. L. Jones and P. W. Atkins. W. H. Freeman and Company, 2000.)

gold jewelry sold in North America is 14-carat gold. We can see what fraction of the alloy is actually gold by rearranging the above formula:

$$\text{fraction that is gold} = \text{carats}/24 = 14/24 = 0.58$$

In other words, a 14K gold broach is 58% gold. The other components are 26% copper, 13% silver, and 3% zinc. The alloy having the smallest fraction of gold that is allowed in jewelry is 10K in the United States, 9K in Great Britain and 18K in France.

Exercise 1.4

a) Calculate the number of carats in alloys that contain 50% and 33% gold by mass respectively.
b) Calculate the fraction of gold in alloys that are called 10-carat and 22-carat gold respectively.

Summarizing the Main Ideas

Matter is anything that has mass and occupies space. About 100 substances are elements. Elements cannot be separated into two or more components and are the fundamental components of all materials. Most elements are metals; the remainder are collectively known as nonmetals. Each element is assigned its own symbol, usually an abbreviation of its English name.

At the smallest level of subdivision, matter is found to consist of individual atoms, each of which is about 10^{-10} meters, or a tenth of a nanometer, in diameter. Atoms are indestructible by ordinary means.

Compounds are substances composed of two or more elements in a fixed ratio. Compounds cannot be easily separated into their component elements. Compounds usually behave very differently from their constituent elements. Together, elements and compounds comprise the class of materials known as pure substances. Many materials consist of two or more pure substances that are simply mixed together. The mixtures can be separated into components without using drastic conditions. A homogeneous mixture, also called a solution, has a uniform composition throughout. A mixture without uniform properties is called a heterogeneous mixture. Mixtures that to the unaided eye appear homogeneous but that actually consist of microscopic particles of one substance dispersed in another are called colloids.

Mixtures of metals are called alloys; some are homogeneous and others are heterogeneous. The characteristic properties of metals are that they are shiny, opaque, malleable, ductile, and good conductors of heat and electricity.

In liquid and solid states of matter, neighboring particles touch, and they move relatively little. In the gaseous state, the individual particles are on average very distant from one another. Thus gases are mainly empty space in which the particles move.

Many pure substances consist of small numbers of atoms held tightly together in independent particles called molecules. The number of atoms of each type is the same in all the molecules of a substance. The formula for a molecule lists the symbol for each of its atoms, followed in each case by a subscript that indicates the number of them that are present. The alternative to the occurrence of pure substances as molecules is an extended network, in which a gigantic number of atoms are held together in a single particle.

The terms *element* and *compound* are used to describe the composition of the two types of pure substances. In contrast, the terms *molecule* and *extended network* describe the structure of substances at the atomic level and can apply to either elements or compounds.

Key Terms

matter	amorphous	insoluble	diatomic
element	compound	heterogeneous mixture	polyatomic
metal	pure substance	colloid	molecular formula
nonmetal	metalloid	emulsion	alloy
atom	mixture	emulsifying agent	interstitial alloy
nearest neighbor	homogeneous mixture	condensed state of matter	substitutional alloy
extended network	solution	pressure	carat
crystal	soluble	molecule	

Web Sites of Interest

To link to Web sites of interest, go to www.whfreeman.com/ciyl2e, Chapter 1, and select the site you want.

For Further Reading

M. D. Lemonick, "Will Tiny Robots Build Diamonds One Atom at a Time?" *Time,* June 19, 2000, pp. 64–67. A look at nanotechnology and the potential for molecular-level manipulation of materials.

"Why Doesn't Stainless Steel Rust?" *Scientific American,* August 2001, p. 96. A metallurgical engineer answers the question in a short article. The complete text can also be accessed by going to www.whfreeman.com/ciyl2e and choosing Scientific American: Ask the Experts.

Review Questions

1. What is the primary concern of the subject of chemistry?

2. Which of the following properties are characteristic of metals?
   a) They are brittle.
   b) They conduct heat.
   c) They are ductile.

3. What are three other characteristic properties of metals, as listed by scientists?

4. Name five metallic elements and five nonmetallic ones, and give the symbol for each.

5. What does *nanotechnology* deal with?

6. What is a *pure substance*?

7. Are the characteristics of a compound the same as those of its elemental components? Provide an example to support your answer.

8. What is a *metalloid*?

9. How does a homogeneous mixture differ from a heterogeneous mixture?

10. Explain what a *colloid* is, and give an example of one.

11. Explain the difference between an *emulsion* and a *gel*. From your experience, provide an example of each.

12. What does an *emulsifying agent* do? Name one.

13. Identify the three common states of matter and provide an everyday example of each.

14. How are solids similar to liquids? How are they different?

15. How does a gas differ from solids and liquids?

16. What is a *condensed state*? What are the two common condensed states of matter?

17. Describe the scientific model of gases.

18. What is *pressure*?

19. How is a molecule different from an atom?

20. Define the term *diatomic molecule*. Give two examples, including their formulas.

21. What is a *polyatomic molecule*?

22. What information does the *chemical formula* of a molecule contain?

23. Explain what each of the chemical formulas below tells you:
    a) H_2 b) O_2 c) H_2O

24. What can you say about the composition of the compound called glucose from knowing that the formula for a molecule of it is $C_6H_{12}O_6$?

25. Where is mercury found in the environment?

26. In what form is mercury most toxic? What are some symptoms of mercury poisoning?

27. What is an *alloy*?

28. What is an *interstitial alloy*?

29. What is a *substitutional alloy*? Give two examples, including the names of the elements that they contain.

30. What are *amalgams*? What metal elements can be found in dental amalgams?

31. Identify the following as elements or alloys:
    a) gold
    b) bronze
    c) silver
    d) brass

32. How are compounds, mixtures, and alloys different from one another? How are they the same?

33. From your experience, provide an example of:
    a) a compound
    b) a mixture
    c) an alloy

34. Explain what is meant by the term *carat* as applied to gold alloys.

35. What are the other elements commonly included in alloys of gold?

Understanding Concepts

36. What accounts for the ductility and malleability of metals?

37. Explain the difference between:
    a) an *element* and a *compound*
    b) a *compound* and a *mixture*

38. Classify each of the following as a likely element, compound, solution, or heterogeneous mixture based upon what you know about the substance. Specify the reasoning you use to decide on the classification in each case:
    a) white sugar   b) your skin   c) wood
    d) nickel   e) salty water
    f) a well-stirred mixture of salt and pepper

39. Pure alcohol and water can be mixed in any proportion to produce a liquid that is homogeneous. Would the resulting liquid be classified as a compound or as a solution?

40. Is air a pure substance, a homogeneous mixture, or a heterogeneous mixture? Explain your reasoning.

41. Classify each of the following as a likely element, compound, solution, or heterogeneous mixture based upon what you know about the substance. Specify the reasoning you use to decide on the classification in each case:
    a) smog   b) helium gas   c) soup
    d) cream   e) copper wire   f) soil
    g) oil-and-vinegar salad dressing

42. Write the molecular formula for the pure substances whose molecules each contain
    a) two atoms of hydrogen and two of oxygen
    b) one atom of carbon and four atoms of hydrogen

c) two atoms of bromine

d) two atoms of carbon, six atoms of hydrogen, and one atom of oxygen

43. Sucrose is a pure substance whose molecules each contain 12 atoms of carbon, 22 of hydrogen, and 11 of oxygen. What is its formula?

44. Is a colloid the same as an emulsion? Explain.

45. What is the difference in structure at the atomic level between a material that forms a crystal and one that is amorphous?

46. Using circles, draw a diagram to represent each of the following:

a) an element having an extended network

b) a pure substance composed of diatomic molecules

c) a pure substance composed of polyatomic molecules

d) a solution of two molecular substances

e) a heterogeneous mixture of two molecular substances

Synthesizing Ideas

52. Which do you think is heavier, 100 aluminum atoms or 100 lead atoms? Explain your answer using both information in the text and what you know about each of these elements from your own experience.

53. Read the section in the text about dental amalgams. If you were designing a material for dental fillings, what properties would you want it to have?

47. Compare the characteristics of a liquid crystal to the liquid state and to the solid state.

48. Use the web resources at the end of the chapter to determine the type of mercury produced by the following activities:

a) burning coal

b) burning garbage

c) medical and dental waste

d) agricultural practices

e) industrial activities

49. Use the web resources at the end of the chapter to determine what forms of mercury are found most commonly in:

a) air   b) water   c) food

50. If a gold alloy is 70% gold and 30% silver, what is its carat rating?

51. Determine the percent gold in each of the following:

a) 24-carat gold   b) 20-carat gold

c) 12-carat gold

54. Use the Web sites listed at www.whfreeman.com/ciyl2e or other resources to answer the following questions: What form of mercury is most likely to be found in sport fish? Based upon the recommendations made by government agencies, how much fish from the Great Lakes could *you* safely eat in a 1-year period?

■ Group Activity: The effect of temperatures on the size of balloons

As we have discussed in section 1.11, temperature has a large effect on the amount of space a gas occupies. This relationship can be easily demonstrated through a simple experiment using balloons (that will contain the gas, which in this case is air), a hair dryer (the heat source), and either a refrigerator/freezer or a large container filled with ice water (the cold source).

For this experiment you will need:

■ enough balloons so each group has two balloons to inflate, and string to tie them

■ hair dryers (one per group or a few for the class to share)

■ a refrigerator/freezer *or* a container, filled with ice water, large enough to fit an inflated balloon inside of it and be sealed (such as an ice cooler)

■ cloth tape measures (or similar materials) for measuring the circumference of the balloons

There are two parts to this experiment. Divide the group so half the group works on the one part and the other half on the other part.

I. The first part of the experiment involves heating the balloon. Have one person blow up the balloon so that it is inflated, but *not* to its maximum possible extent. Tie off the end with string so it does not leak. Using the tape measure, measure the circumference of the balloon and record that value. Next, while holding the balloon, one person should blow hot air from the hair dryer onto the balloon. Keep the hair dryer about 6 inches away from the balloon. The other group member should keep time; the balloon should be heated for between $1\frac{1}{2}$ to 2 minutes. (NOTE: If necessary to obtain a change in size, heat the balloon for longer than 2 minutes.) Now measure the circumference of the balloon. Record that value.

II. The second part of the experiment involves cooling the balloon. Have one person from the second group blow up their balloon so that it is inflated, but not all the way, and tied off with string so it won't leak. Again, you do not want the balloon to be completely filled. Using the tape measure, measure the circumference of the balloon and record that value. Next, place the balloon in the freezer, refrigerator or container of ice water. Have one person keep track of the time and leave the balloon in the freezer/container for 3 minutes. (NOTE: If necessary to obtain a change in size, try leaving the balloon in the cooler longer.) Now measure the circumference of the balloon. Record that value.

Once both parts of the experiment have been completed, the two subgroups should rejoin and reconvene to discuss the results. Note that the volume of the balloon, and hence that occupied by the gas, is proportional to its circumference cubed.

1. What happened to its size, and hence to the volume of air, when the balloon was heated? Is it what you expected?

2. What happened to its size, and hence the air volume when the balloon was cooled? Is it what you expected?

3. Are the results, taken together, consistent with what is stated in the text about the effect of temperature on gas volume?

Taking It Further with Math I

Any sample of matter big enough to see contains a huge number of atoms

We have stated that atoms have sizes of about 10^{-10} meters, or 0.1 nanometers, each. From this information, we can obtain an approximate idea of the number of atoms in a sample of matter that is large enough to be seen with the naked eye. Let's estimate that you can see a speck of matter that is about one-tenth of a millimeter in each dimension. Since 1 millimeter (mm) is one-thousandth (10^{-3}) of a meter, it follows that 0.1 mm is equal to 0.1×10^{-3} meters $= 10^{-4}$ meters. Because each atom is about 10^{-10} meter wide, the number of atoms that can be lined up along the 10^{-4}-meter width of the speck is given by the ratio $10^{-4} / 10^{-10} = 10^{6}$. In other words, there are about one million atoms along each row of the speck of matter.

If we assume that the atoms on the speck are aligned like those on the gold surface illustrated schematically in Figure 1.2c, there are about a million rows a million atoms long on the surface of the speck. The number of atoms on the surface, therefore, is a million million, or $10^{6} \times 10^{6} = 10^{12}$. Since there are a million atoms along each vertical row, our speck is a million rows wide and a million rows deep, with each row containing a million atoms. The total number of atoms in the speck is a million million million, or $10^{6} \times 10^{12} = 10^{18}$. In other words,

The distance 0.1 mm is about a tenth of the width of a one-sixteenth-inch separation on a ruler.

The number 10^{18} is a billion billion, since one billion equals 10^{9}.

the number of individual atoms present in a speck of matter just barely big enough to be visible is about 10^{18}, or 1,000,000,000,000,000,000!

Exercise 1.5

Determine the number of atoms in a sugar cube that has dimensions of 1 cm × 1 cm × 1 cm (1 centimeter = 0.01 meter). Assume that each atom has dimensions of 10^{-10} meter each and that the atoms are lined up simply in rows, as we assumed for gold, above.

One cm³ of water weighs 1 gram.

Figure 1.20 Samples of various elements, each containing the same number of atoms. Clockwise, from top left, are carbon, sulfur, mercury, lead, and copper. (Chip Clark)

According to the results of Exercise 1.6, a sugar cube contains about 10^{24} atoms. We know that a cube having a volume of 1 cm³, and consisting of material such as sugar or water, has a mass of about one gram. It follows then that the mass of each atom is about $1/10^{24} = 10^{-24}$ grams. Thus we see how tiny in size and mass an atom truly is! Because of the different masses of the atoms and the differences in the distances between adjacent atoms, the mass and volume associated with a given number of atoms varies from one element to another, as illustrated by the examples in Figure 1.20.

1. What are the typical dimensions of an atom?
2. How many atoms would be present along the edge of a cubic speck of dust just large enough to be visible?

Exercise 1.6

By definition, a liter has dimensions of 0.1 meter × 0.1 meter × 0.1 meter. Determine the number of atoms in one liter of a metal in the shape of a cube. You should assume that each metal atom has a diameter of 10^{-10} meters and that the atoms are lined up simply in rows, as discussed in the chapter. Calculate the volume of matter, with atoms of this same size and arrangement, that contains about 10^{24} atoms.

Taking It Further with Math II

The Gas Laws show mathematically how the pressure, volume, and temperature of a gas are interrelated

As was stated earlier in this chapter, it is found experimentally that the volume of any enclosed gas decreases in proportion to the pressure that is applied to it and that the enclosed gas exerts back on its container. This result can be expressed as an equation using the symbol P to represent the gas pressure and V its volume. **Boyle's Law** states that—provided the gas temperature does not change when P and V are changed—the product of the pressure times the volume is a constant, that is, it has the same numerical value after the change as before it:

$$PV = \text{a constant}$$

$$P \propto \frac{1}{V}$$

Thus, if we double the pressure exerted on and by a gas sample, the volume is cut in half, and vice-versa. A graph showing the relationship of pressure to volume for a gas sample is shown in Figure 1.21.

Pressure is often expressed as the height of a column of mercury that can be raised into a vacuum by the pressure of the gas. Alternatively, gas pressure can be expressed in atmospheres (atm), kilopascals (kPa), or pounds per square inch (lb / in²). For example, the average pressure exerted by air at sea level can be variously expressed as 1.0 atm, 101.3 kPa, 14.7 lb / in², 29.9 in, or 760 mm of mercury. Commonly, gas volume is expressed in centimeters (cm^3), cubic meters (m^3) or liters (L). For example, consider a cardboard container of volume 0.946 L (equivalent to 1 quart) full of air at a pressure of 1.04 atm. For this sample, the value of the constant is

$$PV = 1.04 \text{ atm} \times 0.946 \text{ L} = 0.984 \text{ atm L}$$

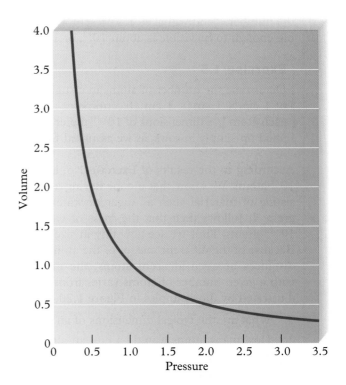

Figure 1.21

If no air is allowed to leak from the container, then to reduce its volume to 0.750 L the pressure that has to be exerted on the walls of the container is

$$P = \frac{0.984 \text{ atm L}}{0.750 \text{ L}} = 1.31 \text{ atm}$$

Exercise 1.7

If a gas sample initially occupies a volume of 45 cubic inches when the gas pressure is 0.95 atm, what will be its volume if the pressure is increased to 1.10 atm?

We also noted earlier in the chapter that the volume occupied by a gas sample increases linearly with its temperature, provided that its pressure stays constant. Thus, the volume of a constant-pressure gas sample becomes smaller and smaller as the temperature decreases, as illustrated in Figure 1.22. If we extrapolate the straight-line relationship between volume and temperature back, the temperature at zero volume turns out to be about $-273°C$ ($-459°F$) regardless of the pressure, sample size, or nature of the gas in the sample. Since the property of volume cannot be less than zero, this temperature is the lowest that a gas—or, as it turns out, any other form of matter—can attain, and for that reason it is called **absolute zero.** The **Kelvin temperature scale** uses this temperature of $-273°C$ as its starting point; thus, $0 K = -273°C$. Since the size of each degree is the same on the Kelvin and Celsius scales, $0°C$ is equivalent to 273 K.

Using temperature as measured on the Kelvin scale, which we symbolize by T, it is found that the volume of a gas sample (if its pressure is held constant) increases in direct proportion to T. Thus, **Charles's law** states that (provided the gas pressure does not change):

$$V = (\text{constant}) T$$

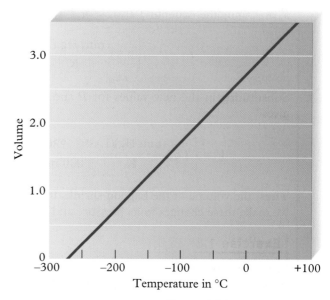

Figure 1.22

For example, heating a gas sample from 100 K to 200 K results in a doubling of its volume since its Kelvin temperature is doubled.

Boyle's and Charles's laws can be combined to allow us to predict what happens when the pressure, volume, and temperature of a gas sample are all changed at once. We find that the value of pressure times volume divided by the Kelvin temperature is the same for the gas sample before and after the change:

$$\frac{PV}{T} = \text{constant}$$

This is known as the **combined gas law.** As an example, let's consider a balloon whose volume V is known to be 1.00 L when its internal gas pressure P is 760 mm Hg and the temperature is 27°C, so $T = 27 + 273 = 300$ K. Thus, the value of the constant ratio of PV/T for this sample is equal to

$$\frac{(760 \text{ mm Hg})(1.00 \text{ L})}{300 \text{ K}} = 2.53 \frac{(\text{mm Hg})(\text{L})}{\text{K}}$$

The value of PV/T is the same under all pressure, volume, and temperature conditions provided that no gas escapes from or enters the balloon. Thus, if we increase the internal pressure to 800 mm Hg and reduce the temperature to 7°C, which is equivalent to 280 K, we can predict the new volume of the balloon. We have

$$\frac{PV}{T} = 2.53 \frac{(\text{mm Hg})(\text{L})}{\text{K}}$$

Rearranging the equation to solve for V, we obtain

$$V = \left[2.53 \frac{(\text{mm Hg})(\text{L})}{\text{K}} \right] \frac{T}{P}$$

Substituting in the new values for P and T and evaluating the ratio gives

$$V = \left[2.53 \frac{(\text{mm Hg})(\text{L})}{\text{K}} \right] \frac{280 \text{ K}}{800 \text{ mm Hg}} = 0.89 \text{ L}$$

Thus, the volume of the balloon shrinks from 1.00 L to 0.89 L when we make these changes to its pressure and temperature.

Exercise 1.8

For the balloon discussed above, determine what pressure it will exert if its volume is increased to 2.0 L and its temperature to 37°C.

Taking It Further III

Liquid crystal is an intermediate state for molecules, between liquid and solid

Looking up from your laptop computer, you glance at the digital clock and see you have two hours before meeting friends for dinner. Deciding that you need some stimulation for the task at hand, you heat a cup of coffee in the microwave and set the CD player to your favorite track. As you use each of these devices, you see what is happening on an LCD, or *liquid crystal display*. LCDs are all around us in electronic devices that display information. LCDs have advantages over other display technologies in that they are thinner, lighter, and use much less power.

The substance used in the screen is a molecular substance in the **liquid crystal state,** a form of matter that is intermediate between the liquid and solid states and that occurs for a few materials. In the liquid crystal state, the substance flows like a viscous liquid—one that flows somewhat but not very readily—even though its molecules exist in highly ordered patterns, as in a crystalline solid. The molecules of substances that form liquid crystals are usually long and rod shaped. The rod shape makes them stack together so that they lie in a parallel fashion (see Figure 1.23), but they are able to slide past each other (like freight trains on parallel tracks).

Because all the molecules in a liquid crystal are aligned in one direction, some of the properties of the substance depend upon the direction in which they are aligned. For example, light can be either transmitted through the substance or reflected by it, depending on

Figure 1.23 Liquid crystal. (Chip Clark, from L. L. Jones and P. W. Atkins, Chemistry, 4th ed. © 2000 by L. L. Jones and P. W. Atkins. W. H. Freeman and Company, 2000.)

whether the light beam is directed parallel to or perpendicular to the direction of the molecules. Furthermore, the orientation of the molecules changes direction if an electric field is applied to the substance. Consequently, the pattern of spots—whether transparent or opaque—that liquid crystals display on a screen depends upon the orientation of the molecules, and therefore upon whether or not a field is applied, at each position (see Figure 1.24). Using this phenomenon, letters and numbers and pictures can be displayed on a screen.

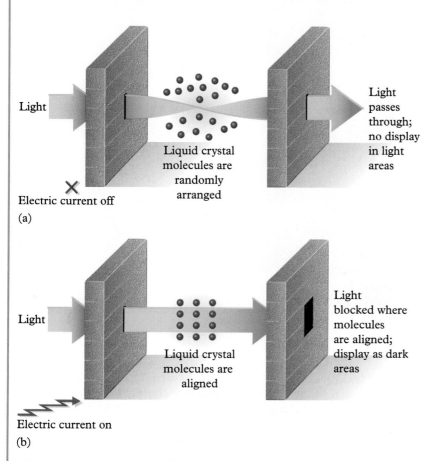

Figure 1.24 Rearrangement of molecules in a liquid crystal when an electric field is applied.

In this chapter you will learn:

- about physical versus chemical change, including how odors are produced and detected;
- about chemical reactions, including how your body oxidizes food for energy;
- why carbon monoxide is so common but also very deadly;
- how metals tarnish or rust;
- how to apply the scientific method.

Have you ever considered how much chemistry is involved in snowboarding?

In fact, this snowboarder is surrounded by both chemical and physical change. Physical processes convert water to snow and ice. In a chemical change, the body "burns" nutrients to produce energy, much like burning fuel produces heat for cooking. (Digital Vision)

New Identities?
Physical and Chemical Change

You put water in a tray in the freezer, and ice cubes form. You put water in a pot on the stove, turn on the burner, and after a while the liquid boils. Water can also be absorbed by foods such as pasta, and is taken in and together with other substances is converted to leaves by growing green plants. Water is produced from other substances when they burn—think of the clouds of water that rise from smokestacks. Though you may not have thought of these everyday occurrences in this way, they are evidence that *matter can undergo change.*

In this chapter, we explore the types of changes that matter can undergo and try to understand what occurs at the atomic level when they happen. At the end of the chapter, a case study that applies the scientific method to a familiar phenomenon—the burning of a candle—illustrates many of the points about change. In later chapters, we shall find that this knowledge allows us not only to understand more fully the natural world but also to exploit this understanding to create new materials and processes.

Before we begin our study of how materials change into new substances, however, we have to understand simpler processes in which the material does not change its constitution. So, we turn first to the physical properties and physical changes of matter.

Physical Properties and Physical Changes

2.1 Physical properties are characteristics of a substance

Many of the characteristics of a material are aspects of the substance itself and can be observed and specified *without* involving reference to its transformation to another substance; these are called its **physical properties.** For example, the physical properties of a metal such as gold include its color, its melting point, and its hardness. The physical properties of a substance can be determined without reference to its interaction with any other materials. We can often, although not always, see a difference in physical properties in different physical states of a substance (see Figure 2.1).

People commonly characterize materials such as aluminum foil and cotton candy as substances that are "light," whereas lead and iron are said to be "heavy." In making these comparisons, people are contrasting the masses of objects that occupy the same volume. They are, in fact, comparing the characteristic physical property of materials called density. In general, the **density** of a material is defined

Whenever you see this icon in this chapter, go to
www.whfreeman.com/ciyl2e

Figure 2.1 Ice cream has different physical properties in the solid and the melted (liquid) states. (Royalty-Free/Corbis)

Since a volume of 1 cm³ is identical to 1 milliliter (1 mL), densities expressed in grams per milliliter have the same values as those expressed in grams per cubic centimeter.

as *the ratio of the mass to the volume* for a sample of it. For example, 1 cubic centimeter of mercury has a mass of 13.5 grams. Mercury's density is therefore said to be 13.5 grams per cubic centimeter, or 13.5 g/cm³. Although the mass and volume of different samples of mercury vary widely, the density is the same for all samples of the element (at 25°C and under standard atmospheric pressure). (Strategies and examples of problem solving using density are discussed in *Taking It Further I* at the end of this chapter.)

In contrast to the value for mercury, the density of aluminum is only 2.7 g/cm³. Indeed, mercury and some other elements such as cadmium and lead are called heavy metals because their densities are large compared to most other metals and indeed to most other substances.

The density of water is only 1.00 g/cm³. Substances having densities greater than this will sink when placed in water. Substances having lower densities will float at the surface of liquid water. From the observation that ice floats in liquid water, we can conclude that ice is less dense than liquid water; measurements indicate that the density of ice is 0.92 g/cm³, which is less than the 1.00 g/cm³ value for liquid water. Water is unique because its solid phase is less dense than its liquid phase. Most oils, including olive oil and others used in cooking, have lower densities than water. Thus in oil-and-vinegar salad dressings, the oil layer lies on top of the vinegar-containing water layer.

Exercise 2.1

One gram of the metallic element magnesium occupies a volume of 0.59 cm³. What is the density of magnesium? Would you call it a "heavy metal" or not?

2.2 Changes of state are physical changes

Processes in which the composition (identity) of a pure substance is *not* altered are called **physical changes.** An important category of processes of this type is the **change of state** or change of the phase—whether gas, liquid, or solid—in which a substance happens to exist at a given time. These possible changes are:

- The *melting* of a solid, or its opposite, the *freezing* of a liquid
- The *boiling* of a liquid, or its opposite, the *condensation* of a gas
- The *sublimation* of a solid directly to a gas, or the corresponding *deposition* of a gas as a solid

The transitions between the various states are summarized in Figure 2.2. For pure substances, these changes of state usually occur at a characteristic, precise temperature rather than gradually over a large temperature range (see Figure 2.3). For example, we are familiar with the fact that ice will remain solid indefinitely at −1°C, but at exactly 0°C it melts. Even at +1°C, liquid water will not freeze to a solid no matter how long we wait, but it will freeze completely at 0°C. The

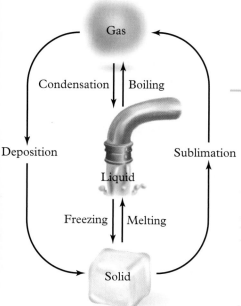

Figure 2.2 The states of matter and their interconversions. (Note that sublimation and deposition do not occur for most substances under normal conditions of pressure and temperature.)

constitution of water in all three phases is the same, namely molecules having the formula H_2O, so its identity does not change during transitions between the phases.

■ Chemistry in Your Home: Chocolate that melts in your mouth

Place a piece of chocolate in your mouth and wait. What happens to the chocolate? What can you say about the temperature of the melting point of chocolate from this evidence?

When water boils, which occurs at 100°C if the liquid is located at sea level, it is converted into a *colorless* gas. The average distance between molecules greatly increases in going from the liquid to the gaseous form, as we saw in Chapter 1. The white puffs of steam that we commonly think of as water vapor are not actually gaseous water but consist of a large number of tiny droplets of liquid water dispersed

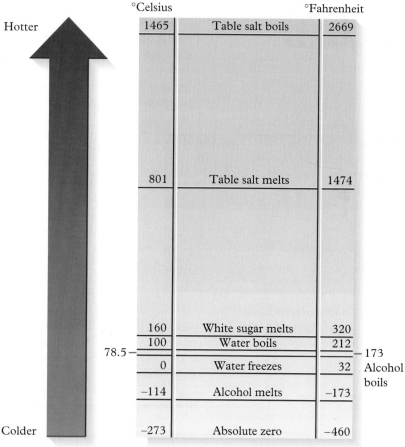

Figure 2.3 Temperature ranges for boiling and melting of some common substances.

in air, which reflect light so that they appear to be white—as occurs in clouds. These droplets form temporarily when hot, gaseous water molecules are quickly cooled by the air into which they are discharged and their temperature drops below 100°C, and so they condense back to the liquid state. Usually the steam puffs dissipate once the droplets have had a chance to evaporate, and the water returns to the gas phase. (We discuss evaporation in section 2.4.) The same phenomenon occurs when jet planes emit moist exhaust gases. Once the emissions cool, the water vapor condenses to droplets, producing the contrails we see in the sky. The noticeable gap between the back of the plane and the contrail exists because some time is required for the gases to cool sufficiently.

Of the various changes of state, you are probably least familiar with the direct interconversions between the solid and gas states. One example of sublimation that you may have seen involves solid carbon dioxide, commonly known as dry ice. When the surface of a piece of dry ice becomes warmer than $-78°C$, molecules of CO_2 vaporize to the gaseous state. Over time, the entire solid gradually warms to $-78°C$ and sublimes to become a gas without ever passing through an intermediate liquid state.

As a consequence of their vaporization, very cold molecules of carbon dioxide occur in the air above the dry ice, and they rapidly cool the air in the surrounding region. Because this nearby air is quite cold, water vapor in it condenses into small liquid droplets which form swirls of clouds that appear to emanate from the solid CO_2. Dramatic stage productions can achieve the effects of fog by placing dry ice on the floor of the stage and allowing it to warm (see Figure 2.4).

Figure 2.4 Dry ice sublimating is used to produce dramatic fog effects. (Robert Goldwitz/Photo Researchers)

The reverse process, the deposition of gaseous CO_2 to the solid state, occurs if the temperature of the gas falls to $-78°C$ or below. For those of us who live in cold climates, a more familiar example of deposition occurs in winter when warm indoor air meets a cold window, and ice crystals are deposited directly on the glass. The same phenomenon occurs when humid outdoor air cools at night to less than 0°C and ice crystals, known as "white frost," are deposited on cold surfaces. In the same way, ice forms on the walls of a refrigerator freezer compartment by deposition of water vapor. Some of the gaseous water in the freezer results from evaporation from ice cubes, which is why the cubes shrink over time. The word **sublime** is sometimes used to describe the evaporation of a solid. A more rapid version of this process is used in preparing "freeze-dried" foods. They first are frozen, and then the surrounding air is removed using a vacuum pump. The water evaporates from the frozen food because there is a vacuum. The freeze-dried solid food is left behind and can be rehydrated by liquid water when it is time to eat it.

Clouds in the sky above us and fog at ground level are visible accumulations of fine droplets of water and/or tiny ice crystals, suspended in air. In contrast, haze is caused by solid particles, often composed of pollutants, the origin of which we'll discuss in Chapter 14.

The liquid or solid droplets in clouds often become sufficiently large that they fall to the ground as rain or snow, or what is known in general as *precipitation*. The phenomenon of *freezing rain,* so dangerous to drivers, occurs when liquid droplets that have been "supercooled" below the freezing point, but that have not yet had a chance to form crystals, freeze on contact with solid objects such as pavement or car bodies. On cold days, some low clouds are composed of small, super-cooled water droplets, which freeze to the wings of aircraft flying through them. This deposition sometimes fatally affects the ability of the aircraft to fly, especially small planes. Another uncomfortable form of precipitation encountered in cold climates is *sleet,* which consists of small ice crystals that partially melted during their descent and then refroze near Earth's surface. *Hail,* which is composed of lumps of ice of substantial size, is often associated with violent thunderstorms.

2.3 Dissolving is different from melting

View the processes of dissolving and melting at Chapter 2: Visualizations: Media Link 1.

The dissolving of sugar crystals in water to produce a solution is another example of a physical process. In the resulting liquid, the sugar molecules and water molecules remain individually intact, as we discussed for solutions in Chapter 1. Indeed, you may have observed that if the water is allowed to evaporate from the solution, crystals of sugar remain, having been reconstituted from sugar molecules.

Notice that dissolving is a different phenomenon from melting. When sugar melts, which it does when it is heated to 186°C, its molecules move around and past each other in a medium that contains only other molecules of the same type (see Figure 2.5a). However, when sugar dissolves in water, even at room temperature, its molecules move among the water molecules as well as among other sugar molecules (see Figure 2.5b).

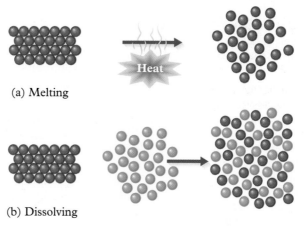

(a) Melting

(b) Dissolving

Figure 2.5 The difference between (a) melting and (b) dissolving at the molecular level.

■ Chemistry in Your Home: Is salt incorporated into ice when you freeze salt water?

As salty water cools and starts to freeze, does the salt remain behind in the unfrozen water or does it become part of the ice? (Write your prediction of the answer to this question before beginning the activity.) You can answer this question by dissolving about half a teaspoon of salt in a glass of water and stirring thoroughly until it is all dissolved. (It should not be cloudy.) Pour the solution into some ice-cube trays, and place the trays in the freezer section of a refrigerator for an hour or two. After the solution has *partially* frozen, remove some of the ice and wash it off with clean water. Taste bits of the ice, allowing them to melt in your mouth. Is the water from the ice salty or not? (Compare with the taste of the unfrozen liquid.) What is your answer to the question at the beginning of this activity? From your result, predict whether or not icebergs are a source of fresh water.

View the evaporation of water at Chapter 2: Visualizations: Media Link 2.

The Evaporation and Boiling of Liquids

2.4 In evaporation, molecules escape from the liquid's surface

Our common experience tells us that if a liquid is left to stand in an open container, it will slowly evaporate, eventually to dryness. Chemists interpret the phenomenon of evaporation as the escape of molecules from the surface of the liquid. The escaped molecules become gaseous and a component of the air that surrounds the liquid. The molecules that leave the liquid's surface are those that happen to be traveling in an upward direction when they meet the surface *and* happen to have sufficient velocity to overcome the weak attractive forces that hold molecules closely together in the liquid state. We know that not all the molecules in a sample of matter have exactly the same energy of motion—at any instant, some are traveling faster than average and some slower. It is the faster ones that escape the surface.

As an analogy to evaporation of molecules, think of a rocket ship that is traveling upward from Earth's surface fast enough to overcome the attractive force of gravity and therefore can escape from the planet. If the rocket were traveling slowly, it would have insufficient velocity to overcome gravity and would fall back to Earth. In the case of molecules, the attractive force is not gravity, but arises from the electrical interactions among the constituents of neighboring molecules.

Even metals evaporate, given a high enough temperature. The tungsten metal wire in a lightbulb slowly evaporates during its use, since it is heated to such a high temperature, over 2000°C, by the electrical current passing through it that it glows. The dark coating you see on the inside of a lightbulb that has been used extensively is condensed tungsten metal. Eventually the lightbulb "burns out" because the tungsten filament becomes so thin that it breaks or melts.

Figure 2.6 Molecules of perfume evaporate readily into the air above the liquid. (Photo from Royalty-Free/Corbis)

2.5 We smell substances by responding to evaporated molecules

Another common experience involving evaporation lies with our sense of smell. We know that we need not bring liquid perfume right into our nostrils to smell it. When we sniff a liquid, we deliberately inhale air that lies just above the surface, so we take into our noses some of the molecules of the liquid that have evaporated and that now exist as gas molecules in that region (see Figure 2.6).

We become aware of the odor of substances when their gaseous molecules reach our nose. Receptors that are located in an area of the nose just below the eyes temporarily pick up specific types of gas molecules. These receptors are themselves large molecules that have regions into which gaseous molecules of specific sizes and characteristics can fit. We have about 1000 different types of odorant receptors. When a gas molecule fits into a receptor, an electrical signal conveys information about the type of molecule directly to the brain. A given substance, such as a flower, emits many types of odor molecules, and they elicit responses from several receptors simultaneously. The brain integrates these signals and interprets this information as a smell (see Figure 2.7). (Information concerning the molecules used in perfumes, etc., that have specific odors are discussed in Chapter 6.)

The area of the nose that contains receptors for gas molecules is directly linked to areas of the brain that deal with emotion, memory, and sexual and maternal behavior. This helps us to understand why certain scents sometimes evoke a flood of memories and emotions in a person. Scientists are currently trying to understand the link between choosing a sexual partner and scent substances called *pheromones*.

Dogs have many more odor receptors than humans do, and consequently they can detect a substance when many fewer molecules of it are present. For this reason, dogs are often used to detect drugs and explosives and to follow the path taken by criminals or missing people. Someday, it may be possible to somehow enhance the ability of humans

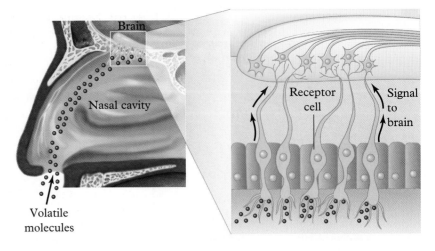

Figure 2.7 The human olfactory system. Volatile gas molecules enter the nose and fit into receptor cells, which then send signals to the brain. (Adapted from N. Campbell, L. Mitchell, and J. Reece, *Biology: Concepts and Connections*. Benjamin/Cummings, 1994.)

to discern scents. In the book *The Man Who Mistook His Wife For a Hat and Other Clinical Tales,* the author, Oliver Sacks, reports the case of a young medical student who, after having taken mind-altering drugs, dreamed he became a dog and found himself in a world surrounded by smells. The effects lingered for several weeks after the dream, during which time the young man reported recognizing people, shops, and streets by their smells. He described the scent-filled world as ". . . a world of pure perception, rich, alive, self-sufficient, and full . . . I now see what we give up in being civilized and human."

In order for us (and other animals) to smell them, molecules must be able to dissolve in water, to a slight extent at least. The reason is that the odor receptors are covered with a thin layer of water, which the molecules must penetrate before they can be detected. Thus, substances that are very insoluble in water do not have an odor. Artificial noses are being developed that contain sensors that overcome this difficulty and respond to contact with gaseous molecules. Such devices may one day be useful, for example, in fire-fighting equipment to distinguish between different types of fires (for example, chemical and wood fires).

2.6 Evaporation and condensation can eventually reach equilibrium

Recall from Chapter 1 that nitrogen and oxygen gases are the top two components of air.

If, rather than being open to the environment, the liquid in a container is enclosed so that there is only a finite air space into which molecules from the liquid can escape—think of a jar, with a tight-fitting lid, that is only partially full of water—our experience tells us that complete evaporation, to dryness, does *not* occur. Indeed, keeping the liquid from evaporating is one of the main reasons that we place lids on containers!

Consider an experiment in which we mark the level of liquid in a jar before closing the lid of the container. If we chose a jar with an ample air space above the liquid, after some time had elapsed, we would find that the level of the liquid would have fallen a tiny bit because *some* of the liquid would have evaporated. After a while longer, the level of the liquid would not decrease further. Sophisticated experiments indicate, however, that the evaporation process has *not* ended once the liquid's

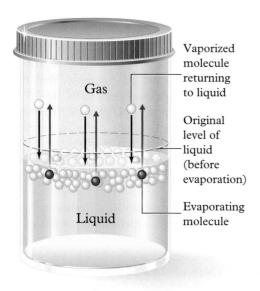

Figure 2.8 Evaporation and condensation of a liquid in a closed container.

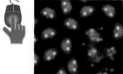

View a system in dynamic equilibrium at Chapter 2: Visualizations: Media Link 3.

level has become constant. Instead, some vaporized molecules in the enclosed air space above the liquid again become part of the liquid when they encounter the surface of the liquid in their random motions. Their return to the liquid counters the continuing evaporation of other molecules from the surface. After a while, the rates of evaporation and of return to the liquid become equal, and we say that **equilibrium** has then been achieved in this situation (see Figure 2.8). Scientists use the term *equilibrium* to signify the equality in rate between two opposing processes of any type, not just evaporation and condensation.

As a consequence of their motion in air, the evaporated molecules of a liquid exert a specific, usually small, amount of pressure—known as the **vapor pressure**—on the surfaces of the container in which the gas is enclosed. The magnitude of this vapor pressure varies substantially between different liquids, in agreement with our common experience that some liquids evaporate more readily than others. Liquids that evaporate readily are said to be **volatile.** For example, nail polish remover is quite volatile and evaporates quickly, whereas liquid mercury is not very volatile and is slow to evaporate unless heated. Generally speaking, liquids with high vapor pressures are also quick to evaporate.

These ideas are probably familiar to you if you've ever experienced a very humid day. The ratio of the actual amount of water in air to the maximum amount it is capable of holding is called the relative humidity when expressed on a percentage scale. For example, if the relative humidity outdoors during the daytime is 40%, then it contains only 0.4 times the maximum amount of water vapor it could *at that temperature.* Although the air inside heated buildings is thought to be "dry," the air in an average-sized room contains several liters of evaporated water. Even desert air contains some water vapor. In fact, technology is being developed to condense this air to obtain a small supply of fresh water.

When an outside air mass cools at night, its relative humidity value rises—even though the amount of water vapor in it has not increased—since the maximum amount of water vapor decreases swiftly with decreasing temperature (see Figure 2.9). When the temperature is lowered so that the relative humidity just equals 100%, the *dew point* temperature is reached, and condensed water ("dew") starts to form on cold solid surfaces, such as cars, grass, trees, buildings, etc.

■ Chemistry in Your Home: Liquids with different volatilities

You can compare the volatility of two different highly aromatic liquids (for example, perfume, vanilla, vinegar, alcohol) by opening a bottle of each at an equal distance far from your nose. Open the first liquid and move it slowly toward your nose, stopping when you just smell it. Take note of the distance between your nose and the liquid. Repeat for the second liquid. Is there a difference? Does one liquid need to be closer to your nose in order to smell it? Which liquid has the higher vapor pressure? Explain.

The Evaporation and Boiling of Liquids

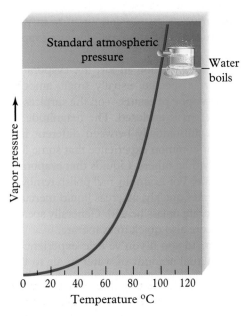

Figure 2.9 The increase in the vapor pressure of liquid water with temperature.

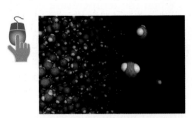

View the process of boiling from inside a bubble at Chapter 2: Visualizations: Media Link 4.

2.7 When does a liquid boil?

For any liquid, the magnitude of the vapor pressure increases sharply with increasing temperature. The behavior of water in this regard is illustrated by the curve in Figure 2.9. Notice, for example, that the increase in vapor pressure as the temperature rises from 60°C to 70°C is much greater than the pressure increase from 10°C to 20°C. Because the amount of water that evaporates into air varies so dramatically with temperature, the concentration of water vapor in air varies widely, from 0 to 5%, depending on the temperature and the level of humidity. Water vapor is usually the third most predominant gas in air.

Using the concept of vapor pressure of a liquid, we can explore the phenomenon of **boiling,** which is the *complete* and rapid conversion, at a specific temperature, of a liquid into the gas state. We know from experience that boiling is associated with the production and rising of bubbles in a liquid—think of water boiling in a pot on a stove. When a liquid boils, the bubbles that are formed consist almost entirely of vaporized molecules of the liquid, *not* of air.

For the purposes of analyzing boiling, think of a bubble consisting of vaporized molecules that happens to form spontaneously at the bottom of a glass container filled with the liquid—as illustrated in Figure 2.10. First, consider what happens at temperatures that are *lower* than the liquid's boiling point, as shown at the lower left of Figure 2.10. The vapor pressure exerted by the gas-phase molecules in the bubble on the liquid water located at the surface of the bubble is *less* than the pressure exerted by the atmosphere which is transmitted through the liquid and experienced by the bubble. Consequently, the bubble collapses—and the vaporized molecules again become part of the liquid.

Let us repeat this "thought experiment" at a higher temperature at the lower right of Figure 2.10. You have learned that the vapor pressure of any liquid increases with temperature. Consequently, as the liquid is heated, more and more molecules escape from it and enter a gas bubble as shown at the bottom of the liquid. Since more molecules are present in the gas bubble at the higher temperature, they exert a greater overall pressure on the surrounding liquid. Eventually a temperature is reached at which the liquid's vapor pressure is just equal to the external, atmospheric pressure. Under these conditions, a bubble is stable and does not collapse, since the pressure that the vaporized molecules exert outward on the liquid equals the atmospheric pressure exerted on them through the liquid. Furthermore, since the bubble is much less dense than the liquid because the gases in it consist mainly of empty space, the bubble will rise through the liquid to its surface. By this mechanism, boiling liquid in an open container is converted to vapor dispersed in air.

In general, *the boiling point of a liquid is the temperature at which its vapor pressure is equal to the atmospheric pressure it experiences.* Thus, water normally boils at 100°C because not until this temperature is reached

Figure 2.10 The stability of gas bubbles near the boiling point of the liquid.

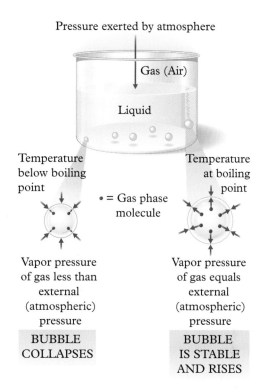

Pressure exerted by atmosphere

Gas (Air)

Liquid

Temperature below boiling point

Temperature at boiling point

• = Gas phase molecule

Vapor pressure of gas less than external (atmospheric) pressure

BUBBLE COLLAPSES

Vapor pressure of gas equals external (atmospheric) pressure

BUBBLE IS STABLE AND RISES

does its vapor pressure equal the pressure of the atmosphere. At elevations that are much higher than sea level, the pressure exerted by the atmosphere is considerably less than that at sea level since there is less air above to exert pressure. At the elevation of Salt Lake City, for example, water boils at a temperature of only 95°C, since at this lower temperature the vapor pressure of water equals the lower atmospheric pressure.

When water is boiled to produce a gas, the H_2O molecules that were present in the liquid remain intact in the gaseous state. For this reason, water vapor in the air can condense back to the liquid state if the temperature falls below the boiling point and a sufficient number of molecules join together to form liquid water droplets. Thus, it is possible to condense steam on a cold surface and recover the water as a liquid. You have no doubt seen this phenomenon when you use a pot with a top on it to cook a moist food. If you lift the pot's top to check the condition of the food, you can see liquid water present on the underside of the top. This liquid forms by the condensation of water vapor that boiled off the liquid in the pot and then encountered the cooler surface at the top. Since the condensed liquid on the lid is cooler than the boiling liquid, it does not reboil.

An interesting variation of the boiling process occurs when water is heated in a microwave oven. It often does not boil spontaneously until the container is moved, and then all of a sudden vigorous boiling begins. This occurs because it is difficult for bubbles to form in the middle of the liquid, where the temperature is the highest. Once bubble formation starts—for example, by placing a teabag or a spoon into the liquid and thereby providing surfaces on which bubble formation can occur—the process occurs rapidly and sometimes explosively because the liquid water is "superheated" above its boiling point.

Before leaving this topic, we should clarify the difference between evaporation and boiling. If the space above a liquid is enclosed, then evaporation—at temperatures below the boiling point—will result in an equilibrium in which only a small amount of the substance is in the gas state. If the space is not enclosed, then eventually all the liquid will evaporate since the vaporized molecules escape the immediate area and cannot recondense into the liquid. However, when the liquid is heated to its boiling point, all of it can quickly be converted to the gaseous form.

Chemical Change

Most elements and compounds are stable indefinitely if they are kept in isolation, out of contact with other substances, and not subject to extreme heat or pressure. However, this stability generally disappears if they are heated excessively or exposed to strong light, and/or brought into contact with certain other substances.

2.8 Chemical change alters pure substances

The **chemical properties** of a pure substance describe how it can change into other substances. Perhaps the most familiar examples of such change involve the phenomenon of the burning of a substance during the self-sustaining process called *fire* that occurs with a flame and that gives off heat and light. We all are aware that the butane liquid in a lighter, or the wax in a candle, is consumed in the fire—it becomes "burned up" and no longer exists as such once the reaction is complete. Simple experiments indicate that in both cases molecules of the fuel, which are composed of carbon and hydrogen atoms, join together in the flame with oxygen molecules, O_2, from the air, and are collectively destroyed. It is important to realize that the *atoms* of these molecules are *not* destroyed, nor are new atoms created. The atoms are simply rearranged in the burning process so that they now exist as molecules of carbon dioxide, CO_2, and of water, H_2O (see Figure 2.11). Thus, one of the chemical properties of wax is that, in the presence of oxygen, it will burn to produce carbon dioxide and water.

Burning is an example of a **chemical change,** *a process in which the identity of pure substances is altered as a result of the rearrangement of atoms.* In chemical changes involving molecules, atoms are interchanged between molecules, so that the identity of the molecules after the change is quite different from their identity before it occurs. In the terminology that is used by chemists to describe such chemical changes:

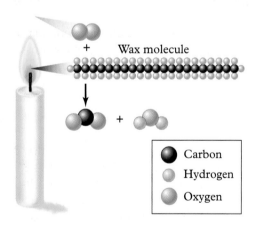

Carbon
Hydrogen
Oxygen

Figure 2.11 The burning of carbon and hydrogen in a wax candle to produce carbon dioxide and water.

Chapter 2: New Identities?

- The process itself is called a **chemical reaction.**
- The substances which are present before the change occurs are called the **reactants.**
- The substances that are present after the change occurs are called the **products.**

In experiments, the total mass of the products of a chemical reaction is always found to be exactly equal to the total mass of the reactants. There is no change in mass when either a chemical or a physical change occurs. This result is known as the **law of conservation of mass.** It also is true that no atoms are destroyed, formed, or changed during a reaction or physical change. Thus, the mass of carbon dioxide and water produced when wax is burned is equal to the combined mass of the wax and the oxygen consumed in the flame.

Chemists usually refer to the self-sustaining process of burning in air, which always involves the rapid reaction of oxygen with a fuel, as a **combustion** reaction. Combustion releases energy, some of which is then used to heat up unburned fuel. This increases the rate of the reaction and sustains it. Combustion reactions you have seen include the burning of wood—whether in a campfire or a forest fire—and the burning of natural gas in gas stoves, water heaters, or furnaces. Many of the common fuels that we burn, such as gasoline and natural gas, are mixtures of molecular compounds called **hydrocarbons,** which are defined as those that consist entirely of carbon and hydrogen atoms and which are discussed in detail in Chapter 4. In general, substances that will burn—including almost all hydrocarbons—are said to be **flammable** or the older term *inflammable*. Substances that will not burn, such as asbestos, are called **nonflammable.**

In general, chemical changes that involve the consumption of O_2 by its interaction with another substance are called **oxidation** reactions, whether or not they occur as combustion with a flame or at room or body temperatures.

The burning of a wax candle is perhaps one of the most familiar of all chemical reactions. A more detailed analysis of the processes that occur in this phenomenon is presented later in this chapter.

2.9 Chemical reaction equations summarize the interchange of atoms

Often the nature of a chemical change is summarized in a shorthand manner by a **chemical reaction equation,** in which the formulas for the reactant(s) are listed, followed by an arrow, and completed by a listing of the formulas for the product(s). If more than one reactant or product is involved in the reaction, their formulas are joined by + signs. In other words,

$$\text{reactant \#1} + \text{reactant \#2} \rightarrow \text{product \#1} + \text{product \#2}$$

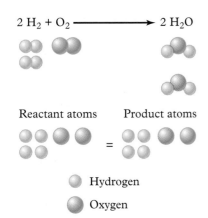

2 H$_2$ + O$_2$ $\longrightarrow$ 2 H$_2$O

Reactant atoms   Product atoms

=

Hydrogen

Oxygen

Figure 2.12 The conservation of mass.

A *reaction equation summarizes the interchange of atoms that occurs between pure substances.* It is important to keep in mind that the equation summarizes only the rearrangement of a particular set of atoms. Because the law of conservation of mass tells us that atoms are neither created nor destroyed in a chemical reaction, *the number of atoms of each type in the products of a reaction must equal the number of atoms of each type in the reactants* (see Figure 2.12). When a reaction equation is written in such a way that it recognizes this principle, it is said to be *balanced.*

Consider, for example, the process of burning the gas called propane, a fuel often used in portable barbeques. The molecular formula for propane is C$_3$H$_8$ and the chemical equation for its combustion is written as:

$$C_3H_8 + O_2 \rightarrow CO_2 + H_2O$$

In the reaction equation above, all the substances are molecules. Since three atoms of carbon are present in the molecule of propane (C$_3$H$_8$), there must also be three carbon atoms in the product. Since each carbon dioxide molecule contains only one carbon atom, it follows that each propane molecule must produce *three* molecules of CO$_2$. We signify this in the equation by placing the number 3 in front of the CO$_2$ formula. Numbers placed in front of chemical formulas in equations are called **coefficients.** Note that these coefficients apply to all atoms and subscripts within the formula that directly follows them. Thus, 3 CO$_2$ means three carbon atoms and *six* (3 × 2) oxygen atoms.

Similarly, one molecule of propane contains eight hydrogen atoms (C$_3$H$_8$), and all of these hydrogens end up as components of water molecules. Since each water molecule possesses two hydrogen atoms, it follows that there must be four water molecules produced for each molecule of propane that burns, and so we introduce this coefficient into the reaction equation. Thus far, we have the following "partially balanced" reaction equation:

$$C_3H_8 + O_2 \rightarrow 3\ CO_2 + 4\ H_2O$$

When we look closely at the products in this revised equation, we see that there are six oxygen atoms in the three molecules of CO$_2$ product, since each coefficient applies to *each* of the atoms in the formula that follows it, and four oxygen atoms in the four water molecules, for a total of ten oxygen atoms. Clearly more than one molecule of O$_2$ must be involved in the reaction with one C$_3$H$_8$ molecule. In fact, it requires five molecules of O$_2$ to provide ten oxygen atoms, and so the coefficient of O$_2$ in the equation must be 5:

$$C_3H_8 + 5\ O_2 \rightarrow 3\ CO_2 + 4\ H_2O$$

The equation above is balanced in terms of the number of C, H, and O atoms in the reactants and products. Notice that we balanced carbon

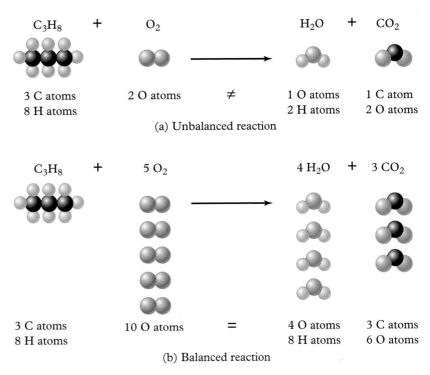

C₃H₈ + **O₂** → **H₂O** + **CO₂**

3 C atoms   2 O atoms   ≠   1 O atoms   1 C atom
8 H atoms   2 H atoms   2 O atoms

(a) Unbalanced reaction

C₃H₈ + **5 O₂** → **4 H₂O** + **3 CO₂**

3 C atoms   10 O atoms   =   4 O atoms   3 C atoms
8 H atoms   8 H atoms   6 O atoms

(b) Balanced reaction

Figure 2.13 (a) Unbalanced and (b) balanced chemical reactions of the complete combustion of propane.

and hydrogen, which each appear in only one substance on each side of the equation, before balancing oxygen, which appears in two substances on the right. In general, beginning by balancing elements that only appear once leads to a final result more quickly.

The final equation above states that, in the burning reaction, one propane molecule combines with five oxygen molecules and that, at the completion of the process, these molecules no longer exist. Instead, their constituent atoms have rearranged themselves as three molecules of CO_2 and four of H_2O. A symbolic representation of the balanced reaction is shown in Figure 2.13.

2.10 A systematic procedure is used to balance reaction equations

In some cases, an initial attempt at balancing an equation leads to a situation in which fractions of molecules appear to be involved. Consider the combustion, using the oxygen in air, of butane, C_4H_{10}. If we balance the number of carbon and hydrogen atoms, we obtain the following equation:

$$C_4H_{10} + O_2 \rightarrow 4\ CO_2 + 5\ H_2O$$

Butane is the main component of the liquid in cigarette lighters.

Chemical Change

Since there are 13 oxygen atoms ($4 \times 2 = 8$ plus 5×1) in the product, we require 6.5 molecules of O_2 to balance them:

$$C_4H_{10} + 6.5\ O_2 \rightarrow 4\ CO_2 + 5\ H_2O$$

However, this equation, while "balanced" in all elements, is unrealistic if we interpret it in terms of actual molecules, since we cannot have half molecules of O_2 or of any other substance. In order to clear the equation of fractions, we multiply *each* coefficient by the same lowest factor that clears the fractions. In this case multiplication by 2 works, so we obtain:

$$2\ C_4H_{10} + 13\ O_2 \rightarrow 8\ CO_2 + 10\ H_2O$$

The above reaction is the simplest **balanced equation** using integer numbers of molecules that we can write for the burning of butane.

We can check that an equation is indeed balanced by adding up the number of atoms on each side of the arrow and ensuring equality for each atom type. For the butane reaction equation, we have:

$2 \times 4 = 8$ atoms of C in the reactants, and
$8 \times 1 = 8$ atoms of C in the products

$2 \times 10 = 20$ atoms of H in the reactants, and
$10 \times 2 = 20$ atoms of H in the products

$13 \times 2 = 26$ atoms of O in the reactants, and
$(8 \times 2) + (10 \times 1) = 26$ atoms of O in the products

This shows the equation is indeed balanced for all three types of atoms.

Notice that the number of *molecules* does not necessarily stay constant as the result of a chemical change. In our balanced butane reaction equation, there are 15 molecules before the reaction occurs and 18 after it is complete.

When students first encounter the exercise of balancing equations, they often try to modify the subscripts in one or more of the chemical formulas. This is *not* correct, because it changes the nature of the reactant or product molecules from those that actually are involved. We cannot write 26 O to balance the above equation in oxygen, since O represents solitary oxygen atoms—which are not involved in this reaction as reactants or products—rather than O_2 molecules. Similarly you cannot write O_{26} to balance oxygen, since no real oxygen molecules contain 26 atoms.

Of course, as you know from Chapter 1, no real sample of butane or any other substance contains only two molecules of it. However, we can easily scale up the balanced equation to whatever number of butane molecules are present in a sample, and the *ratio* of butane molecules to O_2 molecules in the process will still be 2:13.

In balancing chemical equations, you may find it useful to keep a "running score" of the number of atoms of each type in the equation.

For our butane reaction, the *initial* numbers of atoms of each type are summarized in the table below:

Element	No. of atoms in reactants (collectively)	No. of atoms in products (collectively)
C	4	1
H	10	2
O	2	3

As you work at balancing an equation, you should update the table each time you insert or change a coefficient. When the two numbers in each row are equal, the equation is balanced.

Exercise 2.2

Starting with the unbalanced equation for the butane combustion reaction, balance it first for carbon, then for hydrogen, then for oxygen, and finally to clear fractions. At each step, update the above table so that you always have the current number of atoms of each type showing. Check your results with those in the table below:

Element	No. of atoms in reactants (collectively)	No. of atoms in products (collectively)
C	~~4~~ 8	~~1~~ ~~4~~ 8
H	~~10~~ 20	~~2~~ ~~10~~ 20
O	~~2~~ ~~13~~ 26	~~3~~ ~~13~~ 26

Exercise 2.3

Determine balanced reaction equations for the combustion of each of the following molecules with oxygen, O_2, to produce carbon dioxide and water. In each case, use a table to keep a running score of the atoms of each type: **a)** methane, CH_4 **b)** ethane, C_2H_6.

■ Chemistry in Your Home: Building a model of molecules

Use marshmallows, raisins, and pieces of dry cereal to build models of the methane molecule discussed in Exercise 2.3a above. Use the marshmallow for carbon, the cereal for oxygen, and the raisins for hydrogen. Build an oxygen molecule by attaching two pieces of cereal with a toothpick. Build a methane molecule by attaching four raisins to a marshmallow, one on each side.

The reaction between oxygen and methane involves breaking old links and re-forming new links between atoms. Pull the two molecules apart and rebuild the components into water and carbon dioxide: one cereal piece on each side of the marshmallow for carbon dioxide and one raisin on each side of a cereal piece for water. Do you have any leftover pieces? Are some molecules incomplete? How many water and carbon dioxide molecules can you make? Look at the balanced equation you wrote for the combustion of methane in Exercise 2.3 and build any additional molecules. Start over with the new molecules and try recombining the components again. Are your results different now?

Exercise 2.4

Consider the following recipe for baking a loaf of bread:

2 tablespoons of honey
1 package of yeast, dissolved in a cup of warm water
3/4 pound of whole wheat flour

Add the honey and yeast to flour. Stir thoroughly for 3 minutes. Add a handful more flour if necessary to make a smooth, firm dough. This dough does not have to be kneaded. Put the dough in a long bread pan and place in a warm spot (not the oven) for 45 minutes. Bake in a hot oven for one hour.

What are the "reactants" in the recipe? What is the "product"? How many loaves of bread does the recipe produce? In words, write a "chemical reaction" describing the formation of the product from the reactants. What is analogous to the coefficient of each reactant? How would you change these if you wanted to double the number of products?

2.11 Your body generates energy by oxidizing your food

A process that is entirely analogous to combustion, but that occurs without a flame and without raising the temperature of the reactants, occurs continuously in your body. Your blood carries oxygen, O_2, absorbed from the air that you inhale, to the various cells of your body. The cells use the oxygen by combining it with the glucose molecules, $C_6H_{12}O_6$, that are derived from the sugars and starch in the food you eat and that circulate in your bloodstream. The oxidation of glucose is a chemical reaction that, like combustion, produces carbon dioxide, water, and energy. The balanced reaction equation for this process is:

$$C_6H_{12}O_6 + 6\ O_2 \rightarrow 6\ CO_2 + 6\ H_2O + \text{energy}$$

Exercise 2.5

Check that this equation is balanced. Do this by adding the number of atoms of each element that appear on the two sides of the equation.

Another biological oxidation reaction is the rotting of plant material such as vegetables. Every year, a number of North Americans die when they enter confined areas, such as root cellars, in which the air has been depleted of oxygen, partially as a result of such oxidation reactions.

The carbon dioxide you exhale is produced from this process. The overall effect of this remarkable reaction is equivalent to burning the food in air. But, because it occurs gradually, without a flame, you are able to capture and use the energy that the process produces. All the energy you use to walk, run, think, and so on is derived from this process. Indeed, the fact that life occurs is due to chemical reactions, most of which are facilitated by the energy that is supplied in this way.

Complete and Incomplete Combustion

2.12 Complete combustion produces CO_2

When a substance containing carbon and hydrogen burns to carbon dioxide and water, and no other carbon-containing substance is also produced, the reaction is called a **complete combustion** (see Figure 2.14a). In the reaction of butane with oxygen, we have seen that 13 molecules of oxygen are required for each pair of butane molecules present in such a process. What will happen if the ratio of oxygen to butane that actually is available is *less* than 13:2, or if the butane molecules do not remain in the high-temperature flame long enough for the complete reaction to occur? There are two possibilities:

- Not all the butane will react, but that which does react will produce CO_2 and H_2O.
- Most of the butane will react with O_2 to produce CO_2 and H_2O, but some of it will react to give different products.

In reality, some combination of these two possibilities will probably occur. One possible alternative product of the reaction between butane and oxygen is the gas carbon monoxide, CO, a poisonous substance that forms whenever burning occurs and insufficient oxygen is available to convert all the fuel to CO_2. Its formation under these conditions is not unreasonable, given that the production of CO requires fewer oxygen

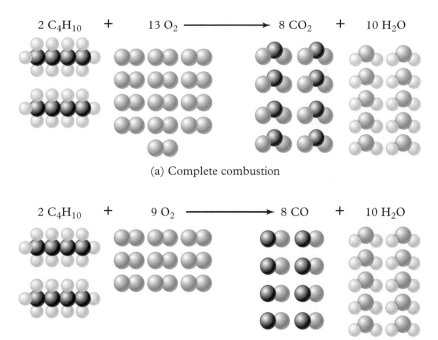

$$2\ C_4H_{10} \quad + \quad 13\ O_2 \longrightarrow 8\ CO_2 \quad + \quad 10\ H_2O$$

(a) Complete combustion

$$2\ C_4H_{10} \quad + \quad 9\ O_2 \longrightarrow 8\ CO \quad + \quad 10\ H_2O$$

(b) Incomplete combustion

Figure 2.14 (a) Complete and (b) incomplete combustion.

atoms than CO_2. Such a process is called **incomplete combustion** because it does not generate the product, CO_2, that contains the maximum amount of oxygen (see Figure 2.14b). The unbalanced reaction equation for this type of incomplete combustion of butane is:

$$C_4H_{10} + O_2 \rightarrow CO + H_2O$$

Exercise 2.6

Balance the reaction for the incomplete combustion of butane.

In summary, when combustion occurs and there is insufficient oxygen or reaction time to change all the carbon to carbon dioxide, some of the hydrocarbon molecules are converted to carbon monoxide, in accordance with the reaction shown above and balanced in Exercise 2.6. The majority of the butane molecules still react to produce carbon dioxide, according to the balanced equation discussed previously. Thus, two different chemical reactions proceed simultaneously and compete for the butane that is available. In this instance, the carbon monoxide is called a **by-product** of the main reaction, which is a general term used by chemists for small amounts of alternative substances produced when a main reaction is not the exclusive process that the reactants undergo.

It should be clear from our butane example that the term *incomplete combustion,* as used by chemists, does *not* mean that some of the fuel is necessarily left over, unreacted when the reaction ceases. Rather, as Figure 2.14 illustrates, it means that the substances obtained are a mixture of those from two or more alternative combustion reactions.

2.13 Soot is a common by-product of incomplete combustion

Another common product of the incomplete combustion of carbon-containing materials is soot, which is an impure form of the element carbon, C. (The composition of soot is discussed in more detail in Chapter 5.) The formation of soot from butane is described by the following equation:

$$2\ C_4H_{10} + 5\ O_2 \rightarrow 8\ C + 10\ H_2O$$

Soot is black and is often found in chimneys and on other solid surfaces that are in the region of a flame undergoing incomplete combustion (see Figure 2.15). Thus, like CO, soot is a by-product of the combustion of a hydrocarbon. In the flame, some fuel burns completely to carbon dioxide, some to carbon monoxide, and some just to carbon. The black color of smoke emitted from diesel-powered trucks, buses, and trains, from industries, and even from bonfires and backyard barbecues, is due to soot. Although pictures of the white or black smoke emanating from industrial smokestacks is often used by the media to draw attention to carbon dioxide emissions into the atmosphere, none of

Figure 2.15 Wood burning in a fireplace. (Corbis/Punchstock)

the visible smoke is in fact CO_2 since CO_2 is a colorless gas (as is CO). In reality, the light color of smoke is due to droplets of water and the dark color to small particles of soot.

2.14 Concentrations of gases are expressed as parts per million

Many of the carbon monoxide detectors found in homes and industries display, as a readout, the fraction of the molecules in the tested air that are molecules of carbon monoxide. This parameter is an example of a **concentration scale,** which chemists use to state *the fraction of a mixture that corresponds to a single component.* Sometimes the word *level* is used instead of *concentration.* The carat scale, discussed in Chapter 1, is an example of a concentration scale, since it indicates what fraction (using a denominator of 24) of an alloy is gold.

Since the fraction of many pollutant gases—CO for example—in air is very small, the number corresponding to the concentration is much less than one and so it contains many zeros, for example 0.000002. In order to avoid using such inconvenient numbers, scientists use concentration scales other than simple fractions for such mixtures. Thus carbon monoxide concentrations in air are often reported on the **parts per million (ppm) scale,** which gives the *number of molecules of a substance that are present in one million molecules of an air sample.* The concentration of carbon monoxide gas in average indoor and outdoor air is about two parts per million, often abbreviated as 2 ppm. This means that, on average, in every 1 million molecules of air, two molecules are CO, and so the fraction of CO in air is 0.000002.

Worked Example: Converting concentrations to numbers of molecules

If the concentration of CO in polluted air is reported to be 10 ppm, how many molecules of CO are present in 10 million molecules of air?

Solution: First we recall that ppm stands for parts per million, so 10 ppm means that there are 10 molecules of CO in every 1 million molecules of air. Since 10 million is 10 times 1 million, the amount of CO in this larger sample must also be 10 times as large, namely $10 \times 10 = 100$ molecules.

Exercise 2.7

If the concentration of a gas in air is 25 ppm, how many molecules of the gas are present in 1 million molecules of air? How many are there in 1 billion (that is, 1000 million) molecules of air?

2.15 Carbon monoxide gas can severely degrade your health

Carbon monoxide concentrations that are much higher than a few ppm can have significant effects on human health. The effects of short-term exposures to high levels of CO are listed in Table 2.1. There is usually a lag after exposure to CO before symptoms are observed, because it takes time for the carbon monoxide in the air to gradually enter the bloodstream. The first symptoms of CO poisoning, which occur at the level of several hundred ppm, are drowsiness and then headache. Since these conditions have many other causes, people may not immediately connect them with the presence of elevated concentrations of CO.

Carbon monoxide is a health hazard because it interferes with the efficient transport of oxygen molecules, O_2, from your lungs through your bloodstream to the various cells of your body. The component of the blood that binds to the oxygen and transports it is called hemoglobin (discussed in greater detail in Chapter 9). Carbon monoxide molecules in your lungs can attach themselves even more strongly than oxygen to hemoglobin, thereby reducing the number of hemoglobin molecules available to carry oxygen to your cells. As a result, your heart must work harder to supply cells with oxygen when you are exposed to CO.

About 70,000 visits to hospital emergency rooms due to CO poisoning occur annually in North America. The condition is treated by administering a gas mixture very high in oxygen, which eventually displaces the CO from the blood. However, brain damage occurs in about 30% of cases because the replacement process is slow.

Medical authorities believe that the role of smoking in heart disease is due mainly to the carbon monoxide that the smoker inhales in cigarette smoke. A portion of the tobacco undergoes incomplete combus-

Cigarette smokers have two to three times as much of their hemoglobin tied up with CO as nonsmokers.

Table 2.1 Effects on healthy adults of exposure to various levels of carbon monoxide

Air concentration of CO in ppm	Symptoms	Death occurs after exposure of
50	Maximum allowable 8-hour concentration (adults)	
200	Slight headache, fatigue, dizziness, nausea after 2–3 hours of exposure	
400	Frontal headache after 1–2 hours of exposure	> 3 hours
800	Dizziness, nausea, convulsions after 45 minutes of exposure; unconsciousness in < 2 hours	2–3 hours
1600	Same as above after 20 minutes of exposure	1 hour
3200	Same as above after 5–10 minutes of exposure	25–30 minutes
6400	Same as above after 1–2 minutes of exposure	10–15 minutes
12,800		1–3 minutes

tion and so produces CO. Indeed, a smoldering cigarette undergoes even less complete combustion than one from which smoke is currently being drawn, and consequently considerable carbon monoxide is present in the "secondhand smoke" to which nonsmokers are exposed.

2.16 Carbon monoxide is a common air pollutant

All carbon-containing substances are first converted in flames—the zone in which the reaction occurs—to CO, which then reacts with more oxygen to form CO_2. Incomplete combustion results if the carbon monoxide does not remain in the flame long enough, or if there is insufficient oxygen available. Thus, incomplete combustion results in the emission to air of measurable amounts of carbon monoxide gas when gas or oil furnaces malfunction. It also commonly occurs when gasoline is combusted in automobile engines, and even when wood is burned in a fireplace. Because carbon monoxide is poisonous to humans, it is important that the exhaust gases from furnaces, vehicles, and fireplaces be efficiently vented to the outdoors so that occupants of buildings or cars are not unduly exposed to this gas. It is also important not to use charcoal grills or unvented kerosene heaters inside buildings since you would be directly exposed to their exhaust gases. Unfortunately, carbon monoxide is odorless, tasteless, and colorless, and so you cannot easily detect it. For this reason, many people are installing carbon monoxide detectors in their homes, much as they do smoke detectors (Figure 2.16).

Figure 2.16 A carbon monoxide detector used in homes. The CO concentration in ppm is displayed if it is elevated. (B. Moscatelli for W. H. Freeman)

The U.S. government has set 9 ppm as the maximum concentration of carbon monoxide in outdoor air to which people should be exposed over any eight-hour period. Some cities regularly exceed this standard, due principally to the CO that is exhausted into the air from motor vehicles. In order to minimize the levels of carbon monoxide, and those of other pollutants, in outdoor air, catalytic converters are required to treat the output gases from motor vehicles (and wood stoves in some cases) before they are released into the air. Within these devices, the reaction of CO with oxygen to give CO_2 is completed *without* the use of a flame. The hot gases CO and O_2 in the engine's exhaust pass over finely divided grains of a metal such as platinum, Pt, which have the ability to speed up their conversion to CO_2 (see Figure 2.17). *In general, a substance that can speed up a reaction without itself being consumed is called a* **catalyst,** hence the name *catalytic* converter. Here the platinum metal is the catalyst, since it facilitates the rapid conversion of one CO molecule after another to CO_2.

The net reaction here is

$$2\ CO + O_2 \rightarrow 2\ CO_2$$

The catalyst is not permanently altered by the reaction, but it does participate in it, so it could be shown as both a reactant and a product:

$$2\ CO + O_2 + Pt \rightarrow 2\ CO_2 + Pt$$

Figure 2.17 A catalytic converter. Gases from the engine pass through the catalytic converter before exiting as exhaust.

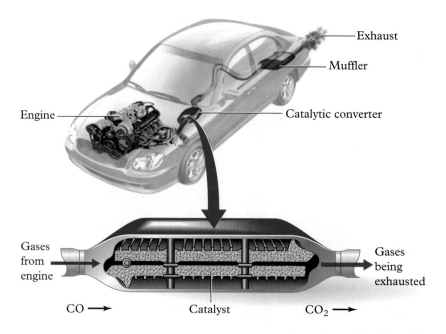

More conventionally, however, its presence is indication by an annotation above the arrow:

$$2\ CO + O_2 \xrightarrow{\text{Pt}} 2\ CO_2$$

Because of the above reaction, very little of the CO produced in the engine is emitted from the catalytic converter and enters the atmosphere. The catalytic converter must itself be hot to operate effectively, so it does not work when the engine has just been started. At those times, the exhaust gases pass through the converter unchanged and enter the air as such.

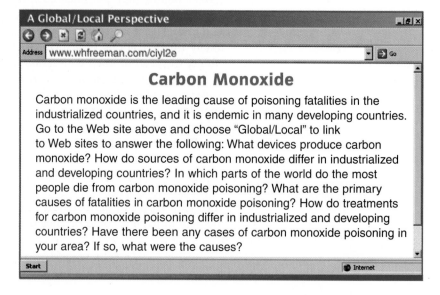

A Global/Local Perspective

Address www.whfreeman.com/ciyl2e

Carbon Monoxide

Carbon monoxide is the leading cause of poisoning fatalities in the industrialized countries, and it is endemic in many developing countries. Go to the Web site above and choose "Global/Local" to link to Web sites to answer the following: What devices produce carbon monoxide? How do sources of carbon monoxide differ in industrialized and developing countries? In which parts of the world do the most people die from carbon monoxide poisoning? What are the primary causes of fatalities in carbon monoxide poisoning? How do treatments for carbon monoxide poisoning differ in industrialized and developing countries? Have there been any cases of carbon monoxide poisoning in your area? If so, what were the causes?

Examples of Chemical Reactions That Are Not Combustions

All of the examples of chemical reactions discussed so far involve oxygen as a reactant in a combustion process. In this section, we illustrate some interesting reactions that do not fall into this category.

2.17 Tarnishing and rusting are chemical reactions

Many households have objects such as jewelry or cutlery that are made of alloys of silver. For example, *sterling silver* consists of 92.5% silver and 7.5% copper. It is well known that when such silver objects are exposed to air, the surface gradually dulls and eventually becomes brownish or blackened—what we call "tarnished." (An analogous reaction does not occur with gold.) The process that occurs to alter the surface in this way is a chemical reaction. It occurs between atoms of silver on the surface of the object and molecules of the gas hydrogen sulfide, H_2S, that exist even in very small concentration in air that comes into contact with the silver.

The source of the hydrogen sulfide is often the cooking of foods such as eggs, a process that naturally releases small quantities of H_2S gas. However, there are many other sources of small amounts of hydrogen sulfide in air and it is therefore a common air pollutant.

When silver atoms react with molecules of H_2S gas that collide with them, they are converted into the grayish-black compound silver sulfide, Ag_2S. Silver sulfide is what gives the surface its dull, gray appearance. The balanced chemical reaction equation for the process is:

$$2\,Ag + H_2S \rightarrow Ag_2S + H_2$$

The thin coating of silver sulfide that clings to the surface of the silver can be cleaned off by a "silver polish" and the shine of silver reestablished. However, this results in a net loss of silver to the object, which can be noticeable in time if the cleaning is repeated over and over. Alternatively, the Ag_2S can be converted back to metallic silver by a chemical reaction. This can be achieved by placing the silver object in contact with a piece of aluminum, such as aluminum foil, and immersing the two solids in a solution of sodium bicarbonate (baking soda) in water. The aluminum extracts the sulfur atoms from the tarnished silver and temporarily forms the compound aluminum sulfide, Al_2S_3:

$$2\,Al + 3\,Ag_2S \rightarrow 6\,Ag + Al_2S_3$$

■ Chemistry in Your Home: Cleaning silver with aluminum foil

As stated in the text, many silver polishes remove the silver contained in the tarnish. However, there is a simple (and relatively inexpensive) method for polishing silver that retains the metal. The procedure uses aluminum foil, baking soda, and water. We suggest you perform this experiment in a well-ventilated (or preferably outdoor) space as a strong odor results from the reaction. Place the silver object into full contact with the foil, and immerse the two into a solution of baking

Figure 2.18 In deoxygenated water *(left)*, dissolved O_2 gas has been removed and therefore is not available to react with the iron in the nail. In normal water *(right)*, dissolved O_2 combines with the iron and forms rust. (Ken Karp for W. H. Freeman and Company. From L. L. Jones and P. W. Atkins, *Chemistry*, 4th ed. © 2000 by L. L. Jones and P. W. Atkins.)

soda (sodium bicarbonate) that has been dissolved in warm water. If you detect characteristic "rotten egg" odor of hydrogen sulfide gas originating from the solution, it is due to a complicated reaction involving the decomposition of Al_2S_3.

In a process analogous to the tarnishing of silver, the atoms at the surface of the element iron and its alloys such as steel react with atmospheric oxygen dissolved in water in a complicated process called rusting. A mixture of compounds—including iron oxide, Fe_2O_3—that is collectively known as rust is formed at the surface:

$$4\ Fe + 3\ O_2 \rightarrow 2\ Fe_2O_3$$

Rust does not adhere well to the underlying metal and eventually flakes off, weakening the solid structure (see Figure 2.18). Commonly, iron and steel are protected from rusting by galvanizing their surfaces. This means that the iron or steel is coated with a thin surface layer of zinc, a metal whose oxide does not easily flake off.

2.18 Explosions result from some chemical reactions

Some chemical reactions occur so quickly, and release so much energy, that explosions result. One common example is the explosive decomposition of ammonium nitrate, NH_4NO_3, a solid substance that is used as a fertilizer and therefore is easily obtained in bulk. The chemical reaction that constitutes the explosion is

$$2\ NH_4NO_3 \rightarrow 2\ N_2 + O_2 + 4\ H_2O$$
$$\text{solid} \qquad \text{gas} \quad \text{gas} \quad \text{gas}$$

The energy released by the reaction initially heats the product gases to high temperatures. Since gas pressure is proportional to temperature (see *Taking It Further II* at the end of Chapter 1), the pressure of these gases is much greater than the pressure of the air. This pressure is quickly reduced by expanding their volume. The rapid expansion of the gases produces a large force and causes much destruction to objects that it encounters.

Clay is used to coat the particles of ammonium nitrate in commercial supplies so it will not spontaneously explode. However, combining it with fuel oil and igniting the mixture results in explosions. This mixture was probably used in the 1995 Oklahoma City bombing.

A controlled explosion of practical importance occurs when airbags in modern automobiles are activated to expand. The nylon airbag fills explosively with nitrogen gas, N_2, after a collision has occurred, thereby protecting front seat occupants from smashing into the glass windshield as they lurch forward. The nitrogen gas is produced quickly by the decomposition of the solid compound sodium azide, NaN_3:

$$2\ NaN_3 \rightarrow 2\ Na + 3\ N_2$$

Although this reaction has been known for some time, chemists and engineers devised a way to initiate it electrically so that it occurs in an extremely short period of time—a few hundredths of a second—once a signal is received from the car's computer that a collision has been detected.

(a)

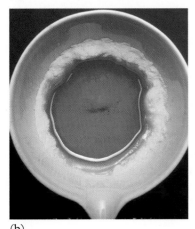
(b)

(c)

(d)

Figure 2.19 (a) Solid elemental sulfur. (b) When solid sulfur is heated to the melting point, it forms an orange liquid (physical change). (c) Upon further heating, the liquid turns brown and becomes viscous (chemical change). (d) At even higher temperatures, the liquid becomes red and flows readily again (chemical change).

(Part a, Gary Retherford/Photo Researchers; parts b, c, and d, Chip Clark)

● **Tying Concepts Together:** Molten sulfur involves both physical and chemical change

Though only a physical change, not a chemical one, occurs when a substance melts, the heating of elemental sulfur produces changes of both types within a rather short period of time.

Solid sulfur (see Figure 2.19a) is a yellow solid that consists of S_8 molecules. When it is heated to 113°C, it melts to form an orange liquid (see Figure 2.19b). Since the orange liquid also consists exclusively of S_8 molecules, this transition is a physical change. If the molten sulfur is heated further, to about 159°C, the liquid turns brown and becomes more viscous (see Figure 2.19c) because at this temperature many of the S_8 molecules react together to form long chains of sulfur atoms. Further heating to higher temperatures results in the decomposition of many of these long chains into small molecules, producing a red liquid that is less viscous and flows more readily (see Figure 2.19d). The processes that occur *above* sulfur's melting point (113°C) are *chemical* changes, because the nature of the molecules changes when they occur, whereas the transition from solid to liquid at the melting point is a *physical* change since the molecules remain S_8.

In the case of sulfur, decomposition occurs when the substance is heated moderately beyond its melting point. There are some substances for which melting and decomposition occur simultaneously—an example is sugar, for which the browning of the solid at the melting point signifies that a chemical change is also occurring.

Examples of Chemical Reactions That Are Not Combustions

The Burning of a Candle: The Scientific Method in Action

The scientific method is the procedure that is used by all scientists to investigate nature. Only by using this technique is knowledge gained in science so that we can explain and manipulate nature. Rather than discussing the method in abstract terms, we shall illustrate its workings by means of a specific example.

One clear-cut example of a chemical reaction with which we are all familiar is the burning of a wax candle. We will use everyday knowledge, supplemented with the results of simple additional experiments, to illustrate how the scientific method can help us investigate the burning of a candle.

In using the scientific method, we first gather **observations** about a natural phenomenon—in this case the burning of a candle—and then form a **hypothesis,** an attempt to explain these observations in terms of materials and processes that we already understand. We then test the hypothesis, and perhaps refine and extend it, by performing experiments that are devised to discover whether or not various aspects of the original hypothesis are valid. The experiments are normally duplicated a number of times to make sure the same result is obtained each time. Only then would we be confident that the findings were significant.

■ Chemistry in Your Home: What do you see when a candle burns?

Light a candle and, as you watch it burn, make a list of all of your observations about the candle and what happens when it burns. Restrict yourself to things you can observe with your own eyes and other senses, rather than trying to interpret the phenomenon in terms of what specific chemicals are consumed or produced.

2.19 Initial observations

One simple observation that we make about the burning of a candle is that the process generates both heat and light. Indeed, it is to produce these phenomena that candles are burned! We also observe that, over time, the candle itself gradually "disappears" and that no significant amount of any other liquid or solid material is produced by its transformation. We know that both the light and the heat are generated by the flame, which is present in a region located *above* the top of the wax which encloses the wick.

If you have ever tried to light solid wax—for example, that at the side of a candle, or even the liquid wax that falls from the flame area—you know that neither material will burn. Indeed, the flame of a candle reaches down to the surface of the liquid puddle that forms at the top of the candle, but the flame does not spread across the liquid. Careful observation of the flame indicates that it is not uniform throughout. The inner portion of the flame, near the wick, is not as bright as the outer portion (see Figure 2.20).

Figure 2.20 The inner portion of the flame is darker than the outer portion. (Photodisc/Punchstock)

2.20 We can form an initial hypothesis and test it

Given our initial observations, we might form the hypothesis that it is the *vapor* of the wax that undergoes combustion, rather than its liquid or solid phases. We would imagine that the heat produced by the combustion initiated by the flame of a match melts some of the wax at the top of the candle to form the puddle that we observe there, and that some of this liquid wax makes its way up the wick where it is evaporated by the flame's heat and then, as a gas, becomes ignited.

We can test this simple hypothesis by an experiment in which we determine the nature of the substance that lies closest to the wick of the flame. If we insert a glass tube into the dark part of the flame—that is, the part that lies closest to the wick—and trap the material that flows through the tube into a flask that is cooled (for example by ice water), we discover that liquid and solid wax are deposited at the bottom of the flask (see Figure 2.21). This tells us that the gas that was tapped from the dark part of the flame and that traveled through the tube and into the flask must have consisted of wax vapor, which condensed back to the liquid and solid states once it was cooled at the bottom of the flask. We can confirm this by an additional experiment in which, rather than leading the tube from the flame into the flask, we place a lighted match at the end of the tube. The gas leaving the tube burns brightly, supporting our contention that it is wax vapor.

Thus our simple additional experiments have supported our hypothesis that what burns in the flame is the vapor phase of wax, not

The wicks of some candles are treated with lead to stiffen them and to give a more even burn to the wax. However, the lead in the wicks becomes an airborne pollutant when the candle is burned.

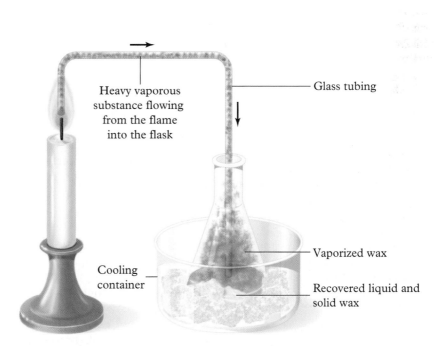

Heavy vaporous substance flowing from the flame into the flask

Glass tubing

Cooling container

Vaporized wax

Recovered liquid and solid wax

Figure 2.21 Extraction of the substance in the darker portion of the flame.

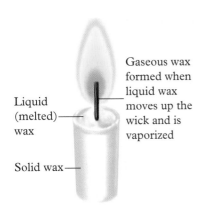

Figure 2.22 Gaseous, liquid, and solid components of a burning wax candle.

Liquid (melted) wax

Gaseous wax formed when liquid wax moves up the wick and is vaporized

Solid wax

Figure 2.23 Testing of the gases emitted from the top of the candle's flame. (Dave Wrobel/Visuals Unlimited)

the liquid or solid forms (see Figure 2.22). Further support for this hypothesis is provided by the fact that when we extinguish a candle by snuffing out the flame, the odor of the candle's materials becomes much more apparent to us. Presumably, the wax that had been vaporized but not yet combusted simply becomes part of the air which we smell. We can perform a chemical test of this point. If immediately after snuffing the candle we bring a lighted match to a position a few centimeters above the wick, we observe a flame traveling through the air to the wick (see Figure 2.23). We interpret this phenomenon as the ignition of vaporized wax in the air, which is consistent with our hypothesis.

2.21 Further experiments can identify the products of the combustion

One of our observations was that the wax of the candle seems to just "disappear" gradually over time, and it is not replaced by any liquid or solid product. We could form two alternative hypotheses to explain this phenomenon:

1. This process is unique in that matter is not conserved and that the atoms are destroyed.

2. Matter is indeed conserved in the reaction, but the products are gases whose existence we do not observe because they are colorless and odorless.

The experiments described below helped scientists determine which of these two possibilities is correct.

Over a period of decades and centuries, chemists have developed tests that can be used to determine the nature of the gases that are produced when a candle burns. One of the simplest tests is that for water vapor. We know that water vapor condenses to liquid water when its temperature falls to 100°C. If we place a cold, solid object—such as a metal spoon that has been cooled in a refrigerator—over the top of the flame, we find that a drop of water gradually forms on the bottom of the spoon. Thus, water must be a product of the burning process.

Numerous other tests for specific substances show that the only other gas emanating from the burning flame is carbon dioxide. Thus experiments confirm that gases are indeed produced by the burning of the wax candle and that these gases are water, H_2O, and carbon dioxide, CO_2.

2.22 A simple experiment identifies the oxygen-containing reactant

The tests indicate that the products of combustion contain carbon, hydrogen, and oxygen. We know from other information that the wax we used is composed exclusively of carbon and hydrogen. So where does the oxygen found in the products come from? Since the flame burns in air, we could hypothesize that it comes from the oxygen gas that is a major component of air. We can test this hypothesis by burning the candle in a closed container. If oxygen is indeed a reactant, the candle ought to be extinguished once the oxygen concentration in the container becomes low. Indeed, when we burn the candle in a closed container, we find that the flame gradually becomes smaller and smaller, and dies long before the candle is completely burned.

2.23 Where is the elemental carbon present?

Candles often produce some soot—which we know to be elemental carbon—when they burn. Careful observation of a burning candle indicates that some soot almost always escapes from the flame, especially when it is starved for air and thus undergoes incomplete combustion. You can sometimes observe the soot on a ceiling directly above the flame of a candle left burning for some time. The mass of soot obtained is variable, depending upon how the candle burns, and corresponds to only a small fraction of the mass of the candle, so it is only a by-product of the main reaction. Nothing in our hypothesis accounts for this by-product of combustion.

We can gain some insight into the production of soot from a simple experiment involving the luminous part of the candle's flame. If we use our glass tubing to extract some of the gas from the *brightest* part of the flame, we obtain a gas that contains black soot (see Figure 2.24). Indeed, it is the soot that produces the brightest light in the flame, since particles of elemental carbon glow brightly when they are heated to a high temperature. We can conclude that the combustion of the wax must be occurring in the bright part of the flame and that at least a portion of the wax's carbon must be converted temporarily to elemental form as the reactant migrates outward from the wick.

Confirmation that the main reaction occurs in the outer, more luminous portion of the flame is provided by an additional simple

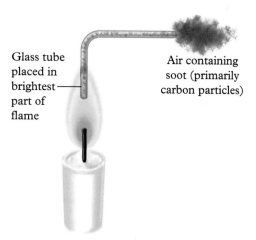

Glass tube placed in brightest part of flame

Air containing soot (primarily carbon particles)

Figure 2.24 Extraction of the substance in the brightest part of the candle's flame.

experiment. If we place a piece of white cardboard in the lower part of the flame for a few seconds, and remove it before it bursts into flame, we observe that only a small circular region has been scorched by the flame's heat. The central portion of the cardboard, inside the scorched ring, was in the cooler part of the flame, where the wax had been vaporized but had not yet reached the outer part where it comes into contact with oxygen and burns. Since the flame's heat is given off from the bright part of the flame, we confirm that the reaction must occur there.

2.24 The mass of chemicals is collectively unchanged

Is the total mass of the products of our candle-burning reaction greater than, less than, or equal to the total for the reactants? We can investigate this question by modifying the experiment so that the combustion is carried out in a closed container. We place the airtight container on a sensitive scale and watch to see if the mass of the apparatus changes *during* the course of burning the candle. In fact, we find that the mass of the apparatus does not change *at all* during the burning. Thus we confirm the law of conservation of mass for our experiment.

2.25 Summary of the expanded and tested hypothesis

We can conclude from our observations and previous considerations that fire, such as undergone by a burning candle, is a process in which an energy-releasing, self-sustaining combustion reaction consumes more and more fuel over a period of time. This confirms the statement in Chapter 1 that fire is a process, not a substance.

Let us now see how our hypothesis, as revised and expanded by experiments, explains the phenomenon of the burning of a wax candle. The solid candle, consisting of a wax that contains carbon and hydrogen, does not ignite on its own. Under the influence of heat given out from the flame (or the match used to ignite it), a small portion of the solid wax melts to liquid wax around the wick of the candle, a physical change. The liquid itself does not burn, but travels up the wick where the heat evaporates it and releases the vapor into the dark area of the flame surrounding the wick. Oxygen from the air combines with evaporated wax in the luminous area of the flame in a chemical reaction that produces water and carbon dioxide, though at least some of the wax is first converted to elemental carbon. Both light and heat are produced when the reaction occurs in the bright area; much of the heat is carried away from the flame by the product gases that are given off. The operation of different parts of the candle and flame are summarized in the diagram in Figure 2.25.

■ Chemistry in Your Home: Using the scientific method in everyday life

You can apply the scientific method by determining the shortest route to your favorite store. First, based upon your previous experience (observations), determine which route you believe to be the shortest to the store. This is your hypothesis. Test your hypothesis by driving (or biking or walking) to the store using your route and alternative routes. Keep track of the mileage on your odometer or the time taken to bike or walk. This is data gathering and experimental activity. Compare the results of your experiments to your hypothesis. Was your initial route indeed the shortest?

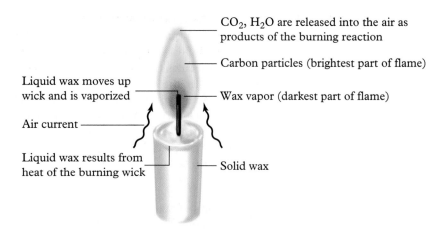

Figure 2.25 Summary of processes in a burning candle.

Summarizing the Main Ideas

Physical properties are those such as color, melting point, or density that do not involve the transformation of one substance to another and that are inherent characteristics of a substance. Physical changes are those in which the nature of a substance does not change. Common examples of physical changes are the changes of state that a substance undergoes as it is transformed from a gas to a liquid or a solid and so on.

Liquids commonly evaporate over time, and receptors in our nasal passages detect certain evaporated molecules as having an odor. Evaporation of a liquid in a closed container ultimately yields an equilibrium when the vaporized molecules return to the liquid at the same rate that other molecules evaporate from the liquid. When the temperature of the liquid is raised so that the pressure exerted by the evaporated molecules is equal to the pressure exerted by the atmosphere, the liquid boils.

In chemical change, an element or compound is converted into a different pure substance. According to the law of conservation of mass, the total mass of chemicals does not change during any chemical or physical process. Combustion of substances by burning them in fires is a common example of a chemical change. Reactions involving O_2 as a reactant are called oxidations.

In general, processes of chemical change are called chemical reactions. The initial substances are the reactants and the final ones obtained are called the products. Chemists write reaction equations to symbolize the transformations that occur in a reaction. In a balanced equation, the number of atoms of each type is the same in the reactants and in the products. This is achieved by placing coefficients in front of the formulas for the various pure substances that are involved.

Complete combustion of a carbon-containing substance in air produces carbon dioxide and water as products. If some of the carbon is converted instead to other carbon-containing substances such as carbon monoxide and/or soot, the process is called incomplete combustion.

Carbon monoxide can be very damaging to health and can pollute the air both indoors and outdoors. The concentrations of air pollutants such as CO are often given as parts per million, the number of pollutant molecules that occur in every million molecules of air.

In the scientific method, scientists gather observations and then form a hypothesis that is a tentative explanation of them. The hypothesis is then tested by performing experiments. The experiments will perhaps cause the hypothesis to be modified or rejected.

Key Terms

physical property	boiling	flammable	incomplete combustion
density	chemical property	nonflammable	by-product
physical change	chemical change	oxidation	concentration scale
change of state	chemical reaction	chemical reaction	parts per million scale
sublime	reactant	equation	catalyst
equilibrium	product	coefficient	observation
vapor pressure	combustion	balanced equation	hypothesis
volatile	hydrocarbon	complete combustion	

Web Sites of Interest

To link to Web sites of interest, go to www.whfreeman.com/ciyl2e, Chapter 2, and select the site you want.

Review Questions

1. Define the terms *physical properties* and *physical changes*.

2. Identify three physical changes you have encountered.

3. Define *density*.

4. On separate lines, write the three common states (phases) of matter. Name the two transitions that occur between each combination of them.

5. What is *dry ice*? Why do "clouds" form above dry ice?

6. Compare the formation and composition of freezing rain, sleet, and hail.

7. Explain how the process of dissolving a solid differs from that of melting it.

8. Explain, on a molecular level, how we smell substances.

9. What is meant by the terms *equilibrium* and *volatile*?

10. Define *vapor pressure*.

11. Compare the events, on a molecular level, of boiling to those of evaporation. How are they the same? How are they different?

12. Why does the boiling point of water decrease when elevation increases?

13. Define the terms *chemical properties* and *chemical change*.

14. Identify three chemical changes you have encountered.

15. In a chemical reaction, what is meant by the terms *reactants* and *products*?

16. What is the *law of conservation of mass*?

17. What is *combustion*?

18. What is meant by the term *oxidation*?

19. What is meant by a *balanced* chemical reaction equation?

20. What is the significance of the coefficients placed in front of the molecular formulas in a chemical equation?

21. Explain the meaning of each of the following:
    a) 3 O_2 b) 2 $C_6H_{12}O_6$ c) H_2O d) Ar

22. What is the difference between *complete* and *incomplete combustion*? What are the characteristic carbon-containing reaction products in the two cases?

23. What is *soot*?

24. What is meant by the term *ppm*?

25. What is meant by the statement that the CO concentration in a room is 1 ppm?

26. What does the term *concentration* describe?

27. How does carbon monoxide affect oxygen transport? Why is it dangerous to health?

28. What does a *catalytic converter* do?

29. What is a *catalyst*?

30. How is *tarnish* produced?

31. How is rust produced?

32. What substances are produced when ammonium nitrate, NH_4NO_3, decomposes?

33. What substances are produced when sodium azide, NaN_3, decomposes?

34. Outline the steps of the scientific method.

35. What substances are produced when wax undergoes complete combustion? What substance produces the light that emanates from the flame? What substance can be extracted from the dark portion of the flame?

Understanding Concepts

36. Using your past observations and logical reasoning, classify each of the following processes as a physical or a chemical change. Specify the logic that you used in each case.
    a) the melting of an ice cube in a glass of water
    b) the rusting of a piece of steel in a car body
    c) the ripening of fruit (a process in which it becomes sweeter)
    d) the evaporation of water from a bowl into air over time
    e) the souring of milk

37. Many substances that consist of extended networks, such as gold, do not have scents detectable to humans. Explain why this might be.

38. Explain how the vapor pressure of a liquid is related to its boiling point. Provide an example to support your answer.

39. Which do you think has a higher vapor pressure, rubbing alcohol or water? Explain your answer.

40. How do you know if a liquid is volatile?

41. Is gasoline a volatile liquid? Explain.

42. Why is the vapor pressure of a substance dependent upon the temperature?

43. Is combustion an oxidation reaction? Explain.

44. Write the chemical reaction equation for the combustion of carbon in oxygen to produce carbon dioxide. What does the arrow in the equation imply?

45. Balance each of the following chemical reaction equations:
    a) $SO_2 + O_2 \rightarrow SO_3$
    b) $C_2H_6 + O_2 \rightarrow CO + H_2O$
    c) $C_2H_4 + O_2 \rightarrow CO_2 + H_2O$
    d) $CO_2 + H_2O \rightarrow O_2 + C_{12}H_{22}O_{11}$

46. Write the chemical equations for the reactions described below:
    a) One molecule of diatomic oxygen gas reacts with two molecules of diatomic hydrogen gas to produce two molecules of liquid water.
    b) One atom of elemental carbon reacts with one molecule of gaseous water to produce one molecule of carbon monoxide gas and one molecule of diatomic hydrogen gas.
    c) One molecule of glucose, $C_6H_{12}O_6$, produces two molecules of ethanol, C_2H_5OH, and two molecules of carbon dioxide, CO_2.

47. What chemical reaction generates energy in the body? Write the balanced chemical equation for the reaction. What type of reaction is this? What is the fuel in the reaction?

48. Describe, in general terms, how catalytic converters change carbon monoxide, CO, into a harmless substance.

49. Describe, using balanced chemical equations, how hydrogen sulfide can tarnish silver and how the tarnish can be converted back to silver.

50. If the concentration of CO in a sample of polluted air is reported as 200 ppm, how many molecules of CO are present in 100,000 molecules of air?

51. How are the processes of breathing, burning, and rusting related?

52. Compare the products produced when solid sulfur is heated to 113°C and to 159°C. How are they different? How are they the same? What process produces the sulfur product at 113°C? What process produces the sulfur product at 159°C?

Synthesizing Ideas

53. Devise a way to determine the relative densities of two gold rings.

54. Helium gas is less dense ("lighter") than air. Xenon gas is more dense ("heavier") than air. If you filled two balloons, one with each gas, would both rise in air? Would one fall?

55. Consider how a liquid boils and the relationship between its vapor pressure and the atmosphere. Use your understanding of these concepts to explain how you think the process of water boiling is different at higher altitudes. *Hint*: Think about the difference in atmospheric pressure at higher and lower altitudes.

56. Carbon monoxide molecules bind to hemoglobin about 200 times more strongly than do O_2 molecules. Calculate the concentration, in parts per million, of CO in air that would lead to half the hemoglobin in your body binding to CO and half to O_2, given that the concentration of O_2 in air is about 200,000 ppm. Does your answer correspond to a fatal level of CO according to the data in Table 2.1?

57. The concentration of water vapor in humid, hot air is about 5%. Convert this concentration into parts per million.

58. One of the warning signs of possible CO contamination of a house is a buildup of black soot on various surfaces. Why do you think the presence of soot should be taken as a sign of possible CO contamination?

59. Examine activities in which you engage every day. Do you use the scientific method in any way?

60. A familiar example of a somewhat complicated series of physical changes is the popping of popcorn. Make a list of what you have observed through the years about this phenomenon. Then formulate a hypothesis concerning what happens inside a kernel of corn that causes it to pop.

61. The process of sweating cools the body by removing heat. Use the information in this chapter, your own experience, and the fact that sweat is mostly water to explain how sweating works to cool the body.

62. The superior scenting abilities of dogs have been used in many ways. Use the Internet to answer the following questions: How are dogs used by police, fire, and rescue agencies? Why are bloodhounds so good at tracking scent trails? What concentration, in ppm, of scent molecules is a bloodhound able to detect? How does this compare to moths and to humans? What other animals have been used to detect scents? For what purposes?

■ Group Activity: The Speed of Chemical Reactions
Adapted from an experiment provided by Professor Bob Perkins, Kwantlen University College

Alka-Seltzer tablets contain three ingredients, two of which react vigorously with each other when they come into contact with water. The fizz that the reaction produces is caused by bubbles of carbon dioxide gas, which quickly rise and escape from the surface of the water. In this activity, you will observe how the temperature of the water affects the speed of the reaction. Assign one member of your group to be the timekeeper while the other members watch carefully to see when the reaction is over. Note: If you have a thermometer, take an exact reading of the temperature of the water you use in each experiment before the addition of the Alka-Seltzer tablet and after the reaction is complete.

Directions:

1. Fill a cup about three-quarters full with cold tap water (approximately 70°F/20°C).

   a. Have one group member hold the cup in their hand and then describe the temperature of the cup (for example, cool, warm, cold, etc.).

   b. Drop an Alka-Seltzer tablet into the water and determine the number of seconds it takes for the solid tablet to completely disappear. (Note: Ignore any nonreacting solid that forms on the bottom of the cup.)

   c. Have the same group member who held the cup before hold it again and record their description of the temperature of the cup.

2. Repeat part b. of the experiment using hot tap water (approximately 130°F/ 55°C) and a new tablet.

3. Repeat part b. of the experiment using ice-cold water (approximately 32°F/0°C).

Results

1. Based upon your results, do you conclude that the reaction speed increases with temperature, decreases, or stays the same?

2. In the first experiment, did the temperature of the cup differ before and after the Alka-Seltzer tablet was added? If so, how did the temperature differ and what is a possible explanation for this difference?

3. The reaction speed is the inverse of the number of seconds recorded. What is the ratio of the reaction speed with hot water to that with cold water?

4. Using a piece of graph paper or a computer, plot the reaction speed against the change in temperature using your recorded temperatures. If you do not have a thermometer, plot the speed of the reaction using 32°F/0°C, 70°F/ 20°C, and 130°F/55°C as approximations of your actual temperatures.

Additional Experiment: Using a cup of cold tap water, add another tablet of Alka-Seltzer. When the reaction has been going for a few seconds, place the flaming tip of a lighted match about 1 centimeter (half an inch) above the water surface. From your observations, decide whether or not carbon dioxide can support combustion—that is, whether flames can occur in an atmosphere of carbon dioxide gas.

Taking It Further with Math I

Density and Dimensional Analysis

As discussed earlier in the chapter, the density of a sample of matter is defined as its mass divided by its volume. This relationship can be stated mathematically by symbolizing density as d, mass as m, and volume as V:

$$d = \frac{m}{V}$$

The formula for density is useful in problem solving. For example, knowing that the density of mercury is 13.5 grams per cubic centimeter allows us to determine the volume for any amount of mercury. To do this, we first isolate the variable V by multiplying both sides of the above equation by V and dividing both sides by d. We obtain the result:

$$V = \frac{m}{d}$$

Thus, if the mass of a sample of mercury is 100 grams, its volume is:

$$V = \frac{100 \text{ g}}{13.5 \text{ g/cm}^3} = 7.41 \text{ cm}^3$$

Similarly, the mass can be deduced for any specified volume of mercury. For example, to determine the mass of mercury that will fit into a container of volume 25.0 cm³, we first isolate the variable m by multiplying both sides of the equation by d:

$$dV = m$$

By substituting the values for d and V, we obtain the mass:

$$m = (13.5 \text{ g/cm}^3) \times 25.0 \text{ cm}^3 = 338 \text{ g}$$

We can use the formula for density to illustrate the problem-solving technique called *dimensional analysis* or the *conversion factor method*. The center of attention in this technique is the *units* for the property of interest. To see how these techniques work, consider the density of mercury: 13.5 g/cm³. This can be expressed both by the statement that

13.5 g of mercury occupies a volume of 1 cm³

and by the alternative statement that

a volume of 1 cm³ of mercury has a mass of 13.5 g

These statements can be expressed as ratios, called **conversion factors,** which are valid for any sample of mercury:

$$\frac{13.5 \text{ g}}{1 \text{ cm}^3} \qquad \frac{1 \text{ cm}^3}{13.5 \text{ g}}$$

The use of the conversion factor ratios in problem-solving can be illustrated by redoing the problems we did above. First, to determine the volume associated with 100 g of mercury, we note that the units of the answer should be in cm³, whereas the units of the starting information are in g. To convert the information from units of g to cm³, we need to multiply by a conversion factor that has g in its denominator. Of the two possible conversion factors, one has units of g/cm³ and the other cm³/g. We use the factor that has units of cm³/g so that the g is cancelled and the unit that remains is cm³. Thus, we convert mass to volume by multiplying 100 g by this conversion factor ratio:

$$100 \text{ g} \quad \times \quad \frac{1 \text{ cm}^3}{13.5 \text{ g}} \quad = 7.41 \text{ cm}^3$$

Similarly, to convert a volume of 25.0 cm³ of mercury to mass, we want to change units from cm³ to g, so we multiply by the conversion-factor ratio that has g in the numerator and cm³ in the denominator:

$$25.0 \text{ cm}^3 \quad \times \quad \frac{13.5 \text{ g}}{1 \text{ cm}^3} \quad = 338 \text{ g}$$

The dimensional analysis technique also allows us to easily convert units by using conversion factors. For example, knowing that 1 kilogram is 1000 grams, we can write the two conversion-factor ratios as

$$\frac{1 \text{ kg}}{1000 \text{ g}} \qquad \frac{1000 \text{ g}}{1 \text{ kg}}$$

To change, for example, 240 grams to kilograms, we multiply by the factor that has kg as numerator (since this is the desired unit) and g as denominator:

$$240 \text{ g} \quad \times \quad \frac{1 \text{ kg}}{1000 \text{ g}} \quad = 0.240 \text{ kg}$$

Exercise 2.8

The element lead has a density of 11.3 g/cm³. Determine a) the volume occupied by 100 grams of lead, and b) the mass of lead in a sample of it that has a volume of one liter, that is, 1000 cm³.

Exercise 2.9

A cup of water has a volume of about 250 cm³. Given that the densities of liquid water and of ice are 1.00 g/cm³ and 0.92 g/cm³, respectively, determine the volume occupied by a cup of liquid water when it freezes to ice.

Taking It Further with Math II

Calculating Masses of Substances in Chemical Reactions

The coefficients for the substances in the balanced equation for a reaction can be used to determine the mass of one substance that reacts with, or is produced by, a specified mass of another substance in the reaction. We can do these calculations using the atomic masses shown for each element in the table of elements inside the back cover.

To calculate the masses involved in a given chemical reaction, first calculate the **molar mass** for each substance that is of interest. (The concept of the mole is discussed in Chapter 11.) The molar mass of a substance equals the sum of the atomic masses for the individual atoms times the subscript for that element in the substance, and has units of grams. Thus, the molar mass for butane, C_4H_{10}, equals four times the atomic mass of carbon plus ten times the atomic mass of hydrogen. We have molar mass of:

$$C_4H_{10} = (4 \times \text{atomic mass of C}) + (10 \times \text{atomic mass of H})$$
$$= (4 \times 12.01 \text{ g}) + (10 \times 1.0079 \text{ g})$$
$$= 58.12 \text{ g}$$

Exercise 2.10

Show that the molar masses for O_2, CO_2, and H_2O are 32.00 g, 44.01 g, and 18.02 g respectively.

Now consider the balanced equation, discussed earlier, for the combustion reaction between butane and oxygen that produces carbon dioxide and water:

$$2 \text{ C}_4\text{H}_{10} \quad + \quad 13 \text{ O}_2 \quad \rightarrow \quad 8 \text{ CO}_2 \quad + \quad 10 \text{ H}_2\text{O}$$

The coefficients in the equation specify the number of molar masses of each substance in the process. In other words, 2 molar masses of C_4H_{10} combine with 13 molar masses of O_2 to produce 8 molar masses of CO_2 and 10 molar masses of H_2O. Let's write the molar masses obtained above and coefficients for each substance under the formulas in the balanced equation:

$2 \text{ C}_4\text{H}_{10}$	$+$	13 O_2	$\rightarrow$	8 CO_2	$+$	$10 \text{ H}_2\text{O}$
2×58.12 g		13×32.00 g		8×44.01 g		10×18.02 g
116.24 g		416.00 g		352.08 g		180.20 g

Thus, we know that 116.24 grams of butane is the amount that will combine with 416.00 grams of oxygen, and together they will produce 352.08 grams of carbon dioxide and 180.20 grams of water. Notice that the total mass of reactants, 532.2 grams, is exactly equal to that of the products.

The relative amount of reactants and products consumed and produced in an actual chemical reaction can be determined by using ratios obtained from balanced chemical equations and atomic masses. For example, if we wish to know the amount of oxygen that is required to react completely with 5.00 grams of butane, we first note that the ratio of butane mass to the value listed under the equation is

$$\frac{5.00 \text{ g}}{116.24 \text{ g}} = 0.0430$$

The amount of oxygen that will react with the 5.00 grams of butane has the same ratio, 0.0430, to the value for O_2 listed in the reaction equation, 416.00 g. Using this ratio, we find that the mass of oxygen required in this case is

$$0.0430 \times 416.00 \text{ g} = 17.89 \text{ g}$$

Similarly, the amount of carbon dioxide produced from the complete reaction of 5.00 g of butane is $0.0430 \times 352.08 \text{ g} = 15.14$ grams.

Exercise 2.11

Using the method described above, calculate the amount of butane that will react with 10.00 grams of oxygen. What mass of water is produced?

Notice that the amounts of reactants and products depend on the coefficients in the balanced reaction equation. Consider the reaction of butane with oxygen to produce water and carbon monoxide (rather than carbon dioxide) that was discussed in Section 2.12. The balanced equation is

$$2 \text{ C}_4\text{H}_{10} + 9 \text{ O}_2 \rightarrow 8 \text{ CO} + 10 \text{ H}_2\text{O}$$

For this reaction, the mass of 116.24 g, corresponding to two molar masses of butane, reacts with nine molar masses of O_2 rather than 13 molar masses when the product was CO_2 rather than CO.

Exercise 2.12

For the reaction above, calculate the mass of carbon monoxide produced by the combustion of two molar masses of butane. How much CO would be produced by the combustion of 1.00 gram of butane?

Exercise 2.13

Using the balanced reaction you obtained in the text Exercise 2.3b, calculate the mass of ethane, C_2H_6, that would have to be burned to produce 100 grams of carbon dioxide. How much water is also produced?

Exercise 2.14

As discussed in the text, the rusting of iron in air corresponds to the balanced reaction

$$4 \text{ Fe} + 3 \text{ O}_2 \rightarrow 2 \text{ Fe}_2\text{O}_3$$

Calculate the mass of oxygen required to react with 1.00 gram of iron, and the mass of Fe_2O_3 that is produced. Is the amount of Fe_2O_3 you calculate equal to the sum of the masses of iron and oxygen that reacted? Why or why not?

In this chapter you will learn:

- about atoms and their electrons, including how neon signs work;
- why hydrogen has a bad reputation for exploding and how hydrogen peroxide can be used to bleach your hair;
- about a type of atomic interaction called covalent chemical bonding;
- about ions and ionic bonding, including why we need sodium, potassium, calcium, zinc, and iron in our diets;
- how deicers melt ice.

When you look at the bright colors of the neon lights on this Las Vegas strip of casinos, do you know what makes them glow?

The famous bright neon lights of Las Vegas, New York City, and Hong Kong are all produced when energy interacts with atoms. An understanding of how energy reacts with specific types of atoms allows manufacturers of neon lights to design signs that glow with a desired color or brightness. (Corbis/Punchstock)

An Insider's Perspective
The Internal Workings of Atoms and Molecules

When you use any of the myriad consumer products in our modern world, whether they be plastics, food, clothing, gasoline, medications, or many other substances, you probably don't think about their chemical nature. In fact, almost all of the articles you deal with every day are molecular in form. During the last two centuries, chemists discovered that atoms obey simple rules of combination when they form molecules. Knowledge of these principles has led to the creation of many new molecular compounds with useful properties and consequently to many consumer products that we take for granted. In this chapter we will learn the principles of combination that atoms follow when they form molecules and other networks. In order to understand the origin of the rules, we shall first explore the internal workings of atoms and see how they determine chemical behavior.

The Components of Atoms

3.1 Atoms contain electrons and a nucleus

In Chapter 1, we implied that atoms are indestructible. In actual fact, atoms *can* be decomposed into simpler components but only by the application of large amounts of energy. What results in this case is a gaseous mixture called a **plasma** which consists of electrically charged particles. Let's take the element argon (symbol Ar) as an example. When a very large amount of energy is applied to the atoms in a gaseous sample of argon, the result is a plasma made up of a mixture of two types of particles, as illustrated in Figure 3.1.

- The most numerous particles are **electrons,** each of which has a very small mass, about 0.001% of the whole argon atom. Each electron carries a *negative* electrical charge. The size of an electron's charge is the same regardless of the atom from which it is obtained. Since in chemistry all other charges are found to be a multiple of that of the electron, its charge is taken to be a fundamental quantity. By convention, the numerical magnitude of the charge is given the symbol e. Each electron's charge is written as $-1e$ (or $-e$) since it has that amount of negative charge. Eighteen electrons are associated with each argon atom, so the total negative electrical charge of each argon atom is $-18e$.

- The other type of particle in an atom is called the **nucleus.** One nucleus is associated with each argon atom—in fact, with any atom. The nucleus is much, much heavier than all the electrons collectively. The argon nucleus,

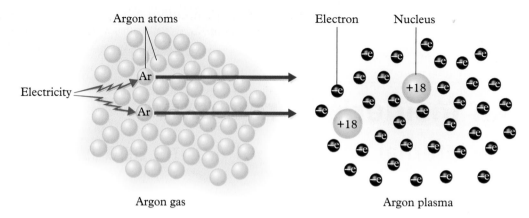

Figure 3.1 Decomposition of atoms of argon into a plasma, showing the positively charged nuclei and the negatively charged electrons. The sizes of the particles are not to scale.

Plasmas created from neon and xenon gases are now used in some TV picture tubes. When an electrical current is sent to one of the hundreds of thousands of tiny cells in the picture tube, a plasma is instantaneously created. The collision of one of the plasma's electrons with a positively charged particle stimulates the release of ultraviolet light, which then reacts with the phosphor material coated on the cell to produce visible light (either red, blue, or green).

for example, possesses a mass that is 99.98% of that of the entire atom. Each nucleus of an argon atom carries a positive electrical charge of +18e, which is exactly equal and opposite to that of the 18 electrons that the argon atom contained (see Figure 3.1).

If a similar experiment were performed on the atoms of gold in a gaseous sample of the pure metal, the result would be qualitatively similar. A large number of electrons—each identical to those found in argon—would be found in the plasma for each gold atom that was decomposed, and only one nucleus per atom would be obtained. For gold, the nucleus is even more massive than for argon, and the positive charge is even greater than that of argon. But again, for each gold atom nucleus, the number of electrons per atom—in this case, 79—is equal to the charge on the nucleus. For gold atoms, the charge carried by the nucleus is 79 times the charge of each electron.

The findings are qualitatively similar for atoms of every element. Therefore, we conclude that *each atom of any element consists of one or more electrons and a nucleus. The nucleus possesses almost all the mass of the atom and carries an amount of positive charge that is equal to the total charge of the negatively charged electrons.* Since the positive and negative charges exactly balance, *each atom is electrically neutral overall.* Every atom of a given element has the same number of electrons and possesses a nucleus with the same quantity of electrical charge, which is a characteristic of that element.

3.2 The atomic number identifies elements

The magnitude of the positive charge of the nucleus, expressed as a multiple of the size of the electron's charge, e, is called the **atomic**

number of the element. *The atomic number is the fundamental quantity that distinguishes atoms of one element from those of another.* Thus all nuclei of hydrogen atoms have a +1e charge; consequently their atomic number is 1. No other atom's nuclei have a charge of +1e. The atomic number of helium is 2 since all helium atom nuclei, and only the nuclei of atoms that are helium, have a +2e charge.

All the elements having atomic numbers ranging from 1 to 118 are known. The **periodic table** of the elements, shown in Figure 3.2 (and repeated on the inside front cover of this book), is an arrangement of all these known elements, in order of their atomic numbers. (In places, the sequencing of elements may look a little odd, but it has a rationale that we will explain later.) For example, carbon is the element with the atomic number 6, chlorine is 17, uranium is 92, and so on. We will be looking at the periodic table in more detail as we go along.

None of the elements with atomic numbers greater than 92 occur naturally. These elements have only been produced synthetically, as is

The plural of the word *nucleus* is *nuclei*.

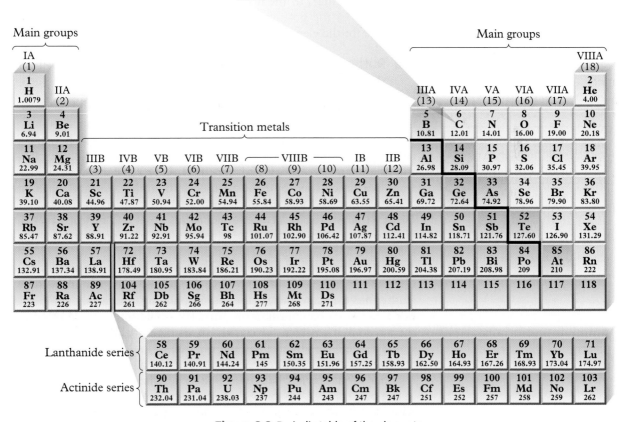

Figure 3.2 Periodic table of the elements.

also the case for elements with the atomic numbers of 43 and 91. The most famous synthetic element is plutonium, atomic number 94, which is generated in nuclear reactors and which is used in some atomic bombs.

Worked Example: Deducing the charges in an atom of an element

Locate the element oxygen in the periodic table. From the information shown for it, determine **a)** the magnitude of the positive charge on its nucleus and **b)** the number of electrons that one atom of oxygen contains.

Solution: Given the name oxygen, we expect that the element's symbol should be O or Ox. Indeed, we find an element called O as the eighth element in the periodic table, with an atomic number of 8. **a)** Because, by definition, the atomic number equals the magnitude of the nuclear charge, the positive charge on the nucleus of each O atom is $+8e$. **b)** Because the number of electrons is equal to the magnitude, in e units, of the nuclear charge, the number of electrons must be equal to 8.

Exercise 3.1

Locate the element boron in the periodic table. From the information shown, determine **a)** the number of electrons that one atom of it contains and **b)** the positive charge on each boron nucleus. Repeat the exercise for the element sulfur.

3.3 Rutherford's experiments led to the nuclear model of the atom

The modern conceptual model of an atom is that of a sphere consisting mostly of empty space. At the center of the sphere lies a nucleus that is tiny even by atomic standards—its diameter is approximately of 10^{-15} meters, or 0.000001 nanometers. The electrons travel constantly at extremely high speeds around the nucleus. This picture was deduced from experiments led by Ernest Rutherford, a New Zealand–born scientist who worked in the early part of the 20th century in Canada and Great Britain. Prior to his experiments, scientists had believed that the positive charge was evenly distributed through an atom, with electrons suspended in it like raisins in a pudding.

In Rutherford's experiments small, positively charged particles were directed at a piece of gold foil so thin that it was only a few atoms thick. Rutherford expected that, if the "raisin pudding" model shown in Figure 3.3a was correct, most of the particles should travel straight through the material since the positive charge is evenly dispersed. A few particles would be deflected by a slight amount due to the strong repulsion that occurs between charges of the same sign when they move very close together. What actually happened was that, although most particles did in fact pass through the foil with their direction unchanged or only modified slightly, about 1 in every 20,000 of them was deflected by huge angles. In essence, these exceptional particles bounced back in the direc-

The principles of naming discussed in Chapter 1, section 1.3, will help you to locate specific elements in the periodic table. Also see the table titled "The Elements" at the back of the book.

You may want to review Appendix A on scientific notation to remind yourself of just how small 10^{-15} is.

tion from which they were fired, as shown in Figure 3.3b. As Rutherford later stated, "It was almost as incredible as if you had fired a 15-inch shell at a piece of tissue paper and it had come back and hit you."

The only explanation for Rutherford's results is a model of the atom in which the positive charge, rather than being distributed evenly over the whole atom, is *concentrated* in a tiny volume of what is mainly empty space sparsely populated by even tinier electrons. Most of the particles fired by Rutherford into the foil would pass through the empty space undeflected, or they would be slightly deflected if they came close to the nucleus. The occasional particle that was aimed right at the nucleus would experience a huge repulsive force when it almost "hit" the nucleus, and would bounce back.

As a consequence of these experiments, scientists changed their model of the atom from one in which the positive charge was evenly distributed throughout to one in which it was highly concentrated in a tiny central region. Rutherford named this central region the nucleus, from the Latin word for kernel.

The nuclear model of the atom—with negatively charged electrons in constant motion in what is mainly empty space around a positively charged nucleus—yields a simple answer to the question of what holds

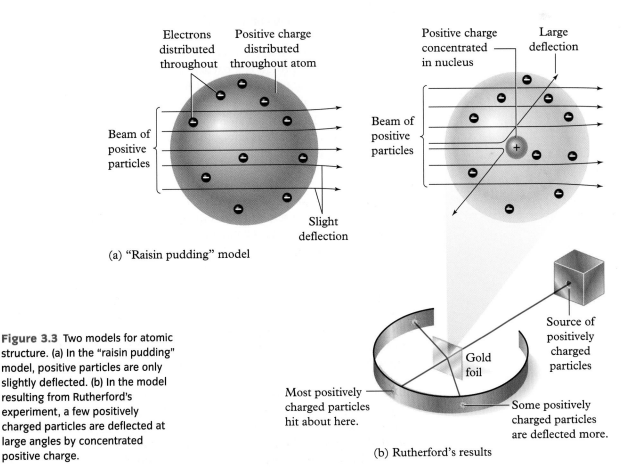

Figure 3.3 Two models for atomic structure. (a) In the "raisin pudding" model, positive particles are only slightly deflected. (b) In the model resulting from Rutherford's experiment, a few positively charged particles are deflected at large angles by concentrated positive charge.

(a) "Raisin pudding" model

(b) Rutherford's results

The Components of Atoms

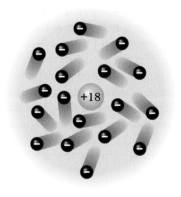

Figure 3.4 Distribution of charged particles in an argon atom.

the atom together. The electrons are held within the volume of the atom by **electrostatic forces** of attraction—the attraction between opposite electrical charges—in this case between the negatively charged electrons and the positively charged nucleus. Thus each electron, of charge $-1e$, is attracted to the oppositely charged nucleus.

Opposing these attractions are the forces of repulsion that occur within each atom between each electron and every other electron, since these particles have charges of the same sign. Just as opposite charges attract each other, charges of the same sign repel each other. Furthermore, the electrons must be in constant motion. The energy associated with this motion of the electrons provides a counterbalance to the force of attraction to the nucleus and keeps the electrons from falling into it, just as the motion of planets in our solar system keeps them from falling into the Sun.

The distribution of charged particles in an argon atom is illustrated in Figure 3.4. Notice that for convenience in this and future diagrams, we show the charges in units of e, so the charge on the nucleus is shown as $+18$ and that for each electron is -1, or simply $-$.

3.4 The nucleus contains protons and neutrons

Subsequent experiments on the nature of the atom showed the following:

- The positive charge on the nucleus results from the presence of **protons.** A proton is a particle that carries an electrical charge equal to that of the electron; the proton's charge, however, is positive rather than negative.

- The nucleus of most atoms also contains **neutrons.** A neutron is a particle that has about the same mass as that of a proton, but no electrical charge.

Since protons account for all the positive charge of the nucleus, *the number of protons present must equal the atomic number.* Thus, since carbon is the element of atomic number 6, its nuclei must each contain 6 protons, each of charge $+1e$, giving a total charge of $+6e$.

Neutrons do not influence the chemical properties of an atom. Indeed, different atoms associated with the same element often have different numbers of neutrons. Consequently, the nuclei of atoms of an element can have different masses. *The **mass number** of a nucleus equals the sum of the number of protons and neutrons it contains.* Nuclei of a given element with different mass numbers, due to the varying numbers of neutrons, are called **isotopes.** Mass numbers are sometimes displayed as leading superscripts to the atom symbol. For example, most nuclei of hydrogen contain no neutrons and consist entirely of a proton, so the mass number for this isotope is 1, and its *isotopic symbol* is ^{1}H. However, some hydrogen nuclei, about 0.02% of those on Earth, contain one neutron. These nuclei therefore contain one proton and one neutron. The mass number for such nuclei is 2, and their isotopic symbol is ^{2}H. The third, very rare, isotope of hydrogen has nuclei with two

We will look further at isotopes and some of their uses in Chapter 17.

neutrons and one proton, and so has a mass number of 3. Its isotopic symbol is ^{3}H. Hydrogen with mass number of 2 is sometimes called *deuterium,* and that of mass number three is sometimes called *tritium.*

Exercise 3.2

The element phosphorus, symbol P, has an atomic number of 15. For the isotope of phosphorus that has 16 neutrons in its nucleus, determine its mass number and write its atomic symbol to display this value. How many protons does an atom of phosphorus have?

Electron Configurations

Once the nuclear model of the atom had been deduced, scientists wondered what paths the electrons followed in their motions around the nucleus. An early 20th-century model, proposed by the Danish scientist Niels Bohr, envisioned circular or elliptical orbits, much like the orbits in which planets revolve about our Sun. We now know that this model is too simplistic for electrons in atoms. In fact, it is not possible even in principle to know or predict the exact path that an electron follows. Due to the inherent uncertainty in its position and path, all we can state is the *probability* that an electron will be found in a particular region of space around the nucleus.

3.5 Electrons are held to their atoms by varying forces

In any given atom some of the electrons travel on average closer to the nucleus than others do. The consequence is that different groups of electrons are held to the atom by different amounts of force. Although the model of an orbit is not accurate for atomic structure, just as it would require energy to remove a planet from its orbit around the Sun, it requires energy to remove an electron from an atom. If we were to attempt to remove an electron from a particular atom, the required amount of energy would depend on the electron's distance from the nucleus.

The energies that are required to remove different electrons from an atom can be determined by bombarding a gas sample consisting exclusively of atoms of any one element with fast-moving, small particles, all of which have a uniform speed, and hence the same energy. In the experiment, we vary the energy of the particles over a wide range and observe whether or not an electron is dislodged from the atom at each particular energy value.

The experiment reveals that for argon atoms there are three distinct groups of electrons in each atom. The minimum energies required to remove electrons of each of the three groups stand in the ratio of about 1:14:170. In other words, as diagrammed in Figure 3.5, it requires about 170 times as much energy to remove one electron from the group that is held most firmly to the argon atom as it does to remove an electron from the group that is attached by the weakest attractive force. Electrons in the intermediate group require 14 times as much energy as is necessary for those held by the weakest attraction.

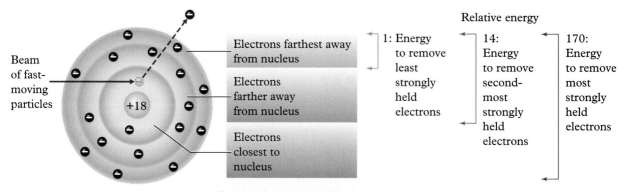

Beam of fast-moving particles

+18

Electrons farthest away from nucleus

Electrons farther away from nucleus

Electrons closest to nucleus

Relative energy

1: Energy to remove least strongly held electrons

14: Energy to remove second-most strongly held electrons

170: Energy to remove most strongly held electrons

Figure 3.5 Removal of electrons from an atom by particles of varying energies. Electrons closest to the nucleus are most strongly held and require the most energy to be removed.

View the first, second, and third electron shells of an atom at Chapter 3: Visualizations: Media Link 1.

Similar evidence for atoms of all the elements leads us to a model known as the **shell structure** for the electrons in atoms. We know that the electrostatic energy of attraction between oppositely charged particles is greater the smaller the distance between them. Therefore, we can conclude that the electrons that require the *greatest* amount of energy to be liberated from the atom must on average lie *closer* to the nucleus than do the other groups of electrons. Since the energy that holds the second-most strongly held electrons is only a fraction of that holding the first, we conclude that the second group must on average be located farther from the nucleus than is the first group. Similarly, the least-strongly-held electrons must lie even farther from the nucleus than do those in the second group. Thus, electrons can be thought of as occupying one of several concentric shells centered on the nucleus. Figure 3.5 shows one way to visualize this structure. Keep in mind that the shells are simply *regions of space* in which the electrons spend the majority of their time during their complicated motions inside the atom; they are not orbits.

We can appreciate shell structure by considering the electron arrangement in atoms having small atomic numbers. Helium, with the atomic number of 2, has two electrons. Experiments indicate that both electrons are held to the atom by the same amount of attractive force, so it follows that they must be present in the same shell. However, in lithium, which contains three electrons since it has the atomic number 3, one electron is found to be much less strongly held to the nucleus than are the other two, indicating that this electron occupies a different shell, more distant from the nucleus. From this and other evidence, we hypothesize that only *two* electrons can be present in the stablest, inner-most electron shell of a lithium atom. In fact, this hypothesis turns out to be valid for atoms of *all* the elements, not just lithium.

What happens when we encounter atoms with even more electrons? Let's consider atoms of the elements neon and sodium as examples. Neon atoms have ten electrons, eight of which are held to the atom by energies that are only about 3% of that required to remove an electron from the first shell. Therefore we conclude that neon has two shells, one

Figure 3.6 The distribution of electrons among shells in lithium and sodium. Note that both have only two electrons in their inner shells.

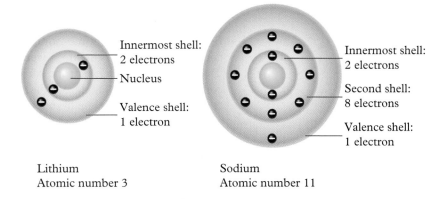

Innermost shell: 2 electrons

Nucleus

Valence shell: 1 electron

Lithium
Atomic number 3

Innermost shell: 2 electrons

Second shell: 8 electrons

Valence shell: 1 electron

Sodium
Atomic number 11

with two electrons and one with eight. An atom of sodium has just one more electron than neon does. However, the energy holding the additional electron to the atom is only about one-tenth as great as that holding the eight electrons in the second shell, so clearly the additional electron must occupy a third shell. Following the reasoning we used to define the first shell, we deduce that the second most stable shell of an atom is capable of containing a *maximum of eight* electrons (see Figure 3.6). This deduction is confirmed by similar findings for other atoms.

Evidence of this type leads to the conclusion that the third shell can contain up to 18 electrons. There is a definite mathematical pattern to this trend of increasing capacity with increasing shell number, n. In particular, it is found that the maximum number of electrons that shell n can hold is $2n^2$. Thus, the fourth shell, corresponding to $n = 4$, can have up to $2 \times (4)^2 = 32$ electrons. The maximum values for the first few shells are listed in Table 3.1 and illustrated in Figure 3.7. Note that although we won't be concerned with n values greater than 5, there is no upper limit to the value that n can possess, since it can be any positive integer.

Table 3.1	Maximum number of electrons per shell in atoms

Shell number	Maximum no. of electrons
1	2
2	8
3	18
4	32
5	50

3.6 Electron configurations of atoms indicate electron arrangements

The allocation of electrons to the various shells in an atom is called its **electron configuration.** The highest shell in a particular atom that has any electrons in it is called its **valence shell,** which we shall

Figure 3.7 The shell structure for the location of electrons in an atom. (a) The innermost shell can hold only two electrons. (b) The second shell can hold a maximum of eight electrons. (c) The third shell can hold a maximum of eighteen electrons. The shells are numbered from the closest to the nucleus outward.

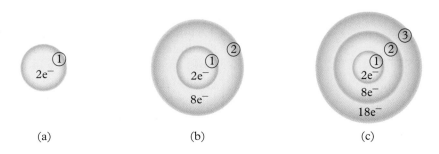

(a) 2e⁻

(b) 2e⁻ 8e⁻

(c) 2e⁻ 8e⁻ 18e⁻

sometimes refer to as the "outermost occupied" shell. The other shells that contain electrons in a given atom are known as its **inner shells.** Every atom potentially has some shells that contain no electrons but could be occupied if the atom were to acquire many extra electrons or if one of its existing electrons were to become repositioned. Electron configuration is important because, as we shall see, the number of electrons in the valence shell determines how atoms react with other atoms, which is ultimately the basis of all chemical behavior. Furthermore, the outermost occupied shell defines the size of the atom, in much the same way that the outermost planet defines the size of a solar system.

The allocation of electrons to different shells in atoms of low atomic number—up to 18—follows the principle that the *shells fill in order of their energetic stability*. In other words, as the number of electrons increases from hydrogen to helium to lithium, and so on, the first and second electrons can be assigned to the first shell, the third through the tenth electrons to the second shell, and so on. Thus, for example, in atoms of the element magnesium, with a total of 12 electrons, there are two electrons in the first shell, eight in the second, and two in the third.

An atom's electron configuration is sometimes summarized in a shorthand notation. For example, we could summarize the electron configuration for magnesium as 2, 8, 2. The electron shells are sometimes represented by the integers 1, 2, 3, . . . with the number of electrons in each shell shown as superscripts, giving 1^2, 2^8, 3^2 for magnesium.

Beyond the first 18 elements, electrons do *not* fill successive shells in strict order. We shall ignore this complication, however, since we don't need to know the detailed electron configurations for the heavier elements that we'll consider in this book.

Worked Example: Deducing electron configurations for atoms

What would be the electron configuration for atoms of the element whose atomic number is **a)** 6 and **b)** 14?

Solution: **a)** If the atomic number is 6, then the atom must have 6 electrons and must be carbon. Since electron shells fill in order for atoms with low atomic numbers, and since the first shell has a maximum of 2 electrons, we conclude that the first shell will have 2 electrons in it. This leaves $6-2 = 4$ electrons to occupy other shells. Since the maximum number of electrons in the second shell is 8, all of these 4 will occupy the second shell. Hence, the full electron configuration for this atom is 2 electrons in the first shell and 4 in the second shell. **b)** In an atom with a total of 14 electrons (silicon), the first shell will have 2 electrons, leaving 12 to distribute. Since the maximum for the second shell is 8 electrons, the remaining 4 must be in the third shell. Hence the electron configuration is 2 in the first, 8 in the second, and 4 in the third shell. The shorthand notations would be 2, 8, 4 or $1^2 2^8 3^4$.

Exercise 3.3

a) What is the electron configuration of an atom whose atomic number is 17?

b) Find the element nitrogen (symbol N) in the periodic table. From its atomic number deduce the total number of electrons that an atom of nitrogen possesses, and also deduce its electron configuration.

3.7 Atoms with eight electrons in their valence shell are unreactive

A special chemical stability occurs when atoms have exactly *eight* electrons in their valence shell. Such atoms tend not to react or combine with other atoms, either of the same element or of other elements. Because of its importance, we often refer to such an arrangement as an **octet of electrons.** For example, argon, which we met at the beginning of this chapter, is the element whose atoms have eight electrons in the third shell (see Figure 3.8a). The element neon also has eight electrons in its valence shell, in this case the second (see Figure 3.8b). Both neon and argon are so stable that they exist in the gas phase as single atoms that do not form diatomic or polyatomic molecules. Note that in the case of argon, the valence (third) shell is *not* filled to its maximum capacity of 18 electrons; in the case of the larger shells, stability is associated with *eight* electrons, not necessarily with a filled shell. Elements such as iron and nickel have more than eight electrons in the third shell, but they are not unreactive as is argon.

All the atoms that have an octet in their valence shell are *noble gases* (sometime called *inert* gases), the name used for elements that exist as single atoms and that resist entering into chemical combination. (Noble here is used in a sense analogous to that employed in the terms noblemen and noblewomen, individuals who prefer not to associate with anyone else!) Argon is the most abundant noble gas on Earth, constituting almost 1% of air. It finds use in ordinary incandescent lightbulbs, where its inertness can be an advantage—its presence can extend the lifetime of the filament that is heated by the electrical current and that emits the light we see. Krypton and xenon are other noble gases that, like argon, have an octet of electrons in the valence shell that is far from being filled. And like argon, they consist of atoms that show no tendency to combine with each other. Much smaller concentrations of the denser noble gases—krypton, xenon, and radon—are also found in air. Although they occur in air and in the pure state as monatomic gases, these three elements do form chemical compounds under certain conditions.

Figure 3.8 An octet of valence electrons. (a) Argon has eight valence electrons in its third shell. (b) Neon has eight valence electrons in its second shell.

(a) Argon

(b) Neon

3.8 Helium is the only noble gas without eight valence electrons

The noble gas helium is the second-most-abundant element in the universe. It is the exception to the rule that noble gas atoms have eight valence shell electrons: the atoms of helium have only two electrons in their valence shell (the first). Nevertheless, helium is completely inert and cannot explode under any circumstances. Helium gas is "light"—by which in this context we mean that its density is much less than that of air—and so it rises in air. Because helium's density is so different from that of air, and the density of the gas surrounding the vocal cords affects the way in which sound is transmitted, our voices sound quite different when we inhale helium and then speak.

Though it exists mainly in stars, helium is also found on Earth and has many important uses in our world, all of which exploit its low density and/or its chemical inertness. For both these reasons, helium is the gas used to fill party balloons and dirigible airships such as the blimps used for airborne advertising. Helium has the lowest melting point of any substance—element or compound—and so it is used when the lowest possible temperatures are needed. Its very low solubility in water and in water-based liquids such as blood is exploited by deep-sea divers, who breathe mixtures of helium and oxygen. If a diver uses regular air, the nitrogen gas in it dissolves in blood under the high pressures of the deep. When the diver returns to the lower pressure at the surface, the gas bubbles out of the blood, causing considerable pain ("the bends"). Because helium does not dissolve in blood, the problem is prevented.

Divers can use regular air to breathe if they ascend slowly, allowing time for the nitrogen bubbles to be exhaled as the blood circulates.

3.9 Noble gases are used commercially for lighting

Now that we know about neon's electron configuration and existence as a gaseous collection of individual, unreactive atoms, we can understand how it is used to emit light in commercial signs. Electrical energy is passed through a tube containing neon gas at low pressure. The neon atoms absorb the energy and use it to transfer—or "excite"—one electron from the (filled) second shell to a less stable, previously empty shell lying much farther from the nucleus. The electron doesn't stay in the far shell for long, however. It cascades down through the intermediate, empty shells back to the second shell, in each transition releasing the extra energy as light (see Figure 3.9a). In this way, electrical energy is transformed into the energy of light by the intermediacy of neon atoms. The light is emitted all along the tube by the atoms, not at the filament as occurs in ordinary lightbulbs.

The predominant color of the light emitted by neon is orange-red. "Neon" lights having other colors use a different gas or colored glass (see Figure 3.9b). In particular, argon, krypton, and xenon are also used in some "neon" lights, since excitation of an electron from their valence shells also results in the emission of visible light of various colors when the electron returns to its proper shell. The predominant color for argon emissions, for example, is blue.

Chapter 3: An Insider's Perspective

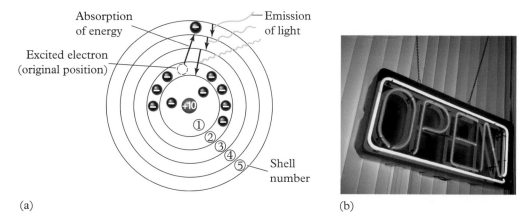

(a)

(b)

Figure 3.9 Neon as a source of lighting. (a) Excitation of valence electrons results in the emission of light. (b) Neon light. (Part b, Digital Vision/Getty Images)

The same principle of converting electricity into light is used in mercury and sodium streetlights, and in mercury-based fluorescent lamps (see Figure 3.10). Here, vaporized metal atoms are used to accept the electrical energy by exciting one of the valence shell electrons to a less stable shell, thereby producing visible light when the electron returns to its original shell.

Light of different characteristic colors is also emitted by electrons in atoms when they are placed in flames and absorb the energy of the flames, become excited, and subsequently lose the excess energy when the electrons return to their normal shells. For example, compounds of strontium produce a red light when they are placed in flames, and sodium compounds produce yellow light. The colors of fireworks are based upon these principles.

Figure 3.10 Mercury vapor streetlights. (Matthias Clamer/Stone)

The Periodic Table: Elements Grouped by Similar Properties

In the original periodic tables, elements were ordered by their relative masses rather than by their nuclear charges. Since the two properties correlate strongly, however, there is little difference in the orderings obtained.

In listing the chemical elements in order of their atomic number, chemists noticed that it was possible to construct a diagram in which elements of similar properties are grouped in columns. For example, in the partial periodic table shown in Figure 3.11 (which shows only what are called the main groups), the noble gas elements He, Ne, Ar, etc., are all found in the same column at the extreme right side. All elements in this last column of the periodic table have the same number of electrons in their valence shell, namely eight, except for helium, which has the maximum number (two) that is allowed for the first shell.

The 19th-century Russian chemist Dimitri Mendeleev was the first to construct the table of elements according to their repeating—or *periodic*—properties in the manner of the modern periodic table. He was so convinced that all elements would behave according to this pattern that he left gaps in the table for undiscovered elements having properties that were consistent with this concept. The gaps have been filled in over the years as more elements were discovered, providing solid support for Mendeleev's theory.

3.10 Elements that behave similarly occur periodically

The connection between the properties of an element and the number of electrons in its valence shell also applies to the other columns of elements, all of which are called **groups.** The so-called main groups are shown in the partial periodic table of Figure 3.11. The atoms of each

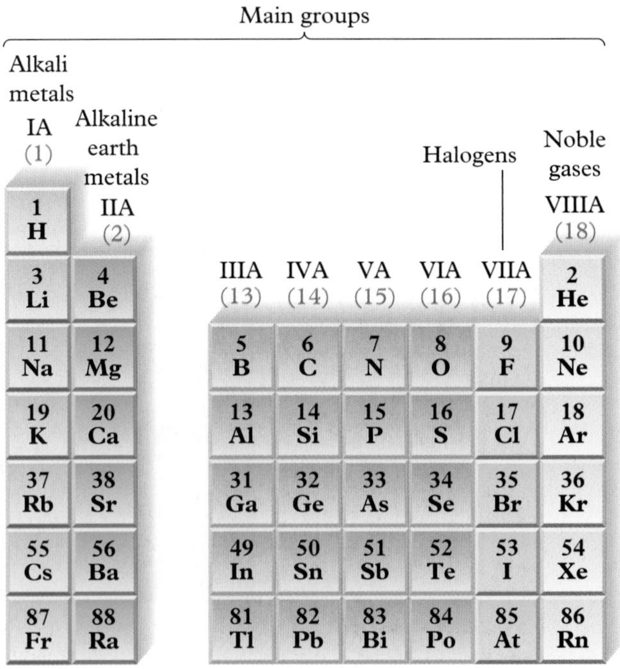

Figure 3.11 A partial periodic table, showing main group elements.

Figure 3.12 Some halogen elements: chlorine, bromine, and iodine. (Chip Clark)

element in the group, labeled VIIA, that lies to the immediate left of the noble gases have an electron configuration that is one electron short of that for the adjacent noble gas. All the elements in Group VII behave similarly, though not identically, in forming chemical substances. For example, each of these elements—which we will refer to generically as X—exists in stablest form as a diatomic molecule of the formula X_2 (for example F_2, Cl_2, . . .), which is rather reactive toward other substances. The elements of Group VII are known as the *halogens*. Some of the halogens are illustrated in Figure 3.12. Notice that similar chemical reactivity doesn't necessarily mean that the elements look alike. Elemental chlorine exists as a yellow-green gas, elemental bromine as a red-brown liquid, and elemental iodine as a blue-black solid.

The elements of the group that lies at the extreme left side of the periodic table, Group I, have atoms that all possess only one electron in their valence shell. These elements (Li, Na, K, . . .), known collectively as the *alkali metals*, are all soft, highly reactive metals. (You saw an example of their softness in Figure 1.24.) To their immediate right in the periodic table lie the *alkaline earths*, Group II, a set of metals whose atoms contain two electrons in the valence shell. The alkaline earth elements are somewhat less reactive than the alkali metals, but the two groups have many similarities (see Figure 3.13). For example, each alkaline earth metal, abbreviated M, forms a compound of formula MX_2 with each of the halogens (X). The formulas of the compounds formed by the alkali metals with halogens are all MX.

The elements that are placed on the same horizontal level of the periodic table are said to constitute a **row,** or period, of it. In general, as we proceed from left to right across a row of the so-called main groups of the periodic table (those shown in Figure 3.11), the number of electrons in the valence shell of the atoms increases by one in each transition to a new group. A new group numbering scheme, illustrated in blue on the partial periodic table in Figure 3.11, designates the groups by the

(a)

(b)

(c)

Figure 3.13 Some elements of Group **II**, the alkaline earths. (a) Magnesium. (b) Calcium. (c) Strontium. (Parts a, b, and c, Ken Karp for W. H. Freeman and Company. From L. L. Jones and P. W. Atkins, *Chemistry,* 4th ed. © 2000 by L. L. Jones and P. W. Atkins)

The Periodic Table: Elements Grouped by Similar Properties

Arabic numerals 1–18. In the new scheme, if the group number exceeds 10, simply subtract 10 from it (that is, ignore the first digit) to obtain the number of valence shell electrons.

Using the generalizations above, we can conclude that atoms of the elements in

- Group III or 13 (B, Al, . . .) have three valence shell electrons
- Group IV or 14 (C, Si, . . .) have four valence shell electrons
- Group V or 15 (N, P, . . .) have five valence shell electrons
- Group VI or 16 (O, S, . . .) have six valence shell electrons

And as we already have seen, the halogens have seven valence shell electrons and the noble gases (Group VIII) from neon onward have eight. Thus the periodic table provides a handy guide to valence electron shell configurations.

The position of the element hydrogen in the periodic table is somewhat ambiguous. Given that it has but one electron in its valence shell, the case could be made that it belongs at the top of the alkali metal group, where it is shown in most versions of the periodic table. However, it is also only one electron short of a full shell and of a noble gas electron (helium) configuration, so it could be argued that it belongs at the top of the halogen group.

> Recall from Chapter 1 that under extremely high pressure, hydrogen is indeed a metal.

Worked Example: Deducing the number of valence shell electrons

By inspecting the periodic table, deduce the number of electrons in the valence shell of atoms of arsenic (As) without determining its entire electron configuration.

Solution: Arsenic is found below P in Group V (15) of the periodic table. Thus it has five electrons in its valence shell.

Exercise 3.4

By inspecting the periodic table and without working out the full electron configurations, deduce the number of electrons in the valence shells of atoms of the elements **a)** calcium (Ca) and **b)** selenium (Se).

The full periodic table of the elements (see Figure 3.2) has a rather complicated structure. This feature arises because, as we noted earlier, starting with the nineteenth element, the shells do not fill strictly in order. Rather, before a given shell is complete with its maximum number of electrons, a few electrons enter the next higher shell, after which the lower shell again resumes filling. The gradual completion of these inner shells gives rise to the large number of elements known as the transition, lanthanide, and actinide metals in the full periodic table. The relationship between the properties of these metals and their electron configurations is complicated, and in this book, we shall not need to worry about being able to deduce their configurations. (The numbers listed under each elemental symbol in the periodic table in Figure 3.2 and given at the front of this book show the relative masses of the atoms, which we will discuss later.)

3.11 Chemical properties are determined by valence shell electrons

On the basis of the evidence we have discussed, and much more supporting data, chemists have concluded that *the chemical properties of an element are primarily determined by the number of electrons that its atoms possess in their valence shell.* Consequently, *elements having atoms with the same number of valence shell electrons behave in a similar fashion.* Of course, no two elements behave *exactly* alike in all respects, but the similarity between two elements in the same group is greater than, say, two elements in the same row even though the latter have almost the same atomic number and almost the same total number of electrons. Because the number of electrons in the valence shell is a property that varies periodically, so too do the chemical properties of elements. It is for this reason that our table of elements is called a *periodic* table.

Covalent Bonding and the Formation of Molecules

As we have seen, the configuration of an octet of electrons in the valence shell is apparently the basis for the stability of the noble gases beyond helium. Because the nonmetal elements (located to the right of the zigzag black line in the periodic table, Figure 3.2) lack this configuration, they do not exist as simple monatomic gases as do neon, argon, and so on. Instead, atoms of these elements achieve stability by *sharing* one or more of their valence shell electrons with other atoms in order to acquire the electron configuration of a noble gas. In the process, they form molecules or extended networks of atoms.

Let us see how the process of molecule formation works, first using hydrogen as an example. This is a particularly simple example, because the noble gas whose valence shell configuration hydrogen seeks to achieve is helium, which has only two electrons rather than eight in its valence shell (since that is the maximum allowed for the first shell). Once we understand how two hydrogen atoms form a molecule, we will consider the bonding of hydrogen to atoms which need to acquire an octet of valence electrons.

3.12 A hydrogen molecule is held together by the sharing of electrons

As we shall see in Chapter 4, there are also chemical bonds that consist of four or six electrons.

When two unbonded hydrogen atoms combine to form a molecule of H_2, the single electrons of each atom are said to form a **bonding pair of electrons** (see Figure 3.14). The pair of electrons travels around both of the nuclei—they are *shared* by the two atoms. The sharing of two electrons

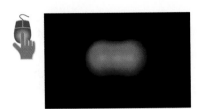

View a representation of a covalent bond at Chapter 3: Visualizations: Media Link 2.

between a pair of adjacent atoms is called a **covalent chemical bond.** Because the two atoms are very close together—their nuclei are separated by only 0.07 nanometers in the H_2 molecule—the volumes in space that were associated with the individual electron shells in the nonbonded, isolated atoms overlap each other to a considerable extent, as shown in Figure 3.14. Thus each shell is in effect occupied by two electrons, the same number as occurs in the noble gas helium, for much of the time.

This situation, in which each hydrogen atom has access to two electrons, is sufficiently stabilizing that the atoms make no attempt to share electrons with additional hydrogen atoms. Consequently, elemental hydrogen exists as a collection of H_2 molecules, rather than, for example, H_3, H_4, or some other such combination, or as unbonded hydrogen atoms. If more than two hydrogen atoms were to combine by sharing electrons, each would have more than two electrons, a destabilizing situation since the maximum number in the first shell is two. Therefore H_3 does not exist as a stable molecule; any H_3 that does form immediately splits off one hydrogen atom.

As a form of shorthand, chemists often represent the electrons in the valence shells in atoms by simple symbols, such as dots placed around the elemental symbol. For convenience, different colors can be used to distinguish the origin of electrons that become shared, although in fact electrons are all identical and indistinguishable particles. Thus for H_2, the electron from one atom can be symbolized by a red dot, and the electron from the other atom by a black dot. The fact that these electrons are shared in the molecule H_2 is indicated by placing both symbols *between* the atoms in a "bonding diagram" for the molecule:

$$\boxed{H \cdot\cdot H}$$

Because covalent bonds, formed by the sharing of two electrons, are so common in molecules, often a simple dash between the atoms is used to represent each bond, rather than showing the corresponding electron pair explicitly:

$$\boxed{H-H}$$

As you go through this book, keep in mind that the above figures are equivalent representations.

3.13 Hydrogen is a stable gas, but it can burn and explode

Because the bond between the hydrogen atoms stabilizes the two atoms, hydrogen normally exists as H_2 molecules, whether the element is a gas, a liquid, or a solid. There are attractive forces that operate *between* H_2 molecules, but these are very weak compared to those holding the atoms of each molecule together. Because the attractive forces between

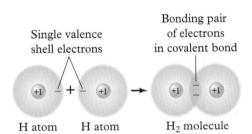

Single valence shell electrons

Bonding pair of electrons in covalent bond

H atom   H atom H_2 molecule

Figure 3.14 Formation of a hydrogen molecule. The two H atoms are held together by a covalent bond since two electrons are shared.

Chapter 3: An Insider's Perspective

$$2\,H_2 + O_2 \longrightarrow 2\,H_2O + \text{energy}$$

Figure 3.15 Destruction by fire of the hydrogen-filled airship *Hindenburg* in New Jersey in 1937, after its inaugural transatlantic flight. (Archive Photo Royalty/ The Image Bank)

the molecules are very weak, hydrogen molecules do not remain close together in fixed positions—in other words as a solid—above the very low temperature of $-259°C$ (which is 14 degrees on the Kelvin scale). Above this temperature, the molecules have sufficient energy to move around each other; that is, they form a liquid. At a slightly higher temperature, $-253°C$, the molecules separate from each other entirely to become a gas. To remain in a liquid or solid form, therefore, hydrogen must be kept at temperatures only slightly above the lowest temperature that can be achieved, $-273°C$, or *absolute zero.*

Achieving and maintaining such low temperatures requires the expenditure of much energy and consequently is expensive, so the practical applications of liquid and solid hydrogen are limited. On the other hand, as we shall see in Chapter 12, gaseous hydrogen is considered to be the fuel of the future, since it burns cleanly in oxygen to produce only water as its product and in the process releases a considerable amount of heat:

$$2\,H_2 + O_2 \rightarrow 2\,H_2O + \text{heat}$$

Unfortunately, this reaction can often occur in an explosive manner. When car batteries malfunction internally, they can produce hydrogen gas that can react explosively with the oxygen in air if there is a short circuit in the battery that produces a spark to initiate the reaction. It was this same reaction that produced a catastrophic hydrogen fire and destruction of the *Hindenberg* airship in the early part of the 20th century when its aluminum skin was ignited (see Figure 3.15). As a result of the *Hindenberg* disaster, hydrogen was replaced by helium as the "light" gas used to lift dirigibles.

Although achieving a noble gas electron configuration stabilizes the hydrogen atoms and greatly reduces their reactivity, the molecules that form are not completely inert and will react under certain conditions.

3.14 Hydrogen forms molecules with other atoms by sharing electrons

In addition to sharing electrons with other atoms of the same element, atoms can share electrons with atoms of *different* elements. As our example, let us consider the formation of a molecule between an atom of hydrogen and one of fluorine.

As we have seen, a hydrogen atom needs to acquire a share in a second electron in order to obtain a stable, noble gas electron configuration. A fluorine atom has seven electrons of its own in its valence shell, and thus it also needs one more electron to acquire the octet of electrons associated with its nearest noble gas, neon. Consequently, in bonding to a hydrogen atom, a fluorine atom will contribute *one* of its seven valence shell electrons to be shared. The sharing of one electron from fluorine with one from hydrogen, to form a covalent bond, serves to increase the number of valence shell electrons accessible by the H atom to two, and by the F atom to eight (see Figure 3.16). In this way, fluorine forms the molecule HF, hydrogen fluoride, with hydrogen. As in the case of hydrogen molecules, and indeed for every solid composed of molecules, the forces holding adjacent HF molecules to each other are much weaker than those of the covalent bond holding the H and F atoms together.

Just as fluorine atoms form HF, chlorine atoms form HCl (hydrogen chloride) molecules, bromine atoms form HBr (hydrogen bromide) molecules, and iodine atoms form HI (hydrogen iodide) molecules when combining with hydrogen atoms. All the Group VII halogen atoms require one additional electron to acquire an octet of electrons in the valence shell, as occurs in atoms of the noble gas that follows them in the periodic table. This means the compounds HF, HCl, HBr, and HI all exist as **diatomic molecules,** that is, molecules that each consist of two atoms.

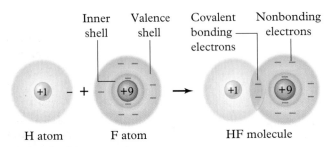

Figure 3.16 Formation of a molecule of hydrogen fluoride, HF. Notice that the F atom has access to eight outer electrons, six nonbonding and two shared.

3.15 Lewis structures show electron distributions

The six valence shell electrons that fluorine does *not* use for bonding purposes are called **nonbonding electrons.** They occupy regions of space which do not much overlap the electron shell of hydrogen. They are sometimes shown as three pairs of dots around the F atom in a bonding diagram for a molecule. Here we show them as blue dots to distinguish them from the two electrons doing the bonding:

$$ \text{H} \cdot \cdot \ddot{\underset{..}{\text{F}}} \colon $$

Notice again that the electron pair that is shared can be symbolized by two electron symbols or by a dash. In some representations used by chemists, *only* the bonds in a molecule—not the nonbonding electrons—are displayed:

$$ \text{H} - \text{F} $$

Chapter 3: An Insider's Perspective

Lewis diagrams were named for G. N. Lewis, the American chemist who studied bond formation and developed this form of notation.

Diagrams that show bonds and nonbonding electrons in molecules are known as **Lewis structures.** Note that *only* electrons from the valence shell of each atom, not from its inner shells, are shown in such diagrams. Since even nonbonding electrons occupy separate regions of space from each other, they are usually shown as distinct, separate pairs around the element symbol in Lewis structures.

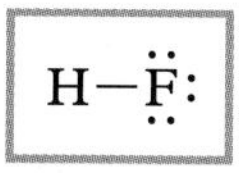

$$H-\ddot{\underset{\cdot\cdot}{F}}:$$

When chemists want to picture the shape of a molecule, they use a different form of representation, which we shall learn about in Chapter 4.

3.16 Some atoms form bonds to more than one hydrogen atom

Free atoms of oxygen, sulfur, and the other Group VI elements each possess six electrons in their valence shell. This means each element needs *two* additional electrons to complete the octet of electrons associated with the noble gas that follows it in the periodic table. The sharing of one electron from an oxygen atom with one electron from a hydrogen atom to form a covalent bond between the oxygen and hydrogen atoms increases the number of electrons about the O nucleus beyond six—but only to seven. Since a hydrogen atom can share an electron and bond with only one atom, an oxygen atom needs to bond to *two* individual hydrogen atoms, sharing one of its six electrons with one H atom and another electron with the second H atom, in order to attain an octet of valence shell electrons. The Lewis structure for this molecule—H_2O—would be written as:

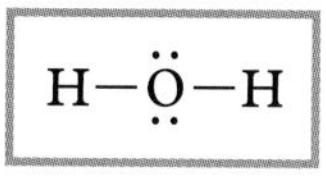

$$H-\underset{\cdot\cdot}{\ddot{O}}-H$$

We conclude that oxygen atoms, and atoms of the other Group VI elements, will form molecules with two hydrogen atoms each, giving molecules of formula H_2O, H_2S, and so on. Since these molecules each contain three atoms, they are triatomic molecules.

Worked Example: Predicting molecular formulas and drawing Lewis structures for molecules

By analogy with the analyses given above, **a)** predict the formula of the molecule formed by atoms of hydrogen with one atom of nitrogen, and **b)** draw the Lewis structure for the molecule.

Solution: **a)** Hydrogen has one electron in its valence shell, so it will form one covalent bond, as we have shown. Nitrogen is found in Group V of the periodic table. Therefore, its atoms each have five valence shell electrons. Consequently, a nitrogen atom needs to acquire access to three more electrons to achieve an octet in its valence shell, as in the noble gas neon. Therefore, nitrogen will share three of its five valence shell electrons. Since each H can share only one electron and form one covalent bond, three hydrogen atoms are required to bond to one nitrogen atom. Consequently the formula of the molecule formed by nitrogen and hydrogen is NH_3. **b)** Since the nitrogen atom forms three bonds, one to each of three hydrogens, we show in the Lewis structure the symbol N connected by dashes to each of the three H atoms. We also show a pair of dots on the nitrogen to denote the presence of its two nonbonding valence shell electrons, corresponding to the $5 - 3 = 2$ electrons not used for bonding:

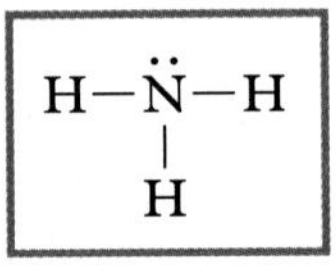

$$
\begin{array}{c}
\overset{\displaystyle ..}{\text{N}} \\
\text{H} - \text{N} - \text{H} \\
| \\
\text{H}
\end{array}
$$

Note that in Lewis structures the positioning of the electrons and bonds about any particular atom is arbitrary and is not intended to illustrate the positions of the particles in space. In other words, we could also diagram NH_3 as, for instance:

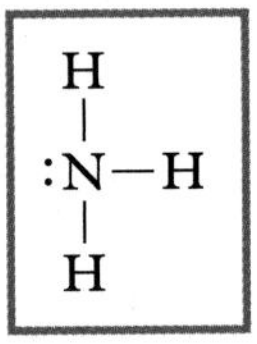

$$
\begin{array}{c}
\text{H} \\
| \\
:\text{N} - \text{H} \\
| \\
\text{H}
\end{array}
$$

Exercise 3.5

What would be the formulas and Lewis structures of molecules formed by hydrogen atoms with **a)** one atom of carbon, **b)** one atom of sulfur, and **c)** one atom of phosphorus?

3.17 The position of bonds can be deduced from electron configurations

From the analyses for Ar, HF, H_2O, and NH_3 discussed above, we can draw a number of general conclusions:

- Since atoms of Group V (15) elements (N, P, As, Sb, and Bi) each have five valence shell electrons, they need to form three bonds in order to acquire an octet of electrons.

- Since atoms of Group VI (16) elements (O, S, Se, Te, and Po) each have six valence shell electrons, they need to form two bonds in order to acquire an octet of electrons.

- Since atoms of Group VII (17) elements (F, Cl, Br, I, and At) each have seven valence shell electrons, they need to form one bond in order to acquire an octet of electrons.

- Since atoms of Group VIII (18) elements with the exception of helium (Ne, Ar, Kr, Xe, and Rn) each have eight valence shell electrons, they have no need to form bonds in order to acquire an octet of electrons.

- Since hydrogen atoms have but one electron each, and need only one more to fill their valence shell, they always form one bond in molecules. A helium atom's only shell is filled with two electrons, and so, like the other Group VIII atoms, it forms no bonds.

All the examples we discussed in earlier sections involved the bonding of an atom to one or more hydrogen atoms. However, many types of covalent bonds do not involve this element. For example, pairs of halogen atoms combine together to form diatomic molecules in which the atoms are joined by a covalent bond. An example is chlorine, which forms Cl_2 molecules with the following Lewis structure:

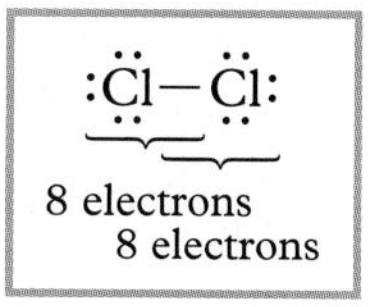

$$:\ddot{C}l - \ddot{C}l:$$

8 electrons
8 electrons

Notice that each chlorine atom contributes only one of its seven valence shell electrons to be shared, since it needs to add only one more electron to achieve an octet. Since they also are only one electron short of an octet, the other halogens also exist as diatomic molecules: F_2, Br_2, and I_2. We shall see in Chapter 4 that in some cases, two bonded atoms need to share two or three pairs of electrons to achieve an octet.

The number of covalent bonds formed by an atom is called its **valence.** Thus the valences of carbon, nitrogen, oxygen, fluorine, and argon are 4, 3, 2, 1, and 0 respectively.

Exercise 3.6

Deduce the appropriate Lewis structures for the molecules **a)** F_2 and **b)** FCl.

Because the halogen atoms F, Cl, Br, and I are similar to hydrogen atoms since they generally form only one bond, the formulas of many halogen-containing molecules are analogous to those containing hydrogen. For example, chlorine combines with carbon to form CCl_4, analogous to the carbon–hydrogen molecule CH_4 (encountered in Exercise 3.5a). Carbon–halogen compounds are discussed further in Chapter 5.

In forming covalent molecules, individual nonmetal atoms do not achieve the same complete "ownership" of an octet of electrons as the noble gas atoms do. Rather, each atom has *complete* control over only its *nonbonding* electrons and partial control over ("access to") the electrons that it shares with atom(s). In most molecules that exist as stable entities, each constituent atom shares enough electrons to obtain an octet.

Larger stable molecules contain a network of covalent bonds holding each atom to the overall structure.

It is important to realize that the idea that atoms share specific electrons in molecules is a *concept*, not something that can be directly observed, since we cannot follow the paths of individual electrons with any form of instrumentation. However, the concept is consistent with observations that can be made by sophisticated techniques that measure the average location of electrons in space.

> **Worked Example:** Deducing Lewis structures for more complicated molecules
>
> Using the generalizations above regarding the number of bonds formed by the atoms of different elements, deduce whether there would be a continuous series of bonds joining the atoms in the molecule HOOH.
>
> **Solution:** Starting from the left side of HOOH, the hydrogen can form one bond to the oxygen atom adjacent to it, thus fulfilling its bonding requirements:
>
> H—O
>
> This oxygen will form one additional bond, to the second oxygen atom, since we know that all oxygen atoms form two bonds:
>
> H—O—O
>
> The second oxygen atom can form a second bond to the right-hand hydrogen, thus fulfilling its requirements for two bonds. Thus HOOH should correspond to a real molecule, since it is connected by the continuous bond network:
>
> H—O—O—H

> **Exercise 3.7**
>
> Using the generalizations regarding the bonds formed by the atoms of different elements, deduce whether there would be a continuous series of bonds joining the atoms in the potential molecule HFFH. Start at the left side of the molecule, as in the Worked Example. From your results, predict whether or not it would be a stable molecule.

3.18 Hydrogen peroxide is a chemically unstable compound

The substance HOOH that we encountered in the Worked Example above is hydrogen peroxide, whose formula we write as H_2O_2 if we are simply showing its elemental composition rather than the relative location of the atoms in its molecules. As the pure compound, hydrogen peroxide is similar to water in that it is a colorless liquid that freezes at about 0°C, though it boils at a higher temperature (150°C) than water.

Hydrogen peroxide differs from water in that as a pure compound it is chemically unstable. Molecules of the substance react with each other, so it decomposes, yielding water and molecular oxygen as products:

$$2 \ H_2O_2 \rightarrow 2 \ H_2O + O_2$$

This reaction occurs rapidly, and sometimes explosively, when the liquid is heated.

For disinfectant uses, hydrogen peroxide is usually encountered as a (nonexplosive) 3% solution in water. As a somewhat more concentrated solution (6–12%), it is used as a *bleaching agent,* by which we mean that it removes the color from materials such as fabrics and human hair. Simple bleaching of hair does convert brunettes and redheads into blondes of a sort, but usually an alternative process that is less damaging to the hair is employed whereby a dye is used in combination with the hydrogen peroxide in order to achieve a particular color and shade. In the bleaching process, the hydrogen peroxide chemically converts substances that have color to those that do not by changing the nature of the bonds within the substance's molecules.

Hydrogen peroxide is also produced in your body. When you eat foods that contain fat and protein, they are broken down so that your body can use the component molecules. A minor by-product of this breakdown is hydrogen peroxide. Since hydrogen peroxide is toxic to cells, your body has special catalyst molecules called *catalase* that convert hydrogen peroxide into water and oxygen, through the chemical reaction shown above. This destruction of H_2O_2 and of other molecules that contain the —O—O— bond, which in general are called peroxides, is important, since their buildup could result in reactions that destroy important biological molecules in cells, in turn leading to the death or at least the dysfunction of the cells. For example, peroxide damage to the membranes that surround cells can disrupt the flow of materials in and out of the cell. The inability of the body to rid itself efficiently of harmful peroxides has been linked to aging and to diseases such as *amytropic lateral sclerosis,* or *Lou Gehrig's disease.* A diet high in vitamins C and E can help limit the damage done by peroxides.

????????????????? **?** ????????????
Fact or Fiction **:**

Honey can kill germs.

Honey has been used historically to dress wounds. The presence of a small amount of hydrogen peroxide in honey is thought to be responsible in part for its antibiotic properties.

■ Chemistry in Your Home: Catalysts in potatoes

Potatoes contain a substance that speeds the reaction that converts peroxides into water and oxygen. If you add a slice of raw potato to a little hydrogen peroxide (a bottle of which is available in a local drugstore), you can see oxygen bubbles forming. You can collect some of this oxygen by inverting a small drinking glass over the potato for a few minutes, then quickly removing the glass and covering it. Then light a match or candle's wick, blow it out, and quickly insert it into the glass. The oxygen makes the match or wick glow and may even cause it to relight.

● Tying Concepts Together: Water

Water, the compound at the basis of life on Earth, is a good example to let us review and extend the various concepts about matter we have developed so far in this book. First consider that, individually, hydrogen atoms and oxygen atoms are very tiny spheres of matter that have, at their centers, nuclei of charges $+1e$ and $+8e$, respectively. Around these nuclei swiftly travel electrons, tiny particles each having a charge of $-1e$ but very little mass. The electrons of oxygen atoms are divided between two shells, the inner shell having two electrons that play no role in bonding. The valence shell of each oxygen atom, before the oxygen forms a bond, contains six electrons. When it joins with hydrogen to form water molecules, each oxygen atom shares two of its six valence shell electrons and acquires a share in the electron associated with each of two hydrogen atoms. Because they share electrons in electron shells which must partially overlap each other, the two hydrogen atoms in every H_2O molecule are strongly and closely held to the oxygen atom by covalent bonds (see Figure 3.17). All three atoms travel together as an intact particle—whether in the gas, liquid, or solid phase of the compound called water (Figure 3.18a).

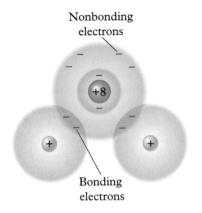

Nonbonding electrons

Bonding electrons

Figure 3.17 Covalent bonding in a water molecule.

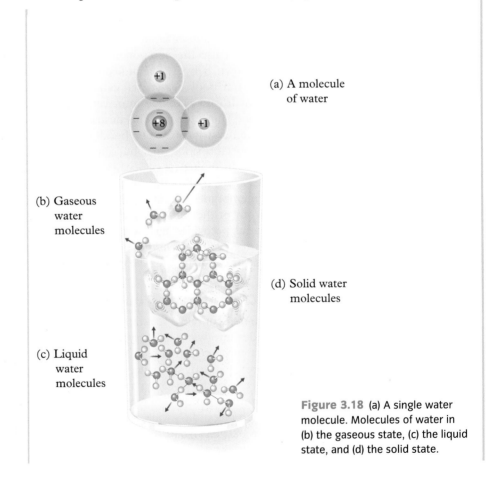

(a) A molecule of water

(b) Gaseous water molecules

(d) Solid water molecules

(c) Liquid water molecules

Figure 3.18 (a) A single water molecule. Molecules of water in (b) the gaseous state, (c) the liquid state, and (d) the solid state.

In the gas phase, individual H_2O molecules travel in random motion and are largely independent of each other since they are separated on average by long distances (Figure 3.18b). Their direction and speed of motion can change as a result of frequent collisions with other molecules, be they additional H_2O molecules, those of N_2 or O_2, or any other atoms or molecules. When water molecules are a component of a gas, as when water evaporates or boils, they are invisible. If they are present in very large numbers, and the temperature drops below 100°C, water molecules can collect together to form droplets of liquid water. If the droplets are suspended together in air, we can see them—think of fog or steam.

When water is in the liquid state, the individual H_2O molecules remain as intact units, and undergo some motion relative to each other (Figure 3.18c). If the temperature drops below 0°C, liquid water freezes to ice, in which the individual H_2O molecules largely remain in fixed positions relative to each other (Figure 3.18d). The closest water molecules are separated from each other by a few tenths of a nanometer in ice and indeed in liquid water.

We know even more about the way hydrogen and oxygen atoms are linked to form water molecules than we have discussed so far. By measuring the average positions of the electrons and/or locating the nuclei, chemists can determine the relative positions of the atoms in individual molecules. They have determined that in H_2O molecules each of the two hydrogen nuclei is equidistant from the nuclei of the oxygen atom, and that the distance between the O and H nuclei is about 0.1 nanometer, whether the molecules are in the gas, liquid, or solid phase (Figure 3.19a). They have also found that the H-to-O-to-H angle that is defined by the nuclei averages about 105°, which is slightly more than a right angle (90°) (Figure 3.19b). In other words, the three atoms are *not* aligned in a straight line, for which the angle would be 180°, but rather are present in a bent orientation. Thus we say that the water molecule has a bent shape. Figure 3.19c shows a "ball-and-stick" model of the water molecule. In such models, the circles are meant to represent the atoms, and the sticks represent the bonds.

Chemists are interested in the shapes of molecules since the properties of substances are strongly influenced by the 3-D arrangement of atoms in space as defined by what are called **bond angles**—the angles that the bonds joining two atoms to a common third atom make with each other. Thus the H-O-H bond angle in water molecules is 105°. Bond angle and molecular shape are very important concepts in chemistry, and we will devote considerable attention to them in the next chapter.

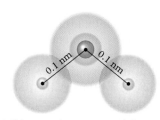

(a) Distance between nuclei

(b) Bond angle

(c) Ball-and-stick model

Figure 3.19 Three visualizations of a water molecule.

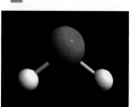

View a representation of a ball-and-stick model of a water molecule at Chapter 3: Visualizations: Media Link 3.

View a representation of a cation and an anion at Chapter 3: Visualizations: Media Link 4.

View representations of sodium, potassium, and magnesium ions at Chapter 3: Visualizations: Media Link 5.

Ions and Ionic Bonding

3.19 Ions are electrically charged atoms

Atoms do not *always* exist in compounds as neutral atoms joined by covalent bonds. This is particularly true when one or more of the atoms is a metal. Such atoms do not have enough valence-shell electrons to form a large number of covalent bonds in order to achieve an octet.

In some compounds, the constituent atoms are missing one or more of their electrons or have added one or more electrons. This makes the atoms electrically charged in either a positive or a negative sense, respectively. In chemistry, an atom or molecule, or indeed any group of atoms bound together, that has gained or lost electrons and has become electrically charged is called an **ion.** The electrical charge on the ion, in units of e, is shown as a superscript following the elemental symbol(s), as discussed below.

Metal atoms usually occur in compounds as positively charged ions, that is, atoms that are missing electrons. Positive ions are called **cations.** Since the positive charge of the protons in the nucleus is greater than the negative charge of the remaining electrons, the net charge shown as a subscript is positive for cations. Thus, a sodium ion that has formed a cation by losing one electron is symbolized as Na^+.

The ions formed by nonmetal atoms are usually negatively charged, that is, they carry extra electrons. Negatively charged ions are called **anions.** Since the negative charge of the electrons is greater than the positive charge of the protons in the nucleus, the net charge shown as a superscript is negative for anions. Thus, a chlorine atom that has formed an anion by gaining one electron is symbolized as Cl^-.

When atoms of the Group I or II elements form cations, they do so by losing *all* the electrons from their valence shell. These electrons are transferred to other types of atoms in the compounds that are formed. By losing these electrons, the Group I and II atoms achieve the same, highly unreactive electron configuration as the noble gas that just *precedes* them in atomic number. For example, atoms of sodium, atomic number 11, have a 2,8,1 electron configuration, that is, $1^2 2^8 3^1$. Loss of the lone electron in the third shell gives them the 2,8,0 configuration of argon, the element one lower (10) in atomic number. In general, then, the Group I atoms H, Li, etc., form +1 ions, whereas the Group II atoms Be, Mg, etc., form +2 ions. Consequently the common ions formed by Group I metals are Li^+, Na^+, K^+, . . . , and those formed by Group II elements are Be^{2+}, Mg^{2+}, Ca^{2+}, . . . Note that the charges shown for these ions signify that the atoms overall have a greater positive charge (in the nucleus) than the total negative charge, since negatively charged electrons have been lost. (Do not interpret the + to mean that an atom has gained electrons!)

The chemical properties of atoms and their ions are very different. For example, sodium, Na, as an atom in its elemental form is highly reactive with water. However, Na^+, a very important biological ion, is not reactive at all with water, including that in your body.

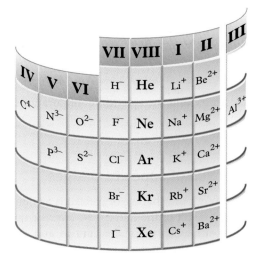

Figure 3.20 Common ions formed by elements of the main groups.

When nonmetal atoms form anions in compounds, they usually *add* sufficient electrons to attain the same number of valence shell electrons that the next heavier noble gas atoms possess. With the exception of hydrogen, which adds one electron to attain the full first shell of two electrons as in helium, the anions of nonmetals, therefore, have an octet of valence shell electrons. For example, the halogen atoms (Group VII) each commonly add one electron to the seven that they already possess as neutral atoms, thereby forming the ions F^-, Cl^-, Br^-, and I^-. Similarly, Group VI atoms add two electrons to their existing six to give the anions O^{2-}, S^{2-}, etc.; and Group V atoms add three electrons to their existing five to give N^{3-}, P^{3-}, etc. Carbon atoms form anions in a few compounds. As expected from the position of the element in Group IV, its atoms add four electrons to yield C^{4-}. The charges on the ions formed by elements of the main groups are summarized in Figure 3.20. The figure shows the periodic table as if it were formed into a cylinder so you can see more easily how the ions of each group relate to the noble gas nearest to them.

Although cations formed by single atoms are simply named using the identity of the element, anions are specified somewhat differently. The suffix *-ide* is applied to the root of the name of the element; for example, in the case of anions consisting of charged single atoms, F^- is called fluoride, O^{2-} is oxide, and S^{2-} is sulfide. Thus, it follows that the compound formed by Na^+ and Cl^- ions is called sodium chloride, that formed by Al^{3+} and S^{2-} ions is called aluminum sulfide, etc.

Exercise 3.8:

a) What is the name of the ionic compound of magnesium and oxygen? of lithium and bromine?

b) What is the name of the ionic compound KF? of CaS?

Worked Example: Deducing the electron configurations in ions

Deduce the total number of electrons, and the electron configurations, of the following ions:
a) N^{3-} **b)** Al^{3+}

Solution: a) From the periodic table, we determine that the atomic number of nitrogen is 7. This means a *neutral* atom of N contains seven electrons. Since the charge, in units of electrons, on the ion is −3, it contains three *additional* (negatively charged) electrons, so the total number of electrons in N^{3-} is $7 + 3 = 10$. Since a maximum of two electrons can enter the first shell, and eight can enter the second, and no electrons are available to enter the third, the ion's electron configuration is 2,8,0. **b)** Since the atomic number of Al is 13, and since the charge on the ion is +3, indicating that three electrons have been lost, the total number of electrons in Al^{3+} is $13 - 3 = 10$. Thus, like N^{3-}, it has the electron configuration 2,8,0. (Displaying the final zero is optional, but is useful in some instances.)

Exercise 3.9

Deduce the total number of electrons, and the electron configurations, of the following ions:

a) C^{4-} **b)** Mg^{2+} **c)** Cl^-

3.20 Ions bond by the attraction of opposite charges

Ions do not bond to each other through the sharing of electrons as neutral atoms do. Rather they bond as a result of the *electrostatic attraction* of opposite charges, that is, of the positive charge of the cations and the negative charge of the anions. Consider the interaction between a particle having a positive ($+1$) charge and one having a negative (-1) charge:

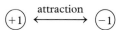

These particles *attract* one another since their charges are *opposite* in sign ("opposites attract"). The *smaller* the distance separating the particles, the greater the attraction. *The attractive interaction between a pair of adjacent oppositely charged ions is called an* **ionic bond.**

In contrast to the forces of covalent bonding, which operate only between the two atoms sharing a particular pair of electrons, an electrostatic attraction of a given particle for another particle does *not* preclude similar interactions with yet other particles. Consider what would happen if a *second* pair of oppositely charged particles came close to the first pair. This action introduces four additional electrostatic interactions:

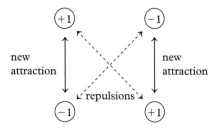

New attractions occur between each cation and the anion of the other pair. In addition, the proximity of the two pairs introduces electrostatic *repulsions* between the two cations and between the two anions, since charges having the same sign repel each other.

In order to decide whether the additional attractions outweigh the repulsions, we must consider the distances between the particles. In the orientation of the particles that is shown above, the distance in the new attractions between the cations and anions is *smaller* than that between the two cations or between the two anions. Therefore, the additional attractions are *greater* in magnitude than are the repulsions, since, as we mentioned, the electrostatic interaction between particles is larger the smaller the distance between them. The *opposite* would be true if the pairs of ions were in the orientation shown below, in which the two cations and the two anions are close together.

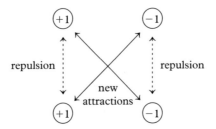

Following this reasoning, additional pairs of ions would experience a net attraction to the existing set of four ions, *provided* that their cations were positioned next to one of the anions of the existing set, and vice versa. Overall, the ions would form two lines:

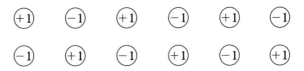

Indeed, because the additional attractions would always outweigh the additional repulsions, it would be energetically stabilizing if pairs of ions continued to join this network in the same manner and extend it indefinitely in two dimensions, forming a plane:

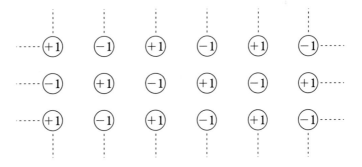

If you focus your attention on any one cation in the body of a layer of such a structure, it should be clear that the closest particles to each cation are four equally distant anions, and vice versa. Recall from Chapter 1 that the closest particles surrounding a given one all at equal distances are called its *nearest neighbors,* so in the structure above, each cation and each anion has four nearest neighbors in the plane.

View representations of ionic lattices at Chapter 3: Visualizations: Media Link 6.

The mineral part of bone, for example, is a complex ionic lattice that includes Ca^{2+} and Mg^{2+} ions.

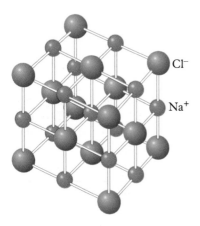

Figure 3.21 A sodium chloride crystal. NaCl is composed of Na^+ and Cl^- ions in an ionic lattice that extends in three dimensions.

3.21 Ions form compounds with three-dimensional networks

Layers of ions can occur on top of and below each other, thereby increasing the number of nearest neighbors for each ion and creating a three-dimensional network called an **ionic lattice** that is held together by electrostatic forces. As a result of such interactions, ions generally exist *not* as isolated pairs of particles (that is, molecules) but rather in three-dimensional extended networks as crystals in the solid state. Each ion in the structure is highly attracted to several other nearest neighbors of opposite charge, all of which lie at the same distance from it. Consequently, no unique individual pairs of ions can be identified in the network.

3.22 Sodium and potassium chloride are common salts

The most common material that has an ionic lattice is sodium chloride, NaCl. Here the cations are all sodium ions, Na^+, and the anions are all chloride ions, Cl^- (see Figure 3.21). Huge deposits of solid sodium chloride are found in many locations on Earth, from which the material is mined and then purified. The salt that you sprinkle on food consists of a large number of small particles, each of which is a crystal of sodium chloride. Another common salt that is found in deposits is potassium chloride, KCl, in which the cations are potassium ions, K^+, and the anions are chloride ions, Cl^-.

The origin of both the NaCl and KCl deposits is salty water from inland seas that over the ages gradually dried up and deposited their salt content on what was the ocean floor. The Great Salt Lake and the Dead Sea are examples of what such waters must have been like in the distant past. Because NaCl is less soluble in water than KCl, as a sea dries up, the sodium salt is deposited first, followed by KCl when even less water remains. As a consequence, sodium chloride deposits occur at greater depths than potassium chloride deposits.

Although all compounds that contain a lattice of ions can properly be called ionic compounds, chemists also use the term **salts** for compounds that consist of ions *other* than H^+ or OH^-. The name "salt" is *not* reserved in chemistry for NaCl, which we shall refer to as "table salt" or "common salt."

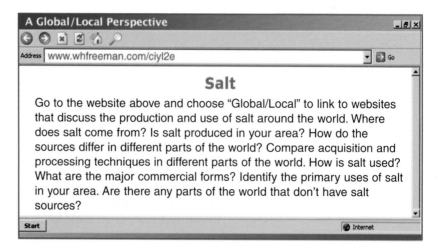

A Global/Local Perspective

Address www.whfreeman.com/ciyl2e

Salt

Go to the website above and choose "Global/Local" to link to websites that discuss the production and use of salt around the world. Where does salt come from? Is salt produced in your area? How do the sources differ in different parts of the world? Compare acquisition and processing techniques in different parts of the world. How is salt used? What are the major commercial forms? Identify the primary uses of salt in your area. Are there any parts of the world that don't have salt sources?

3.23 Minerals, the ions in your diet

From a nutritional point of view, **minerals** are defined as inorganic compounds that occur as ions and that are required for good health. Calcium, magnesium, sodium, potassium, and the nonmetals phosphorus, sulfur, and chlorine are called the **major minerals** because each is required in quantities of 0.1 gram (100 milligrams) or more per day. The **trace minerals,** substances needed in amounts of only a few *milli*grams per day, include iron, zinc, selenium, and several others. The *ultratrace minerals* As, B, Cd, Co, Cr, F, Li, Mn, Mo, Ni, Pb, Si, Sn, and V are so called because they are necessary in even smaller amounts.

Even arsenic (As) is essential; in tiny concentrations, it can act as a tonic to speed metabolism. Too much arsenic, of course, is deadly. In fact, excesses of many minerals can be detrimental to your health. Table 3.2 lists the **Recommended Dietary Allowances,** RDAs, for some selected minerals. RDA values are set by each country and are defined as the level of intake of essential nutrients judged to be adequate to meet the known nutritional needs of most healthy persons.

Table 3.2 Recommended Dietary Allowances of selected minerals

Element	Males age 15–24, in mg per day	Females age 15–24, in mg per day	Males age 25+, in mg per day	Females age 25+, in mg per day
Calcium	1200	1200	800	800
Phosphorus	1200	1200	800	800
Magnesium	350–400*	280–300*	350	280
Selenium	50–70**	50–55**	70	55
Zinc	15	12	15	12
Iron	10–12*	15	10	10–15***
Iodine	0.15	0.15	0.15	0.15

* The higher value is for the subgroup aged 15–18 years.
** The lower value is for the subgroup aged 15–18 years.
*** The lower value is for the subgroup aged 51+ years.

The source is the Food and Nutrition Board, U.S. National Academy of Sciences. Values recommended by the Canadian government differ slightly.

3.24 Sodium and potassium ions are important to your health

Sodium and potassium ions are the most important cations in your body fluids. Like the oceans that surround us, our blood is primarily water, and sodium chloride is the principal ionic component. In fact, the human body is 0.9% sodium chloride.

In contrast to your blood plasma, in which there are 16 times as many sodium ions as potassium ions, K^+ greatly predominates (by a

You can always tell if a food contains high levels of sodium because it will have a salty taste.

False. It is the chloride ion in common salt, sodium chloride, that imparts the salty taste, not the sodium. Many foods and medicines contain sodium but not chloride, and so don't taste salty. Conversely, compounds such as potassium chloride and calcium chloride taste salty, but contain no sodium.

9:1 ratio) over Na^+ in your body's cells. Your body uses energy to maintain these different concentrations of Na^+ and K^+ inside and outside your cells. For example, the differential concentration is used to generate the electrical signals that regulate heartbeat. Your body contains molecules that have cavities that fit ions of a specific size (that exclude ions of any other size) and transport them in and out of cells; since potassium ion is larger than sodium ion, the two ions can be distinguished and selected by such molecules. Some antibiotic medications operate by selectively transporting specific ions into or out of cells.

Excess sodium in the diet has been linked to high blood pressure, which arises in part because kidneys require a high pressure in the arteries in order to excrete the excess sodium from the blood. On average, Americans ingest 4–5 grams of sodium per day, compared to the range of 1.1–3.3 grams recommended for adults by the National Academy of Sciences. One teaspoonful of salt weighs about 6 grams; since NaCl is 39% sodium by mass, it contains about 2.3 grams of sodium, the maximum daily intake recommended for people with high blood pressure. Processed meats (bacon, sausage, ham) and canned soups and vegetables almost all contain added sodium—partially for taste and partially because it acts as a preservative—adding to the amount from the common salt that we add to our food for cooking and flavoring. Sodium also occurs in the flavor enhancer MSG (it's what the S stands for).

Overall, by a ratio of 5:3, the human body needs more potassium than sodium. An improper ratio can lead to cardiac problems. Potassium must be constantly resupplied as part of your diet, because the body excretes it as a component of urine. A deficiency of potassium is not uncommon among older people and those with chronic diseases. Many people with high blood pressure/hypertension are prescribed drugs called *diuretics*. These drugs work by inducing urination so as to reduce the body's volume of retained water. Thus, users of diuretics often require a potassium supplement to replace the loss of potassium due to the increased urination. Potassium loss occurs in healthy people due to vomiting, diarrhea, and other gastrointestinal problems. Potassium deficiency also occurs when you drink too much water and are heavily perspiring—for example, when you try to prevent dehydration during sports. Fatigue is the common symptom of potassium depletion.

Although there is no recommended daily allowance for potassium, experts recommend 2.0–2.5 grams. The average American diet supplies 2–6 grams of potassium (K) per day. Good sources of potassium include bananas, apples, and citrus fruit and juices; vegetables such as broccoli, beans, and peas; bran cereals; meat and fish (such as salmon); and dairy products. Some people use salt that consists of crystalline KCl, or a mixture of NaCl and KCl, to give food a salty taste while reducing their sodium intake since it is Cl^-, not Na^+, that is "salty." Blackstrap molasses, which in the old days was often consumed as part of a healthy diet, contains in 4 tablespoons the recommended daily intake of potassium! Your kidney normally can prevent too high a potassium level in your blood.

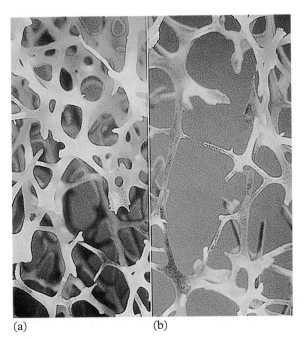

(a)   (b)

Figure 3.22 (a) Normal and (b) osteoporotic human bone.

(IOF/Visuals Unlimited)

3.25 Calcium and magnesium are important for bones and teeth

Calcium, in the form of its ion Ca^{2+}, is the most abundant mineral in your body. About 99% of the approximately one kilogram of calcium in your body is found in your bones and teeth. The remaining 1% occurs in the blood and tissues, where it has many vital functions. Indeed, it is so important there that in order to maintain appropriate calcium levels in the blood, your body will deplete the element from your bones if other sources aren't adequate. Thus a deficiency of calcium in the body can lead to *osteoporosis*, a disease that thins bones so that they fracture easily (see Figure 3.22). Deficiency can result from insufficient calcium in the diet or from poor absorption into the body of the calcium that is in the food that is eaten. Poor absorption of calcium can result from too low a body level of vitamin D. Postmenopausal women are at particularly high risk for osteoporosis. In order to minimize their bone loss, they may be placed on estrogen hormone replacement therapy and/or advised to ensure that their levels of exercise and calcium intake are high.

The calcium RDA for young adults is 1.2 grams (1200 milligrams) per day. Milk products are the best sources of the element, and milk is often fortified with vitamin D, which helps the body absorb calcium. Calcium is also present in dark-green leafy vegetables, though it is difficult for the body to extract and absorb the mineral because it is strongly bonded to ions formed from the oxalic acid in the vegetables. For this reason, you should not count on these foods as your sole source of calcium.

Magnesium is another essential mineral, required in the formation of bone and in certain essential enzymes. The RDA for magnesium in adults is 0.35 gram for males and 0.28 gram for females. Sources of this metal include vegetables and nuts as well as "hard" water, that is, water that contains significant quantities of calcium and magnesium ions. People who drink *only* soft or softened water may have a magnesium deficiency. There exist statistical studies that suggest that magnesium in drinking water protects against death from heart disease.

3.26 Iron and zinc are essential for basic body functions

Iron is a trace mineral that is essential to your well-being. There are two common ionic forms of this element: Fe^{2+} and Fe^{3+}. The two ions behave quite differently because they differ in the number of the electrons that are present. Iron is a necessary component in your diet since it is required as part of hemoglobin, the substance in your blood that transports oxygen, O_2, from your lungs to your cells. Many foods, including meats (especially organ meats), fish, dark-green leafy vegetables, raisins, and whole grain cereals contain iron. Usually your body can absorb and use only a small fraction of the iron you consume in foods.

Menstruating women lose iron as part of the blood loss in their monthly cycle, and some suffer occasionally from *iron deficiency anemia*. Symptoms of anemia include fatigue, weakness, and an inability to concentrate. Due to the lack of iron, the red blood cells contain insufficient hemoglobin to carry the necessary amount of oxygen efficiently (see Figure 3.23). For this reason, many women take iron supplements. Indeed, the RDA of iron of 15 milligrams for young women is greater than that of 10 milligrams for young men. Some men suffer from an *excess* of iron, since their intake—from diet and supplements—contains more than enough of it and, once absorbed, it is only slowly eliminated by their bodies. Too high a dose of iron can be dangerous to your health

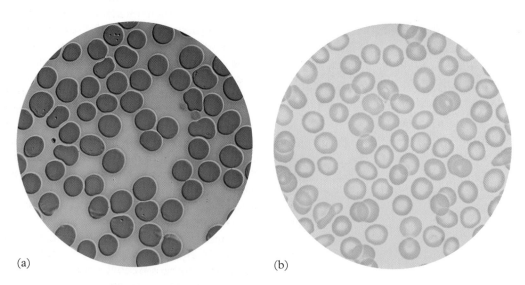

(a)   (b)

Figure 3.23 (a) Normal and (b) anemic red blood cells. The color of the cells shows the difference between adequate and inadequate amounts of iron in the diet. (Part a, David M. Phillips/ Visuals Unlimited; part b, Joaquin Carrillo Farga/Photo Researchers)

because excess iron can damage your liver and pancreas. The average human body contains about 4 grams of iron.

■ Chemistry in Your Home: Pulling iron out of cereal

Some fortified cereals use metallic iron as an ingredient. If this is the case, the word *iron* will appear alone on the nutritional label. (By contrast, iron sulfate is an iron salt, not the metallic form.) You can "pull" the iron out of cereal that has the metallic form by the following procedure. Crush about a cup of cereal until it is fine. Put the cereal into a clear glass bowl or drinking glass. Add water and stir or shake for five minutes. Let the cereal stand for a few minutes. Hold a magnet against the bottom of the bowl or glass for several minutes. You should see a small grouping of black specks form as you keep the magnet against the glass. This is the iron that was in the cereal.

The metal zinc is required by the body for use in many of its enzymes, in amounts comparable to those of iron. Zinc occurs as the cation Zn^{2+}. Good sources of zinc are meat, shellfish, and dairy products. Strict vegetarians often need to supplement their diet with this element.

3.27 Iodine and table salt

Iodine is required by the human body to produce hormones in the thyroid gland. **Hormones** are "chemical messenger" molecules that are carried in the bloodstream from one part of the body to another and regulate the rates of biochemical reactions at a target site. If the diet is low in iodine, the thyroid gland enlarges in a futile attempt to produce more hormones, and as a consequence, the person suffers from low energy, low blood pressure, and weight gain. Extreme cases can lead to *goiter,* in which the gland is grossly enlarged, and in infancy to *cretinism* (extreme mental retardation). On the other hand, excess iodine in the diet can lead to an overactive metabolism.

Iodine is present in small amounts in seawater, so an excellent source of iodine is seafood. It is also found in the air and soil of coastal areas, so it is present in food grown there. Although the RDA of the element iodine is only 0.15 milligram, many people who eat little seafood or consume only food grown in inland regions could be chronically deficient in this substance. For this reason, a small amount (0.01%) of iodine in the form of *potassium iodide,* KI, an ionic compound containing the *iodide ion,* I^-, is added to most common table salt sold for domestic consumption.

Although more than 99% of the salt in the package you buy is indeed sodium chloride, NaCl, other substances are deliberately added in order to improve its properties. Crystals of 100% pure sodium chloride tend to cake together after absorbing a small amount of moisture from the air, preventing the salt from flowing smoothly. For that reason, table salt usually contains about 0.5% of a "drying agent" such as an ionic compound containing the carbonate or silicate anion.

Bromide ion is also present in seawater. In the past, sodium or potassium bromide was used as a sedative and to treat tension headaches.

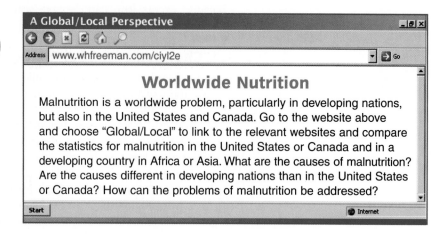

3.28 Many ionic compounds do not have a 1:1 ratio of atoms

Not every ionic lattice has a 1:1 ratio of cations to anions like NaCl. For example, both calcium ions, Ca^{2+}, and magnesium ions, Mg^{2+}, form ionic compounds with chloride ions in a 1:2 ratio. The formulas for ionic compounds use as subscripts the integers that correspond to the *simplest* ratio of the ion types in the substance. The symbol for the cation, usually a metal, is listed before that of the anion, usually a non-metal. Thus the formula for calcium chloride is $CaCl_2$.

The appropriate formula for the compound that is formed by any cation with any anion can be deduced in a simple manner. Because, as you learned earlier, the electrical charge on the structure as a whole must be zero, *the total positive charge contributed by all the cations must exactly equal the total negative charge of all the anions.* The charges are created, after all, by the flow of electrons from what will become the cations to what will become the anions; no electrons are created or destroyed in the process. Since the subscripts in the formula give the *relative* number of ions of both types, it follows that *the total electrical charge associated with the various ions listed in the formula must add up to zero.* For example, calcium ions have twice the charge of chloride ions, so in the structure of the ionic compound formed by Ca^{2+} and Cl^- ions, there must be twice as many anions, each of charge -1, as there are cations, each of charge $+2$. Therefore the formula for the substance is $CaCl_2$. In general, the formulas for ionic compounds of two elements can be determined by using the magnitude of the anions' charge for the cation subscript, and vice versa, as we did for calcium chloride.

Because the formulas for ionic compounds look like formulas for molecular compounds, it is easy to make the mistake of thinking that ionic compounds contain molecules. However, the use of formulas for ionic compounds is *not* meant to imply that they contain individual $CaCl_2$ molecules.

Worked Example: Deducing the formulas and names of simple ionic compounds

What is the formula and name for the ionic compound formed by **a)** the cation Mg^{2+} and the anion N^{3-}? **b)** the cation Ca^{2+} and the anion O^{2-}?

Solution: a) Since the cation charge is $+2$ and the anion is -3, the subscripts for the cation and anion will be 3 and 2, respectively. Thus the compound's formula is Mg_3N_2. Its name is magnesium nitride, since the root of the name *nitrogen* is *nitr*. **b)** Since both cation and anion charges are 2, we initially conclude the compound's formula is Ca_2O_2. However, remember that ionic formulas use the *simplest* ratio of the ion types. Here, we can divide both subscripts by a common factor, 2, and thereby obtain the simplest formula, CaO. The compound's name is calcium oxide, since the root of the name *oxygen* is *ox*.

Exercise 3.10

What is the formula and name for the ionic compound formed by **a)** the cation K^+ and the anion S^{2-}? **b)** the cation Al^{3+} and the anion O^{2-}? c) the cation Al^{3+} and the anion N^{3-}?

3.29 Solutions of ions in water have lowered freezing points

As we all know, water freezes at $0°C$ ($32°F$). However, if the water contains either molecules or ions dissolved in it, it does not freeze until the temperature is lowered further. For example, a liter of water containing about 17 grams of common salt dissolved in it freezes at $-1°C$. The more salt that is dissolved, the colder the solution must be before it freezes. In general, the extent of the decrease in freezing point is proportional to the total number of ions (or molecules) that dissolve in a given amount of water.

One of the main uses for both sodium chloride and calcium chloride is to "melt" the winter ice that forms on roads and sidewalks in northern climates. Spreading one of these salts on ice melts the top layer; the salty water does not easily refreeze until temperatures significantly below $0°C$ are reached. Sodium chloride used for this purpose is obtained from salt mines. Calcium chloride is also readily available since it is produced in huge quantities as a by-product in the chemical industry. A drawback to the use of ionic compounds for road deicing is that the ions promote the rusting of steel in car bodies and bridges, and can be toxic to shrubs and plants adjacent to roadways and sidewalks.

■ **Chemistry in Your Home:** Does salt water freeze faster than fresh water?

Examine the effect of salt on water freezing by dissolving three heaping tablespoons of table salt in 1/2 cup of water. Fill two ice cube trays, one

???????????????????? **?** ????????????
Fact or Fiction :

Does adding salt to boiling water shorten cooking time?

Adding salt to water not only lowers its freezing point, it also increases its boiling point. The difference between salted and unsalted water is small, however. It takes about 30 grams (one ounce) of salt to increase the boiling point of 4 liters (about 4 quarts) of water by about one-half a Celsius degree (1° Fahrenheit). The amount of salt needed to raise the boiling point enough to affect the cooking time would render the food inedible.

with salted water and one with unsalted water, and place them in the freezer. Check the ice cubes every hour to monitor their progress. Which ice cubes freeze first? Is there a very big difference between the two?

3.30 Polyatomic ions have two or more types of atoms

In all the examples mentioned so far, ions have consisted of charged single atoms. However, there also are common ions that consist of *several* covalently bonded atoms, with the extra positive or negative charge shared among them. A simple example of such a **polyatomic ion** is the hydroxide ion, OH^-. Here the oxygen and hydrogen atoms are joined by a single covalent bond. The existence of this bond serves to increase the number of electrons accessible to hydrogen to the required number of two, but it raises that of oxygen only from six to seven, not to the required eight. However, there is an additional electron associated with the negative charge on the ion: this serves as the eighth valence shell electron (shown in blue on the Lewis diagram below) on the oxygen:

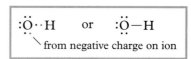

Notice that no attempt is made in the ion's formula to specify the location of the extra electron as being on any particular atom.

Exercise 3.11

By analogy with the case of the hydroxide ion, write the Lewis structure, showing bonds and nonbonding electrons, for the ion SH^-.

Many polyatomic anions contain a central atom to which a number of oxygen atoms are covalently bonded. Boron, carbon, and nitrogen, for example, each form a stable anion with three oxygen atoms, which are called borate, carboronate, and nitrate respectively:

$$BO_3^{3-} \qquad CO_3^{2-} \qquad NO_3^-$$
borate ion   carbonate ion   nitrate ion

Don't be confused by the presence of both a subscript and superscript after the oxygen symbol! The subscript tells you the number of oxygen atoms in the ion, whereas the superscript tells the charge on the ion as a whole. The *-ate* ending to the name denotes that the polyatomic anion contains oxygen. Atoms from the next row of the periodic table form their stablest anions with four oxygens; again the *-ate* ending is used, giving silicate, phosphate, and sulfate when oxygen combines with Si, P, and S, respectively:

$$SiO_4^{4-} \qquad PO_4^{3-} \qquad SO_4^{2-}$$
silicate ion   phosphate ion   sulfate ion

View a representation of a hydroxide ion at Chapter 3: Visualizations: Media Link 7.

The ionic compound sodium carbonate, Na_2CO_3, is one of the main components of salt deposits left when fresh water evaporates. It readily incorporates water into its structure. The ancient Egyptians used such salt deposits to extract water from tissues—and hence prevent their decay—in the process of embalming dead bodies.

The traditional medication called *Epsom salts* is magnesium sulfate. It is still sometimes used as a laxative, and when dissolved in warm water, as a footbath.

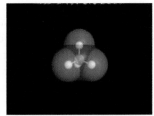

View a representation of an ammonium ion at Chapter 3: Visualizations: Media Link 8.

Some of these nonmetals also form anions with one fewer oxygen atom than those shown above; examples are the nitrite ion NO_2^-, the phosphite ion PO_3^{3-}, and the sulfite ion SO_3^{2-}. The *-ite* ending is used for anions containing fewer oxygen atoms, but the same central atom, as those having the *-ate* ending.

Among cations, the only common polyatomic one is the ammonium ion, NH_4^+. The most common compound containing this cation is ammonium chloride, NH_4Cl.

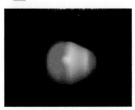

View a representation of a polar covalent bond at Chapter 3: Visualizations: Media Link 9.

● Tying Concepts Together: Bonds that are partially covalent, partially ionic

We have seen that in the formation of a covalent bond between two identical atoms, there is an exactly equal sharing of an electron pair. In contrast, an ionic bond between two atoms can be thought of as a very unequal sharing of an electron pair: one atom donates an electron to pair with an electron of the other atom. These two electrons spend all their time traveling around one atom and no time around the other.

Although we have so far presented only these two bond types, many bonds are in fact intermediate between a covalent and an ionic bond. For example, in the hydrogen–oxygen bond in water (and some other molecules), the shared electron pair actually spends more of its time traveling close to the oxygen atom than it does close to the hydrogen atom. Such bonds, where the electron pair is closer to one atom, are called **polar covalent bonds.** As a result of the polarization, the hydrogen atom in the O—H bond has less than a half share in the two electrons. Consequently, the average number of electrons traveling around the hydrogen atom is less than one, and hence it carries a partial positive charge, symbolized δ^+. Similarly, because the oxygen atom attracts more of the shared pair, it effectively carries a partial negative charge, symbolized δ^-. These charges are sometimes displayed in the Lewis structure for the molecule:

The polarization of the covalent O—H bond occurs because oxygen has a greater ability to attract bonding electrons than does hydrogen. The sharing of electrons between two atoms is exactly equal *only* when the atoms are *identical*. The *relative ability of an atom to attract electrons in bonds to itself is known as its* **electronegativity.** Nitrogen and the halogens also have electronegativities that exceed that of hydrogen, and so the bonds they form with it are polarized in the sense that hydrogen carries a partial positive charge. The electronegativity of carbon is similiar to that of hydrogen, so C—H bonds are not very polarized.

We can see now that the ionic bond is simply the extreme case of a polar covalent bond, in which the electronegativity difference between the atoms is so large that the "shared" electron pair spends essentially all its time traveling around the atom with the much higher electronegativity.

Summarizing the Main Ideas

All atoms are composed of one or more electrons, each of which has a charge of $-e$, and one nucleus, which is positively charged and much more massive than an electron. The charge on the nucleus, in electron units, is called the atomic number for the element and is equal to the number of protons in the nucleus. Free atoms—that is, those that are not parts of molecules or extended networks—have a number of electrons equal to the number of protons in the nucleus, so consequently they are electrically neutral overall. Most nuclei also contain neutrons, particles which are about as massive as protons but which carry no electrical charge.

It is the atomic number that distinguishes one element from another. All atoms of a given element have the same atomic number, and no two elements have the same atomic number. Elements with atomic numbers ranging from 1 to 113, and some with higher numbers, are known. The elements are arranged by atomic number in the periodic table.

Ernest Rutherford's experiments showed that the atom consists of an almost empty sphere, with the nucleus at its center and electrons traveling about it. The electrostatic force of attraction of the electrons for the nucleus holds them within the atom. Electrons do not fall into the nucleus because they are in constant motion about it.

All electrons in a given atom are not held to the atom with the same amount of force, since they occupy shells at different distances from the nucleus. The most energetically stable shell is that closest to the nucleus, which can hold a maximum of two electrons. Successive shells can hold a maximum of 8, 18, 32, 50, and 72 electrons, respectively. Atoms with relatively few electrons fill their electron shells in strict order of stability. Beyond the second shell, however, there are deviations from the simple filling sequence.

Elements whose atoms have eight electrons—an octet—in their valence shell are particularly unreactive and exist as uncombined atoms. These so-called noble gases are neon, argon, krypton, and radon, plus helium, which has only two electrons and so has a filled valence shell.

The chemical behavior of elements is determined mainly by the number of electrons the elements possess in their outermost occupied, or valence, shell. Each column, or group, of the periodic table consists of elements whose atoms have the same valence shell configuration and therefore have many similar properties. The elements shown on the same horizontal level in the periodic table constitute a row or period.

Atoms of nonmetals combine to form stable molecules and extended networks by the sharing of some of their valence shell electrons. Each atom shares sufficient electrons so that it acquires a noble gas electron configuration—usually an octet of electrons in its outermost occupied shell—for a portion of the time. The sharing of two electrons between a pair of adjacent atoms is called a covalent chemical bond.

Chemists indicate the sharing of a pair of electrons by two atoms by placing a line between their atomic symbols. The nonbonding electrons of the valence shell of an atom are depicted by dots placed around the atomic symbol. The complete diagram showing these bonds and dots for a molecule is called its Lewis structure.

Ions are atoms, or groups of covalently bonded atoms, that have gained or lost electrons and so are electrically charged. Positive ions are called cations, whereas negative ions are anions. Metal atoms of Groups I and II generally lose all their valence shell electrons in forming cations to achieve the configuration of the next lighter noble gas. Nonmetal atoms gain sufficient electrons to achieve the configuration of the next heavier noble gas.

Oppositely charged ions are attracted to each other by electrostatic forces and thereby form ionic bonds. Ions arrange themselves in 3-D extended networks called ionic lattices rather than forming small molecules, since it is energetically favorable to do so.

The ratio of cations to anions in an ionic compound need not be 1:1. However, it must be such that the structure overall is electrically neutral, a characteristic which is reflected in the formula for the compound.

The electronegativity of an atom is a measure of its ability to attract bonding electrons to itself. The atom of greater electronegativity carries a partial negative charge when it forms a polar covalent bond, since the shared electron pair spends more time traveling near it.

Key Terms

plasma	shell structure	nonbonding electron	mineral
electron	electron configuration	Lewis structure	major mineral
nucleus	valence shell	valence	trace mineral
atomic number	inner shell	bond angle	Recommended Dietary
periodic table	octet of electrons	ion	Allowance (RDA)
electrostatic force	group	cation	hormone
proton	row	anion	polyatomic ion
neutron	bonding pair of	ionic bond	polar covalent bond
mass number	electrons	ionic lattice	electronegativity
isotope	covalent chemical bond	salt	

Web Sites of Interest

To link to Web sites of interest, go to www.whfreeman.com/ciyl2e, Chapter 3, and select the site you want.

Review Questions

1. What is a *plasma*? What were the components of the plasma obtained from energizing the atoms in argon gas?

2. What is an *electron*? What is its charge?

3. What is a *nucleus*? What is the sign of a nuclear charge?

4. What is meant by the *atomic number* for an element? How does it relate to the number of protons and electrons that an atom has?

5. Describe how the experiments by Ernest Rutherford led to the concept of the nuclear atom.

6. What are *electrostatic forces*?

7. What is the difference between a neutron and a proton?

8. What is a *mass number*?

9. What is an *isotope*? How is an isotope's mass number related to the number of nuclear particles?

10. What are the *isotopic symbols* and common names for the three isotopes of hydrogen?

11. What is meant by the *shell structure* for electrons in atoms?

12. What is the relationship between the energy required to remove an electron from an atom and the electron distance from the nucleus?

13. What is the maximum number of electrons in the first shell? second shell? third shell?

14. What is meant by the terms *inner* and *valence shells*? For an atom with an electron configuration 2, 8, 3, which are the inner shell electrons and which are the valence shell electrons?

15. What electrons define atomic size?

16. What is the characteristic number of valence shell electrons for noble gases other than helium? What is it for helium?

17. Identify two uses of the noble gases.

18. What are vertical columns of the periodic table called?

19. What characteristic of electron configuration is shared by the elements in any given column?

20. Which column is known as the alkali metals? the halogens? the alkaline earths?

21. What does the term *periodic* mean when referring to elements in the periodic table?

22. How do elements change as you move along a row in the periodic table?

23. Identify at least three pieces of information about an element that can be obtained from the periodic table.

24. What determines the chemical properties of an element?

25. Why do elements in the same group exhibit similar chemical behavior?

26. What is meant by *sharing of electrons*?

27. What is meant by the phrase *covalent chemical bond*? Describe the bonding in the hydrogen molecule.

28. Why does hydrogen exist as a gas rather than as a solid at room temperature?

29. Identify one commercial use of hydrogen.

30. What drawbacks are associated with the use of hydrogen as a fuel?

31. What is meant by the *Lewis structure* for a molecule? What are *nonbonding electrons,* and how are they displayed in a Lewis structure?

32. With how many hydrogen atoms will an atom from each of the following groups bond?
    a) Group IV   b) Group V
    c) Group VI

33. What is the molecular formula for hydrogen peroxide?

34. Why is hydrogen peroxide harmful when produced inside your body?

35. How can hydrogen peroxide be useful when applied to the outside of your body?

36. What is meant by the *bond angle* in a molecule?

37. What is the difference between a *cation* and an *anion*? How are the charges displayed on atomic symbols for an ion?

38. How does the group number tell you about an element's ion type and charge?

39. What types of ions do metals form? What types of ions do nonmetals form?

40. What is an *ionic bond*? How does an ionic bond differ from a covalent bond?

41. What is an *ionic lattice*?

42. What is a *salt*?

43. What roles do sodium and potassium play in the body?

44. What is a *diuretic*?

45. What does the term *RDA* mean?

46. What is a *hormone*?

47. What roles does calcium play in the body?

48. What roles does magnesium play in the body?

49. What is a *trace mineral*? Provide an example.

50. Match the minerals below to their sources:
    a) calcium   1) meats and fish
    b) magnesium   2) milk products and leafy
    c) iron   green vegetables
    d) zinc   3) meat, shellfish, and dairy
                         products
                      4) hard water, vegetables,
                         and nuts

51. How does a trace mineral differ from a major mineral?

52. What is *iron deficiency anemia*?

53. Match the minerals below with their roles in the body:
    a) zinc
    b) iron
    c) calcium
    d) magnesium

    1) important in oxygen transport by hemoglobin
    2) used in many enzymes
    3) major component of bones and teeth
    4) important for bone formation and some enzymes

54. How does iodized table salt differ from uniodized table salt? How are they alike?

55. How can you predict the ratio of anions and cations in an ionic compound?

56. What is the net electrical charge on any ionic compound?

57. How does the addition of salt to water affect the water's freezing point?

58. Provide four examples of polyatomic anions that contain oxygen.

59. What is *electronegativity*?

Understanding Concepts

60. Locate the following elements in the periodic table. In each case, from the information shown for the element, determine the magnitude of the positive charge on the nucleus of one of its atoms and the number of electrons and protons that one atom of it contains:
    a) bromine (symbol Br)   b) aluminum
    c) iron (symbol Fe)

61. The element carbon, symbol C, has an atomic number of 6. For the isotopes of carbon that have 6, 7, and 8 neutrons respectively in their nuclei, determine their mass number and write the atomic symbols to display these values.

62. Complete the table below for atoms of the various isotopes. You can use the periodic table to correlate atomic numbers with elemental symbols:

Isotopic symbol	No. of protons	No. of neutrons	No. of electrons	Mass number	Atomic number
^{11}B					
	6	7			
				16	32
		10	9		
^{238}U					

63. Deduce the shorthand electron configuration for atoms of each of the following elements; underline the electrons that belong to the valence shell:
    a) He (helium)   b) Mg (magnesium)
    c) Cl (chlorine)   d) Ar (argon)

64. Which of the following electron configurations is *incorrect*? Why? What is the *correct* configuration for that number of electrons? Underline the valence shell electrons in the correct configurations.
    a) 2, 6, 0
    b) 2, 9, 1
    c) 1, 1

65. Using only the position of the element in the periodic table (and not its atomic number) as information, deduce how many valence shell electrons are associated with atoms of each of the following:
    a) Kr (krypton)   b) S (sulfur)
    c) Sr (strontium)   d) Cs (cesium)

66. Why does the evidence indicate that an octet of valence electrons is a chemically stable situation?

67. Why is helium stable with only two valence electrons?

68. Explain how electron excitation can produce light.

69. Examine the elemental block below and describe the information conveyed by each symbol:

V
(15)
7
N

70. Why doesn't hydrogen form molecules of three or four atoms: H_3 or H_4?

71. Attempt to draw the Lewis structure (using dots for electrons) for a molecule of H_3 and a molecule of H_4. Do these molecules follow the rules?

72. Using only their position in the periodic table, deduce the number of valence shell electrons in atoms of each of the following elements:
a) P (phosphorus)   b) Ge (germanium)

From your answers, deduce the formula of the molecules which one atom of P and one atom of Ge form with hydrogen. Write the Lewis structure for each molecule.

73. What is the relationship between the number of electrons in an atom's valence shell and the number of covalent bonds the atom can form?

74. Determine the positions of the bonds and of the nonbonding electrons in each of the following molecules, and write the appropriate Lewis structure for each:
a) silicon hydride, SiH_4
b) hydrogen bromide, HBr
c) bromine, Br_2

75. If an atom has two electrons in its valence shell, what type of ion will it form? What will the charge of the ion be? What if the atom has five valence shell electrons?

76. Based upon the positions of the following elements in the periodic table, predict the ion that each will form and write the symbol for the ion:
a) P   b) Ca   c) Al   d) O   e) F

77. Deduce the number of electrons, the number of protons, and the charge on the nucleus for each of the following atoms and ions:
a) H^+ b) N^{3-} c) Ca^{2+} d) F^- e) Xe

78. Write Lewis structures for each of the atoms and ions in Exercise 77.

79. Explain why the chemical behavior of a neutral atom is different from that of its ion.

80. Describe the nature of bonding in an ionic lattice. Why would a pair of oppositely charged ions exist as part of such a lattice rather than as a diatomic molecule?

81. Predict the formula and give the name for the ionic compound that would be formed by combining the following pairs of ions:
a) Na^+ and Br^- b) Na^+ and N^{3-}
c) Mg^{2+} and S^{2-} d) Mg^{2+} and Cl^-
e) Sr^{2+} and O^{2-} f) Sr^{2+} and P^{3-}

82. Predict the formulas of the ionic compounds formed by the following pairs of elements. Note that you must first deduce the likely charge on the ion that each forms.
a) lithium and oxygen
b) magnesium and nitrogen
c) aluminum and sulfur
d) strontium (Sr) and fluorine

83. How are the following forms of chlorine the same? How are they different?
a) Cl   b) Cl^- c) Cl_2 d) ^{35}Cl

84. How are the following forms of sodium the same? How are they different?
a) Na   b) Na^+ c) ^{23}Na

85. Deduce the diagram showing the position of the bonding and nonbonding electrons in the anion NH_2^-. Note that both hydrogen atoms are connected to the nitrogen atom but not to each other.

86. Deduce the Lewis structure, showing bonds and any nonbonding electrons, for the following ions, in which all the hydrogen atoms are bonded to the non-hydrogen atom:
a) NH_4^+ b) H_3O^+

87. If atom X has a greater electronegativity than atom Y, does the shared electron pair in the X—Y bond spend more of its time near X or near Y? Which atom carries a partial negative charge and which a partial positive charge?

Synthesizing Ideas

88. Based upon the discussion in this chapter regarding the appropriate position in the periodic table for hydrogen, in which group other than VIII could one place the element helium?

89. Below are given the formulas for two groupings of atoms that may constitute molecules. Determine the positions of the bonds and of the nonbonding electrons, and write the appropriate Lewis structure for each

grouping. In each case, predict whether all the components are bonded together in a single molecule or whether they constitute several separate molecules:

a) N_2H_4, where two H atoms are connected to each N atom, and the N atoms are situated next to each other

b) N_2H_6, where three H atoms are connected to each N atom, and the N atoms are situated next to each other

90. Look at labels for two different processed breakfast cereals. Using the information about compound names in this chapter, can you identify any ionic compounds in the cereals? Are they the same or are they different in the two cereals?

91. Explain why one of the symptoms of iron deficiency anemia is lack of energy.

■ Group Activity: Really Cold Water

Adapted from an experiment provided by Professor Rudy Luck, Michigan Technological University

For this activity, you will need:

- A Styrofoam drinking cup
- A thermometer for liquids that registers at least to −20°C (0°F)
- Snow or ice chipped into small pieces
- Several tablespoons of salt

Procedure

1. Fill the cup with the snow or chipped ice and record its temperature once a little water has formed.

2. Add 1 tablespoon of salt to the cup and stir the mixture for a few minutes. The snow or ice should slowly melt and the salt should dissolve in the resulting liquid.

3. Once no further melting or dissolving seems to be happening, record the temperature of the "brine" liquid.

4. Repeat the experiment with a clean cup, fresh snow or ice, and 2 tablespoons of salt. Record the temperature when no further melting or dissolving seems to occur.

Results

1. For each experiment, calculate the change in temperature between the brine and the pure snow or ice. Was the temperature difference in the second experiment about twice that found in the first?

2. The process by which salt dissolves in cold water is endothermic, that is, it absorbs heat as it proceeds. Use this information to explain how the liquid in the cup could become colder than 0°C when the external temperature is much higher.

3. From your results, what do you conclude about the freezing point of brine relative to that of pure water?

4. Using your results, explain why it is effective to throw salt on icy roads to melt the ice.

In this chapter you will learn:

- about simple hydrocarbons, including methane, butane, benzene and those in candle wax, asphalt;
- about fossil fuels, including natural gas, LPG, and petroleum;
- where fossil fuels are found, their chemical properties, and the pollution caused by them;
- how gasoline is made and rated (octane numbers);
- how oil spills and lead pollute the environment;
- about coal as a fuel, including the environmental issues concerning the mining and burning of coal.

You are probably aware that recreational vehicles such as jet skis and snowmobiles are powered by gasoline.

But what is the chemistry that allows jet skis to speed over the surface of an ocean or lake? The energy stored in carbon-rich fossil fuels allows us to ride a snowmobile, commute to work or school, mow our lawns, and heat our homes. This fuel source does, however, often come at a cost. (Digital Vision/Getty Images)

Powering the Planet

Hydrocarbons and Fossil Fuels

Whether or not we are aware of it, we all use considerable amounts of carbon each day, both in the form of its compounds and sometimes as the element itself. Carbon and hydrogen are the predominant elements in the food that we eat, the clothes that we wear, the medications and vitamins that we take, and the plastics that we use. Indeed, as we shall see in Chapters 8 and 9, carbon also is arguably the most important element in our bodies.

In the next few chapters we explore the chemistry of carbon, first from the point of view of the element and its simple compounds, including alcohol and substances such as caffeine and cocaine used as stimulants. We then turn to carbon as a component in modern materials, in our nutritional needs, in our bodies, and especially in our DNA.

The specialization within chemistry that is concerned with the compounds of carbon, many of which are at the basis of life as we know it, is called **organic chemistry.** In this chapter we will introduce some of the essential concepts of organic chemistry and apply them to some compounds that contain carbon and hydrogen, including gasoline and other fuels that are composed of these two elements. We begin this exploration by considering the carbon-based substances that we burn to fuel our cars, our homes, and cooking appliances, and to produce much of our electricity. As we shall see, there is inevitably some pollution—usually of the air—produced whenever we burn these fuels.

> The term *organic* as used by chemists differs from that used in the phrase *organic food,* by which is meant food grown without the use of chemically produced fertilizers or other substances that have not previously been living matter.

Hydrocarbons That Contain Only Single Bonds

4.1 Hydrocarbons are compounds of carbon and hydrogen

Of the atoms of all the elements, those of carbon and hydrogen are the best at forming covalent bonds. The huge number of compounds made up *exclusively* of carbon and hydrogen are collectively known as **hydrocarbons.** The reason that so many different hydrocarbons exist is that carbon atoms readily form covalent bonds with other carbon atoms, thereby producing molecules that can have short or long networks of carbon atoms—aligned like pearls on a chain, in a strand, or in even more complicated arrangements. Bonded to the carbon atoms are enough hydrogen atoms to bring the total number of bonds formed by each carbon up to four. In forming long networks in which atoms of the same type are bonded to each other, carbon is quite unlike most other elements, few of which form stable chains longer than three atoms. As a consequence, there are more known compounds that contain both carbon and hydrogen than all the compounds put together that do not!

 Whenever you see this icon in this chapter, go to www.whfreeman.com/ciyl2e

As we have seen, hydrogen shares its (only) electron by forming one bond, thereby achieving the pair of electrons associated with helium. Unbonded carbon atoms have four electrons in their outermost electron shell, and so they require four more to achieve the noble gas configuration of eight valence shell electrons. Thus a carbon atom will share each of these four electrons to form *four* covalent bonds in order to complete the octet.

To summarize, in all cases:

- Hydrocarbon compounds consist of molecules.
- Each hydrogen atom is bonded to a carbon atom.
- Each carbon atom forms a total of four bonds.

4.2 Methane molecules contain one carbon and four hydrogen atoms

The simplest molecule containing carbon and hydrogen that fulfills the requirements of four bonds for each carbon and one for each hydrogen is methane, CH_4, which has the following Lewis structure:

$$\begin{array}{cc}
\text{H} & \text{H} \\
\vdots & | \\
\text{H}\cdot\cdot\text{C}\cdot\cdot\text{H} \qquad \text{H}-\text{C}-\text{H} \\
\vdots & | \\
\text{H} & \text{H}
\end{array}$$

Methane

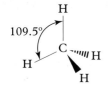

109.5°

(a) Structural diagram

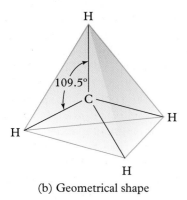

109.5°

(b) Geometrical shape

Figure 4.1 The three-dimensional shape of methane.

Because CH_4 and most organic molecules have three-dimensional structures (rather than being flat), it is not simple to portray their actual geometry on a two-dimensional surface such as a sheet of paper. To visualize the actual 3-D structure of a methane molecule, assume that two hydrogen atoms and the carbon atom are located *in* the plane of the paper, as is the case for the left-side pair of H atoms and the C atom in the structural diagram in Figure 4.1a. The two remaining H atoms lie outside this plane. One C—H bond projects *out* from the paper toward you. The convention for depicting such an arrangement is the solid wedge shown in the diagram. The other C—H bond lies *in back* of the plane. The convention for this is the dashed wedge. All the six H—C—H bond angles in CH_4 molecules are 109.5°, not the four 90° and two 180° angles suggested by the Lewis structure. You may find it hard to visualize the six bond angles; the activity below will help you to do this. The geometrical structure of methane is a **tetrahedron** (see Figure 4.1b), the characteristic geometry of all carbon atoms that form bonds to four atoms in all organic molecules, not just methane. The tetrahedral geometry occurs because it minimizes the total repulsion among the four pairs of valence-shell electrons of the carbon atom. (For the same reason, the water molecule is bent rather than linear because this shape minimizes the repulsion among the four electron pairs—two bonding and two nonbonding—of the oxygen atom.)

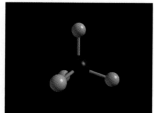

View different representations of a methane molecule at Chapter 4: Visualizations: Media Link 1.

■ Chemistry in Your Home: Building a model of methane

To help visualize the 3-D structure of methane, gather together four toothpicks, four raisins, and a marshmallow. Stick two toothpicks into the marshmallow so that they make an angle slightly greater than a 90° angle. Place raisins at the ends of these two toothpicks. Now place the third toothpick into the marshmallow above the plane of the existing pair, such that it makes an angle slightly greater than 90° with the other toothpicks. Finally, place the fourth toothpick in the marshmallow below the plane of the pair that have raisins attached, again at an angle of slightly more than 90° with all three other toothpicks. Place raisins on the ends of the third and fourth toothpicks. You have built a model of methane, with the marshmallow representing the carbon atom and the four raisins representing the hydrogen atoms. You may have to modify slightly the positions of the toothpicks to generate a structure in which all six angles defined by the toothpick pairs are indeed equal. Rotate the structure around and convince yourself that all C—H bonds are equivalent. By comparing your model to Figure 4.1b, clarify the meaning of the diagram.

Clearly, a diagram in which the 3-D structure of CH_4 is shown is more complex than its Lewis structure, which shows only which atom is bonded to which. Unfortunately, simple dashes are used in chemistry in two different ways: in Lewis structures, dashes are used to represent bonds, whether or not they lie in a plane, whereas in the representations of 3-D structures, dashes are used to mean bonds of any type that lie within a plane. To help differentiate the two uses, we shall use longer, thinner lines when we are trying to specify bonds in a plane (compare the lines in Figure 4.1a to those in the Lewis structure above for methane).

4.3 Methane is a component of air

As a compound composed of CH_4 molecules, methane is colorless and odorless, and is a gas under normal conditions of temperature and pressure. Methane molecules do not attract each other very strongly, so it is not until the gas is cooled to −164°C that its molecules condense to form a liquid. Upon further cooling, it freezes to a solid at −182°C.

Methane is a minor component of our atmosphere, to the extent of 2 ppm; that is, there are two molecules of it for every million molecules of air. Methane is released into the atmosphere when living matter undergoes decomposition in the absence of air, as occurs in landfills, rice paddies, swamps, etc., or when belching, flatulence, or excretion occurs. The latter are not only human phenomena. Methane is also produced in the stomachs of ruminant animals, such as cows, and in the guts of termites when the action of bacteria in the intestine digests the fiber cellulose. The spontaneous combustion of methane in swamp gas has been linked to reports of unidentified flying objects (UFOs) (see Figure 4.2).

As we will see later in this chapter, methane occurs naturally in underground deposits as the main component of the mixture called *natural gas*. Methane is also produced and released into the air as a byproduct of coal mining. Under the right conditions, methane and coal

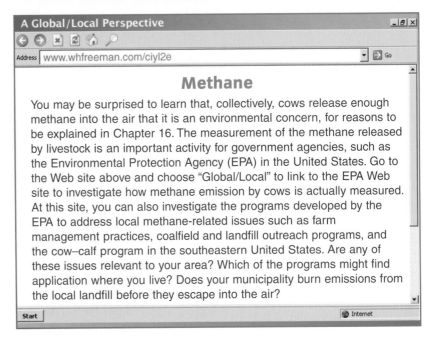

Methane

You may be surprised to learn that, collectively, cows release enough methane into the air that it is an environmental concern, for reasons to be explained in Chapter 16. The measurement of the methane released by livestock is an important activity for government agencies, such as the Environmental Protection Agency (EPA) in the United States. Go to the Web site above and choose "Global/Local" to link to the EPA Web site to investigate how methane emission by cows is actually measured. At this site, you can also investigate the programs developed by the EPA to address local methane-related issues such as farm management practices, coalfield and landfill outreach programs, and the cow–calf program in the southeastern United States. Are any of these issues relevant to your area? Which of the programs might find application where you live? Does your municipality burn emissions from the local landfill before they escape into the air?

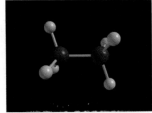

View different representations of an ethane molecule at Chapter 4: Visualizations: Media Link 2.

dust in the mine's air can ignite, with explosive results. You can read about such a disaster in the Westray coal mine in Nova Scotia by visiting the Web site for this book.

4.4 Ethane molecules have two carbons joined by a C—C bond

The simplest hydrocarbon that contains a carbon–carbon bond is ethane, which has the molecular formula C_2H_6. Like methane, ethane is a gas under normal conditions. The carbon atoms here are joined by a covalent C—C bond. Since each carbon atom must form four bonds, three hydrogen atoms are attached to each carbon atom:

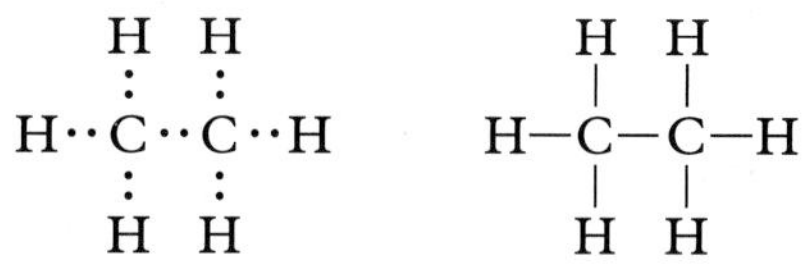

$$
\begin{array}{ccccc}
 & H & H & & H & H \\
 & \cdot\cdot & \cdot\cdot & & | & | \\
H\cdot\cdot & C & \cdot\cdot C & \cdot\cdot H \qquad H- & C-C & -H \\
 & \cdot\cdot & \cdot\cdot & & | & | \\
 & H & H & & H & H
\end{array}
$$

Ethane

To understand the 3-D structure of ethane, try the following activity.

■ Chemistry in Your Home: Making a model of ethane

Gather together six raisins, two marshmallows, and seven toothpicks. Join the marshmallows by one toothpick, and insert another toothpick (with a raisin at its end) in each marshmallow at an angle of slightly more than 90°. Then add two more toothpicks (with attached raisins) to each marshmallow, following the instructions of generating a tetrahedral shape given in the previous activity. You have now constructed a model for ethane. Notice that you can rotate the group of three toothpicks at one end relative to those at the other by twisting the group about the toothpick joining the marshmallows.

In order to indicate the internal bonding structure of molecules such as ethane, chemists sometimes write their formulas carbon atom by carbon atom, with the number of hydrogen atoms attached to each shown after the C symbol. Thus ethane would be written as CH_3CH_3, rather than C_2H_6. Such representations are known as **condensed formulas.** A common variation on condensed formulas is to show the bonds that exist between the carbon atoms, but not the bonds to different atoms such as hydrogen. For example, for ethane the formula would be shown as CH_3-CH_3. Note that in the latter representation, the dash refers to the bond formed between the carbon at the left and the carbon at the right, not between the hydrogens at the left to the carbon at the right, even though the H symbol is placed directly before the dash! As we'll see over the next few chapters, it is very useful to represent the structure of organic molecules by condensed formulas since they are often large and contain many carbon atoms at which reaction could occur.

4.5 Alkanes have the general formula C_mH_{2m+2}

In addition to methane and ethane, carbon and hydrogen form a huge number of other types of molecules. In some of these molecules carbon atoms are bonded to each other in an extended chain in which the only bond types are C—C and C—H. In all cases, the carbon forms four bonds. Conceptually, the simplest such series has a continuous chain of carbon atoms bonded to each other:

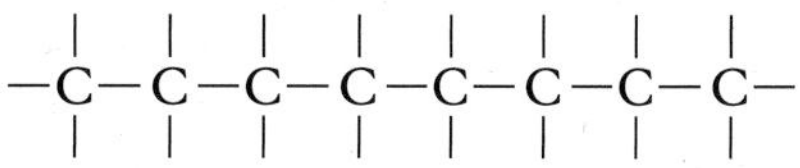

$$
\begin{array}{cccccccc}
| & | & | & | & | & | & | & | \\
-C- & C- & C- & C- & C- & C- & C- & C- \\
| & | & | & | & | & | & | & |
\end{array}
$$

Other than the two end carbons, all carbons in such a chain form two bonds to other carbons. Each "internal" carbon atom must also be bonded to two hydrogen atoms in order for each carbon atom to achieve a total of four bonds. In contrast, both *end* (terminal) carbons require three hydrogen atoms in order to form a total of four bonds. Thus, in general terms, if the chain has m carbon atoms, where m is any positive integer, then the total number of hydrogen atoms required is $2m+2$, since the two terminal carbons each require one more hydrogen than the two associated with the internal carbons. Therefore the general formula for such molecules is C_mH_{2m+2}. The many hydrocarbons with this formula are collectively called the alkanes. The names and formulas for the alkanes from $m=1$ to $m=10$ are shown in Table 4.1.

In the alkane series, methane is the molecule that corresponds to $m=1$, and ethane corresponds to $m=2$. The $m=3$ member of the series, named propane, has $2m+2 = 2 \times 3+2 = 8$ hydrogen atoms, and so has the formula C_3H_8. As expected, the structure of a propane molecule is $CH_3—CH_2—CH_3$, since it contains the C—C—C chain of carbon atoms.

Propane occurs as a minor component of natural gas and is also present in petroleum. Like most hydrocarbons, propane burns easily and controllably, and since it is plentiful, it finds extensive use as a fuel. Some of it is also used to make plastics, as we shall see in Chapter 5.

Exercise 4.1

Draw the Lewis structure, and the simplest condensed formula, for a molecule of propane. *Hint:* Write the three carbons in a line and place hydrogen atoms around them according to the condensed structure given above. Then work from left to right, forming the usual number of bonds for each atom and thereby connecting the molecule together.

Table 4.1 Names, prefixes and condensed formulas of alkanes

Number of carbon atoms	Name with prefix in bold	Condensed Formula
1	**meth**ane	CH_4
2	**eth**ane	$CH_3—CH_3$
3	**prop**ane	$CH_3—CH_2—CH_3$
4	**but**ane	$CH_3—CH_2—CH_2—CH_3$
5	**pent**ane	$CH_3—CH_2—CH_2—CH_2—CH_3$
6	**hex**ane	$CH_3—CH_2—CH_2—CH_2—CH_2—CH_3$
7	**hept**ane	$CH_3—CH_2—CH_2—CH_2—CH_2—CH_2—CH_3$
8	**oct**ane	$CH_3—CH_2—CH_2—CH_2—CH_2—CH_2—CH_2—CH_3$
9	**non**ane	$CH_3—CH_2—CH_2—CH_2—CH_2—CH_2—CH_2—CH_2—CH_3$
10	**dec**ane	$CH_3—CH_2—CH_2—CH_2—CH_2—CH_2—CH_2—CH_2—CH_2—CH_3$

4.6 Longer alkanes have several nonequivalent forms

The next molecule in this series—that is, the alkane with $m=4$—is butane, C_4H_{10}. The Lewis structure for this molecule is:

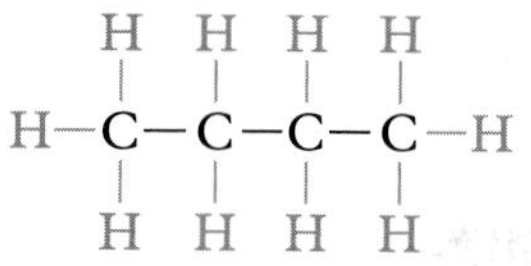

$$
\begin{array}{ccccccccc}
 & H & & H & & H & & H & \\
 & | & & | & & | & & | & \\
H- & C & - & C & - & C & - & C & -H \\
 & | & & | & & | & & | & \\
 & H & & H & & H & & H &
\end{array}
$$

Butane

In some structures, hydrogen atoms are displayed in a lighter shade to emphasize the carbon network.

Chemists sometimes write Lewis structures for alkane molecules with some carbons shown at 90° angles to the line defined by the chain of the other carbons, as illustrated below for butane:

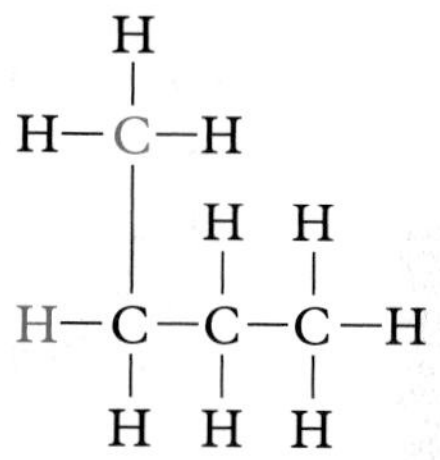

$$
\begin{array}{c}
H \\
| \\
H-C-H \\
| \\
\begin{array}{cccccc}
 & & H & & H & \\
 & & | & & | & \\
H-C & - & C & - & C & -H \\
| & & | & & | & \\
H & & H & & H &
\end{array}
\end{array}
$$

It is important to realize that, chemically, this structure is in fact equivalent to the one in which all the carbon atoms are shown in a straight line. This is because, as we noted earlier, Lewis structures show only the bonds in a molecule, and not its 3-D geometry. Remember that, as we saw for methane, the four bonds radiating from any one carbon are equivalent and define a tetrahedron. As a consequence, any two atoms or groups bonded *to a given carbon atom* can be interchanged in the Lewis structure without changing its meaning. For example, in the butane structure above, we can interchange the position of the red carbon atom (bonded to the leftmost C) with the blue hydrogen atom because both are bonded to the same carbon atom. Then it becomes apparent that the structure below is in fact equivalent to the one in which all four carbons are shown along one line:

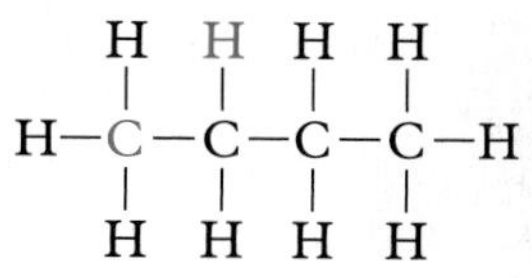

$$
\begin{array}{ccccccccc}
 & H & & H & & H & & H & \\
 & | & & | & & | & & | & \\
H- & C & - & C & - & C & - & C & -H \\
 & | & & | & & | & & | & \\
 & H & & H & & H & & H &
\end{array}
$$

However, if you interchange atoms or groups bonded to *different* carbons, you *do* change the molecule that you are representing. You can reduce confusion in writing Lewis structures if you always *write out the Lewis structure of organic molecules with the longest possible carbon chain displayed in the horizontal direction.* Remember that you can interchange carbon groups or hydrogen atoms bonded to the *same* carbon, but *not* those bonded to *different* carbons. Thus the structure shown on the left below is best symbolically transformed to the linear chain on the right by interchanging the red carbons with the blue hydrogens at both ends of the molecule:

$$
\begin{array}{c}
\quad\;\; \overset{\displaystyle |}{\underset{\displaystyle |}{C}} \\
H-\overset{\displaystyle |}{\underset{\displaystyle |}{C}}-\overset{\displaystyle |}{\underset{\displaystyle |}{C}}-H \\
\quad\;\; \overset{\displaystyle |}{\underset{\displaystyle |}{C}}
\end{array}
\;\rightarrow\;
-\overset{\displaystyle |}{\underset{\displaystyle |}{C}}-\overset{\displaystyle |}{\underset{\displaystyle |}{\underset{\displaystyle H}{C}}}-\overset{\displaystyle |}{\underset{\displaystyle |}{\underset{\displaystyle H}{C}}}-\overset{\displaystyle |}{\underset{\displaystyle |}{C}}-
$$

Alkanes such as the butane molecule discussed above, in which *all* the carbon atoms are along one continuous chain (that is, which you can draw as a straight-line path in the Lewis structure), are called straight-chain or **unbranched hydrocarbons.** There is, however, another way to join four carbon atoms together that is indeed different from the CCCC chain form discussed above. You can think of the C—C—C—C chain as a C—C—C chain with the fourth carbon atom connected to one of its end atoms. If the fourth carbon is instead connected to the *central* carbon of the C—C—C unit, a chemically different molecule will result. The Lewis structure for this form of C_4H_{10} is:

$$
\begin{array}{ccc}
H & H & H \\
| & | & | \\
H-C-&C-&C-H \\
| & | & | \\
H & | & H \\
& H-C-H & \\
& | & \\
& H &
\end{array}
$$

Isobutane

This type of butane is actually different—not simply written differently—because three of the carbon atoms have three hydrogen atoms bonded to them and the other carbon has but one, compared to the "3223" hydrogen atom distribution for the C—C—C—C structure. In general, carbon chains that are attached to the main continuous network of carbon atoms are called *branches,* and hydrocarbons having one or more such units are said to be **branched.**

Molecules having the same formula but different internal bonding structures are called **structural isomers.** Thus we have established that there are two structural isomers of C_4H_{10}. By convention, the name of the unbranched alkane isomer is given the prefix *n-* (for *normal*); thus the compound whose molecules correspond to $CH_3CH_2CH_2CH_3$ is called *n*-butane. The second, branched, isomer of butane is called isobutane.

Because isobutane has a different bonding network than does the C—C—C—C form, it has properties that also differ slightly both at the

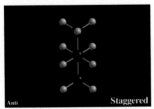

Anti   Staggered

View different representations of a butane molecule at Chapter 4: Visualizations: Media Link 3.

Chapter 4: Powering the Planet

level of individual molecules and also when there is a macroscopic collection of these molecules. Indeed, there are two distinctly different compounds having the C_4H_{10} formula, corresponding to molecules with the two different carbon networks we have described. Under standard atmospheric pressure, liquid isobutane boils at a temperature 10°C lower than for its unbranched isomer, which boils at 0°C. Under higher pressure, both isomers remain liquid at room temperature, as occurs in butane lighters (until the pressure is released). The freezing points of the two isomers also differ somewhat; that of isobutane is 22°C lower than that of *n*-butane.

■ Chemistry in Your Home: Making models for the two isomers of butane

Gather eight marshmallows and six toothpicks. Construct a model for the carbon atom chain in the two isomers of butane, using four marshmallows and three toothpicks in each case. Note that all angles made by the toothpicks with each other around each marshmallow should be a little greater than a right angle. Show that the unbranched structure does not convert to the branched one by rotating it about the axis.

Exercise 4.2

Convince yourself that the following diagram does *not* represent a third isomer of butane but in fact is equivalent to *n*-butane or isobutane. *Hint:* Rewrite the structure, by interchanging groups bonded to the same carbon atom, until it is identical to that shown previously for one of the isomers.

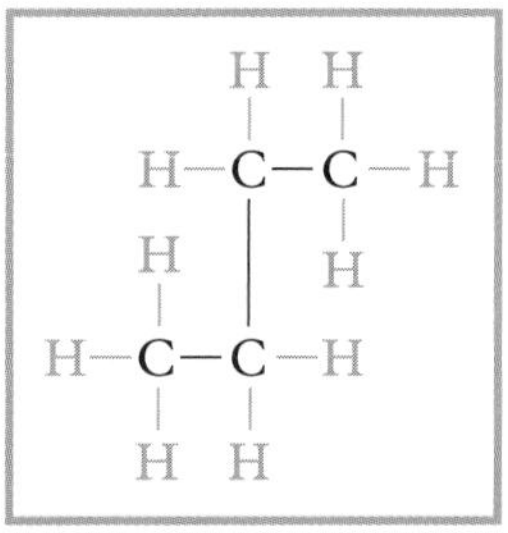

$$
\begin{array}{c}
\overset{\displaystyle H}{\underset{\displaystyle}{|}} \quad \overset{\displaystyle H}{\underset{\displaystyle}{|}} \\
H-C-C-H \\
\overset{\displaystyle H}{\underset{\displaystyle}{|}} \quad\quad \overset{\displaystyle H}{\underset{\displaystyle}{|}} \\
H-C-C-H \\
\overset{\displaystyle H}{\underset{\displaystyle}{|}} \quad \overset{\displaystyle H}{\underset{\displaystyle}{|}}
\end{array}
$$

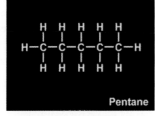

Pentane

View different representations of the alkane series from one carbon (methane) to five carbons (pentane) at Chapter 4: Visualizations: Media Link 4.

Branches, or side chains, of several carbon atoms are often connected to continuous chains of carbon atoms in organic compounds. Commonly these branches correspond to alkanes in which one carbon is missing one of its hydrogen atoms; the branch is connected to the main chain at that carbon. Branches have names that indicate their length. In general, the name of any side chain is that of the corresponding hydrocarbon with the *-ane* ending replaced by *-yl*. The roots of the alkane names, to which *-yl* is appended for the side chain, are shown in bold in Table 4.1. Thus, the branch containing one carbon, namely CH_3—, is called *methyl*. Similarly, the two-carbon branch CH_3CH_2— is named *ethyl*, since CH_3CH_3 is ethane. Using the so-called systematic approach, such designations are used to name hydrocarbons containing branches. For example, another name for isobutane (page 134) is methylpropane, since it can be thought of as a propane molecule to which a methyl group has replaced one hydrogen of the central carbon. The branches are often called *groups*; thus CH_3— is the methyl group. Although systematic names generally are preferred, chemists often use common names that employ prefixes such as *iso-* and *neo-* to name branched isomers of common hydrocarbons.

4.7 Alkanes with longer carbon chains are liquids and solids

After butane, the next formula in the C_mH_{2m+2} alkane series is C_5H_{12}, corresponding to $m=5$; compounds with this formula are called the pentanes. Starting with pentane and for all higher values of m, the alkane names correspond to the Roman- or Greek-based number for the number of carbons (*pent* for 5, *hex* for 6, *hept* for 7, *oct* for 8, and so on, without limit), followed by the suffix *-ane*.

The alkanes from C_5H_{12} through $C_{12}H_{26}$ are all liquids and are prime components of gasoline, which is a solution mainly of hydrocarbons. Liquid alkanes such as hexane are also used as **solvents,** which are substances that readily dissolve many other substances. Alkanes are good solvents for many organic compounds, especially those that are insoluble in water, which is why they are found in paint and varnish removers and grease cutters. If a solid—for example, a solid alkane whose molecules have very long chains of carbon atoms—is stirred with a solvent such as hexane, a homogeneous liquid solution is formed in which the molecules of hexane and the longer-chain alkane are mixed. As discussed in section 2.3, this is *not* the same phenomenon as melting the solid alkane compound by heating it, although in both cases the molecules of the long-chain alkane are liberated from their fixed positions in the solid.

Alkane hydrocarbons having up to several hundreds of carbon atoms linked together in chains by single bonds are known; many of them occur naturally in petroleum. The wax used in candles (the focus of our discussion of combustion in Chapter 2) is a homogeneous solution consisting predominantly of many different alkanes, each having several dozen carbon atoms on average. When isolated as pure substances, each of these alkanes is a waxy solid. Chemists have also been able to prepare mixtures of even longer chains containing thousands of carbon atoms. Some such materials, such as the waxy plastic known as "polyethylene," have many commercial and industrial uses and are discussed at length in Chapter 5.

4.8 Cycloalkanes are formed when carbon atoms form rings

In addition to forming chains of almost any length, carbon atoms also form structures in which the carbon atoms and their C—C bonds define a closed ring. The smallest possible ring contains three carbons, and the most commonly encountered rings contain five or six carbon atoms. The **cyclic hydrocarbons** are named by placing the prefix *cyclo-* in front of the name for the relevant chain alkane. Notice that the "rings" shown here are not literally circular:

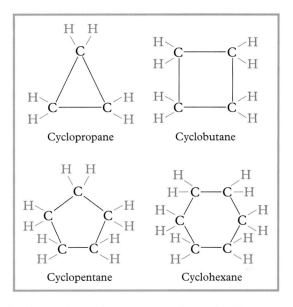

Cyclopropane   Cyclobutane

Cyclopentane   Cyclohexane

View different representations of a cyclohexane molecule at Chapter 4: Visualizations: Media Link 5.

Simple cycloalkanes have the same number of carbon atoms but *two fewer hydrogen atoms* than their noncyclic equivalents. In the ring compounds, fewer hydrogen atoms are required for each carbon to form four bonds because there is one more C—C bond than in the corresponding *n*-alkane since the "ends" of the chain are connected. This difference is illustrated below for *n*-hexane and cyclohexane, both of which contain six carbon atoms:

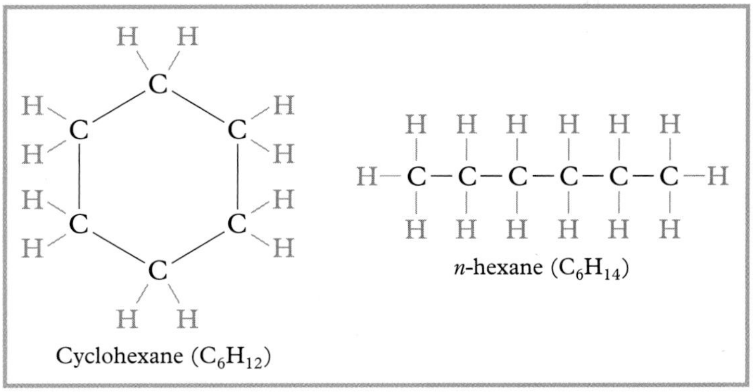

Cyclohexane (C$_6$H$_{12}$)

n-hexane (C$_6$H$_{14}$)

4.9 Shorthand representations are used for carbon chains and rings

In order to display the all-important carbon network in cycloalkanes and indeed in all organic molecules, chemists have devised a simple shorthand representation for structural formulas. In this shorthand, the hydrogen atoms are *not* displayed at all. Indeed, the carbon atoms are not shown either; *only the C—C bonds* (and, in organic chemistry generally, all bonds except C—H ones) *are shown*. Thus the shorthand diagrams for each of the three cycloalkanes displayed earlier are shown below their carbon–carbon networks:

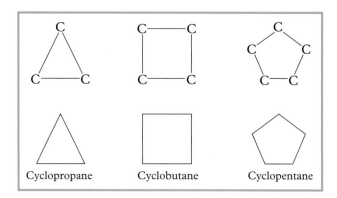

Cyclopropane   Cyclobutane   Cyclopentane

Exercise 4.3

Deduce the Lewis structure, the shorthand representation, and the name for the cycloalkane that contains a ring of six carbons. How many hydrogen atoms does the molecule have?

In shorthand representations, carbon chains are shown not as a continuous line of C—C bonds, but rather as a zigzag pattern, in order that each separate bond is obvious. Thus the shorthand representation for propane is:

Propane

It is understood that in such structures, *there is a carbon atom at the junction of any two lines, as well as at the ends of the lines.* As usual, we can deduce the number of hydrogen atoms at any point from the requirement that a carbon atom always forms four bonds. Thus the central carbon in propane is bonded to two hydrogens, since it is shown as forming two C—C bonds. The two terminal carbons each must be bonded to three hydrogens, since they are shown as forming only one C—C bond each.

Worked Example: Writing Lewis structures from shorthand structures

Draw the Lewis structure for the hydrocarbon whose shorthand structure follows:

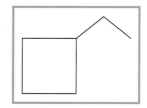

Solution: The square represents a ring of four carbons, since there are four corners to it. The side chain consists of two carbons, one at the "kink" and one at the terminus. Thus the carbon network in the hydrocarbon is cyclobutane with an ethyl group attached to one ring carbon:

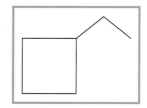

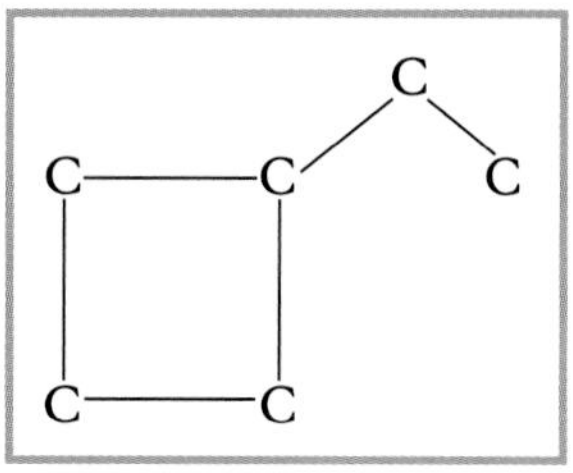

For each ring carbon except the one bonded to the side chain, there must be two hydrogen atoms attached to those carbon atoms that only form two bonds to other carbons. The carbon attached to the side chain forms three C—C bonds, so it has one hydrogen bonded to it. The central carbon of the side chain forms two C—C bonds, so it has two hydrogens. The terminal carbon forms only one C—C bond, so it must have three hydrogen atoms attached:

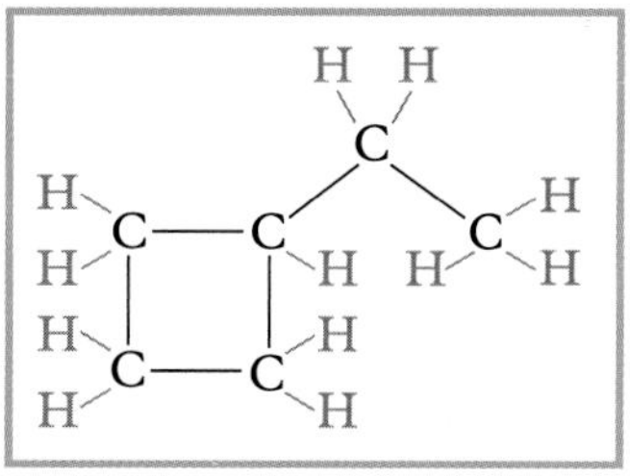

Exercise 4.4

Draw the Lewis structures for each of the three following shorthand structures:

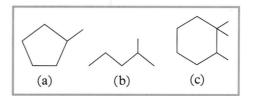

(a)   (b)   (c)

Draw the shorthand diagrams for the two isomers of butane.

Solution: Start by drawing as many carbon atoms as possible in a continuous line. The *n*-butane isomer has all of its carbons in a continuous line: C—C—C—C. Since it contains three C—C bonds, we draw three lines, one after another. We orient these lines alternately forward and backward so that we can see that three carbon atoms are present, where the lines join:

represents   C—C—C—C

The isobutane isomer has only two C—C bonds in a row. The third one is attached to the middle carbon, whose position corresponds to the junction of the other two bonds. Thus we draw the third line emanating from this junction:

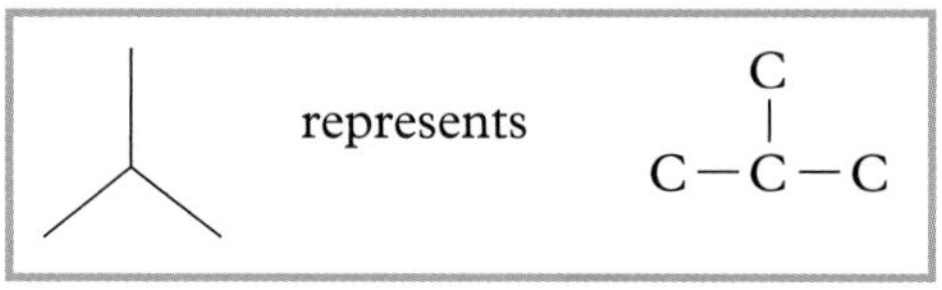

represents

$$\begin{array}{c} C \\ | \\ C-C-C \end{array}$$

Exercise 4.5

Draw the shorthand structure for the isomer of pentane whose Lewis structure is:

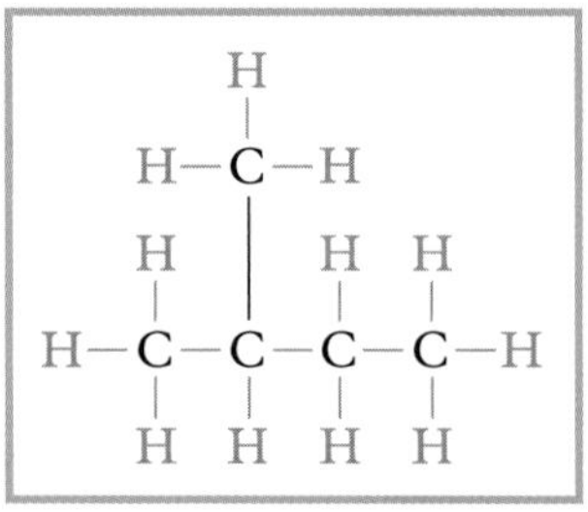

Molecules That Have Double and Triple Bonds

4.10 Some atoms can form double bonds with each other

If a sample of ethane gas, C_2H_6, is heated to a high temperature, a chemical reaction occurs. The net result is that a hydrogen molecule, H_2, is expelled from each C_2H_6 unit:

$$C_2H_6 \rightarrow C_2H_4 + H_2$$

Each of the pair of carbon atoms in the carbon-containing product, C_2H_4, forms only *two* C—H bonds, since one hydrogen atom from each ethane carbon is removed in the reaction. In order to form a total of four bonds each, then, the carbon atoms in C_2H_4 must share *two pairs* of electrons:

$$
\begin{array}{cc}
H & H \\
\vdots & \vdots \\
C & :: & C \\
\vdots & \vdots \\
H & H
\end{array}
$$

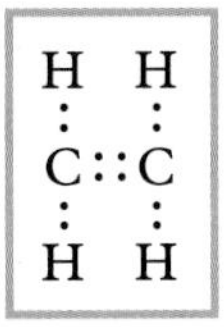

When a pair of atoms share *two* pairs of electrons, normally with two electrons being contributed from each atom, there is said to be a **double bond** between them. A double bond is represented by *two* lines joining the atoms, one on top of the other:

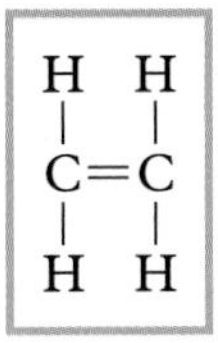

$$
\begin{array}{cc}
H & H \\
| & | \\
C & = & C \\
| & | \\
H & H
\end{array}
$$

To differentiate them clearly from double bonds, those such as C—H and C—C that involve only one electron pair are called **single bonds.**

Some of the properties of molecular oxygen are not consistent with this simple picture of its bonding, but we shall ignore this complication.

Atoms of some elements other than carbon also form double bonds. Consider, for example, the O_2 molecule. Before forming bonds, each oxygen atom has six valence shell electrons, since oxygen lies in Group VI of the periodic table. In order to acquire an octet of electrons, each oxygen must share two of its electrons with the other, forming a double bond:

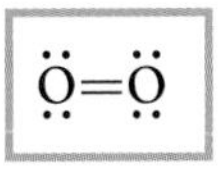

Exercise 4.6

Determine the Lewis structure for the molecule having the atom network HNNH, and thereby show that the two nitrogen atoms are connected by a double bond.

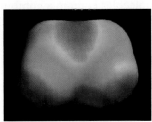

View different representations of an ethene molecule at Chapter 4: Visualizations: Media Link 6.

4.11 Alkenes contain a C=C bond

As a class, hydrocarbons containing a C=C bond are called alkenes. Formally, alkenes are given the same name as the alkane with the same carbon network but with all C—C single bonds, except that the *-ane* ending is changed to *-ene*. Thus C_2H_4 is properly called ethene, since C_2H_6 is called ethane. Ethene also has a traditional name, ethylene, that is still commonly used for it.

Double bonds are stronger than single bonds, approximately doubly so. Thus it would take much more energy to pry apart the CH_2 units in ethene than the CH_3 units in ethane. Another important difference between carbon atoms forming double bonds compared to those forming only single bonds is the geometry about the carbon. When a carbon atom forms one double and two single bonds, it and the atoms to which it is attached all lie in the *same plane*—in other words, the molecule is flat, or *planar*, around the atom. Because the sum of the angles about any point in a plane adds to 360°, the three bond angles about the two carbon atoms in a double bond are all approximately 120°:

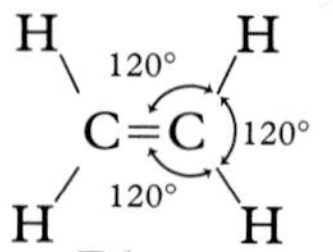

Ethene

In contrast to C—C single bonds, around which twisting occurs rapidly, the two groups at one end of a double bond can *not* rotate (revolve) relative to the pair at the other end.

Ethene can be obtained commercially from natural gas and petroleum. The process of forming ethene by heating the ethane in natural gas, described in section 4.10, is an important one in modern industry. Since hydrocarbons containing one or more C=C bonds are much more reactive than those with only C—C bonds, ethene can be converted into many interesting and useful materials such as plastics, as we'll see in Chapter 5.

Ethene also occurs naturally and is present in the air in small concentrations. It is emitted from fruit such as apples and lemons as they ripen. The presence of the ethene enhances ripening and hence further emission of the gas, so ethene is often intentionally used commercially to speed up the ripening process. However, rapid ripening can be set off unintentionally. If there is a piece of damaged fruit in a container of unripened pieces, it will quickly begin to ripen, emitting ethene which will prematurely accelerate ripening and rotting of the rest—thus the old saying "one rotten apple can spoil the barrel" is based in fact.

4.12 Some carbon chains have both double and single bonds

Many hydrocarbons and other organic molecules contain a combination of both C=C and C—C bonds, interspersed along a chain or within a ring. The simplest example is propene, also known as propylene. Both Lewis and shorthand structures for it are shown below:

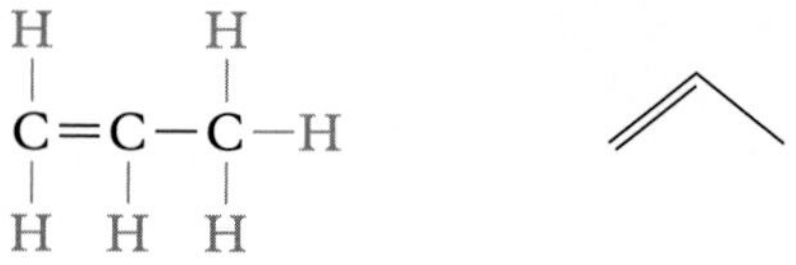

Propene

Propene is obtained by heating propane from natural gas or petroleum, producing it and hydrogen gas, in a reaction similar to that used to produce ethene from ethane, as explained in section 4.10.

A more complicated example is butadiene, molecules of which possess two carbon–carbon double bonds separated by a single one:

Butadiene

The Greek letter β—"beta"—here is used to denote the specific isomer of carotene that has this structure. This substance sometimes is written as "beta-carotene."

An example of a much larger hydrocarbon molecule, shown below, in which double and single bonds alternate along the chain is β-carotene, the substance that is responsible for the orange color of carrots and other vegetables (see Figure 4.3):

Beta-carotene

Beta-carotene is the precursor molecule to vitamin A, which is essential for bone development, vision, and maintenance of healthy skin. There is some evidence that a diet high in beta-carotene can increase resistance to certain types of cancer.

Benzene Because a carbon atom can form three bonds in a plane with angles of 120°, as in ethene, some interesting structures can result. For example, each of the angles in a (planar) regular hexagon is 120°. Thus a stable ring structure should exist with carbons at each of the six corners of a hexagon. This is possible provided that each carbon forms one double bond to one of its two adjacent carbons in the ring and a single bond to the other, since each of the resulting $C-C=C$ angles is 120°:

Figure 4.3 Foods containing beta-carotene. Note that not all foods containing this molecule are orange.
(George Semple for W. H. Freeman and Company)

In order to fulfill the requirement that it form four bonds, each carbon atom is also bonded to one hydrogen atom, which also lies in the C_6 plane:

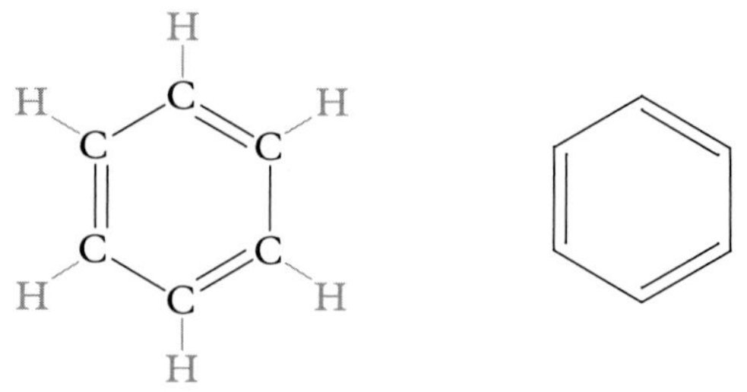

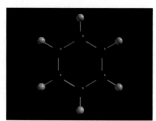

Benzene

The molecule shown above is benzene, C_6H_6. Notice that, in a benzene molecule, the single and double bonds *alternate* with each other. Such an alternation provides extra stability to a molecule if there are a total of exactly *three* double bonds and *three* single ones in one ring, as occurs in benzene. For this reason, benzene and other molecules that contain its C_6 ring display a special stability: they are quite resistant to participation in chemical reactions.

As a compound, benzene exists as a liquid, with a freezing point (5.5°C) just above that of water and a boiling point (80°C) somewhat below that of water. Benzene and some other organic compounds with the same characteristic ring structure have strong odors and are therefore called aromatic compounds.

Large quantities of benzene are obtained from petroleum. For many years, benzene was used both industrially and domestically as a solvent—for example, as a dry-cleaning agent. However, it eventually became clear that benzene is **carcinogenic**—in other words, it causes cancer, in this case human leukemia—and consequently many of its uses have been phased out. In modern times, most of our exposure to benzene comes from cigarette smoke, automobile exhaust, and the evaporation of gasoline. Because of its toxicity, the benzene content of gasoline has been restricted to no more than 1% in many jurisdictions.

The discovery of the cyclic character of benzene is alleged to have come to the German chemist Friedrich Kekulé in a dream, when he visualized a chain of carbon atoms as a snake seizing its own tail.

View different representations of a benzene molecule at Chapter 4: Visualizations: Media Link 7.

4.13 The atoms of a few elements form triple bonds

Nitrogen occurs in Group V of the periodic table and therefore has five valence-shell electrons. As we saw in section 3.16, a nitrogen atom needs to form three covalent bonds with another atom in order to achieve an octet. In N_2, the nitrogen atoms form all three bonds with each other, producing a molecule with the following Lewis structure:

$$:N{\equiv}N:$$

The sharing of three pairs of electrons by two atoms is called a **triple bond.**

Two carbon atoms can also form a triple bond. Since each carbon forms a total of four bonds, one other atom is connected by a single bond to each of these carbons:

$$-C\equiv C-$$

The simplest hydrocarbon that contains a carbon–carbon triple bond has both carbons bonded to a hydrogen atom:

$$H-C\equiv C-H$$
Ethyne

The compound consisting of such C_2H_2 molecules is called ethyne, more commonly known as acetylene. Like other hydrocarbons, acetylene burns in air. Since acetylene burns with a particularly hot flame, it is the gas used to fuel welding torches.

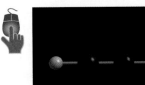

View different representations of an ethyne molecule at Chapter 4: Visualizations: Media Link 8.

Exercise 4.7

Determine the Lewis structure for a molecule of the poisonous gas hydrogen cyanide, HCN. Note that the hydrogen is bonded to the carbon, and the carbon to the nitrogen.

Fossil Fuels: Natural Gas and Petroleum

Modern society is reliant on organic compounds, especially for the generation of energy and for the production of synthetic materials, notably plastics and pharmaceuticals. Three carbon sources satisfy our enormous appetite for this element: oil (petroleum), coal, and natural gas. These three substances are known as **fossil fuels,** since they all burn in air to release energy in the form of heat, and since they are the residual by-products of organisms that lived hundreds of millions of years ago. Then, as now, the majority of plants and animals decayed after death and were ultimately transformed into carbon dioxide and water, in a process whose *net effect* is similar to the combustion of hydrocarbons. However, a small fraction of the dead matter was buried by natural forces before it had completely decayed. Because this material was cut off from the oxygen that was required for its oxidation, it was unable to complete its decomposition to carbon dioxide and water. This unoxidized carbon material is what we know today as fossil fuel. It occurs in huge deposits in many places around the world.

4.14 Supplies of fossil fuels are limited

Because the global quantities of energy used annually are so large, we use the energy unit of exajoules, EJ, where 1 EJ equals 10^{18} joules, to discuss them. The total amount of commercial energy that is consumed annually by humans worldwide amounts currently to about 400 EJ.

Table 4.2 World reserves of fossil fuels

Fuel type	Proven reserves	Years of supply
Coal	28,000 EJ	200
Petroleum	5,700 EJ	30–40
Natural gas*	5,600 EJ	60

Source: Original reserves data (as of 2000) obained from U.S. Government Web site. www.eia.doe.gov/emeu/iea, and converted to joules.

*Does not include methane hydrates.

Although it is commonly said that we are "running out" of fossil fuels, there is little evidence to support this contention globally in the short-to-medium term, that is, the next few decades. Estimates of the world's fossil fuel energy reserves are listed in Table 4.2. Proven reserves are those which we are sure are available and which can definitely be exploited. Proven reserves of natural gas continue to climb, even though the use of natural gas is growing faster than the use of either coal or oil. Continuing improvements in technology allow greater and greater proportions of the oil in a given deposit to be extracted. Consequently, it appears that there is enough oil and gas to last us until about mid-century (see Table 4.2), though perhaps at a higher cost than we have been accustomed to paying. The reserves of coal appear to be sufficient for several centuries. In addition to these reserves, an enormous quantity of natural gas is held in methane hydrates (methane held in weak combination with water), buried deeply in ocean sediments as ice crystals and also present in permafrost (see Figure 4.4). These sources would double the total fossil fuel reserves if they could be tapped.

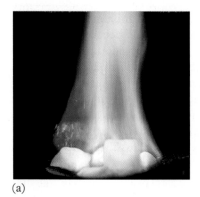

(a)

Figure 4.4 (a) Flaming methane hydrate. (b) Methane hydrate crystals. (Part a, L. Stern and J. Pinkston, U.S. Geological Survey; part b, from video by Dr. P. Aharon, funded by NOAA Ocean Exploration Program)

(b)

Chapter 4: Powering the Planet

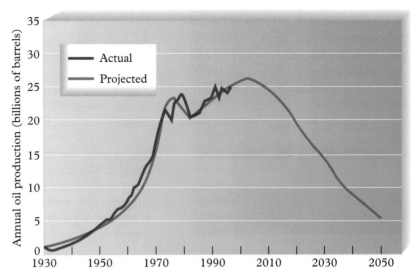

Figure 4.5 Worldwide oil production, 1930–2050. (Adapted from "The End of Cheap Oil" by Colin J. Campbell and Jean H. Laherrere. *Scientific American,* March 1998, vol. 278, no. 3, p. 81. © 1998 by Scientific American, Inc. All rights reserved.)

Taking It Further

For information and problems on clean coal, go to Taking It Further at www.whfreeman.com/ciyl2e.

Although it took nature about half a billion years to create the world's supply of petroleum, humans probably will have used almost all of it during the 200-year period that started just before the end of the 19th century. Indeed, the production of petroleum in the lower 48 states of the United States has already peaked. Many, though not all, experts predict that world production will reach its maximum, and we will begin to "run out" of oil by about 2010, give or take a decade. The rise and fall of oil production is predicted by one popular model to follow a descending curve as shown in the red line of Figure 4.5. In commerce, petroleum is measured in barrels, each of which is equivalent to 159 liters, or 42 U.S. gallons. Current annual world oil production amounts to about four *trillion* (4×10^{12}) liters, or 27 billion barrels. Much of the proven oil reserves are found in the Middle East.

At current rates of consumption of oil, the total proven reserves correspond to about a 45-year supply. Very long chain hydrocarbons from *oil shales,* a type of sedimentary rock, and *tar sands,* which are sandstone or porous rock impregnated with very viscous crude oil (see Figure 4.6), could extend the supply. Indeed, the potential for oil from tar sands in Alberta, Canada, exceeds the oil reserves of Saudi Arabia! Extracting crude oil from these sources is expensive. However, as the more accessible supplies dry up and the price of oil rises, such extraction processes are becoming more economically viable. The same is true of oil shales, such as those in Wyoming, Colorado, and Utah. The shales contain the material *kerogen,* which upon heating in the absence of air releases its bound oxygen, leaving a petroleum-like oil.

Figure 4.6 Tar sands. (Courtesy of Syncrude Canada Ltd.)

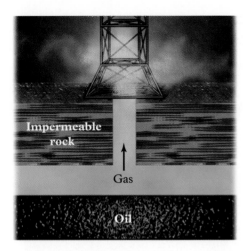

Figure 4.7 Releasing natural gas deposits from the earth.

4.15 Natural gas consists of short-chain alkanes

We have mentioned the common fuel called natural gas a few times in our discussion of hydrocarbons, but what does this name actually refer to? In terms of its hydrocarbon content, natural gas as it exits from the ground consists predominantly (60–90%) of methane, CH_4. The other component alkanes—ethane, propane, and the two butanes—all of which are gases that consist of small molecules, are present to varying extents depending upon the geographic origin of the deposit. Natural gas deposits are found in geological formations in which the gas mixture has been trapped by a mass of impermeable rock. Drilling a hole down through the rock releases the gas in a steady flow to the surface, as shown in Figure 4.7.

Methane Methane's boiling point is so low ($-164°C$) that it does not readily condense into a liquid, even at moderately high pressures. The other alkanes all possess boiling points that are substantially higher than that of methane. This makes it possible to remove the other alkanes from natural gas by lowering the mixture's temperature and thereby condensing these other hydrocarbons to liquids. After processing, the natural gas, which now is mainly methane, is transported by pipelines to consumers. Methane burns readily, producing substantial amounts of heat as it does:

$$CH_4 + 2\ O_2 \rightarrow CO_2 + 2\ H_2O + \text{heat}$$

As a domestic fuel, natural gas is used especially to heat homes (about half of those in North America) and other buildings, and to fire gas stoves and hot-water heaters.

In some respects, methane's very low boiling point is unfortunate, because as a liquid it would be much more convenient to use and more efficient to transport. A tankful of a noncompressed methane gas would contain such a small mass of the substance that it would possess little fuel value. Methane is usually sold as *compressed natural gas* (CNG) in heavy tanks that have been constructed to withstand high gas pressures. As a highly compressed gas, methane is used as the fuel for motor vehicles that are specially designed or adapted for its use (see Figure 4.8).

Methane is also used in the production of commercial quantities of hydrogen gas. Currently the cheapest and most efficient method of producing hydrogen is to chemically convert methane into a mixture of hydrogen and carbon dioxide by reacting it with water in the form of steam:

$$CH_4 + 2\ H_2O \rightarrow CO_2 + 4\ H_2$$

Propane Propane, C_3H_8, obtained from natural gas or from petroleum refining (described in *Taking It Further* at the end of the chapter), has achieved more popularity than methane as an alternative fuel for motor

Figure 4.8 Vehicles powered by natural gas *(left)* and propane *(right)*. (Courtesy of the U.S. Department of Energy.)

vehicles adapted to its use. Propane condenses at $-42°C$ and exists as a liquid even at ambient temperatures if high pressure is applied to it, so it can be readily liquefied. The volume required to store a given amount of energy is considerably less for propane than for methane since liquids are more dense than even compressed gases. Propane fuel is liquid, but it is the vaporized gas that exits the fuel tank when it is drawn to burn.

Propane and/or butane are sold individually, or as a mixture called *liquefied petroleum gas (LPG)*, in small tanks as a liquid fuel that can be used for heating and cooking purposes where natural gas pipelines do not exist. Like methane, propane is an odorless gas. A tiny amount of a sulfur-containing organic molecule with a very strong, unpleasant odor is added to both CNG and LPG so that any leakage of the gases from the containers or pipelines can be readily detected by the consumer.

4.16 The hydrogen sulfide in natural gas is removed before combustion

Natural gas, as it exits from the ground, also contains molecular substances other than hydrocarbons. Although some of these constituents, such as N_2, CO_2, and noble gases, are innocuous, the inevitable presence of hydrogen sulfide is problematic since it is poisonous. "Sour" natural gas wells are those in which the concentration of hydrogen sulfide is particularly high.

Hydrogen sulfide, H_2S, a gas having the odor of rotting eggs, is produced wherever living matter decays in the absence of air—for example, in swamps and sewers. In high concentrations, around 500 ppm, the gas is *asphyxiating;* that is to say, it produces unconsciousness and possibly even death by paralyzing the respiratory center of the brain, thus depriving the individual of oxygen. Even in developed countries, there are regular reports of death from hydrogen sulfide poisoning when natural gas employees work in deep holes at drilling sites or in small rooms at refineries. Exposure to lower levels can cause brain damage and spontaneous abortions. At the much lower concentrations likely to be commonly encountered in everyday life, hydrogen sulfide is an irritant to the eyes, respiratory tract, and skin.

Upon combustion in air, H_2S produces sulfur dioxide gas. The reaction is analogous to that for methane when it burns:

$$2 H_2S + 3 O_2 \rightarrow 2 SO_2 + 2 H_2O$$

This reaction occurs when natural gas from a well that cannot be immediately sold or processed is "flared"—that is, combusted on the spot. Unfortunately the flaring often leaves a significant fraction of the hydrogen sulfide (and the methane) unburnt, and thus H_2S levels in the air surrounding the well can become elevated and hazardous to local residents.

The hydrogen sulfide in natural gas that is destined for market can be readily removed by converting it to elemental sulfur, a yellow solid. This conversion is accomplished by exposing the gas to some sulfur dioxide gas:

$$2 H_2S + SO_2 \rightarrow 3 S + 2 H_2O$$

The sulfur dioxide required for this reaction can be produced by burning sulfur or H_2S in air.

4.17 Petroleum is a mixture of hydrocarbons

Like natural gas, the sticky, viscous liquid called **petroleum**—also called crude oil—is a complex mixture consisting mainly of alkane hydrocarbons. The exact composition varies with the location of the source. In general, the boiling points of alkanes increase with the number of carbon atoms that they contain, as illustrated in Figure 4.9. Thus, since the average chain length of the molecules is much greater in petroleum than in natural gas, petroleum is a liquid rather than a gas.

Petroleum corresponds to the fossil remains of microscopic animals that lived in the seas in ancient times. The fat content of these animals, like that of our own bodies, consisted of molecules having long unbranched chains of carbon atoms that survive relatively intact as the alkanes found in oil. Thus, the alkane molecules in oil are mainly unbranched. They range from methane to those containing more than 40 carbon atoms which, if isolated as pure compounds, would be solids rather than liquids. In addition to alkane chains, petroleum contains some cyclic hydrocarbons, including those containing benzene rings. The fraction of the mixture that is cyclic compounds depends strongly on the geographic origin of the oil.

Petroleum, found in certain rock formations in the ground, is pumped to the surface in oil wells. As it exits from the ground, crude oil is not a very useful substance because it is a mixture of so many compounds. To gain utility, it must first be separated into components, each of which has several particular uses. For example, the molecules contain-

Figure 4.9 Straight-chain alkane boiling points versus chain length.

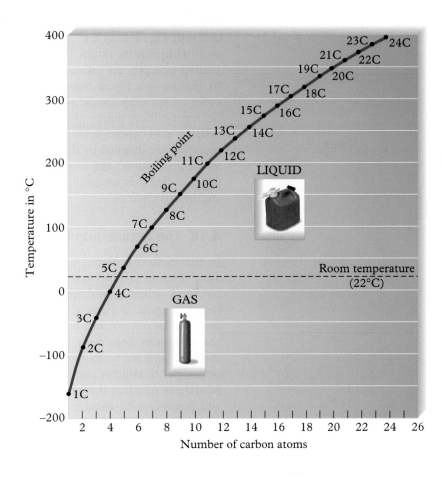

ing one to four carbon atoms each are gases (methane through butane) that are dissolved in the liquid oil. As a group, they can be separated from each other and employed as discussed earlier.

Similarly, hydrocarbon solids are dissolved in the liquid oil and as a group can be separated as described in *Taking It Further* at the end of this chapter. The solids with carbon chains having about 20 to 40 carbon atoms are useful as *paraffin wax*. The remainder of the solid component contains from 40 to 200 carbon atoms, and this component collectively can be separated out and used as asphalt for paving roads and as tar for roofs. Although the bulk of the atoms in asphalt are carbon and hydrogen, its constituent molecules contain some sulfur and small concentrations of other elements, including metals. Like plastic, asphalt softens when it is heated, and it even flows at certain temperatures. Asphalt used in paved roads hardens and begins to crack and crumble when its more volatile components are eventually lost.

The liquid compounds that are present in crude oil consist of hydrocarbons containing from 5 to about 20 carbon atoms each. Although no attempt is usually made to isolate individual compounds from the mixture, crude oil is separated into a number of *fractions*—different liquid solutions whose components all boil within a relatively small temperature range. This separation of oil into fractions is accomplished by a process called **distillation,** which involves the vaporization by boiling of a liquid mixture, followed by the cooling of the vapor in order to cause its condensation back to the liquid state. Because of the different boiling points of the compounds, it is possible to separate the mixture into components. Each day, a total of about 10 billion liters of crude oil are distilled by this procedure in hundreds of *petroleum refineries* located around the world. The procedure, fractional distillation, by which refining is accomplished is described in *Taking It Further* at the end of this chapter.

4.18 The octane number of gasoline

In the modern internal combustion engines present in our cars and vans, ignition is supposed to occur, and maximum power can only be obtained, when a mixture of gasoline and air is highly compressed before a spark initiates the combustion reaction. Gasoline consisting primarily of unbranched alkanes and cycloalkanes has poor combustion characteristics because a mixture consisting of vaporized gasoline of this type and air tends to ignite spontaneously in the engine cylinders before the mixture is completely compressed. As a consequence of this premature ignition, the engine "knocks," with a resulting loss of power and possibly with damage to the engine. To overcome such problems, all modern gasoline is formulated to contain substances that will prevent knocking.

In contrast to unbranched alkanes, highly branched alkanes—such as the octane isomer called isooctane—have excellent burning characteristics for internal combustion engines. In particular, they are much less prone to premature ignition. The carbon network of isooctane is:

> Octane is the name for alkanes that contain eight carbon atoms.

$$
\begin{array}{ccccc}
 & C & & C & \\
 & | & & | & \\
C{-}C & {-}C{-}C & {-}C \\
 & | & & & \\
 & C & & &
\end{array}
$$

Isooctane's carbon skeleton

Unfortunately, significant amounts of highly branched alkanes do not occur naturally in crude oil. To overcome this deficiency, it is common for oil refineries to chemically convert some petroleum into highly branched alkanes and to blend this product with the "straight-run" gasoline, which is that extracted directly from petroleum.

The ability of a gasoline to generate power without engine knocking is measured by its **octane number.** On this scale, isooctane is assigned an octane number of 100, and *n*-heptane, the straight-chain alkane with seven carbons, is arbitrarily assigned a value of zero. The octane number of other substances is determined by matching their knocking performance with various blends of *n*-heptane and isooctane. Numerical values

Table 4.3 **Octane numbers for selected hydrocarbons**

Formula	Hydrocarbon	Symbolic structure	Octane number
C_4H_{10}	n-butane		94
C_5H_{12}	n-pentane		62
C_6H_{14}	n-hexane		25
C_7H_{16}	n-heptane		0
	2-methylhexane*		42
	2,3-dimethylpentane*		90
C_8H_{18}	isooctane (2,2,4-trimethylpentane)*		100
C_6H_6	benzene		106
C_7H_8	toluene		118
C_8H_{10}	p-xylene		116

* The numbers here refer to the positions of the methyl group branches along the chain, where *2* designates the second carbon, *3* the third carbon, and so on, from the beginning.

of the octane number for a few hydrocarbons are given in Table 4.3. Notice in the C_7H_{16} series how the value increases with the extent of branching. Notice also that short-chain alkanes such as n-butane have high octane numbers and that these decrease with increasing chain length. Straight-run gasoline has an octane number of about 55, which is much too low for use in modern vehicles. Blending in some branched alkanes raises the octane number somewhat, but not up to the high 80s value that is required in modern car engines.

4.19 Lead compounds raise gasoline's octane number

When hydrocarbon groups are bonded to a central atom, the molecular formulas are often written using parentheses to illustrate the nature of the group.

It was discovered decades ago that, when dissolved in gasoline in small amounts, the compounds tetramethyllead, $Pb(CH_3)_4$, and tetraethyl lead, $Pb(C_2H_5)_4$, prevent engine knocking and hence greatly boost the octane number of gasoline. For decades, these "octane number enhancers" were added worldwide to gasoline consisting predominantly of unbranched alkanes and cycloalkanes. However, not only were lead compounds emitted into the air when the gasoline was combusted, but lead interfered with the operation of catalytic converters, so its use indirectly increased the emissions of other pollutants as well.

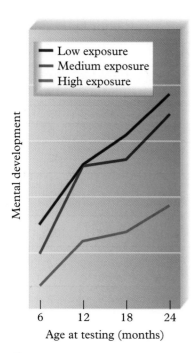

Figure 4.10 Mental development after different levels of prenatal lead exposure. (Data from *New England Journal of Medicine*, 136 (1987): 1037–1043.)

As a consequence of the emission of lead compounds from automobiles into the atmosphere, the air, the vegetation, and the top level of the soil in urban areas and those close to highways all became polluted with lead. Indeed, the lead from gasoline became the leading source of the element for people in many urban areas, and remains so today where leaded gasoline is still sold. Lead additives have largely been phased out now in most developed countries, resulting in a substantial reduction in environmental lead levels. For example, European scientists have been able to monitor the rise and fall of lead by analyzing for the element in different vintages of a French red wine (Chateauneuf du Pape) that was produced for many years from grapes that were grown near two busy freeways. They found that lead levels rose steadily to a maximum in the mid-1970s, a period that was followed by a steady decline in the level to about one-tenth its former peak by the early 1990s.

Humans are most at risk from lead as fetuses and as young children, because they absorb more of the element as their brains are growing. Several studies have established a small but consistent neuropsychological impairment in young children who absorbed lead before or after their birth. Lead has a deleterious effect on children's behavior and attentiveness, and probably also on their intelligence. This latter effect is illustrated in Figure 4.10, where the mental development index of groups of children who were inadvertently exposed before birth to different amounts of lead is plotted against their age. Clearly, the higher the level of exposure, the more their mental development was retarded. Once Americans switched to unleaded gasolines, children's blood lead levels declined remarkably. Figure 4.11 contrasts the distribution of levels in children born in the 1980s with those born in the 1970s, before lead was removed from gas.

4.20 Lead compounds have been replaced in gasoline

In some European countries and in Canada, lead compounds in gasoline were replaced by small quantities of a carbon-containing compound of manganese that acts as an octane enhancer but also pollutes the environment. The alternative to using lead or manganese additives to boost octane ratings is to blend into gasoline significant quantities of oxygen-containing substances such as ethanol or MTBE, both of which will be discussed in Chapter 6. Also used to boost octane numbers are compounds related to benzene, but in which one or two of the six hydrogen atoms have been replaced by a methyl group, CH_3. The replacement of one such hydrogen gives the hydrocarbon named toluene, and the replacement of two gives the three isomers of xylene:

Toluene   Xylenes

Figure 4.11 Blood lead levels in U.S. children for two different time periods. (Data from R. A. Goyer, 1996. Results of lead research: Prenatal exposure and neurological consequences. *Environmental Health Perspectives*, 104: 1050–1054.)

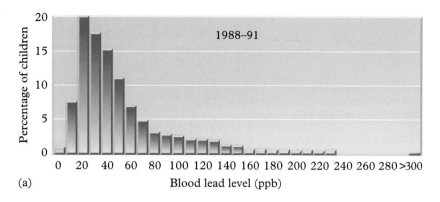

(a)

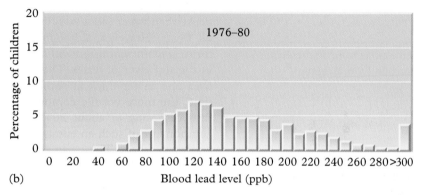

(b)

Taken as a group, benzene, toluene, and xylene are known as the *BTX* fraction of gasoline, or sometimes its "aromatic" content.

The BTX content of gasoline sold in the United States was as high as 40% in the past. Unfortunately, the BTX hydrocarbons are more reactive than the alkanes that they replace in causing air pollution, so the lead pollution has been reduced at the price of producing more smog. Benzene itself is a worrisome air pollutant since, as we saw earlier, at high levels it has been linked to increases in the incidence of leukemia. The reformulated gasoline used in the second half of the 1990s in North America contained a maximum of 1% benzene and 25% aromatics in total, with a minimum of 2% oxygen content, which is achieved by incorporating ethanol or MTBE. The second phase of reformulated gasoline reduced the benzene and BTX components even further. This phase of the program, implemented in 2000, mandated that reformulated gas be used in the nine cities in the United States with the worst smog problem: Los Angeles, San Diego, Chicago, Houston, Milwaukee, Baltimore, Philadelphia, Hartford, and New York City.

4.21 Oil spills pollute waterways

The majority of the world's petroleum production lies far from the regions of its greatest consumption, so most oil is transported over long distances—by pipeline or by ocean tanker. Even the normal operation of the tankers results collectively in the discharge of substantial amounts of oil into natural waters each year. Accidental spills of large amounts of oil produce huge concentrations at one location, often near a coast, and result in dramatic effects on the plants and animals that live in the waters and the nearby shoreline.

The most famous oil spill in recent times was that from the tanker called *Exxon Valdez*, in March 1989. The ship had just left the port of Valdez, Alaska, when, in an attempt to avoid sea ice, it ran aground in shallow water in Prince William Sound and began to leak its oil cargo. About one-fifth of the 200 million liters of oil it was carrying leaked into the waters. The rest was pumped into other ships sent to help.

Since the hydrocarbon mixture that constitutes crude oil is less dense than water (about 0.8 versus 1.0 grams per cubic centimeter), the oil floats. However, although the oil is almost insoluble in water, it forms a foamy emulsion with it—tiny droplets of oil become suspended in the top layer of water. The emulsified "mousse" of oil in water was over one meter thick at some places in Prince William Sound! Over time, the more volatile components in the oil evaporated into the air, thus disposing of about 20% of the spill. Unfortunately, the heavier components of such an emulsion gradually form tar balls, which persist and wash ashore. These can also sink and cover the habitats of shallow-water organisms or threaten coral reefs. Also, small amounts of the lighter components dissolve in water and prove toxic to sea creatures.

The seriousness of an oil spill depends upon its size, the type of oil spilled, where the oil finally settles, and environmental factors such as weather and time of year. Short-term harm to shellfish from oil washed onto a rocky or sandy shore is contrasted to long-term effects such as those seen in stands of mangrove trees. Mangroves exposed to an oil spill in Panama in 1986 still held toxic oil after five years. Contaminated marshes can ooze liquid oil up to ten years after a spill. Bottom-dwelling fish exposed to compounds released after oil spills can develop diseases and may have growth problems or be unable to reproduce.

The majority of oil from the *Valdez* spill, and indeed most of the oil that enters the ocean waters from routine tanker operations and even from natural processes, is eventually ingested and oxidized by bacteria to carbon dioxide and water. Indeed, the most effective treatment found for the oil that washed up on the shores in Alaska was to fertilize the bacteria, using nitrogen and phosphorus compounds, to encourage their growth and ability to consume the hydrocarbons. About half the oil was oxidized by bacteria; about one-eighth was recovered by cleanup crews (see Figure 4.12).

You may have seen television pictures of efforts to clean up oil spills around the world. The best technique is to contain the floating oil in a small surface area using booms, and then to skim the oil from the surface. Although this technique works in calm waters, the preferred method in rough seas is to add chemical dispersants that break the oil into tiny droplets that subsequently decompose more quickly in the

Emulsions were discussed in section 1.9 of Chapter 1.

Figure 4.12 Cleanup from the *Exxon Valdez* oil spill. (Vanessa Vick/Photo Researchers)

Figure 4.13 Oil spill containment and dispersal. (Bernd Wittich/Visuals Unlimited)

water (see Figure 4.13). Another technique is to ignite the oil on the surface and burn it away, though this usually generates great clouds of black smoke since the combustion is very incomplete.

As dramatic and serious as they are, oil spills account for only 37 million of the approximately 700 million gallons of oil put into the oceans worldwide every year. The majority, about 360 million gallons, enters the oceans from land, industrial, and municipal runoff.

● Tying Concepts Together: The volatility of gasoline

Recall from Chapter 2 that the vapor pressure of every liquid has a specific value at a given temperature. In general, the vapor pressure of an alkane hydrocarbon is greater the fewer the carbons it possesses. For example, pentanes (five C atoms) are more volatile than heptanes (seven C atoms), so a larger fraction of the pentane molecules in a gasoline mixture will vaporize and exist as a gas above the surface of a liquid gasoline mixture.

The issue of gasoline volatility is important to the operation of motor vehicles. Some components of the gasoline must be quite volatile, so that some fuel vapor is available when the engine is first started. This is usually provided by the evaporation of short-chain alkanes, such as butanes and pentanes. In cold climates, the butane component of gasoline is increased in winter in order to provide sufficient gaseous fuel, since, as we saw in Chapter 2, vapor pressure decreases rapidly with decreasing temperature.

In warm weather, the most volatile components evaporate from the fuel tank into the air, where they contribute to air pollution. For this reason, summer gasoline formulated for the southern U.S. states must be less volatile than that for northern areas.

Fossil Fuels: Natural Gas and Petroleum

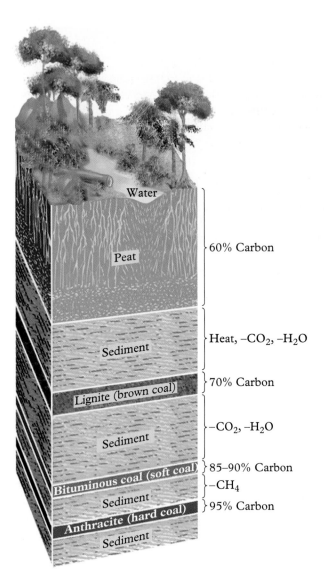

Water

Peat

60% Carbon

Sediment

Heat, $-CO_2$, $-H_2O$

Lignite (brown coal)

70% Carbon

Sediment

$-CO_2$, $-H_2O$

Bituminous coal (soft coal)

85–90% Carbon

$-CH_4$

Sediment

Anthracite (hard coal)

95% Carbon

Sediment

Figure 4.14 Different types of coal and their compositions.

Fossil Fuels: Coal

As indicated in Table 4.2, coal is by far the most abundant of the fossil fuels. It is available in great amounts in many regions of the world, including developing countries, and is quite cheap—in monetary terms—to mine and transport. North America has about one-third of the known coal reserves.

4.22 The composition of coal varies with its age

The solid that we call coal is largely elemental carbon, C, mixed with some mineral material. It was formed over the last half billion years from the tiny proportion of ancient plant matter that was covered over by mud or water and that as a consequence could not be oxidized back to CO_2. The organic component of coal was formed from the remains of land-based woody plants that grew and accumulated in swamps. Initially the submerged plant matter decomposed to some extent through the action of microorganisms, which converted it into the material known as peat. Eventually the peat was covered by water and sediments; biological transformations then ceased, but geochemical ones replaced them and in time transformed the peat, through various stages, into coal.

The plant material that formed peat was the lignin component of the original plants. At the atomic level, lignin consists of very large molecules consisting mainly of carbon, hydrogen, and oxygen. Over long periods of time, as the material was buried deeper and deeper into the ground, it was subjected to elevated temperatures. Oxygen was gradually lost—as water and carbon dioxide—from the buried material. In the process, even more extensive networks, mainly of carbon atoms, were formed, eventually yielding the very carbon-rich, hard material known as coal (see Figure 4.14). Not all the coal available today has undergone all the stages of transformation. For example, much of the coal in eastern Europe is brown coal.

In the final transition of coal, to anthracite, the hydrogen component decreases due to the elimination of methane, CH_4, from the solid. The methane gas does not escape the solid completely but is released when the coal is mined. The buildup of methane in underground mines generates conditions in which fires and explosions can occur, so the gas is often extracted from coal seams before they are mined. The methane is then used to supplement other sources of natural gas.

4.23 Coal mining creates environmental problems

The seams of coal in commercially viable deposits are generally several meters thick. Coal is often obtained from underground mines, which is expensive in human lives. Over 100,000 miners were killed by accidents, explosions, cave-ins, etc., in American mines alone during the 20th century. Coal is also obtained from *strip mines,* which are safer for the workers but which often produce serious environmental degradation. In strip mining, a large area of the surface soil, rock, and vegetation is removed to reveal the coal seam, which is then removed using heavy machinery. The large holes in the ground, and the great piles of soil, that result are visually ugly. Perhaps more significantly, the exposed soil is readily washed away by rainfall. Modern regulations for strip mining in the United States require that once coal is removed from an area, the land must be restored by filling the gaps and by planting trees and vegetation over the restored landscape. Strip mining now is not permitted on steep slopes where such restoration is impractical.

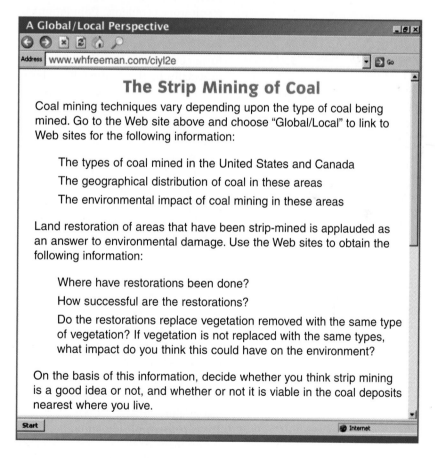

A Global/Local Perspective

Address www.whfreeman.com/ciyl2e

The Strip Mining of Coal

Coal mining techniques vary depending upon the type of coal being mined. Go to the Web site above and choose "Global/Local" to link to Web sites for the following information:

The types of coal mined in the United States and Canada

The geographical distribution of coal in these areas

The environmental impact of coal mining in these areas

Land restoration of areas that have been strip-mined is applauded as an answer to environmental damage. Use the Web sites to obtain the following information:

Where have restorations been done?

How successful are the restorations?

Do the restorations replace vegetation removed with the same type of vegetation? If vegetation is not replaced with the same types, what impact do you think this could have on the environment?

On the basis of this information, decide whether you think strip mining is a good idea or not, and whether or not it is viable in the coal deposits nearest where you live.

Start — Internet

4.24 Coal is called a "dirty" fuel

The problem for the next few hundred years is not the *supply* of coal, but rather the pollution caused when it is mined and when it is burned.

Although the mineral component of coal is largely clay, it unfortunately also contains measurable quantities of virtually every naturally occurring element. Thus coal has a reputation for being a "dirty fuel." Consequently, when coal is burnt, it emits not only CO_2 and H_2O but also substantial quantities of many air pollutants, notably sulfur dioxide, fluoride compounds, uranium and other radioactive metals, and heavy metals, especially mercury. Coal-fired power plants and municipal and medical waste incinerators are the biggest sources of mercury emissions to the atmosphere in North America. The vaporized mercury is eventually oxidized and returns in rain and snow, often falling far from the site of the original emissions. The U.S. EPA had formulated regulations requiring all coal-fired power plants to install pollution equipment that would remove mercury from exhaust gases before they are emitted into the air. However, an alternative program that would reduce total emissions, rather than ones from each plant has been adopted instead.

Soot released into the air from the burning of coal is a respiratory irritant associated with asthma attacks, decreased lung function, and respiratory problems. Populations that are especially vulnerable to these effects include young children, the elderly, and people with cardiopulmonary disease. Sooty smoke from coal burning in domestic stoves and furnaces has been an air pollution problem for many hundreds of years, especially in the United Kingdom. John Evelyn wrote in his January, 1684 diary that "London by reason of the excessive coldness of the air, hindering the ascent of the smoke, was so filled with the fuliginous [sooty] steam of sea-coal, that hardly could one see across the street, and this filling the lungs with its gross particles exceedingly obstructed the breast, so as one would scarce breathe." Indeed, unsuccessful attempts to control coal burning and punish offenders had begun in Britain as early as the 13th century. Domestic coal burning has been largely discontinued in developed countries.

Since it can be burned cleanly in large-scale facilities and is relatively inexpensive, coal is still used in most developed and developing countries for electric power production. The heat from its combustion is used to generate steam that in turn is used to turn turbines, the action of which generates an electrical current. Indeed, almost all the coal that is mined in North America is used to produce electricity. Unfortunately, the ratio of CO_2 emitted to energy produced is substantially greater for coal than for the other fossil fuels. Carbon dioxide is an important "greenhouse gas" that produces global warming of the atmosphere; this topic is discussed intensively in Chapter 16.

The combustion of coal is the main human-derived source of atmospheric SO_2, since the fuel contains 1 to 6% sulfur, depending upon the geographic area from which it is mined. About 0.5 to 1.5% of coal's mass consists of sulfur that is covalently bonded to carbon in the organic component of the solid and that cannot be removed without expensive processing. However, usually half or more of the sulfur is present in the mineral component of the coal. High-sulfur coal arises when there is a marine origin to the substance because sulfur compounds that were dissolved in the salt water became deposited in the mineral sediments.

If coal is pulverized before combustion, the sulfur associated with the mineral component can be mechanically removed by a technique that relies on the difference in density between the pulverized particles that are mainly mineral and those that are mainly carbon. The coal is mixed with a liquid that has a density intermediate between those of coal's components. The carbon particles float to the top, where they can be skimmed off and subsequently used as low-sulfur solid coal. The mineral particles, containing most of the sulfur, sink to the bottom and can be disposed of as a waste solid.

Alternatively, sulfur dioxide emissions from power plants can be reduced by substituting oil, natural gas, or low-sulfur coal for the uncleaned coal, but these fuels usually are more expensive than high-sulfur coal.

> ▶ **Discussion Point:** Deciding whether to remain a gasoline-based society

Currently we are reliant upon gasoline derived from petroleum to fuel our vehicles. No matter how we reformulate this liquid, there are flaws with its use. In the coming decades, as petroleum runs out, society will face several choices regarding fuel for our vehicles, including (1) to remain gasoline-based by converting our ample supply of coal into synthetic gasoline, though, as we've seen, the use of coal involves environmental problems, or (2) to switch to fuels such as hydrogen, which could eventually be obtained using solar energy but which would be more expensive than gasoline for some time to come.

Develop arguments for and against each of these two options by using material in this chapter and available at some of the Web sites listed on the Web site for this book, as well as your own opinions regarding these issues.

Summarizing the Main Ideas

Organic compounds are those that contain carbon and other elements; hydrocarbons contain carbon and hydrogen only. Carbon forms four covalent bonds, and hydrogen one bond, in organic molecules. The geometry about a carbon atom that forms bonds to four separate atoms is tetrahedral. Carbon atoms readily form covalent bonds with other carbon atoms and thus can form rings and long chains. Condensed formulas show the internal structure of molecules by listing the carbon atoms in sequence, with the hydrogen atoms displayed after each one.

Alkane hydrocarbons consist of chains of single-bonded carbon atoms to which hydrogen atoms are also attached. The general formula for an alkane hydrocarbon is C_mH_{2m+2}. Alkanes having four or more carbon atoms exist in several structural isomeric forms, depending upon how the carbon atoms are arranged. Straight-chain or unbranched alkanes have all the carbons in one continuous bonding network. Branched alkanes have carbon chains leading off the main chain. Cyclic hydrocarbons have some or all the carbon atoms in a continuous ring.

The topic of "Clean coal" is discussed in detail in a Taking It Further section of the Web site for this textbook.

In shorthand diagram representations of organic molecules, the symbols for carbon and hydrogen atoms are not shown. Only the bonds are shown, though C—H bonds are not displayed at all. The carbon atoms are assumed to be located at the junction of two lines or at the terminus of a line.

A double bond between two atoms requires them to share two pairs of electrons, four electrons in total, between them. Hydrocarbons containing a C=C bond are called alkenes. Double bonds are approximately twice as strong as single ones. The geometry about a carbon atom that forms one double and two single bonds is planar, with angles of about 120°. The groups attached to the carbon atoms of a double bond do not rotate relative to each other.

In some hydrocarbons, carbon–carbon double bonds alternate with single C—C bonds. The presence of three sets of alternating single and double bonds, as in benzene, produces a molecule having extra stability and resistance to chemical reaction. Some hydrocarbons contain a triple carbon–carbon bond.

Fossil fuels are predominantly carbon-containing substances that are the residues of living matter from the ancient past. The main fossil fuels are coal, oil, and natural gas. The proven reserves of oil and natural gas are sufficient to last for decades and those of coal for several centuries. Natural gas and, to a lesser extent, petroleum contain hydrogen sulfide, a poisonous gas.

Petroleum, or crude oil, is a mixture of mostly alkane hydrocarbons, especially those with unbranched chains but some of which also consist of rings. Different fractions of petroleum have different boiling point ranges and are separated from each other by distillation. Hydrocarbons having a small number of carbon atoms are gases under normal conditions, whereas those with a large number of carbons are solids. The remainder are liquids.

The burning of gasoline in internal combustion engines can produce engine knocking, which reduces output power and can harm the engine. Unbranched alkanes such as *n*-heptane produce the most knocking and are assigned low octane numbers. Highly branched alkanes such as isooctane burn without knocking and are assigned high octane numbers. Small amounts of compounds in whose molecules an atom of lead is bonded to short chains of carbon atoms greatly increase the octane number of gasoline, and they were used as additives for many years. These lead compounds were replaced in gasoline sold in the United States by increasing the content of hydrocarbons whose molecules contain the benzene ring.

Coal is principally carbon. It is formed over long periods of time from the residue of organisms as high temperatures within Earth convert it, mainly by loss of oxygen, from peat through various stages, eventually to hard coal. The various pollutants in coal can be removed in a variety of ways, including pulverizing, liquefying, or gasifying the coal.

Key Terms

organic chemistry
hydrocarbon
tetrahedron
condensed formula

unbranched hydrocarbon
branched hydrocarbon
structural isomer
solvent

cyclic hydrocarbon
double bond
single bond
carcinogenic

triple bond
fossil fuel
petroleum
distillation

Web Sites of Interest

To link to Web sites of interest, go to www.whfreeman.com/ciyl2e, Chapter 4, and select the site you want.

For Further Reading

C. J. Campbell and J. H. Laherre, "The End of Cheap Oil," *Scientific American,* March 1998, pp. 78–83. The article discusses the coming decline in global oil production which will be coupled with increased demands in this century.

E. Suess et al., "Flammable Ice," *Scientific American,* November 1999, pp. 76–83. The article discusses the occurrence, extraction, and potential use of methane hydrates.

Review Questions

1. What is a *hydrocarbon*?
2. How many bonds can a carbon atom form?
3. How many bonds can a hydrogen atom form?
4. Sketch the three-dimensional shape of a methane molecule and an ethane molecule.
5. In what physical state is methane at room temperature? Explain why.
6. Give the general molecular formula for an alkane and an alkene.
7. How does a *straight-chain* carbon molecule differ from a *branched* carbon chain?
8. What is a *structural isomer*? Provide an example.
9. Sketch the possible arrangements for five carbon atoms joined together in different ways.
10. What is a *solvent*?
11. What is a *cycloalkane*? Provide an example.
12. Sketch the shorthand representation for the following molecules:
    a) C_5H_{10} (cyclic)
    b) C_4H_8 (cyclic)
    c) C_7H_{16} (straight chain)

13. Identify two sources of each of the hydrocarbons below:
    a) methane   b) ethane   c) propane
    d) butane
14. How many electrons are involved in a double bond? Write the Lewis structure for ethene.
15. How do *alkanes* differ from *alkenes*?
16. What is benzene? Why is it called an *aromatic compound*?
17. Identify a source of benzene emissions into the air.
18. Benzene is a *carcinogenic* substance. Explain what that means.
19. What structural characteristics do *aromatic* compounds share?
20. What are two elements that form triple bonds?
21. Why are hydrocarbons called *fossil fuels*?
22. What is a *proven reserve* of fuel?
23. How long can we expect proven fuel reserves to provide for global energy needs?
24. What is a *methane hydrate*?

25. What is *natural gas*?

26. How do the uses of methane and propane differ?

27. What does the term *sour* mean when referring to a natural gas well?

28. What is *crude oil*?

29. Why is crude oil not useful as a fuel?

30. What property of hydrocarbons allows them to be separated into fractions?

31. What is an *octane number*?

32. What environmental and health problems were associated with the use of lead in gasoline?

33. How do oil spills occur?

34. Explain why oil spills are difficult to clean and to contain. Outline the approaches used to clean them up.

35. Use the Web sites listed on the Web site for this book to identify some short- and long-term environmental effects associated with oil spills. What impacts do spills have on organisms and habitats, including plants and microorganisms?

36. Describe how coal was formed.

37. What is *peat*? How is peat associated with coal?

38. What is *strip mining*?

39. Why is coal a "dirty fuel"?

Understanding Concepts

40. Methane, ethane, propane, and butane are all hydrocarbon gases. What does this tell you about the attractive forces between these molecules?

41. Write the condensed formulas for the molecules below:

a)

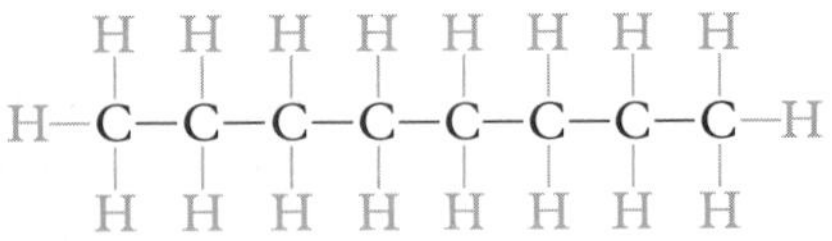

b)

c)

d)

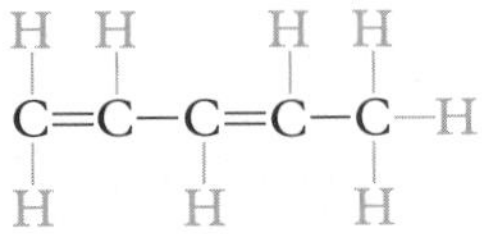

42. Predict the formulas of the alkane and alkene hydrocarbons with the following number of carbon atoms:
a) four   b) six   c) eight

43. Draw symbolic shorthand structures for
a) cycloheptane (the cyclic alkane containing seven carbon atoms) b) *n*-pentane

44. If you are feeling brave, try identifying the carbon networks of the five isomers of hexane, C_6H_{14}, and of the nine isomers of heptane, C_7H_{16}.

45. Which of the following pairs of symbolic diagrams correspond to carbon networks of unique isomers and which are duplicates?

a)

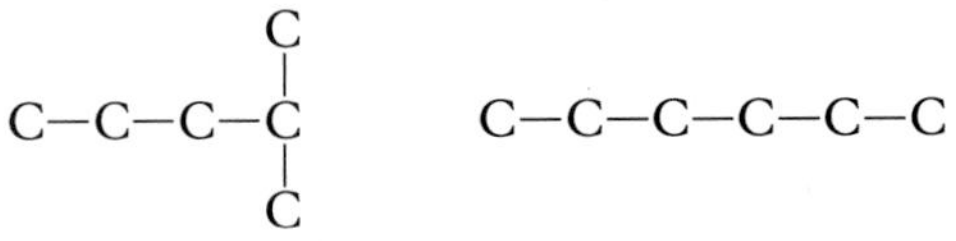

b)

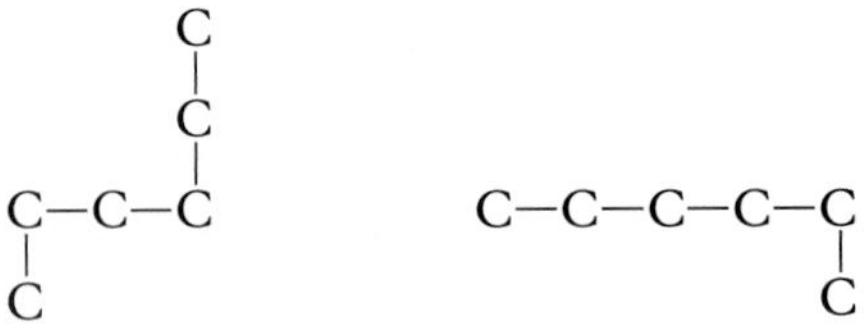

c)

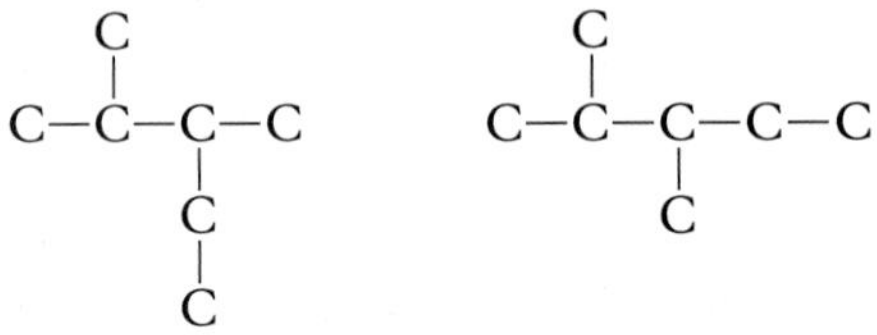

46. Ethane and ethene are both hydrocarbons. How are they the same? How are they different? Describe the way they differ in their geometry around the carbon atoms.

47. Why do you think that hydrocarbons make good solvents?

48. Why are hydrocarbons with more than six carbon atoms liquids or solids at room temperature?

49. Compare the sources of oil and of natural gas. Which of these fossil fuels is more easily extracted?

50. Why are hydrocarbons useful as fuels?

51. Alkanes of more than 12 carbons cannot be used in gasoline. Explain why not.

52. Explain how a gasoline with an octane number of 100 is different from a gasoline with an octane number of 55.

53. In a closed container having hydrocarbons of both fewer and more than five carbon atoms, which hydrocarbons will occupy the greatest percentage of the space above the liquid? Which hydrocarbons will vaporize to the greatest extent? Which will vaporize to a relatively lesser extent?

Synthesizing Ideas

54. Draw a 3-D representation of silicon hydride, SiH_4, a molecule that has the same tetrahedral shape as methane.

55. Using solid and/or dashed wedges as necessary, draw a representation of the 3-D structure of the NH_3 molecule (ammonia), assuming that the HNH bond angles are all equal to the tetrahedral value (109.5°).

56. When a carbon atom is bonded to four hydrogen atoms, it has tetrahedral geometry. Another geometry that could result from four atoms surrounding a central atom is called *square planar*: a central atom bonded to four atoms, each at the corner of a square and all of them in the same plane. Draw a tetrahedral arrangement and a square planar arrangement. Describe how the arrangement of atoms in a tetrahedron is different from the arrangement of atoms in a flat square.

57. Do you think that a molecule of formula C_2H_6 could exist in which four hydrogen atoms were bonded to one carbon and two hydrogen atoms to the other carbon? Explain why or why not. Try to draw a Lewis structure for such a molecule to test your hypothesis.

58. Shown below are two different representations used by chemists for a molecule of hydrogen peroxide. For each diagram, state what each line or geometric symbol represents:

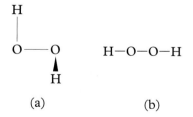

(a)   (b)

59. Given the general formula for alkanes, determine how many hydrogen atoms would be required to bond to an *n*-alkane having 100 carbon atoms.

60. Recall that the general formula for alkanes that exist entirely as chains is C_mH_{2m+2}. What would be the general formula for cycloalkanes that have all their m carbon atoms in one continuous ring?

61. On a piece of paper, make a diagram showing the relative positions of the four carbon atoms in butadiene, given that all the CCC angles are about 120° and that all four atoms lie flat in the same plane. You should be able to draw two unique structures, depending upon how you position the fourth carbon atom relative to the first three. If you formulate more than two structures, show by cutting them out and rotating them that there are only two unique structures.

62. There are 1.1×10^{21} molecules of the *n*-alkane $C_{30}H_{62}$ in each gram of this compound, and 0.775 gram of it occupies 1 cm³. Calculate the volume of one molecule of $C_{30}H_{62}$. From your answer, and assuming that each molecule occupies a perfect cube (whose volume equals its height times its width times its depth), determine the dimension of the cube. *Hint:* Mathematically, the volume of a cube is d^3, where d is the length of each of its three dimensions.

63. Consider the following chemical reactions:
a) H_2S burns in oxygen to produce SO_2 and H_2O:

$$H_2S + O_2 \rightarrow SO_2 + H_2O$$

b) H_2S and SO_2 combine to produce sulfur:

$$H_2S + SO_2 \rightarrow S + H_2O$$

(i) Balance both equations.
(ii) How would you have to change the coefficients of the balanced equations so that all the SO_2 produced in reaction (a) is consumed to produce S in reaction (b)?
(iii) Add the left sides of the two equations together, and then do the same for the right sides, in order to deduce the *net* effects of these reactions proceeding in sequence. Cancel any terms that appear on both sides of your resulting reaction equation. Divide all the coefficients by any common factor to obtain the final result.

64. Based upon the discussion in this and previous chapters, explain why a flame source can ignite gasoline without being near the liquid.

65. Compare the four types of coal. Which has the highest oxygen content? Based upon your answer, and recalling the discussions of combustion reactions in previous chapters, explain which type of coal has the highest fuel value and why.

Taking It Further

Petroleum refining: Fractional distillation

As we noted in section 4.17, fractional distillation separates crude oil into a number of fractions having molecules of similar sizes. The crude oil mixture is first continuously fed through pipes that pass through a furnace that heats the oil to 360°–400°C. Even higher temperatures are not employed because of the tendency of the oil to decompose under such conditions. At the temperatures that are used, most of the oil is converted to a gas. The portion of the oil that is *not* vaporized is a hot liquid, called the *bottoms* or the *residuum*, which contains the heaviest molecules found in oil. It is drawn off and subsequently separated into components that are used as solid products such as waxes and asphalt, or is used for making the form of carbon called *coke* that is employed in steel production (discussed further in Chapter 5). It is also possible, by reducing the pressure to almost a vacuum, to boil the bottoms fraction of crude oil and, by various techniques, to split the vaporized long molecules of this fraction into shorter ones, for use in gasoline and diesel fuel.

The vaporized oil is injected into a vertical distillation or *fractionating* tower, which is several meters in diameter and up to 30 meters high. The temperatures in the tower *decrease* as the hot gases move higher and higher; thus the vapor cools as it rises (see Figure 4.15). Since they correspond to compounds with high boiling points and hence high condensation temperatures, the first gaseous molecules to recondense to liquids as the vapor rises through the tower are those with 17 or 18 or more carbon atoms. By means of a series of collection trays situated in the tower at positions where the temperature falls to an average of about about 350°C, this liquid fraction of the oil can be collected and drained off, thereby separating it from the remainder,

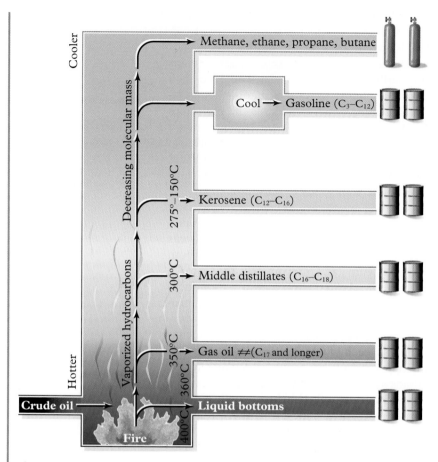

Figure 4.15 Petroleum fractions in a refining tower.

which continues to rise in the tower. This first fraction of the petroleum, called *gas oil,* is a rather viscous liquid when cooled to room temperatures and finds commercial use as lubricating oils.

Another series of collection trays and an output pipe are located somewhat higher in the tower, where the temperature has cooled sufficiently, to about 300°C, to allow hydrocarbons in the C_{16} to C_{18} range to condense and be collected. This second fraction, used as diesel fuel and industrial heating oil, is called *middle distillates.* The final fraction, called *kerosene* or *heavy naphtha,* which contains primarily hydrocarbons with 12 to 16 carbons, is collected by trays near the top of the tower, where temperatures have cooled to about 150°–275°C. The kerosene fraction is used as diesel and jet fuels and as oil for home heating.

There is no particular reason why the fractions described above, with these particular boiling point ranges, should be the *only* ones collected. In fact, different petroleum distillation towers collect different

Taking It Further

fractions by using different collection temperatures, not just those we have described. The decision as to exactly which fractions are to be collected is made by considering the end uses for the various products.

At the top of the tower, the remaining uncondensed vapor contains hydrocarbons consisting primarily of molecules having 1 to 12 carbon atoms each. This vapor is cooled almost to normal outdoor temperatures in a separate unit, a procedure which condenses the molecules with 5–12 carbons to the liquid known as *straight-run gasoline* or *light naphtha*. This fraction, which constitutes about one-fifth of the original oil, is the basis of the gasoline used to power motor vehicles. Alkanes having more than about 12 carbons cannot be used in gasoline since they do not evaporate in the engine sufficiently to burn properly.

The C_1 to C_4 gases—namely methane, ethane, propane, and butane—that remain uncondensed at the top of the tower can be collected and used for the purposes previously described for components of natural gas. The C_4 alkanes (butanes) are used as components both of gasolines and of liquefied petroleum gas. Unfortunately, the C_1 to C_3 gases are sometimes ignited and simply "flared off" into the air if facilities for their condensation or transportation do not exist at the petroleum refining site (see Figure 4.16).

In summary, the fractionating tower separates crude petroleum into a number of useful materials, each of which is a mixture of hydrocarbons in which the different constituents have approximately the same number of carbon atoms and which all boil within a small range relative to the larger range of the crude oil. For many applications, a further separation of a fraction into subfractions, each consisting of a smaller set of hydrocarbons, is accomplished subsequently.

Figure 4.16 **A petroleum gas flare-off.** (Royalty-Free/Corbis)

In addition to hydrocarbons, crude oil also contains small quantities of compounds that contain other elements. The most predominant of these elements is sulfur, which occurs in oil to the extent of 0.5–6%, depending upon the origin of the material. Metals such as vanadium, nickel, and iron also are present, at a total concentration of more than 1000 ppm in some cases, but they usually are found mainly in the bottoms.

1. Explain the process by which crude oil is converted to usable fractions.
2. How are gasoline and kerosene different? How are they alike?
3. Why can both gasoline and kerosene be used as fuels?

In this chapter you will learn:

- how useful solvents, as well as some banned ones (including CFCs and chloroform), are derived from hydrocarbons that contain halogen atoms;
- how polymers are used to create commercial plastics and fibers, including polyethylene, PVC, Styrofoam, Gore-Tex, and Teflon;
- how the physical properties of polymers, such as whether or not they conduct electricity or if they are like rubber or like glass, determine their uses;
- about the pros and cons of recycling of plastic;
- how diamonds and graphite are the result of intricate carbon networks;
- about the issues concerning the disposal of rubber tires.

What would an outdoor adventure be like without synthetic materials and plastics?

Waterproof, windproof, tough, and light—the properties we take for granted in camping equipment are due to the composition and structures of large carbon-based molecules called polymers, which are made of the same hydrocarbons that make up crude oil and natural gas. In this chapter we examine how these hydrocarbons become consumer products. (Courtesy of the Coleman Company, Inc.)

From Diamonds to Plastics

Carbon's Elemental Forms, Addition Polymers, and Substituted Hydrocarbons

In the 1967 movie *The Graduate,* Benjamin (played by Dustin Hoffman, in one of his first roles) was given one word of advice by a family friend about his possible career plans. That word was "plastics." The friend was accurate in predicting the direction in which society was headed for its materials. (Manufacturing plastics as a means of becoming rich was mentioned in films as far back as 1946, in Frank Capra's classic film, *It's a Wonderful Life.*) Indeed, it is sometimes said that we are now living in the age of plastics.

The materials we generally call "plastics" and "synthetic fibers" start as hydrocarbons, the molecules we studied as energy sources in Chapter 4. Much of the clothing and footwear you have on now, the CDs and DVDs you listen to and watch, many parts of your computer and your car, the packaging for your food and other products, including cosmetics and even pharmaceuticals—all are built starting with crude oil and natural gas. Chemistry re-forms these substances into the very long molecules that constitute so many of the materials that are an integral part of your everyday life.

▪ Chemistry in Your Home: The synthetic products you use

Consider the clothes you are now wearing. What sort of fibers or plastics are used to make the synthetic ones? What clothes do you own that are made of natural fibers (wool, cotton, leather, silk)? Identify five plastic objects you have used already today. What materials do you think these objects would have been made from a century ago?

In this chapter, we shall see how a variety of common synthetic materials, both fibers and plastics, are made starting from simple hydrocarbons. Carbon is the structural backbone of many natural and synthetic materials. Indeed, it is the structural diversity of carbon, its ability to combine with other elements, and its availability that combine to produce the plethora of carbon-based materials that we see today. Though much of our focus in the chapter is on synthesis, we shall also investigate two natural materials—diamond and graphite—that are nothing but the element carbon in the form of extended networks. We'll also open a window on exciting developments underway in producing synthetic molecules of elemental carbon that someday may form the basis of nano-scale computers and other forms of technology.

Whenever you see this icon in this chapter, go to
www.whfreeman.com/ciyl2e

Although synthetic organic materials are primarily carbon and hydrogen, chemists have incorporated a small proportion of other elements, especially nitrogen, oxygen, fluorine and chlorine, into organic compounds in order to impart characteristics such as fire resistance to the final materials. For this reason, we begin our discussion by introducing some simple organic compounds in which halogen atoms are incorporated. We then investigate the well-known synthetic materials made from halogens as well as carbon and hydrogen. The analogous materials containing oxygen and/or nitrogen are discussed in Chapters 6–8.

Organic Molecules Containing Halogen Atoms

Halogen atoms form a single bond to carbon atoms, so they can replace hydrogen atoms in hydrocarbons on a one-for-one basis. A large number of such halogenated organic molecules are known. Some of them occur naturally, but the majority were first synthesized by chemists in their quest for substances with interesting and useful properties. In this section, we investigate some of these substances, especially those used to form plastics.

5.1 Halogen atoms can substitute for hydrogen

Taking It Further

For more information and problems on isomers of multihalogentated alkanes, go to Taking It Further at www.whfreeman.com/ciyl2e

One halogen-containing organic compound is *methyl bromide*, CH_3Br, which is formed naturally in chemical reactions that occur in the ocean. After its formation, much of the methyl bromide slowly escapes from the surface water into the atmosphere. This compound is also produced synthetically by the chemical industry for agricultural use to sterilize soil, especially to prevent crops such as strawberries from being destroyed by microscopic worms that eat the roots of the plants. However, although methyl bromide is agriculturally useful, its production and use are controversial because it is a threat to the protective ozone layer that lies far above us in the sky, which we shall discuss in Chapter 15.

The structure of a methyl bromide molecule is the same as that of methane, CH_4, with one of the four hydrogen atoms replaced by a bromine atom:

Lewis structure of
CH_3Br

3-D structure of
CH_3Br

All four of the hydrogen atoms in methane are equivalent, so only one isomer of methyl bromide exists. Like methane, the 3-D structure of methyl bromide is tetrahedral. Stable molecules similar to methyl bromide also exist in which an atom of fluorine or chlorine or iodine, three other types of halogen atoms, rather than bromine replaces one hydrogen atom of methane. Indeed, both methyl chloride, CH_3Cl, and methyl iodide, CH_3I, are also present in the air due to their formation in, and release from, the oceans.

Chemists have been able to produce compounds consisting of molecules in which *several* halogen atoms have replaced hydrogen atoms in various hydrocarbons, including methane. For example, the liquid *methylene chloride,* a common industrial and domestic solvent used as, among other things, a paint stripper, consists of molecules of CH_2Cl_2. Each such molecule corresponds to one of methane but with two of its four hydrogen atoms replaced by those of chlorine. Note that although there are several ways of *writing* the Lewis structure for methylene chloride molecules, there is only one unique molecule, since all pairs of the four bond directions in a tetrahedron are identical:

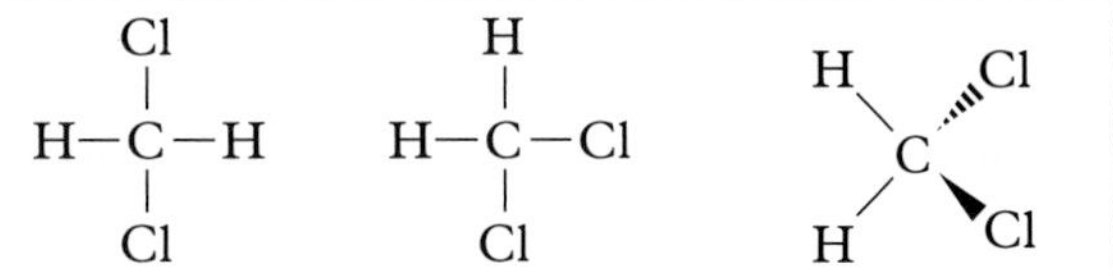

Two possible Lewis structures for CH_2Cl_2

3-D structure of CH_2Cl_2

■ Chemistry in Your Home: There is only one molecular structure for CH_2Cl_2

Construct a model for CH_2Cl_2 using marshmallows and raisins (to represent H and Cl, respectively) and toothpicks. Show by rotating the model about a toothpick that the apparently different Lewis structures shown above are interconverted by this motion.

Molecules containing halogen atoms in addition to carbon and hydrogen atoms are not always produced, either in nature or in the laboratory, with the corresponding hydrocarbon molecule as the starting material. Nevertheless, it is *conceptually* useful to think of such molecules as being hydrocarbons in which one or more of the hydrogen atoms have been "substituted" by halogen atoms. Thus, we can think of molecules of methylene chloride, CH_2Cl_2, as methane molecules in which two of the four hydrogen atoms have been replaced or "substituted" by chlorine atoms.

Compounds with the formulas $CHCl_3$ and CCl_4 consist of molecules corresponding to methane molecules in which three, and all four, hydrogen atoms, respectively, have been substituted by chlorine atoms:

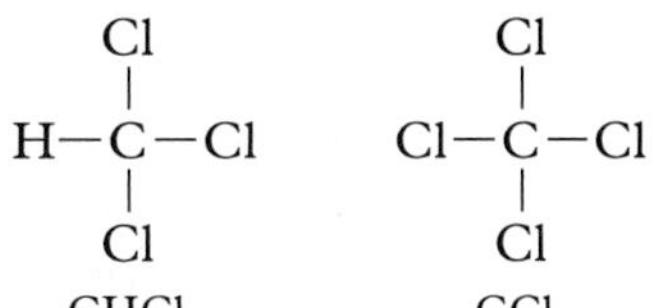

$CHCl_3$
Chloroform

CCl_4
Carbon tetrachloride

These two compounds were well-known liquid solvents for many organic materials. Although its use has been phased out for environmental reasons, *carbon tetrachloride*, CCl_4, was formerly used as a common solvent, especially for dry cleaning, and as the liquid in some fire extinguishers. This latter application was feasible because of the general tendency of organic compounds containing halogens to resist reaction in general and combustion in particular. The lack of reactivity arises from the unreactive nature of carbon–halogen bonds and increases as the proportion of halogen atoms to carbon atoms increases. Consequently, carbon tetrachloride is very unreactive indeed.

Many gases, or vapors of highly volatile liquids including many organic compounds, behave as *anesthetics*. An anesthetic works by dissolving in cell membranes and subsequently blocking nerve impulses from reaching the brain. Much of a membrane behaves like a hydrocarbon, and consequently small amounts of many organic compounds, including solvents, are readily soluble in it. *Chloroform*, $CHCl_3$, was one of the world's first medical anesthetics, though it is rarely used nowadays because of the damage that it causes to the liver and because a small overdose can be fatal. Another multi-halogenated alkane that was used more recently as an anesthetic is CF_3—$CHBrCl$, which is known as *halothane*.

Many organic compounds are soluble in each other. Here chloroform is the solute rather than the solvent since it is present in the smaller amount.

5.2 CFCs damaged the atmospheric ozone layer

Chemists have been able to create compounds in which atoms of two *different* halogens substitute for hydrogen atoms in the same molecule. The most famous—or infamous—of these compounds are those in which both chlorine and fluorine are present; their general name is *chlorofluorocarbons*, or CFCs. Two of them, in which all four hydrogens of methane have been substituted, are CF_2Cl_2 and $CFCl_3$, the trade names of which are CFC-12, or *Freon-12*, and CFC-11, or *Freon-11*, respectively:

$$
\begin{array}{cc}
\mathrm{Cl} & \mathrm{Cl} \\
| & | \\
\mathrm{F-C-F} & \mathrm{Cl-C-F} \\
| & | \\
\mathrm{Cl} & \mathrm{Cl} \\
CF_2Cl_2 & CFCl_3
\end{array}
$$

Because of their properties as stable, nontoxic, nonflammable liquids that boil below room temperature ($-30°C$ and $-81°C$, respectively), such CFCs were used extensively in the 20th century as the "working" (circulating) fluid in refrigerators and air conditioners, the propellant in spray cans, and the substances by which holes could be created in the formation of foam products made from liquids. Unfortunately, all of these uses eventually result in the dispersion of the CFC compounds into the air, where they wreak havoc with the protective ozone layer in the upper atmosphere. Equally damaging are molecules of the compound *halon*, CF_3Br, in which fluorine and bromine replace all of the hydrogen atoms in methane. Halon was used in the fire extinguishers designed for safe use in computer centers and jet aircraft. Because of their environmental destructiveness, CFCs are no longer commonly used in developed countries, as we shall see in Chapter 15.

5.3 Chlorine-substituted ethenes are used in plastics

Molecules that correspond to ethene, $H_2C{=}CH_2$, with one or more of its hydrogen atoms substituted by halogen atoms, also constitute stable compounds that resist reaction if their halogen-to-carbon ratio is high. One of the simplest, but most important, examples substitutes a chlorine atom for one hydrogen atom. The resulting molecule, *chloroethene*, is better known by its common name *vinyl chloride:*

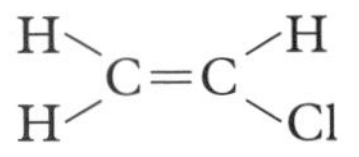

$$
\begin{array}{cc}
H & H \\
_{\diagdown} C {=} C _{\diagup} & \\
H ^{\diagup} ^{\diagdown} Cl &
\end{array}
$$

Vinyl chloride

Since all four hydrogen atoms in ethene are equivalent, substitution of any one of them by a Cl atom yields the same molecule. As we shall see later in this chapter, vinyl chloride is the starting point for the preparation of one of the world's most common plastics, PVC. Because of this use, it is the most important organic chemical industrially, as ranked by the quantity produced. Unfortunately, vinyl chloride gas is a human carcinogen, a fact that was discovered when several workers in a vinyl chloride plant were stricken with the same form of a rare liver cancer after chronic exposure to relatively high levels of the compound. Exposure of workers at PVC plants to the gas is now much more tightly controlled and monitored.

Substitution of three of the four hydrogens by chlorine gives trichloroethene:

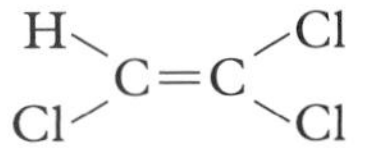

$$
\begin{array}{cc}
H & Cl \\
_{\diagdown} C {=} C _{\diagup} & \\
Cl ^{\diagup} ^{\diagdown} Cl &
\end{array}
$$

Trichloroethene (TCE)

Trichloroethene, or TCE as it is commonly known, is a colorless, oily liquid that is widely used as a solvent for dissolving oily and greasy substances. In the past, it was used as a dry-cleaning solvent, to clean grease from metals, to extract caffeine from coffee, and as a paint stripper, though many of these uses have largely been phased out for safety and environmental reasons.

Complete substitution of hydrogen by chlorine in ethene gives C_2Cl_4, the compound tetrachloroethene, more commonly called *perchloroethene* (nicknamed "perc"), abbreviated PCE.

$$\begin{array}{c} \text{Cl} \diagdown \diagup \text{Cl} \\ \text{C}=\text{C} \\ \text{Cl} \diagup \diagdown \text{Cl} \end{array}$$

Perchloroethene (PCE)

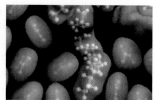

View liquid tetrachloroethene dissolving grease at Chapter 5: Visualizations: Media Link 1.

The *per*- prefix indicates that *all* the hydrogen atoms in a given molecule have been substituted by the same atom or group. Like TCE, PCE is an excellent organic solvent and is used extensively as a dry-cleaning liquid to remove grease spots and other stains on clothing. Because they have been widely dispersed in the environment and they are such stable molecules, both TCE and PCE eventually work their way through the layers of soil down into the water table; thus, they are now common pollutants in well water.

Addition Polymers

In Chapter 4, we mentioned that some alkane hydrocarbon molecules found naturally in crude oil contain 20 or more carbon atoms—for example, $n\text{-}C_{20}H_{42}$:

$$\text{CH}_3-\text{CH}_2-\text{CH}_2-\text{CH}_2-\text{CH}_2-\text{CH}_2-\text{CH}_2-\text{CH}_2-\text{CH}_2-\text{CH}_2-\text{CH}_2-$$
$$\text{CH}_2-\text{CH}_2-\text{CH}_2-\text{CH}_2-\text{CH}_2-\text{CH}_2-\text{CH}_2-\text{CH}_2-\text{CH}_3$$

Recall that condensed formulas were explained in section 4.4.

For simplicity, we could rewrite this formula in condensed form as $\text{CH}_3(\text{CH}_2)_{18}\text{CH}_3$, where the unit in parentheses, $-\text{CH}_2-$, is repeated in a row the number of times indicated by its subscript, 18.

In the 20th century, chemists devised methods whereby molecules with *very* long chains—thousands of carbon atoms rather than just a dozen or two—could be made synthetically in the lab and later produced on a huge scale in chemical factories. Substances that contain such very long chains *with the same structural unit, such as* $-CH_2-$, *repeating over and over again* are known as **polymers.** The impetus for polymer research was the finding that these substances could be used to create the useful products we now know as plastics and synthetic fibers. The dictionary definition of **plastic** is *material that can be molded*. Indeed, the main superiority of plastics to natural materials such as wood is that plactics are much more readily molded into the desired shape. Plastics also have many other advantages over natural materials: they don't rot, rust, corrode, or break easily, and they can be permanently colored. **Thermoplastics** are those which become melted, and therefore softened, but not decomposed by heating, and that consequently can be molded—over and over again if necessary, for example, during recycling. When they are cooled, they become hard again.

The prefix *poly-* is derived from the Greek word for *many* and the suffix *-mer* is derived from the Greek word for *part*.

5.4 Polyethylene is the simplest organic polymer

The simplest polymer, *polyethylene*, also known as *polyethene* and *polythene*, is one that can be made into a plastic. The name is derived from the fact that this polymer is made starting with the gas ethene, also

Figure 5.1 High-density polyethylene (HDPE), the polymer used in plastic milk containers. (Photo by George Semple for W. H. Freeman and Company)

In fact, the density of HDPE is slightly less than that of water, though it is higher than those of other forms of polyethylene and of most hydrocarbons.

known as ethylene. In general, a polymer is named after its monomer. The molecular structure of the simplest form of polyethylene is similar to that shown for $C_{20}H_{42}$, except that the carbon chain is about 20 *thousand*, rather than 20, carbon atoms long!

Since the chains in the polyethylene polymer described above are unbranched, they pack together rather well and consequently form a solid that is relatively high in density compared to some other forms of polyethylene that we'll describe later. Consequently, this type of polyethylene is known as *high-density polyethylene*, abbreviated HDPE.

Although HDPE molecules differ enormously in size from those of the long-chain alkanes that are found in petroleum, nevertheless the two solid substances they form have some similar properties. For example, HDPE plastic looks and feels "waxy" in the same way that the compounds of long-chain alkanes derived from petroleum do. Both substances also are opaque; that is, you cannot see through them although some light passes through. However, HDPE possesses considerable structural strength, toughness, and rigidity. It is the plastic that is used to make objects such as bottles, toys, pipes, containers for fluids such as milk or gasoline, and even opaque plastic bags (see Figure 5.1).

5.5 Addition polymers consist of repeating molecules

Any collection of plastics that is used domestically and/or industrially usually includes some plastics made from polymers similar to polyethylene, in that their molecules contain long chains of carbon atoms singly bonded to each other. However, in these polymers, *some* or all of the hydrogen atoms have been substituted by other atoms or groups of atoms. In several cases, *at every second carbon* along the chain one hydrogen is replaced by another group that we'll symbolize in general as X. Any small portion of the long chain—except that near either end—looks like the following:

$$
\begin{array}{cccccccc}
H & H & H & H & H & H & H & H \\
| & | & | & | & | & | & | & | \\
-C & -C & -C & -C & -C & -C & -C & -C- \\
| & | & | & | & | & | & | & | \\
H & X & H & X & H & X & H & X
\end{array}
$$

Given the regular repetition in the long chain, we could reformulate the representation of the structure as follows:

$$-M-M-M-M-M-M-$$

Here —M— stands for the two-carbon unit

$$
\begin{array}{cc}
H & H \\
| & | \\
-C & -C- \\
| & | \\
H & X
\end{array}
$$

which is the **repeating unit** of the polymer. Since these polymers consist of intact molecules that combine together, they (and polyethylene itself) are known as **addition polymers.** The simplest way to make molecules of a specific addition polymer would presumably be for individual CH_2CHX molecules to link together successively to form a very long chain; in fact, this is what happens. Since each carbon forms four bonds, the *original,* independent CH_2CHX molecules must contain a carbon–carbon *double* bond:

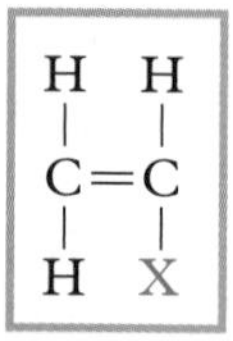

$$
\begin{array}{ccc}
H & & H \\
| & & | \\
C & = & C \\
| & & | \\
H & & X \\
\end{array}
$$

Each small molecule that successively combines to form part of the very long chain in the polymer is called a **monomer.** Here the monomers are appropriately substituted ethene molecules.

Notice that whereas the carbon–carbon bonds within the unconnected monomers are *double* bonds, once the monomer unit is part of the polymer its carbons are joined by *single* bonds. Indeed, the formation of addition polymers of this type depends upon two of the four electrons originally used in the monomer C=C bonds now creating new C—C single bonds that join the monomer units. In Figure 5.2, we depict these two electrons in each monomer by dots and illustrate how

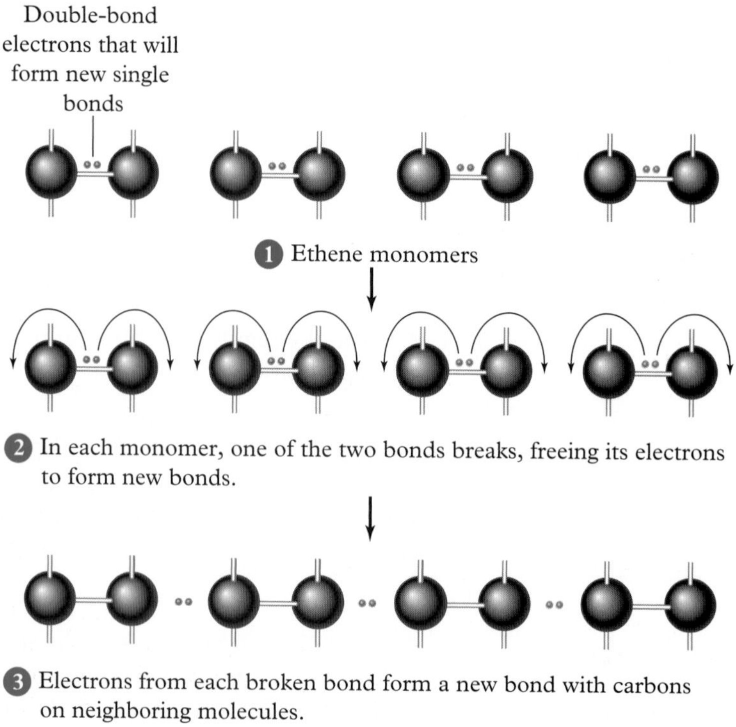

Although the simplest repeating unit in polyethylene is —CH_2—, it is made by reacting together monomers containing a C=C bond since the unit CH_2 is not a stable entity and therefore cannot be used as the starting substance.

Double-bond electrons that will form new single bonds

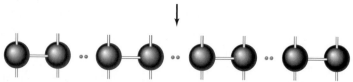

① Ethene monomers

② In each monomer, one of the two bonds breaks, freeing its electrons to form new bonds.

③ Electrons from each broken bond form a new bond with carbons on neighboring molecules.

Figure 5.2 Step by Step How ethene monomers form a polymer.

Chapter 5: From Diamonds to Plastics

one of them is used at each end of the monomer to link to another monomer. We should emphasize that this linking of thousands of monomer units together does *not* happen all at once, since it is extremely unlikely that the monomer units would ever spontaneously line up in order! Rather the chain lengthens itself sequentially, adding one monomer unit at a time.

Generally speaking, for addition polymers, the repeating unit has the same arrangement of atoms as does the monomer, though the bond types—whether single or double—differ. For example, since the repeating unit in the polymer $\cdots -CH_2-CHF-CH_2-CHF-CH_2-CHF- \cdots$ is $-CH_2-CHF-$, the monomer from which it is made is $CH_2=CHF$.

Exercise 5.1

Identify the repeating unit in the polymer $-CH_2-CHBr-CH_2-CHBr-CH_2-CHBr-$. What is the monomer that could be polymerized by an addition process to obtain this polymer?

Figure 5.3 Polyvinyl chloride is the polymer used to make garden hoses. (Photo by George Semple for W. H. Freeman and Company)

PVC The original name for molecules containing C=C bonds was *olefins;* hence the wide variety of addition polymers formed by monomers containing the C=C unit are, as a class, called *polyolefins.* In one of the most useful polyolefin polymers with the CH_2CHX repeating unit, X is the atom chlorine. Consequently, the monomer in this case is vinyl chloride, $CH_2=CHCl$:

$$
\begin{array}{ccc}
H & & H \\
| & & | \\
C & = & C \\
| & & | \\
H & & Cl
\end{array}
$$

The polymer it forms is called *polyvinyl chloride,* PVC. Small objects made using the plastic formed from this polymer carry the number 3 inside the triangle. PVC plastics are rigid and strong, and are often used for structural purposes to replace metals—for example, in water pipes and siding for houses—but PVC is also used to make floor tiles, electrical insulation, garden hoses, raincoats, surgical gloves, shower curtains, and simulated leather apparel and upholstery (see Figure 5.3). Much more PVC is used in building materials than in consumer products.

You may wonder how an inflexible material such as the PVC that is used for pipes and siding can also be flexible enough to be used for hoses and clothing. This is accomplished by forming a mixture (up to 50%) of the PVC with a **plasticizer,** which is *a liquid that blends easily with a polymer and effectively acts as a lubricant between individual polymer chains,*

The molecular structure of typical plasticizers will be discussed in Chapter 7.

Taking It Further

For information and problems on other addition polymers, go to Taking It Further at www.whfreeman.com/ciyl2e

thereby softening the plastic and imparting flexibility to the mixture as a whole. Unfortunately, plasticizers eventually leak out of plastics as they age. The result is both environmental pollution by molecules of the liquid and the deterioration of the plastic, which becomes brittle and easily broken as a result of the lack of plasticizer.

Polypropylene In some polymers used to produce fibers and plastics, X is a group that itself contains carbon atoms. The simplest example is when X is a methyl group, CH_3. The monomer for this case is *propene*, $CH_2=CHCH_3$, also called *propylene,* and thus the name of the polymer is *polypropylene* (plastic triangle number 5):

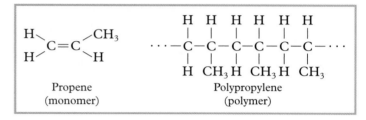

Propene (monomer)   Polypropylene (polymer)

The monomer is derived from propane obtained from natural gas or petroleum refining.

As you might expect from the similarity in its atomic constitution and structure, polypropylene has properties that are similar to those of HDPE; for example, it also is opaque. However, polypropylene is stronger and more rigid than HDPE. It can be heated to a higher temperature than HDPE before it softens. Consequently, objects such as syringes made from it can withstand the high temperatures that are required for sterilization, and domestic objects made from it can be placed in dishwashers without distorting. Because of its toughness and the fact that it can be drawn into fibers, polypropylene is used to make ropes, indoor–outdoor carpeting, artificial turf, and a replacement for canvas.

All **fibers,** including those of polypropylene, are composed of threadlike strands of solid material. Synthetic fibers are formed when a melted polymer, or one that is dissolved in a liquid, is squeezed through a small hole in a "spinneret"—much as toothpaste exits a tube when you squeeze it. It forms a long, thin stream of highly viscous liquid, which then is solidified (see Figure 5.4a). Stretching the fibers during the process of solidification makes them particularly strong. One new form of polyethylene has molecules that are much longer than in conventional HDPE and can be drawn into fibers that have a tensile strength 20 times that of steel (see Figure 5.4b)! Plastic films such as those used to wrap food are made in a similar way to fibers but are extruded through slots rather than holes.

Polystyrene and Styrofoam In the other important $CH_2=CHX$ polymer that involves a hydrocarbon group, X is a benzene ring molecule minus one of its hydrogen atoms:

(a)

(b)

Figure 5.4 Polypropylene. (a) Thread production. (b) Astroturf. (Part a, courtesy of CarolMac Corp.; part b, Photodisc Red/Getty Images)

$$X = \quad \begin{array}{c} H \qquad H \\ C - C \\ -C \qquad C-H \\ C = C \\ H \qquad H \end{array}$$

Polystyrene

The monomer in this case is the molecule called styrene, so the polymer is named *polystyrene*, which corresponds to plastic triangle number 6.

Polystyrene is a transparent, hard, brittle polymer that is thermoplastic and can readily be molded into different shapes. Consequently, it is used for objects such as plastic cutlery, clear drinking glasses, medicinal vials, cases for CDs and audiotapes, etc.

Polystyrene is also commonly encountered as a foam such as *Styrofoam*. Recall from Chapter 1 that a foam is a mixture of two substances. One, consisting of small bubbles of a gas, is trapped within the other, which can be a liquid or a solid. The production of polystryrene foams starts with small beads of polystyrene that have previously been impregnated with a small percentage of a liquid, commonly pentane, that has a boiling point which lies not too far above room temperature. The beads are then heated by hot air or steam to a temperature at which the incorporated liquid boils to produce a gas. At this point, the polystyrene softens somewhat but does not itself boil. The gas expands, puffing up the beads, and the resulting foam acquires the shape of whatever mold

Addition Polymers

Figure 5.5 Polystyrene foam. (Photo by George Semple for W. H. Freeman and Company)

encloses the bead. Upon cooling, the polystyrene foam hardens and retains its shape. In this manner, polystyrene foam products such as coffee cups, egg cartons, packaging material—including "peanuts"—and insulation material is produced (see Figure 5.5).

As the environmental impact of nondegrading polymers such as polystyrene mounts, new materials are being developed to take their place. Carbohydrate-based packing "peanuts," for example, have many of the properties of Styrofoam but dissolve in water, thus reducing landfill burdens. Research is also underway to develop polymers that degrade upon exposure to sunlight, so those that are carelessly discarded on streets, in vacant lots, etc., would eventually disintegrate.

Teflon and Gore-Tex Polymerization of monomers having the formula $CX_2{=}CX_2$ produces a polymer with the structure. $\cdots{-}CX_2{-}CX_2{-}CX_2{-}CX_2{-}\cdots$. The important example occurs when X is fluorine, so the polymer is $\cdots{-}CF_2{-}CF_2{-}CF_2{-}CF_2{-}\cdots$, *poly-tetrafluoroethylene*, or PTFE. Like other fully halogenated substances, this polymer has a very high resistance to both chemical attack and to thermal decomposition. As a plastic, PTFE is known as *Teflon.* Because of its resistance to reaction it is used to produce materials such as gaskets, plumber's tape, and laboratory equipment that could come into contact with chemicals. And because of its stability when heated and the lack of frictional resistance to motion on its surface, it is used as a coating on frying and baking pans.

Fibers made from the PTFE polymer are used to make *Gore-Tex* brand clothing (see Figure 5.6a). The membrane of fabric that characterizes such apparel consists of PTFE polymer strands that are so close together and are so effective at repelling water that they do not allow *liquid* water, even fine droplets of it, to pass between them, and consequently the material is waterproof. However, in contrast to the case for many

Figure 5.6 (a) Polytetrafluoroethylene (PTFE) fibers are used in Gore-Tex clothing. (b) Waterproof Gore-Tex fabric "breathes" because vaporized perspiration can move out of pores too small to allow water drops in.

(Part a, photo courtesy of Gore-Tex)

(a)

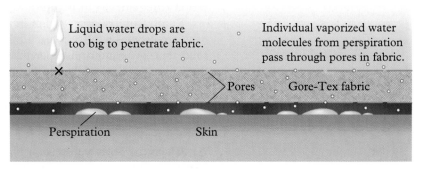

Liquid water drops are too big to penetrate fabric.

Individual vaporized water molecules from perspiration pass through pores in fabric.

Pores   Gore-Tex fabric

Perspiration   Skin

(b)

other waterproof fabrics, *gaseous* water molecules *can* pass between the fibers, and hence the material "breathes" in the sense that vaporized perspiration from the wearer's body can escape (see Figure 5.6b). The direction of flow of water vapor is from inside the material to the outside air *provided* that the outdoor temperature is cooler than the temperature inside the clothing. The direction of spontaneous flow is from the warmer to the cooler air.

Properties of More Complex Polymers

5.6 Polymers exist in two different solid states

We are accustomed to encountering matter in one of three states—gas, liquid, or solid. Plastics, however, have *two* solidlike states. Indeed,

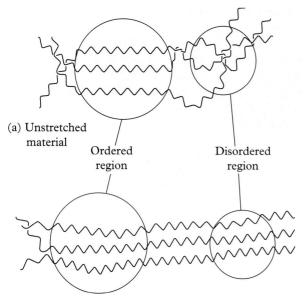

(a) Unstretched material

Ordered region

Disordered region

(b) Stretched material

Figure 5.7 Ordered and disordered regions of a polymer when the material is (a) unstretched and (b) stretched.

The incorporation of plasticizers significantly changes the glass transition temperature for a polymer.

it is a common experience to find that some solid plastics adopt different characteristics, such as becoming brittle, when cooled, which indicates that the nature of the solid state has changed. Plastics also exist in a liquid state. Like other extended networks, however, plastics cannot exist in the gas phase. When they are strongly heated, they decompose to a mixture of smaller molecules.

When most nonpolymeric liquid materials cool, their molecules can align and pack together in a regular structure without much difficulty, forming a *crystalline* solid. However, it is quite difficult for many polymers to achieve a crystalline solid state. First of all, the chance of having *all* the very long chains perfectly aligned is slight. To complicate matters further, the chains are of unequal length. It is common, then, for a solid polymer to contain finite *regions* of regular, crystal-like order and other regions that are very disordered, with the molecular strands mixed together like a plate of cooked spaghetti. These *disordered* regions are analogous to the arrangement of atoms in amorphous solids, which, as we learned in Chapter 1, are solids without overall order. The two regions in polymers are illustrated schematically in Figure 5.7.

Recall from Chapter 1 that in the solid state, there is no movement of the molecules relative to each other. Like amorphous solids, glass is a solid that lacks uniform order throughout its structure. Consequently, the corresponding solid state of polymeric materials is called the **glass state.** Polymers in the glass state are often brittle and will break readily when force is applied to them. For example, the plastic knives, forks, and spoons made with polystyrene fall into this category, since at normal temperatures, polystyrene is in the glass state.

When the temperature of a plastic in the glass state is raised sufficiently, molecular motion begins, but at first only in the *disordered* (amorphous) regions. Thus the plastic material is essentially a liquid in the disordered regions but is still a solid in its ordered regions. Overall the substance is still a "solid," but it has properties that are quite different from those in its glass state. This new solid state is called the **rubber state,** since natural rubber normally exists in this same state. Material in its rubber state is usually more flexible—like rubber—than it is in the glass state, since the liquidlike random regions in the rubber state provide little resistance to distortion of their shape.

The transition from the glass state to the rubber state occurs at a characteristic temperature for each plastic. The so-called **glass transition temperature,** T_g, for polystyrene, for example, is about 100°C. If you have ever put polystyrene plastic utensils in a dishwasher, you have probably seen the effects of the transition from the glass state to the rubber state. The utensils will undergo this transition at the hot temperatures reached in dishwashers, at which point the plastic can flow slightly and assume a new shape. Once the plastic cools, the material is "frozen" into its new, distorted shape.

Many of the plastic materials that we commonly encounter—such as polyethylene, polypropylene, and rubber—exist in the rubber state at normal temperatures, so we are often unaware that their properties will change at a lower temperature. One well-known application of this transition occurs when people try to remove chewing gum that has become stuck to clothes or carpets. The main ingredient in gum is a polymer that normally occurs in the rubber state. It sticks to other materials because of the existence of the liquidlike disordered areas. If ice is applied to the gum, the polymer material falls below its glass transition temperature and enters the glass state, where it is much less sticky and can be removed.

5.7 One form of polyethylene is a branched polymer

All the plastics that we have discussed so far consist of polymeric molecules that are long, continuous chains. However, other plastics exist in which the chains are *branched*.

The commercially most important polymer having side chains (branches) is *low-density polyethylene*, LDPE. In this substance, a small percentage of the carbon atoms along each long chain are not simple $-CH_2-$ units, but rather are $-CHR-$, where R stands for an unbranched carbon chain about four carbon atoms long:

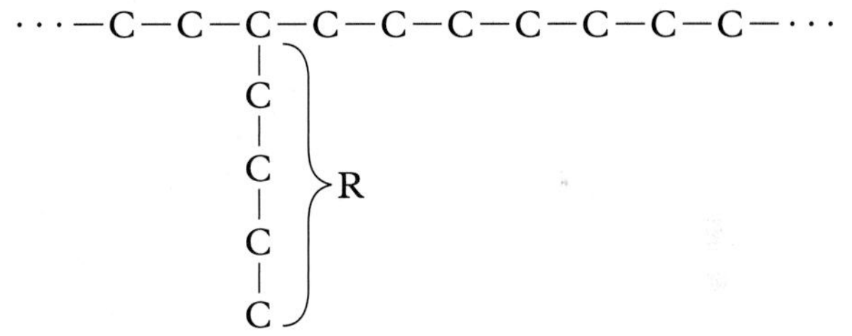

Carbon backbone of low-density polyethylene (LDPE)

When the chains branch, they cannot pack together in the solid state as efficiently as they do in HDPE, and consequently the material has a lower density (0.92 vs. 0.97 gram per cubic centimeter). The individual chains in low-density material are only about 500 carbon atoms long, much shorter than the chains of tens of thousands of atoms found in HDPE.

LDPE plastics are more waxy, more flexible, and softer than HDPE because they are less crystalline at the molecular level. As a consequence of the difference in packing, LDPE melts to form a liquid at a lower temperature (108°C) than does HDPE (133°C). Even in boiling water and in the hot water used in dishwashers, objects made from LDPE will deform whereas those made from HDPE will not. LDPE is used to make sandwich, grocery, and trash bags;

LDPE was first formulated into a usable plastic in the Second World War, when it was employed to insulate electric cables for radar equipment in aircraft, thereby helping the British air force to fend off attack.

Figure 5.8 Some low-density polyethylene (LDPE) consumer items. (George Semple for W. H. Freeman and Company)

food-wrapping film; squeeze bottles; insulation for electrical wires and cables; and even plastic flowers (Figure 5.8). Its plastic triangle number is 4.

5.8 Rubber is a polymer chain that contains double bonds

The long carbon chains of all the polymers discussed above consist of carbon atoms that are all singly bonded to each other. However, both nature and chemists have produced some addition polymers in which carbon–carbon *double bonds* appear at regular intervals along the chain. Such polymers are useful in producing rubber and similar products.

In many of these polymers, there are C=C units separated from each other by *three* C—C bonds (that is, two —CH$_2$— units) on either side. Thus the carbon backbone of the chain itself looks like

$$\cdots -C=C-C-C-C=C-C-C-C=C-C-C-C=C-\cdots$$

In the diagram below, we also show the hydrogen atoms in the chain, and indicate that a nonhydrogen group X, bonded to one of the C=C carbons, is present in some polymers:

$$
\begin{array}{c}
\text{H} \quad \text{X} \quad \text{H} \quad \text{H} \quad \text{H} \quad \text{X} \quad \text{H} \quad \text{H} \quad \text{H} \quad \text{X} \quad \text{H} \quad \text{H} \quad \text{H} \quad \text{X} \\
| \quad\ | \quad\ | \quad\ | \quad\ | \quad\ | \quad\ | \quad\ | \quad\ | \quad\ | \quad\ | \quad\ | \quad\ | \quad\ | \\
\cdots -\text{C}=\text{C}-\text{C}-\text{C}-\text{C}=\text{C}-\text{C}-\text{C}-\text{C}=\text{C}-\text{C}-\text{C}-\text{C}=\text{C}-\cdots \\
| \quad\ | \qquad\qquad\ \ | \quad\ | \qquad\qquad\ \ | \quad\ | \\
\text{H} \quad \text{H} \qquad\quad\ \text{H} \quad \text{H} \qquad\quad\ \text{H} \quad \text{H}
\end{array}
$$

Exercise 5.2

Bracket the repeating unit in the structure above.

Natural rubber and vulcanization Natural rubber consists of the polymer shown above when X is a methyl group, CH_3. It is obtained from latex, a substance that oozes from rubber trees when they are slashed and that consists of a mixture of small particles of rubber present in water. Separating the rubber particles from the water yields a solid that is sticky but elastic, especially if it is warmed, since it then becomes even stickier and loses much of its resilience. It also becomes rigid when cold. In this condition, the material has rather limited use.

The American inventor Charles Goodyear in 1839 discovered a modification to natural rubber that overcame its loss of resilience upon heating. By accident, he spilled a mixture of rubber and elemental sulfur onto the surface of a hot stove—and discovered that the resulting product retained its elasticity even when warm! We now understand what Goodyear's modification did at the molecular level. As an element, sulfur exists as cyclic S_8 molecules. When sulfur is heated above its melting point, the rings open to produce sulfur atom chains of varying lengths. The two terminal sulfur atoms in the chains are reactive, since each such atom forms only one bond, not the two that are required to provide a stable octet of electrons around it:

$$\cdot\ddot{S}-\ddot{S}-\ddot{S}-\ddot{S}-\ddot{S}-\ddot{S}-\ddot{S}-\ddot{S}\cdot$$

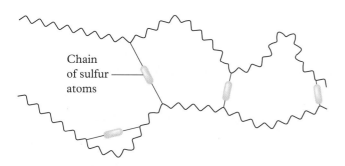

Chain of sulfur atoms

Figure 5.9 Sulfur cross-links between polymers.

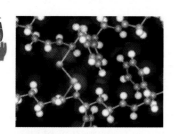

View the effects of compression on rubber molecules at Chapter 5: Visualizations: Media Link 2.

Each terminal sulfur atom subsequently forms a bond to a carbon atom that was present in a C=C bond, thereby destroying the double bond character of that carbon–carbon linkage. Since each sulfur chain has two reactive ends, the net result is to cross-link nearby polymeric carbon chains—or even two parts of the same chain—by short sulfur chains of 1–4 atoms, as illustrated in Figure 5.9. Goodyear went on to develop this process, known as **vulcanization,** commercially.

Rubber that has been vulcanized is harder and stronger and resists flowing because the sulfur connections keep the carbon chains from moving relative to each other very much. As a result, it is longer-wearing than nonvulcanized rubber. It is also more elastic because the sulfur chains help to pull the polymer chains back to their original positions once stress is released after the rubber is stretched, even when it is warm. Substances that have this "elastic" property are called **elastomers.** Vulcanized natural rubber found much use in the fabrication of automobile and bicycle tires, and it still represents about one-third of world rubber production. Even in vulcanized rubber, the fraction of C=C bonds that are destroyed by adding sulfur chains is small. If too much sulfur is added in the process, the cross-linking by sulfur chains becomes too extensive, and the resulting material is very hard and inflexible—but can be used to construct bowling balls!

Properties of More Complex Polymers

Synthetic rubber Synthetic rubber can be produced by polymerizing certain hydrocarbons to obtain the characteristic pattern of double bonds and single bonds in the general polymer on page 186. Synthetic rubber with polymer chains in which X=H is called *polybutadiene rubber*, since the monomers are *butadiene* molecules:

$$
\begin{array}{cccc}
\text{H} & \text{H} & \text{H} & \text{H} \\
| & | & | & | \\
\text{H}-\text{C}=\text{C}-\text{C}=\text{C}-\text{H}
\end{array}
$$

Butadiene

Another commonly used synthetic rubber is *neoprene*, in which X = chlorine in the general structure on page 186. Because (after vulcanization) it is resistant to attack by gasoline, oil, and grease, neoprene is used to make hoses for gasoline pumps and rubber components for car engines and chemical labs.

5.9 Some polymers can conduct electricity

Polymers, and indeed most organic chemicals, in general do not conduct electricity. However, in recent decades scientists have devised several plastics that can conduct an electrical current. Such materials are an exciting development, since plastics are more lightweight and flexible than metals, which are used as the electrical conductors in all of our current technology.

The first polymer to be synthesized that displays electrical conduction was *polyacetylene*. Since acetylene, HCCH, contains a triple carbon–carbon bond, when it forms an addition polymer by the same method used for ethylenes, one of the three bonds is destroyed. The two electrons so liberated instead form C—C bonds that connect the monomers. In this way, a polymer is formed in which single and double bonds alternate with each other along the chain:

$$
\cdots - \text{C}=\text{C}-\text{C}=\text{C}-\text{C}=\text{C}-\text{C}=\text{C}-\cdots
$$

Extended networks in which each carbon forms a double bond are known to conduct electricity and therefore behave like metals. Indeed, solid polyacetylene film is silvery and shines like a metal. However, the individual polymer molecules are not so long that they form wires of macroscopic size. Since an electrical current cannot readily flow between polymer molecules, electricity is not conducted across solid pure polyacetylene. However, by "doping" the material with small amounts of other substances that assist in the transmission of an electrical current between polymer molecules, polyacetylene can be transformed into an electrical conductor.

Research in the synthesis of polyacetylene and the early development of its metallic properties resulted in a Nobel Prize in Chemistry in 2000 for its chief investigators: Alan MacDiarmid of the University of Pennsylvania, Hideki Shirakawa of the University of Tsukuba in Japan, and Alan Heeger of the University of California at Santa Barbara.

Since the discovery of polyacetylene, several other conducting polymers have been synthesized. Currently, technologists are developing the field of "plastic electronics" using various conducting polymers. These materials can also be used to absorb light and convert it into electricity, as in solar cells, or transform electrical energy into light, as used in computer technology. Over the next decade, you will see more and more applications in which plastics replace metals.

● Tying Concepts Together: Variations on the theme of a polymer molecule

Our tour of common polymers has revealed many subtle variations on the general theme. We have seen that, in contrast to molecular compounds, a sample of a polymer contains molecules of many different sizes, all of them much longer than non-polymer molecules.

The simplest addition polymers have a long, continuous chain of monomers joined together in an unbranched fashion: $\cdots$—M—M—M—M—M—M—$\cdots$. The monomer M is a simple ethylenic unit in many cases. Variations in the polymer properties are dependent upon the nature of the atoms or groups bonded to the two carbon atoms of this unit.

The properties of the plastic or fiber formed from the polymer are also modified to advantage in some instances by introducing an occasional branch off the main polymer backbone:

Connecting adjacent polymer chains together by cross-linking them through a small chain of atoms sometimes leads to an improvement in the quality of the final product:

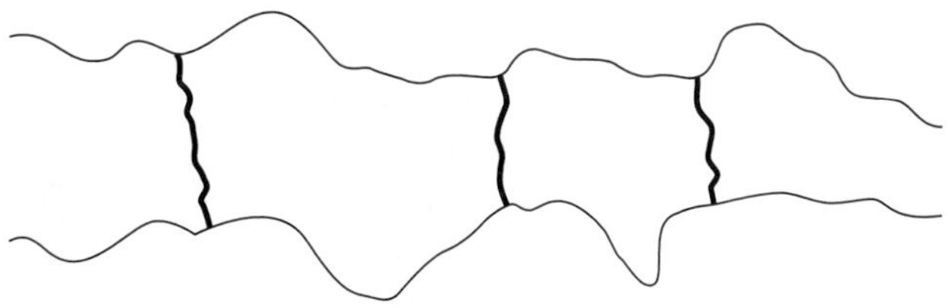

We'll see in Chapters 8 and 9 that more complicated types of polymers also exist, both in nature and in the chemical laboratory.

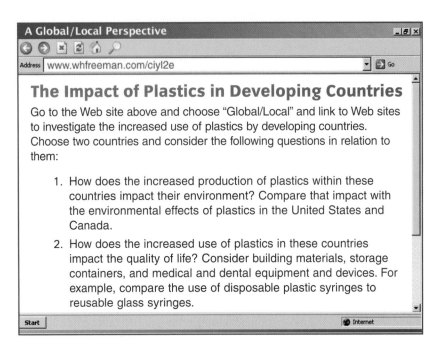

A Global/Local Perspective

Address www.whfreeman.com/ciyl2e

The Impact of Plastics in Developing Countries

Go to the Web site above and choose "Global/Local" and link to Web sites to investigate the increased use of plastics by developing countries. Choose two countries and consider the following questions in relation to them:

1. How does the increased production of plastics within these countries impact their environment? Compare that impact with the environmental effects of plastics in the United States and Canada.

2. How does the increased use of plastics in these countries impact the quality of life? Consider building materials, storage containers, and medical and dental equipment and devices. For example, compare the use of disposable plastic syringes to reusable glass syringes.

Start   Internet

The Recycling of Plastics

In the last quarter of the 20th century, plastics became the symbol of the "throwaway society," since many products—especially those used in packaging—were designed to be used once and then quickly discarded. Many environmentalists believed that waste plastic was a major culprit in the "garbage crisis." Indeed, plastics are the second-most-common constituent of municipal garbage, following paper and cardboard, though they follow by a large margin. Molded plastics take up more room in landfills—otherwise known as garbage dumps—than is indicated by their percentage by mass because they are not very dense. These plastics do, however, become compressed by the weight of the materials placed on top of them, as well as by compacting machinery before placement in the landfill (see Figure 5.10).

The per capita annual use of plastics in North America is approximately 30 kilograms.

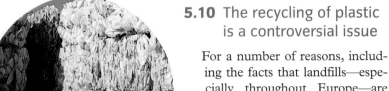

(a)

Figure 5.10 (a) Uncompressed plastics and (b) compressed plastics in a landfill. (Part a, Mark E. Gibson/ Visuals Unlimited; part b, Ken Lucas/ Visuals Unlimited)

(b)

5.10 The recycling of plastic is a controversial issue

For a number of reasons, including the facts that landfills—especially throughout Europe—are reaching their capacity and that many citizens in developed countries are opposed to garbage incineration, many plastics now are collected from consumers and recycled. Some countries, such as Sweden and Germany, have made manufacturers legally responsible for the collection and recycling of the packaging used in their products.

There has been resistance to the recycling of plastics in some quarters, including much of the plastics industry. Their argument is that *virgin plastic*—that is, the material yet unused and just synthesized from fossil fuels—is a low-cost material that is made from relatively low-cost raw materials, namely natural gas and crude oil. As a result, the amount of energy used to make plastics is tiny compared to the amount used to produce aluminum or steel from its raw materials. The cost of cleaning used plastic and converting it back into its monomers so it can again be polymerized is substantial, compared to the current cost of oil. Some in the plastics industry argue that the natural disposal method for plastics is simply to burn them and utilize the heat energy provided, especially since there is little objection by the public to burning the oil produced for vehicles, domestic furnaces, and power plants. Furthermore, experiments indicate that the presence of plastics makes the other materials in domestic garbage burn more cleanly and reduces the need for fossil fuel to be added. Although plastics account for less than 10% of the mass of garbage, they make up more than one-third of its energy content.

Environmentalists counter these arguments by pointing out that if environmental impacts were to be included in determining the cost of virgin materials, recycled plastic would be the cheaper choice. In addition, the combustion of some plastics, notably PVC, produces some toxic organic compounds and releases hydrogen chloride gas, HCl. Because HCl dissolves easily in water and is corrosive in the presence of moisture, it attacks mucous membranes (very moist areas) and damages the lungs when it is inhaled. HCl reactions with metals cause corrosion and the release of hydrogen gas.

Notwithstanding this debate, there is little doubt that the public in many developed countries has embraced recycling of plastics. By the late 1990s, for example, about half the urban communities in the United States had curbside recycling programs that included plastics. The numbers stamped inside of the three-arrow triangle on each plastic object help consumers identify the types of plastic collected in their area. In the mid-1990s, the great majority of the plastics recycled in the United States were HDPE and LDPE, plus the more complex plastic called PET (discussed in Chapter 8). However, the recycling rate for PET soda bottles in 1996 had fallen to 34%, considerably less than the 45% rate achieved in 1994.

5.11 There are several ways to recycle plastics

There are four basic ways to recycle plastics:

- Reprocess it by remelting or reshaping. Often the plastics are washed, shredded, and ground up, so that clean, new products can be made from them.

- Depolymerize it back to its component monomers by a chemical or thermal process so that it can be polymerized again to form new products.

- Transform it into a lower-quality (that is, lower value, less pure) substance from which other materials can be made.
- Burn it to obtain energy, that is, "energy recycling."

Examples of the reprocess option include the production of plastic trash cans, grocery bags, etc., from recycled HDPE, and of CD cases and office accessories such as trays and rulers from recycled polystyrene. Further examples are listed for each category of packaging plastic in the last column of Table 5.1.

The depolymerization of plastics has a number of difficulties. Often, small amounts of organic and inorganic compounds are added to the original polymer in forming the plastic to modify its physical properties. In many cases these compounds must be removed from the plastic or the monomers before they can be reused. It is also difficult,

Table 5.1   Recycling categories for common plastics

Plastic recycling number	Acronym and name of polymer	Original uses	Recycle uses
1	PET Poly(ethylene terephthalate)	Beverage bottles, food and cleanser bottles	Carpet fibers, fiberfill insulation, nonfood containers
2	HDPE High-density polyethylene	Milk, juice, water bottles, grocery bags (crinkly)	Oil and soap bottles, trash cans, grocery bags, pipes
3	PVC (or V) Polyvinyl chloride	Food and water bottles, food wraps, blister packs, construction materials	Drainage pipes, flooring tile, traffic cones
4	LDPE Low-density polyethylene	Flexible bags for trash, bread, milk, groceries; flexible wraps and containers	Bags for trash, groceries; irrigation pipes; oil bottles
5	PP Polypropylene	Handles, bottle caps, lids, wraps, bottles	Auto parts, fibers, pails, refuse containers
6	PS Polystyrene	Foam cups, packaging; cutlery; furniture; appliances	Insulation, toys, trays, packaging "peanuts"
7	Other	Various	Plastic "timber," posts, fencing, pallets

Reprinted from C. Baird, *Environmental Chemistry*, 2nd edition, W. H. Freeman, 1999, p. 527.

Chapter 5: From Diamonds to Plastics

for many addition polymers, to devise a process for reforming the original monomers. The monomer yield in the thermal depolymerization of polystryene is about 40%. However, it is close to zero for polyethylene because the chain can be broken with equal chance at almost any position rather than exclusively to produce two-carbon units.

Examples of the transformation option are:

- Heating of plastics to a high temperature, to "crack" the polymeric molecules and thereby produce synthetic crude oil that consists of hydrocarbon chains having a dozen or less C atoms each. It can be done even with mixed plastics.

- Plastics are reacted with oxygen and steam in order to produce *synthesis gas,* which as we saw in Chapter 4 is a mixture of hydrogen and carbon monoxide that can be used as a fuel or transformed into useful chemicals.

Exercise 5.3

Write the Lewis structure of the monomer that would be obtained if the following polymer were to be decomposed by heat back into two-carbon units:

$$\cdots\!-\!CHF\!-\!CF_2\!-\!CHF\!-\!CF_2\!-\!CHF\!-\!CF_2\!-\!\cdots$$

Exercise 5.4

Classify each of the four basic processes for recycling plastics as a physical or a chemical change.

▶ Discussion Point: Should plastics be recycled?

Take one side of the issue of whether or not plastics should be recycled, and prepare a list of points in favor of your position that you could use in a debate on this issue. Then prepare a list of points your debate opponent would likely raise, and prepare rebuttals to them. Be sure to consider the question of whether or not consumers should subsidize plastic recycling through their taxes, or whether manufacturers of plastics should bear the recycling costs. Do you agree that burning plastics for energy is a form of "recycling"?

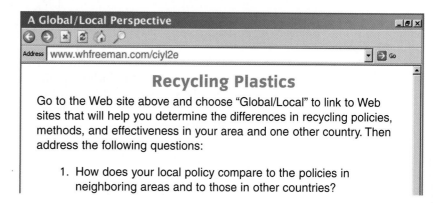

A Global/Local Perspective

Address www.whfreeman.com/ciyl2e

Recycling Plastics

Go to the Web site above and choose "Global/Local" to link to Web sites that will help you determine the differences in recycling policies, methods, and effectiveness in your area and one other country. Then address the following questions:

1. How does your local policy compare to the policies in neighboring areas and to those in other countries?

2. How effective are the recycling policies of the various areas?
3. Are recycling statistics available for your area? How do they compare with the statistics from other countries?
4. What recycling methods are used in your area? How do these compare with methods in other countries?

Start | Internet

Elemental Carbon

In Chapter 4, we saw that carbon can form a wide variety of molecular structures when it is bonded to hydrogen, and in the previous section we found that very long chains of carbon atoms could be formed synthetically. With this background, we are ready to investigate the even more extensive networks of carbon atoms in elemental carbon. Until recently, the only stable forms known of this substance were diamond and graphite, both of which consist of extended networks of carbon atoms. In diamond, only C—C bonds are present, whereas in graphite there is an alternation of single and double carbon–carbon bonds. Modern research has revealed that carbon can also form molecules such as C_{36} and C_{60}, as well as tiny tubes of carbon atoms that may someday form the basis of a new electronics industry. In the material that follows, we explore these various **allotropes** of carbon, the term used for different forms of a pure element.

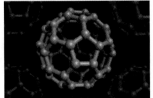

View the molecular structures of carbon allotropes at Chapter 5: Visualizations: Media Link 3.

5.12 Diamond: Its structure and bonding

Just as gold has traditionally been the world's most valuable metal, diamond has been the most valuable gemstone. These materials owe their popularity to their inherent beauty, rarity, and great resistance to destruction or corrosion. Both also are elements: diamond is one form of the element carbon.

We can understand the structure of diamond by considering first a carbon atom that is attached by single C—C bonds to four other carbon atoms:

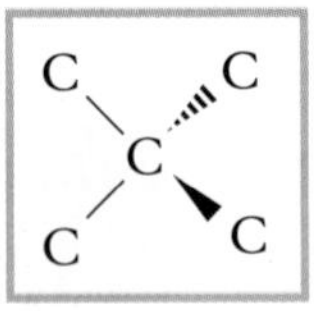

In diamond, *all* the carbons atoms are bonded to four others in this way, thereby eliminating the need for any hydrogen atoms to complete carbon's electron octet. If you choose any one carbon atom in the middle of the diamond structure in Figure 5.11a and follow the bonding network from it to the next carbon and then to one bonded to the second atom, and so on, you will discover that the structure consists of six-membered rings. Furthermore, each ring is "fused" to several neighboring ones by the sharing of two adjacent carbon atoms (see Figure 5.11b).

Figure 5.11 (a) A portion of a diamond network and (b) close-up of carbon bonding to form six-membered rings.

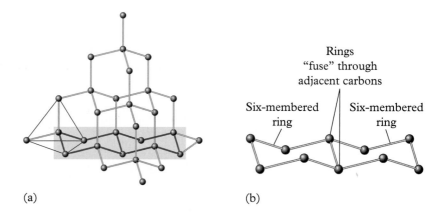

Since each carbon in diamond forms four bonds, then as in methane and as illustrated above, the geometry at each carbon is tetrahedral. As a consequence of this geometry, these rings are three-dimensional. Thus you might consider diamond to be one giant molecule of fused, 3-D rings consisting of six carbon atoms each. However, diamond is *not* a molecule; each diamond is a single crystal composed of a network of carbon atoms. Thus we see that covalent bonds are not restricted to small molecules but also exist in some solids composed of an extended network of atoms.

Because its bonds are individually strong and the entire network is almost "welded" together as a unit, diamond is an extremely hard material—the hardest substance known, in fact. Thus, diamonds will scratch most other materials but will not be scratched by them. Indeed, diamond dust is the only practical material for use in polishing larger diamonds! The hardness of the material leads to great industrial importance for diamonds. Because diamonds will scratch other materials, they are used to engrave stone and metals. Diamonds are insoluble in water, oil, or any other solvent, and they are quite unreactive, although under extreme conditions they can be ignited and then burn in air to produce carbon dioxide.

Fact or Fiction?

The carat scale for diamonds is the same as the carat scale for gold.

False. The carat scale for diamonds is based upon mass rather than concentration. One carat is defined as 1/150 of a Troy ounce, which is itself 1.097 times the mass of a regular ounce. Originally this carat scale was based on carob seeds, about 144 of which constitute one ounce.

5.13 The formation and color of diamonds

Diamonds are formed 150–200 kilometers deep in Earth's mantle from carbon-containing material that is subjected to high temperatures and pressures. The diamonds are ejected from the depths, along with rock, by erupting volcanoes. Synthetic diamonds are produced by subjecting carbon-containing substances to very high pressures and temperatures in a molten metal such as nickel or iron. The diamond crystallizes in the molten metal. These techniques work because although graphite, the other common form of elemental carbon (which is discussed later in this chapter), is stabler than diamond under moderate conditions, diamond is the stabler form at high temperatures and pressures. The two forms of carbon do not easily interconvert because they both contain

(a)

(b)

Figure 5.12 (a) Pure and colored diamonds and (b) a microdiamond. (Part a, Jody Dole/The Image Bank; part b, David Scharf)

strong bonds that must first be broken in order for the transition to occur.

Pure diamond is colorless. Any permanent color that a diamond displays is due to impurities that were trapped in the crystal when it was forming. For example, a diamond will have a yellowish tinge if it contains some iron oxide as an impurity. The inherent color of a diamond should be distinguished from the flashes of color that it displays when light strikes it at certain angles. A diamond surface reflects about one-sixth of the light that hits it. By contrast, glass reflects only about 4%. Furthermore, some of the remainder of the light is reflected internally in a diamond and dispersed into different hues, thereby giving rise to multicolor flashes of light (see Figure 5.12).

Diamonds do not conduct electricity well, though they are good heat conductors. Diamonds feel cool to the touch since they conduct heat away from your fingers when you touch them. Experts can distinguish a real diamond from a fake one by placing it on their tongue: the real diamond feels cool, like a metal.

5.14 Polycyclic aromatic hydrocarbons (PAHs) fuse benzene rings

Before we discuss the structure of the other forms of elemental carbon, it is useful to discuss a group of hydrocarbon molecules that contain similar

Chapter 5: From Diamonds to Plastics

structural characteristics. By proceeding in this way, we shall gradually build up the level of complexity in the structure.

Recall from Chapter 4 that the hydrocarbon benzene consists of a six-membered ring in which C=C and C—C bonds alternate. The hexagonal ring structure of benzene is also present in compounds whose molecules contain several six-membered rings, each of which *shares* 2 carbon atoms with one or more adjacent rings. The simplest example is naphthalene, $C_{10}H_8$, which has two six-membered rings fused to each other—that is, they share 2 adjacent carbon atoms. Notice that the total number of carbon atoms in naphthalene is only 10, not 12, since 2 carbons atoms are shared, and that it has only 8 hydrogen atoms, since neither of the 2 shared carbons is bonded to hydrogen:

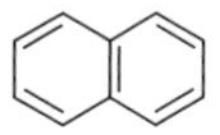

Naphthalene

Naphthalene is a solid compound with a melting point of 90°C. However, this solid is quite volatile—that is, it loses surface molecules to the gas phase by evaporation over time. Thus, the air in a small closed container in which a sample of solid naphthalene is present will contain a measurable concentration of the gaseous molecules. For this reason, and because its vapor is toxic to certain insects, including moths, it is used as one form of *moth balls* or *moth flakes*.

Exercise 5.5

Draw the complete Lewis structure for the molecule naphthalene, showing all atoms and bonds explicitly.

A number of hydrocarbons consist of planar molecules having several six-membered benzene rings fused to each other; collectively, they are called *polycyclic* (or polynuclear) *aromatic hydrocarbons*, or PAHs for short. They are commonly produced as by-products of incomplete combustion processes and subsequently appear as common pollutants in air, water, and soil.

5.15 PAHs are carcinogenic environmental pollutants

PAHs enter the environment from a number of combustion sources: in the exhaust of gasoline and especially diesel combustion engines, as the "tar" of cigarette smoke, on the surface of charred or burnt food, in the smoke from burning wood or coal, and from other combustion processes in which the carbon of the fuel is not completely converted to CO_2. Elevated PAH concentrations in indoor air are typically due to the smoking of tobacco and the burning of wood and coal.

PAHs are of concern as pollutants because many of them are known to be carcinogenic, at least in test animals. The most notorious and common carcinogenic PAH is *benzo[a]pyrene*, BaP, which contains five fused benzene rings:

Benzo[a]pyrene, BaP

Benzo[a]pyrene is a common by-product of the incomplete combustion of fossil fuels, organic matter (including garbage), tobacco, and wood (see Figure 5.13). It is a carcinogen in test animals and is classified as a "probable human carcinogen."

Has exposure to PAHs been demonstrated to actually produce cancer in humans? The answer to this question is both yes and no. For over 200 years, it has been known that prolonged exposure in occupational settings to very high levels of *coal tar*, the liquid product of heating coal in the absence of air, leads to cancer in humans. The principal toxic ingredient of coal tar is benzo[a]pyrene. In 1775, the occurrence of scrotal cancer in chimney sweeps was associated with the soot, produced during the burning of coal, that became lodged in the crevices of the skin of their genitalia.

PAHs have also been implicated as the carcinogen that led to high rates of skin cancer among workers in the coal tar and oil refining industries in the late 19th century. Modern workers in coke and gas production plants likewise experience increased levels of lung and kidney cancer due to benzo[a]pyrene.

The evidence that these compounds can induce cancer in the general public is less clear cut. General exposure to PAHs is at levels many powers of 10 lower than in the occupational environments discussed above. Though inhalation of cigarette smoke is the main cause of lung cancer, the smoke contains many carcinogenic compounds besides

Figure 5.13 A forest fire in Kakadu National Park, Australia. (Royalty-Free/Corbis)

Figure 5.14 Beluga whale in the St. Lawrence River. (Art Wolfe/The Image Bank)

PAHs. It is difficult to extract from public health statistics the much smaller influences of pollutants such as PAHs when other dominant causes are present.

Nevertheless, as common air pollutants PAHs are strongly implicated in the degradation of human health in some urban areas. Many cities in developing countries have chronic problems with carbon-based air pollution. The serious indoor and outdoor air pollution arises primarily from the unvented burning of coal and natural materials such as wood, crop residues, and animal dung used to produce energy for cooking and heating. These emissions consist primarily of PAHs, sulfur dioxide gas, and tiny, suspended soot particles; they are reputed to be responsible for over one million deaths annually in China.

Polycyclic aromatic hydrocarbons are also serious water pollutants in some locales. For example, PAHs are generated in substantial amounts in the production of such coal tar derivatives as *creosote,* a wood preservative. The leaching of PAHs from the creosote that was used to preserve the immersed lumber of fishing docks and the like represents a significant source of pollution to crustaceans such as lobsters. There is also some evidence that PAHs may be the component of pulp mill effluent, which has a very detrimental effect on fish. In addition, PAHs enter the aquatic environment as a result of oil spills from tankers, refineries, and offshore oil drilling sites, and are thought to play a role in the devastation of the populations of beluga whales in the St. Lawrence River (see Figure 5.14). In drinking water, the PAH levels are typically very small, and consequently this usually is not an important source of these compounds to humans.

For most nonsmokers in developed countries, by far the greatest exposure to carcinogenic PAHs arises from their diet, especially from charcoal-broiled and smoked meat and fish, which contain some of the highest levels of PAHs found in food. Apparently most of the PAHs associated with barbecuing meat arise as by-products when fat drops onto hot coals and is partially decomposed as a consequence.

5.16 Graphite consists of weakly linked giant planes of carbon atoms

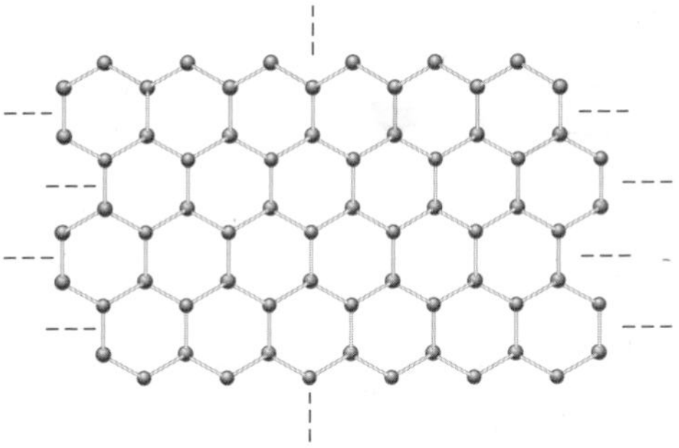

It is not difficult to imagine a network of fused six-carbon, benzenelike rings that extends indefinitely in both the horizontal and vertical directions within a plane. Such a network, pictured in Figure 5.15, corresponds to a giant PAH molecule. The ratio of carbon to hydrogen in such a structure becomes smaller and smaller as the network size increases, since hydrogen atoms are present only at the peripheries (edges) of the structure. There are millions of carbon atoms along any of the two directions and a negligible number of hydrogen atoms, so the substance is really just a form of elemental carbon. Its structure can be compared to that of chicken wire.

Figure 5.15 Graphite layer, top view.

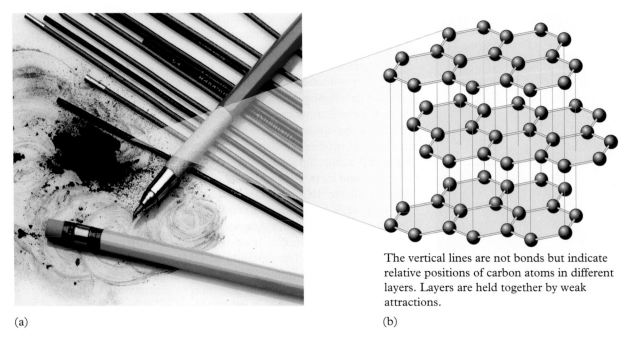

The vertical lines are not bonds but indicate relative positions of carbon atoms in different layers. Layers are held together by weak attractions.

(a)   (b)

Figure 5.16 (a) Powdered and solid forms of graphite and (b) graphite structure, showing the stacking of layers. (Part a, Peticolas & Megna/Fundamental Photographs)

Figure 5.17 A tennis racket made from graphite composite material. (Courtesy of Wilson Sporting Goods Company)

The black solid graphite, a form of elemental carbon, consists of many such planar layers, or "sheets," stacked one upon another (see Figure 5.16). Since each carbon atom forms four bonds *within* its plane, there are no covalent chemical bonds joining atoms that lie in *different* planes. Consequently, a given plane is held only weakly to the adjacent ones, by the same sort of weak forces that hold hydrocarbon molecules together in the solid state. Consequently, the carbon planes can slip over each other fairly readily—since no chemical bonds need be broken to do so—and thus graphite feels slippery and can be used as a lubricant. The "lead" in pencils is graphite mixed with some clay to bind it, and the marks that a pencil makes on paper correspond to a few layers of the graphite-containing mixture which rubs off the surface of the black solid when it is pressed on paper.

Since every carbon atom in a graphite plane forms a double bond, then like polyacetylene it can conduct an electrical current. Consequently, graphite is an electrical conductor in the direction along the planes, but not perpendicular to the planes since they are not connected by bonds.

High-strength fibers of graphite can be made. When encased in a plastic medium to protect them, they form a strong yet somewhat flexible solid material that can be used in consumer products such as the frames of tennis rackets (see Figure 5.17). In general, solid mixtures of nonmetallic materials such as this are called **composite materials;** other examples of fiber-reinforced plastics include the materials used for the bodies of some sports cars and boats.

5.17 Impure forms of elemental carbon include soot and carbon black

There exist several *impure* forms of carbon that are variations on the theme of graphite. Soot consists of tiny crystals whose structure is like graphite but in which about 10% of the atoms are hydrogen. Soot particles are of micrometer size and consist of chainlike clusters of tiny, spherical particles each measuring about 30 nanometers across. These particles each consist of several, usually 5 or 6, carbon layers having the hexagonal graphite structure (see Figure 5.15). The outside layers can readily **adsorb,** or hold on their surfaces, organic molecules such as PAHs. Carbon black is a pure form of soot that consists of tiny spheres and aggregates of the spheres. Because of its intense color, carbon black is used as a pigment in ink (such as the ink on this page) and paint, and also to reinforce and color tires for vehicles. Coke, which is made by heating coal in the absence of air, contains up to 98% carbon. It is often used in industry as a cheap but reactive form of carbon, for example, in the production of metals such as steel.

Do not confuse *adsorption* with *absorption;* the latter means incorporation into the body of the structure, not just onto the surface.

5.18 Activated carbon is a form of charcoal

Charcoal is an impure form of carbon that is made by heating wood or other organic materials in the absence of air. The resulting black material feels light when lifted—it has a low density compared to other common solids. Charcoal's low density is due to its highly porous nature. Its structure is like that of a sponge, though its holes are too small to be seen with the naked eye. The many holes and internal channels near the surface give charcoal a very large surface area, which allows it to adsorb considerable amounts of other substances. This property is exploited in activated charcoal, sometimes called activated carbon, which is charcoal that has been cleaned of any extraneous materials with steam. Activated charcoal has many environmental applications, such as the removal of pollutant materials from vapors and liquids, including drinking water, as we shall discuss in Chapter 13.

5.19 Fullerenes are synthetic molecules containing many carbon atoms

All the forms of elemental carbon described so far, including diamond, are network solids. Given its versatility in forming structures, you might wonder whether there could be stable *molecular* forms of elemental carbon. Although none were known before 1985, chemists have recently synthesized many intriguing carbon molecules whose potential uses are just beginning to be explored. All of them are based upon the structure of a graphite sheet in that they contain (mostly) six-membered, fused rings in which each carbon forms one double bond and two single bonds.

Nanotubes of carbon We can understand the formation of the molecular forms of elemental carbon by returning to the analogy between sheets of graphite and chicken wire. Imagine if the sheet of such wire mesh were to be cut along two parallel lines so that a long, thin strip of wire was obtained. If each of the junctions of the wire represents a carbon atom, those lying along the two edges are missing one bond each. In many of the molecular forms of carbon, this deficiency is overcome by curling the mesh around in a circle so that the two ends meet, with the carbons on one edge bonding to those of the other, and the structure as a whole forming a cylinder (see Figure 5.18). In this way, **nanotubes** of carbon atoms are formed, so called because they are only a few nanometers in diameter. The length of a cylindrical nanotube can be thousands or millions of times its diameter, and so usually falls in the micrometer range. Although the geometry at any carbon is not *exactly* planar, the distortion from planarity is not so extreme that the structure is destabilized, provided that the circle's diameter is not too small. Winding about a dozen six-membered rings around the circumference is the average, with about ten being the minimum.

Of course, creating a tube from a strip of chicken wire still leaves the ends dangling. In order for each carbon atom at these ends to form four bonds, an oval "cap" of carbons forms at each end of the tube. The geometry at the carbon atoms in five-membered rings is easier to distort from planarity than in six-membered ones. Since the curvature required to accomplish the closure of the tubes is greater than in the bulk of the tube, both five-membered rings of carbon atoms as well as six-membered ones are present in the caps.

Strands of aligned nanotubes up to 15 centimeters in length have been made. The long strands act like flexible ropes. Although single-walled nanotubes are still expensive to produce, bulk quantities of multiwalled nanotubes are relatively cheap to make and already are in use as additives to certain plastics. In multiwalled nanonotubes, about a dozen nanotubes of successively smaller diameters are embedded within another like layers of an onion, or like wooden Russian dolls in which each doll contains an even smaller version of itself. The spacing between the layers is similar to that between adjacent layers of graphite (0.34 nm). Some nanotubes conduct electricity and heat very well, are up to 100 times stronger than steel, and can withstand repeated attempts to distort them. Nanotubes open at one end may also be an excellent storage medium for hydrogen gas. In the future, "tweezers"

Recall that 1 nanometer = 1 nm = 10^{-9} m

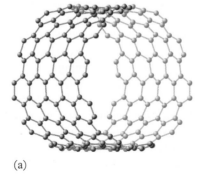 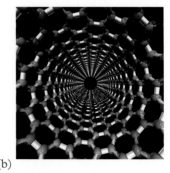

Figure 5.18 Carbon nanotubes.
(a) Cross-sectional view of cylinder.
(b) Computer image of a nanotube.
(Part b by Chris Ewels, www.ewels.info)

(a)

(b)

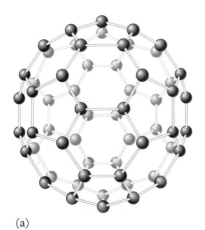

(a)

(b)

Figure 5.19 (a) Structure of buckminsterfullerene, C_{60}. (b) The structure of C_{60} resembles that of a soccer ball. (Part b: Photodisc Blue/Getty Images)

made from nanotubes may be used to move individual tiny particles of matter from one place to another.

Since their initial discovery in the early 1990s, many different nanotubes of carbon have been prepared, and there is now intense interest in them since they may be central characters in the developing **nanotechnology**—the construction of machinery so small that its components have dimensions of nanometers. Some futurists believe that nanotubes will eventually be used as the cornerstones of the electronics industry, used in computers in the same ways that metal wires and silicon chips are employed today. This usage would allow even further miniaturization than currently exists. Carbon nanotubes may also find application in combination with other synthetic materials to add strength without adding much weight.

Buckyballs Recall that fibers are composed of threadlike strands of material. Carbon nanotube fibers of many different lengths have been prepared. In the limiting case of a very short one, there is no "middle" section to the tube at all, but just two "caps" connected together. Although it is energetically less stable on a per-atom basis, than fibrous nanotube molecules, such a molecule exists and indeed was the first *molecular* form of carbon to be discovered. This ball-shaped C_{60} molecule is illustrated in Figure 5.19. Sports fans will notice its resemblance to a soccer ball. The bonds correspond to the positions of the seams in a soccer ball, with the carbon atoms corresponding to the points at which three seams intersect.

Nanotubes and the balls of carbon are collectively called *fullerenes*, named after the American architect Buckminster Fuller, whose "geodesic domes" the balls of carbon resemble. Indeed, the C_{60} molecule is officially called *buckminsterfullerene*. Unofficially, it and similar small carbon molecules are often called *buckyballs*. They were first intentionally synthesized, and their structures determined, in 1985, during research initiated by the British chemist Harold Kroto, who had set out to synthesize carbon chains of the type he had speculated were present in stars. The research he undertook in 1985 in collaboration with Richard Smalley and Robert Curl of Rice University in Houston, Texas, quickly changed direction in order to concentrate on the buckyballs once it was determined that buckyballs were the more stable form of carbon formed under the highly energetic conditions of the experiments. Kroto, Smalley, and Curl were awarded the Nobel Prize in Chemistry in 1996 for their researches in synthesizing C_{60} and other "great balls of carbon." In the late 1990s, an even smaller stable molecule of carbon, C_{36}, was synthesized and described.

Recycling of Tires

We conclude this chapter by considering a problem that is becoming increasingly troublesome in modern society—what to do with used rubber tires. We are able to tackle this subject because of our background in addition polymers—of which rubber is one—and our knowledge concerning the nature of soot and PAHs.

Why do old tires present a special recycling challenge? See what happens to discarded tires at Chemistry in Action 5.1.

Like plastic, rubber tires represent consumer commodity that is a waste management headache. In North America, an average of about one 10-kilogram rubber tire per person per year is discarded—about one-third of a *billion* tires are added annually to the supply of approximately 1 billion tires presently stored in mountainous piles, awaiting ultimate disposal. Because tires are made primarily from oil and so are flammable, "tire fires"—often set by vandals—in these huge piles are not uncommon. These fires produce tremendous amounts of smoke, carbon monoxide, and toxins such as PAHs (see Figure 5.20). The fires are difficult to extinguish because of air pockets present in and between the tires. There have been efforts to use the tires either as fuel or as a filler for asphalt, but currently such applications consume only about 10% of the tires that are discarded annually. Recently, chemists at the Goodyear Tire and Rubber Company have found a way to remove the sulfur added to rubber in vulcanization, so that the rubber can be reused; it is not yet known whether the process is commercially viable.

A number of attempts have been made to reprocess shredded scrap tires by **pyrolysis**—the process of thermal degradation of a material in the absence of oxygen. The resulting products are low-grade gaseous and liquid fuels, along with a char, a mixture containing minerals and a low-grade version of carbon black, which can be further treated and converted into activated carbon. It may be possible to convert the liquid component from pyrolysis into high-grade—that is, a more uniform, purer, and therefore more useful—carbon black, thereby making the process economically profitable. The rubber in tires consists of about 62% of a hydrocarbon polymer and 31% of carbon black, and there is a ready market for the carbon black. Using the liquid component of the char instead as a fuel is problematic because of its high content of aromatic hydrocarbons.

Recycled rubber tires have been used to build homes. These structures use recycled tires that are filled with dirt and covered with stucco or adobe. Recycled rubber tires are also making an appearance in consumer products, including doormats and footwear.

Figure 5.20 A tire fire near Tracy, California. (AP/*The Tri Valley Herald*, Matthew Stanndard)

Halogen atoms can substitute for any or all of the hydrogen atoms in hydrocarbons. Chlorine-substituted hydrocarbons are often produced and used commercially as solvents for organic compounds.

Addition polymers consist of very long chains of carbon atoms in which the same basic structural unit, the monomer, repeats over and over. Addition polymers having a backbone of carbon atoms all singly bonded can be made by reacting together molecules of ethene having the appropriate substituents.

Polymers are often made into plastics or fibers. Plasticizers are sometimes added to the polymer in order to impart flexibility to the resulting plastic. Fibers are made by squeezing a melted polymer through the small hole of a spinneret and drawing the resulting long, thin material into a fiber. Foams are made by impregnating the polymer with beads of a material that becomes a gas when heated, thereby producing gas bubbles trapped inside the polymer.

Depending upon temperature, polymers exist in two different solid-like states. At low temperatures, polymers exist in the glass state, in which all movement of molecules relative to each other has ceased. Heating the polymer above the glass transition temperature produces the rubber state, in which regions of random orientation of polymer molecules behave as a liquid, whereas the more ordered regions still behave like a solid.

In some polymer chains, including those of rubber, double bonds link some of the carbon atoms in the carbon backbone chain. In the vulcanization process, some of these carbon atoms in nearby chains become linked by chains of sulfur atoms.

Plastics can be recycled, though this is a controversial practice. The four ways plastics can be recycled are by reprocessing, depolymerizing, transforming, and burning.

Elemental carbon can exist in several allotropic forms. The two forms that have been known for ages are both extended networks. In diamond, each carbon is connected tetrahedrally by single bonds to four others. In graphite, each carbon occurs in one of the many parallel planes, and forms single bonds to two carbon atoms and a double bond to the third, within the same plane. The individual planes of graphite are not connected by covalent bonds and can move over each other without much force being applied. Small molecules called polycyclic aromatic hydrocarbons contain the same type of fused benzenelike six-carbon rings as does graphite.

A number of impure forms of elemental carbon exist. The most useful of these is activated carbon (activated charcoal), in which the total surface area is so large it can be used to adsorb pollutant molecules and thereby purify liquids and gases.

Molecular forms of elemental carbon are called fullerenes. Nanotubes consist of relatively long cylinders of graphitelike carbon atoms, with the ends terminated by an oval cap of carbon atoms. In buckyballs, all of the carbon atoms occur in a round structure.

Key Terms

polymer	plasticizer	vulcanization	adsorb
plastic	fiber	elastomer	nanotube
thermoplastic	glass state	depolymerization	nanotechnology
repeating unit	rubber state	allotrope	pyrolysis
addition polymer	glass transition	composite material	
monomer	temperature		

Web Sites of Interest

To link to Web sites of interest, go to www.whfreeman.com/ciyl2e, Chapter 5, and select the site you want.

For Further Reading

W. Rathje and C. Murphy, "Five Myths About Garbage and Why They Are Wrong," *Smithsonian*, July 1992, pp. 113–122. The results of a 20-year study of landfills in the United States; examines the true rates of biodegradation, the kinds and amounts of garbage in landfills, and real garbage production rates.

R. Trautman et al., "Microdiamonds," *Scientific American*, August 1998, p. 82. Article discusses how tiny diamonds, less than a millimeter in size, can be used in industry and provides a model of how diamond grows.

J. Winters, "Tomorrow's Tubes," *Discover*, January 1999, p. 38. Discussion of the state of nanotube technology and of potential applications.

Review Questions

1. Match each carbon-containing molecule in the left column with the appropriate description of it in the right column:
   a) CFC
   b) carbon tetrachloride
   c) methylene chloride
   d) vinyl chloride
   e) TCE
   f) PCE

   1. dry-cleaning solvent in which all carbons are bonded to chlorine
   2. solvent used to extract caffeine from coffee
   3. the starting material for PVC
   4. solvent used as a paint stripper
   5. dry-cleaning solvent also used in fire extinguishers
   6. both chlorine and fluorine are bonded to carbons in these materials

2. Name two halogenated hydrocarbons that act as anesthetics.

3. What are *CFCs*? Why are they an environmental hazard?

4. What is a *plastic*?

5. What are some of the advantages of using plastics instead of natural materials?

6. What is a *polymer*? Provide an example.

7. How is a plastic similar to a polymer?

8. What is a *thermoplastic*?

9. What is the simplest organic polymer? What is the starting material for this polymer?

10. Why is HDPE denser than other forms of polyethylene?

11. What is a *monomer*?

12. What is an *addition polymer*? Provide an example and identify the monomer unit.

13. Identify the monomer unit in the sequences below:
    a) —C—C—C—C—X—X—C—C—C—C—X—X—
    b) —C=C—C—C=C—C—C=C—C—C=C—
    c) —C=C—C=C—C=C—C=C—

14. What is a *plasticizer* and how is it used in polymer technology?

15. Why do polymers such as PVC become brittle with age?

16. What is a *fiber*? How is a synthetic fiber made?

17. Explain why Styrofoam is a foam.

18. What happens to a polymer when the temperature goes above its glass transition temperature?

19. How does the structure of a branched polymer differ from that of a straight-chain polymer?

20. What is the substance called *rubber*?

21. How is rubber different from latex?

22. Describe the process of *vulcanization*.

23. How does vulcanization affect the properties of rubber?

24. What is an *elastomer*?

25. How does a *cross-linked polymer* differ from a *branched polymer*?

26. What is the meaning of the triangular symbol found on plastic products?

27. Describe the methods by which plastics can be recycled.

28. Identify the problems or shortcomings associated with each process in question 28.

29. Why is diamond such a hard substance?

30. How are diamonds formed? What gives them color?

31. Identify three physical properties of diamond.

32. What is a *PAH*? Provide an example.

33. How do PAHs find their way into the environment?

34. Why are PAHs pollutants? What health problems are associated with them?

35. What is *creosote*? How is it used?

36. Compare the structures of diamond and graphite. How are they the same? How are they different?

37. What is a *composite material*?

38. What is an *allotrope*? Identify three allotropes of carbon.

39. What is *soot*?

40. What is *activated charcoal* and how is it used?

41. Describe the structure of a carbon nanotube.

42. Describe the structure of a carbon buckyball.

43. What is *nanotechnology*?

44. What problems are associated with the use and recycling of rubber tires?

Understanding Concepts

45. Explain why the structures of CH_2Cl_2 on page 173 are actually all the same molecule.

46. Draw three different representations of the 3-D structure for a molecule of chloroform, with the lone hydrogen atom a) in the plane, b) projecting forward from the plane, and c) projecting in back of the plane of the paper. Are there three different isomers of chloroform, or are all three representations in fact equivalent?

47. Generate the formulas of the three molecules that contain one carbon atom, plus at least one atom each of hydrogen, fluorine, and chlorine.

48. Explain what you see in the following properties as the ratio of halogen to carbon atoms in a molecule increases:
   a) boiling point of molecule
   b) state of molecule

49. Write out the structural formula, showing about a dozen carbons atoms, for the polymer that would be formed starting from the monomer $CH_2{=}CCl_2$.

50. How does an anesthetic work?

51. What is the repeating unit in the polymer $-CCl_2-CHCl-CCl_2-CHCl-CCl_2-CHCl-$? What is the monomer that could be polymerized by an addition process to obtain this polymer?

52. Compare the structures and properties of long-chain alkanes and HDPE. How are they alike? How are they different?

53. Why are polymers generally more flexible in their rubber state?

54. Why is polystyrene a good choice for use in insulation?

55. Are all polymers plastics? Are all plastics polymers? Explain.

56. What properties would you want in a polymer that is used to keep food from spoiling?

57. Compare the molecular-level structures of high-density and low-density polymers. What factors determine the density?

58. Do the structures of natural and synthetic rubber differ? Explain.

59. Explain how the structures of some polymers confer the ability to conduct electricity. Why is this a directional property?

60. What properties would you want in a polymer that was to be recycled?

61. Describe the similarities in structure between a layer of graphite and a carbon nanotube.

62. Diamond and naphthalene both contain networks of carbon atoms. Describe the differences in their structures and compare their properties.

Synthesizing Ideas

63. Construct a table of structural monomers and properties for the following polymers: HDPE, PVC, polypropylene, polystyrene, and PTFE. What observations can you make about the composition of a polymer and its properties?

64. For each of the polymers below, explain how its properties make it appropriate for the use described:
a) Teflon used in cooking equipment
b) polystyrene used in CD cases
c) PVC used in water pipes

65. Describe and compare the structures and properties of a carbon nanotube and a carbon buckyball. How do they differ? How are they the same? Based upon your answer and information from the Web site that accompanies this book, what uses could you envision for nanotubes and buckyballs?

66. Deduce the structural formulas for the five unique isomers of $C_{18}H_{12}$ which contain four fused benzene rings.

67. Use the resources listed at the Web site for this book to investigate the potential danger of using polymers in medical equipment such as intravenous tubes and bags.
a) Which polymers are used in medical equipment?
b) What properties are essential for materials used in intravenous therapy (IV bags, tubes, tube connectors, etc.)?
c) Is there evidence that the use of these polymers in IV bags can be harmful?
d) Are alternatives available?

68. Use the resources listed at the Web site for this book to investigate the use of buckyballs as drug delivery systems.
a) What properties of buckyballs make them good candidates for use in drug delivery?
b) How would drug delivery via buckyballs work?

What are the advantages of using buckyballs for drug delivery systems? How do they compare with pills or injections?

■ Group Activity: The plastics in your life

Before meeting as a group, have each member make an inventory of the consumer plastic materials in his/her bedroom, kitchen, and bathroom. Check each container and packaging material for the recycling number found inside a triangle of arrows stamped (usually on the bottom) of the object. Make a list or table containing the name of the object (e.g., liquid soap bottle), the recycling number, a short description of the extent of its flexibility (for example, "flexes upon squeezing"), and whether or not it is transparent. Have one member of the group determine the recycling numbers of the plastics recycled in your community.

When the group meets, consider each recycling number in turn and make a master list of the objects and their properties and answer the following questions:

1. What are the uses and properties common to the objects with the same recycling number?
2. Do some types of objects appear under more than one recycling number?
3. Can you see why some types of plastics would be inappropriate for certain uses?
4. Would you be able to tell people which types of plastics in your community they should recycle without checking for the recycling number?

In this chapter, you will learn:

- about organic compounds that contain oxygen atoms, and the variety of everyday materials they are used for;

- about an organic class of compounds called ethers, used as anesthetics, and as gasoline additives such as MTBE;

- about an organic class of compounds called alcohols, used as fuel in racing cars, and found in alcoholic beverages;

- why water and alcohols have unusual physical properties;

- about the many pleasant-tasting and sweet-smelling organic compounds called aldehydes and ketones, often used in foods and perfumes;

- about a class of organic compounds called carboxylic acids, which are responsible for body odor and hangovers;

- what the alpha-hydroxy acids found in body lotions are;

- how an organic class of sweet smelling compounds called esters, are often used in the production of contact lenses and waxes for lipsticks.

Can you smell and taste the foods in this picnic scene?

The flavors of juicy ripe fruit, the fragrance of flowers, and the effects of wine are due to the presence of small carbon-based molecules that contain oxygen atoms. In this chapter, we will learn more about these oxygen-containing organic molecules. (Photodisc/Getty Images)

The Flavor of Our World

The Oxygen-Containing Organic Compounds We Drink, Smell, and Taste

When we smell a rose or perfume, our noses are actually detecting specific organic molecules that contain oxygen. The same is true when we taste the tartness of a lemon or a citrus fruit drink. Other oxygen-containing organic molecules have very different effects. The alcohols, for instance, produce intoxication and are later transformed in our bodies into substances that can give us hangovers.

In this chapter, we explore the formation of covalent bonds between carbon atoms and oxygen atoms, and the many different types of molecules that form as a consequence. These include many of those that produce the odors, tastes, and effects associated with many common foodstuffs, flowers, beverages, etc. In the following chapter, we shall continue this exploration by considering the chemistry of some natural components of our diet, namely carbohydrates and fats. In Chapters 8 and 9, we shall explore some of the chemistry of life, including disease and heredity.

We begin this chapter with a look at ethers and alcohols, two groups of organic compounds whose molecules contain carbon–oxygen single bonds, C—O. However, carbon can also form a double bond with oxygen. There exist several different types of molecules in which a carbon atom forms a double bond to only one oxygen atom—these are the aldehydes, ketones, carboxylic acids, and esters, which we consider in this chapter after ethers and alcohols. As we shall see, organic compounds containing oxygen are important industrially as well as biologically and aesthetically.

As well as exploring the bonding within molecules, in this chapter we shall also discover how special weak interactions between molecules profoundly affect properties such as the water solubility and the boiling point of some organic compounds of oxygen. Another of the general concepts to be developed is the toxicity of substances and how they are reported. Thus, the chapter is a balance between general principles and descriptive information about specific types of chemicals.

Ethers

Carbon atoms form stable, strong single covalent bonds with oxygen atoms: C—O. However, in contrast to hydrogen and halogen atoms, the oxygen atom forms an *additional* bond, since as an isolated atom it is *two* electrons short of a stable octet. One way in which it can achieve two bonds is by forming two carbon–oxygen single bonds; that is, linking the oxygen to carbon atoms in two chains or rings:

$$C—O—C$$

 Whenever you see this icon in this chapter, go to
www.whfreeman.com/ciyl2e

Compounds whose molecules contain such a three-atom unit, but no other types of carbon–oxygen bonds, are called **ethers.** These compounds are named informally after the carbon chains or rings attached to the oxygen. For example, if both chains bonded to the oxygen atom contain just one carbon atom and three associated hydrogen atoms—in other words, the methyl group—the molecule is called *dimethyl ether:*

Methyl group · Methyl group

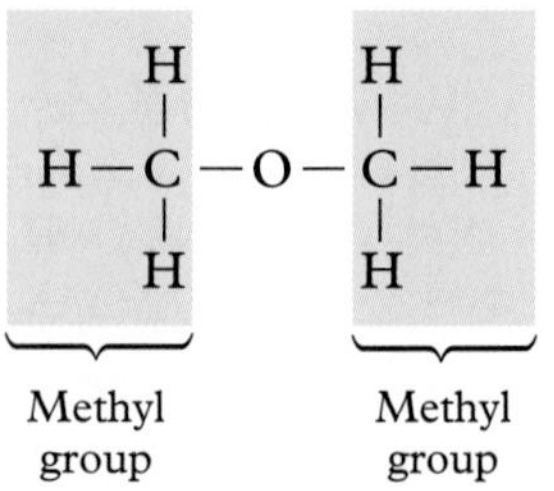

Dimethyl ether

If both chains contain two bonded carbon atoms—the ethyl group—the molecule is called *diethyl ether:*

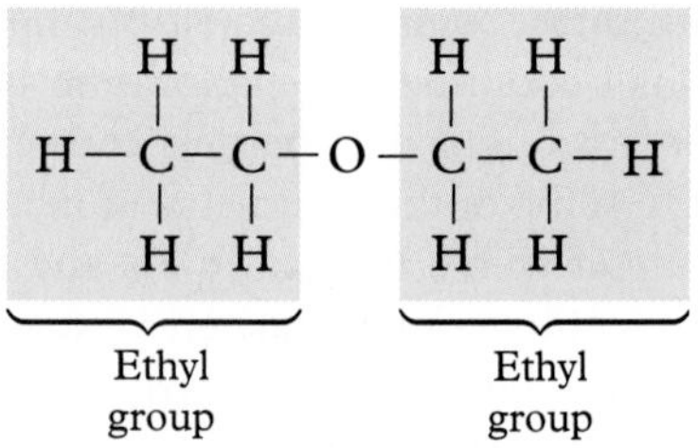

Ethyl group · Ethyl group

Diethyl ether

The ether $C-O-C$ unit is an example of what chemists call a **functional group,** which in general is an atom or a characteristic bonded collection of atoms that occur in organic molecules.

Diethyl ether, whose condensed formula we can write as $CH_3CH_2-O-CH_2CH_3$ or as $(CH_3CH_2)_2O$, is often referred to simply as "ether." It is used in chemistry laboratories as a solvent for organic compounds. Like hydrocarbons, simple ethers including diethyl ether are flammable, and so safety precautions must be taken in laboratories to prevent fires.

Diethyl ether was used for many years (starting in the mid-19th century) as an anesthetic. It has fallen out of favor in developed countries for a number of reasons, including its high flammability; the fact that it takes patients hours to regain consciousness; and its potential to produce negative side effects (such as vomiting) when the patient has awakened. Related ethers, such as enflurane, CHF_2-O-CF_2CHFCl, and isoflurane, $CH_2F-O-CHClCF_3$, which have halogen atoms and are consequently much less flammable, are now popular anesthetics.

Dimethyl ether has been tested as a replacement for, and an additive in, diesel fuel because of its clean-burning tendencies.

6.1 The fuel additive MTBE is an ether

The two carbon chains that are bonded to the oxygen in ethers are not necessarily identical, nor must they be unbranched. For example, in the ether known as *MTBE*, one group is simply methyl whereas the other is the *tertiary-butyl* group, giving rise to its full name, *methyl tert-butyl ether* (hydrogen atoms are not shown in this structure):

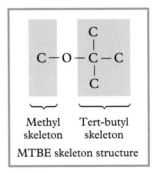

Notice that there are four carbon atoms in the tertiary-butyl group—and hence the name butyl, derived from butane, the four-carbon alkane—but that the network is branched and the carbon atom used to bond to the oxygen is from the middle, not the end, of the chain.

As we noted in Chapter 4, in the 1990s, MTBE found use as a replacement for some of the hydrocarbon content of gasoline. Because it introduces oxygen into the fuel mixture, it makes the gasoline burn more completely and so vehicles emit less carbon monoxide. Furthermore, since MTBE has a high octane rating (116), it can replace some of the aromatic hydrocarbons, thereby lessening the emission of benzene.

The use of MTBE in gasoline has been quite controversial, however. Its strong odor—at least when it is mixed with certain types of gasoline—is sickening to some people. In addition, it is slow to decompose naturally and so, as a result of gasoline spills, it has seeped down through the soil to the water table and contaminated the well water in some areas. For these reasons, and because of other unresolved questions concerning its toxicity, its use in gasoline is being phased out, by law, in the United States. For example, the state of California banned the use of MTBE in 2003.

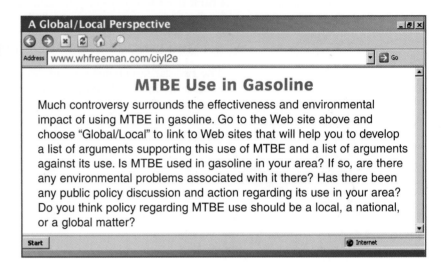

MTBE Use in Gasoline

Much controversy surrounds the effectiveness and environmental impact of using MTBE in gasoline. Go to the Web site above and choose "Global/Local" to link to Web sites that will help you to develop a list of arguments supporting this use of MTBE and a list of arguments against its use. Is MTBE used in gasoline in your area? If so, are there any environmental problems associated with it there? Has there been any public policy discussion and action regarding its use in your area? Do you think policy regarding MTBE use should be a local, a national, or a global matter?

Alcohols

The term *alcohol* is widely used for a variety of different substances. In addition to the alcohol that is used for drinking, you may have heard of *rubbing alcohol, wood alcohol, fuel alcohol, grain alcohol, renewable alcohol, denatured alcohol, methyl alcohol,* and so on. There are indeed many different kinds of alcohol, which is a category of organic compounds. In this section, we will sort out the meaning of many of these terms and discuss the chemical structure, uses, and abuses of the various alcohols.

6.2 Alcohol molecules contain a C—O—H unit

In an ether molecule, an oxygen atom forms two single bonds to carbon atoms. In a water molecule, the oxygen forms two single bonds to hydrogen atoms. In **alcohols,** the oxygen atoms form one single bond to a carbon atom and the other to a hydrogen atom. Thus, alcohols characteristically contain the three-atom unit C—O—H, which is another example of a functional group. This structural unit, intermediate as it is between water and ethers, leads to some intermediate properties, as we shall see.

The scientific name for each particular alcohol terminates in *-ol,* with the root based upon the name of the hydrocarbon unit bonded to the —OH. Thus CH_3OH is called *methanol* because it contains a methyl unit. However, alternative names in which the hydrocarbon group is named, and followed by the word *alcohol,* are quite common; thus CH_3OH is also called *methyl alcohol.* The formulas for alcohols are usually written with the OH unit shown explicitly so that the substance can

Methanol is not actually made commercially by replacing one —H by —OH in a chemical reaction. But it is convenient for us to imagine that —OH replaces —H.

A method for estimating the quantity of heat released by burning methanol (or by other chemical reactions involving simple organic compounds) is discussed in *Taking It Further* at the end of this chapter.

Figure 6.1 Methanol. A fuel source that can provide heat for cooking.
(George Semple for W. H. Freeman and Company)

be easily identified as an alcohol. Usage of the molecular formulas, such as CH_4O for methanol, is less common.

6.3 Methanol is often used as a fuel

Methanol is the simplest and one of the most important of the alcohols. Only one carbon atom is present in methanol molecules, so it is bonded to three hydrogen atoms, forming a methyl group:

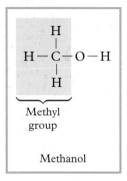

Methyl group

Methanol

Conceptually, the —OH group essentially replaces one of the —H atoms in methane. The —OH unit is sometimes called the *hydroxyl group*, and is incorporated as *hydroxy* in compound names.

As a compound, methanol is a clear, colorless liquid that boils at 65°C. It is also known as "wood alcohol," since it can be obtained in bulk by heating wood in the absence of air until it decomposes. This so-called destructive distillation process was the commercial source of methanol for hundreds of years until the 1930s; since that time it has been produced primarily from fossil fuels, particularly natural gas.

In contrast to water, but in common with ethers, methanol and other alcohols burn readily in air. As in the case of hydrocarbons, carbon dioxide and water are the products of the complete combustion:

$$2\ CH_3OH + 3\ O_2 \rightarrow 2\ CO_2 + 4\ H_2O + heat$$

Methanol is used as a fuel in a number of situations. It is a common fuel for small indoor cooking appliances, such as fondue pots, and for outdoor camp stoves (see Figure 6.1).

Given its convenient form as a liquid, its ability to undergo smooth combustion, and its high octane number (107), methanol can be used as a fuel for vehicles whose engines have been constructed or adapted to overcome its somewhat corrosive properties. For example, all the cars at the Indianapolis 500 and other races use methanol rather than gasoline as fuel, not only because of its high octane number but also because, in the

Alcohols

event of a crash, it is less likely to produce a fatal fireball. Domestic vehicles fueled by pure methanol are now appearing on the market.

A blend of about 5–10% methanol in gasoline improves the burning characteristics and octane number of the liquid. Gasoline–methanol fuel mixtures are characterized by an M symbol with a subscript indicating the percent, by volume, of methanol that is present. For example, M_{10} is a solution that is 90% gasoline and 10% methanol.

Most methanol is produced by reacting methane obtained from natural gas with steam and with carbon dioxide in the presence of a catalyst. The reaction involves several sequential steps, with the overall result:

$$3 \; CH_4 + 2 \; H_2O + CO_2 \xrightarrow{\text{catalyst}} 4 \; CH_3OH$$

The fuel energy content of the natural gas is thereby converted to that of a liquid fuel, methanol. Methanol can also be made from coal by a similar reaction, and this technique will presumably increase in importance in the future as natural gas reserves become depleted. However, CO_2 emissions are no lower when methanol is produced in these ways, and then combusted, than when natural gas or coal is simply burned directly as a fuel.

In addition to its growing use as a fuel, much of the methanol currently manufactured is converted to the fuel additive MTBE and to other organic compounds. Some of it finds use as a solvent—for example, mixed with water as windshield wiper fluid—and in industrial processes.

6.4 Methanol is not for drinking!

Don't drink methanol! It is much more toxic than ethanol, the alcohol in commercial alcoholic beverages. In fact, methanol is the most toxic of all the alcohols. Consuming even a relatively small amount (30 mL, or about 1 ounce) can lead to blindness, as its decomposition products in the body attack the optic nerve. Drinking greater quantities of it can lead to death. You may wonder why anyone would drink methanol, but more than 100 people were killed in two incidents in recent decades, one in Italy and the other in India, when homemade "liquor" that had been formulated with methanol was consumed at parties. So don't confuse the alcohols!

The above warnings notwithstanding, the human body *can* cope successfully with the *small* amounts of methanol that are found naturally in the juices of fruits such as pineapple. Small amounts of methanol are also found in normal alcoholic beverages and are produced in your body when the artificial sweetener *Aspartame* is metabolized.

6.5 Ethanol is found in alcoholic beverages

Ethanol, or ethyl alcohol, C_2H_5OH, is sometimes called "grain alcohol" since it can be made from that source. Ethanol molecules contain two carbon atoms singly bonded to each other in the ethyl group, which itself is singly bonded to the —O—H unit:

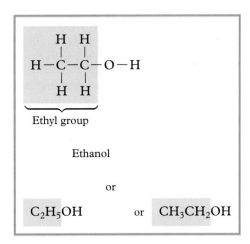

Like methanol, ethanol is a clear, colorless liquid. Its boiling point is 78.5°C, and it is soluble in water in any proportion. Indeed, pure ethanol absorbs water vapor from air. Ethanol's density is 0.79 gram per cubic centimeter. It is the least toxic of all the alcohols.

Ethanol is the alcohol that is the principal ingredient, along with water, in alcoholic beverages, including beer, wine, "hard liquor" (such as bourbon, whiskey, rum, gin, vodka, etc.), and liqueurs (cordials). The flavors of these beverages are due to other, minor components, since the alcohol itself is almost tasteless.

The concentration scale that is used commercially for alcoholic beverages that are more potent than beer or wine is the **proof value,** which in North America is two times the percent by volume of ethanol in the ethanol–water solution. Thus if a whiskey is 45% ethanol by volume, it is labeled *90 proof.* Other types of whiskeys will have slightly different proof values, since they are manufactured by slightly different processes. Pure alcohol is 200 proof. The term *proof* is based upon 17th-century techniques that checked the quality of alcoholic spirits. The proof test involved moistening gunpowder with the spirits and lighting it. If the gunpowder burned slowly, the spirits were at proof; if it did not ignite, the spirits were under proof; if it burst into flames, the spirits were over proof.

Taking It Further

For information and problems on Ethanol, go to Taking It Further at www.whfreeman.com/ciyl2e.

The proof scale in the United Kingdom differs somewhat from the North American scale. In the United Kingdom scale, 100 proof corresponds to 57% alcohol by volume.

Worked Example: Deducing alcohol content from proof values

What is the percentage of alcohol by volume in 100 proof vodka?

Solution: Since the value of the proof is twice the percentage of alcohol, 100 proof corresponds to 100/2 = 50% of alcohol by volume.

Exercise 6.3

a) What would be the proof value for a beer whose alcohol content is 5%?
b) What is the alcohol content of 80 proof whiskey?

Alcohols

6.6 Ethanol has many effects on human health

Is drinking small quantities of alcohol—about one glass of wine per day—good for a person's health, as some medical studies suggest? Or is it bad for you since ethanol is a poisonous substance that can kill you if too much of it is ingested? Seeming contradictions of this type are not unusual for many substances, including most drugs. Indeed, it appears that consuming large amounts of almost *anything* is harmful—and in some cases fatal—to your health, but it does *not* necessarily follow that consuming small amounts of the same substance is harmful. In most situations, it is the dose that makes the poison!

Numerous studies have established that people who drink moderate amounts of alcohol—up to one drink per day—have a significantly lower death rate than do nondrinkers; that is, nondrinkers die on average at an earlier age. Most of the positive benefits seem to be independent of whether the alcohol is in the form of wine, beer, or spirits, and the benefits are mostly due to a reduction in the rate of coronary heart disease. Unfortunately, alcohol also brings a significant *increase* in the risk of death from cancer, especially at high levels of consumption, though this may not be true for wine. "Binge drinking" is also known to result in brain damage, especially among young people. Drunk driving is also a leading cause of traffic accidents and deaths. On average, only middle-aged and older people have been found to increase their life spans with moderate alcohol consumption when all these factors, positive and negative, are taken into account.

Of course, the amounts of *some* materials that one needs to ingest before harmful effects occur are quite small; these are commonly called **toxic substances.** The toxicity of a substance is measured by its lethal dose, which is the amount of it that is required to kill an organism. Generally speaking, the heavier the organism, the greater the dose of a given substance that is required to be lethal. Thus, lethal doses are usually quoted as the mass of the minimum amount of the substance that is lethal *divided* by the mass of the organism that is killed by it.

For many people, the lethal dose of ethanol is probably about 10 grams of ethanol per kilogram of body weight. For example, if you weigh 70 kilograms (155 pounds), the lethal dose would be $10 \times 70 = 700$ grams. This is *not* to say that in a group of people all weighing 70 kilograms, ingestion of less than 700 grams of ethanol will not kill *any* of them, nor that ingestion of 700 grams or more of ethanol will prove fatal for *all* members of the group. For any substance there are always a few members of any group that are particularly susceptible to a toxin and some others that are quite resistant to it.

The most commonly quoted toxicity value for substances is the **LD_{50},** the minimum dose that is lethal to 50% of the population. The LD_{50} values for a number of alcohols are listed in Table 6.1. Note that the lower its LD_{50} value, the more toxic a substance is—less of it is needed to be poisonous. LD_{50} values are based on animal studies. Though there is evidence that the values for many substances are approximately transferable to humans, they are inaccurate in certain cases. However, in the

Your weight in kilograms is that in pounds divided by 2.2.

Taking It Further
with Math

For information and problems on converting units, go to Taking It Further with Math at www.whfreeman.com/ciyl2e.

Table 6.1	LD$_{50}$ values for various alcohols	
Alcohol	Name	LD$_{50}$ in grams / kilogram
C$_2$—OH	ethanol	7–10
C$_3$—OH	1-propanol	1.9
C$_4$—OH	1-butanol	4.4
C$_5$—OH	1-pentanol	3.0
C$_6$—OH	1-hexanol	4.6

absence of any other information, LD$_{50}$ values are a good approximate guide as to the toxicity to humans of various substances.

Exercise 6.4

Calculate the mass of 1-propanol, listed as C$_3$—OH in Table 6.1, that would prove fatal to you if you were of average susceptibility to this substance. Note that your mass in kilograms equals your weight in pounds divided by 2.2.

Many detrimental effects of drinking too much alcohol are well known. They include ulcers, cancers, accidents, and cirrhosis of the liver. *Cirrhosis* is a condition in which healthy liver cells are replaced by hard scar tissue. This scar tissue builds up and prevents blood from reaching the surviving cells in the liver, leading to an accumulation of the toxic substances the liver would normally filter out and resulting in a permanent loss of function. Cirrhosis causes up to 26,000 deaths annually in the United States. Because a cirrhotic liver has an orange-like color to it, it was named cirrhosis, which means "orange color" in Greek. Although the disease is commonly seen in alcoholics, other conditions—such as chronic hepatitis C, B, or D, and autoimmune hepatitis, can also lead to cirrhosis of the liver. In males alcohol lowers the blood level of *testosterone*, a substance required to induce erections. Ethanol and its metabolism products also interfere with the transmission of the nerve impulses that are required to produce and maintain erections. Thus it is not an uncommon experience for a male to find that his sexual performance is affected negatively during bouts of heavy drinking and chronically as a consequence of high alcohol consumption.

The MRI image on the right in Figure 6.2 shows the brain of a child diagnosed with *fetal alcohol syndrome* (abbreviated FAS). FAS describes a host of physical and psychological abnormalities that result from the consumption of alcohol by a pregnant woman. If you compare the healthy brain (Figure 6.2a) with that of the brain affected by FAS (Figure 6.2b), you will see that the brain tissue of the image on the right is shrunken. This is most obvious in the center of each image, where all

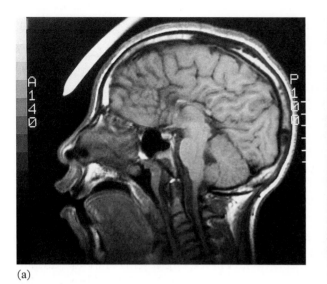

(a)

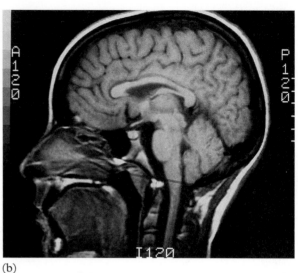

(b)

Figure 6.2 MRI image showing brain activity in (a) a healthy child's brain and (b) a child's brain affected by fetal alcohol syndrome. (Courtesy of Edward Riley, San Diego State University)

the parts of the middle of the brain are seen to be smaller and more spread out in the image of the alcohol-affected brain as compared to the healthy brain. This occurs because the development, function, movement, and survival of certain nerve cells are disrupted by exposure to alcohol. Indeed, research shows that ethanol concentrations equivalent to a single glass of wine can inhibit the formation of connections between developing nerve cells. The physical changes in the brain that result from the poorly connected nerve cells are believed to be responsible for the physical and psychological problems commonly seen with FAS children, which include attention disorders, hyperactivity, aggression, psychiatric problems, mental retardation (in severe cases), and deformation of the limbs, face, and other body parts. Children with FAS may display any combination of these abnormalities. The likelihood of a fetus developing FAS increases as the amount of alcohol the mother consumes during the pregnancy increases. In the United States, about 5000 babies are born with FAS each year, many of them to alcoholic mothers.

6.7 The history of a can of beer in the body

To see how alcohol is metabolized and how it produces its effects, let us take an imaginary journey with the contents of a can of beer as it makes its way through the human body. You can follow the steps in Figure 6.3. Beer in North America comes in containers usually having a volume of about 340 mL (12 ounces). In "regular"—as opposed to "light"—beer, about 5% by volume of the liquid, or 0.05×340 mL = 17 mL, is actually alcohol. Since ethanol has a density of about 0.8 gram per mL, the mass of alcohol present is 0.8 g/mL $\times 17$ mL = 14 grams (about 1/2 ounce).

    When beer is swallowed, the liquid is diluted by stomach juices. Some of the alcohol almost immediately enters the bloodstream—it does

The journey of a glass of wine, a shot of liquor, or a serving of any other alcoholic beverage would be analogous, though the percentage of alcohol in the original drink would be different.

A shot of tequila, a glass of wine, and a glass of beer all contain about the same amount of alcohol.

True—at least in the U.S. The amount of alcohol that is equal to a standard drink in the U.S., 14 grams, is about the same as that contained in a 12-oz. can of regular beer, a glass of wine (about 12% alcohol by volume), or in a 1¼ ounce (38 mL) shot of hard liquor such as whiskey (about 40% alcohol). However, outside of the U.S., standards are different. For example, a standard drink in Great Britain contains only 8 grams of alcohol, whereas in Japan it contains 19.75 grams!

not need to undergo prior digestion or transformation in the stomach (Figure 6.3, step ❶). The rest of the beer passes to the small intestine, where the remainder of the alcohol is transferred rather quickly into the bloodstream (step ❷). For the beer to move from the stomach to the small intestine, the *pyloric valve* must open, an action that is delayed if the stomach contains fats or sugar that are being digested. Thus, eating a high-fat meal while drinking alcohol will slow down the absorption of alcohol into the bloodstream. In contrast, consuming a carbonated drink will result in even faster absorption of alcohol into the bloodstream. The alcohol dissolves readily in the blood and then is largely transferred to tissues that have high blood flow. Within one minute of absorption, alcohol is delivered to the brain and the changes associated with inebriation begin. The lungs, kidneys, and liver also receive absorbed alcohol within minutes of absorption.

As soon as there is some alcohol in the blood that passes through the liver, zinc-containing enzymes in that organ set to work to oxidize it (step ❸). Thus the concentration of alcohol in the blood begins to fall about half an hour after the beer is consumed and is reduced to about one-quarter of its peak level at the end of an hour or so.

Fifteen milliliters of ethanol, a little less than the amount in one beer, is as much as the liver can handle in one hour; if a person drinks two cans in a short interval, it will take two hours or more for the alcohol level in the blood to fall substantially. The extent to which a person becomes inebriated is the result of the race between ethanol absorption into the blood and its breakdown in the liver. Figure 6.4 plots the percent of alcohol in blood against time for an average adult male after he has quickly consumed one or more beers. The difference in blood alcohol resulting from the consumption of seven 12-ounce beers all at once, or gradually over a period of four hours, is illustrated in Figure 6.5. Notice that the peak level is much higher when the alcohol is consumed all at once.

At whatever rate it is consumed, over 90% of the alcohol is disposed of by **metabolic processes,** chemical reactions that occur in the body and that alter the nature of chemicals present there. The metabolism of alcohol occurs mainly in the liver. A small proportion of the alcohol is excreted in sweat, and a small amount passes through the kidneys and enters the urine (Figure 6.3, step ❹). Only a very small fraction of it is exhaled as a gas

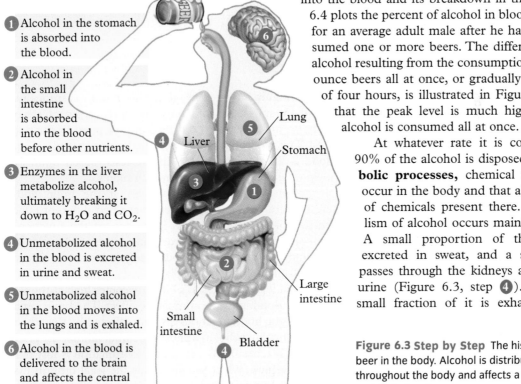

❶ Alcohol in the stomach is absorbed into the blood.

❷ Alcohol in the small intestine is absorbed into the blood before other nutrients.

❸ Enzymes in the liver metabolize alcohol, ultimately breaking it down to H_2O and CO_2.

❹ Unmetabolized alcohol in the blood is excreted in urine and sweat.

❺ Unmetabolized alcohol in the blood moves into the lungs and is exhaled.

❻ Alcohol in the blood is delivered to the brain and affects the central nervous system.

Figure 6.3 Step by Step The history of a can of beer in the body. Alcohol is distributed quickly throughout the body and affects a variety of metabolic functions.

Alcohols

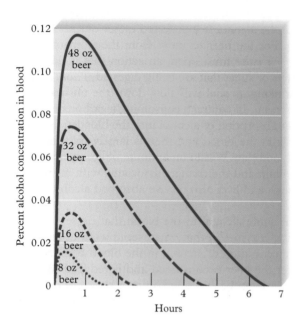

Figure 6.4 The relationship between the amount of alcohol consumed and the time it remains in the blood. Data are for an average adult male. (Adapted from *The New Encyclopaedia Britannica*, 15th ed., Chicago, 1995.)

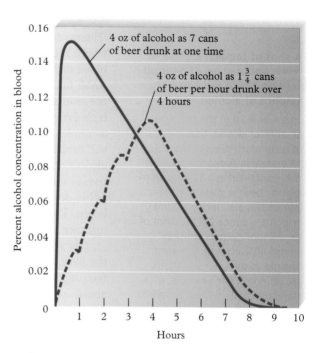

Figure 6.5 The difference in blood alcohol levels after consuming an amount of alcohol all at once and over a period of time. When alcohol is consumed over a period of time, blood alcohol concentration does not rise to the same levels. (Adapted from *The New Encyclopaedia Britannica*, 15th ed., Chicago, 1995.)

through the lungs (Figure 6.3, step ❺), but this amount is sufficient to be used in Breathalyzer tests to indirectly determine a person's blood alcohol level if he or she is driving a vehicle that is stopped by the police (see Figure 6.6). The ratio of ethanol in exhaled air to that in an equal volume of blood is taken to be 1:2100 in the tests, based upon both chemical principles and experiments on human blood.

When alcohol in the bloodstream arrives at the brain and is absorbed there (Figure 6.3, step ❻), it can have two different effects. Initially, when its concentration is low, it suppresses the function of some of the inhibitory brain centers, producing some exhilaration and loss of tension and perhaps some mood changes, loss of social constraint, and uncontrolled emotional displays.

When the alcohol level in a person's blood becomes higher, the sedative effect of alcohol dominates the stimulative effect. The brain centers that control some of the body functions also are affected by the ethanol. If one consumes so much alcohol in a short time interval that the level of alcohol in the blood reaches 0.08%, so much coordination is lost that it is unsafe to drive a vehicle, which is why this is the legal limit in many states and provinces. At higher levels, the brain centers are so affected that speech is slurred, gait is unsteady, and reaction time is appreciably increased. At the molecular level, these effects have occurred because ethanol has replaced some of the water around the nerve cells, and this change interferes with the transmission of information pulses along nerve fibers. The stagger in the walk of an inebriated person is caused by attempts to compensate for the loss of some of the sense of balance. Balance is controlled by the density of the tissue and fluid in the ears, both of which are changed by alcohol, which has a lower density than water.

People usually urinate more than usual in the hours after drinking—not just the amount corresponding to the volume of drinks consumed, but at least an additional equal volume. This occurs because the ethanol suppresses the normal recycling of water in the body, so more of the water simply passes into the bladder, promoting increased urination and leading to some dehydration. If a person consumes a fair amount of alcohol, it is a good idea to drink water before going to bed.

If the level of alcohol in the blood increases, to about 0.3 to 0.4%, the depressant effect of ethanol becomes dominant, leading to sedation and stupor. This is when people "pass out." Coma and even death can result if the level becomes higher still because at such concentrations the breathing center in the brain and/or the action of the heart can be anesthetized.

Eventually, the alcohol in the bloodstream is all oxidized in a series of metabolism steps to carbon dioxide, which is exhaled, and to water. During the oxidation of a specific weight of alcohol, even more energy or fat is produced than if the same weight of sugar or starchy food had been eaten.

Figure 6.6 (a) A Breathalyzer. (b) This device detects unmetabolized ethanol that is exhaled by the lungs by reacting it with an aqueous solution of intensely yellow dichromate ion, $Cr_2O_7^{2-}$, to produce green chromate ion, Cr^{3+}. (R. Pearce for W. H. Freeman & Company.)

(a)

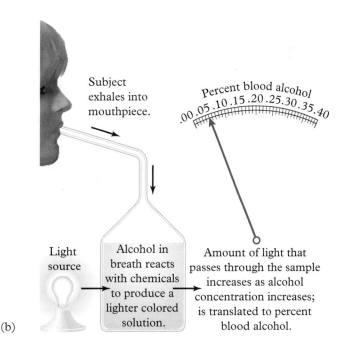

(b)

Subject exhales into mouthpiece.

Percent blood alcohol

.00 .05 .10 .15 .20 .25 .30 .35 .40

Light source

Alcohol in breath reacts with chemicals to produce a lighter colored solution.

Amount of light that passes through the sample increases as alcohol concentration increases; is translated to percent blood alcohol.

■ Chemistry in Your Home: Two tests for sobriety

One of the field tests used by police officers to determine sobriety involves having the subject follow a finger or light with his or her eyes. The basis for this is a phenomenon called *nystagmus*. Nystagmus is an involuntary jerking of the eye as it moves to the side. The more intoxicated a person is, the less the eye has to move toward the side before jerking begins. You can observe nystagmus by having a friend focus on your finger in front of his or her eyes. Move your finger very smoothly and slowly back and forth and observe the eye movements. Then move your finger smoothly but quickly and observe the eye movements. What observations can you make? Do eye movements differ when the object is moving slowly instead of quickly?

Another field test for sobriety requires that a subject walk in a straight line touching heel to toe for nine steps, turn around on the line, and repeat the process. This is a divided-attention task that tests the subject's ability to understand and execute two or more simple instructions at one time. Impaired subjects have trouble performing the tasks. Try the test yourself. Some people have a problem doing this even when they are sober!

6.8 Ethanol can be used as a fuel or as an additive

You may know that some alcoholic products such as brandy and hard liquor, in which the ethanol content exceeds about 20%, will burn if a lighted match is put to their surface—this property is often exploited in flambé desserts. In such reactions, the alcohol is combusted by oxygen in the air into carbon dioxide and water:

$$C_2H_5OH + 3\ O_2 \rightarrow 2\ CO_2 + 3\ H_2O + heat$$

Since this reaction generates heat, ethanol can be used as a fuel in the same manner that was discussed earlier for methanol.

Indeed, large quantities of ethanol currently are produced in North America from crops such as corn, and in warmer countries such as Brazil and Zimbabwe from sugarcane, for use as a gasoline additive (a 10% ethanol and 90% gasoline solution, for example, called *gasohol*) or as an automotive fuel in itself. In the latter case, the alcohol is usually mixed with some gasoline to improve the ease of starting the engine. The mixed fuel is characterized by a symbol such as E_{85}, where the subscript denotes the percentage (by volume) of ethanol that is present, analogous to the M scale for methanol–gasoline mixtures. This use of ethanol promotes clean burning of gasolines, since less carbon monoxide is emitted. It also improves their octane number (since that of ethanol is 108) and reduces the amount of petroleum that a country needs to import. Brazil has led the way in using ethanol as a fuel, in a major effort to reduce its reliance on foreign oil. In North America, ethanol is viewed as a replacement for MTBE in gasoline since it also contains oxygen.

Exercise 6.5

What would be the appropriate "E" symbols for gasohol (10% ethanol) and for the 85% ethanol–15% gasoline solutions used as automobile fuels?

In principle, ethanol can be continuously produced from carbohydrate crops, which during their growing absorb from the air as much CO_2 as is emitted when the alcohol is burned (see Figure 6.7). Ethanol is thus potentially an example of a **renewable fuel,** which is defined as one whose supply can be continuously renewed and in which there is no *net* carbon dioxide release during its production/combustion cycle. However, agricultural ethanol is much less advantageous as an oil replacement if, as in North America, natural gas or coal is burned to distill the ethanol. This step, plus the natural gas and petroleum requirements of producing fertilizer and operating machinery, together consume almost as much fossil fuel energy as is stored in the alcohol. Another objection to the production of crops for fuel is that the land could instead be used to grow food. The biotechnology industry is developing enzymes that will allow ethanol to be produced directly from waste paper and other waste wood products. In the future, methanol obtained from wood may be a more acceptable renewable fuel than ethanol.

Exercise 6.6

Like methanol, ethanol can be used to make an ether for use as a gasoline additive. Draw the structure of the analog to MTBE that would be obtained starting with ethanol rather than methanol. Suggest a name and four-letter abbreviation for it.

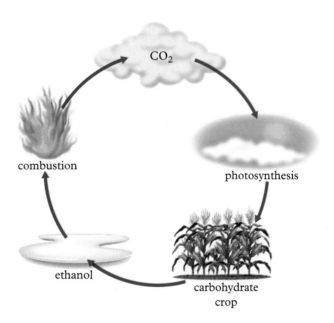

Figure 6.7 Cycle of CO_2 absorption and release during agricultural production and its use as a fuel.

Alcohols

▷ **Discussion Point:** Should the use of agricultural ethanol in gasoline be mandatory?

We have seen that oxygen-containing compounds can be used as gasoline additives to make the fuel cleaner burning and to displace some of the crude oil that otherwise would have to be imported. Some people, especially those with an interest in producing the material, believe that ethanol from agricultural production should be used for this purpose, rather than methanol or ethanol or other compounds produced synthetically by the chemical industry. They also believe that the ethanol should be taxed less than the hydrocarbon components of gasoline.

Use the Web site that accompanies this book and other resources to develop a list of arguments supporting the use of agricultural ethanol in gasoline and a list of arguments against its use. Is ethanol used in gasoline in your area? Are there any environmental problems associated with ethanol in your area? What are the economic advantages of using ethanol? Who benefits economically by using ethanol? Are there economic disadvantages to using ethanol? Do you agree or disagree with legislation forcing some or all of the oxygen-containing additives in gasoline to be agricultural ethanol?

In addition to its use as a fuel and as a component of beverages, ethanol has a variety of other uses. It is used as a solvent, both industrially and to dissolve active ingredients in products such as cough medicines, and to prepare other compounds. Highly concentrated solutions (70% ethanol in water) act as antiseptics to bacteria, and somewhat lower concentrations are used in mouthwashes. Alcohol that is not intended for human consumption is taxed at a much lower rate. To ensure that "industrial alcohol" is not drunk, a small amount of *Bitrex*, the world's most bitter-tasting substance, is added to it. Bitrex is also added to some household chemicals to keep young children from drinking them.

6.9 Hydrogen bonding is a weak attraction

Dimethyl ether is a gas at room temperature, whereas its isomer, ethanol, is a liquid which does not boil until 79°C, a full 104°C higher than the ether. Similarly, the alcohol butanol, $CH_3CH_2CH_2CH_2OH$, boils at a temperature that is 82°C higher than for its isomeric diethyl ether.

These differences in boiling point are surprising given that the boiling point is typically a property that is determined for most organic molecules by their size. For example, propane and dimethyl ether, compounds whose molecules have the same number of electrons and almost the same mass and size, are both gases at room temperature and have similar boiling points of −42°C and −25°C respectively. Higher boiling points are usually associated with heavier molecules, since the weak attractions that exist between adjacent molecules in the liquid (and solid) increase with the number of electrons that each possesses. Thus, the boiling points of both the *n*-alkanes and the ethers increase regularly as the length of the carbon chain (or carbon-oxygen-carbon chain) increases.

However, alcohols generally have much higher boiling points than do their isomeric ethers. What can be the source of the extra attraction between alcohol molecules that is responsible for increasing their boiling points so dramatically? Scientists ascribe the attraction to the phenomenon known as **hydrogen bonding.** To understand hydrogen bonding, we must remember that oxygen and hydrogen atoms have very different electronegativities, as discussed in Chapter 3. Because the electronegativity of oxygen is greater than that of hydrogen, oxygen–hydrogen bonds are polar: the oxygen atom receives *more* than an equal share of the two electrons that are shared between the atoms. Consequently, the oxygen carries a partial negative charge, δ^-, and the hydrogen a partial positive charge, δ^+.

Since opposite charges attract one another, alcohol molecules can attract each other if the hydrogen atom of the —OH group on one molecule is closely aligned to the oxygen atom of an adjacent alcohol molecule ROH, where R stands for any alkyl group:

Hydrogen bond

The partial positive charge of the hydrogen is particularly attracted to the nonbonding electrons of the oxygen, since these electrons represent regions of high negative electron density. Symbolically, we represent the hydrogen bond as a long dashed line: O---H. It sometimes is useful to indicate the atom to which the hydrogen is covalently bonded, so we use representations such as O---H—O.

Hydrogen bonds are not bonds in the conventional sense. The attraction *between* alcohol molecules amounts only to about 5% of the energy of each *internal,* covalent O—H bond. Nevertheless, it is sufficient to hold the molecules to each other more tightly than molecules of this size which are not hydrogen bonded. Since the hydrogen-bonded molecules are held together more tightly, more energy is required to separate them and form a gas. It is for this reason that the boiling points of alcohols are significantly higher than those of isomeric compounds in which hydrogen bonding cannot occur, such as the hydrocarbons and ethers.

An extensive 3-D network of hydrogen bonds weakly connects the adjacent molecules of an alcohol together. Because the molecules in a liquid are in constant motion, each hydrogen bond is transient, lasting only a tiny fraction of a second—but new bonds are constantly being formed as the molecules move around. The hydrogen bonds become permanent when the liquid freezes to a solid.

Exercise 6.7

Methanol, CH_3OH, is a liquid at room temperature (boiling point: $+65°C$). Ethane, CH_3CH_3, is a gas whose molecules have the same number of electrons as methanol but which boils far below room temperature (at $-89°C$). Explain this trend by drawing a hydrogen-bonding network for molecules of the alcohol. *Hint:* Refer to the diagram below to view the analogous hydrogen-bonding system in a generic alcohol.

Because each water molecule possesses two O—H bonds, the extent of hydrogen bonding in liquid and solid water is even more extensive than in alcohols. As a consequence, H_2O not only has much higher melting and boiling points than expected from those of molecules having the same number of electrons, but it also displays a number of other unusual properties.

Hydrogen bonding is a phenomenon that can occur between any adjacent pair of suitable molecules, not just between identical ones. In general, hydrogen bonding is the special stabilizing interaction that occurs under the following conditions:

- There is a hydrogen atom with a partial positive charge.
- There is a nearby atom with a partial (or a net) charge that is negative and that also possesses at least one pair of nonbonding electrons.

As shown below, hydrogen bonding can occur between a molecule of water and a molecule of alcohol. Can you see how the two conditions necessary for hydrogen bonding to occur have been satisfied here?

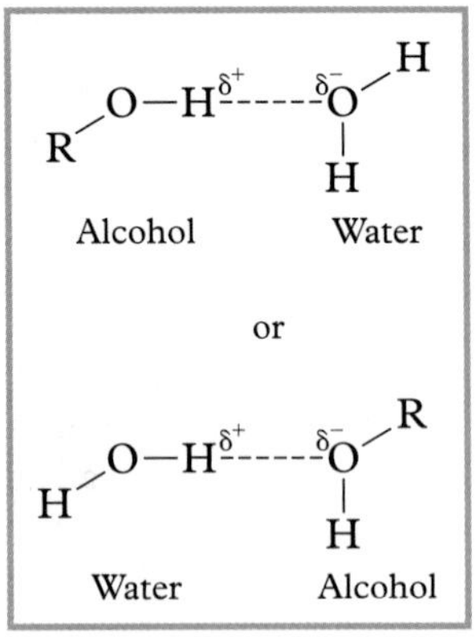

It is partially because of such hydrogen bonding that alcohols—at least those with short chains—are readily soluble in water. Thus methanol, ethanol, and isopropyl alcohol are all **miscible** with water; that is, *one*

In reality, the atom carrying the negative partial charge must be nitrogen, oxygen, or fluorine in order for the hydrogen bond to have significant strength.

Chapter 6: The Flavor of Our World

will dissolve in the other in any proportions to form (homogeneous) solutions. This behavior is unusual, because hydrocarbons and indeed most other organic compounds are practically insoluble in water, since they cannot form hydrogen bonds to it.

A 70% (by volume) solution of *isopropyl alcohol*, $(CH_3)_2CH$—OH, in water is known as "rubbing alcohol." After it is applied to skin, the alcohol rapidly evaporates by absorbing heat from the skin, thereby cooling it. This cooling effect is sometimes used to reduce perspiration and medically to reduce a fever. Since it is an effective bactericide, rubbing alcohol is also used as a topical antiseptic and to sterilize medical instruments such as stethoscopes between uses. (Some brands of rubbing alcohol use ethanol rather than isopropyl alcohol.)

The —OH portion of an alcohol molecule is termed **hydrophilic,** or water-loving, since it can form hydrogen bonds with H_2O molecules. The hydrocarbon portion, however, cannot form hydrogen bonds because C—H bonds are not appreciably polar; hence the carbon portion is **hydrophobic,** or water-hating. In alcohols having long carbon chains, the hydrophobic fraction of the molecule overwhelms the hydrophilic, and consequently the alcohol is not soluble in water. You can see the gradually decreasing water solubility with increasing chain length illustrated in Figure 6.8.

A *bactericide* is a substance that kills bacteria.

The suffixes *-philic* and *-phobic* are derived from Greek words meaning "loving" and "fearing," respectively.

 **Taking It Further** *with Math*

For information and problems on how to interpret graphs, go to Taking It Further with Math at www.whfreeman.com/ciyl2e.

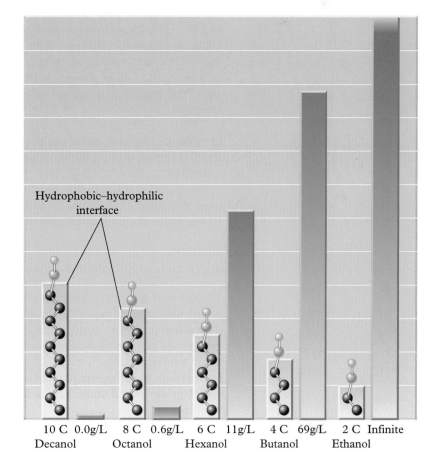

Hydrophobic–hydrophilic interface

| 10 C 0.0g/L | 8 C 0.6g/L | 6 C 11g/L | 4 C 69g/L | 2 C Infinite |
| Decanol | Octanol | Hexanol | Butanol | Ethanol |

Figure 6.8 The relationship between the solubility of an alcohol (green) and the length (yellow) of the carbon chain. As the number of carbons in an alcohol chain increases, the amount of the alcohol that can dissolve in a liter of water decreases.

Alcohols

The hydrogen-bonding attraction between the single —OH units in pure, liquid alcohols becomes progressively less dominant than the weak interactions between the carbon chains in adjacent molecules as the chains become longer and longer. Thus, the alcohols with long chains behave like alkanes and ethers since most of the molecule is hydrophobic and hydrocarbonlike. The boiling points of alcohols come closer and closer to those of their isomeric ethers as the chains lengthen.

■ Chemistry in Your Home: Determining which substances are hydrophobic

Try this simple experiment to determine which substances are hydrophilic and which are hydrophobic. First, mix together about 25 mL (1/4 of a cup) of water and the same volume of colored oil (such as olive oil). Do they form a solution, as they would if the oil were hydrophilic? Now add 25 mL of ethanol to the mixture and stir. After the liquids have separated, notice which layer—the water or the oil—has increased in volume by forming a solution with the alcohol. What do you conclude about the hydrophilic or hydrophobic nature of alcohol?

Although we have mentioned only the O---H interaction, N---H and F---H hydrogen bonds also occur. All three types are weaker than covalent bonds but stronger than other forces between organic molecules. Thus the presence of hydrogen bonds between adjacent molecules in liquids and solids strongly affects the properties of the substances, since the molecules are held together more strongly than otherwise expected.

Both O---H and N---H hydrogen bonds are of crucial importance in biological systems, as we shall see in Chapters 8 and 9. The hydrogen bond is responsible for the structures of proteins in the body, the three-dimensional shape of DNA, and the dissolution of important nutrients in the blood. Hydrophobic and hydrophilic interactions between proteins and water in the body determine overall three-dimensional protein shape, as we shall see in Chapter 9.

6.10 Some alcohols have complex structures

Glycols and glycerine A number of alcohols have more than one —OH group within each molecule. In all such alcohols, each —OH group is on a different carbon atom, since molecules in which two or more —OH groups are positioned on the same carbon spontaneously react to form other substances. Many alcohols with multiple —OH groups are sweet to the taste, as are sugars (which, as discussed below, contain numerous —OH groups), and some have a very pleasant odor, resulting in commercial use.

Ethylene glycol is the alcohol whose molecules contain a two-carbon atom chain, with an —OH group bonded to each carbon:

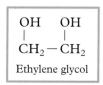

OH   OH
| |
CH$_2$ — CH$_2$
Ethylene glycol

Due to its more extensive hydrogen bonding, its boiling point of 187°C is even higher than that of ethanol (78.5°C). The syrupy liquid compound is added to water as *permanent antifreeze* for vehicles in cold climates. Like other substances dissolved in a solvent, it lowers the freezing point—thus preventing the radiator water from freezing. Unlike most of the other candidates that are suitable for this function, such as methanol or isopropyl alcohol, it boils at such a high temperature that it won't evaporate away in summer months and so is called "permanent." Ethylene glycol is also used in other liquid products and in the formulation of certain plastics, as we shall see below. It is both sweet and rather toxic, an unfortunate combination. *Propylene glycol,* which is propane in which two hydrogen atoms have been substituted by —OH groups, has replaced ethylene glycol in such antifreeze formulations since it is much less toxic. A concentrated aqueous solution of propylene glycol is used as *aircraft deicer:* it is sprayed hot and under pressure onto aircraft surfaces before takeoff to melt ice, snow, and frost.

It is interesting that the antidote to poisoning by swallowing of methanol or ethylene glycol is ethanol. Because the body preferentially metabolizes the ethanol, it is kept from quickly metabolizing more toxic alcohols and thereby converting them into deadly substances. Due to the delay, some of these other alcohols would be eliminated from the blood by the kidneys; sometimes *dialysis,* under medical supervision, is used to remove them from the blood.

Another syrupy liquid is obtained if each of the three carbon atoms in propane has one —OH group bonded to it:

OH OH OH
| | |
H — C — C — C — H
| | |
H   H   H
Glycerol

This substance, *glycerol,* or *glycerine* as it is sometimes called, has been obtained for many years as a by-product of soap manufacturing. Its capability for extensive hydrogen bonding with water prevents the evaporation of moisture from a substance. Since its toxicity is very low, it can be widely used in cooking, for example to keep cakes moist. It is also used in soaps to keep hands moist, and therefore soft, and in toothpastes to sweeten them. Glycerol also is used to prepare various chemicals including *nitroglycerin,* a substance used to relieve the pain of the heart ailment *angina* by dilating (expanding) the blood vessels. In commerce, nitroglycerine is used as a component of dynamite.

PVA: Polyvinyl alcohol Another important example of a polyolefin containing —OH side groups is *polyvinyl alcohol,* or PVA, which is the CH_2=CHX polymer in which X=OH:

$$\left(CH_2-\underset{\underset{OH}{|}}{CH}\right)$$

Polyvinyl alcohol

Because it contains so many —OH groups, it is one of the few water-soluble polymers. It finds use as a glue and a thickening agent, and as a water-soluble packaging material for medicines, vitamins, bath oils, and so on (see Figure 6.9). Once a PVA capsule containing medicine is swallowed, the plastic slowly dissolves, releasing the medicine.

THC: Tetrahydrocannabinol The active ingredient in marijuana that produces its effects on the human brain is the alcohol known as *tetrahydrocannabinol,* or THC. It has quite a complicated structure:

Tetrahydrocannabinol (THC)

Notice that THC also contains an ether linkage (also shown in blue) within one of its rings. Different strains of the plant contain quite different levels of THC. The psychoactivity of THC is apparently due to its

Figure 6.9 Some commercial products that use polyvinyl alcohol **(PVA).** (H. Thonet for W. H. Freeman and Company.)

similarity to a natural brain chemical, since it can attach itself to the receptor site in the brain intended for molecules of the other substance. Although this continues to be a controversial issue, marijuana is now legally used in some places in the world for certain medical purposes, since smoking it reduces the nausea associated with chemotherapy and the pain of diseases such as cancer.

6.11 Alcohols can be converted to ethers

Finally, in our consideration of alcohols we note that they can be converted into ethers. This occurs in the presence of a **dehydrating agent,** a substance, such as sulfuric acid, that can remove water. The hydrogen atom of the —OH group on one alcohol molecule and the entire —OH group of another are extracted from their respective molecules, and combine to form a water molecule, H—OH. Simultaneously, the oxygen atom that remains on the first alcohol attaches to the carbon atom of the alcohol molecule that lost its —OH, thus forming an ether C—O—C link:

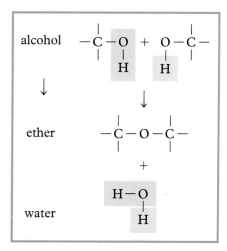

As we shall see in Chapter 8, this reaction is carried out by enzymes in living systems on larger molecules that have multiple —OH groups attached to them.

■ Chemistry in Your Home: Finding the alcohols in your home

Alcohols are found in many household and consumer products. Conduct a survey of labels on household cleaners, mouthwashes, deodorants, lotions, aftershaves, etc., and identify those products that contain alcohols. Note that alcohols have names that end in -*ol*.

Aldehydes and Ketones

The most prominent molecule having carbon–oxygen double bonds is carbon dioxide, whose bonding structure is O=C=O. Conceptually, the simplest organic molecules in which a carbon atom is doubly bonded to only one oxygen atom are aldehydes and ketones:

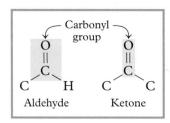

The $\overset{\diagdown}{C}=O$ unit, if bonded only to hydrogen atoms and/or carbon systems, is known as the **carbonyl group.** Aldehydes and ketones are commonly the products of the reaction of alcohols with atmospheric oxygen. Consequently they are said to be "more oxidized" than are alcohols. As we shall see, aldehydes and ketones affect us in diverse ways—some are toxic, whereas others provide some of the most sublime odors our noses can detect.

6.12 Formaldehyde is toxic

In **aldehydes,** at least one of the two groups singly bonded to the $\overset{\diagdown}{C}=O$ unit is a hydrogen atom. The molecule *formaldehyde,* $H_2C=O$, results when two hydrogen atoms are connected to the carbon:

$$\begin{array}{c} H \diagdown \\ \diagup C=O \\ H \end{array}$$
Formaldehyde

Formaldehyde exists as a gaseous compound under normal temperatures and pressures, and is produced in large quantities primarily because it is used to prepare plastics. As formaldehyde is toxic to bacteria, it is also employed to preserve biological specimens in laboratories, as an embalming fluid, and as a domestic disinfectant. In these applications, it is conveniently used as a 40% solution in water known as *formalin.* In embalming, it not only disinfects but also alters the skin to make it less prone to decomposition.

Formaldehyde is naturally present in the smoke of wood fires and assists in the preservation of smoked meats. However, gaseous formaldehyde is somewhat toxic to humans, so it is partially responsible for the eye tearing you may have experienced from exposure to smoke. Formaldehyde is an air pollutant that is produced during episodes of photochemical smog. It has a pungent odor, which humans can detect at concentrations of about 0.1 part per million or higher. This is the smell you may have noticed in stores that sell carpets and synthetic fabrics, since these materials slowly release the gas. At levels higher than 0.1 ppm, many people report problems of irritation to their eyes, especially if they wear contact lenses, and to their noses, throats, and skin.

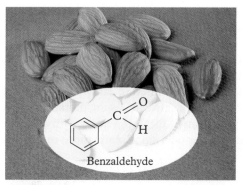

(a) Almond

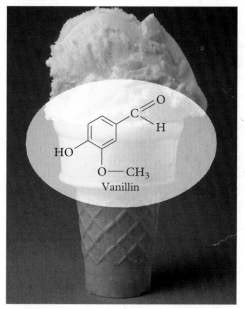

(b) Vanilla

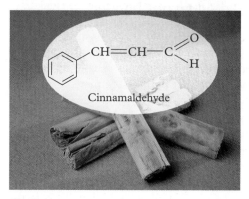

(c) Cinnamon

Figure 6.10 Some familiar aldehydes with pleasing flavors and fragrances. (Photos by George Semple for W. H. Freeman and Company)

The role of formaldehyde as an indoor air pollutant is discussed in Chapter 14.

Commercially, formaldehyde is produced by the *partial oxidation* of methanol:

$$2\ CH_3OH + O_2 \rightarrow 2\ H_2CO + 2\ H_2O$$

This is the same reaction that occurs in the human body when it metabolizes methanol. Formaldehyde is probably the compound responsible for the attack on proteins in the optic nerve following methanol consumption; ultimately death occurs if too high a concentration is present in the body. The above reaction also occurs in automobile engines when methanol is used as a fuel or fuel additive if it is not combusted completely. Some of the concerns about using methanol for fuel purposes lie with the increased atmospheric concentrations of formaldehyde that would result. In addition, in warm, sunny air, formaldehyde will decompose to increase the production of other substances associated with smog.

6.13 Aldehydes are used in foods and perfumes

Aldehydes generally have characteristic odors and tastes, which in some cases are quite pleasant. These properties give rise to their use in perfumes and foods. Aldehydes often occur naturally in fruits and other plants. For example, attachment of a benzene ring to the C=O carbon gives molecules of *benzaldehyde*, also called "oil of almond" (see Figure 6.10a). This substance contributes to the natural odor of cherries and almonds, and is used by food manufacturers to flavor maraschino cherries. If two of the hydrogens of the benzene ring in benzaldehyde molecules are substituted, one by an —OCH₃ group and one by —OH, the result is *vanillin,* the tasty ingredient of vanilla (see Figure 6.10b). Other benzene-based aldehydes produce the characteristic odors associated with cinnamon (see Figure 6.10c) and with hyacinths, for example.

Aldehydes synthesized in chemical laboratories are often used in perfumes. For example, the long-chain aldehyde shown below is a prime component of *Chanel No. 5:*

$$CH_3\,(CH_2)_8 - \overset{\overset{\displaystyle H}{|}}{\underset{\underset{\displaystyle CH_3}{|}}{C}} - C \overset{\displaystyle O}{\underset{\displaystyle H}{\diagup}}$$

Recall from Chapter 2 that the detection of odor involves the interaction of its molecules with a receptor in our noses. Recent research on test animals has shown, for example, that one specific nasal receptor detects aldehydes in which the unbranched carbon chain bonded to C=O has six to nine

Although many aldehydes taste and smell very good, some do not. For example, the aldehyde with the formula $CH_3(CH_2)_5CH{=}CH{-}CHO$ causes the characteristic cardboard taste of stale beer. Another example is the characteristic odor of overheated fat, which is due to acrolein, $CH_2{=}CH{-}CHO$.

carbons. Since the odorous molecules must be in the gas phase and must travel to one of approximately 1000 receptors in your nose, the smell of a liquid (or solid) can only be detected if a reasonable number of molecules evaporate from the liquid's surface. As you learned in Chapter 2, every liquid or solid has a characteristic vapor pressure, the magnitude of which is quite different for different substances. Thus if different components of a perfume have different vapor pressures, then all other things being equal you will smell the component with a higher vapor pressure before that with a lower one, since many more molecules of the former would be in the air sooner.

Although perfumes are solutions generally made up of dozens to hundreds of compounds, they evaporate in three stages, called the *top, middle,* and *base notes.* Aldehydes are used as the one of the top notes, the first scents we experience when we smell a perfume. The middle notes are often a mixture of floral scents; they constitute the heart of the perfume and linger for hours. Aldehydes make the floral notes smell fresher and more brilliant. Base notes, usually with a musky smell, can linger for a day or more. Chemists discover many new perfume components by analyzing the volatile chemicals that are released from flowers.

■ Chemistry in Your Home: Extracting the scents from flowers

The essence of aromatic flower scents can be captured in water or alcohol. Floral waters can be made by steeping flowers such as lavender, rose, or orange blossoms in a mixture of water and ethyl alcohol (2 fluid ounces of alcohol to 8 fluid ounces of water) or vodka (3 fluid ounces of vodka to 7 fluid ounces of water) for several days. Put the flowers into a jar and cover with the water/alcohol mixture. Let steep for seven days, shaking the mixture every other day. To obtain the scented liquid, strain the mixture through a cheesecloth or coffee filter and store the liquid in a dark glass jar or bottle. Are the scent molecules in the liquid hydrophobic or hydrophilic?

6.14 Ketones have two carbon groups bonded to the C=O carbon

Cholesterol and other related molecules containing alcohol and/or ketone groups are discussed in Chapter 9.

Compounds whose molecules have two carbon chains or rings connected to the $C{=}O$ group are called **ketones;** their names usually end in *-one.* The simplest example is *acetone,* in which both attachments are methyl groups:

Methyl group

H_3C
$\searrow C{=}O$
H_3C

Methyl group

Acetone

View different representations of an acetone molecule at Chapter 6: Visualizations: Media Link 1.

Acetone, which is an excellent solvent for many organic substances, may be familiar to you as the major component in many nail polish removers. Acetone is often used to clean lab glassware and equipment because it is a good, cheap solvent and because its residue after rinsing evaporates easily. *Methyl ethyl ketone,* known as MEK, is also a liquid solvent that, like acetone, is used in paint and varnish removers as well as in nail polish remover.

Some ketones with long carbon chains and rings, such as those associated with cloves, raspberries, and spearmint, have more pleasant odors than acetone and MEK. Other ketones have a heavier smell—such as *muscone,* the compound responsible for much of the smell of musk. The ketone called *2-heptanone* gives blue cheese its characteristic aroma and flavor, and *carvone,* the main component of oil of spearmint, is used to flavor spearmint gum. Some complex ketones have a bitter flavor—for example, the ones derived from hops and used to counter the sweetness in beer.

$$CH_2 \overset{CH_3}{\underset{CH_3}{\diagdown}} \quad \overset{O}{\overset{\|}{\diagdown}} - CH_3$$

Carvone

Exercise 6.8

Write the Lewis structure for methyl ethyl ketone, the components of which you can deduce from its name.

Carboxylic Acids

We have seen that, in alcohols and ethers, a carbon atom forms a single bond to oxygen, whereas in aldehydes and ketones, a carbon–oxygen double bond is present. Imagine now a carbon atom that forms a double bond to one oxygen atom *and* a single bond to another oxygen. This

$$\overset{O}{\underset{}{\overset{\|}{-C}}} - O - $$

structural unit is present in the molecules of many important organic compounds, namely those called *carboxylic acids* and *esters.*

When the singly bonded oxygen is also joined to a hydrogen atom,

Alcohols differ from carboxylic acids in that in alcohols, the carbon of the C—O group does not also form a double bond to a second oxygen.

as occurs also in alcohols, the compounds contain the $\overset{O}{\underset{}{\overset{\|}{-C}}} - O - H$ functional group in their molecules and are called **carboxylic acids.** There are many such acids, since the fourth bond formed by the carbonyl carbon atom can join either a hydrogen atom or one of the many carbon chains or rings. In condensed formulas for acids, the $\overset{O}{\underset{}{\overset{\|}{-C}}} - O - H$ group often is represented simply as —COOH; however, this version is not meant to imply that the second oxygen atom is bonded to the other oxygen rather than to the carbon.

Carboxylic acids are produced by the reaction of alcohols or of aldehydes with atmospheric oxygen. Thus they are "more oxidized" than are aldehydes and ketones. The highest stage of oxidation of carbon compounds is carbon dioxide. We can list organic substances in order of increasing oxidation as

hydrocarbons < alcohols < aldehydes and ketones
< carboxylic acids and esters < CO_2

As we shall see in detail in Chapter 11, the term *acid* is used to refer to substances that release H^+ ions when they dissolve in water. For carboxylic acids, it is the hydrogen atom of the $O=C-O-H$ group that forms this ion. Like other acids, most carboxylic acids have a sour taste.

6.15 Formic and acetic acids are common carboxylic acids

The simplest carboxylic acid is *formic acid*, HCOOH. The additional group bonded to the carbon is just a hydrogen atom:

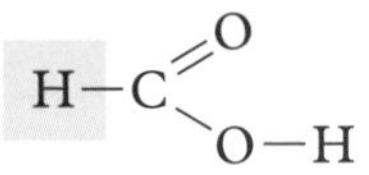

$$H-C\underset{O-H}{\overset{O}{\lessgtr}} \quad \text{or} \quad HCOOH$$

Formic acid

The original source of this acid was the liquid employed as a sting by certain types of ants; indeed, the root of the acid's name is derived from *formica*, the Latin word for "ants." Formic acid is also a component of the sting of bees and wasps. Formic acid is irritating, and if you are stung by one of the insects that produce it, your skin in that area often swells with water, as your body tries to dilute the acid that has been injected into you.

If a methyl group is bonded to the $-COOH$ unit, molecules of *acetic acid*, CH_3COOH, are obtained:

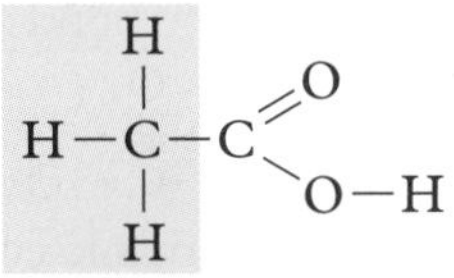

$$H-\underset{\underset{H}{|}}{\overset{\overset{H}{|}}{C}}-C\underset{O-H}{\overset{O}{\lessgtr}}$$

Acetic acid

Pure acetic acid is a somewhat toxic liquid that is used in many chemical preparations. More familiar is the 5% water solution of acetic acid that is known as vinegar, a substance that has been known for millennia since it forms easily when ethanol is oxidized by the oxygen in air:

$$CH_3CH_2OH + O_2 \rightarrow CH_3COOH + H_2O$$

The odor of vinegar is due principally to the acetic acid that it contains. The sharp, sour taste that vinegar imparts to foods is due not only to the acetic acid but to other compounds that also are present in small amounts. Though the sour taste of acetic acid is often desirable, it can have its disadvantages. For example, once a bottle of wine is opened, it becomes progressively more sour in taste as the ethanol in it is slowly oxidized by atmospheric oxygen to acetic acid; the reaction is speeded up by the presence of bacteria. To prevent this oxidation from occurring, some wine makers create an oxygen-free environment—for example, by using nitrogen rather than air as the gas enclosed between the liquid and the cork in wine bottles.

6.16 Hangovers are produced by aldehydes

In the human body, ethanol is eventually oxidized to acetic acid, which is ultimately oxidized to carbon dioxide and water. However, as we noted in our discussion of ethanol, this alcohol is first converted by the liver to acetaldehyde. The acetaldehyde in turn is oxidized to acetic acid in a long series of steps, all of which are catalyzed by enzymes:

$$\text{ethanol} \rightarrow \text{acetaldehyde} \rightarrow \text{acetic acid} \rightarrow \rightarrow \rightarrow CO_2 + H_2O$$

Acetaldehyde is transformed into acetic acid and carbon dioxide relatively slowly. While it is in the bloodstream, acetaldehyde produces the painful effects associated with the common "hangover." In particular, it dilates (expands) the blood vessels in the scalp and around the brain, producing a headache. Acetaldehyde in the body can also produce nausea. Thus, the uncomfortable "morning-after" recovery from a night of drinking is actually a recovery from the acetaldehyde produced from the alcohol. Alcoholics undergoing treatment are sometimes given a drug (*Antabuse*) that inhibits the enzyme that converts acetaldehyde into acetic acid. This intervention produces severe, long-lasting hangovers. The anticipated effect of the drug obviously discourages the consumption of alcohol. Some people naturally are missing the enzyme that metabolizes acetaldehyde, and they, too, find drinking alcohol a very painful experience.

When humans drink methanol, the liver oxidizes it first to formaldehyde and then to formic acid in a process analogous to that described above for ethanol:

$$\text{methanol} \rightarrow \text{formaldehyde} \rightarrow \text{formic acid} \rightarrow \rightarrow \rightarrow CO_2 + H_2O$$

The acid severely disrupts the normal chemistry in the bloodstream; this disruption, as we have noted, can result in death.

In addition to ethanol, alcoholic beverages also contain tiny concentrations of other organic compounds, including other alcohols, aldehydes and ketones, and carboxylic acids. Some of these substances are components of the original plant material used to make the alcohol; some arise from the fermentation process; and some are introduced during the aging process when the liquid is in contact with containers. These other substances, and the products they form during metabolism in the body, also can contribute to a hangover. In particular, fruit-derived alcoholic beverages, such as brandy and red wine, can contain up to 2% methanol, in addition to larger quantities of ethanol. The formaldehyde and formic acid produced when this methanol is metabolized may well produce hangover effects that are worse than those from drinks such as vodka that contain little or no methanol.

6.17 Some carboxylic acids have awful smells

Owing to extensive hydrogen bonding involving their —OH units, even the shortest carboxylic acids are liquids rather than gases at room temperature. However, enough of the liquid acids evaporate to give these substances their characteristic sharp odors. Acetic acid and *propionic acid,* CH_3CH_2COOH, which contributes to the odor and taste of Swiss cheese, are the least offensive. The addition of one more carbon to the chain gives *butyric acid,* $CH_3CH_2CH_2COOH$, which gives rancid butter and vomit their appalling smells. (The acid name is derived from the Latin word *butyrum,* for "butter.") Tiny concentrations of this acid are even present in your body odor, and it is the high sensitivity of dogs to its smell that allows them to track people's travels by their odor.

Butyric acid also helps give some strong cheeses their characteristic odors.

Exercise 6.9

What is the formula for the hydrocarbon whose molecules each have the same number of valence-shell electrons as molecules of formic acid, HCOOH? Explain why formic acid is a liquid (with a boiling point that is essentially the same as that of water), whereas the hydrocarbon is a gas at room temperature.

The carboxylic acids containing 6, 8, and 10 carbon atoms in unbranched chains have various names derived from the Latin word for "goats" (*caper*), since the odors of all of these acids contribute to the unpleasant smell of these animals. For example, the acid with 6 carbons is called *caproic acid.*

The sometimes offensive smell of human armpits is due to the presence of the volatile carboxylic acid shown below dissolved in sweat, which is mainly water:

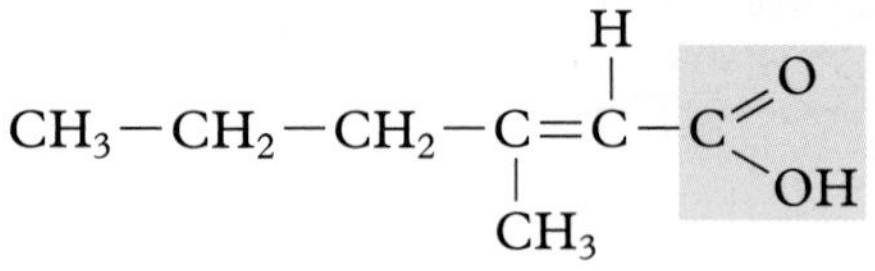

$$CH_3-CH_2-CH_2-\overset{\displaystyle H}{\underset{\displaystyle CH_3}{C}}=C-C\overset{\displaystyle O}{\underset{\displaystyle OH}{\diagup}}$$

3-methyl-2-hexanoic acid

We do not emit this substance as such. Rather, bacteria attack odorless, large compounds that are released from our glands and convert them to the acid. In order to prevent this reaction from happening, some people use deodorants that contain a substance that kills the bacteria before they have a chance to produce the acid. The most common bactericide in deodorants is *triclosan:*

Triclosan is also used as a bactericide in many toothpastes. Check the ingredients in yours!

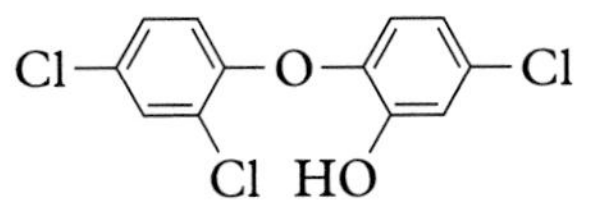

Triclosan

Fact or Fiction?

The metals often found in antiperspirants work by absorbing the odor from sweat.

False. The particles of aluminum salts found in most antiperspirants actually block sweat glands, thereby reducing underarm wetness. The aluminum salts also have some antibacterial properties.

Deodorants usually also contain a fragrance to mask any body odors that do eventually arise; a gelling agent; and aqueous ethanol or a glycol to act as solvent for the components. Simply washing the armpits is also effective for a short period, since it removes the bacteria.

Chemistry in Your Home: Finding the carboxylic acids in your home

Use the names listed for carboxylic acids in this chapter to carry out the following exercise. Carboxylic acids are used as food additives to impart sourness to a product. Examine the labels of various sour candies. What carboxylic acids can you identify? Carboxylic acids are also used in cosmetic products that claim to reduce the effects of aging. Examine labels for various skin products. What carboxylic acids can you identify? Are the same carboxylic acids used in both food and cosmetic products?

6.18 Carboxylic acids are a common component of foods

A number of other carboxylic acids occur naturally in foods and help give them their characteristic tastes. For example, medium- and short-chain carboxylic acids give cheese much of its flavor and odor. The acids are largely produced from the breakdown of protein in milk; thus as cheese ages, its flavor intensifies. Goat cheeses are smellier than most since the carboxylic acids in their milk fat are shorter-chain acids, and hence more volatile, than are those from cow's milk. Other components of cheese odor arise from aldehydes and ketones produced by natural oxidation of carboxylic acids.

In many cases, a single —COOH unit at the end of a chain is not the only functional group present in the carboxylic acids in foods. One group of naturally occurring acids has a —COOH group at *both* ends of a chain. These are called **dicarboxylic acids.** The simplest of these is *oxalic acid,* which is just two —COOH units bonded together:

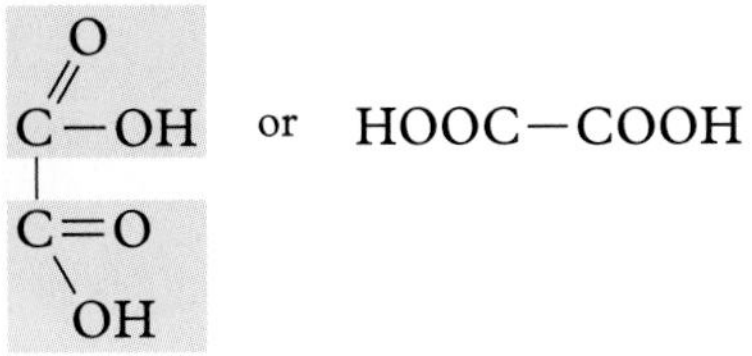

Oxalic acid

Oxalic acid is toxic, though the body has little difficulty in processing the small amounts present in natural foods such as spinach, cabbage, tomatoes, sorrell, and cooked rhubarb stalks. The leaves of rhubarb plants and of houseplants such as *philodendron* and *dieffenbachia* contain much higher concentrations of the acid and should not be eaten. As we

shall see in Chapter 8, dicarboxylic acids are also important in the preparation of certain plastics.

Some other carboxylic acids in foods also contain alcoholic —OH groups on one or more carbon atoms of the chain:

- *Tartaric acid,* found in grapes, is used as a domestic food ingredient in some recipes. Residues of tartaric acid are used as proof that ancient cultures made wine from grapes.
- *Malic acid,* the predominant acid in apples and peaches, is also present in grapes, rhubarb, and pineapples.
- *Lactic acid,* found in sour milk, causes the tart flavor in dill pickles and sauerkraut.
- *Citric acid* gives the characteristic sour "bite" to citrus fruits and juices, tomatoes, and many popular fruit drinks.

The structures of these four acids are illustrated in Figure 6.11.

The tartness, or "bite," in cola drinks is produced by phosphoric acid, H_3PO_4, rather than by the sour taste of a carboxylic acid.

Tartaric acid

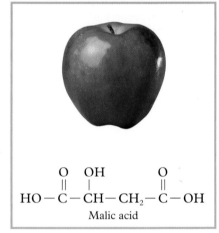

Malic acid

Lactic acid

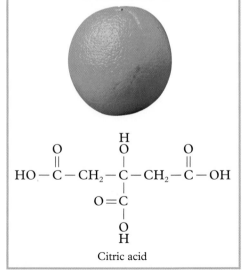

Citric acid

Figure 6.11 Some carboxylic acids in common foods.
(Photos by George Semple for W. H. Freeman and Company)

Chapter 6: The Flavor of Our World

6.19 Lactic, malic, citric, and tartaric acids are alpha-hydroxy acids

Alpha-hydroxy acids all contain a hydroxyl group, $-OH$, bonded to the first carbon atom attached to the $-COOH$ group. Lactic, malic, citric, and tartaric acids are all alpha-hydroxy acids, but the simplest example is *glycolic acid,* $CH_2OH-COOH$. These acids are often included in skin cream products. Their function there is to absorb moisture from the air and hence make the skin softer and more flexible, and to accelerate the natural *exfoliation* of the skin, the removal of the layer of dead cells at the surface of the skin. The acids are also used at higher concentrations by cosmetic dermatologists and skin-care salons to peel off damaged or aged skin. The positive effects of alpha-hydroxy acids on skin have been exploited unknowingly through history—for example, by Egyptian women who bathed in sour milk, a good source of lactic acid.

Lactic acid is formed naturally in our bodies as an **intermediate** in the oxidation of glucose. An intermediate is a substance that is produced during the initial stages of a sequence of reactions, survives for a short time, and is consumed in a later step, so that none of it normally remains when the overall reaction sequence is completed. During vigorous exercise, lactic acid is produced faster than it can be consumed; indeed, its odor can often be detected in human sweat.

Esters

When the carbon of a $C=O$ group is singly bonded to an oxygen that is itself bonded to another carbon (forming an etherlike $C-O-C$ unit), the

$$\overset{\displaystyle O}{\overset{\displaystyle \|}{-C}}-O-C$$

resulting $-C-O-C$ functional group is called an **ester.** The word is a component of the term *polyester;* these polymeric esters, often used to make fibers, are discussed in Chapter 7.

The simplest ester is:

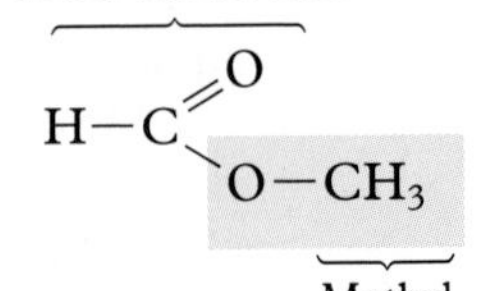

From formic acid

$$H-C\overset{\displaystyle O}{\underset{\displaystyle O-CH_3}{<}}$$

Methyl

Methyl formate

This structure is almost identical to that of formic acid, but the $-OH$ group is replaced by $-OCH_3$. Esters are named after the carboxylic acids from which they are derived, so all those containing the

$$\overset{\displaystyle O}{\overset{\displaystyle \|}{H-C}}-O-$$

unit are called *formates.* This word is preceded by the name of the carbon chain that replaces the hydrogen of the $-OH$ group, and so the molecule above is *methyl formate,* $HCOOCH_3$.

As we saw previously, *acetic* acid is one of the important carboxylic acids. In accordance with the naming rule above, its esters are called *acetates*. If the chain attached to the singly bonded oxygen is the two-carbon ethyl group, then the compound whose molecules have the structural formula $CH_3COOCH_2CH_3$ would be called *ethyl acetate*.

Exercise 6.10

Given that CH_3CH_2COOH is called propionic acid, and that $-CH_2CH_2CH_2CH_3$ is the butyl group, what would be the name of the ester $CH_3CH_2COOCH_2CH_2CH_2CH_3$?

Since ester molecules cannot engage in hydrogen bonding with each other (though carboxylic acids can do so), esters become gases at temperatures that are considerably lower than those for the acids. For example, methyl formate boils at 32°C, so it is a gas rather than a liquid in hot climates, whereas the isomeric compound acetic acid, CH_3COOH, remains a liquid until it reaches 118°C; that is, a temperature that lies above the boiling point of water.

Exercise 6.11

What is the carboxylic acid that is isomeric with the ester methyl acetate, CH_3COOCH_3? The boiling point of this ester is 57°C. What do you predict the boiling point of the isomeric acid to be, if the ester–acid boiling point difference in this case is the same value as the boiling point difference between methyl formate and acetic acid, as discussed above?

View different representations of a methyl acetate molecule at Chapter 6: Visualizations: Media Link 2.

6.20 That nice smell in the air may be an ester

Most esters are rather volatile liquids, which means they evaporate readily and thereby become minor components of the air surrounding the original liquid. In contrast to carboxylic acids, whose odors often are obnoxious, the corresponding esters often have pleasant smells. Indeed, when we smell fruits and flowers, the molecules that our noses detect are usually esters!

Most of the pleasant odors we associate with fruits stem from esters containing zero to four carbons in the chain connected to the $C=O$ carbon, and one to eight carbons in the chain connected to the $-O-$ oxygen (see Table 6.2). For example, a formate ester is a component of the odor of raspberries, and acetate esters contribute to the odors of bananas, pears, oranges, and jasmine. The overall smell that we experience when we detect a fruit smell is usually due to a combination of esters (and other compounds), as you can see by the multiple listings for some fruits in Table 6.2. On the other hand, the main odor of certain flowers, such as roses, is due to a combination of multicarbon alcohols, rather than of esters. However, not all esters have a pleasant smell: ethyl acetate is familiar as the powerful smell in some nail polish removers.

Table 6.2 Some esters and their odors

Name	Structure	Odor
ethyl formate	$HCOO-CH_2CH_3$	rum
isobutyl formate	$HCOO-CH_2CH(CH_3)_2$	raspberry
ethyl acetate	$CH_3COO-CH_2CH_3$	floral
propyl acetate	$CH_3COO-CH_2CH_2CH_3$	pear
pentyl acetate	$CH_3COO-CH_2CH_2CH_2CH_2CH_3$	banana
isopentyl acetate	$CH_3COO-CH_2CH_2CH(CH_3)_2$	banana
octyl acetate	$CH_3COO-CH_2CH_2CH_2CH_2CH_2CH_2CH_2CH_3$	orange
benzyl acetate	$CH_3COO-CH_2C_6H_5$	jasmine
isobutyl propionate	$CH_3CH_2COO-CH_2CH(CH_3)_2$	apple
methyl butyrate	$CH_3CH_2CH_2COO-CH_3$	rum
ethyl butyrate	$CH_3CH_2CH_2COO-CH_2CH_3$	pineapple
butyl butyrate	$CH_3CH_2CH_2COO-CH_2CH_2CH_2CH_3$	pineapple
amyl butyrate	$CH_3CH_2CH_2COO-CH_2CH_2CH_2CH_2CH_3$	apricot
isoamyl pentanoate	$CH_3CH_2CH_2CH_2COO-CH_2CH_2CH(CH_3)_2$	apple

As a consequence of their pleasant odors, esters find extensive use in perfumes. As a consequence of their pleasing tastes, they are used as flavorings in candies and baked goods. Natural fragrances are extracted from plants by simple chemical techniques, and the components of the extracts are used in perfumes. Fragrances that are not destroyed by heat can be extracted by distillation. In this method, the plant's components, such as flower petals, that contain a desired fragrance are exposed to steam, which often has been superheated beyond 100°C. The volatile oils containing the esters are carried off by the steam. Condensation of the steam produces a liquid mixture having two separate layers—oil and water. Since the oil is largely insoluble in liquid water, it rises to the surface, where it can be readily skimmed off. Some separation of the oil into individual components can be achieved by collecting separate fractions early in the distillation process and not mixing them with later ones.

Fragrances which would deteriorate in steam because of the high temperature of steam must be extracted using techniques that are carried out at room temperature. The oils containing the fragrances can be extracted by placing the plant's components in contact with an organic medium in which they are highly soluble. For example, many plants can be crushed and placed in ethanol, into which the oils dissolve. In the ancient art of *enfleurage*, delicate scents were captured by placing intact flower petals onto lard, which extracted the oils. Later, ethanol was used to extract the oils from the lard (see a similar process in the Chemistry in Your Home on page 236).

6.21 Some useful plastics contain esters

If you wear contact lenses, or glasses with plastic lenses, you may be using a plastic made from an ethylenic polymer that has ester side groups. In the most important of these plastics, the repeat unit is $-CH_2-CXY-$, where X is a methyl group, CH_3, and Y is a small ester group. In poly(methyl methacrylate), PMMA, Y is simply $-\overset{\overset{\displaystyle O}{\|}}{C}-O-CH_3$; that is, a methyl ester:

PMMA (H atoms not shown)

You can find out more about contact lenses in the *Science News* article, "Vision Quest," listed in this chapter's *For Further Reading*.

Figure 6.12 Some consumer items made of poly(methyl methacrylate) (PMMA). (George Semple for W. H. Freeman and Company)

PMMA is a hard, clear, colorless plastic whose optical clarity rivals that of glass but in a less brittle, more impact-resistant, and much "lighter" material (density of 1.1 grams per cubic centimeter, compared to 2.5 for glass). It is used to produce the plastic lenses in lightweight eyeglasses and cameras, fiber optics material, and the many structures—including windows in aircraft, bathtubs, hospital incubators, and even some sculptures—that are made with *Plexiglass*, *Lucite*, and *Perspex* (see Figure 6.12). PMMA is the main polymer used to make hard contact lenses and even dentures.

Soft contact lenses, on the other hand, consist mainly of a variation of PMMA in which one hydrogen of the ester group is replaced by an $-OH$ group. Hydrogen bonding by this $-OH$ permits the lens to absorb water, so that the plastic behaves like a film of water and causes little irritation to the eye.

6.22 Esters form the waxes that are used in lipstick

Lipstick is a heterogeneous mixture of several ingredients, the most prominent of which are natural waxes, oils, emulsifiers such as lanolin, and a synthetic dye.

The wax gives the lipstick its shape and its hardness, and makes it easy to apply. Waxes found in nature are mixtures of long-chain esters. For example, the beeswax commonly found in lipstick consists of esters that were originally formed by the reaction of alcohols

Chapter 6: The Flavor of Our World

having 24 to 36 carbon atoms with carboxylic acids having up to 36 carbons. Another ester-containing substance found commonly in lipstick is *carnauba oil.*

The oil in lipstick is largely castor oil! Its presence allows the lipstick to remain a solid in its container but to flow when it is warmed slightly and when pressure is applied; that is, when it touches the lips. But it then forms a tough, shiny film when it dries after being applied. The oils and waxes also help keep the lips moist and soft.

The dye in lipstick is not water-soluble, or it would be licked off in no time. Instead, the dye, usually a red or orange one, is oil-soluble. In color-change lipsticks, the dye changes color upon application since it interacts with protein that is present in the skin.

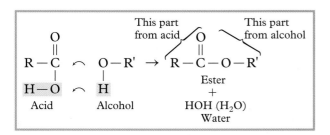

Figure 6.13 The reaction between an alcohol and a carboxylic acid to form an ester. R and R′ are used to designate the portion of the molecule other than the functional group of interest.

6.23 Alcohols react with carboxylic acids to produce esters

We are now set to introduce a chemical reaction that we will see over and over again in Chapter 7, since it is used so extensively both by chemists and by nature in the making of many different polymers. In the present context, however, the reaction involves just two molecules as reactants. In particular, it transforms a carboxylic acid into an ester by reacting the acid with an alcohol that contains the alkyl group that will ultimately be bonded in the C—O—C link. The reaction is illustrated in Figure 6.13. In this diagram, we have oriented the reactant molecules in a specific way in order to illustrate what happens during the process. In particular, the —OH unit of the acid combines with the H atom of the alcohol's OH group to form a molecule of water, H—OH (or H_2O). Simultaneously, the C=O carbon of the acid becomes attached to the oxygen atom remaining on the alcohol.

This process is known as a **condensation reaction,** since water is condensed out from the two original reactants. It is really no different in concept from the reaction we considered in section 6.11, in which an ether molecule was formed by reacting two alcohol molecules and which we called dehydration. Note that by the condensation reaction, we can prepare any ester that we wish by simply combining the appropriate acid and alcohol.

Worked Example: Deducing which alcohol and which acid to combine to make a specific ester

A food chemist wishes to prepare the ester having the formula $CH_3CH_2COOCH_3$, since she suspects that it may have a pleasant smell. Which acid and which alcohol should she combine together?

Solution: The problem can be solved by conceptually *reversing* the ester formation reaction. Thus we mentally add a hydrogen atom to the oxygen of the C—O—C ester linkage, producing the alcohol, and add an —OH to the C=O carbon, producing the acid:

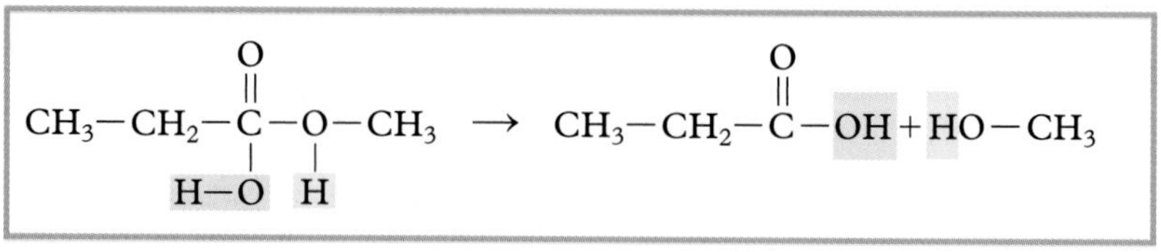

$$CH_3-CH_2-\overset{\overset{O}{\|}}{\underset{\underset{H-O\ \ H}{|\ \ |}}{C}}-O-CH_3 \ \rightarrow \ CH_3-CH_2-\overset{\overset{O}{\|}}{C}-OH + HO-CH_3$$

Thus the appropriate carboxylic acid to use is CH_3CH_2COOH, propionic acid, and the alcohol is CH_3OH, methanol.

Exercise 6.12

Which acid and which alcohol would a food chemist need to combine in order to make the ester $CH_3COOCH_2CH_2CH_3$, one of the substances responsible for the odor of ripe pears?

● Tying Concepts Together: Oxygen-containing functional groups

We have seen that five distinct categories of organic compounds containing oxygen exist: ethers, alcohols, aldehydes/ketones, carboxylic acids, and esters. The structural components of each category are summarized in Table 6.3. There are similarities between different categories, based on the three bond types C—O, C=O, and O—H. The C=O bond occurs on its own in aldehydes and ketones and in combination with the C—O bond in carboxylic acids and esters. The C—O bond occurs also in ethers and alcohols, and the O—H bond occurs in alcohols and acids.

Table 6.3 Characteristic bond networks in oxygen-containing organic molecules

Category	Bond types	Category	Bond types
Alcohol	$-\overset{|}{\underset{|}{C}}-O-H$	Ketone	$-\overset{|}{\underset{|}{C}}-\overset{\overset{O}{\|}}{C}-\overset{|}{\underset{|}{C}}-$
Ether	$-\overset{|}{\underset{|}{C}}-O-\overset{|}{\underset{|}{C}}-$	Carboxylic acid	$-\overset{\overset{O}{\|}}{C}-O-H$
Aldehyde	$-\overset{\overset{O}{\|}}{C}-H$	Ester	$-\overset{\overset{O}{\|}}{C}-O-\overset{|}{\underset{|}{C}}-$

Summarizing the Main Ideas

Ethers are compounds whose molecules contain a $C-O-C$ bond, whereas alcohols contain a $C-O-H$ unit. Like alkanes, ethers are flammable, and some have been used as gasoline additives since they have high octane numbers.

Alcohols undergo combustion efficiently, and short-chain ones are also useful as gasoline additives or replacements. However, ethanol is better known for its use as a component of alcoholic beverages. The "proof" of alcoholic beverages is equal to twice the percentage by volume of ethanol, the remainder being mainly water.

Toxic substances, such as the alcohols, can be harmful to humans even in small amounts. There is a lethal dose of each substance for every subject. The LD_{50} value of a substance, the measure of its toxicity, is the dose required to kill half a given population. It is expressed as the mass of the toxic substance per kilogram of body weight of the subject. Its value is usually determined by experiments on test animals.

Ethanol is produced from carbohydrates in agricultural crops by fermentation, a process that is catalyzed by the enzymes in yeast. Since this reaction produces a dilute solution of alcohol in water, distillation must be used to produce highly concentrated ethanol.

Some ethanol is used as a fuel additive in gasoline. Ethanol potentially is a renewable fuel because it can be produced continuously and because the amount of carbon dioxide emitted by the combustion of ethanol from one crop is balanced by the amount absorbed by the growing crop for the following year. However, fossil fuels are used extensively in the raising of the crop and the distillation of water from the alcohol, making ethanol less environmentally advantageous as a fuel.

Alcohols have much higher boiling points than their isomeric ethers because their molecules are attracted to each other by hydrogen bonding. In this phenomenon, the hydrogen atom of the $-OH$ bond on one alcohol molecule is attracted to the oxygen atom of an adjoining molecule since they carry partial electrical charges that are opposite in sign. The partial charges result from the polarity of the $O-H$ bond: the shared electrons in it are more attracted to the oxygen than to the hydrogen atom, since the former has a higher electronegativity value than does the latter. It is also for this reason that alcohols are more soluble in water than are the isomeric ethers.

Alcohols can be converted into ethers by dehydrating agents, which abstract a hydrogen atom from one alcohol molecule and an $-OH$ unit from another, forming water and joining the remainders of the alcohols by a $C-O-C$ bond in the process.

Many organic molecules contain a carbon–oxygen double bond, $\diagdown C=O$, the carbonyl unit. If the carbon is bonded to one or two hydrogen atoms, an aldehyde is formed, whereas if it is bonded to two carbon atoms, the resultant molecules are called ketones. Aldehydes are produced when alcohols undergo partial oxidation, whether in the body or in an internal combustion engine. Some aldehydes have pleasant odors and are used in perfumes, and other aldehydes have pleasing tastes and are used in cooking. Acetaldehyde is largely responsible for the discomfort of alcoholic hangovers.

$$\overset{\displaystyle O}{\underset{\displaystyle \|}{}}$$

Carboxylic acids contain the $-\overset{O}{\overset{\|}{C}}-OH$ group. The acids have unpleasant odors if they contain an intermediate number of carbon atoms. Some carboxylic acids, especially those containing another $-OH$ group or two acid groupings, are common constituents of foods.

Esters contain the $-\overset{O}{\overset{\|}{C}}-O-C$ unit. Most esters have pleasant smells and are components of the odors we associate with flowers and fruits. Esters are produced from carboxylic acids by reacting them with an alcohol. The alcohol loses the hydrogen of its $-OH$ group and, together with the acid's $-OH$ unit, forms water molecules. This process is known as a condensation reaction.

Key Terms

ether	metabolic process	dehydrating agent	intermediate
functional group	renewable fuel	carbonyl group	ester
alcohol	hydrogen bonding	aldehyde	condensation reaction
proof value	miscible	ketone	
toxic substance	hydrophilic	carboxylic acid	
LD_{50}	hydrophobic	dicarboxylic acid	

Web Sites of Interest

To link to Web sites of interest, go to www.whfreeman.com/ciyl2e, Chapter 6, and select the site you want.

For Further Reading

D. Christensen, "Sobering Work," *Science News, 158,* July 8, 2000, pp. 28–29. An examination of fetal alcohol syndrome and its impact on childhood development.

"Vision Quest," *Science News, 157,* February 5, 2000, pp. 88–89. Interesting article about the development, structures, and properties of a new generation of contact lenses.

Review Questions

1. How is diethyl ether used today? What was a historical use of this compound?

2. What is *MTBE*?

3. Why is MTBE added to gasoline?

4. What is a *functional group*?

5. Look at the structures for the molecules below and classify them according to their functional groups:
   a) $CH_3OCH_2CH_3$
   b) CH_3CH_2CHO
   c) $CH_3CH_2CH_2COOH$

6. What is the relationship between the proof value of an alcoholic beverage and its alcohol content?

7. What does the term LD_{50} signify?

8. How would LD_{50} differ from LD_{10}?

9. What are two negative physiological effects of drinking too much ethanol for a long period of time?

10. What is *fetal alcohol syndrome*?

11. Define the term *metabolic process*.

12. What metabolic processes are affected by drinking alcoholic beverages?

13. What does it mean when alcohol is *denatured*?

14. What are the common names for the following alcohols:
    a) methanol   b) ethanol

15. Why can ethanol be used as a fuel?

16. What are the products of the combustion of ethanol?

17. Why is ethanol called a *renewable fuel*?

18. What is *hydrogen bonding*?

19. Explain why ethanol and water are miscible.

20. Explain the difference between *hydrophilic* and *hydrophobic*.

21. Why is 70% ethanol used as a disinfectant?

22. What is *THC*? What effect does it have on the body?

23. Why are phenol, glycerol, and THC all classified as alcohols?

24. What is a *dehydrating agent*?

25. Why is formaldehyde harmful?

26. How does a *dicarboxylic acid* differ from a carboxylic acid?

27. What is the relationship between ethanol, acetaldehyde, and a hangover?

28. What does the term *intermediate* mean in regard to a chemical reaction?

29. Match the substances below with their occurrences/uses.

a) formic acid	1. an intermediate in the oxidation of glucose
b) benzaldehyde	2. found in ant stings
c) lactic acid	3. preserves biological specimens
d) citric acid	4. present in oranges
e) formaldehyde	5. oil of almond

30. What types of molecules are combined to make esters?

31. What is a *condensation reaction*?

Understanding Concepts

32. Draw the structure of the ether and the alcohol that are isomers of the chemical formula C_2H_6O.

33. What chemical properties of MTBE make it attractive as an additive to boost gasoline efficiency?

34. Draw Lewis structures for the reactants and products in the dehydration reaction in which two ethanol molecules combine to form an ether and water.

35. Show by use of molecular structures why methanol and ethanol dissolve completely in water.

36. Explain how methanol and ethanol differ in their biological activity. How are they similar?

37. Draw the Lewis structure for the following compounds:
    a) methyl ethyl ether
    b) an alcohol isomer of the ether in (a)

38. Write the condensed formula for an ether that is the isomer of *n*-butanol, $CH_3CH_2CH_2CH_2OH$.

39. What would be the appropriate "M" symbols for a solution that is 8% methanol and 92% gasoline, and for one that is 90% methanol and 10% gasoline?

40. What would be the "proof" value for pure ethanol? What is the alcohol content, by volume, of 90 proof whiskey?

41. What is the proof value of a wine that has a 9% alcohol content? What is the alcohol content of a strong ale that has a proof value of 14?

42. Look at the LD_{50} values for the following substances and rank them in order of most toxic to least toxic. Explain how you established the order.

Substance	LD_{50} (mg/kg body weight)
Acetaminophen (Tylenol)	338
Nicotine	230
Ibuprofen (Advil, Nuprin, Motrin)	1,255
Naproxen (Aleve)	1,234

43. How does alcohol production by fermentation differ from alcohol production by distillation?

44. From the list of esters below, identify the original carboxylic acid and alcohol that formed the ester:
   a) methyl acetate   b) ethyl butyrate
   c) pentyl acetate

45. What does the temperature at which a liquid boils tell you about the attractive forces between molecules of that liquid?

46. How do the structures of aldehydes and ketones differ?

47. Draw the Lewis structure for the aldehyde that is an isomer of acetone.

48. What ester would result from the condensation reaction between ethanol and propanoic acid?

49. Which acid and which alcohol would a food chemist need to combine in order to synthesize the ester called butyl butyrate?

50. What property of some fragrant oils makes it possible to collect them in ethanol?

51. How does the water content of a soft contact lens differ from that of a hard contact lens?

Synthesizing Ideas

52. Use the following information to calculate the dose of caffeine and the dose of aspirin that could be lethal to you. (See question 58 for data.)

Substance	LD_{50}
Caffeine	127 mg/kg body weight
Aspirin	1,100 mg/kg body weight

53. How could you determine the relative vapor pressures of different perfumes?

54. Use the information at the Web site of Intoximeters, Inc. to calculate the amount of ethanol that would produce impairment in you. How many glasses of table wine would you have to consume to reach that level?

55. Describe the properties and characteristics a material must possess to be renewable fuel. Make a list of all the substances you can identify as potential renewable fuels.

56. Draw the structures of a two-carbon alcohol (ethanol), a two-carbon aldehyde (acetaldehyde), and a two-carbon carboxylic acid (acetic acid). How many hydrogen bonds can each molecule form? Which molecule would make up the liquid that would have the highest boiling point? Explain.

57. What is the concentration in milligrams per kilogram of 0.01% alcohol in blood?

58. Convert your body weight into kilograms (given that 1 kg = 2.2 lb), and determine the LD_{50} value of ethanol for you. How many bottles of beer does this amount of alcohol correspond to? Would drinking one bottle less than this be a guarantee that you would not kill yourself, and one more guarantee that you would? Explain your answer.

59. Draw the Lewis structure for the isomer, containing two —OH groups, of the commercial product called propylene glycol.

60. Consider an unbranched chain of five carbon atoms connected by C—C bonds, as is present in *n*-pentane. Draw the Lewis structures for all the different aldehydes and ketones that could be formed by doubly bonding one of the carbon atoms to an oxygen atom and adjusting the necessary number of hydrogen atoms.

61. Based upon the behavior of methanol and ethanol, what do you expect to be the metabolism product that contains C=O of $CH_3CH_2CH_2OH$ in your body? Is it an aldehyde or a ketone?

62. The family of weak acids called *alpha-hydroxy acids* has been widely used in cosmetic items claiming to reverse or prevent the effects of aging on facial skin. Visit the *Food and Drug Administration* Web site to find a list of alpha-hydroxy acids in cosmetics. What is the source of most alpha-hydroxy acids? How do they work? Why does the use of alpha-hydroxy products make users more sensitive to ultraviolet radiation?

Taking It Further with Math

Calculating the heat released or absorbed during chemical reactions

The great majority of chemical reactions either release energy or absorb energy when they occur. Most reactions are **exothermic.** This means they release energy, usually in the form of heat, and sometimes in part as light or electricity. Processes that absorb energy as they occur are termed **endothermic.** These terms can also be applied to physical processes: thus the boiling of water is an endothermic process since it requires heat, whereas its opposite, the condensation of water, is exothermic.

Chemical reactions release or absorb energy because the collective strengths of the bonds in the products are different from those in the reactants. Thus, for example, when methane burns in oxygen (in air) to produce carbon dioxide and water, heat energy is given off because the collective strength of the bonds in CO_2 and H_2O molecules is greater than those in CH_4 and O_2.

$$CH_4 + 2\ O_2 \rightarrow CO_2 + 2\ H_2O$$

Scientists usually report the energy released or absorbed by reactions in units of joules, abbreviated J, though the older unit *calorie* is still widely used in North America.

When one gram of methane gas undergoes complete combustion to produce carbon dioxide gas and water vapor according to the reaction above, 50,000 joules of heat energy are released. Because reactions usually involve thousands of joules of energy, the unit of kilojoules, abbreviated kJ, is usually used. Thus, burning one gram of methane releases 50 kJ of heat. We say that the **heat of combustion** of methane is 50 kJ per gram. In general, the amount of heat released (or absorbed) by any reaction is in exact proportion to the amount of reactants consumed, so burning 10 grams of methane will produce 500 kJ of heat. Knowing this, how much heat will be produced by burning 50 grams of methane?

Although it is useful to know the heats of reaction for various substances per gram, mass is not a convenient basis for computations involving energy. Instead, we calculate reaction heats from data using the mole as the unit of amount for the substances involved. Recall from *Taking it Further II* in Chapter 2 that the molar mass of any element or compound is equal to the sum of the atomic masses for its constituent elements in grams. Therefore, the molar mass of methane, CH_4, is equal to the atomic mass of carbon plus four times that of hydrogen:

$$12.01 \text{ g} + (4 \times 1.008 \text{ g}) = 16.05 \text{ g}$$

From this information, we can calculate the heat of combustion for methane in units of moles:

$$50.0 \text{ kJ/g} \times 16.05 \text{ g/mole} = 802 \text{ kJ/mole}$$

By analyzing information about reaction energies, chemists have found interesting and useful consistencies. In particular, they have established that the strength of a given type of bond (for example, C—H or C=C) is remarkably similar in all the molecules in which it occurs. For each bond type, chemists have defined the **bond energy,** which is the amount of heat energy that must be absorbed to break the bond when the molecule is decomposed into free, nonbonded atoms. For example, since each methane molecule contains 4 C—H bonds, the heat energy required to decompose a methane molecule into a carbon atom and four hydrogen atoms must be equal to four times the energy of a carbon–hydrogen bond. Bond energies are normally expressed in kilojoules *per mole of bonds.*

The energies for the common bonds in molecules are collected in Table 6.4. The value for the C—H bond is at the H column and the C row, that is, 416 kJ/mole. From it, we calculate that the heat energy required to decompose one mole of methane into carbon and hydrogen atoms is 4×416 kJ/mole $= 1664$ kJ/mole.

Bond energies are useful for calculating the amount of heat energy that is released or absorbed when a reaction occurs. The results of such calculations are useful, for example, to engineers in determining the amount of power expected when a fuel is burned in a car's engine, and to dietitians and consumers in determining the energy values of different foods. To calculate the heat associated with a process, we make use of the following relationship:

Table 6.4 Bond energies in kJ/mole

					Single bonds				
Element	H	C	N	O	S	F	Cl	Br	I
H	436								
C	416	356							
N	391	285	160						
O	467	336	201	146					
S	347	272	—	—	226				
F	566	485	272	190	326	158			
Cl	431	327	193	205	255	255	242		
Br	366	285	—	234	213	—	217	193	
I	299	213	—	201	—	—	209	180	151

Multiple bonds

C=C 598	C=N 616	C=O 803 in CO_2
C≡C 813	C≡N 866	C≡O 1073
N=N 418	C=N 616	
N≡N 946	O=O 398	

Source: C.L. Stantinski et al., *Chemistry in Context*, 4th ed., McGraw-Hill, New York, 2003.

heat of a reaction = sum of the bond energies of the reactants
− sum of the bond energies of the products

For exothermic processes, the heat of reaction is a negative number, whereas for endothermic processes, the heat of reaction is a positive number.

To see how this equation works in practice, let us determine the heat of combustion for methanol, CH_3OH.

- First we write the unbalanced equation for the reaction in which methanol reacts with oxygen to produce carbon dioxide and water

$$CH_3OH + O_2 \rightarrow CO_2 + H_2O$$

- Next we balance the equation so that only integers, not fractions, are involved:

$$2\ CH_3OH + 3\ O_2 \rightarrow 2\ CO_2 + 4\ H_2O$$

- We now write Lewis structures for all the molecules involved to identify all the types of bonds:

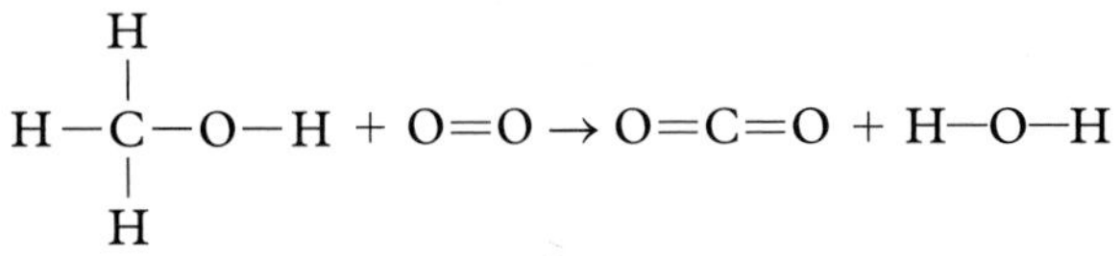

$$H-\underset{\underset{H}{|}}{\overset{\overset{H}{|}}{C}}-O-H + O{=}O \rightarrow O{=}C{=}O + H-O-H$$

- Let us now count up the number of bonds of each type:

$$6\ C{-}H + 2\ C{-}O + 2\ O{-}H + 3\ O{=}O \rightarrow 4\ C{=}O + 8\ O{-}H$$

- Using the equation above for the heat of reaction and letting E(bond) represent the energy of the bond, we can now write:

$$\begin{aligned}\text{heat of reaction} = &\ 6\ E(C{-}H) + 2\ E(C{-}O) + 2\ E(O{-}H)\\ &+ 3\ E(O{=}O) - 4\ E(C{=}O) - 8\ E(O{-}H)\end{aligned}$$

- The values for each bond type can be obtained from Table 6.4 and substituted into this equation:

$$\begin{aligned}\text{heat of reaction} = &\ 6 \times 416\ \text{kJ/mole} + 2 \times 336\ \text{kJ/mole}\\ &+ 2 \times 467\ \text{kJ/mole} + 3 \times 398\ \text{kJ/mole}\\ &- 4 \times 803\ \text{kJ/mole} - 8 \times 467\ \text{kJ/mole}\\ = &\ -1352\ \text{kJ/mole}\end{aligned}$$

When two moles of methanol burn in air, the reaction releases 1352 kJ of energy as heat, so the heat of combustion of methanol is 1423 kJ per mole. Because the molar mass of methanol is 32.05 grams, its heat of combustion per gram is:

$$(1423\ \text{kJ/mole})/(32.05\ \text{g/mole}) = 44\ \text{kJ/g}$$

It should be noted that all bond energies, and the heats of reaction calculated from them, refer to gases only; thus the value we have calculated for methanol refers to the combustion of its vapor. However, the values for the liquid and solid states do not vary much from those for the gas state.

Exercise 6.13

Using the bond energies listed in Table 6.4, calculate the heat of combustion of a) ethanol, and b) dimethyl ether. Which of these two isomers releases the most heat upon combustion? Which would make the better fuel in terms of energy release? (*Hint:* Use the structural diagrams in the text to determine the number of bonds of each kind in the molecule.)

Using the bond energies listed in Table 6.4, calculate the heat associated with the reaction in which one mole of formaldehyde reacts with one mole of hydrogen gas, H_2, to form methanol. According to the sign of your answer, is the reaction exothermic or endothermic?

In this chapter, you will learn:

- about the different kinds of sugar in our diet, including sucrose, fructose, and lactose;

- why sugar is sweet and why some people are lactose intolerant;

- about complex carbohydrates, such as starch from plants;

- why your body cannot digest roughage (called cellulose) and why our bodies need fiber;

- why some fats are solid and some fats are liquid;

- the difference between unsaturated fats, trans fats, and omega-3 fatty acids;

- the scientific basis of the low-carbohydrate diet;

- about cholesterol and its sources.

Do you see chemicals when you look at the colorful produce in the stalls of this farmers' market?

All the calories and nutrition in food comes from relatively simple molecules. In this chapter we will see how these chemicals provide energy and nutritional value, and influence your body's chemistry.

(Photodisc/Punchstock)

Health and Energy
Carbohydrates, Fats, and Oils

We read and hear so much about "natural" foods that you may have the impression that it is possible for food to be free of chemicals. But like everything on Earth, all food—"natural" or not—is made of chemicals. The chemicals and chemical processes that are part of biological systems, such as our food and our bodies, are the subject of the subdiscipline known as **biochemistry.** With the necessary background of organic chemistry from Chapter 6, we turn now to biochemistry and nutrition.

In this chapter, we'll explore the carbohydrate and fat components of our diet and see that they are all composed of relatively simple chemicals whose structure we can readily understand. We'll see that some carbohydrates are polymers made by nature, though they are polymers of a slightly different type than the addition polymers we've previously studied. While we're on the subject, we'll also discuss a nonnutritional, but commercially useful, polymer of this same type, cellulose.

The mineral component of diet was discussed in Chapter 3.

Carbohydrates: Food, Fuel, and Plant Structure

Athletes "load up" on carbohydrates in the days that precede sporting events that will challenge their endurance. Those of us with a "sweet tooth" love almost any food with sugar in it. Hospital patients are often fed aqueous solutions of glucose directly into their veins. We are told by dietitians to increase the fraction of "complex" carbohydrates in our diet for health reasons, and also to increase our intake of "fiber" or "roughage."

Each of the substances mentioned above consists of molecules containing only carbon, hydrogen, and oxygen, in a roughly 1:2:1 atomic ratio. Such substances in general are called **carbohydrates** (from "carbon hydrates"). This name implies that the molecules consist of water molecules attached to carbon atoms, but this is not literally the case. Instead, *most* of the carbon atoms in carbohydrate molecules are bonded to one —OH group and to one hydrogen atom, thereby yielding the characteristic 1:2:1 ratio of elements. Indeed, one definition of carbohydrates is that they are polyhydroxyl organic compounds.

7.1 Glucose and fructose are simple sugars

Our first example of a carbohydrate is *glucose*, the formula of which is $C_6H_{12}O_6$, and whose molecular structure follows:

Whenever you see this icon in this chapter, go to
www.whfreeman.com/ciyl2e

View two forms of a glucose molecule at Chapter 7: Visualizations: Media Link 1.

$$
\begin{array}{c}
\text{HO} \quad \text{CH}_2\text{OH} \\
\text{O} \\
\text{HO} \quad\quad \text{OH} \\
\text{OH} \\
\text{Glucose}
\end{array}
$$

Notice that five of the six carbon atoms are contained in a six-membered ring whose other atom is an oxygen atom attached by ether linkages to two of the carbons. Only single bonds are present in the molecule. Each ring carbon in glucose is bonded to one —OH group and one hydrogen atom, *except* the carbon that is bonded to the carbon atom external to the ring. This external carbon is itself connected to one —OH group and two hydrogen atoms.

Glucose occurs in most fruits, but in especially high concentrations in grapes, figs, and dates. When you eat, your body turns all the other various carbohydrates you digest into glucose. Regardless of its source, the glucose then circulates in the blood, delivering fuel to cells. Glucose is also known as *dextrose* and as "blood sugar." The latter name is used because glucose circulates in your blood, at a concentration of about 0.1%, as a way of delivering it to your cells. Like other sugars, glucose has many —OH groups that form hydrogen bonds to H_2O molecules, making it highly water soluble. Since blood is largely water, the solubility of glucose is assured. When the glucose arrives at a cell, it can react there with oxygen—also carried by your blood—to produce the energy required, for example, by your muscles to do work or by your brain to think. In a cell, the six-carbon glucose is gradually oxidized to carbon dioxide and water. The net reaction is the same as if the glucose had undergone high-temperature complete combustion (compare Chapter 2, section 2.12), even though the whole process occurs at body temperature and without the presence of a flame:

$$
C_6H_{12}O_6 + 6\,O_2 \xrightarrow{\text{enzymes}} 6\,CO_2 + 6\,H_2O + \text{energy}
$$

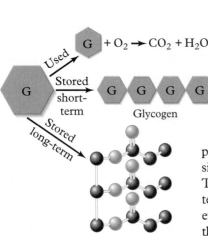

Figure 7.1 The fate of glucose in the body.

Figure 7.1 summarizes the fate of glucose in the body. At this point you will recognize only the first step, the immediate use of the substance for energy. As you learn about the other steps later, you will have occasion to refer back to this figure.

Glucose is the most important member of the **sugar** subgroup of carbohydrates, all of which are molecules containing one or two rings similar to that in glucose and all of which are sweet to the taste. Glucose is a moderately sweet sugar.

Sugars, such as glucose, that consist of molecules with only one ring are called "simple" or "single" sugars. More scientifically, sugars are the smaller members of a group known as **saccharides** (from the Latin

Taking It Further

For information on how sugars determine the human blood groups, go to Taking It Further at www.whfreeman.com/ciyl2e.

saccharum for "sugar"). Those with one ring are called *mono*saccharides, those with two are *di*saccharides, and so on. By convention, all sugars have names that end in *-ose*. Common monosaccharides have five or six carbon atoms in each molecule, though ones with three, four, and seven carbons also are known.

Another important monosaccharide is *fructose*, also called *levulose*. Like glucose, fructose has the formula $C_6H_{12}O_6$, so the two compounds are isomers. Unlike glucose, the fructose molecules consist of a five-membered ring and two single-carbon side chains, rather than a six-membered ring with one side chain:

$$HO \quad \overset{CH_2OH}{\underset{O}{\diagup}}$$

Fructose

Fructose is the sweetest of the monosaccharides, about twice as sweet per gram as glucose. Fructose is also twice as sweet as sucrose (table sugar). For this reason, fructose is used in some low-calorie diets since less of it achieves the same sweet taste. A common consumer source of this sugar is *high-fructose corn syrup*, which is used in many processed foods. Fructose occurs in abundance in many fruits (for example, at the 5% level in apples, at 8% in grapes); hence its name, which means "fruit sugar."

7.2 In diabetes, glucose is not available to body cells

Diabetes is a chronic disease that affects 21 million Americans. There are three major types of diabetes: *insulin-dependent* (type 1), *non-insulin-dependent* (type 2), and *gestational* diabetes. While each type differs in the disease process, in all cases blood glucose levels are elevated because the body cannot maintain a healthy range. When glucose builds up in the blood, body cells can't access it and are literally starved for energy. Normally, the hormone insulin, which is produced by the pancreas, assists in the movement of glucose from the blood into the cells (see Figure 7.2). In diabetes, there may be no insulin available to do the job (type 1) or there may be insulin, but the cells can't use it properly (type 2). Gestational diabetes can occur when a woman is pregnant, but usually goes away after her baby is born. If uncontrolled, the woman's risk of having high blood pressure during pregnancy, and of having a large baby requiring a cesarean section at delivery, are increased. The baby may also experience low blood glucose levels right after birth and have breathing problems.

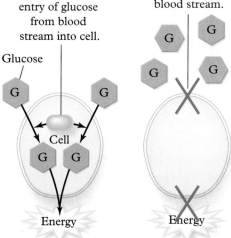

Insulin assists entry of glucose from blood stream into cell.

Without the aid of insulin, glucose can't enter cells and builds up in blood stream.

(a) Normal cell uses glucose to produce energy.

(b) Diabetic cell can't use glucose to produce energy.

Figure 7.2 The role of insulin in the body: cells can't use glucose without insulin.

Carbohydrates: Food, Fuel, and Plant Structure

The treatment for diabetes depends upon the type. Patients who don't produce insulin must inject themselves daily with the hormone and monitor their diets and blood sugar levels very closely. Patients who produce insulin, but aren't able to use it properly, can control their diabetes by eating a diet that is low in fat, has only moderate amounts of protein, and is high in "complex carbohydrates," substances which will be discussed later in this chapter. Research has shown that complex carbohydrates are absorbed more slowly and thus prevent surges in blood glucose levels. Diabetics must monitor their blood glucose levels several times per day, either by sticking themselves to obtain a drop of blood for testing or with "stickless" monitoring devices that take the discomfort out of the procedure.

7.3 Sucrose is common table sugar

In discussing alcohols in Chapter 6, we noted that two molecules could react and combine to form an ether by the collective elimination of a water molecule. Although in the laboratory this condensation reaction requires a dehydrating agent such as sulfuric acid, enzymes (biological catalysts) accelerate and control the same process in some plants. In particular, one specific —OH group of a glucose molecule and one specific —OH of a fructose molecule react together to eliminate H_2O and thereby to join the two ring systems by an etherlike C—O—C unit. The combination of glucose and fructose is the disaccharide molecule called *sucrose,* known more commonly as table sugar:

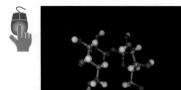

View a sucrose molecule at Chapter 7: Visualizations: Media Link 2.

Glucose   +   Fructose   =   Sucrose

We will often depict such complex structures by a simpler symbolism in which most parts of the original monosaccharide molecules are represented by blue-shaded hexagons, with the oxygen atom that joins them shown explicitly. The specific monosaccharides are signified, if necessary, by a letter within the hexagon: in this case, G for glucose and F for fructose. Note that we use a hexagon symbol for fructose even though its ring has five, not six, atoms bonded together around the ring:

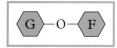

Figure 7.3 Various types of sugar.
(C Squared Studios/PhotoDisc/Getty)

Sucrose is found in many fruits (at the 5% level in ripe oranges, 7% in bananas) and to a lesser extent in vegetables. Thus, we add carbohydrates to our diet when we eat fruit since they contain sugar. Sucrose occurs in higher concentration (about 15%) in sugar cane and sugar beet plants, from which it is commercially extracted. Sucrose is the pure compound that is produced in the greatest amount in the world. The cane or beet is first crushed or sliced, respectively, and then treated with water to extract the sugar. Impurities in the raw sugar are removed by chemical means, and the solution is then boiled to remove water. This results in a mixture of sugar crystals and a syrup, which we know as *molasses*. The molasses is removed by several sequences of spinning the mixture in a centrifuge. A final quick wash with water leaves pale yellow sugar crystals. "Natural," or "organic" sugar does not undergo the initial chemical processing. The noncarbohydrate nutrients, such as iron, calcium, and potassium, that are present initially in unrefined sugar are concentrated in the molasses, and so they are absent in the "refined" white sugar that we buy. Brown sugar is usually produced by adding some molasses back into white sugar. Molasses is also used in cooking and is fermented to produce rum (see Figure 7.3).

Sucrose itself cannot be used by the body. Instead, the condensation reaction that produced it originally is reversed by enzymes in your digestive system, recreating glucose and fructose:

$$\text{sucrose} + \text{water} \xrightarrow{\text{enzymes}} \text{glucose} + \text{fructose}$$

The sucrose is said to be *hydrolyzed* by this reaction, since water is added to the C—O—C link. This process is the reverse of the condensation reaction (depicted on the previous page) by which two C—OH groups react by the formation of an ether bond and a molecule of water. In the body, glucose is easily metabolized, as we have seen. The fructose is usually converted in your liver to its isomer glucose before it is used. Consuming glucose directly provides a faster source of energy for the body than eating sucrose, since the hydrolysis step is not required.

The term **hydrolysis reaction** in general means the reaction of a substance with water. The same hydrolysis reaction for sucrose that occurs in the body can be executed commercially using an acid or the enzyme called *invertase* as a catalyst. The resulting mixture of glucose and fructose is called "invert sugar." Bees collect sucrose from the aqueous nectar of flowers, and invert it into fructose and glucose using enzymes in their saliva. Consequently, honey is a 70% solution of fructose and

The hydrolysis of sucrose begins with the saliva in your mouth and continues in your stomach.

glucose, plus water and small amounts of other compounds, including a small amount of unreacted sucrose.

Exercise 7.1

Explain the difference between a mixture of glucose and fructose, as found in honey and invert sugar, and the sucrose combination of these two monosaccharides.

7.4 The relative sweetness of molecules

Why are sugars, and indeed other compounds composed of small molecules containing many hydroxyl groups, sweet to the taste whereas most other substances are not? According to the theory developed to explain this phenomenon, a molecule triggers the sensation of sweetness when it fits into and binds to a specific type of submicroscopic receptor site in the taste buds on the tongue.

A schematic diagram of the sweetness receptor site is shown in Figure 7.4. According to the current theory of sweetness, a number of conditions make for optimal binding between the receptor and the sugar. First, the O—H unit of the sweet molecule forms an X---H—O hydrogen bond with the atom X, which could be oxygen or nitrogen, on the receptor (see Figure 7.4, step ❶). Second, another atom of the sweet molecule, here oxygen, forms a hydrogen bond to a nearby site on the receptor having the Y—H unit, giving a Y—H---O hydrogen bond, where Y is O or N (step ❷). The third requirement is that the region (R group) behind these two oxygen atoms must be hydrophobic (step ❸). Furthermore, the two distances shown in the sweet molecule must both be about 0.3 nanometers. The better the molecule fulfills these requirements, the sweeter it is detected to be by the receptor. Glucose fits quite well. The groups on glucose, fructose, and sucrose that fit the two 0.3-nanometer distance requirements are shaded in the structures in Figure 7.5.

Virtually all molecules that we sense as sweet share the characteristics described above. This information lets scientists deduce what the requirements for the receptor site must be. In most artificial sweeteners, which per molecule are sweeter than sugars, the H—O bond attached to the X unit is replaced by an H—N bond, which has essentially the same length and which participates even better than does H—O in hydrogen bonds. Another substance that is much sweeter than sucrose is Sucralose, whose molecules correspond to those of sucrose but with three of its —OH groups replaced by chlorine atoms (Figure 7.5). Its characteristic sweetness was accidentally discovered by a researcher who misunderstood a verbal request from management to "test" a sample of the new compound as a request to "taste" it—which he promptly did.

In Chapter 6 we mentioned that glycerol and even the toxic ethylene glycol are sweet to the taste. Using Figure 7.4, you can now understand why many molecules with at least two hydroxyl groups on

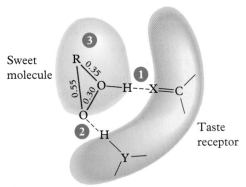

Sweet molecule

Taste receptor

❶ H in OH group on sweet molecule forms hydrogen bond with X on receptor.

❷ O on sweet molecule forms hydrogen bond with H in HY group on receptor.

❸ Hydrophobic area behind O atoms on sweet molecule.

Figure 7.4 Step by Step How sugars bind to receptor sites on taste buds.

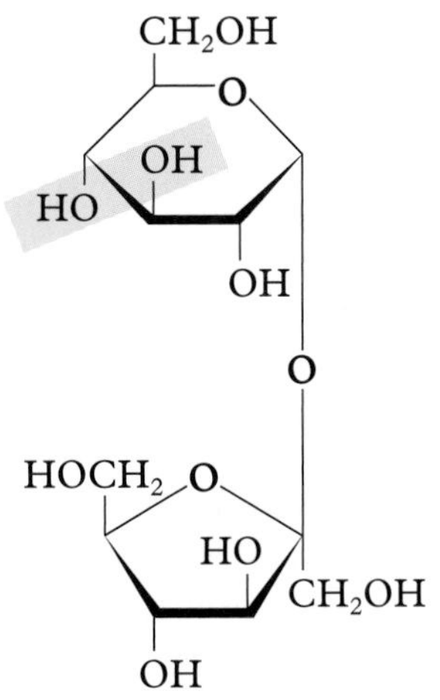

Figure 7.5 Structures of several molecules and their relative sweetness. Sweetness is expressed relative to sucrose (table sugar), which is arbitrarily assigned a value of 1.0. In the structures for sucrose and fructose, the heavy black bond lines indicate that part of the molecule extends "out" from the plane, toward you. The other molecules are shown as two-dimensional figures.

Sucrose (1.0)

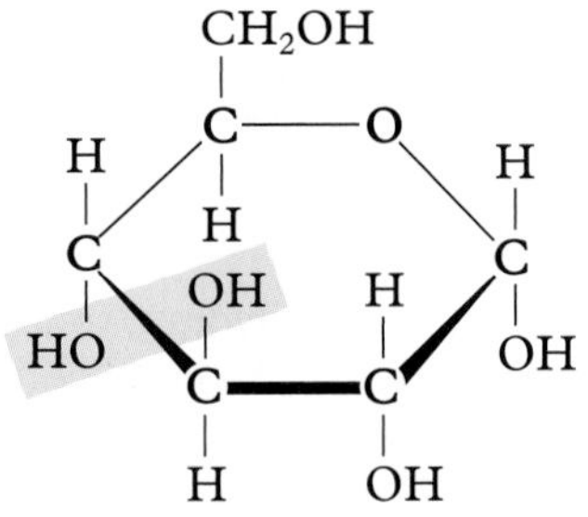

Glucose (0.74)

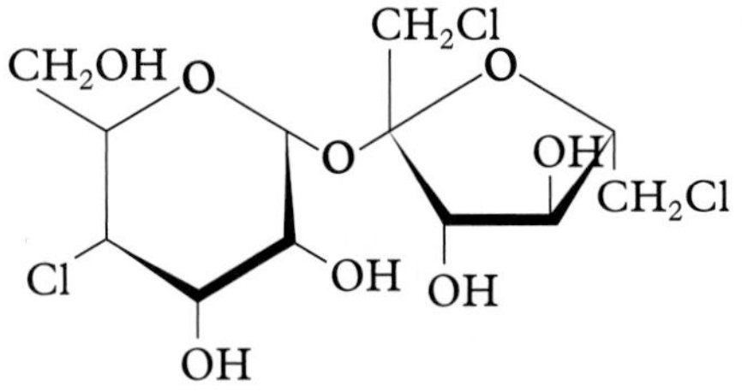

Sucralose (600.0)

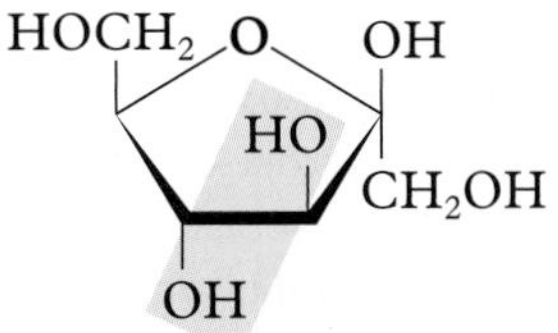

Fructose (1.73)

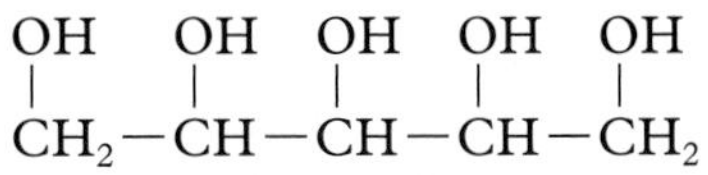

Xylitol (1.0)

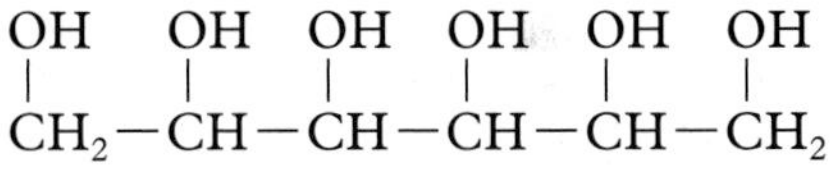

Sorbitol (0.6)

adjacent carbon atoms possess this property. Thus the molecules *xylitol* (five C atoms, each with an —OH) and *sorbitol* (six C atoms and —OH's) are also sweet (see Figure 7.5). These substances are used in "sugarless" gum because they are not broken down by bacteria in the mouth into the carboxylic acids that cause tooth decay. Both these compounds are, however, metabolized by the body in the same way as sugars and thus provide energy.

Their multiple —OH groups give sorbitol, glycerol, and propylene glycol other uses as well. Because the —OH groups can hydrogen-bond to water molecules, these liquid compounds attract water from the air to themselves and are used as components of commercial skin moisturizers.

■ Chemistry in Your Home: Diet versus regular

For this activity, you need two unopened cans of soft drinks (soda), preferably of the same brand (for example, Coke and Diet Coke, Sprite and Diet Sprite). One can of soda should be the *regular* version (containing sugar) and one should be the *diet* version (containing an artificial sweetener). The regular version of the drink contains many spoonfuls of sugar, whereas the diet version contains a much smaller amount of a more powerful sweetener. Place the cans of pop in a sink or plastic tub full of water and observe what happens. Keep in mind that they both have the same volume, so if one floats and the other sinks, it is because one weighs more than the other. Based on your observations, can you conclude that sugar has a greater density than water?

Carbohydrates: Food, Fuel, and Plant Structure

7.5 Lactose is a problematic sugar for many people

"Milk sugar," formally called *lactose,* is a disaccharide that combines glucose with another of its isomers, *galactose* (Ga):

Glucose + Galactose = Lactose

Lactose is the only carbohydrate found in milk (see Figure 7.6). During digestion, it is hydrolyzed to glucose and galactose. Unfortunately, many adult humans—in fact, most people not derived from Northern European stock—and a small fraction of babies lack the specific enzyme, lactase, that is required to perform this hydrolysis reaction. The lactose they ingest passes undigested to the large intestine. There it is consumed by bacteria, which convert it to carbon dioxide and hydrogen gases, which produce bloating, and lactic acid, which produces diarrhea.

Because some lactose is converted by processing and cooking into its component monosaccharides, milk products such as cheese and yogurt and cooked food containing milk usually pose less of a problem to individuals who suffer from this *lactose intolerance.* Lactose-intolerant adults who wish to eat milk products directly can ingest lactase enzyme in tablet form with their meals or purchase lactose-free milk, which has been pretreated with lactase.

A phenomenon similar to lactose intolerance occurs when we consume beans, one component of which is *raffinose,* a trisaccharide (three sugar rings) that humans have no enzyme to process. Like lactose, the raffinose is consumed by intestinal bacteria, producing the gases that eventually escape from our bodies one way or another!

Figure 7.6 Typical dairy products.
(Comstock Images/Getty Images)

Exercise 7.2

a) Given that raffinose is a trisaccharide, draw the symbolic representation of its structure analogous to that for sucrose on page 263.
b) Beano is a tablet that people can take to avoid the development of gas when they eat beans. By analogy with the action of lactase, suggest the likely mechanism by which Beano accomplishes its objective.

Polysaccharides

As you have learned, when two monosaccharide molecules combine to form a disaccharide by collectively eliminating an H_2O molecule, each unit still has some —OH groups unreacted. It follows, then, that a disaccharide could add another monosaccharide molecule to form a trisaccharide, such as the raffinose mentioned earlier:

G—OH + HO—G—O—F → G—O—G—O—F

Monosaccharide + Disaccharide   =   Trisaccharide

This process of addition of monosaccharides by consecutive condensation reactions can continue indefinitely beyond this structure, because, as noted in the symbolic structures, the terminal units always have one (or more) unused —OH groups. Eventually a **condensation polymer** of several thousand units can be obtained:

CHEMISTRY IN ACTION

Why are food guidelines arranged in a pyramid? Learn how to use the food pyramid at Chemistry in Action 7.1.

Condensation polymers differ from addition polymers in that the monomer units are joined together by the operation of a condensation reaction, which splits out a small molecule, H_2O in most reactions. Polymers that are made from saccharide monomers are called **polysaccharides.**

In nature, enzymes force glucose molecules to combine together in large numbers to form several different types of condensation polymers, which are also known as **complex carbohydrates.** Carbohydrates—particularly complex carbohydrates—are a major food group in human diets. Indeed, there are those anthropologists who believe that the discovery that carbohydrate-rich tubers could be cultivated and made more palatable by cooking was a pivotal event in human history. Even today, some cheap grain, such as rice or wheat, constitutes a main dietary staple in most countries. Indeed, in developing countries, a single carbohydrate source may provide 90% of the energy in the whole diet. By contrast, carbohydrates supply 50–60% of the energy in the average North American's diet, though nutritionists recommend that this be increased to 65–80%. In many Western countries, *per capita* bread consumption has decreased considerably in the past few decades and been supplanted by an increased amount of sugar. Nutritionists call sugar "empty calories" since it supplies energy but no other essential nutrients such as the vitamins and minerals found in grains, potatoes, and so on (see Figure 7.7).

Nutritionists classify foods by the **glycemic index,** which measures how fast the component carbohydrates are broken down into simple sugars and transferred to the bloodstream. Some researchers believe that foods with a high glycemic index produce hunger a few hours later. This occurs because the high blood sugars they immediately produce causes the body to produce excess insulin, which results in the storage of glucose in cells, making the sugar inaccessible later. There

Figure 7.7 The U.S. Food Pyramid showing the five food groups.

Polysaccharides

is evidence that the pancreas can eventually "burn out" from chronically producing excess insulin to cope with excess glucose, leading to diabetes.

Foods with a high glycemic index include simple sugars, white rice and pasta, whereas low index foods are fruits, vegetables, legumes, and unprocessed starches. The sugars in fruits, vegetables, and whole grains are released only when the body digests the plant cells that contain them, a process that requires time.

7.6 Starch is a complex carbohydrate

The polysaccharide known as *starch* is the major food reserve found in plants. Grain crops such as wheat, oats, rye, and barley are all high in starch content, as are corn (70% starch), rice (80% starch), and potatoes. Since flour is made primarily from grains, it is also about 70% starch. Unlike glucose, starch itself is not sweet. Unripened fruit is high in starch, which is mainly hydrolyzed to sugar as ripening takes place.

In our digestive systems, starch is gradually broken down by successive hydrolysis reactions with the assistance of enzymes, principally in the small intestine. In chemical terms the polymer starch is *depolymerized* right back to its monomers, namely glucose molecules. The process begins in your mouth. You can prove this if you keep a piece of raw potato or bread in your mouth for a while. Soon you will experience a sweet taste due to the glucose that is produced from the starch.

There are in fact two types of polymer molecules in starch. About 20% of the molecules consist of unbranched chains formed by the condensation of about 100 glucose monomers. This water-soluble material is known as *amylose*.

Grains are called *cereals* and corn is called *maize* in the United Kingdom and elsewhere.

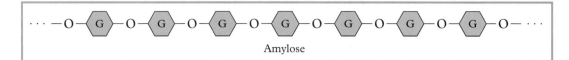

Amylose

The presence of many —OH groups on an amylose molecule enable it to form hydrogen bonds with water. However, some starches have different arrangements and don't form hydrogen bonds as easily. These starches are insoluble in water.

However, most of the starch molecules are much larger—usually containing thousands of glucose units—and consist of highly branched chains (see Figure 7.8). The existence of more than two —OH groups on each glucose monomer naturally allows for this branching, an example of which is shown in the structure on top of page 269. Notice that the branches themselves form additional branches. The water-insoluble material containing these large, branched chains is called *amylopectin*.

View the structure of starch at Chapter 7: Visualizations: Media Link 3.

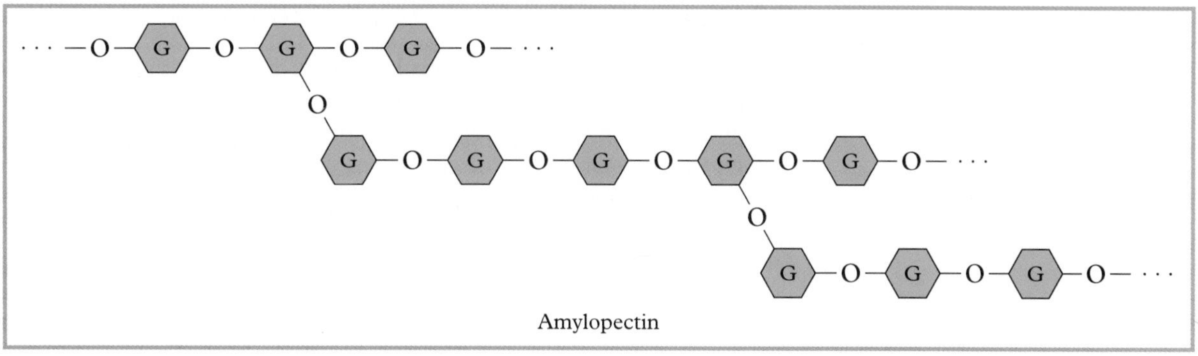

Amylopectin

Exercise 7.3

Identify the repeating unit in the amylose structure shown on page 268.

Recall that 1 micrometer is 10^{-6} meters.

Polylactic acid is a biodegradeble polymer used for packaging. It is made by polymerizing lactic acid (Chapter 6) derived from starch.

The starch in plants is contained in granules whose size lies in the 1- to 200-micrometer range. The granules are covered in protein and are insoluble in cold water. Cooking a starchy vegetable such as a potato in water produces "gelatinization" of the starch. Upon heating, water penetrates the outer layers of the starch granules, and they begin to swell. Upon further heating, the starch granules burst, and their contents are dispersed into the surrounding water. On cooling, the starch molecules attempt to recombine and, at 50°C–70°C, form a network mesh with each other. However, small droplets of water become trapped inside the mesh as H_2O molecules form hydrogen bonds with the numerous —OH groups of the starch. The resulting semisolid is a gel (Chapter 1). Such colloids form in general whenever a liquid solvent becomes trapped and immobilized within the 3-D network formed by polymer chains. The rigidity of a gel depends upon the strengths of the chains. Cooking also reduces the lengths of the starch molecules by the usual hydrolysis reaction involving the addition of H_2O molecules.

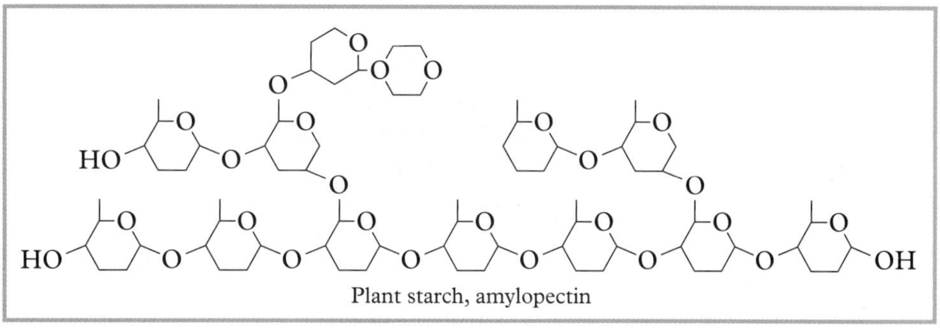

Plant starch, amylopectin

Figure 7.8 Starch. Plant starch differs from animal starch only in the degree of branching in the chains. Plant starch molecules have fewer branches per chain than animal starch molecules.

Polysaccharides

7.7 Glucose is stored temporarily in the body

As you have learned, starch is broken down by depolymerization reactions into glucose in the body, and in this way it becomes available for use to generate energy in cells. If all your immediate energy needs are met and there is an excess of glucose in your body, a small amount of it is temporarily polymerized to the highly branched material known as *glycogen*, sometimes called "animal starch," which is stored in muscles and the liver (see Figure 7.9 and Figure 7.1). Glycogen can quickly be depolymerized back to glucose in order to supply energy when you need it. In fact, your supply of glucose stored as glycogen—amounting to about one day's energy requirements—greatly exceeds the amount of free glucose flowing in your blood.

Endurance athletes have found a way, called "carbohydrate loading," to temporarily double or triple the usual amount—about 300 grams—of glycogen that is available. They first deplete their glycogen supply by a regime of extensive exercise and/or a low-carbohydrate diet for several days. Then, a few days before their competitive event, they switch to a high-carbohydrate diet, which generates much more glucose than they can use. The excess glucose is in turn converted to the high levels of glycogen they require for enhanced performance.

The beverages consumed by athletes during competitive events typically consist of some form of soluble carbohydrate, such as glucose or one of its shorter polymers and sometimes sucrose, along with water and sodium and potassium ions. The carbohydrates are required since blood glucose levels start to decline after an hour or so of exercise. Generally athletes do not consume carbohydrates immediately before competitions because the extra glucose generated is quickly countered by the release of insulin. Insulin promotes the movement of glucose from the blood and into the cells, which can overcompensate and even produce *hypoglycemia* (low blood sugar) temporarily. Excesses of glucose beyond what can be stored as glycogen are converted to fat (see Figure 7.1) and stored as potential sources of energy, as we will see later in this chapter.

> Losing weight by low-carbohydrate, high-fat diets is discussed in section 7.14.

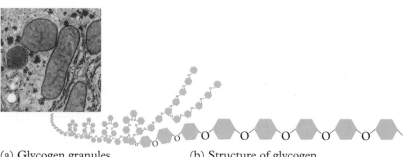

(a) Glycogen granules in human liver cells

(b) Structure of glycogen

Figure 7.9 Glycogen, or animal starch in a human liver cell. Glucose units are linked together by bonds called alpha linkages that differ from the bonds in cellulose. (Part a by Dr. Dennis Kunkel/Phototake)

Chapter 7: Health and Energy

7.8 Polysaccharides have complex 3-D structures

Although the representations of glucose molecules displayed so far are two-dimensional, in fact the requirements for tetrahedral angles at all the carbons (as in methane) make it a nonplanar molecule. The carbon atom that participates in the C—O—C unit joining two rings also forms bonds to two nonring atoms. One of these bonds is to oxygen and the other is to hydrogen. As a consequence of the tetrahedral structure at the carbon atom, the two nonring bonds point in two nonequivalent directions. In particular, one bond points outward, more or less parallel to the rings, and the other points above (or below) the ring. Substances with somewhat different properties are produced depending on whether the oxygen joining the rings occupies the in-plane or the out-of-plane bond position. One way for two ring systems to be joined uses an oxygen atom that lies *below* one of them and *in* the plane of the other:

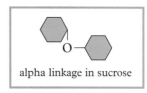

alpha linkage in sucrose

This "stepped" linking of rings is called the *alpha* form of the C—O—C linkage. It is present at each ring-joining C—O—C unit in starch. The alpha linkage also joins the glucose and fructose rings in sucrose. In general, humans have enzymes that can facilitate the addition of H_2O molecules (the hydrolysis reaction) at C—O—C units having the alpha geometry. This is why sucrose and starch are substances that we can digest.

The alternative geometry at the C—O—C unit involves in-plane bonds from *both* rings to the external oxygen atom:

beta linkage in cellulose

This is called a *beta* linkage of rings. In an extended network of beta connections, the rings follow each other in a more linear sequence than that of starch. In general, humans do *not* possess enzymes to speed the addition of water to beta C—O—C units, which occur in lactose sugar, the trisaccharide raffinose, and some other natural substances, though, as we've seen, babies and some adults do possess a special enzyme for lactose.

7.9 Cellulose is not digestible by humans

The C—O—C units in the unbranched polymer called *cellulose* have the beta linkage between glucose units, and consequently humans and other mammals lack any enzyme that can hydrolyze cellulose to glucose units

View the structure of cellulose at Chapter 7: Visualizations: Media Link 4.

and thereby digest it and use its energy. We do not even possess intestinal bacteria that can metabolize cellulose, so the parts of plants that contain it pass through our systems undigested. Ruminant animals such as cows, sheep, goats, and camels—and insects such as termites—do have bacteria in their digestive tracts whose enzymes can degrade cellulose to small units. This process does not occur by simple hydrolysis reactions to produce glucose, however. Rather, fragments of cellulose are produced that contain two to four carbons, as well as methane and hydrogen gases. Thus ruminant animals can eat and digest grass, hay, and so on, and termites can eat wood, though all these organisms emit phenomenal amounts of methane as they process the cellulose.

Recall that we discussed such natural sources of methane in Chapter 4.

Molecules of cellulose contain several hundred to several thousand rings all joined together by beta-type C—O—C units; the structure of cellulose is illustrated in Figure 7.10. The arrangement of adjacent rings produces hydrogen bonding *internal* to each chain, between adjacent glucose rings and between adjacent chains. Because this leaves little hydrogen-bonding ability available for external molecules such as H_2O, cellulose is insoluble in water. The hydrogen bonding also allows adjacent molecules to lie parallel to each other and pack efficiently, resulting in a rigid, strong material. Plants use cellulose as the major structural material in both stems and leaves. Indeed, *cell*ulose is the main component by mass of plant *cell*s and is named after them. The cellulose in plant cells consists of bundles of parallel chains called *fibrils,* which in turn lie parallel to each other.

Unusual, weak C—H---O hydrogen bonds also help cellulose form a strong structure.

Cotton, the hairy protective covering for cotton seeds, is a rather pure form of cellulose (see Figure 7.11). The quality of cotton cloth depends on the length of the fibers, and cotton clothes owe their ability to withstand repeated washings to the fact that the material is stronger when wet than dry. Wood is primarily cellulose, with other organic polymers present as well. *Cellophane* is a film made from regenerated cellulose.

Due to its crystalline, rigid structure, cellulose itself is not moldable and cannot be directly made into plastics. However, the first synthetic plastic produced, *celluloid,* was made from a form of cellulose in which some of the —OH groups were reacted with nitric acid, and the product was mixed with camphor and heated. These reactions of necessity elim-

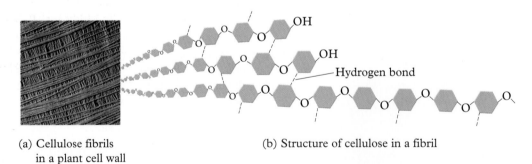

(a) Cellulose fibrils in a plant cell wall

(b) Structure of cellulose in a fibril

Figure 7.10 Cellulose in the cell wall of a marine alga. Glucose units are linked together by bonds called beta linkages that produce a molecule with different three-dimensional characteristics than glycogen or starch. (Part a, BioPhoto Associates/Photo Researchers)

Chapter 7: Health and Energy

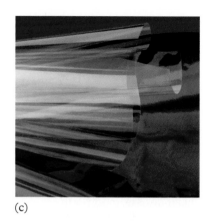

(a)   (b)   (c)

Figure 7.11 Different forms of cellulose. (a) Cotton. (b) Wood. (c) Cellophane. (part a, Photodisc Green/Getty Images; part b, Brand X Pictures/Getty Images; part c, BCA & D Photo Illustration/Alamy)

inated some of the hydrogen bonding, so that the material became moldable. Celluloid was used to produce the original film used for movies and snapshots, and the stiff collars for men's shirts, and is still used to make billiard balls. Celluloid derivatives also form the silklike fibers known as rayon and viscose.

Plants produce cellulose by the process of **photosynthesis,** which is the reverse of the combustion reaction with which we are familiar. Carbon dioxide in the air and water in the plant are combined with an *input* of energy provided by sunlight to produce oxygen and a carbohydrate:

$$CO_2 + H_2O + \text{energy (from sunlight)} \rightarrow O_2 + \underbrace{CH_2O}_{\text{Carbohydrate}}$$

For convenience we write the formula CH_2O to indicate a carbohydrate, with the understanding that we do not mean CH_2O molecules, but some carbohydrate with a C:H:O ratio of about 1:2:1.

Cellulose is produced in greater abundance than any other substance on Earth. It is unfortunate that we cannot readily depolymerize cellulose to the glucose that we could use to formulate foods. At present, to produce glucose from cellulose we must rely primarily on the ability of acids to break the bonds and allow an H_2O molecule to add the H to the oxygen on one carbon and the OH to the other carbon to produce two —OH groups from a C—O—C unit. This process is not an economical source of food. Perhaps biotechnology research will someday provide us with a cheap, ready, efficient set of enzymes by which we can convert cellulose into food.

7.10 Dietary fiber and resistant starch are nondigestible

Although we cannot digest cellulose, it forms an important part of our diet in that it supplies much of our dietary *fiber* (also known as "roughage" or "bulk"). As it passes through our digestive system, cellulose absorbs water—though it is insoluble in it—and thus increases in volume. This "bulk" promotes the rapid passage of solid wastes through our body, which may be an advantage in reducing the risk of developing colon cancer and other diseases since it minimizes the time that toxic substances, produced by the digestion of other food, remain in contact with our intestinal tissue. It may also be that the more fiber we consume, the less of other foods we will eat—and thus produce fewer of the other toxic substances. Foods that are rich in fiber include the *bran* component—that is, the outer coat—of grains, especially wheat, and the skins and fibrous parts of fruits and vegetables, especially celery, lettuce, and cabbage. Cooking a vegetable induces the softening or even complete breakdown of cell walls composed of cellulose.

When certain foods, including rice, are cooked, some of their starch is converted into a form called *resistant starch* (or "hidden fiber"), which is resistant to hydrolysis. The combination of a high cooking temperature and cooking water makes some of the starch molecules bind so closely together that our digestive enzymes cannot force them to react, and they pass through our bodies intact. Corn flakes are high in resistant starch.

Soluble fiber consists of polymers of various monosaccharides, not just glucose, and it is very good at binding to water. Oat bran is especially high in soluble fiber. Soluble fiber forms a gel in the stomach, slowing the rate of digestion and absorption. This prevents large, rapid increases in blood sugar. Food rich in soluble fiber can also help to reduce elevated blood cholesterol levels. Many other polymers of similar constitution form *gums,* which are water-soluble sticky substances that are exuded from certain plants and that often thicken and harden when exposed to air.

Fats and Oils

Our bodies obtain much of their energy from the carbohydrates in our diet. However, storing energy as carbohydrates for long periods is inefficient for organisms such as ourselves, which, in contrast to plants, often are in motion. In particular, the heat energy that is ultimately generated by combustion of 1 gram of carbohydrate is about 17 kilojoules, or 4 Calories, which is only about one-third of the energy obtained from a comparable amount of a hydrocarbon fuel such as heptane. From a chemical viewpoint, carbohydrates are an inefficient way to store energy because of their high oxygen content. Compared to hydrocarbons, carbohydrates are partially oxidized already, and thus they produce less energy per gram when burned.

This analysis predicts that animals could store fuel for energy much more efficiently in molecules that contained much less oxygen and so were more similar to hydrocarbons. Indeed, nature has developed such an efficient energy storage system: it is the solid called **fat.** In this section, we investigate fats and their liquid counterparts, **oils,** as components of the diet. Fats and oils, which store about 9 Calories of energy per gram, are thus much more efficient in this respect than carbohydrates. These substances are not only energy storage systems but have other functions as well.

Foods that are high in fat include nuts (cashews are 45% fat), cheese (cheddar is 35%), and avocados (25%), as well as the many foods cooked in fat or oil since they absorb the substance upon heating (Figure 7.12). Because fats and oils are much higher in energy content

The calculation of the amounts of heat given off by reactions was covered in a Taking It Further section in Chapter 6.

Figure 7.12 Different types of nuts. (Photodisc Green/Getty Images)

than are carbohydrates (and proteins), people who want to lose weight are usually advised to reduce their intake of fatty foods.

7.11 Fatty acids and glycerol combine to form fats and oils

The even number of carbon atoms occurs because in nature fatty acids are constructed sequentially from two-carbon units.

To begin our study of fats and oils, we return to the subject of carboxylic acids, the structures of which are conveniently summarized by the compact formula RCOOH. A **fatty acid** is a carboxylic acid in which the R group is an unbranched carbon chain of medium length, about a dozen or so carbon atoms. A common example of a fatty acid is *stearic acid,* whose compact condensed formula is $CH_3(CH_2)_{16}COOH$ and whose overall structure is:

$$CH_3-CH_2-CH_2-CH_2-CH_2-CH_2-CH_2-CH_2-CH_2-CH_2-CH_2$$
$$-CH_2-CH_2-CH_2-CH_2-CH_2-CH_2-COOH$$

Notice that the ratio of carbon to oxygen atoms in such molecules, here 9:1, is quite high compared to the approximately 1:1 ratio found in carbohydrates, and that there is overall an even number of carbon atoms present. Both these factors are characteristic of natural fatty acid molecules. Notice also that all the carbon–carbon bonds in stearic acid are single bonds; fatty acids containing only single C—C bonds are called **saturated.** Diets that are high in saturated fatty acids can be unhealthy since they lead to higher levels of cholesterol, as explained later in this chapter.

One of the properties of fatty acids that is pivotal in determining their performance is their melting points. Recall that the temperatures at which alkane hydrocarbons melt, and the temperatures at which they boil, both generally *increase* as the size of the molecule increases. The same is true for saturated fatty acids: the longer the hydrocarbon chain R, the higher the temperature at which they melt from solid to liquid. Figure 7.13 shows the melting points for saturated fatty acids plotted against the number of carbon atoms they contain. Notice that the shorter-chain saturated fatty acids (up to C_8) are liquids at room temperature (about 22°C), but the medium-sized and longer-chain ones are all solids.

Fats and oils are not simply fatty acids, but rather are **triesters**—that is, molecules having three ester functional groups each—formed by fatty acids with glycerol, which is the alcohol with an —OH group on each of its three carbons (see Chapter 6, section 6.11). We can imagine forming such a molecule by the

Figure 7.13 Melting point vs. carbon chain length in fatty acids. The melting point increases as the length of the carbon chain increases.

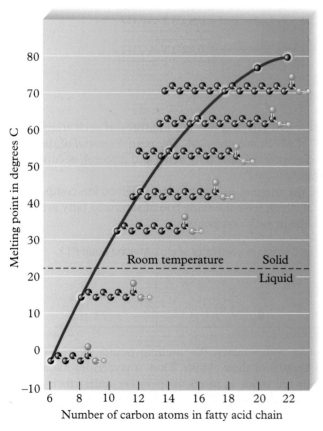

Melting point in degrees C

Room temperature — Solid

Liquid

Number of carbon atoms in fatty acid chain

successive reaction of the three —OH groups in a glycerol molecule with the —OH component of the —COOH group of three fatty acid molecules. An H_2O molecule is eliminated in each case, to form three esters:

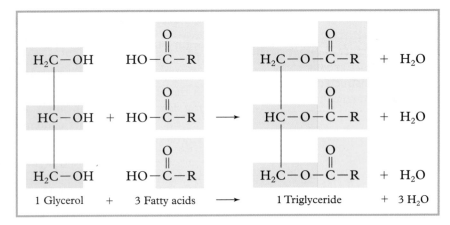

1 Glycerol + 3 Fatty acids ⟶ 1 Triglyceride + 3 H_2O

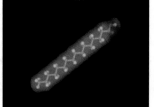

The resulting triester molecule is a **triglyceride** which we represent symbolically as:

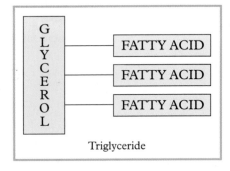

Triglyceride

View structures of triglycerides and saturated fats at Chapter 7: Visualizations: Media Link 5.

What we commonly call fats and oils are mixtures of such triglyceride molecules, not all of which are identical. The partial *reversal* of the reaction above, to produce the original fatty acids and glycerol, occurs when a fat or oil becomes "rancid."

The three fatty acids used to form any one triglyceride molecule are not usually all identical.

Triglyceride levels in blood are often obtained in routine medical tests.

Following are the melting points of three triglyceride compounds, all the molecules of which have the same trio of saturated fatty acids but with different lengths in each compound:

Number of C atoms in each fatty acid	Melting point (°C)
4	−75
12	46
18	73

Like their fatty acid components, the melting points of triglycerides increase with the lengths of their chains. Thus triglycerides made from saturated fatty acids are solids at room temperature unless their chains are rather short. They will melt to become liquid as temperatures are

Chapter 7: Health and Energy

increased somewhat—for example, as they are heated when the food that contains them is cooked.

Most fats originate from animals and most oils from plants, though there are exceptions. Fats and oils from natural sources—animals and plants—are *mixtures* of triglycerides formed by different fatty acids with glycerol. A given fat or oil is not a pure compound; its component fatty acids reflect the source of the substance. Thus, for example, "tropical oils" such as coconut oil and palm oil contain a high proportion of the esters of shorter-chain saturated fatty acids. Since their chains are short, tropical oils have a rather low melting point and in fact become liquids just above room temperature (at about 25°C for both oils mentioned above). In contrast, saturated fats such as those obtained from animal fat are solid at room temperature.

Though they are composed of three chains of carbon atoms, fats and oils are *not* polymers. The number of atoms involved is nowhere near the hundreds or thousands of atoms that are joined together in what are called polymer molecules.

Natural *waxes*, produced and used by plants to protect their surfaces from drying or injury, are esters of simple alcohols with fatty acids.

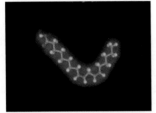

View structures of unsaturated fatty acids and triglycerides at Chapter 7: Visualizations: Media Link 6.

7.12 Unsaturated fats and oils contain C=C bonds

Some of the fatty acids in natural oils and fats have one or more carbon–carbon *double* bonds along their unbranched hydrocarbon chain. Such molecules are called **unsaturated.** If only one C=C unit is present in a molecule, it is said to be *monounsaturated.* The most common monounsaturated fatty acid is *oleic acid,* which has the formula

$$CH_3-CH_2-CH_2-CH_2-CH_2-CH_2-CH_2-CH_2-CH=CH-$$
$$CH_2-CH_2-CH_2-CH_2-CH_2-CH_2-CH_2-COOH$$

For convenience we can abbreviate this to a compact condensed formula:

$$CH_3(CH_2)_7CH=CH(CH_2)_7COOH$$

If more than one C=C unit is present, the fatty acid is said to be *polyunsaturated.* Common dietary examples here are *linoleic acid,* which has two C=C bonds, and *linolenic acid,* which has three:

$$CH_3(CH_2)_4CH=CH-CH_2-CH=CH(CH_2)_7COOH \qquad \text{Linoleic acid}$$

$$CH_3CH_2CH=CH-CH_2-CH=CH-CH_2-CH=CH(CH_2)_7COOH$$
Linolenic acid

An isomer of linolenic acid is found in *oil of primrose,* a substance reputed to have healing power.

Recent research indicates that certain isomers of linoleic acid, in which the two double bonds occur in a "conjugated" or —CH=CH—CH=CH— sequence, also occur naturally in foods and are said to be effective in promoting good health and weight loss.

Fats and Oils

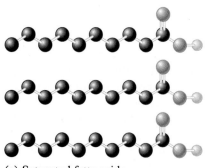

(a) Saturated fatty acids

(b) Unsaturated fatty acids

Figure 7.14 Geometry of saturated and unsaturated fatty acids. The presence or absence of C=C double bonds makes the molecule more or less liquid.

Generally speaking, for different fatty acids having the same chain length, the melting point *decreases* as the number of C=C units *increases*. This phenomenon occurs because the C—C=C bond angle is larger than a C—C—C bond angle, so the presence of C=C double bonds changes the regular geometry of saturated fatty acid molecules to the extent that they do not pack together as well (see Figure 7.14). Consequently, the solid requires less energy to disrupt the weak forces that operate between the molecules. For example, the melting points of the C_{18} fatty acids with various numbers of double bonds follow:

Number of C=C bonds	Melting point (°C)
0	71
1	16
2	−5
3	−11

Since the melting points of triglycerides reflect the melting points of their component fatty acids, the melting points of triglycerides decrease as the number of their C=C bonds becomes greater. For that reason, most natural oils—other than those that contain rather short chains—are high in unsaturated fatty acids. For example, the triglyceride that has three of the monounsaturated "oleic" fatty acids has a melting point of −6°C and so is a liquid even in a domestic refrigerator. By contrast, the triglyceride with three *saturated* chains of the same length has a melting point of +73°C and so remains a solid until it is heated substantially.

7.13 Common fats and oils contain mixtures of fatty acids

The fraction of saturated, monounsaturated, and polyunsaturated fatty acids in a number of common fats and oils is given in Table 7.1. For example, 86% of the fatty acids in the triglyceride molecules of olive oil, a liquid, are unsaturated, mainly being the monounsaturated oleic acid. Sunflower and corn oils have an even higher percentage of unsaturated fatty acids, though almost all of them are polyunsaturated. In contrast, butterfat and beef fat contain about 50% saturated fatty acids and consequently are solids. The typical triglycerides in beef tallow (fat) have a saturated fatty acid chain attached to the middle carbon of the glycerol unit and unsaturated fatty acid chains on the two end carbons. Human fat is about 35% saturated, 52% monounsaturated, and 13% polyunsaturated.

Table 7.1   Composition of common fats and oils

Dietary fat/oil	% Saturated fat	% Monounsaturated fat	% Polyunsaturated fat
Canola oil	6	58	36
Safflower oil	9	13	78
Sunflower oil	11	20	69
Corn oil	13	25	62
Olive oil	14	77	9
Soybean oil	15	24	61
Peanut oil	18	48	34
Cottonseed oil	27	19	54
Lard	41	47	12
Palm oil	51	39	10
Beef tallow	52	44	4
Butterfat	66	30	4
Coconut oil	92	6	2

Exercise 7.4

Predict the order of melting points for triglycerides containing the following combinations of averaged chain lengths and averaged extent of unsaturation for their component fatty acids:

a) short chains and high extent of unsaturation
b) long chains and low extent of unsaturation
c) long chains and high extent of unsaturation

In the diet of people living in developed, Western countries, on average fats and oils collectively provide 34–40% of dietary energy. Health authorities maintain that a level of 30%, of which only one-third is saturated, would be healthier for us. One can't, and shouldn't try, to *completely* eliminate fats and oils from the diet, since three fatty acids are **essential**—that is to say, they are required by the human body but cannot be synthesized by it from other components.

The essential fatty acids are linoleic, α-linolenic, and arachidonic acids. Linoleic acid is essential since from it the body synthesizes certain hormones.

■ Chemistry in Your Home: Calculating your intake of fat from processed foods

Most health experts recommend that you limit your fat intake to about 30% of your daily diet. That doesn't mean that every food you eat should have 30% fat, but that no more than 30% of your overall daily calorie intake should come from fat. In this activity you will calculate how many grams of fat provide 30% of the calories in your diet.

a) Using the nutrition labels on food or the Web sites listed at www.whfreeman.com/ciyl2e, add up how many calories you eat every day. You can keep a log, if you want, of a detailed breakdown of your caloric intake.

Taking It Further

For information and problems on Chocolate, go to Taking It Further at www.whfreeman.com/ciyl2e.

b) Multiply the number of calories you take in every day by 0.30. That gives you the number of calories that should come from fat.

c) Divide the calories from fat (step b) by 9 to get the maximum grams of fat you should have each day. (Each gram of fat provides 9 calories.)

For example, if you eat 2200 calories per day, no more than 660 (2200 × 0.30) calories should come from fat, which is equivalent to 73 grams of fat (660 ÷ 9). This means that if you consume approximately 2200 calories a day, you should have no more than about 73 grams of fat in your daily diet.

7.14 How do low-carbohydrate diets work?

Early in 2004, about 10% of the U.S. population had adopted a low-carbohydrate diet, but this had fallen to 5% by the end of the year.

In recent years, many people have tried to lose weight by adopting a diet that is very low in carbohydrates, such as the Atkins plan. Recent research suggests that the weight loss in the first few weeks of such diets is due to the substantial reduction in calories that people consume when carbohydrates are restricted. The initial weight loss associated with a very low carbohydrate diet is not entirely due to a loss of fat, but also to the excretion of the water that had been stored (in a 3:1 ratio) with the glycogen that the body converts to glucose to replace the missing calories.

In both the induction and long-term stage of this diet, about 60% of the dieter's energy is obtained from fats (versus a maximum of 30% recommended by nutritionists), and the amount of saturated fat consumed in all stages is twice the recommended maximum. The amount of dietary fiber in the diet is considerably less than generally recommended. For these reasons, some scientists fear that the long-term side effects of low-carbohydrate diets could be an increase in cancer (especially of the colon) and heart disease. The most common short-term side effects reported anecdotally include constipation, bad breath, loss of energy, and kidney problems. In the South Beach diet, a low-carbohydrate regime is instituted only in the first stage, and increased consumption of monounsaturated rather than saturated fats is emphasized.

In metric units, your BMI equals your weight in kilograms divided by the square of your height in meters.

Before you go on any sort of diet, you might find out first whether medically you are considered to be underweight, overweight, or just right. Medical scientists use the **body mass index** (BMI) to answer this question. To calculate your BMI, multiply your weight in pounds by 700, then divide the answer by the *square* of your height in inches. (Thus if you are 64 inches tall and weigh 120 pounds, then your BMI is (700 × 120) ÷ (64 × 64) = 20.5.) A BMI of less than 18.5 is typically classified as underweight, whereas normal weight is associated with the 18.5–25.0 range. A person whose BMI lies between 25 and 30 is often classified as overweight; those with BMIs greater than 30 are generally considered obese.

7.15 Omega−3 fatty acids are good for you

Fish oils are rich in polyunsaturated fatty acids, especially acids with five or six C=C units. In such molecules, the last double bond in the chain, counting from the —COOH group, often starts at the fourth carbon from the end of the chain:

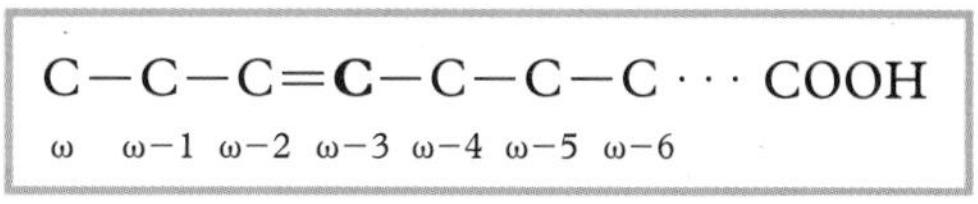

$$C-C-C=C-C-C-C \cdots COOH$$
$$\omega \quad \omega-1 \quad \omega-2 \quad \omega-3 \quad \omega-4 \quad \omega-5 \quad \omega-6$$

This carbon is known as $\omega-3$, or omega minus three, since the final carbon is by convention called ω, or omega.

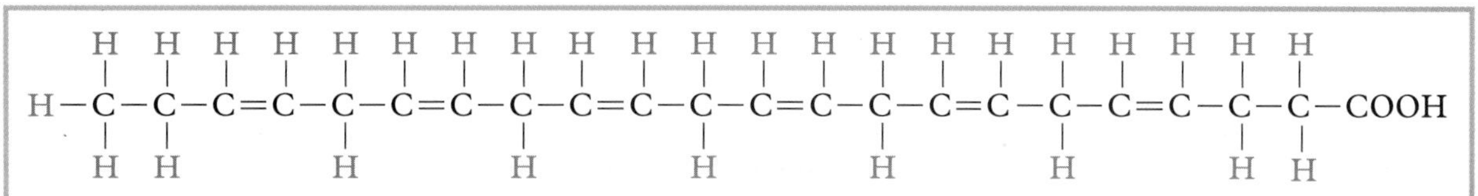

The essential fatty acid DHA (docosahexaenoic acid)

Thinking the minus sign is simply a dash, some people call omega-minus-three fatty acids just "omega three" or just "omega."

The presence of fish such as salmon, mackerel, sardines, and herring whose oils contain a high proportion of these so-called omega−3 fatty acids in your diet may serve to reduce the incidence of heart disease. Recent evidence suggests that this beneficial effect occurs because omega−3 fatty acids are converted in the body into specific lipids that suppress chronic inflammation, a condition that can lead to hardening of the arteries, and hence to an increased likelihood of heart attack and stroke. Some eggs are now produced enriched in omega−3 fatty acids. This enrichment is accomplished by feeding hens with flaxseed, which contains high levels of omega−3 acids. Flaxseed itself can be a direct source for humans if it is used in baking. Most unsaturated fatty acids in regular food products are those classified as $\omega-6$ (omega−6), not $\omega-3$. To give them long shelf lives, packaged foods contain much more $\omega-6$ than $\omega-3$ fatty acids in their oils. As a consequence, our diets now contain much more $\omega-6$ than $\omega-3$ acids, compared to the roughly equal amounts in past centuries. Some scientists believe that low consumption of $\omega-3$ fatty acids causes depression; indeed, DHA (structure shown above) is an important component of our brain cells.

Exercise 7.5

Using the structure for linoleic acid on page 277, deduce the class of acid to which it belongs according to the "omega minus" classification scheme.

7.16 Fats and oils are emulsified and hydrolyzed in the body

The molecules of fat from the food you consume must first undergo hydrolysis to produce the fatty acids and glycerol your body can use. However, the clumps of fat that reach your small intestine are too large to be hydrolyzed by enzymes (see Figure 7.15, step ❶). *Bile* from the gallbladder acts as an emulsifier, breaking up fat clumps into small

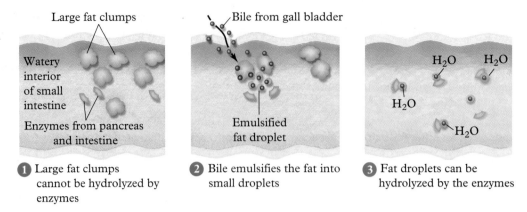

Figure 7.15 **Step by Step** Emulsification of fat by bile.

① Large fat clumps cannot be hydrolyzed by enzymes

② Bile emulsifies the fat into small droplets

③ Fat droplets can be hydrolyzed by the enzymes

droplets (step ②) which can then be hydrolyzed by enzymes from the pancreas and intestinal cells (step ③). If the fatty acids and glycerol produced are not used for other purposes, they are eventually combined to form body fat, which is then stored until needed to supply energy.

Because of the presence of long hydrocarbon chains and the absence of —OH groups, triglycerides are quite insoluble in water—giving rise to the old saying that "oil and water do not mix." Because they have low water solubility, they cannot dissolve in blood to a significant extent. Hence, fats and oils cannot circulate and be used as *immediate* energy sources for cells. Carbohydrates serve this purpose, even though, as we have seen, they are less efficient energy sources. For long-term storage of energy, however, organisms, including ourselves, use fat. Indeed, we would all weigh substantially more if we carried the same energy reserves around as carbohydrates rather than as body fat—though we would not necessarily be much bigger, since fat is about half as dense as carbohydrates.

7.17 Hydrogenation is used to produce some margarines

Margarine largely replaced butter in the diet of many health-conscious people in the last decades of the 20th century, since margarine has a higher ratio of unsaturated to saturated fatty acids due to its origin from plants rather than animals. However, there is some evidence that margarines that have been modified chemically may be as bad or worse for your health as are the saturated fatty acids in butter!

Most of the common, inexpensive plant oils that are used as starting materials for margarine are liquids at room temperature and remain so even under refrigeration, since they are polyunsaturated. In order to produce a semisolid, easily spread product from them, the extent or "degree" of unsaturation must be decreased, as you might expect from the correlation of melting point with unsaturation level. This decrease in unsaturation is accomplished industrially using the chemical reaction called **hydrogenation.** At the molecular level, hydrogenation is the

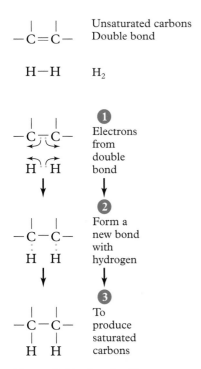

| $-\overset{\mid}{C}=\overset{\mid}{C}-$ | Unsaturated carbons Double bond |
| $H-H$ | H_2 |

1 Electrons from double bond

2 Form a new bond with hydrogen

3 To produce saturated carbons

Figure 7.16 Step by Step Hydrogenation of an unsaturated molecule.

Figure 7.17 A typical dinner basket of food fried with trans fats in the oil. (Royalty-Free/Corbis)

catalyzed reaction of a molecule of hydrogen gas, H_2, with a carbon–carbon double bond in a molecule such as an unsaturated fatty acid. In this reaction, illustrated in Figure 7.16, each of the two atoms of hydrogen from the H_2 molecule (step **1**) *adds to* one of the two carbon atoms of the double bond (step **2**). Since each C atom forms this additional bond to an H atom, it can no longer form a double bond but only a single bond to the other carbon (step **3**). By reacting the substance with more and more hydrogen gas, the number of double bonds in the fatty acid chains of the triglycerides in a plant-derived oil can be gradually decreased. In practice, the reaction is carried on until the oil has reached the desired extent of solidification.

Because a fully hydrogenated margarine would not have the required softness characteristics desired by the consumer, only *partial* hydrogenation of the oil is carried out. Thus, some carbon–carbon double bonds survive in the final margarine. However, due to the heating required for the hydrogenation of other bonds, a change in the molecular geometry occurs even at some C=C bonds that survive the reaction intact. In the original oils—and in fact in all natural fats and oils—both R groups at each C=C unit lie on the *same* side of the bond, as indicated in the left-hand diagram below. This is called the *cis* orientation of R groups about a double bond:

$$
\begin{array}{cc}
R \qquad\qquad R & R \qquad\qquad H \\
\quad C=C & \quad C=C \\
H \qquad\qquad H & H \qquad\qquad R \\
\text{cis orientation} & \text{trans orientation} \\
\text{of R groups} & \text{of R groups}
\end{array}
$$

This orientation does not change with time since under normal conditions there is no rotation about C=C bonds (in contrast to C—C bonds). However, the harsh conditions that are employed in the hydrogenation of *other* C=C sites alter the orientation of some of the surviving C=C bonds such that in the final product, the two R groups are on *different* sides of the C=C bond. This is called the *trans* orientation and is shown in the right-hand diagram above. So-called **trans fatty acids** are inadvertently produced and are present in margarines made in this way. Some brands of margarines are not produced by hydrogenation and therefore contain no trans fatty acids.

Although our emphasis in the discussion above has been on margarine, most North Americans in fact consume the majority of their trans fatty acids from two other sources:

- Fried foods, such as onion rings, french fries, and fried seafood (Figure 7.17). Most restaurants use shortening for frying that is about 25% trans fat.

- Commercial baked goods, such as doughnuts, cookies, and pastries. The trans fats help make the pastries flaky and the cookies fresh-tasting.

In the early 2000s, the average Canadian (and presumably the average American) consumed about 8 grams of trans fats (approximately 10% of total fat intake) daily, though men 18–34 years old were found on average to consume 39 grams!

We shall discuss the health effects of saturated fat and trans fat further in Chapter 9. Ironically, hydrogenated polyunsaturated oils—and hence trans fats—came into use to overcome health concerns about using tropical oils—which as previously discussed have a high proportion of saturated fat.

7.18 "Fake fats" have the mouth-feel of real fat

Many people wish to reduce the amount of fat in their diet for reasons of weight or cholesterol control. However, most of us enjoy eating fat-containing foods because of their "mouth-feel"—a creamy texture, a tendency to melt on the tongue, and the enjoyable flavor of substances dissolved in the fat. For these reasons, the food industry has developed "fake fats" which possess the pleasurable characteristics of fats without their drawbacks.

In some fake fats, the fatty acid chains found in real triglycerides are attached to a substance other than glycerol, producing a substance that tastes and feels like fat but that is not digested in the body. For example, Olestra is a polyester in which six to eight fatty acids are connected to the —OH units of a sucrose molecule. Although the evidence is controversial, Olestra in large amounts may cause abdominal cramps and loose stools in some people. Other similar fat substitutes under development use sugars other than sucrose to attach fatty acid chains, use short fatty acids chains, or replace fatty acids entirely by long-chain alcohols.

A second category of fake fats uses substances that don't contain long-chain hydrocarbon components but happen to have a mouth-feel like fat. For example, Simplesse is a substance that has a creamy feel in the mouth but actually consists of protein, not fat. Emulsions of gelatin or certain starches also have this property, especially when the particle size is uniform, thereby giving a smooth feel in the mouth. They are used to make the "light" versions of foods such as margarines and salad dressings.

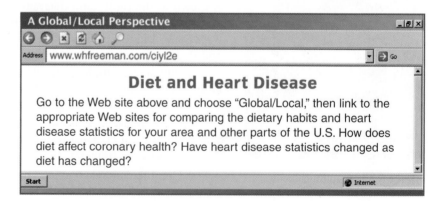

A Global/Local Perspective

Address www.whfreeman.com/ciyl2e

Diet and Heart Disease

Go to the Web site above and choose "Global/Local," then link to the appropriate Web sites for comparing the dietary habits and heart disease statistics for your area and other parts of the U.S. How does diet affect coronary health? Have heart disease statistics changed as diet has changed?

Summarizing the Main Ideas

Carbohydrates are compounds of carbon, hydrogen, and oxygen in which the three atom types are present in a 1:2:1 ratio, or approximately so. Saccharides are carbohydrates, the smaller members of which are called sugars. Saccharides are classified as mono-, di-, or poly-, depending upon whether there are one, two, or many rings joined together by oxygen atoms. Sugars are more sweet or less sweet, depending on how their molecular shape fits the sweetness receptor on the tongue.

Condensation polymers are formed by the successive condensation reactions of molecules, thereby forming long chains such as the polysaccharides starch and glycogen. The reversal of this polymerization process occurs in the body by enzyme-catalyzed hydrolysis reactions. Hydrolysis reactions are those in which water reacts with a substance. In the case of carbohydrates, a water molecule adds to a C—O—C link joining two rings, converting it to two separated C—OH groups and thereby splitting the polymer.

Two types of glucose polymers are known; they differ with respect to the orientation in space of successive glucose rings. In the alpha orientation, the rings are stepped, whereas in the beta form the sequence is approximately linear since in-plane C—OH units become joined together in the condensation process. Human enzymes can attack and speed up the hydrolysis of the alpha linkages but not most beta ones.

Photosynthesis is the creation of carbohydrates and oxygen gas by the reaction of carbon dioxide, water, and the energy from sunlight.

Although carbohydrates are used for energy storage in plants, animals use fats and oils. These substances are triesters of glycerol and three medium- to long-chain carboxylic acids called fatty acids. The fatty acids are called saturated, monounsaturated, or polyunsaturated, depending upon whether their hydrocarbon chains contain no, one, or more than one C=C bond. In general, the melting points of triglycerides increase with the lengths of the fatty acids but decrease with their degree of unsaturation. Common fats and oils are made from a mixture of saturated and unsaturated fatty acids; they are mixtures rather than pure compounds or polymers. Unsaturated fatty acids are sometimes classified according to the position of the last double bond along their carbon chain: the "omega minus" system.

The hydrogenation reaction, by which a molecule of H_2 gas is added to a C=C bond, converting it to C—C, is the process by which polyunsaturated liquid oils are converted to the semisolids useful in margarine. In this process, some of the remaining C=C bonds in the oils are converted from the cis orientation, in which both carbon chains lie on the same side of the C=C unit, to the trans orientation, in which they lie on opposite sides.

Key Terms

carbohydrate	polysaccharide	fatty acid	essential fatty acid
sugar	complex carbohydrate	saturated fatty acid	hydrogenation
saccharide	photosynthesis	triester	trans fatty acid
hydrolysis reaction	fat	triglyceride	
condensation polymer	oil	unsaturated fatty acid	

Web Sites of Interest

To link to Web sites of interest, go to www.whfreeman.com/ciyl2e, Chapter 7, and select the site you want.

For Further Reading

Raloff, J., "Chocolate Hearts: Yummy and Good Medicine?" *Science News, 157,* March 18, 2000, pp. 188–189. Article looks at the recent discovery that chocolate has antioxidant properties and may be beneficial to health.

Travis, J., "Teasing Out a Tongue's Taste," *Science News, 155,* February 27, 1999, p. 132. Article discusses current research results that identify two proteins on the tongue involved in the sense of taste.

Review Questions

1. What components are common to carbohydrates?
2. Why is glucose classified as a carbohydrate?
3. Why is dextrose called *blood sugar*?
4. What characteristic identifies a sugar?
5. How do monosaccharides differ from disaccharides?
6. What is meant by the term gestational diabetes? What are the primary differences between gestational, type 1, and type 2 diabetes?
7. How are the chemical formulas of glucose and fructose related?
8. Match the saccharides below with their sources:
   a) sucrose   1) beans
   b) glucose   2) sugar beets and sugar cane
   c) fructose   3) raisins, grapes, and figs
   d) lactose   4) fruits and honey
   e) raffinose   5) milk
9. What is a *hydrolysis reaction*?
10. What are the products of the hydrolysis of sucrose?
11. Identify three "sugarless" sweeteners.
12. What property of polyhydroxyl molecules makes them useful as components of skin moisturizers?
13. What is *lactose intolerance*?
14. How does a *condensation reaction* produce a polysaccharide?
15. Identify the products of the digestion of starch.
16. How does amylose differ from amylopectin? How are they the same?
17. What is the function of glycogen in the body?
18. How does your body store glucose for long-term needs?
19. How do the structures of alpha-linked carbohydrates differ from those of beta-linked carbohydrates?
20. What is the process by which plants produce cellulose?
21. How is *soluble fiber* different from *insoluble fiber*?
22. How does a *fat* differ from an *oil*?
23. What is *paraffin wax*?
24. Describe the structure of a triglyceride.
25. How are saturated and unsaturated fatty acids different?
26. Identify three plant and three animal sources of fats and of oils.
27. What is an *omega−3* fatty acid?
28. What is a *trans fatty acid*?

Understanding Concepts

29. Use the structure of glucose to explain its solubility in water.

30. Draw the structure for a hypothetical monosaccharide molecule which has only three carbons and oxygen in the ring, plus one —CH_2OH group and two other hydroxyl group.

31. How is the oxidation of glucose in the body similar to combustion?

32. The hydrolysis of sugars requires acid when performed in a laboratory. Why can these reactions be carried out without acid in a human body?

33. What structural characteristics and properties does a molecule need to possess in order to be perceived as sweet by humans?

34. What is the relationship between the structure of a sugar molecule and its sweetness?

35. Explain *carbohydrate loading* and its effect on the human body. Why do athletes engage in carbohydrate loading?

36. Explain how sugarless gum can be sweet if it contains no sugar.

37. Look back in the chapter at the structures of a carbohydrate molecule and a fat molecule. How do the structural features explain why the fat molecule produces more energy than the carbohydrate? *Hint:* Think about the chemical reactions involved in oxidizing food molecules.

38. In what two ways does the structure of amylose differ from that of amylopectin?

39. How does hydrogenation change the structure of an unsaturated fatty acid?

Synthesizing Ideas

40. How is the oxidation of glucose in the body more effective at supplying energy to it than if it were combusted like wood? What would the effect be in your body if glucose were combusted? Could you use the energy contained in the glucose if it were combusted? Why or why not?

41. A diet rich in complex carbohydrates rather than simple carbohydrates is recommended for people suffering from diabetes. Using the information in the chapter and any other resources, explain the basis for this recommendation.

42. The body stores glucose as the polysaccharide glycogen. Look at the structure of glycogen (Figure 7.9b) and explain why it is a more effective storage structure than a straight chain of glucose molecules. In formulating your answer, think about how the body must use glycogen to liberate glucose.

43. Explain why doctors recommend that people lower their intake of meat products in order to maintain a healthy lifestyle.

44. Draw a representation of the 3-D structure of a tetrasaccharide formed by four molecules of glucose in which a) all the C—O—C bonds joining the rings are alpha in character, and

b) all the C—O—C links are of the beta type. Which of these two forms of the tetrasaccharide would you be able to digest?

45. Draw the simplified representation, using hexagons joined by oxygen atoms, for a molecule of a) lactose, and b) a trisaccharide formed by three molecules of glucose.

46. Draw the Lewis structure for the triglyceride in which all three component fatty acids are saturated and each contains six carbon atoms. Based upon the table in the text for the melting points of such triglycerides (page 276), try to predict an upper and a lower limit as to what its melting point would be.

47. Explain the difference in structure between a) starch and glycogen, and b) amylose and amylopectin.

48. Classify a) oleic acid, and b) linolenic acid, according to the "omega minus" system.

49. Consider three triglycerides, each composed of three identical fatty acids. In the first triglyceride the fatty acids are saturated, whereas in the second they are monounsaturated, and in the third they are polyunsaturated. Predict the trend, from lowest to highest, of melting points for these three compounds.

50. How many molecules of H_2 would be required to completely hydrogenate one molecule of linolenic acid? Draw the Lewis structure for the saturated fatty acid that would be obtained after complete hydrogenation has occurred.

51. Write out the balanced reaction for the complete combustion of one molecule of glucose, $C_6H_{12}O_6$, and for one molecule of the alkane hydrocarbon $C_{13}H_{28}$, which weighs almost the same. What is the ratio of the number of molecules of O_2 consumed by the hydrocarbon compared to the sugar in the reactions? Given that the amount of heat generated by burning is proportional to the amount of O_2 used, the ratio of heat produced by the hydrocarbon to that produced by glucose ought to be approximately equal to the ratio of O_2 molecules you just computed.

Taking It Further I

Beer and spirits are alcoholic beverages produced from carbohydrates

Beer and spirits such as whiskey, gin, and vodka are alcoholic beverages made from the starch of various grains. In contrast to wine, in which the sugars in grapes or other fruits are fermented directly, the starch requires depolymerization prior to fermentation.

Beer Beer is usually made from barley (see Figure 7.18). To generate the enzymes that convert the starch in the grain into a sugar, barley grain is

Figure 7.18 The ingredients that produce beer. Clockwise, from top left, are: enzymes, barley, hops, and water. (Center photo by John A. Rizzo/PhotoDisc/ Getty; others by George Semple for W. H. Freeman and Company)

soaked in water for a few days and then left in a warm, damp environment for a week, at which point it begins to germinate. The germination process produces the enzymes required for fermentation but must be halted before the grains sprout by heating the barley at 65–100°C. The higher end of this temperature range is used if a dark beer is to be produced. The very dark color of stout beer is produced by incorporating some roasted barley or caramel. Some other, cheaper sources of starch—corn or rice or wheat—can be added to provide extra starch, if required. The malted barley is then steeped in water and crushed to extract the starch, sugar, and enzymes, and then run through a sieve to remove solids. The starch is mashed by heating gently, and thereby converted to glucose and the related disaccharide maltose by the enzymes from the malt. Because the enzymes cannot cleave starch chains at points where the chain branches, some short-chain carbohydrates called dextrins remain and are not subsequently converted to ethanol.

Once the malting process is complete, *hops*—the blossoms of the hop plant—are added and the mixture is boiled to provide the characteristic bitter hop flavor, which counters the sweetness of the residual carbohydrates. Rather less hops are added if a lager is being made, and none are added if ale is desired. In modern large-scale brewing, an extract from hops is added later in the process rather than at this stage.

The addition of yeast initiates the process of fermentation, in which the original glucose molecules, and those obtained by cleaving molecules of maltose obtained from starch, are converted into carbon dioxide and ethanol (see Figure 7.19). Fermentation is a complex reaction taking about a week and involving a dozen individual steps, each of which has its own enzyme to accelerate it. Small amounts of other organic compounds are also produced, which collectively help give beer its characteristic flavor. *Low-alcohol beer* (<1.2% ethanol) is produced by stopping the fermentation early, or using a yeast (such as that from wine) that cannot convert the maltose to glucose and hence ferments only the existing glucose from the malting step. *Alcohol-free beer* (<0.5%) is obtained by distilling the alcohol from beer after fermentation but leaves most of the beer's flavor intact. *Light beer* contains about 80% of the alcohol of regular beer but many fewer calories, since the dextrin content—which contributes about one-third of the calorie count of normal beer—is reduced by more than half. Regular beer contains about 13 grams of carbohydrates per can or bottle; light beer has only about one-third this amount.

Once fermentation is complete, sediment and residual yeast are removed and the beer is stored for a few months so that it clarifies and additional flavor develops. Carbon dioxide under pressure is added before the beer is bottled. When the bottle is opened and the beer poured, the carbon dioxide is released from the liquid and produces a foam of gas bubbles trapped in liquid. Unlike soft drinks, beer retains the foam due to the presence of dextrin carbohydrates and other ingredients that remain in the beer. Beer in bottles is pasteurized—by heating it for a brief period—to kill the yeast and microbes.

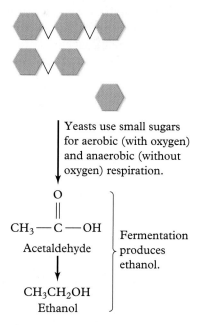

Yeasts use small sugars for aerobic (with oxygen) and anaerobic (without oxygen) respiration.

$$CH_3-\overset{\overset{\displaystyle O}{\|}}{C}-OH$$

Acetaldehyde

$$CH_3CH_2OH$$

Ethanol

Fermentation produces ethanol.

Figure 7.19 The fermentation process.

Spirits Alcoholic "white spirits"—colorless beverages such as gin and vodka—are also made from an inexpensive source of starch such as corn, potatoes, or a grain. Traditionally, the starch was converted to fermentable sugars by adding some malted barley, but nowadays industrial sources for the appropriate enzymes are often used. For these liquors, the idea is to produce alcohol free of color or flavor from the original raw materials, and then add the desired flavor as part of the distillation step. In the manufacturing of gin, for example, flavor is added by distilling the ethanol–water solution to which juniper berries, plus components of other plants, have been added.

Scotch malt whiskey is made by fermenting barley grain and distilling the beerlike product. In blended Scotch, the malt whiskey is combined with whiskeys made with corn. *Canadian whiskey* (simply called *rye* in its native land) is made mainly from rye grain. *Bourbon,* produced in the United States, uses mainly corn as its source of starch. *Rum* is produced from the fermentation of sugar cane or molasses, followed by distillation. Whiskeys and rums are aged in wood barrels; they obtain their color and some of their flavor as extracts from the wood and from previous contents of the barrels, such as molasses or wine.

Taking It Further II

Many common food products are largely fats or oil

The fat content of dairy products varies across an enormous range, from butter (80% fat); through heavy cream (35% fat); to whole milk (3–4% fat); to "fat-free" or skim milk, which has less than 0.5% fat since most of it has been "skimmed off." So-called reduced fat or partially skimmed milk has 1 or 2% fat. Evaporated milk is 6.5% fat, since about 60% of the water of whole milk is evaporated from it before it is canned and heat sterilized. Sweetened condensed milk has up to 8% fat, since it is a milk concentrate to which sucrose has been added.

As you learned in Chapter 1, milk is a colloid, a suspension of tiny droplets of butterfat in water. In general for a colloid, the smaller the particles suspended in the water, the less the tendency for them to join together and form a separate phase. In the *homogenization* of milk, the globules of fat are reduced in size so that they remain suspended in the milk rather than collecting and rising to the top as cream. This is accomplished by forcing the milk through a very fine sieve under high pressure. In contrast to homogenization, churning of the cream from milk forces the butterfat globules to aggregate into larger particles that then coagulate to form butter and a separate watery layer ("buttermilk"), which is removed.

Though butterfat is mainly saturated fat, its component fatty acids vary in chain length from C_4 to C_{18}. Like butter, margarine is about 16% water and 80% fat. For butter and margarine, the "plastic" range, which is the temperature range over which they retain their shape but can be readily deformed and cut with a knife, is important. The upper limit to the range must be just lower than body temperatures so that these products will "melt in your mouth." The lower range for margarines is usually less than for butter; thus butter from the refrigerator often cannot be cut readily, whereas many margarines can.

In butter and other dairy products, microorganisms catalyze the hydrolysis reaction by which water is added to the ester bond, and the fat decomposes, liberating its fatty acids. Because they have short chains, these fatty acids are volatile and we detect their odors easily.

Many oils are extracted from the seeds of plants by crushing. Peanut oil is used in cooking since it does not decompose and "smoke" at high temperatures. Olive oil is obtained from the macerated pulp of the fruit. The first pressing of the pulp is accomplished without heat and yields the virgin olive oil. Subsequent pressings require the use of heat to extract more oil, which is of a lower grade. Unlike other oils meant for human consumption, olive oil is not usually further processed to remove its subtle, desirable flavor, although "light" olive oil may well be so treated.

Salad and cooking oils, and solid shortening, consist almost entirely of triglycerides. Salad dressings are usually 35–50% oil in content, the remainder being vinegar, sugar, and starch (to thicken them). Mayonnaise, on the other hand, is an emulsion consisting of about 80% oil, in which tiny droplets of water are suspended. Both salad dressings and mayonnaise contain about 5% egg yolk, which acts as an emulsifying agent. Egg yolk helps keep the oil and water components from separating into different layers by stabilizing the presence of small droplets of one in the other. Other emulsifying agents such as *monoglycerides* (which are monoesters of glycerol) are added to egg-free versions of mayonnaise and salad dressing.

A naturally occurring emulsifier is *lecithin,* which is a triglyceride in which one fatty acid chain has been replaced by a complicated chain incorporating a water-soluble phosphate group. Lecithin sources include vegetables such as cabbage and cauliflower as well as animal sources such as eggs and liver. It is found throughout the human body.

In this chapter, you will learn:

- about a class of nitrogen-containing organic compounds, called amines;
- about the many drugs and stimulants that contain amines, such as Prozac, nicotine, caffeine, cocaine, Ecstasy, and some painkillers;
- how and why drug levels decline over time in your body;
- about a class of nitrogen and oxygen-containing organic compounds called amides;
- about the commercial products made from condensation polymers, such as nylon, polyester plastics, and polyester fabrics.

What do a cup of coffee and a nylon tent have in common?

The answer is nitrogen. Many small organic molecules that contain nitrogen are biologically active, such as the caffeine in coffee. In this chapter, we will examine large and small nitrogen-containing organic molecules. (Digital Vision/Getty Images)

Condensation Polymers, Especially Those Containing Nitrogen

The Chemistry of Medication and Clothing

What do fibers like nylon and polyester have in common? How do drugs like cocaine and Ecstasy work, and why does a depressive stage always eventually follow the initial elation experienced by users? Why do decomposing fish and meat smell so bad? What's a half-life?

These questions, and many more, are answered in this chapter. Our main theme is the organic substances formed by nitrogen, and we shall find that this encompasses drugs—both legal and illegal—as well as synthetic fibers and plastics such as nylon. Some analogous polymers that do not contain nitrogen are also included. The important biochemical substances of nitrogen are proteins, DNA, and RNA, which have a chapter of their own following this one.

As usual, we begin by considering the simplest type of organic nitrogen compounds and work our way to the more complex.

Important Nitrogen-Containing Organic Molecules

8.1 Amine molecules contain nitrogen atoms

We saw in Chapter 6 that alcohol and ether molecules can be thought of as water molecules in which one or both of the two hydrogen atoms, respectively, have been replaced by chains or rings—symbolized by R in structural formulas—of carbon atoms (and their associated hydrogen atoms):

$$H-\ddot{O}-H \qquad R-\ddot{O}-H \qquad R-\ddot{O}-R$$
Water $\qquad$ Alcohol $\qquad$ Ether

In the same way, molecules of **amines** correspond to molecules of *ammonia*, NH_3, in which one, two, or all three of the hydrogen atoms have been replaced by R groups consisting of chains or rings of carbon atoms (with their associated hydrogen atoms):

Whenever you see this icon in this chapter, go to
www.whfreeman.com/ciyl2e

View the structure of an ammonia molecule at Chapter 8: Visualizations: Media Link 1.

The once-popular diet drug *Fen Phen* consisted of a mixture of two amines, fenfluramine and phentermine. Both compounds are appetite suppressants that consist of molecules with an amino nitrogen bonded to a two-carbon chain that is connected to a benzene ring. Fenfluramine was withdrawn from the U.S. market in 1997 because it was found to significantly increase the risk of heart-valve disease.

$$\begin{array}{cccc} & H & H & H & R \\ & | & | & | & | \\ H-N-H & R-N-H & R-N-R & R-N-R \end{array}$$

Ammonia   Amines

As is the case for the analogous ethers, the various R groups need not all be identical within a molecule. As expected from the fact that it requires three electrons to achieve an octet, each nitrogen atom forms three bonds and possesses one nonbonded pair of electrons. The names for many am*ine*s end in *-ine,* reflecting the name of the general category. For example, CH_3NH_2 is called *methylamine.*

Recall that in Chapters 6 and 7 we found it useful to consider organic molecules as consisting of carbon networks to which are attached functional groups. A nitrogen atom with its two bonded atoms is called the **amine group.** Thus, CH_3NH_2 can be thought of as methane, CH_4, in which the amine functional group $-NH_2$ has replaced one of the hydrogen atoms. Alternatively, methylamine can be considered to be ammonia in which one hydrogen atom has been replaced by the CH_3 group:

$$\begin{array}{cc} H & H \\ | & | \\ H-C-N-H \\ | \\ H \end{array}$$

Methyl group

Amine group

Methylamine

In the liquid and solid states, amine molecules that contain one or two N—H bonds engage in N—H---N hydrogen bonding with neighboring ones:

$$\begin{array}{c} \diagdown \\ N-H \overset{\delta^+}{---} : \overset{\delta^-}{N}- \\ \diagup \end{array}$$

Hydrogen bond

Consequently, such amines have higher boiling points—and evaporate less readily—than the alkane that has the same number of electrons and that cannot exhibit hydrogen bonding:

	CH_3-CH_3	CH_3-NH_2	CH_3-OH
Boiling point	$-89°C$	$-6°C$	$+65°C$

However, since the effect of N—H---N hydrogen bonding in increasing the boiling point of methylamine, CH_3NH_2, above the boiling point of ethane is about half the effect of the O—H---O hydrogen bonding in methanol, we can conclude that N—H---N hydrogen bonds are weaker than O—H---O ones.

8.2 Many amines have odors that disgust us

One of the striking characteristics of amines is their odor. Ammonia's odor is piercing but not too unpleasant in low concentrations. Simple amines have "fishy" smells that are much less appealing. Rotting fish, for example, generates some *trimethylamine*, $(CH_3)_3N$, which we associate as the characteristic smell of such deteriorating material. We readily detect the odors of such amines because they are water soluble and are relatively volatile liquids (or gases), which therefore reach and are absorbed by the odor detectors in our noses.

In the chemical reactions that correspond to the gradual decomposition after death, some of the nitrogen in animal flesh is converted into amines. Our disgust at the odor of rotting flesh is based largely on amines that it releases into the air. The worst such smells come from two **diamines**—that is, amines that contain two amino groups—known by the descriptive names *putrescine* and *cadaverine*, both of which are found in rotting meat and fish (see Figure 8.1). Both of these amines also contribute to the scents of urine, bad breath, and semen. Notice that both putrescine and cadaverine have two $—NH_2$ groups located at the opposite ends of a short unbranched chain of four or five carbon atoms. As we'll

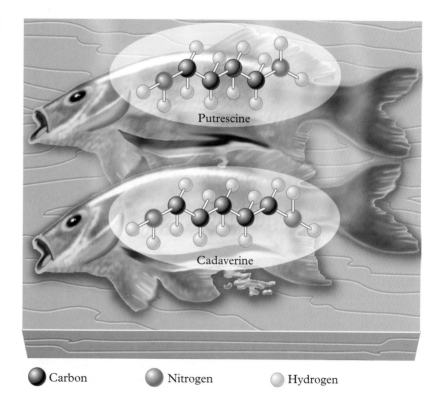

Figure 8.1 Amines. Amines like putrescine and cadaverine smell terrible.

Putrescine

Cadaverine

● Carbon   ● Nitrogen   ● Hydrogen

see later, they are structurally related to a molecule used to make one type of nylon, though happily we can't smell them in nylon fabrics.

8.3 Antidepressants, such as Prozac, are amines

Prozac, or *fluoxetine* to use its generic name, and Paxil, or *paroxetine,* are well known as antidepressant medications. Chemically, they are amines, as the structure for fluoxetine illustrates:

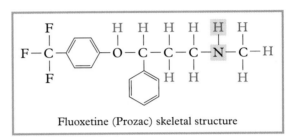

Fluoxetine (Prozac) skeletal structure

To understand how Prozac, Paxil, and other mood-altering drugs work, we must consider the way in which nerve impulses are transmitted within the human brain. *Within* a nerve cell, an impulse is transmitted electrically, by the movement of ions. However, when an impulse is to be transmitted *between* adjacent nerve cells, a different mechanism is used. Molecules called **neurotransmitters** are released by one nerve cell—the one from which the impulse is coming (see Figure 8.2, step ❶) —and then travel for less than a millisecond across a gap, known as *the synaptic area,* toward receptors on the adjacent cell. The arrival and

CHEMISTRY IN ACTION

How do neurotransmitters act as chemical messengers? See how the neurotransmitter acetylcholine generates electrical impulses at Chemistry in Action 8.1.

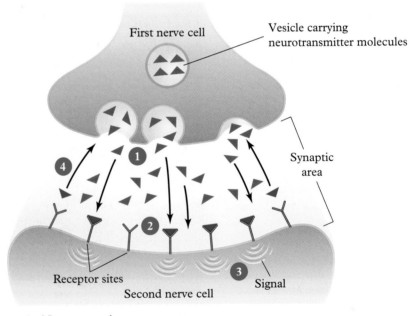

Figure 8.2 Step by Step
The mechanism by which neurotransmitters work. ❶ Upon receiving an impulse, a nerve cell releases neurotransmitters into the synaptic area. ❷ Neurotransmitters migrate to a second nerve cell and bind to receptor sites. ❸ Receptor site binding results in the production of a signal in the second cell. ❹ Neurotransmitters are released from the receptor sites and can travel back to the original cell.

temporary attachment of the neurotransmitter molecule (step ❷) signals an electrical impulse to travel along the second nerve cell (step ❸).

After the signal is generated, the neurotransmitter molecules are quickly released from the receptor site, and most of them travel back across the synapse to the release points on the original nerve cell, ready to be deployed again at a later time (step ❹).

The antidepressant drugs Prozac and Paxil work by modifying the recycling of the neurotransmitter *serotonin* (an amine) in certain nerve cells:

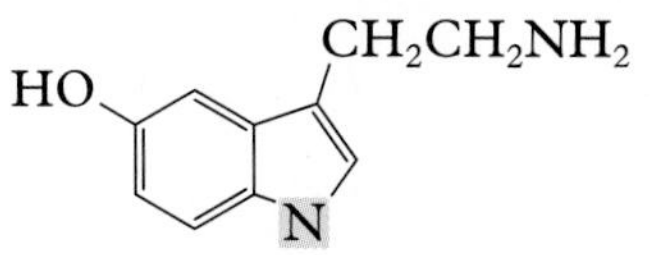

Serotonin

Much of the structure of the hallucinogen LSD, *lysergic acid diethylamide,* is identical to that of serotonin.

One cause of depression is a chronic deficiency of serotonin (and perhaps of other neurotransmitters) in the synapse. Prozac and Paxil increase the concentration of serotonin in the synapse region by preventing the efficient recapture of the neurotransmitter by the original nerve cell. Many such serotonin molecules can then travel again to the receptor site and initiate the transmission of another impulse on that nerve *without* having to be released by nerve impulse excitation at the first cell.

8.4 Nitrogen atoms are found in nicotine and caffeine

Plant-based compounds containing amine-type nitrogen and carbon rings are called *alkaloids*. These compounds are used in medicines such as codeine, quinine, and pseudoephedrine.

In many chemicals that are found in plants, one or more nitrogen atoms are components of five- or six-membered rings, the other ring atoms of which are carbon. The carbon and nitrogen atoms within the rings are joined by single or double bonds, depending upon the specific molecule. Two important examples of the component rings in such molecules follow:

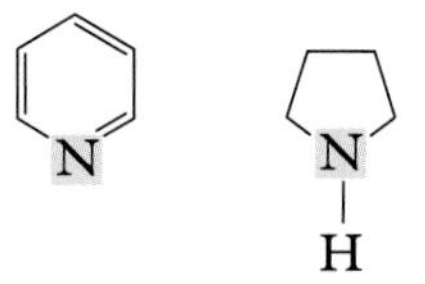

Pyridine   Pyrrolidine

Pyridine is analogous to benzene in that the bonds joining the ring atoms alternate between single and double; notice that the nitrogen here is *not* bonded to a hydrogen atom since it already forms three bonds. In contrast, in *pyrrolidine,* the nitrogen *is* attached to a hydrogen atom since the bonds it forms to the ring carbons are both single.

Nicotine and tobacco *Nicotine,* a substance well known to smokers, consists of molecules of one pyridine ring and one pyrrolidine ring,

Important Nitrogen-Containing Organic Molecules

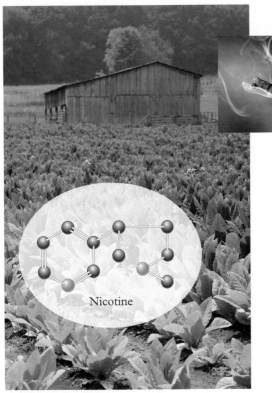

Figure 8.3 Nicotine. This amine is found in tobacco plants. (Larger photo by Royalty-Free/Corbis; inset photo by Brand X Pictures/Getty Images)

joined by a single bond (see Figure 8.3). The nitrogen in the five-membered ring is also attached to a methyl group.

Nicotine is a mild **stimulant,** a substance that stimulates the body's central nervous system and thereby temporarily increases our level of alertness and the speed of our mental processes, improves our mood by providing a sense of euphoria, and decreases our level of fatigue and our appetite. Nicotine also increases the heart rate and blood pressure as well as the body's level of *adrenaline*. Nicotine's ability to induce euphoria is due presumably to its ability to liberate neurotransmitters such as dopamine and serotonin. As indicated by its name, adrenaline is also an amine (see Figure 8.4). Adrenaline increases the rate of glycogen breakdown to glucose in the liver and muscles, thereby increasing the glucose supply in blood and providing a fast energy boost as a response to stress. It also increases the heart and breathing rates, providing more oxygen to the blood.

Nicotine occurs naturally in the leaves of tobacco plants, and the smoking of tobacco is an efficient way to quickly transfer this drug into the bloodstream. Indeed, cigarettes have often been called "nicotine delivery systems." Chewing tobacco releases nicotine directly into the mouth, and from there eventually into the bloodstream.

Unfortunately, nicotine is addictive: a decrease of the level of nicotine in the bloodstream produces withdrawal

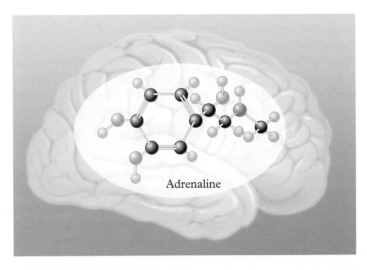

Figure 8.4 Adrenaline. Produced in the brain, adrenaline leads to the "fight-or-flight" response when it circulates in the blood.

Chapter 8: Condensation Polymers, Especially Those Containing Nitrogen

CHEMISTRY IN ACTION

Why is smoking addictive? Examine the action of nicotine in the brain at Chemistry in Action 8.2.

symptoms of irritability, anxiety, difficulty in concentrating, restlessness, etc. The blood level of nicotine decreases relatively quickly, since about one-third of the nicotine absorbed from smoking a cigarette is removed from the blood after one hour. The smoker's solution to the decline is to light up another cigarette and thus to quickly restore the higher nicotine concentration in the bloodstream. Most smokers crave a cigarette immediately upon waking in the morning since their nicotine level has dropped greatly during the night.

Because smoking tobacco produces very serious long-term health problems—including greatly increased rates of lung cancer and other serious lung diseases, increased rates of cancers at other sites, and increased rates of cardiovascular disease—many people have given up the habit or have decided not to start smoking. Quitting smoking can be quite difficult. One method that some people find effective is "the patch," a device that continuously delivers nicotine to the bloodstream by being absorbed through the skin. In this way, an ex-smoker can receive the nicotine he or she craves in ever-decreasing amounts until the addiction is cured, without inhaling the other chemicals in the smoke that produce most of the negative health effects. Nicotine itself is toxic in high doses, which is why individuals wearing the patch must not smoke.

One of the physiological effects of nicotine is the constriction of small blood vessels, including those near the skin. The repeated constriction of these vessels causes the skin of smokers to lose some of its elasticity and thus to become wrinkled. Consequently, long-term smokers often appear to be older than they really are!

The nicotine in tobacco plants is part of the plant's natural defenses, and people commonly use it as an insecticide on other plants. The nicotine in cigarettes would prove even more of a danger to smokers were it not for the fact that some of it is combusted or converted to less toxic substances when tobacco burns. In fact, in highly concentrated doses, nicotine can be fatal to humans.

Caffeine The other common, legal stimulant is caffeine. Caffeine molecules contain four nitrogen atoms distributed between two fused rings, one a six-membered ring and the other five-membered (see Figure 8.5).

The common "delivery systems" for caffeine are coffee, tea, cola drinks, and the over-the-counter "stay awake" drugs (e.g., No-Doz) favored by some students to keep them conscious while studying. One average-sized cup of coffee, two or three cola drinks or cups of tea, or one No-Doz or Excedrin tablet each contains about 100 milligrams of caffeine. However, the amount of caffeine in coffee depends upon its origin and mode of preparation. Instant coffee generally has the lowest caffeine content. Caffeine is also contained in many nonprescription

Caffeine

Figure 8.5 Caffeine. This amine, found in coffee beans, gives the beverage its "jolt." (Larger photo by Punchstock; inset photo by Digital Vision/Getty Images)

> ???????????????? ? ????????????
> **Fact or Fiction?**
>
> **Drinking 100% decaffeinated coffee and tea will cause you no stimulation or other cardiovascular effects.**
>
> False. Scientists have discovered recently that some of the cardiovascular effects of coffee are due to components other than caffeine, which are not removed during decaffeination. In addition to caffeine, tea contains theophylline, whose structure is identical to caffeine except that it lacks the methyl group on the five-membered ring. Theophylline dilates bronchial tubes.

drugs whose actions have unwanted sedative effects or are improved by the presence of a stimulant.

Caffeine remains in the bloodstream and effects its action about twice as long as does nicotine. Thus people generally drink only a few cups of coffee a day whereas many smokers consume 20–40 cigarettes over the same period. Many people avoid drinking coffee in the evening since their sleep is disturbed by the long-lasting stimulant effects of the caffeine. There is some evidence that some people metabolize caffeine more slowly than others, and hence their sleep is more disturbed by it. Over time, people become habituated to some of the effects of caffeine but not necessarily to others, such as its ability to delay and shorten sleep.

■ **Chemistry in Your Home:** How many cups of coffee do you *really* drink per day?

Although coffee is the main source of caffeine for most North Americans, other products can be significant sources, as you can see from the table below. For this activity, record your consumption of coffee, tea, colas, and the other products listed in the table, each day for several days in a row. For each day, add up your total caffeine intake using the approximate value listed. To calculate your real daily coffee cup intake, divide each daily total by 135, which is the caffeine content in standard cups of regular coffee. Is your daily intake lower or higher than you expected? Compare your results to those of other members of the class.

Item	Approximate caffeine content, in milligrams*
Coffee, gourmet (16 oz)	550
Coffee, gourmet (12 oz)	375
Coffee, drip or gourmet (8 oz)	250
Maximum Strength NoDoz (1); Vivarin (1)	200
Weight-control aids	200–275
Coffee, percolated (8 oz)	175
Coffee, nongourmet (8 oz)	135
Excedrin (2)	130
Regular Strength NoDoz (1); Caffedrine (1)	100
Coffee, instant (8 oz)	95**
Caffeinated water (Edge 2-0, 8 oz)	70
Jolt Cola (12 oz)	70
Anacin (2)	65
Colas, caffeinated (12 oz)	50**
Tea, brewed (8 oz)	50**
Milk chocolate (8 oz)	40**
Café au lait (8 oz); cappuccino (8 oz); espresso (1 oz)	35**
Coffee, decaf (8 oz); cocoa (8 oz)	5

* Data from Center for Science in the Public Interest and other sources.
** The content varies significantly between different brands or methods of preparation.

Caffeine is broken down in the liver. Its lethal dose is about 10 grams, so you would have to drink about 100 cups of coffee quickly to achieve the lethal level.

The caffeine is removed from decaffeinated coffee by mixing it with a solvent such as carbon dioxide that has been put under high pressure. The caffeine preferentially dissolves in the solvent, which is then separated from the coffee. The CO_2 evaporates into the air, leaving no residue in the coffee.

8.5 Decomposing substances are characterized by their half-life periods

Most substances, including stimulants, that gradually decompose inside or outside the body by being converted into other compounds do so at a rate that gradually slows over time as the substance is used up. The rate of the decomposition is defined as the number of molecules that decompose per second. Instead of discussing rates, however, it is useful to talk instead about the **half-life period,** or just the half-life. *The half-life period of a substance is the length of time required for half the total amount of a substance to decompose.* The half-life is a constant span of time that is different for every substance.

Note that 100% of a substance does *not* decompose in two half-life periods! Because the speed of decomposition slows as time goes on, another full half-life period is required for *half* the *residual* amount to disappear, so that after two half-life periods have elapsed, one-quarter of the original amount remains. Similarly, one-half of the residual one-quarter, or one-eighth, still remains after the third half-life period has passed. These fractional amounts are summarized in the following table and are plotted in Figure 8.6:

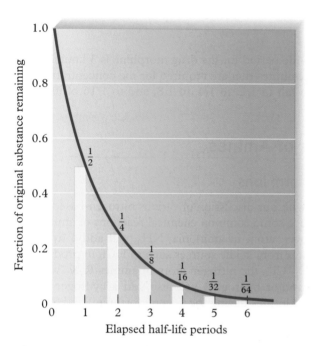

Figure 8.6 Half-life period. Over each half-life period, the amount of a substance declines by one-half.

The half-life period of caffeine is doubled for women taking oral contraceptives, but halved for smokers.

Time elapsed	Fraction of original amount remaining
1 half-life period	1/2
2 half-life periods	1/4 ($\frac{1}{2} \times \frac{1}{2}$)
3 half-life periods	1/8 ($\frac{1}{2} \times \frac{1}{2} \times \frac{1}{2}$)
4 half-life periods	1/16 ($\frac{1}{2} \times \frac{1}{2} \times \frac{1}{2} \times \frac{1}{2}$)

Worked Example: Deducing the remaining fraction of a substance

Caffeine has a half-life period of about 4 hours. What fraction of the caffeine from a cup of coffee still remains in your body 8 hours after drinking the liquid? How much still remains after 12 hours? How long does it take for the concentration to fall to 1/32 of the original amount?

Solution: Since caffeine's half-life is 4 hours, 8 hours corresponds to two half-life periods for it. Half remains after one half-life, and one half of this, or 1/4, remains after the next half-life period has elapsed. We therefore conclude that one-quarter of the caffeine will remain after 8 hours. Since 12 hours is three half-life periods, the fraction remaining is half of 1/4, or 1/8. Finally, since 1/32 equals ($\frac{1}{2} \times \frac{1}{2} \times \frac{1}{2} \times \frac{1}{2} \times \frac{1}{2}$)— that is, one-half five times in a multiplication string—it requires five half-lives, or 5×4 hours = 20 hours, for the level to fall to 1/32 times of the original amount.

Exercise 8.1

Given that the half-life period for the drug morphine is 3 hours, calculate the number of minutes that are required for the concentration of a dose of the drug to fall to 1/2, to 1/4, to 1/8, and to 1/16 of its initial value.

The Reactions of Amines

8.6 Amines can form ions

Before we continue with our discussion of various nitrogen-based drugs, it is useful to introduce some common chemical reactions of amines.

Just as the nitrogen atom in ammonia, NH_3, will add a hydrogen ion, H^+, that originates from acidic sources such as HCl, to produce the ionic compound $NH_4^+Cl^-$, the nitrogen atom in amines R_3N (where R are organic groups and/or hydrogen) will also add a hydrogen ion to produce ionic hydrochlorides having the general formula $R_3NH^+Cl^-$.

The nitrogen forms a covalent single bond to the extracted hydrogen. This bond differs from normal covalent single bonds in that it uses the *pair* of (nitrogen's) electrons that is nonbonding in the neutral molecule in order to supply *both* electrons, since the H^+ itself has no electrons to contribute.

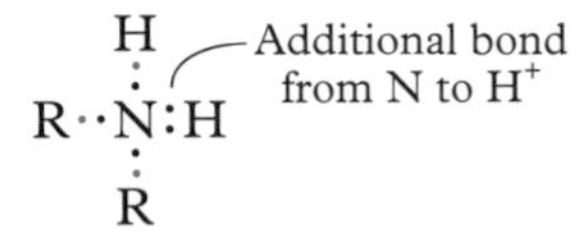

Nitrogen-containing ion carries a +1 charge

The solubility in water of some amine medications is not high enough for them to be available in intravenous form. However, the corresponding ionic compound containing the amine ion R_3NH^+ is generally very water soluble since the ion is strongly stabilized by interactions with the nonbonding electrons on the oxygen atoms of the water molecules that surround it. Consequently, some intravenous medications that consist of nitrogen-containing molecules are supplied as a water solution of a salt that contains the R_3NH^+ ion paired with an anion such as chloride.

Most oral medications that consist of free amines are converted into salts in the stomach, since the concentration there of the H^+ ion is high. However, many amines are liquids, whereas the salts are inevitably solids, so for convenience in packaging some medications are supplied as the salts.

8.7 Cocaine is a highly addictive amine

Cocaine is a powerful nitrogen-containing compound that acts both as a stimulant and as a depressant (see Figure 8.7). This amine is extracted from the crushed leaves of the *Erythroxylum coca* plant by treating the leaves with aqueous hydrochloric acid. The procedure converts the amine, B, which is not very water soluble, into its ionic hydrochloride form, BH^+Cl^-, which is highly water soluble and is therefore transferred efficiently from the leaves into the watery acid. Excess water is then evaporated from the solution, and the solid white residue that remains is the salt *cocaine hydrochloride*.

The ionic form of cocaine is snorted by users because it dissolves in the watery mucous membranes of the nose and quickly enters the bloodstream, where it is transported to the brain. Cocaine hydrochloride can also be dissolved in water and injected ("mainlined") into the bloodstream directly, with a faster and stronger effect because the concentration is increased.

Cocaine

Figure 8.7 Cocaine. Derived from the coca plant, this amine finds its way to the street in powder and rock forms. (Larger photo by Gregory G. Dimijian/Photo Researchers; inset photo by Shahn Kermani/Liaison/Getty)

The Reactions of Amines

Cocaine hydrochloride is an ionic solid and consequently has a high melting point (about 200°C), and so it does not evaporate much upon moderate heating and cannot be smoked; indeed, it decomposes in the burning process. However, in recent years, the smoking of "crack cocaine" has become widespread. Crack consists of the free base of cocaine rather than its hydrochloride salt. The base is produced from the salt by reacting it with a substance such as sodium bicarbonate, $Na^+HCO_3^-$, extracting H^+ ion from it as in the reaction:

$$Na^+HCO_3^- + BH^+Cl^- \rightarrow Na^+ + B + H_2CO_3 + Cl^-$$
$$Na^+ + Cl^- \rightarrow Na^+Cl^-$$

(South American Indians achieve the same result by adding lime to the coca leaves that they chew.) The freebase form of cocaine is a white powder having a relatively low melting point (98°C). Heat from smoking the powder produces a volatile liquid that vaporizes and, upon its inhalation into the lungs, is quickly absorbed into the bloodstream.

Cocaine is highly addictive. It operates biochemically by affecting the level of the neurotransmitter *dopamine*:

Dopamine

Parkinson's disease occurs when the concentration of dopamine in the brain is too low. The dopamine derivative called L-dopa is administered to Parkinson's patients since it is converted to dopamine in the brain.

Many crack users smoke the substance at 15–20 minute intervals for days at a time, without eating or sleeping.

Produced in the brain, dopamine helps regulate movement, mood, and attention. Cocaine blocks the reabsorption of dopamine onto sites at its original transmission nerve fiber, thereby increasing the dopamine concentration in the synapse region. This in turn leads to the frequent attachment of dopamine onto its receptor sites and causes their stimulation. The overstimulation of these receptors is what gives the drug user a "high" sensation. Thus, cocaine is a powerful stimulant of the central nervous system. It gives the user a temporary sense of power, euphoria, and confidence. However, once this effect wears off (within about 20 minutes), severe depression follows (since dopamine production dropped due to the abundance of the substance in the brain) and the addicts crave another dose to reverse these effects. Biochemically, what occurs in the down phase is that the continuing presence of the dopamine molecules in the synapse region allows these molecules to be gradually destroyed by enzymes. The fall of dopamine levels is experienced by the user as a severe depression, relieved only by the consumption of more cocaine. Cocaine's half-life before it is metabolized is only about 45 minutes, so both the dopamine and the drug are destroyed in a short period of time. The cycles of euphoria and depression lead to an addiction to the drug. Long-term users of cocaine often become paranoid and have a high risk of death from heart failure.

Chapter 8: Condensation Polymers, Especially Those Containing Nitrogen

The hallucinogenic drug *mescaline* contains these same structural features.

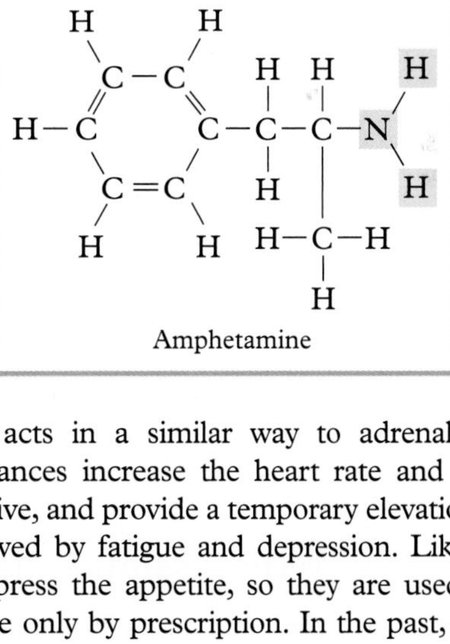

Cocaine has had a long and controversial history. Native people in the countries where the *Erythroxylum coca* plant is indigenous (Bolivia, Peru, Ecuador) have traditionally chewed its leaves for mystical experiences and religious reasons as well as to obtain its stimulant effects. Cocaine also acts as a local (topical) anaesthetic and was used for that purpose not only by indigenous peoples but even by dentists in developed countries during the 19th century. In his early research, Sigmund Freud advocated using cocaine to treat depression in patients, and he used it himself.

Exercise 8.2

Consider the powerful drug *morphine*, which we will discuss later in this chapter. What part of its name tells you that it is an amine? Would you expect it to be more or less soluble in water than its chloride salt? From your answer, explain why morphine as a drug is always supplied as a salt rather than as the free base.

8.8 Amphetamine and its relatives are stimulants

Another common stimulant is *amphetamine,* an amine (note the latter part of the name) in which the —NH$_2$ group is bonded to a short, branched carbon chain that terminates with a benzene ring. The structure of amphetamine is quite similar to that of adrenaline—both have a benzene ring bonded to a two-carbon chain that is connected to an amino group.

Amphetamine

Amphetamine acts in a similar way to adrenaline but is longer-lasting. Both substances increase the heart rate and one's wakefulness, energy level, and drive, and provide a temporary elevation in mood, though this period is followed by fatigue and depression. Like other stimulants, amphetamines suppress the appetite, so they are used in weight-control drugs, now available only by prescription. In the past, Benzedrine (which gave rise to its street name *bennies*) was widely used for this purpose.

Another amine stimulant, *methamphetamine* (*meth* or *speed*) has the same molecule as amphetamine but with a methyl group, —CH$_3$, replacing one of the two hydrogen atoms of the —NH$_2$ unit.

The Reactions of Amines

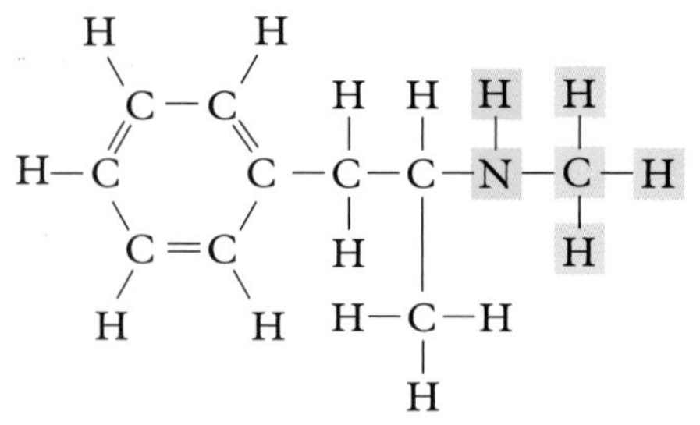

Methamphetamine

Methamphetamine has a greater psychological effect than amphetamine itself. Because it is slowly metabolized, its half-life in the body is about 12 hours, so its effects last much longer than do those of cocaine. Methamphetamine is highly addictive and can have a number of serious effects including anxiety, paranoia, high blood pressure, and breathing problems. *Ecstasy,* a drug popular at all-night "rave" parties, is a derivative of methamphetamine called <u>m</u>ethylene<u>d</u>ioxy<u>m</u>ethamphet<u>a</u>mine, or MDMA. It causes brain cells to release almost all their serotonin at once, thus, the cells reabsorb the serotonin very slowly. As a result of this chemical process, users report feeling serene, calm, and emotionally close to others. In addition to determining one's mood, serotonin also manages the body's temperature, so some users of ecstasy experience dehydration. The high body temperatures associated with some ecstasy users are due both to the effects of the drug itself and to vigorous dancing in hot, humid atmosphere of dance clubs with insufficient intake of liquid. The users are typically unaware that they are overheating. However, a few ecstasy users recognized these potential dangers and drank so much water that they actually over-hydrated, causing their brain tissue to swell. Another danger of ecstasy use is that the substance may contain little or no actual MDMA, but may be contaminated with other drugs or dangerous by-products in its preparation. Of course, these dangers are always present for ecstasy users—they needn't be in a dance club to suffer the ill effects of dehydration or contamination.

Amphetamine, methamphetamine, and ecstasy are normally available as their salts, since in that form they are water soluble and stable. The salt of methamphetamine is sufficiently volatile that it is vaporized upon heating. Illegal use of these substances has become a serious problem in many communities.

▶ **Discussion Point:** Should all stimulant drugs be legalized?

Caffeine and nicotine are mild stimulants that are legal, though their use is frowned upon by some religious groups. However, the stimulant cocaine is now illegal, as are heroin, morphine, and amphetamines. Is it appropriate for governments to outlaw certain stimulants while legalizing other ones? Develop three points supporting each side of this debate.

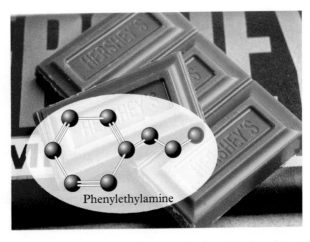

Figure 8.8 Phenylethylamine, an amine naturally found in chocolate. (Photo by George Semple for W. H. Freeman and Company)

Another, but less harmful, relative of amphetamine is *phenylethylamine*. In phenylethylamine molecules, the methyl group bonded to the carbon of the short chain is replaced by a hydrogen atom (see Figure 8.8). This substance is reputed to be a mild stimulant and to occur at small concentrations in chocolate.

8.9 Most painkillers are nitrogen-containing organic compounds

Now that we have covered some of the organic chemistry of nitrogen and oxygen, we can consider the chemistry of the best-known painkillers: acetaminophen, acetylsalicylic acid, and ibuprofen.

Acetaminophen, known commercially as Tylenol, Panadol, and Datril, conceptually combines acetic acid with an amine consisting of a nitrogen group bonded to a benzene ring (see Figure 8.9a). It is an *analgesic*—that is, it relieves pain—and an *antipyretic*—that is, it relieves fever. Acetaminophen is absorbed into the bloodstream via the lining of the stomach and consequently takes about 10 minutes to become effective—faster than aspirin, which is mainly absorbed further along in the digestive system.

Acetylsalicylic acid, ASA, commonly known as *aspirin*, contains both a carboxylic acid group and an ester group of acetic acid bonded to a benzene ring (but no nitrogen) (see Figure 8.9b). In addition to being an analgesic and an antipyretic, aspirin also is an *anti-inflammatory* agent in that it reduces inflammation, such as that which accompanies arthritis. Aspirin relieves pain and inflammation by inhibiting the body's production of *prostaglandins*, a group of compounds that help the brain to experience pain and the body to produce inflammation. Aspirin also reduces the ability of blood to clot, and so it can induce gastrointestinal bleeding in some people and increases the possibility of a type of stroke,

Oil of wintergreen is the methyl ester of salicylic acid, and is used externally to relieve pain.

The Reactions of Amines

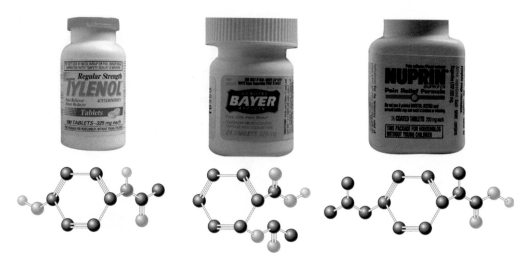

Figure 8.9 Three common over-the-counter painkillers. (Photos by George Semple for W. H. Freeman and Company)

which is excessive bleeding from ruptured blood vessels in the brain. On the other hand, the anticlotting ability of aspirin is used to protect people from heart disease and other types of stroke. Regular small doses of aspirin prevent heart attacks and strokes by suppressing the formation of blood clots in clogged arteries.

Aspirin was developed as a safer version of a natural pain and fever remedy that had been known for millennia. A "tea," prepared by boiling the bark of a white willow tree in water, had been used in many traditional societies to reduce fever. In the 1800s it was analyzed by chemists, and the active ingredient was found to be *salicylic acid*.

$$\underset{\text{Salicylic acid}}{\underset{|}{\overset{O}{\overset{\|}{\underset{OH}{C-OH}}}}}$$

Once isolated, the acid could be prepared synthetically and used to greater effect. Unfortunately, it was so harsh on the stomach that it caused nausea as a side effect. The chemist Felix Hoffmann converted the —OH group on the benzene ring into an ester of acetic acid, giving acetylsalicylic acid, a substance with a much lesser (though not zero) tendency to irritate the stomach. Once aspirin has been ingested and has reached the site of its action, however, the reaction is reversed and the salicyclic acid is re-formed, ready to execute its positive action! In the century since its discovery, people have consumed about one trillion (10^{12}) aspirin tablets and currently consume 50 billion annually.

The third common painkiller is *ibuprofen,* known commercially as Advil, Motrin, Midol, and Nuprin (see Figure 8.9c). Like aspirin, it

Chapter 8: Condensation Polymers, Especially Those Containing Nitrogen

contains a carboxylic acid group and like both ASA and acetamino-phen, a benzene ring. Ibuprofen is an anti-inflammatory as well as an analgesic and an antipyretic agent. It is preferred to aspirin since it is less upsetting to the stomach. For this reason, and because it is very effective against premenstrual and muscle-related pain, it is sometimes used by athletes before races. Regular consumption of ibuprofen, aspirin, and other members of the family called NSAIDs, or non-steroidal anti-inflammatory drugs, have also been found to decrease the incidence of certain cancers, including those of the colon, breast, mouth, and throat, but may increase the incidence of pancreatic cancer.

Exercise 8.3

The ending of the trade names Tylenol and Midol may be taken to mean that acetaminophen and ibuprofen are alcohols. Inspect Figure 8.9 and discover if they are indeed alcohols.

For pain that is not conquered by the three mild analgesics discussed above, physicians often resort to morphine or to substances derived from it. *Morphine,* named after *Morpheus,* the Greek god of dreams, is a complicated amine (see Figure 8.10a) that is a component (~10%) of opium, the substance extracted from unripened seed pods of opium poppies. Morphine acts as a painkiller by blocking nerve receptors that register the sensation of pain. It was easily available, without a prescription, in the 19th century, and was widely used as a painkiller by soldiers in the U.S. Civil War. Unfortunately, morphine and its derivatives are addictive, and many of these soldiers became addicted to it. Since morphine is insoluble in water, it is usually injected for medical reasons as a solution of its hydrochloride salt. Coated tablets containing ionic morphine are now also available.

Scientists have discovered that one or both of the two —OH groups in morphine can be changed without the substance losing all of its analgesic properties. In *codeine* (see Figure 8.10b), one of the —OH groups is converted to —OCH$_3$. This change makes the substance much less addictive but greatly reduces its analgesic power. In *heroin* (see Figure 8.10c), both —OH groups are converted to acetate esters. Since it reaches the brain more quickly, heroin is a more powerful painkiller

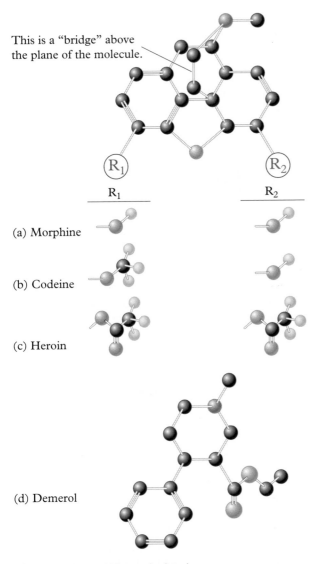

This is a "bridge" above the plane of the molecule.

R$_1$	R$_2$
(a) Morphine	
(b) Codeine	
(c) Heroin	
(d) Demerol	

Figure 8.10 Morphine and related drugs. The painkilling properties of morphine, codeine, heroin, and Demerol are due to similar structural features.

The Reactions of Amines

than morphine. In Europe heroin is used to relieve the suffering of terminal cancer patients, though in the United States it currently has no legally sanctioned medical use. Heroin is even more addictive than is morphine. The drug Demerol (see Figure 8.10d) is an amine that contains a portion of morphine's structure—namely, the nitrogen atom bound on one side to a methyl group and on the other to a ring—and is a painkiller, so we conclude that this structural component gives rise to much of the painkilling ability of all such substances. Because the smaller structure in Demerol is much easier to synthesize than that in morphine, it can be made more cheaply. It is often used to alleviate pain before surgery and childbirth, and has the added bonus of relieving the anxiety of a patient before an operation.

Exercise 8.4

Compare the structure of Demerol (Figure 8.10d) to that of cocaine (Figure 8.7). What structural features do the two molecules have in common?

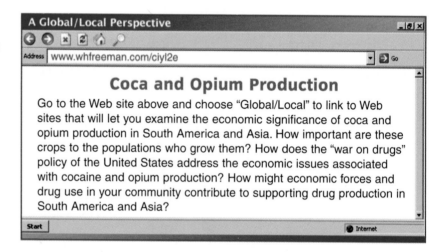

8.10 Amides are analogs of acids and esters

Recall (from Chapter 6) that the presence of an —OH group bonded to the carbon of the carbonyl group (C=O) produces molecules called carboxylic acids and that the presence instead of an —OR group bonded to the carbonyl carbon produces molecules called esters. Analogously, the presence of an amino group bonded to the carbonyl carbon gives rise to molecules known as **amides.** As in amines, the nitrogen can be bonded to two, one, or zero hydrogen atoms, with zero, one, or two carbon chains or rings bringing the number of bonds formed by nitrogen up to a total of three:

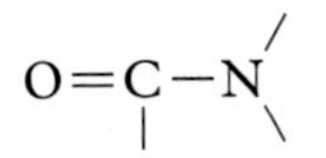

$$O{=}C{-}N\overset{\diagup}{\underset{\diagdown}{\,}}$$

Amide structure

Exercise 8.5

By inspection of its structure in Figure 8.5, decide whether or not caffeine is an amide. What about amphetamine (see page 305) and acetaminophen (Figure 8.9a)?

Again, recall from Chapter 6 that carboxylic acids react with alcohols to produce esters by combining the acid's —OH group with the hydrogen from the alcohol's —OH group to expel water. We could conceive of an analogous reaction in which amides are produced when a carboxylic acid is heated with ammonia or with an amine having at least one H atom. The hydrogen from the amine and the —OH group from the acid would combine to form water, and the carbon of the C=O group and the nitrogen of the amine would become singly bonded:

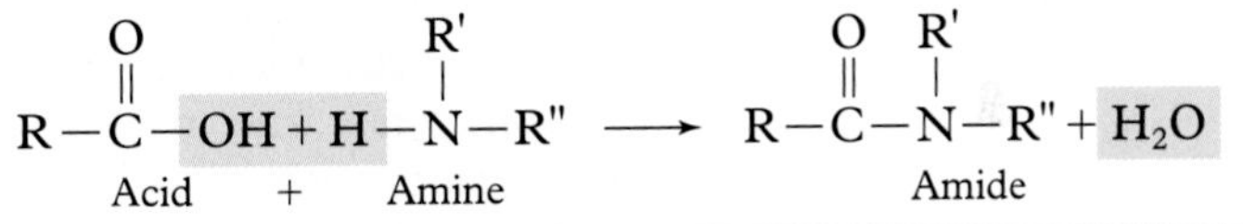

$$
\underset{\text{Acid}}{R-\overset{\overset{\displaystyle O}{\|}}{C}-OH} + \underset{\text{Amine}}{H-\overset{\overset{\displaystyle R'}{|}}{N}-R''} \longrightarrow \underset{\text{Amide}}{R-\overset{\overset{\displaystyle O}{\|}}{C}-\overset{\overset{\displaystyle R'}{|}}{N}-R''} + H_2O
$$

Note that R, R', and R" here can be hydrogen atoms and/or carbon chains or rings; the primes are used to signify the fact that the R groups may differ from one another. (This is not the only reaction that occurs when an acid is heated with an amine, and so it is not a practical way to produce amides.) This reaction can be reversed by adding water to an amide and thereby reconstituting the acid and the amine, but this occurs only in the presence of an acid or a base.

As we'll see in Chapter 9, the nutritional substances called proteins are amides. Several commercially important polymers, such as nylon, as discussed in section 8.11 below.

Condensation Polymers

8.11 Nylon is a polyamide

The ability of a carboxylic acid, which we shall refer to here as A, to combine with an amine, B, to produce the larger amide unit A-B after elimination of H_2O (as explained above) can be exploited to produce long-chain polymers of the type · · · · -A-B-A-B-A-B- · · · · . Such polymers are called **condensation polymers** since a molecule, here water, is eliminated by condensation when they form. Clearly, in order for the unit A to bond to *two* groups B, there must be *two* acid groups originally on A, and so it must be a *diacid*. Similarly for B to bond to two A groups, there must be *two* amino nitrogen groups originally on B, so it must be a *diamine*. The elimination of H_2O lengthens the chain by one A-B combination each time an acid group reacts with an amide group, as illustrated in Figure 8.11.

Exercise 8.6

Identify the repeat unit in the polymer structure shown in Figure 8.11, and enclose it in square brackets.

Polymers of the type described above are called **polyamides** because the internal group, repeated over and over, is an amide:

$$\begin{array}{c} O \\ \parallel \quad \mid \\ -C-N- \end{array}$$

A commercially important polyamide is *nylon*, in which R (see Figure 8.11) is the unbranched four-carbon chain $-CH_2-CH_2-CH_2-CH_2-$ and R' is the unbranched six-carbon chain $-CH_2-CH_2-CH_2-CH_2-CH_2-CH_2-$ which was attached to two $-NH_2$ groups before the condensation reaction. Since both A and B contain six carbons, the polymer is sometimes known as nylon-6,6 to distinguish it from analogous polyamides such as nylon-6,10, which are also useful and have slightly different physical properties. In the nylon numbering system, the number of carbons in the diamine is listed first.

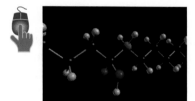

View molecular structures of nylon and nylon chains and the formation of an amide bond in nylon-6,6 at Chapter 8: Visualizations: Media Link 2.

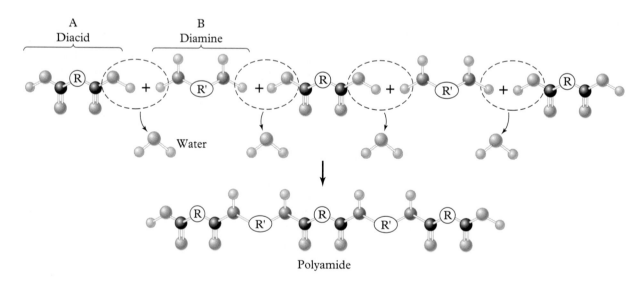

Figure 8.11 Creation of a polyamide. A diacid (A) and a diamine (B) combine to produce a polyamide. R and R' indicate different numbers and types of carbon-containing functional groups or carbon chains.

Chapter 8: Condensation Polymers, Especially Those Containing Nitrogen

Worked Example: Deducing the structure of a nylon

Draw the structure of the polyamide called nylon-6,10 and bracket its repeat unit.

Solution: Since the first number after the word *nylon* gives the number of carbons in the diamine chain, there must be six carbons joining the two amino units. The second number of the pair refers to the *total* number of carbon atoms in the diacid. Since two of these carbons correspond to the

$$\begin{matrix} & O \\ & \| \\ -&C-OH \end{matrix}$$

units, there must be eight carbons in the unbranched chain between the

$$\begin{matrix} & O \\ & \| \\ -&C-OH \end{matrix}$$

units. Thus we write an acid unit, minus the —OH, at each end of an eight-carbon chain and attach it to an —NH— unit with a six-carbon chain terminating in another —NH— unit that is bonded to the —CO— of the next diacid, etc.:

$$\begin{bmatrix} \overset{\displaystyle O}{\overset{\|}{C}}-C-C-C-C-C-C-C-C-\overset{\displaystyle O}{\overset{\|}{C}}-N-C-C-C-C-C-C-N \end{bmatrix}\overset{\displaystyle O}{\overset{\|}{C}}-C-$$

Nylon-6,10 (skeletal)

Exercise 8.7

Draw the skeletal molecular structure of nylon-8,8 and bracket its repeat unit.

In the nylon structures, one hydrogen atom is present on each nitrogen atom in the polyamide chain, since the synthesis begins with molecules containing —NH$_2$ units and only one hydrogen atom is removed from each nitrogen atom. The polymer chains in nylon are oriented in parallel fashion such that the hydrogen atom of each —NH— unit can form a hydrogen bond with the oxygen atom of a C=O bond on an adjacent chain. As a consequence of the extensive hydrogen bonding that occurs between neighboring ones, the chains pack well together and form a strong material (see Figure 8.12).

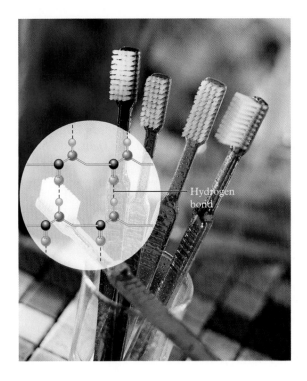
Hydrogen bond

Figure 8.12 The parallel polymer chains of nylon. Nylon is water resistant and strong because of its molecular structure, making it a good choice for toothbrush bristles.
(Photo by Royalty-Free/Corbis)

Condensation Polymers

Kevlar, an even stronger material than nylon, is produced when a benzene ring rather than a carbon chain connects the two —NH₂ groups in the diamine and the two acid carbons in the diacid. Fibers made from this polymer are so strong that they are used to make bulletproof vests and helmets, which have saved the lives of many police officers around the world (see Figure 8.13).

Exercise 8.8

Draw the skeletal structure of Kevlar.

Before leaving the subject of nylon, it is worth noting that a minor variant on nylon-6,6 can be produced by polymerizing an *amino acid,* which is a molecule that contains both an amino group *and* an acid group joined by a carbon chain. For example, if there is a five-carbon chain joining the two groups, the amino acid has the following structure:

Figure 8.13 Garments made of Kevlar. This particularly strong polyamide is used in protective equipment such as this bullet-proof vest worn by a police dog. (Susan Ragan/Reuters NewMedia/Corbis)

If the acid —OH of one molecule reacts with a hydrogen of an amino group on an adjacent molecule, producing water, the remainder of the two molecules bond together through the carbon and nitrogen. The same reaction can occur with still another amino acid molecule using the NH₂ at the left side or the C—OH group at the right side of the joined molecule:

Chapter 8: Condensation Polymers, Especially Those Containing Nitrogen

The polyamide that is produced by polymerizing the above amino acid is called nylon-6 and in fact was the first type of nylon made:

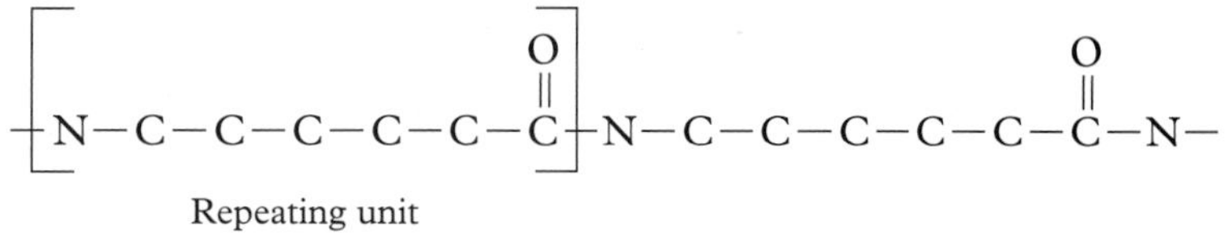

Repeating unit

Nylon-6 (skeletal)

The name is appropriate since it is a single type of six-carbon unit that is repeated over and over again in the polymer chain.

8.12 Polyesters, plastics, and films

Nylon was developed by the research group headed by Wallace Carothers at the DuPont Corporation of Wilmington, Delaware, in the 1930s. Carothers also synthesized A-B-A-B-A condensation copolymers analogous to the polyamides in which a dialcohol, or *diol,* is reacted with the diacid, resulting in the elimination of water and the creation of an ester group $-\overset{\overset{\text{O}}{\|}}{\text{C}}-\text{O}$ at every point of linkage of an A to a B unit. This material, known as polyester, has the following general structure:

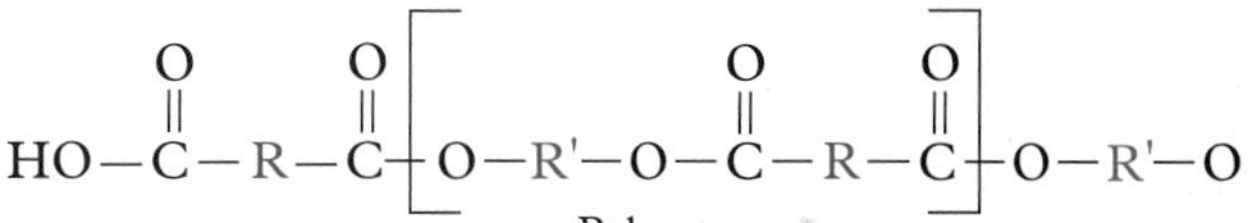

Polyester

The polyesters studied by Carothers were not particularly useful since their glass transition temperatures (discussed in Chapter 5, section 5.6) and melting points were low, and they dissolved readily in dry-cleaning liquids. However, later research in England revealed that when the dialcohol is *ethylene glycol,* $\text{HO}-\text{CH}_2-\text{CH}_2-\text{OH}$, and the diacid has the two COOH groups separated by a benzene ring, corresponding to the substance called *terephthalic acid,* a useful polymer is produced. This polymer, now popularly known as *PET* (or PETE), which stands for *poly(ethylene terephthalate),* is used for a myriad of commercial products. The plastic water bottle that you may have beside you as you read this book is probably made of PET plastic. Plastic PET film is thin and tough, and is used to make some plastic bags that contain foods that will be boiled or frozen. Magnetically coated PET film is Mylar, used to make audio and video recording tapes and computer diskettes.

The repeat units per chain in PET number in the tens of thousands. In contrast to polyamides, there is no possibility of hydrogen bonding

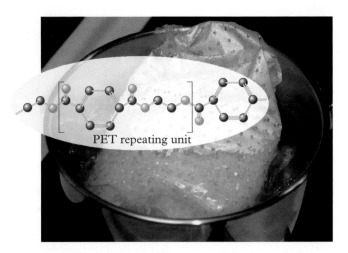

Figure 8.14 PET. The structure of PET in food pouches is not affected by boiling water. (Photo by George Semple for W. H. Freeman and Company)

between adjacent chains in polyesters, since, as you can see in Figure 8.14, there are no hydrogen atoms bonded to nitrogen or oxygen in the latter.

8.13 Polycarbonates are used to make colored plastics

Polycarbonate plastics are a variant on the polyester theme. The diacid in this case is *carbonic acid,* $O=C(OH)_2$, which has two —OH groups on the same carbon atom:

$$
\begin{array}{c}
O \\
\| \\
C \\
H-O \qquad O-H
\end{array}
$$

Carbonic acid

Each of these —OH groups can be combined with the —OH group of a diol to produce two ester linkages with a common C=O group. One diol that is used to produce such plastics is *bis-phenol A,* two phenol units joined by the central carbon of a three-carbon chain:

Recall that phenol is a benzene ring with one —OH substituent. The prefix *bis* means "two."

$$
\text{HO}-\!\!\!\bigcirc\!\!\!-\overset{\overset{\displaystyle CH_3}{|}}{\underset{\underset{\displaystyle CH_3}{|}}{C}}-\!\!\!\bigcirc\!\!\!-\text{OH}
$$

Bis-phenol A

The polycarbonate polymer produced has the following structure:

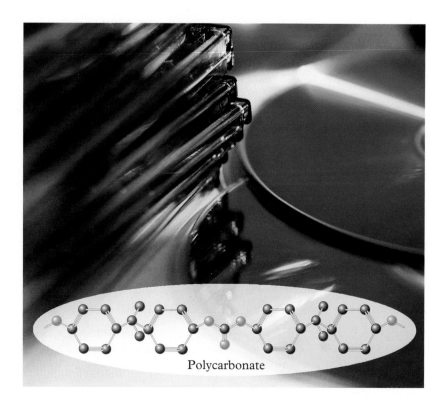

Figure 8.15 Polycarbonate. Polycarbonate polymers produce tough and transparent materials.

(Photo by Royalty-Free/Corbis)

Polycarbonate

Polycarbonate from
bis-phenol A

Polycarbonate is so tough that it is used as a scratch-resistant coating on eyeglasses. Tough, transparent polycarbonate plastics are used to make bottles, CDs, the housings of appliances, protective helmets, and bullet-proof "glass" (see Figure 8.15). Because it is transparent and can easily be brightly colored, polycarbonate has been used for modern products such as the iMac computer.

8.14 Useful fibers are made from condensation polymers

Both polyamides and polyesters find extensive use as fibers. Recall from Chapter 5 that synthetic fibers are formed by squeezing a melted polymer through a small hole in order to form a long, thin stream of viscous liquid. The liquid is stretched as it solidifies, which increases the mechanical strength of the fibers. The fibers formed from nylon have different uses, depending upon their diameter. Very thin ones can be

Figure 8.16 Women waiting in line to buy rationed nylon stockings during World War II. These days, nylon is used not only for hosiery but for many other articles of clothing and other consumer goods. (Bettmann/Corbis)

spun into a thread that is similar to the natural material silk, but is more durable. Nylon stockings were introduced as replacements for the much more expensive silk ones before World War II.

PET plastic can also be extruded into fibers. The fabric produced from PET is called Dacron in North America and Terylene in the United Kingdom, but it is sometimes just called polyester when it is used to make clothes. The polyester fabric is particularly useful in making wash-and-wear clothing. Combining polyester fibers with those of cotton combines the best properties of both fabrics: the comfort of cotton and the easy care of the polyester (see Figure 8.17). Synthetic fibers, including nylon and Dacron, account for about three-quarters of the fabric used currently in North America.

Recently, *polyethylene glycol* has entered the commercial clothing market. Although not actually prepared this way, it corresponds to the condensation polymer obtained by condensing water molecules from ethylene glycol molecules:

Taking It Further

For information and problems on how silicon forms condensation polymers, go to Taking It Further at www.whfreeman.com/ciyl2e

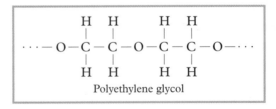

Polyethylene glycol

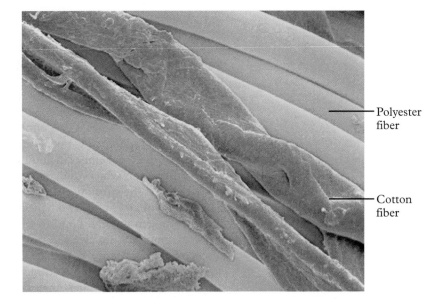

Figure 8.17 A blend of cotton and polyester fibers. In this scanning electron micrograph you can compare the smoothness of the Dacron polyester fibers to the irregular cotton fibers in a blended shirt fabric. [Magnification is 600×.] (Andrew Syred/Science Photo Library/Photo Researchers)

Polyester fiber

Cotton fiber

Because of the ability of the ether oxygen atoms to form extensive hydrogen bonds with water molecules, polyethylene glycol draws moisture and heat away from the wearer by dissolving the water vapor, keeping the person cool and dry. If the weather subsequently turns cool and dry, the material releases the heat and moisture.

Polyurethanes combine structural characteristics of polyesters and polyamides. They are used to create foam for furniture, pillows, etc. Polyurethane fiber is used in carpets, and as Spandex it is used in stretchable clothing such as swim and ski wear and other athletic apparel (see Figure 8.18).

Figure 8.18 Runner in spandex leggings. The polyurethane fibers in spandex clothing make the material lightweight and stretchable. Spandex is always combined with other natural and man-made fibers, and is often used in sports apparel because it is durable and comfortable and enables the body to breathe. (Rubberball/Punchstock)

Condensation Polymers

Summarizing the Main Ideas

Amines correspond to ammonia, NH_3, in which one, two, or all three of the hydrogen atoms have been replaced by chains or rings of carbon atoms. Those having one or two hydrogen atoms attached to the nitrogen can engage in N—H---N hydrogen bonding. Many amines have disgusting odors.

Some amines, such as nicotine, cocaine, and caffeine, are stimulants—they stimulate the body's central nervous system. Such molecules, and other substances including drugs, can be characterized by the half-life period in which they remain active. Only one-half of the original amount of the substance remains active after one half-life period has passed, one-quarter after two half-lives, etc.

A hydrogen ion can bond to the nitrogen of an amine molecule by using the nonbonding electron pair of the latter. The water solubility of the salts formed in this way greatly exceeds that of the amines, so amine-based drugs are often administered in this form.

Amides are the nitrogen analogs of carboxylic acids and esters. They have an amino group attached to the carbon of a C=O group. Amides can be prepared by a condensation reaction. In general, this is a reaction in which a small molecule such as H_2O is condensed out when two other molecules join together. In amides, the hydrogen from an amino group and the OH from a carboxylic acid condense to form water.

Condensation polymers are formed when a condensation reaction joins two units, A and B, together. Owing to the presence of other reactive groups on A and B even after they have combined, an additional condensation reaction occurs at each center and eventually a long polymer chain -A-B-A-B-A-B- is produced. If A is a carboxylic acid, then polyamides such as nylon are produced when B is a diamine, whereas polyesters are obtained when B is a diol. When A is carbonic acid and B is a diol, polycarbonates are produced.

Key Terms

amine	neurotransmitter	amide	polycarbonate
amino group	stimulant	condensation polymer	
diamine	half-life period	polyamide	

Web Sites of Interest

To link to Web sites of interest, go to www.whfreeman.com/ciyl2e, Chapter 8, and select the site you want.

For Further Reading

"Aerial Attack Killing More Than Coca," *The Washington Post,* January 5, 2001, p. A1. This article describes the policy of aerial spraying to kill coca plants in Colombia and its impact on legal crops and local populations.

"Foamy Polymers Hit Goal Right on the Nose," *Science News, 157,* March 4, 2000, p. 149. The article looks at the use of biodegradable polymer foam as scaffolding for cartilage cell growth.

W. Marston, "Wonder Wear," *Discover,* January 2000, pp. 46–48. The article examines and describes the interesting properties of a new generation of synthetic fabrics. Polyethylene glycol fabric is featured prominently in the article, including a look at its antibacterial properties and potential use in medical facilities.

"Now, Nylon Comes in Killer Colors," *Science News, 157,* April 22, 2000, p. 211. The article describes the use of cyclic amines as antibiotic finishes on nylon material.

Review Questions

1. What is an *amine*?

2. How are amines structurally similar to alcohols and ethers?

3. How are the structures of amines and ammonia related?

4. What is a *diamine*?

5. What is a *neurotransmitter*?

6. Identify three neurotransmitters and describe their roles in the body.

7. Describe how nerve impulses are transmitted between nerve cells.

8. Explain how an antidepressant such as Prozac affects the transmission of nerve impulses.

9. How is pyridine structurally similar to benzene?

10. What is a *stimulant*? Explain the effect of a stimulant on the body.

11. What is *nicotine*? Why is it classified as addictive?

12. How does a *nicotine patch* work?

13. What is *caffeine*?

14. Define the term *half-life*.

15. If drug A has a half-life of 1 hour and drug B has a half-life of 10 hours, which drug is 90% eliminated from the body first? Which drug would be taken more frequently? Explain.

16. How does an ammonium ion differ from an ammonia molecule?

17. How do amines bond to H^+?

18. How does *crack* differ from regular cocaine?

19. What is an *analgesic*? An *antipyretic*?

20. Compare the painkilling properties of codeine and aspirin. Which is stronger?

21. What is an *amide*? What is the relationship between an amide and a polyamide?

22. What is a *condensation polymer*? Provide an example.

23. What is a *diacid*? Why can diacids produce polymers when combined with amines?

24. Identify the repeat unit in the nylon polyamide below.

$$\cdots - \overset{\overset{\textstyle O}{\|}}{C} - (CH_2)_5 - \overset{\overset{\textstyle O}{\|}}{C} - \overset{\overset{\textstyle H}{|}}{N} - (CH_2)_{10} - \overset{\overset{\textstyle H}{|}}{N} - \overset{\overset{\textstyle O}{\|}}{C} - (CH_2)_5 - \overset{\overset{\textstyle O}{\|}}{C} - \overset{\overset{\textstyle H}{|}}{N} - (CH_2)_{10} - \overset{\overset{\textstyle H}{|}}{N} -$$

25. What is an *amino acid*?

26. What is a *polyester*?

27. How do interactions between polyester chains differ from interactions between polyamide chains?

28. What kind of chemical reaction produces the polymer PET?

29. How are *polycarbonate* plastics related to polyesters?

30. What properties of polyethylene glycol make it useful for fabrics?

Understanding Concepts

31. Which of the following pair of isomeric amines is likely to have the higher boiling point?
a) $(CH_3CH_2)_3N$ b) $CH_3(CH_2)_5NH_2$

32. Which is likely to have a higher boiling point, an amine with six $-(CH_2)-$ units in the chain or an alcohol with six $-(CH_2)-$ units in the chain? Explain.

33. Why do smokers crave cigarettes more often then coffee drinkers crave coffee?

34. Compare the half-lives of cocaine, caffeine, and nicotine. Which remains in the body longest? Which is 90% eliminated first?

Substance	Half-life (hours)
Cocaine	0.7
Nicotine	2
Caffeine	4

35. If the half-life of a drug is 90 minutes, how long after its administration to a patient would it take for the concentration to fall to 1/4 of its initial value? To 1/16 of its initial value?

36. If the half-life of a substance in your body is 1 hour, what fraction of it still remains in you 2 hours after it is administered? After 3 hours? After 4 hours?

37. The concentration of a drug in a patient had fallen to one-quarter of its initial amount after 3 hours had passed. What is the half-life of the drug?

38. The concentration of a drug in a patient was found to be one-eighth of the original dose some 3 hours after it was administered. What is the half-life period for the drug?

39. Explain how the ratio of carbon atoms to nitrogen atoms in an amine affects its solubility in water.

40. Why is a salt containing a BH^+ ion more soluble in water than the neutral B molecule?

41. Many intravenous (IV) solutions contain dissolved substances that are in their ionic form. Why is this the best way to administer some IV substances?

42. Why can't cocaine hydrochloride be smoked? Why does smoking freebase stimulate the system as quickly as injecting the drug?

43. Compare the biological effects of adrenaline and amphetamines. Why are they similar?

44. Use a molecular perspective to explain why cocaine is addictive.

45. Compare the structures of acetylsalicylic acid, acetaminophen, and ibuprofen. Do you notice any similarities in structures? What properties are associated with any similar areas (e.g., size, polarity, etc.)?

46. Visit the Web site listed at the end of the chapter to compare the biological activities of COX-2 inhibitors with those of the analgesics described in the chapter. Why don't COX-2 inhibitors cause stomach irritation? Are they equally effective in every situation? Explain.

47. Show by a reaction equation why a diacid and a diol can combine to form a polymer, but a monoacid and an alcohol won't polymerize.

48. Draw the structure of the amide formed by polymerization of the following diacid and amine. What is the repeating unit?

$$\underset{\text{Diacid}}{HO-\overset{\overset{\displaystyle O}{\|}}{C}-(CH_2)_4-\overset{\overset{\displaystyle O}{\|}}{C}-OH} \;+\; \underset{\text{Diamine}}{H_2N-(CH_2)_7-NH_2}$$

49. Amino acids can form polymers whereas amines can't. Explain.

Synthesizing Ideas

50. Based upon the different abilities of the isomers $(CH_3)_3N$ and $CH_3CH_2CH_2NH_2$ to form hydrogen bonds to water, which one would you expect to be more water soluble?

51. Can hydrogen bonding occur in pure amines in which all three atoms bonded to the nitrogen are carbon rather than hydrogen? From your answer, predict whether the boiling point of the amine $(CH_3)_3N$ will be significantly higher, significantly lower, or about the same as for its isomer $CH_3CH_2CH_2NH_2$.

52. By reference to Figures 8.7 and 8.9, determine which of the functional groups alcohol, carboxylic acid, ester, amide, and amine are present in a) cocaine and b) in ASA.

53. One of the monomers of the material nylon-6,6 is a six-carbon diamine that is closely related to putrescine and cadaverine, yet we don't smell "death" in nylon fabric. Why do you think we can wear nylon without being surrounded by disgusting odors?

54. Reread the sections in the chapter that discuss Prozac and cocaine and answer the following questions: a) How does each affect neurotransmitters? b) Are their mechanisms of action similar? c) Are they structurally related? d) How does each affect the body? e) Based upon your answers, what features would you look for in a drug that would simulate cocaine activity in the body?

55. Compare the structures of the neurotransmitters and the drugs that block neurotransmitter action shown in the chapter. Identify any feature(s) they have in common. Since each of these molecules either is a neurotransmitter or blocks the action of a neurotransmitter, what do the common features imply about the sites on the nerve cell in which these molecules can bind?

56. Use the structural characteristics of polycarbonates and polyesters to explain the differences in their physical properties.

In this chapter, you will learn:

- how DNA and RNA contain and carry genetic information;
- how RNA translates genetic information from DNA into recipes for proteins;
- about how proteins are produced by your body;
- how proteins are shaped and how their shape effects their function;
- about steroids and sex hormones;
- about dietary proteins, enzymes, lipoproteins, and how your heart is affected by these lipoproteins;
- how genetic mutations happen;
- about genetic diseases and cloning technology;
- how DNA can be used as legal evidence.

What genetic evidence is there that some descendants of the ancient Picts remained in Scotland?

Red hair is an inherited genetic trait, and originated with the ancient Picts. In Roman times, the Picts inhabited Scotland, a country whose redhead population today numbers about 13%, the highest in the world. (Image Source/Getty Images)

The Molecules That Make You What You Are

Nucleic Acids, Proteins, and Hormones

Almost every activity that goes on within your body involves proteins, which are amino acid polymers that can be thousands of units long. The protein molecules in every organism are made according to a coded "instruction book" known as DNA. The DNA instructions are carried by the "messenger" molecule called RNA to sites in the cell where amino acids are assembled into proteins. Plants, animals, people—even bacteria—all have many of the same proteins and carry genetic information coded in the same way on their DNA.

Yet obviously there are enormous differences among the myriad life forms on Earth, so there must also be differences in their DNA to account for them. The diversity of life we see is actually due to small differences in sequences of subunits that make up a part of each DNA molecule. Relatively small chemical differences determine whether an organism is a flower, an ant, or a human being.

In this chapter we will examine both DNA and RNA—their structures and how they work together to translate genetic information and direct protein synthesis. We will then look more closely at the chemical structure of proteins and their role in the body. Finally, we will consider what happens when things go wrong in the genetic process. Let's start by taking a look at the structures and properties of DNA and RNA.

Nucleic Acids: DNA and RNA Structures and Properties

9.1 The basic units of DNA and RNA are nucleotides

DNA and **RNA** are very large molecules called **nucleic acids.** These molecules consist of a sequence of small subunits called **nucleotides.** Nucleotides are made of three simpler components: an acidic phosphate unit, a sugar, and a nitrogen base (see Figure 9.1). Phosphate units, which impart acidity, are derived from phosphoric acid (H_3PO_4) and act as links, joining sugar units via esterlike bonds, which were discussed in Chapter 6. The sugar and phosphate units alternate, forming a long unbranched polymer that is the "backbone" of DNA and RNA (see Figure 9.2). The nitrogen bases carry genetic information. DNA and RNA molecules differ only in the type of sugar and one type of nitrogen base.

Whenever you see this icon in this chapter, go to
www.whfreeman.com/ciyl2e

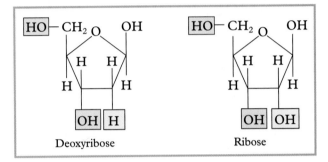

Figure 9.1 An example of a nucleotide.

In DNA, the nucleotide sugar is **deoxyribose,** which has a five-membered ring and in free form has three —OH groups. Two of the —OH groups are used to form ester-like linkages, each with a different phosphoric acid. In RNA, the sugar is **ribose,** which differs from deoxyribose in that it has one more —OH group, which replaces one of the hydrogen atoms in the deoxyribose (shown in blue). In the structures below, the vertical lines represent bonds extending from the carbons, showing the groups that lie above or below the plane of the C_4O ring:

As shown below, each sugar binds to a **nitrogen base** by elimination of an —OH group from the sugar (on the carbon next to the —O— in the ring) and a hydrogen on the nitrogen base to produce a C—N single bond (as occurs in amides and amines):

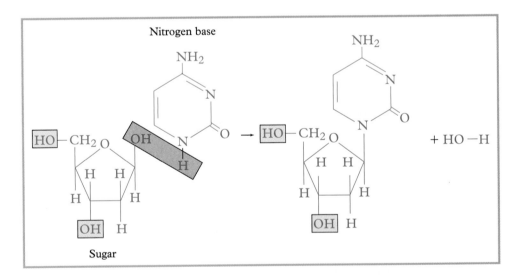

Four different nitrogen bases can occur in each DNA and RNA chain, but they do *not* alternate in any regular, repeating pattern. The specific sequence of the bases in DNA is characteristic of an individual organism.

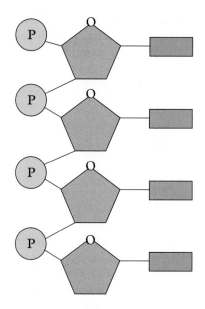

Figure 9.2 A portion of the backbone of DNA and RNA, a phosphate-sugar chain. A blue rectangle represents a nitrogen base, a circle a phosphate group, and a pentagon a sugar.

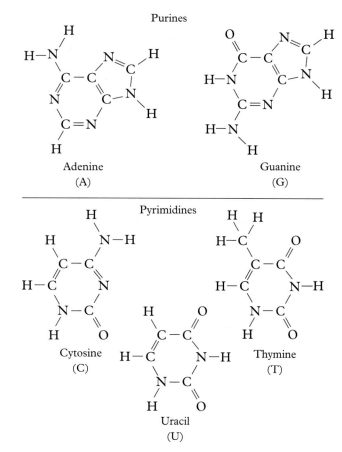

Figure 9.3 DNA and RNA nitrogen bases.

The molecular structures of the nitrogen bases are given in Figure 9.3. Notice that **guanine** and **adenine,** classified as purines, contain two connected rings, whereas **cytosine, thymine,** and **uracil,** classified as pyrimidines, contain only one ring. The bases are generally referred to by one-letter abbreviations:

A for adenine

C for cytosine

G for guanine

T for thymine

U for uracil

Adenine, guanine, and cytosine are found in both DNA and RNA; thymine is found only in DNA and uracil only in RNA. We won't generally be concerned with details of the structure of the nitrogen bases except as they relate to hydrogen bonding between purines and pyrimidines.

Armed with this structural information and the fact that both DNA and RNA are found in the nuclei of cells, we can understand their full names and the source of the three-letter abbreviations commonly used for them:

DeoxyriboNucleic Acid   RiboNucleic Acid

Name of   Name
sugar   of
                                           sugar

9.2 Nucleotides form biological polymers

In discussing polycarbonate plastics in Chapter 8, section 8.13, we saw that carbonic acid could join two other units together in a polymer chain by using both its —OH groups to form ester-like linkages. Nature employs the same technique in the polymers DNA and RNA, where two of the three —OH groups in phosphoric acid, H_3PO_4, join other units into a chain (see Figure 9.4a). One —OH unit of the original acid remains, and so the substance retains some of its acidic properties. The phosphoric acid connects the sugars, which have —OH groups with which the phosphoric acid —OH groups can react to form ester-like linkages by the elimination of H_2O (see Figure 9.4b). Since the nucleotides are joined via condensation reactions, DNA and RNA can be considered biological condensation polymers. Figure 9.5 illustrates a small section of a DNA chain.

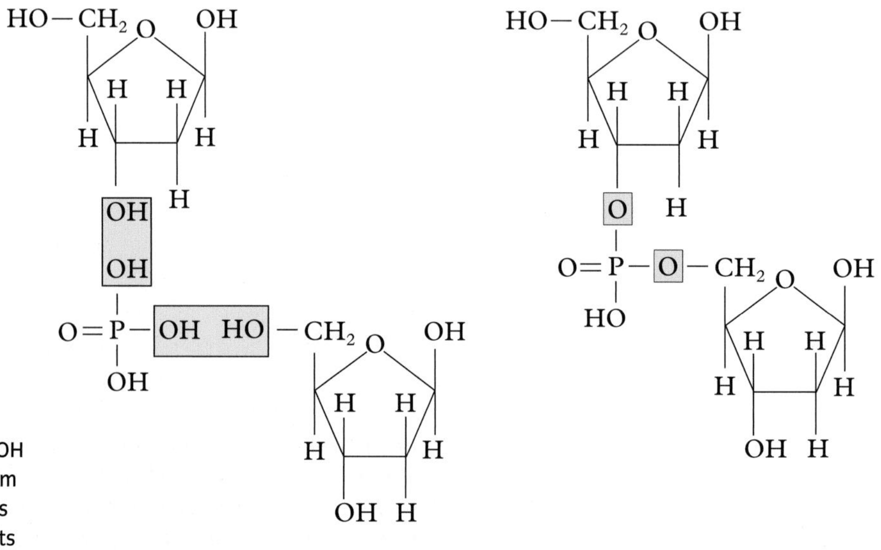

Figure 9.4 Ester-like linkages between nucleotides. (a) The H atoms from H_3PO_4 join with the —OH groups from the sugar units to form water. (b) The remaining —O atoms on the phosphate join the two units together.

(a)   (b)

Figure 9.5 A small section of a DNA strand. (The ATGC sequence of nitrogen bases shown is arbitrary.)

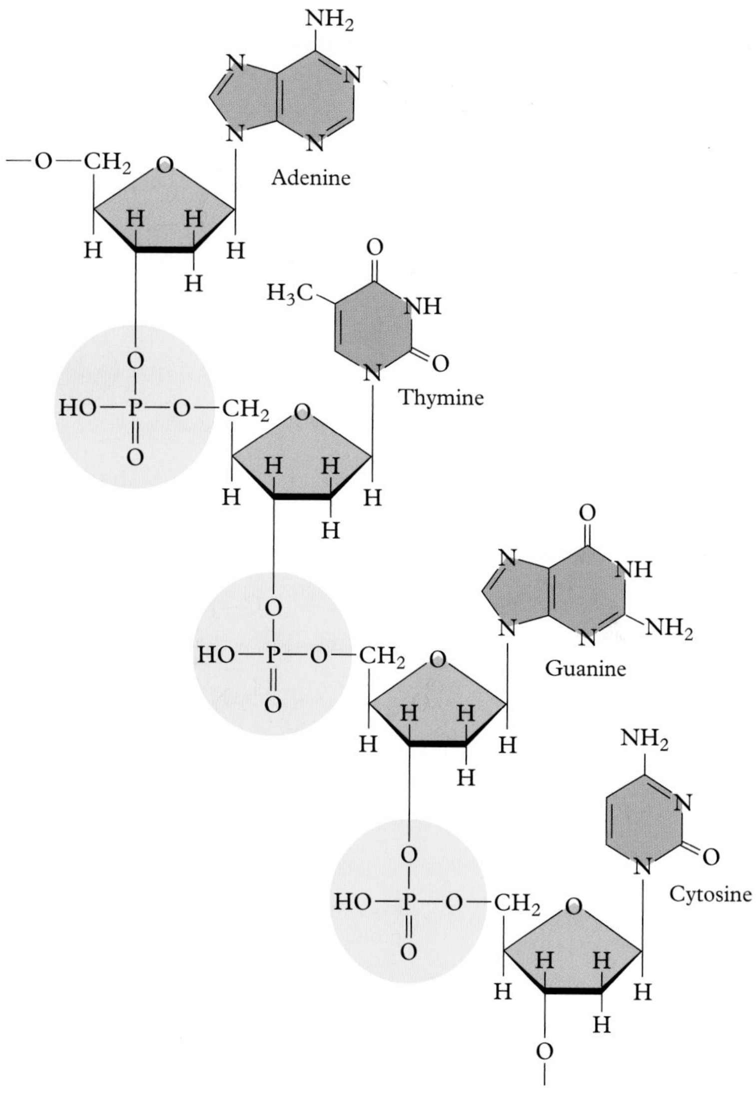

9.3 DNA is a double helix

Thanks to the work done by James Watson, Francis Crick, Maurice Wilkins, and Rosalind Franklin in the 1950s, we know that DNA normally exists not as a single strand of nucleotides, but as a pair of strands wound around each other in a **double helix.** Hydrogen bonding between nitrogen bases that are adjacent to each other on the two strands keeps the structure intact (see Figure 9.6a). The spatial requirements for constructing a stable double helix determine which bases are allowed to bond to one another. *Each of the nitrogen bases can form hydrogen bonds well with only one other type of nitrogen base when two polymer strands are located adjacent to each other and aligned with opposite orientation.* In each hydrogen-bonded combination, one base is a purine and

A helix shape is similar to a coiled spring or Slinky toy.

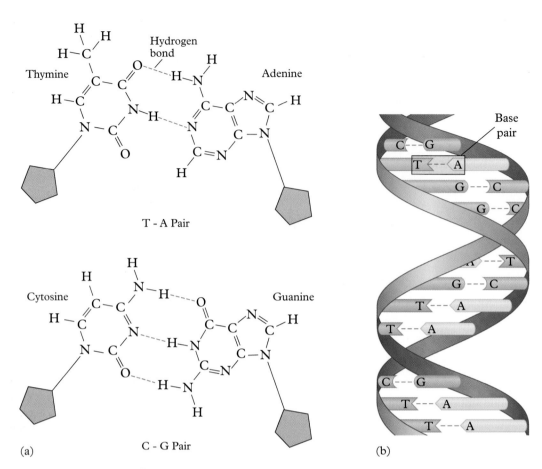

Figure 9.6 (a) Hydrogen bonding (dashed lines) between complementary bases produces (b) the characteristic DNA double helix.

the other is a pyrimidine. For a variety of reasons including the size of the bases and the orientations of the atoms that participate in the hydrogen bonds, guanine only forms strong hydrogen bonds with cytosine, and adenine with thymine or uracil. The two bases in each combination that can hydrogen-bond together are said to be *complementary* to one another and are called base pairs: G and C constitute a base pair, as do A and T, or A and U. This base pairing results in two chains that are complementary to each other in the double-helix configuration (see Figure 9.6b).

The **principle of complementarity** is the basis for the transfer of genetic information. In a process called *replication,* a DNA double helix unwinds and the two strands provide templates along which complementary bases align (A with T, C with G), creating a new DNA double helix. Replication is fast and almost always accurate, assuring that the copied DNA strands contain exactly the same base sequence as the originals (see Figure 9.7).

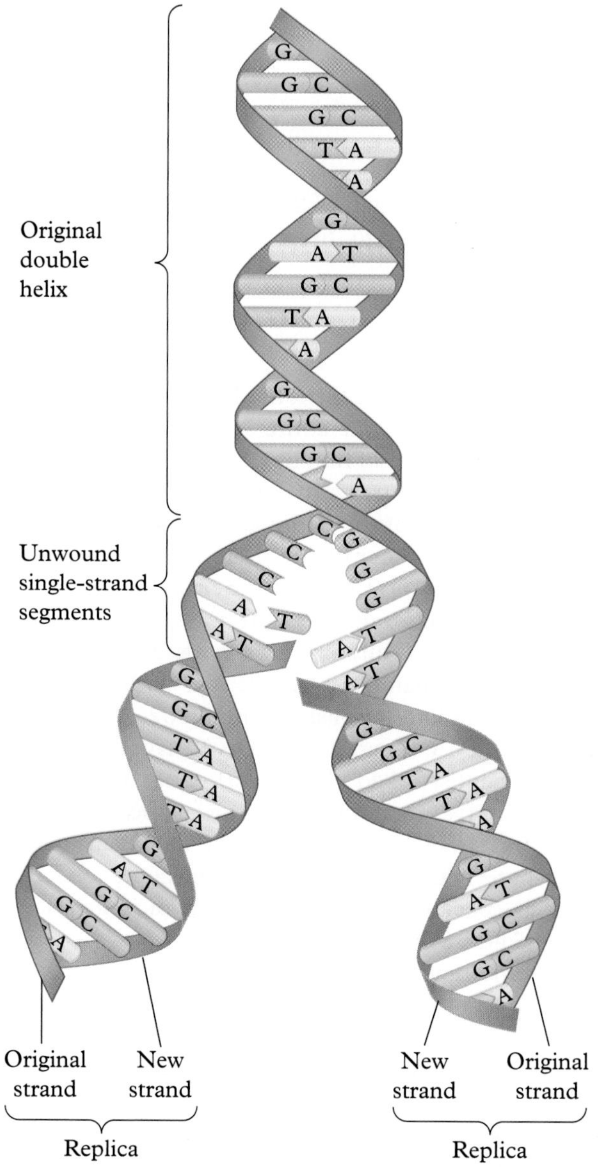

Figure 9.7 Replication. Complementary bases build a new double helix.

Labels in figure:
Original double helix
Unwound single-strand segments
Original strand — New strand — Replica
New strand — Original strand — Replica

DNA: The Genetic Message

9.4 The nucleotide structure of DNA carries information

In English, we use the 26 letters of the alphabet to build words. When words are grouped together into sentences, they convey information as a linear sequence of letters. DNA also has an alphabet that spells words in the same way as human languages, but this alphabet comprises only four letters: A, C, G, T, the four nitrogen bases. However, unlike most written language, there are no spaces or punctuation marks in DNA. All "words" in the DNA language consist of three letters chosen from these four possibilities; there are a total of 64 three-letter combinations—ACG, CGT, GAT, and so on. Information is stored along a strand of DNA in the sequence of triplets of nitrogen bases attached to the sugars. This is the "code" we referred to in the introduction to this chapter.

Of course, the restriction to an alphabet of only four letters and words of only three letters greatly limits the vocabulary of the DNA/RNA language, as compared to English, but on the other hand there are many fewer things to specify. In particular, most of the "words" of the DNA/RNA language each specify a particular amino acid from the set of 20 that are used in living systems. We will look at the way the code works in Section 9.8.

9.5 Specific sequences of nucleotides are genes

Very, very long strands of double-helix DNA are stored within the nucleus of most cells in the form of **chromosomes.** The number of chromosomes in a cell depends upon the species, humans having 46 (23 pairs). The DNA on each chromosome is divided into segments, many, but not all, of which carry specific information. **Genes** are linear sequences of bases that form a major part of the structure of chromosomes and direct the protein-producing activities of the cell. Figure 9.8 summarizes the multilevel structure of the genetic material. Genes are separated from other genes by intergenic regions whose function is as yet unknown. In front of each gene is a regulatory region—a series of bases that switches the gene "on" or "off" (see Figure 9.9). A gene is turned "on" when it is directing the transfer of its base information into RNA and that message is being "translated" into specific proteins. By means of the linear sequence of its nitrogen bases, DNA

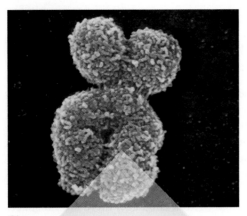

Chromosome

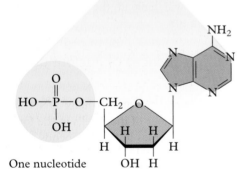

DNA double helix

Gene on a single strand of DNA

GGATATCCAAGC

Nucleotide sequence

One nucleotide

Figure 9.8 The multilevel structure of genetic material. (Biophoto Associates/Photo Researchers)

contains all the genetic coding information for an organism, and in that sense, it is analogous to a library that is full of books. A given gene in DNA specifies the coding for many different types of proteins. An RNA molecule produced from the DNA strand in a specific gene contains information concerning the synthesis of a specific protein, and is thus analogous to a book within the library. About 99.8% of your genes are identical to those of every other human being.

9.6 The Human Genome Project

Begun in 1990, the Human Genome Project is an international effort to determine the entire sequence of bases that make up the complete set of human genes, our genome. The word *genome* combines two words, *gene* and *chromosome*. In 2001, "working drafts" of the entire genome were produced and published, thereby identifying more than 30,000 individual gene sequences and mapping their locations on the 23 pairs of human chromosomes. It was found that the genes make up only about 3% of the base sequence in human DNA, the remainder being in regions between the genes. Including updates published from 2002–2004, about 99% of the gene-rich regions of DNA have now been sequenced. Scientists had been expecting more than 100,000 genes in human DNA. It now appears that all higher vertebrates (such as humans, rats, and even marine mammals) share many of the same sets of genes, and that differences in species arise largely from which genes are switched "on" and which are switched "off."

The genome map will help scientists to identify genes that link to specific physical attributes, behaviors, learning styles, and diseases. Information from the Human Genome Project has the potential to revolutionize medicine by providing insight into the basic chemical processes that underlie many diseases and by pointing the way to new drugs and therapeutic strategies.

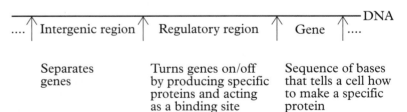

Figure 9.9 Areas on a strand of DNA.

RNA: The Genetic Message Translator

9.7 Genetic information goes from DNA to RNA to protein

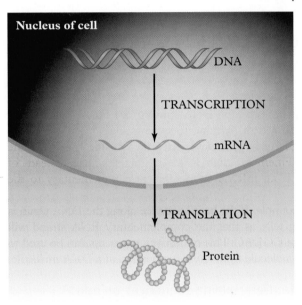

Figure 9.10 Overview of the flow of genetic information.

The DNA in an organism is stored within the nucleus of each of its cells (except red blood cells), not at the site of protein synthesis. Since DNA does not leave the nucleus, all information must be transferred to RNA, which takes the information into the body of the cell. In order to effectively translate the information in a DNA segment into a functional protein, a cell needs the help of three types of RNA, each of which plays a different role. Within the nucleus, **messenger RNA** (mRNA) that is complementary to the DNA sequence to be copied is created by a process called **transcription.** Outside the nucleus the information on the mRNA molecule is used to build a protein with the assistance of **transfer RNA** (tRNA) and **ribosomal RNA** (rRNA) in the process of **translation.** Figure 9.10 summarizes the overall process of genetic information transfer. We will look in more detail at the stages of this process, beginning with transcription.

9.8 Information on DNA is transferred to mRNA

The processes that cause DNA to unwind and the specific molecules involved are beyond the scope of our current discussion.

Base-pairing and the principle of complementarity direct both the construction of mRNA from DNA and the way in which mRNA executes its code of instructions. In order to build a complementary RNA sequence, special proteins first "unwind" the DNA double helix to expose the bases. During transcription, a protein molecule called RNA polymerase assists in bringing unattached nucleotide bases in the nucleus into positions

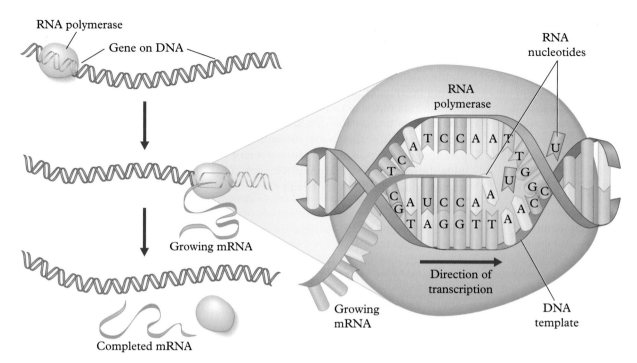

RNA polymerase

Gene on DNA

Growing mRNA

Completed mRNA

RNA nucleotides

RNA polymerase

ATCCAAT

U

T
C

T
G
G
C

G
CGAUCCAA

A
A

TAGGTTA

Direction of transcription

Growing mRNA

DNA template

Figure 9.11 Transcription. mRNA carries a base sequence complementary to the original DNA sequence.

complementary to the bases on the DNA strands (see Figure 9.11). Thus DNA acts as a template, with thymine specifying adenine on the RNA, cytosine specifying guanine, guanine specifying cytosine, and adenine specifying uracil. (Remember that RNA contains uracil instead of thymine.) The result is an mRNA strand that is complementary to the original DNA.

Let's look at an example. If the base sequence along the DNA chain is GCATG, base-pairing tells us that the complementary RNA strand will have the base sequence CGUAC. This transcription process can be used to construct any mRNA molecule from any given portion of a DNA molecule.

Worked Example: Determining sequences on mRNA

If the nitrogen base sequence on part of a DNA molecule is GGTAGCCCT, what is the complementary sequence on the adjacent mRNA strand?

Solution: Since G is complementary to C, and A to U (in RNA), the complementary sequence is obtained by replacing each C in the DNA chain by G, each G by C, each A by U, and each T by A. Thus the complementary mRNA sequence is CCAUCGGGA.

Exercise 9.1

If the nitrogen base sequence on a portion of DNA is ACTTCAGGG, what is the complementary sequence on the adjacent mRNA strand?

9.9 RNA codons dictate a protein's sequence of amino acids

On an mRNA molecule, each sequence of three consecutive nitrogen bases is called a **codon.** These are the "words" of the genetic code. Almost every codon codes for a particular one of the 20 possible amino acids that are the subunits of proteins. For example, the three-letter word, or codon, CAG specifies the amino acid called glycine, whereas AAG specifies lysine. The complete set of 64 codons used by mRNA is given in Table 9.1. (The letter for the first nitrogen base in the codon is shown in the table's first column, that for the second base is given in the table's top row, and that for the third base is given in the table's final column.) Notice that for most amino acids, there is more than one

Table 9.1 Correlation between codons and amino acids

FIRST BASE	SECOND BASE: U		C		A		G		THIRD BASE
U	UUU	Phe	UCU	Ser	UAU	Tyr	UGU	Cys	U
	UUC	Phe	UCC	Ser	UAC	Tyr	UGC	Cys	C
	UUA	Leu	UCA	Ser	UAA	Stop	UGA	Stop	A
	UUG	Leu	UCG	Ser	UAG	Stop	UGG	Trp	G
C	CUU	Leu	CCU	Pro	CAU	His	CGU	Arg	U
	CUC	Leu	CCC	Pro	CAC	His	CGC	Arg	C
	CUA	Leu	CCA	Pro	CAA	Gln	CGA	Arg	A
	CUG	Leu	CCG	Pro	CAG	Gln	CGG	Arg	G
A	AUU	Ile	ACU	Thr	AAU	Asn	AGU	Ser	U
	AUC	Ile	ACC	Thr	AAC	Asn	AGC	Ser	C
	AUA	Ile	ACA	Thr	AAA	Lys	AGA	Arg	A
	AUG	Met or start	ACG	Thr	AAG	Lys	AGG	Arg	G
G	GUU	Val	GCU	Ala	GAU	Asp	GGU	Gly	U
	GUC	Val	GCC	Ala	GAC	Asp	GGC	Gly	C
	GUA	Val	GCA	Ala	GAA	Glu	GGA	Gly	A
	GUG	Val	GCG	Ala	GAG	Glu	GGG	Gly	G

Refer to Figure 9.14 for the full names and abbreviations of the amino acids.

??????????????? ? ????????????
Fact or Fiction?

Taurine, the amino acid in energy drinks such as Red Bull, is one of the 20 amino acids found in human protein.

False. Taurine is naturally present in the body, but does not occur in human proteins. It has several functions, one of which is modulating potassium levels in the body. It has also been synthesized for medicinal uses such as preventing seizures in epileptics, preventing irregular heartbeats (arrhythmia), and lowering cholesterol levels. Taurine is normally produced in our bodies from the consumption of animal-based food products (e.g., dairy and meat). A deficiency of taurine in our bodies can result from extreme physical exertion.

codon that specifies it, although the first two letters of such codons usually are identical. Several three-letter combinations act as "start" or "stop" codons, indicating where a gene begins and ends. In most genes the regions that code for amino acids are interrupted by noncoding sequences that aren't translated into amino acids.

Worked Example: Determining the sequence of amino acids

The first part of an mRNA sequence that directs the beginning of a particular protein reads as follows: AUGCCUAAGAAGGGU. What is the sequence of codons in this list, and what sequence of amino acids does it specify?

Solution: We work from left to right, splitting the sequence of nitrogen bases into successive three-letter codons, thus giving AUG/CCU/AAG/AAG/GGU. Using Table 9.1, we see that AUG specifies the beginning of the sequence and also codes for the amino acid Met (methionine), which, at least temporarily, is the first amino acid in the chain. Similarly, CCU specifies that the second amino acid is Pro (proline). AAG specifies the third amino acid as Lys (lysine), AAG again makes the fourth amino acid Lys, and GGU makes the fifth amino acid Gly (glycine).

Exercise 9.2

The first part of an mRNA sequence that directs the beginning of a particular protein reads as follows: AUGGAUGUCGGGUAU. What is the sequence of codons in this list, and what sequence of amino acids does it specify?

9.10 Ribosomes build polypeptides

The transfer of genetic information from mRNA to the growing amino acid chain that produces a working protein is called translation. This process requires the help of tRNA, rRNA, and the very complex molecules called **ribosomes.** Within their structures ribosomes have specific sites to which both mRNA and tRNA attach when ribosomes are actively translating.

Transfer RNA molecules are single strands of RNA that contain special triplets of bases called **anticodons** (see Figure 9.12, step ❶). Each anticodon, located at one end of the molecule, is complementary to a specific codon on mRNA. At the other end of the molecule is a site to which an amino acid attaches. Each anticodon on a tRNA molecule specifies an amino acid.

Within the ribosome, mRNA codons and tRNA anticodons are brought together. Another tRNA molecule holds the amino acid chain that is built as mRNA's message is read. The ribosome connects each amino acid to a growing chain, as we illustrate in Figure 9.12, steps ❷–❹. A chain of amino acids is called a polypeptide, and long chains of polypeptides combine to form proteins.

Figure 9.12 Step by Step
Translation. Polypeptides are built on ribosomes, with the help of tRNA.

① Transfer RNA (tRNA) carries the amino acid to be added to the polypeptide.

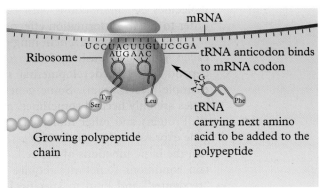

② tRNA anticodon binds to mRNA codon; tRNA, mRNA, and ribosome work together.

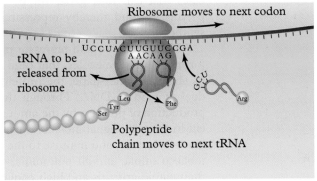

③ After tRNA adds its amino acid to the polypeptide chain, it is released from the ribosome.

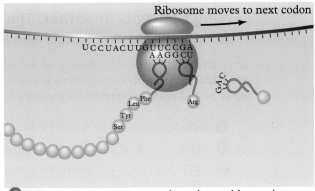

④ Ribosome moves to next codon, along with growing polypeptide chain.

9.11 Cells turn genes on and off

Our understanding of which DNA sequences regulate gene activity, how and why a gene produces (switches "on") or stops producing (switches "off") a protein, and which sequences do not make proteins is still incomplete. Turning genes on and off is vitally important to the cell. Even when genes are identified, their complex relationships with other molecules and interactions with each other need to be understood in order to use the information effectively.

The production of functioning proteins is an example of the process called **gene expression.** Many of the patterns of gene expression depend upon the developmental stage of a cell and its location (for example, heart, brain). Some genes need to be on all the time, some genes are only needed sometimes, and some genes are not turned on at all. Proteins and other molecules act as chemical signals that regulate gene expression. *Housekeeping genes* are turned on all the time and maintain the basic functions of the cell. Genes that are turned on under certain conditions sometimes require the help of *activator proteins,* which "recognize" and bind to specific DNA sequences. Once bound, they help RNA polymerase bind to the DNA, setting the stage for construction of a new protein molecule. *Repressor proteins* are a class of molecules that can turn genes off by preventing the attachment of RNA polymerase and thereby repressing the process that leads to protein synthesis.

Many activator and repressor proteins receive signals, usually in the form of small molecules, that bind to a site on the protein and change its shape. In one shape, the protein binds to DNA and in the other it doesn't bind to DNA. Through this mechanism, the cell controls the production of proteins by the amount and type of signal molecules present in the cell. Signal molecules can be anything from nutrients such as glucose, lactose, and maltose to the environmental carcinogens found in tobacco smoke and air pollutants, or drugs such as those used to treat rheumatoid arthritis and high cholesterol.

● Tying Concepts Together: The base sequence of DNA determines the amino acid sequence of a protein

The (greatly simplified) sequence of events that begins at a section of DNA and ends with a functional protein is summarized in Figure 9.13. Follow that figure as you read the step-by-step descriptions below.

❶ Information from genes on DNA is transcribed to an mRNA molecule.

❷ mRNA leaves the nucleus through nuclear pores.

❸ mRNA binds to a site on ribosomes, the actual protein production centers of the cell.

❹ mRNA and the first tRNA molecules are brought together within the ribosome, initiating construction of a polypeptide.

5 mRNA moves through the ribosome, and a succession of tRNA anticodons "read" the codon information on the mRNA strand. As each codon is read, a new amino acid is added to the growing polypeptide chain.

6 The ribosome recognizes a "stop" codon on mRNA, polypeptide construction ceases, and the chain is released.

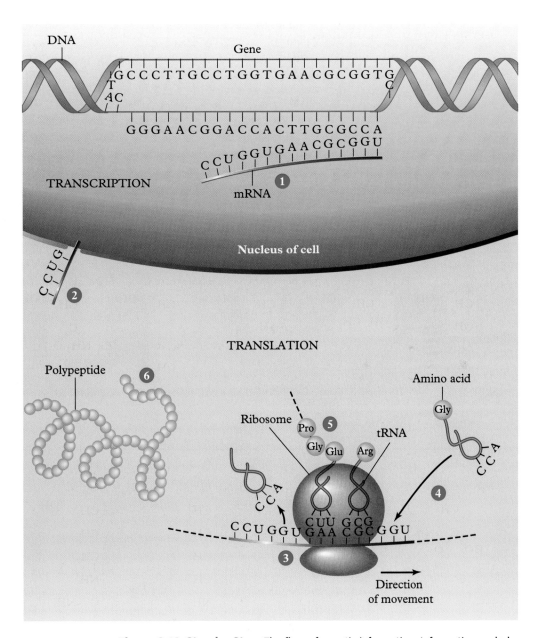

Figure 9.13 Step by Step The flow of genetic information. Information coded on DNA in the nucleus of the cell goes to mRNA, which, with the help of tRNA and ribosomes, directs the synthesis of proteins outside the nucleus.

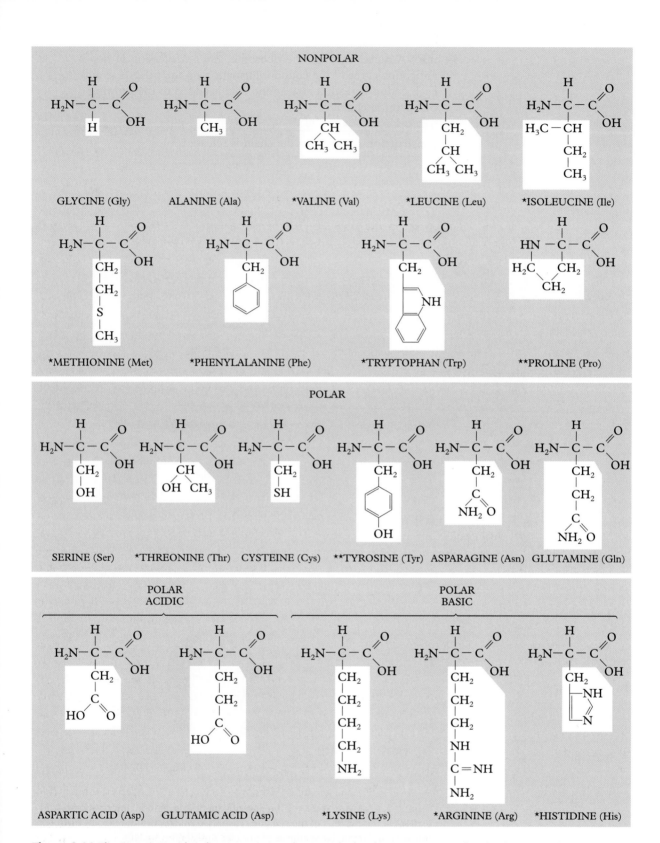

Figure 9.14 The 20 amino acids in human proteins. Those marked with an * are essential in the diet. Some scientists interchange the nonpolar/polar classification of proline and tyrosine.

Proteins are composed of long polypeptide chains. In the next section we explore the expression of the genetic message through proteins, the molecules that determine how you look and what you can do.

The Genetic Message Expressed I: Protein Form

Proteins are among the most important components of all living organisms, including human beings. Our hair, nails, and flesh, the hemoglobin in our blood, and the enzymes that control the rates of chemical reactions in our bodies—all are made of protein. The human body contains tens of thousands of different types of proteins, each of which we synthesize ourselves from our food supply. Any one of the cells in our bodies contains several thousand different proteins.

9.12 Proteins are polyamides

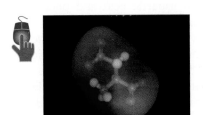

View the structure of the amino acid aspartic acid at Chapter 9: Visualizations: Media Link 1.

Chemically speaking, proteins are polyamides that are similar in structure to the nylon-6 polymers described in Chapter 8. In particular, proteins are the non-branched condensation polymers of **α-amino acids,** where the α designation means that the amine and acid groups are both bonded to the same intermediate carbon atom. In other words, the chain joining the groups is just one carbon atom long.

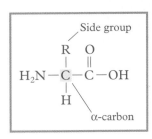

Notice the presence on the central carbon of a hydrogen atom and of a side group, R, which can be either an H atom or a carbon chain or ring. Figure 9.14 shows the various R groups (highlighted in white) associated with the 20 different amino acids in human proteins. Exactly the same set of 20 amino acids are found in all living organisms, though some organisms also use one or two others as well. For example, in some cases the "stop" codon, UGA, is used for the amino acid called selenocysteine.

View the two dipeptides formed from joining aspartic acid and phenylalanine at Chapter 9: Visualizations: Media Link 2.

As we noted above, short chains of amino acids are called polypeptides. Polypeptides that contain long amino acid chains are called proteins. The amino acid sequence is now known for a few tens of thousands of the millions of different proteins found in living organisms.

9.13 Polypeptides are short chains of amino acids

When α-amino acids are polymerized, the —OH group of the acid from one molecule combines with an amino group hydrogen atom of an adjacent amino acid molecule, eliminating a water molecule via a condensation reaction. The two amino acids are joined by what is called a **peptide**

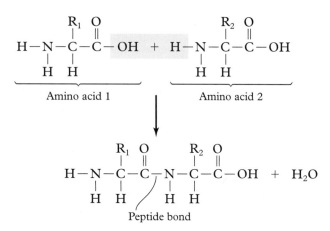

Figure 9.15 Condensation reaction joining amino acids by a peptide bond.

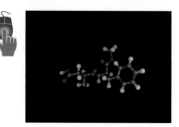

View the structure of aspartame at Chapter 9: Visualizations: Media Link 3.

bond, the single bond between the amino nitrogen and the carbonyl carbon that produces an amide group (see Figure 9.15). If only two amino acid molecules combine together, the result is called a **dipeptide,** three give a **tripeptide,** and more than three form a **polypeptide.** Below we discuss some examples of peptides you may have heard of.

You may be familiar with the artificial sweetener known as aspartame (Equal). This substance is the methyl ester of a simple dipeptide made from aspartic acid and phenylalanine (see Figure 9.16). In the shorthand used by scientists working in the field of proteins, the sequence of amino acids in the dipeptide is described as "Asp-Phe." By convention, the amino acid with the free amino group is listed as the first part of the name. The commercial name for the compound is clearly derived from its aspartic acid origin.

Both of the individual α-amino acids in aspartame are also present in all the proteins in the human body. When most people consume aspartame, their bodies decompose it into its component amino acids for further use. People with a condition called *phenylketonuria* (PKU), however, cannot metabolize the amino acid phenylalanine, which is produced from aspartame, and so the amino acid may build up to toxic levels, causing severe health problems. Those suffering from PKU must avoid consuming

Figure 9.16 The reaction producing the dipeptide aspartame.

Figure 9.17 Phenylalanine warning on a product containing aspartame.
(George Semple for W. H. Freeman and Company)

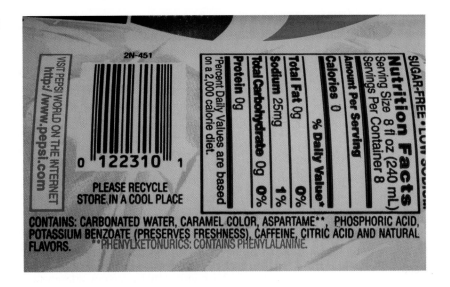

products containing aspartame or other forms of phenylalanine, and packaged foods now carry labels identifying its presence (see Figure 9.17).

Aspartame is 200 times as sweet as sugar.

Aspartame is present in many prepared foods, such as sugar-free fruit yogurt. However, as we will see later in the chapter, polypeptide and protein structures are heat sensitive. For this reason, care must be taken if cooking with aspartame, as it may decompose during preparation.

Polypeptides form the proteins that ultimately determine what your body can do, how you look, and even how you feel. You may have heard, for example, of *endorphins,* substances that occur naturally in the human brain and contribute to reducing pain and to producing the so-called runner's high. Endorphins belong to a class of peptides called opioids, molecules that relieve pain and produce pleasant sensations. One endorphin polypeptide chain has 33 amino acids joined together in a specific sequence. A polypeptide containing any other sequence of the same amino acids would behave somewhat differently and would presumably be less effective in reducing the sensation of pain.

The name *endorphin* is derived from the fact that these substances are produced in the body (endogenous) and behave similarly to morphine in reducing pain.

9.14 Protein shapes are determined by interactions

Proteins are able to fulfill their many biological functions because of their structures. They have specific shapes and regions with particular physical and chemical properties that enable them to interact in different ways with various molecules. Before we see how structure influences function, we will examine how proteins attain their shapes and regional properties. We shall consider some of the characteristics of the individual amino acids and then consider how hydrogen bonding and other interactions between chains or components of a given chain can determine shape.

Primary structure Any particular protein molecule in your body consists of a continuous chain of dozens or hundreds of units, each of which is the main part of a specific amino acid. The sequence of amino acids, as determined by their side groups, R, along the protein chain, or

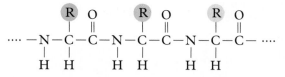

Figure 9.18 Primary structure of a protein.

backbone, is known as its *primary structure* (see Figure 9.18). The type and sequence of amino acids and their side groups in the primary sequence determine the nature of interactions between different parts of a protein chain.

By looking at the side groups (shown in white) in Figure 9.14, you can see that aspartic and glutamic acids behave as acids since the side group includes —COOH. (Recall the discussion of carboxylic acids in Chapter 6, section 6.15.) This is called the *polar acidic* group. Lysine, arginine, and histidine are classified as *polar basic* since their side groups are dominated by amine nitrogens, which makes them accept H^+ ions and therefore act as bases. We expect that both of these side groups will behave as hydrophilic zones in a protein, since their components can form hydrogen bonds with water and even form ions. The six *polar* side groups in Figure 9.14 all contain atoms that can also form hydrogen bonds to water. In contrast, the nine *nonpolar* amino acids all have either hydrocarbon side groups or ones that are dominated by their carbon content, and thus are hydrophobic overall. The combination of amino acids in specific sequences will determine the character (polar or nonpolar, acidic or basic) of particular regions of the polypeptide chain.

Although we can imagine the protein chain as a linear sequence of amino acids, in reality the chain is not straight but adopts a shape called its secondary structure.

The new scientific field of *proteomics* is concerned with the analysis of the structure, functions, and interactions of proteins produced by various genes. Researchers also try to organize this information into databases.

Secondary structure Hydrogen bonds between backbone atoms (see Figure 9.19) form the amino acid sequence into the protein's *secondary structure*, the 3-D shape adopted by its backbone. In most cases the shape of the protein is a regular, repetitive pattern which imparts particular properties to a protein. The two most important types of secondary structures are the **α helix** structure and the **β pleated sheet** structure.

In proteins that have the α helix structure, the backbone chain of the polypeptide winds around an (imaginary) axis, forming a helix (see Figure 9.19a). The backbone adopts this shape because hydrogen bonds form between nearby amino acids on the *same* chain. Specifically, there is a hydrogen bond between the N—H unit of each amino acid with the carbonyl oxygen in the fourth amino acid from that unit along the sequence.

The R groups of the amino acids protrude outward from the axis of the helix. The helix can be stretched a little, since such motion breaks only some of the hydrogen bonds, not the much stronger covalent bonds. When the force stretching the coil is released, it returns to its original length since these broken hydrogen bonds can then re-form.

The α helix is the structural unit in *α-keratin,* which is the substance that makes up animal (including human) hair, wool, skin, and nails. In α-keratin, several helices are wrapped around each other, in the same way and for the same reason that we wrap strings around each other to form rope, namely to produce a stronger fiber.

Proteins that have the β pleated sheet structure have N—H---O hydrogen bonds *between* adjacent polypeptide chains, rather than internal to the chain (see Figure 9.19b). These interactions produce sheets of protein molecules that are pleated rather than flat because the geometry

Figure 9.19 Protein secondary structure. (a) An α helix. (b) A β pleated sheet.

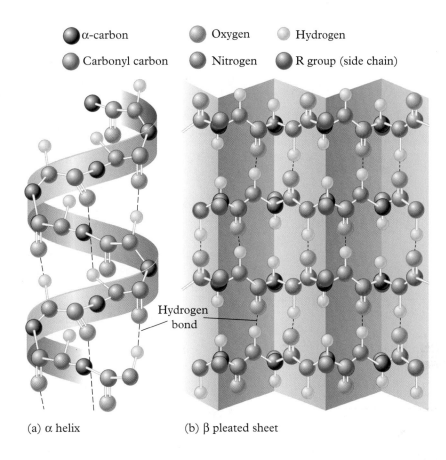

- ● α-carbon
- ● Carbonyl carbon
- ● Oxygen
- ● Nitrogen
- ○ Hydrogen
- ● R group (side chain)

Hydrogen bond

(a) α helix   (b) β pleated sheet

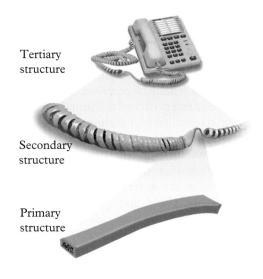

Tertiary structure

Secondary structure

Primary structure

Figure 9.20 Telephone cord analogy for levels of protein structure.

at the carbon that is bonded to the carbonyl carbon and to the nitrogen is tetrahedral. Unlike the coiled helix, pleated sheets are fully extended and impart strength and flexibility to a protein. The silk produced by spiders and certain other insects is made of the protein fibroin, whose structure is mainly the β pleated sheet structure.

Tertiary structure The secondary structure of a protein folds and coils further, into the protein's *tertiary structure*. A good analogy to a tertiary structure is the way a telephone cord coils and folds into a specific form (see Figure 9.20). The folding is determined by the nature of interactions between side groups on the amino acid residues with each other and with the surrounding environment. There are four possible ways in which side groups can interact:

1. Salt bridge attractions between side chain $-COO^-$ and NH_3^+ groups
2. Hydrogen bonds
3. Covalent S—S (disulfide) bonds
4. Weak attractions between hydrophobic (hydrocarbon-like) groups

The Genetic Message Expressed I: Protein Form

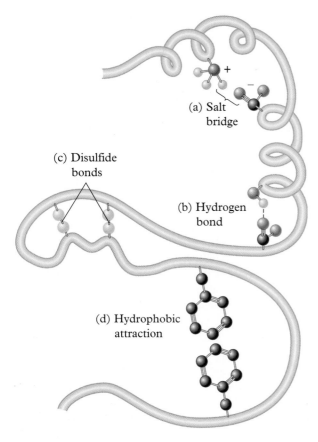

(a) Salt bridge

(c) Disulfide bonds

(b) Hydrogen bond

(d) Hydrophobic attraction

Figure 9.21 The various interactions that can occur between two adjacent protein chains.

Hard animal shells and horn are formed by polypeptides that have high cysteine content and that therefore can form many disulfide linkages.

The —COOH unit in the side group of an acidic amino acid can donate its acidic proton to the nitrogen atom in the side group of an amino acid if they come close together. Once the hydrogen ion transfer is complete, these two groups will probably remain close in space owing to the electrostatic attraction of the positive charge (on the nitrogen) for the negative charge (on the oxygen of the acid group):

$$COOH + :N \rightarrow COO^{-} \cdots {}^{+}HN$$

This interaction is called a **salt bridge** (see Figure 9.21a).

Amino acid side groups on different parts of a polypeptide chain, or on adjoining chains, can also be brought in close contact through hydrogen bonding (see Figure 9.21b). An H atom (attached to N or O) on the one amino acid and an oxygen or nitrogen atom on another amino acid are attracted due to differences in the signs of their partial charges.

A third way of bringing groups in proteins into close contact is through disulfide bonds, S—S (see Figure 9.21c), which are covalent bonds formed by two sulfur atoms, one on each of two cysteine side groups. The sulfur–sulfur bond is created when the —SH units of two cysteines both lose their hydrogen atoms:

$$-S-H \; H-S- \; \rightarrow \; -S-S-$$

An example of such a bond exists in the insulin molecule, where two polypeptide chains, one having 30 and the other 21 amino acids, are joined by two S—S linkages between cysteine groups. Although the S—S link is a covalent bond, its formation can readily be reversed; the breaking and forming of such bonds are an integral part of the process when one receives a "permanent" for hair, as discussed in section 9.15.

The other forces that determine protein shape relate to the hydrophobic (water-hating) and hydrophilic (water-loving) behavior of different side groups. In general, hydrophobic side groups are usually found buried inside a protein, away from the watery environment that normally envelops the molecule as a whole and that repels the hydrophobic groups. In addition, the hydrophobic side groups are weakly attracted to each other, which is similar to what occurs between chains of hydrocarbon molecules and polymers (see Figure 9.21d). Some proteins are folded such that polar side groups are positioned on the outside of the structure, where they can interact via hydrogen bonding or polar attractions with the polar aqueous environment.

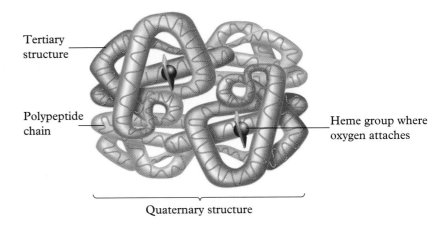

Figure 9.22 Tertiary and quaternary structure of hemoglobin. Two additional heme groups are present but hidden in this representation.

Tertiary structure

Polypeptide chain

Heme group where oxygen attaches

Quaternary structure

In general, the tertiary structure of a protein is determined by a combination of all four types of forces operating at different positions along the chains, producing an elaborately folded form.

Quaternary structure Some proteins consist of more than one type of polypeptide chain and/or contain non-polypeptide units as well. The arrangement of these chains, and of the non-polypeptide units, relative to each other, constitutes the *quaternary structure* of proteins. For example, hemoglobin, the protein that transports oxygen through the bloodstream, consists of four distinct polypeptide chains—two each of two different types—and of four heme groups (see Figure 9.22). The exterior structure of the protein contains many polar R groups, allowing the protein to dissolve in blood, a watery fluid. In contrast, the interior of the protein is nonpolar. The iron ions of the heme groups form weak covalent bonds to intact oxygen molecules, O_2. These weak bonds can be broken readily, allowing the oxygen molecules to be released when they are needed in the body's cells.

Excess iron in the body, beyond that required for hemoglobin synthesis, is stored in a protein called *ferritin,* which is found mainly in the liver.

9.15 Your hair curls due to disulfide bridges and hydrogen bonds

In human hair, two proteins with an α helix structure coil around each other, forming a supercoil, and are held in place by —S—S— bridges and hydrogen bonds between the helices. Numerous supercoil units are packed side by side and staggered lengthwise, joined by some disulfide bonds into the fibers that are eventually thick enough and long enough to constitute a single hair.

The natural shape of your hair—straight, loosely curled, or tightly curled—is determined by the amount and location of cysteine amino acids in your hair's protein chains. Since your DNA specifies the sequence of amino acids in your proteins, this is a hereditary trait—you are stuck with the hair you were born with. Washing your hair breaks some of the hydrogen bonds and allows the hair to straighten somewhat

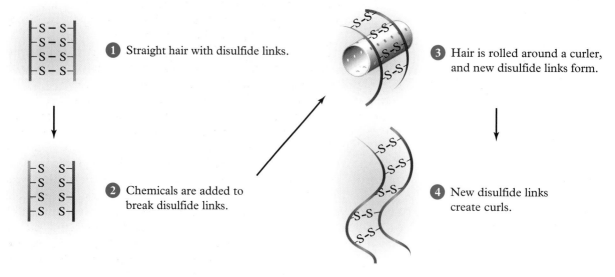

1 Straight hair with disulfide links.

2 Chemicals are added to break disulfide links.

3 Hair is rolled around a curler, and new disulfide links form.

4 New disulfide links create curls.

Figure 9.23 Permanent waving of hair.

because water molecules lodge between the chains and form hydrogen bonds to them. If you shape your hair when it is wet, it will retain some of the new shape temporarily when it dries because of the new hydrogen bonds formed when you arranged your hair. However, this new shape will not survive the next washing of your hair or a severe rainstorm.

In the permanent waving of hair, the —S—S— links between adjacent α helices (Figure 9.23, step **1**) are first broken chemically (step **2**), by the use of an agent that supplies H atoms. The links are temporarily converted to individual —SH groups as the —S—S— covalent bond is broken and the S atoms freed from each other. In addition, the hydrogen bonds *within* the chains are disrupted, and the heat and water that are applied make the helices unwind. Subsequent application of a second chemical causes the unbonded —SH units to form new —S—S— links based upon the shapes and sizes of curlers used in the hair (step **3**). When the hair is subsequently washed and cooled, the α helices reform, but the —S—S— linkages in their new positions exert some force on them and the hair is curled (step **4**). Of course, the hair grows out from the scalp in the genetically determined shape, so permanent wavings must be repeated periodically.

Natural hair color arises from two types of the pigment called *melanin:* eumelanin produces black and brown shades, while phaeomelanin is responsible for red coloring. Hair with only a little melanin is blonde. A total absence of melanin results in gray hair.

Women (and men) have been coloring their hair since ancient Egyptian times. Originally, dyes from plant and animal matter—such as red from henna and cochineal and yellow from chamomile—were used to produce lighter shades. Romans darkened their hair with a solution made by boiling walnut shells and leeks. Today most of the dyes are synthetic, and were developed originally to color textiles.

Melanin is also responsible for skin color. The condition of albinism is the partial or full absence of melanin pigment from skin, hair, and eyes. One version of this genetic condition results from the failure to convert the amino acid tyrosine into melanin.

Chapter 9: The Molecules That Make You What You Are

Hair dying techniques today are much more sophisticated, and consumers have two options for coloring their hair: permanent and temporary hair colors. Temporary dyes just coat intact hair strands, and are removed by washing the hair. For a hair color change to be "permanent," the dye must enter the inner cortex of hair that lies under the outer layer of a hair strand, the cuticle. To achieve that result, a dilute aqueous ammonia solution is applied to the hair, which makes the hair swell and causes the cuticle scales to separate a little. The second chemical in the solution, hydrogen peroxide, reacts over time (about half an hour) with the water-soluble organic molecules that upon oxidation are converted to dyes that can then be incorporated into the hair. (The final dyes are not water-soluble, so cannot be supplied as the original aqueous solutions.) To prevent their premature reaction, the ammonia and the dye precursors initially are in a separate solution from the 6% aqueous solution of hydrogen peroxide. The organic molecules that are supplied in the hair preparation, and that are subsequently oxidized inside the cuticle, are various derivatives of the diamine (called the "primary") and the aminophenol (the "coupler"). The reaction products include some large, highly colored molecules derived by joining diamine and aminophenol units together; these molecules are too large to escape from the cuticle. Different oxidized derivatives of these benzene-based molecules have different colors; the appropriate mixture of the various derivatives is supplied so as to produce the particular shade desired by the consumer. For people with dark hair who wish to dye their hair a much lighter color, the melanin must first be bleached with an oxidizing agent, usually hydrogen peroxide.

The Genetic Message Expressed II: Protein Function

9.16 You need protein in your diet

Protein is essential in your diet; you require about one gram of protein per day per kilogram of your body weight. Your body synthesizes the protein molecules it needs by polymerizing α-amino acids from a pool of them that originally was part of the protein in your food. To use this dietary protein, your body must first decompose it into its component amino acids. The transformation of dietary protein into body protein can be summarized schematically:

$$\text{dietary proteins in humans} \xrightarrow[\text{into}]{\text{broken down}} \text{20 amino acids} \xrightarrow[\text{make}]{\text{essential to}} \text{protein required by body}$$

The process of decomposing proteins into amino acids is accomplished by digestive enzymes, a class of proteins that will be discussed in section 9.19. These enzymes speed up the depolymerization of polypeptides by facilitating the addition of a molecule of water to each peptide bond. The hydrogen from the water adds to the amino group, and the —OH from the water adds to the carbonyl carbon. This hydrolysis reaction reverses the condensation reaction shown in Figure 9.15. DNA and RNA molecules then direct the synthesis of specific body proteins from this pool of amino acids.

There are many sources of protein in the human diet, and they vary from culture to culture. Fish is one common source for many people. It is easily digestible since it contains little connective tissue that is hard for the body to break down. In contrast, some cuts of meat need to be "tenderized" by decomposing some of the longer protein molecules into shorter ones before it becomes pleasurable to consume. Meat tenderizers are natural materials that contain enzymes that can break down proteins into smaller, more tender, and more flavorful components. Some traditional cultures often cook tough cuts of meat together with figs or pineapple, both of which contain natural meat tenderizers.

Of the 20 α-amino acids required to make proteins, only 10 are essential to have in your diet. These **essential amino acids** are marked with an * in Figure 9.14. The other 10, known as nonessential amino acids, are also necessary for your body's protein synthesis. They are called nonessential only because they are not essential in your *diet*. If a nonessential amino acid is in short supply, your body can synthesize it by rearranging the amino acids and other substances that are already present in your cells.

9.17 What's a complete protein?

Dietary protein sources are classified as **complete proteins** if they contain all the essential amino acids in the right ratio for conversion into human protein. Protein found in meat, seafood, poultry, milk, cheese, and eggs is complete protein, as is human milk. Soy protein is one of the few plant-derived proteins that is complete.

Incomplete protein does not contain all the amino acids in the correct proportions necessary to build body proteins. Protein obtained from most plants is incomplete, although by combining different types of plants in the diet, it is possible to achieve the right balance of all essential amino acids required for human growth and maintenance. For example, although grains such as wheat, corn, rice, and oats contain about 10% protein, they are low in the essential α-amino acid lysine and some are low in tryptophan as well. Thus, an all-grain diet does not supply all the essential amino acids in adequate amounts.

On the other hand, legumes such as beans and peas are high in lysine and tryptophan but low in other essential amino acids, such as methionine, that grains possess in good supply. Consequently, *combining* a grain and a legume at a meal provides adequate amounts of all the essential amino acids. Indeed, this pairing is an important part of the traditional diet in many cultures (Figure 9.24):

Corn and beans in Mexico

Rice and black-eyed peas in the U.S. South

Rice and lentils in India

Some powdered "protein sources" available in health food stores are a mixture of amino acids and short-chain peptides.

Figure 9.24 A well-balanced meal. Meat, beans, and rice provide all the amino acids necessary to build body proteins. (Photodisc Green/Getty Images)

Protein synthesis in the body cannot proceed without a supply of all the amino acids, since each one is required in some amount in each protein. Thus if even one essential amino acid is missing from the diet, the creation of protein will cease to some extent.

Any excess amino acids not required for protein synthesis by your body are decomposed, with the nitrogen excreted in your urine in the form of the compound urea. Most of the carbon and hydrogen is converted to glucose for use as energy or stored as fat. Since the body does not maintain storehouses of amino acids (in contrast to fats and carbohydrates), an almost daily intake of protein with the proper ratio of components is necessary. Vegetarians, especially those who do not consume milk or egg products, must take care with their diets to consume all essential amino acids in sufficient quantity.

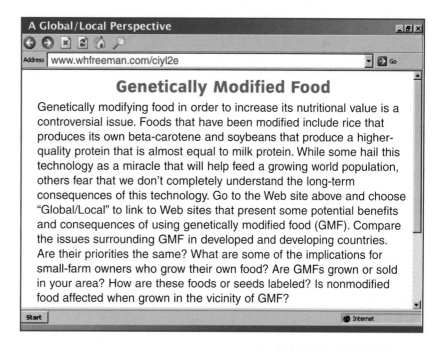

A Global/Local Perspective

Address www.whfreeman.com/ciyl2e

Genetically Modified Food

Genetically modifying food in order to increase its nutritional value is a controversial issue. Foods that have been modified include rice that produces its own beta-carotene and soybeans that produce a higher-quality protein that is almost equal to milk protein. While some hail this technology as a miracle that will help feed a growing world population, others fear that we don't completely understand the long-term consequences of this technology. Go to the Web site above and choose "Global/Local" to link to Web sites that present some potential benefits and consequences of using genetically modified food (GMF). Compare the issues surrounding GMF in developed and developing countries. Are their priorities the same? What are some of the implications for small-farm owners who grow their own food? Are GMFs grown or sold in your area? How are these foods or seeds labeled? Is nonmodified food affected when grown in the vicinity of GMF?

Start   Internet

Some scientists warn against the consumption of soy-based products, including soy milk, by infants and pregnant women since one of its components, genistein, can affect the immune system.

Components of the proteins and starches in certain foods can combine when they are cooked at high temperatures to form a mixture of substances. Some of the products are melanin-like pigments that give fried foods and baked goods their brown outer color. Unfortunately, the compound *acrylamide*, $CH_2{=}CH-C(={O})NH_2$, is formed when the amino acid, asparagine, and the sugar, glucose, combine when baking or frying temperature exceed 120°C (248°F). Based upon animal studies, acrylamide has been classified as a potential human carcinogen (that is, as a cancer-causing agent), and has been found in a wide variety of starch-based foods such as potato chips, French fries, cookies, biscuits, processed cereals, and some breads. However, studies show that it is not associated with human cancers of the breast, bowel, kidney, or bladder. The levels of acryamide in foods are probably too low to be of general concern.

9.18 Proteins have a variety of roles in the body

Since all proteins consist primarily of one or more chains having the same repeating backbone structure, you might think that they would all have pretty much the same physical properties. However, proteins have amazingly different properties and functions, even though they are all formed from the same 20 types of amino acids. Consider the diversity of the proteins in your body in terms of their biological roles:

- *Structural proteins* form much of your physical body. Your hair and fingernails, and the feathers and horns on animals, are composed mainly of the protein α-keratin. The major component of your cartilage and tendons is the protein collagen.
- *Contractile proteins* allow your muscles, such as your heart, to contract, and allow sperm cells to move through space.
- *Regulatory proteins* control cellular activity and constitute certain hormones, such as insulin.
- *Protective (or defense) proteins* help to recognize and ward off invading microorganisms that can make you sick, and help to coagulate blood when your blood vessels suffer an injury.
- *Transport proteins* carry substances such as oxygen from organ to organ in your body.
- *Catalytic proteins* are what we call *enzymes*. They control the rate of production and destruction of substances—including other proteins—in the body. Thousands of different enzymes, each capable of catalyzing a specific reaction, have been discovered.
- *Storage proteins* allow you to store certain nutrients such as iron that are not immediately required.

View the structure of the enzyme amylase at Chapter 9: Visualizations: Media Link 4.

9.19 Enzymes are catalysts for biochemical reactions

Life as we know it would simply not be possible without enzymes. Like the catalysts that are used to speed up reactions in the chemical industry, enzymes greatly speed up the rates of reactions in the body and allow processes to occur that would take almost forever to happen on their own. Enzymes speed up reaction rates by factors of a million times or more.

Some enzymes promote the digestion of food, permitting the breakdown of protein molecules into the component amino acids, and that of carbohydrates into glucose, to occur at body temperature in a

We saw how a catalyst facilitates the reaction that converts peroxide to water and oxygen in Chapter 3, section 3.18.

reasonable time. Other enzymes promote the construction of large molecules such as body proteins and glycogen from simple components. Still other enzymes help to contract muscles, or send nerve signals, or destroy toxins in the body.

The usual reason that many reactions do not occur quickly in the absence of a catalyst is that the reactants need to gather energy in order to partially break some of their component covalent bonds *before* new bonds can begin to form in the products. This means that the reactants must pass through a situation in which they are less stable than they were originally or will be in the future as products. The energy required to achieve this state of lesser stability can be provided by heat, and so most reactions proceed much more rapidly if the reactants are heated. In living systems, however, reactions must proceed *without* high temperatures, since the living organism would be destroyed under such conditions.

Catalysts provide partial, temporary bonds to the reactant molecules as they pass through a transition state in which their own bonds are partially broken. As a consequence of this supplementary bonding, little or no extra heat energy is required to facilitate the reaction. Once the transition state is passed and the new bonds to the products are more fully formed, the partial bonds to the catalyst are gradually broken and the product molecule is released. The enzyme can then catalyze another reaction.

Most enzymes are fairly large, globular-shaped proteins. The secondary structure of a typical enzyme is α helical in some places and β pleated in others. Its specific tertiary, and in many cases quaternary, structure is critical to the enzyme's functioning. Particularly important is a pocketlike region on the enzyme's surface called the **active site,** where the reactant molecule, or **substrate,** fits in, as illustrated in Figure 9.25. The situation is

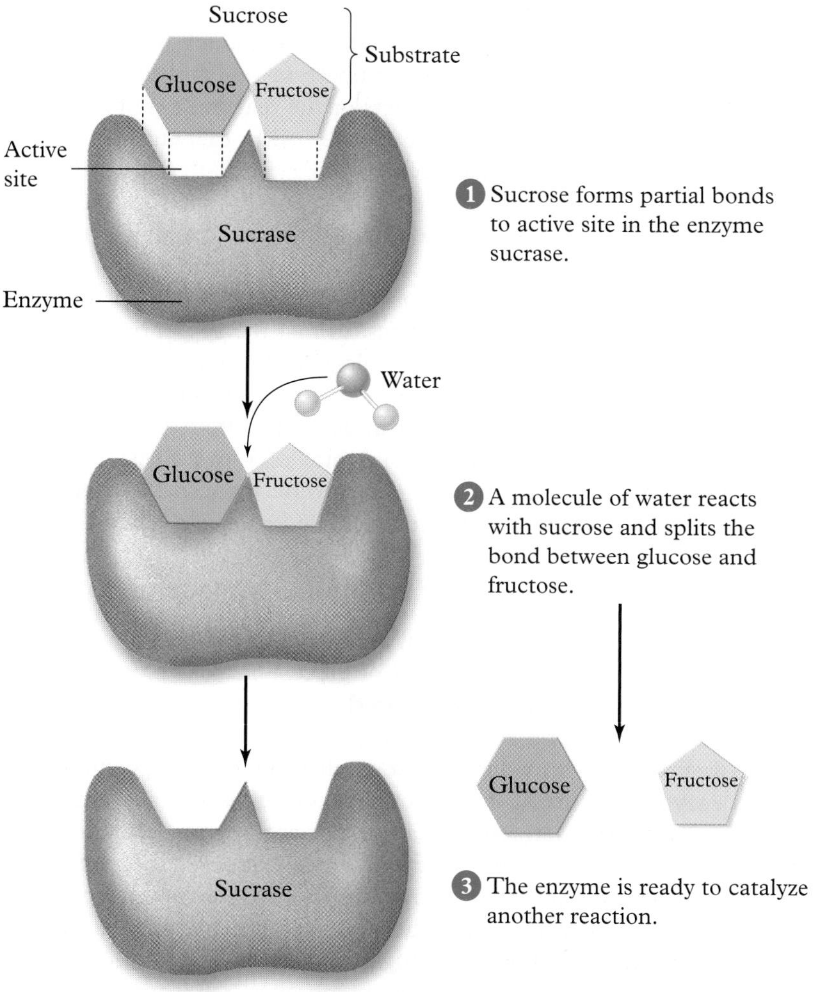

1. Sucrose forms partial bonds to active site in the enzyme sucrase.

2. A molecule of water reacts with sucrose and splits the bond between glucose and fructose.

3. The enzyme is ready to catalyze another reaction.

Figure 9.25 Enzyme action. The enzyme sucrase catalyzes the reaction that splits a molecule of sucrose into a molecule of glucose and a molecule of fructose, that is, glucose—fructose + enzyme + H_2O (sucrose) $\rightarrow$ glucose + fructose + enzyme.

analogous to that of a key (the reactant) fitting into a lock (the enzyme's active site). Since the shape and size of the active site are not very changeable, only a specific molecule, or a specific part of a molecule, having a particular shape and size will fit and become temporarily attached. The attraction between the enzyme and the substrate can be based upon any or all of the following forces:

- Hydrogen bonding
- Hydrophobic interactions
- The attraction between ions and/or partial charges having opposite signs

Attractions between the active site and substrate "hold" the substrate in a specific orientation, facilitating the creation of new bonds or breaking of old bonds.

As an example of how an enzyme in your body works, consider sucrase, whose role is to hydrolyze the ether-like C—O—C bond between the two simple sugars that are joined together in sucrose. The sucrose molecule's shape allows it to fit into the active site of the enzyme that is specifically designed to facilitate the hydrolysis reaction of this type.

The names of enzymes often terminate in -ase and often describe the molecule on which the enzyme acts or the reaction type for which it is effective. You saw an example above, with the enzyme sucrase required to hydrolyze the sugar sucrose.

As you might well imagine, the requirements for the size, shape, and locations of specific atoms within the active site are quite stringent. Enzymes are generally very specific, catalyzing only one type of reaction and then only for a limited number of substrates. Our bodies possess many thousands of different types of enzymes, so that the appropriate reactions can occur with all the various molecules. All the enzymes are created within your body by reactions that themselves require enzymes. The polypeptide component of enzymes can be created from the various dietary α-amino acids. However about one-third of known enzymes require one or more metal ions such as iron, copper, zinc, manganese, magnesium, etc., in order to function. Sucrase, for example, requires cobalt.

9.20 Some proteins require additional molecules

Simple proteins consist entirely of amino acids. An example is insulin, the substance that controls the level of glucose in your blood, which is made of 51 amino acids. **Conjugated proteins** possess some other groups as well as amino acids. These groups can be a metal ion, phosphate groups, a sugar, a lipid, or a small grouping of specific atoms. Some enzymes require complex organic or organometallic structural components called **coenzymes** that are not proteins and that cannot be synthesized in the organism itself from simple components. The function of some of the vitamins that we require in our diet is to supply these structures, or at least ones that the body can modify into coenzymes. Conjugated proteins containing a sugar unit are called *glycoproteins*. Glycoproteins are important parts of cell membranes (see Chapter 7, Figure

7.20) and act as identification markers, such as those that differentiate blood groups. They also act as part of a communication system between the cell and its outside environment. The combination of a protein with a lipid, which is a water-insoluble or fatty molecule, produces a **lipoprotein.** As discussed in section 9.27, lipoproteins have both hydrophilic (protein) and hydrophobic (lipid) sections.

9.21 Mutations: When things go wrong

A **mutation** is a change in the sequence of bases on a section of a DNA molecule. Mutations can affect the final overall shape, and therefore function, of proteins coded by the gene on which they are found. Mutations in human DNA that are passed on to offspring can produce genetic (inherited) diseases or defects. Mutations can arise spontaneously as DNA is replicated in the body or can be induced by environmental factors such as exposure to radiation or toxic chemicals. Agents that induce mutations in DNA are called **mutagens.** *Teratogens* are mutagens that cause mutations that produce abnormal embryonic development. Perhaps the most famous teratogen in recent history is the drug thalidomide. Thalidomide was given to pregnant women from the late 1950s to the early 1960s to prevent miscarriages. In a small percentage of cases, however, the users gave birth to children with malformed or missing limbs. It turns out that thalidomide occurs in a "right"- and "left"-hand form. One form prevents miscarriages and the other form produces birth defects by interfering with the structural development of human embryos.

There are many types of mutations, but not all of them change the shape and function of a protein very much, and relatively few of them have such a major impact as to produce a noticeable defect or disease. Mutations in genes involved in regulating cell processes are very damaging, since cell regulation controls division and growth. Cancer can be a result of damage to cell regulatory genes.

Mutations can alter DNA in several ways. Using a verbal analogy, we can imagine a gene with a normal sequence of bases as a sentence:

Normal sequence   The boy saw the dog.

Changing one or more letters (bases) changes the meaning of the sentence in different ways:

Substitution	The boy saw the **f**og.
Single insertion	The boy saw the d**r**og.
Multiple insertion	The boy saw the dog**gie**.
Single deletion	The boy saw the **dg**.
Multiple deletion	The boy ★★★ ★★★ dog.
Small inversion	The **yob** saw the dog.
Large inversion	**The yob was dog eht**.

In an organism, the various changes may produce effects ranging from nonobservable to lethal. *Silent mutations* do not produce an observable

The pure form of nonteratogenic thalidomide is making a comeback as a treatment for leprosy.

CHEMISTRY IN ACTION

Can defective genes be fixed? Hear what Dr. French Anderson, the father of gene therapy, has to say about this at Chemistry in Action 9.1.

effect and don't impact the operation of a cell. Some mutations replace an amino acid with another of similar size, shape, and charge, so these mutations usually don't alter the function of a protein very much. Mutations that result in very different amino acid substitutions, however, produce low-functioning or nonfunctioning proteins, and these can be devastating to the person. An example of this is found in people with *sickle-cell anemia*. In this inherited disease, a single amino acid change in the sequence results in a red blood cell with the characteristic sickle shape that gives the disease its name. Red blood cells with deformed hemoglobin do not carry oxygen like normal cells and crystallize in the bloodstream, causing great pain and blocking blood flow in capillaries.

Insertions and *deletions* change the way DNA sequences are read and can result in meaningless "sentences." Three bases (or multiples of three), however, might be deleted or inserted with a single amino acid being lost or added and the reading sequence left intact. If the amino acid in question is not in a vital region of the protein, it might not have a lethal effect. Deletions can be very serious if the deletion is of a whole gene. If the gene is not there, the protein cannot be produced, and, if it is essential, the person will not survive. *Inversions* are invariably detrimental, since the resulting segment is meaningless.

Genetic diseases, such as diabetes or sickle-cell anemia, and inherited defects, such as webbed toes or a cleft palate, have their origins in DNA mutations of germ cells. Germ cells are the source of eggs in women and sperm in men. DNA mutations in germ cells can be passed along to offspring through generations. Since humans have two of each chromosome and, therefore, two sets of genes, within a given germ cell the chances of both chromosomes carrying a mutation are slim. There is usually a functioning copy of a gene to take over in the case of mutation in the other gene. When a working copy of the gene hides the defective properties of the mutated gene, the mutation is called *recessive*. Some mutations may confer a benefit which outweighs the disadvantage. For example, people who carry one copy of a gene for cleft-lip mutations may have more resistance to herpes virus, and people with one copy of the sickle-cell gene are more resistant to the tropical disease malaria. Over eons of time, these mutations influence evolution.

If people with similar genetic makeups (close relatives, certain ethnic groups) mate and produce a child, there is a significantly increased risk that the child could receive two defective copies of a gene and thus be afflicted with a disease. This increased risk of what are now recognized as genetic defects may be the basis for long-standing cultural taboos against incest and marriage between close relatives. Maple syrup urine disease among the Amish and hemophilia among European royal families are examples of genetic diseases. Maple syrup urine disease results from the inability of someone's body to metabolize the amino acids leucine, isoleucine, and valine, all of which are present in high-protein foods. These amino acids accumulate in blood, sweat, and urine. Very high levels of these amino acids and their by-products are toxic and can lead to brain injury, mental retardation, and even death. A by-product of isoleucine has a sweet smell, giving the disease its name. In hemophilia, the blood protein that is essential for the clotting of blood following an injury is either missing or is produced at too low a level. This means that a person with hemophilia who gets a cut will continue to bleed for a long period of time (even to the point that a simple wound can become life threatening) because their body's mechanism for halting blood flow (the clotting of the blood) is not functioning.

In *gene therapy*, working versions of defective genes are administered as a drug to the patient. So far, success with this technique has been limited to nonhuman animals. One of the major challenges facing gene therapy is that the body's immune system is "programmed" to destroy foreign genetic material. Another challenge gene therapy has to overcome is that most common diseases are not caused by errors in a single gene, but by errors in several of them. It will probably be several decades before effective gene therapy is widely used.

9.22 Cloning

The term *clone* refers to a group of organisms with exactly the same genetic material. Multiple births—that is, twins, triplets, etc.—that produce genetically identical individuals are examples of natural clones. Despite all the discussion about the cloning of a human embryo, the laboratory production of human clones is a long way off.

The process by which cloned cells are created is simpler, more reliable, and better understood than the process used to clone more complex organisms such as mammals. Scientists are, therefore, more likely to work with cloned cells than with entire organisms. Using techniques that allow them to clone individual cells and create cell colonies, researchers have gained insight into the structure and function of genes and cells, including disease-causing defective genes and cancer cells.

The process by which complex (that is, mammalian) clones are created is complicated, and the details are well beyond the scope of our discussion. Simply outlined, however, an egg cell is obtained from a donor of the same species as the clone. The nucleus of the egg cell is destroyed and replaced with a nucleus taken from a body cell of the animal to be cloned. Since the body cell nucleus contains a copy of the organism's DNA, the egg theoretically will develop into an exact genetic replica of the donor. In reality, the technique of cloning mammals is an imperfect science that is unable to consistently produce healthy clones; in fact, the majority of cloned cells do not survive. Clones that do survive may have inherent problems due to subtle differences in the DNA produced in the cloning process.

Cloning technology may ultimately provide medical benefits or enable us to save endangered species; however, we are not there yet. The possibility of human cloning raises many ethical, religious, and practical questions and has prompted the United States and other countries to consider banning the creation and use of human clones until protocol guidelines can be established.

In 2004, scientists in the United Kingdom cloned monkey embryos and successfully transferred them into mother monkeys. Although the embryos ultimately did not survive, this experiment is the closest scientists have come to cloning primates.

The term *cloning* is also used to describe the production of multiple copies of a gene or other DNA segment. This process, called *DNA cloning,* is useful in *recombinant DNA technology,* where genes from one organism are inserted into the DNA of another.

9.23 Proteins and DNA are used as evidence in legal proceedings

Historically, fingerprints, blood types, and protein profiles have been used to identify or exonerate suspects in criminal investigations. Fingerprints are unique, but are not always available and are rarely perfect. Blood typing identifies the group to which a sample belongs—O, A, B, or AB—but not an individual. Blood types are distributed over a population to differing degrees and can be used to classify a sample as common or uncommon. For example, in the United States 43% of the population has type O blood; 42% has type A, 12% type B, and 3% type AB. A match between a crime scene blood sample and the blood of a suspect would prove only that they *could* be connected, not that they definitely are. Protein profiles are based on the presence or absence of certain cell membrane proteins. Like blood types, these cannot identify an individual, but only classify one as a member of a population with particular statistical parameters. Proteins, including blood, also tend to degrade quickly in the environment, making analysis of all but the freshest samples difficult.

The sequence of DNA bases from even a small biological sample—semen, blood, hair, saliva residues on envelopes and cigarette butts, or cells from fingernail scrapings—can be compared to suspect samples for similarities and differences. DNA evidence has been used both to convict suspects and to exonerate innocent individuals. By contrast, except in the case of identical twins, DNA is unique to a single person and can theoretically be used to identify an individual with a high degree of certainty. It is also very long-lasting, so older samples can be analyzed. Further, only very small samples are needed and these can be taken from most any biological specimen.

Typically, a crime scene sample of DNA is collected and, if necessary, copied using a technique known as polymerase chain reaction, a process that copies a small segment of DNA millions of times to obtain a sample large enough for analysis. Some segments of DNA code for proteins, and are virtually identical in all people. However, a much greater fraction of DNA contains sequences of bases that are repeated over and over—up to 30 times in a row—with no known function. The lengths of most of these repeat segments vary significantly between individuals, and provide forensic scientists a way to distinguish one individual from another. Long strands of DNA are "chopped" into smaller fragments with the help of *restriction enzymes,* special molecules that

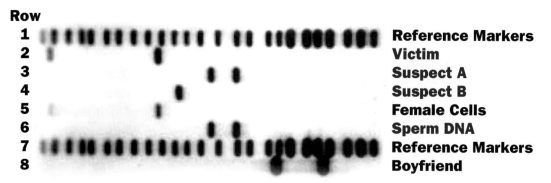

Row	
1	Reference Markers
2	Victim
3	Suspect A
4	Suspect B
5	Female Cells
6	Sperm DNA
7	Reference Markers
8	Boyfriend

Figure 9.26 A DNA "fingerprint." (See text for an explanation.) (© 2001 How Stuff Works, Courtesy Genelex)

seek out specific base sequences on DNA. The fragments are then sorted out by length and electrical charge (originating with ionized phosphate groups). Further treatment and transfer of the separated fragments to a nylon membrane produces a series of bands called a DNA fingerprint that is unique to each individual. DNA bands from suspects and crime scenes or from two or more individuals can be compared directly to determine whether they match.

As an example, consider the set of DNA fingerprints in Figure 9.26 that were used to distinguish between suspects in a sexual assault criminal case. Rows 1 and 7 are simply general reference markers in the test, and correspond to fragments of DNA having various lengths, from longest (left) to shortest (right). The spots in row 2 correspond to the female victim's DNA, whereas those in rows 3 and 4 are those of the two suspects in the case, and that in row 8 is that of her boyfriend. DNA samples from the victim's vaginal canal are shown in rows 5 and 6.

First, it is clear that the sample of row 5 is that of the victim herself, since the pattern of spots coincide with that of her DNA (row 2). Row 6 therefore must be DNA from sperm found in her canal after the assault. Comparing the positions of the dots in this DNA to three possible sources—suspect A, suspect B, and her boyfriend—it is clear that the match is unambiguously that of suspect A.

Although for forensic work it is best to use DNA from the nucleus of cells, DNA found outside the nucleus can be used and is called *mitochondrial DNA*. Mitochondrial DNA can be extracted from hair or bone in "dead" samples. It is derived from the mother alone, is subject to greater variability within a body, has a higher mutation rate than nuclear DNA, and only codes for a few dozen genes. Because of its relatively high mutation rate, it is useful in deducing the evolutionary history of a species. The existence of identical mutations in the mitochondrial DNA taken from different groups of humans is a good indication that the

▶ **Discussion Point:** What are the legal and ethical implications of DNA collection and databases?

Statistics show that a very small number of people are responsible for a very large percentage of criminal activity. In the interest of maintaining a more secure society for everyone, law enforcement agencies are calling for the collection of

DNA samples from every violent criminal. Should every criminal be forced to submit samples of DNA for testing and classification? Who should have access to your genetic information? Where do you stand in the argument over individual liberty versus the right of government agencies to access your genetic information? Use the resources on the Web site for this book to help you develop arguments in support of and against the collection and database storage of DNA information.

groups have been in contact in the past. Extraction and subsequent analysis of the nitrogen base sequence of mitochondrial DNA from bones of several Neanderthals has led to the conclusion that they did not mate with modern humans.

Steroids: Cholesterol, Sex Hormones, and Birth Control Pills

Several aspects of the chemistry of your body are controlled by a set of closely related molecules, called **steroids,** each of which contains alcohol and/or ketone groups connected to a large hydrocarbon structure. The carbon backbone of this structure, which occurs in all plants and animals, is the steroid system, a series of four hydrocarbon rings that are consecutively fused to each other—that is, connected by the sharing of two carbons:

Steroid backbone

Exercise 9.3

Draw the Lewis structure, showing all carbon and hydrogen atoms, for the four-ring steroid shown above.

For identification purposes in discussing particular molecules, we have denoted the six-membered rings in steroids as A, B, and C, and the final, five-membered ring as D. Notice that the centers of the rings are not aligned in a linear fashion, but branched. The specific molecules of interest to us differ only in the nature of the bonds formed by atoms in the first and last rings (i.e., A and D), and so we'll concentrate on these two rings in our discussions.

9.24 Testosterone is one of several anabolic steroids

Recall that hormones are chemical messenger molecules that travel through our bloodstream and signal various parts of our body to undergo change. *Testosterone* is the hormone that promotes muscle growth, the development of male secondary sex characteristics, and the production of sperm. Notice in Figure 9.27a that the A ring for testosterone contains a ketone group (hence the *-one* ending on the name of the substance) and a

Figure 9.27 Steroid structure. Substitutions on the steroid skeleton produce different chemicals.

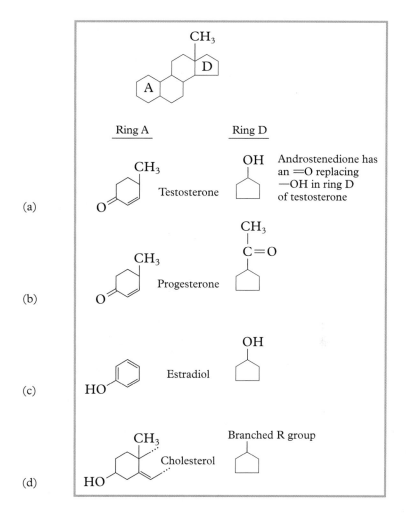

Ring A · Ring D

(a) Testosterone

Androstenedione has an =O replacing —OH in ring D of testosterone

(b) Progesterone

(c) Estradiol

(d) Cholesterol — Branched R group

carbon–carbon double bond. The D ring has an OH group on its top carbon, thereby making that part of the molecule alcohol-like.

Testosterone is an example of an **anabolic steroid,** that is, one that promotes the growth of muscles. People who are recovering from diseases that have resulted in muscle loss, and athletes who want to artificially build up their strength for competitive or cosmetic purposes, often take a synthetic anabolic steroid. The synthetic versions of the hormone have the anabolic characteristics of testosterone but do not rapidly decompose as testosterone itself does if it is taken orally or by injection. One example has a structure identical to that of testosterone except that the alcohol group of its D ring has been replaced by an ester of a long-chain carboxylic acid. In another example, the hydrogen atom bonded to the carbon of the C—OH group on D has been replaced by a short carbon chain. These substitutions not only slow the decomposition of the steroid, but also prevent it from strongly attaching to the body's receptors for **androgen,** the male sex hormones responsible for the development of male sex characteristics.

There can be significant side effects, such as liver damage, to taking anabolic steroids over long periods at high doses. For females, side effects include the development of male sex characteristics such as facial hair and a deeper voice, and some loss of female sex characteristics.

All compounds with the steroid system are highly lipophilic ("fat loving"), so they readily dissolve in fat and consequently remain in the body for some time. For this reason, analysis of a urine sample from an athlete can determine whether or not he or she has taken the substances in the recent past. In order to avoid difficulties with drug enforcement agencies, some athletes and bodybuilders have turned to "Andro-6," the popular name for *androstenedione*, an anabolic steroid that is naturally found in humans. When ingested, it is converted by the liver into testosterone. The structure of the two substances is identical except that Andro-6 has a carbonyl oxygen atom at the top of ring D where testosterone has an —OH group. Since it occurs naturally in the human body—as it does in other animals and some plants—androstenedione is not classified as an anabolic steroid by U.S. government agencies concerned with drugs, so it is sold legally in health food stores.

Cortisone is also a non-anabolic steroid. It is used as an anti-inflammatory agent and to treat rheumatoid arthritis.

> **Discussion Point:** Should athletes be allowed to use Andro-6 and carbohydrate loading?

Athletes can build extra muscle by using the steroid nicknamed Andro-6, and they can build extra energy reserves for competition days by manipulating their glycogen levels through carbohydrate loading. Do you think it should be legal for athletes to use Andro-6 or its equivalent? Or to load carbohydrates to adjust their glycogen? Should the two ways of enhancing performance be treated equally? List at least two points that support your point of view in each case. Do your conclusions depend on the ability of scientists to test athletes for these behaviors?

9.25 Contraceptive pills use female sex hormones

Although there is only one type of androgen, there are two types of female sex hormones: **estrogens** and **progestins.** The main progestin is *progesterone*, which differs structurally from testosterone only in that the —OH group of the D ring is replaced by a short chain that includes another ketone group (Figure 9.27b). When the level of progesterone in a woman's body is low, certain other hormones are released into her bloodstream that provoke the process of ovulation, which is the release of an egg that can subsequently be fertilized by sperm, resulting in pregnancy. When pregnancy does occur, body levels of progesterone rise naturally, promoting the implantation of the fertilized egg in the uterine wall.

High levels of progesterone also prevent subsequent ovulation from occurring. It was this latter characteristic that led to the use of analogs of progesterone in oral contraceptives. Such "birth control pills," which prevent ovulation and therefore pregnancy, do not contain progesterone itself since it is broken down in the body and high doses cause side effects. Instead, they include synthetic variants of progesterone that have minor changes in the groups attached to the A and D rings. These subtle changes allow much smaller doses to be effective and long-lasting contraceptives, without most of the side effects.

The synthetic steroid in the original "morning after" pill, RU-486, is also related chemically to progesterone and to its synthetic variant used in oral contraceptives. RU-486, however, contains an additional large group of atoms bonded to ring C. Its similarity to progesterone allows it to occupy the receptor site in the uterine lining, but it does not provoke the production of proteins that are required to implant the fertilized egg. Since progesterone itself is thereby blocked from the site, implantation of a fertilized egg is prevented, and a spontaneous abortion occurs. The use of RU-486, or one of its alternatives that has the same action, has provoked widespread controversy between pro-choice and antiabortion groups, especially in the United States.

The other constituent of oral contraceptives is a synthetic variant of *estradiol*, the main sex hormone of the estrogen class. This compound consists of molecules that differ from those of testosterone in that the =O bond to the ring carbon is replaced by an —OH group. In addition, the A ring contains three sets of alternating single and double bonds (as in benzene) but is missing the methyl group where rings A and B fuse (see Figure 9.27c). Estradiol and progesterone together regulate the menstrual cycle in females; estradiol also regulates the production of eggs and the development of female sex characteristics.

9.26 Cholesterol is naturally present in animals

The famous—or infamous—compound *cholesterol* is also a steroid. It is an alcohol, as indicated by the suffix *-ol* in its name, with the —OH group in ring A (see Figure 9.27d), as also occurs in estradiol. In contrast to the sex hormones, it contains a double bond in ring B. At the top carbon of ring D lies a branched, eight-carbon alkyl group.

Cholesterol is the most abundant steroid in animals. From it the sex hormones and other important steroids and other substances such as cell membranes are synthesized in our bodies. Don't be too impressed when you see specific plant-based foods advertised as cholesterol-free: this is true of most plants! Although we ingest cholesterol when we eat animal-based food, in fact our own body makes most of the cholesterol that is present in our bodies. We synthesize cholesterol in our livers, the reactants being mainly the saturated fats in our diet. On average, we manufacture about 600 milligrams of cholesterol daily in our livers, and absorb about 80 milligrams more through our diet. Though reducing cholesterol in the diet is promoted as beneficial for the heart, it is usually more important to reduce saturated fat intake in controlling cholesterol levels in our bodies.

9.27 Lipoproteins contribute to heart disease risk

Cholesterol and other lipids are vital parts of cell membranes and certain hormones. Lipids are fatty materials that are insoluble in water, but they must be moved in the body via the bloodstream from the point where they are synthesized to the tissues where they are stored or used. A water-soluble "coat" of lipoproteins makes these hydrophobic molecules temporarily soluble.

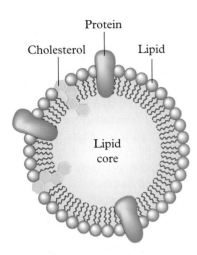

Figure 9.28 The structure of a lipoprotein.

In spite of their nicknames, neither LDL nor HDL is itself cholesterol, though both particles include that molecule.

Lipoproteins are spherical, aggregate particles with hydrophobic lipids at their core, and have hydrophilic side groups of proteins at the surface that allow them to dissolve in blood (see Figure 9.28). Various combinations of proteins and lipids produce lipoprotein particles having different densities. The low-density lipoproteins, LDL, have a smaller mass-to-volume ratio than do those of high-density lipoproteins, HDL.

LDL particles average about 22 nanometers in diameter, and are very rich in cholesterol. They carry the cholesterol from the liver to various tissues, which absorb them. The cholesterol is used in the tissues for membrane construction or hormone synthesis. Unfortunately, if more cholesterol is available in the blood than is needed by the tissues, LDL concentrations increase in blood vessels, and they penetrate the lining of the vessels. Outside the lining, the LDL is taken up by scavenger cells, which accumulate the cholesterol in lipid droplets and eventually form fatty deposits called *plaques* in the walls of the vessels. Recent evidence suggests that it is the number of LDL particles, not their total mass, that is the most important factor in determining plaque formation.

The buildup of plaque on the inner walls of arteries is called *atherosclerosis*. The plaques can cause obstruction of the blood vessels, which can reduce the flow of blood into the heart and may eventually lead to a heart attack. Heart failure arising from blocked coronary arteries causes many deaths annually; in fact, it is the leading cause of death in most industrialized nations.

HDL particles initially are rich in protein and low in cholesterol. As they travel in the bloodstream, HDL particles accumulate cholesterol—from artery walls, for example—and transport it back to the liver, where it is stored or broken down. Thus a *lack* of HDL is correlated with higher incidence of heart disease.

HDL is sometimes called "good cholesterol" whereas LDL is called "bad cholesterol" because of its contribution to plaque buildup. Often the ratio of LDL to HDL is taken as the best indication of whether a person has a "cholesterol problem" or not. A ratio of 5 means that the amount of LDL is five times greater than the amount of HDL. For men, an LDL-to-HDL ratio greater than 5 indicates an increased risk for heart disease. For women, the critical risk ratio is 4.5. It is recommended that the total cholesterol level be 200 milligrams or less per 100 milliliters of blood, with no more than 130 mg/100 mL being LDL. Individuals with total cholesterol levels of 240 mg/100 mL or higher, LDL levels of 160 mg/100 mL or higher, or HDL levels lower than 35 mg/100 mL, are likely to be at high risk for heart disease. People with high levels of cholesterol, which equates to having a higher ratio of LDL to HDL in the bloodstream, are often prescribed drugs called *statins*. These pharmaceuticals block the action of liver enzymes that synthesize cholesterol, and also stimulate the production of the membrane protein in tissues that removes LDL from the blood. Consequently, LDL is removed more quickly and its levels in the blood drop.

As discussed previously, most of your cholesterol is manufactured in your liver rather than being imported by the food you eat. As indicated by the table below, the type and amount of fat you consume plays

Some types of saturated fats, such as those high in stearic acid, apparently have no effect on cholesterol levels. Because they neither raise nor lower the HDL or LDL levels in a person's bloodstream, they are considered neutral.

a large role in determining the type and amount of cholesterol in your bloodstream. Thus, most saturated fats increase both types of cholesterol, whereas monounsaturated fats increase HDL concentrations (the "good" cholesterol) while decreasing the amount of LDL (the "bad" cholesterol). Trans-fatty acids seem even worse than saturated fat since they decrease the level of "good" cholesterol and increase the amount of "bad" cholesterol.

Type of fat	Total cholesterol	HDL	LDL	Net effect on health
Saturated	↑	↑	↑	Bad
Monounsaturated	~	↑	↓	Good
Polyunsaturated	↓	↑		Good
Trans-fatty acids		↓	↑	Very bad

In general, premenopausal women have higher levels of HDL than do men, and as a consequence, they are less prone to heart attacks than men in their early and midlife years. The level of HDL in the blood can be increased by exercise and weight loss.

Summarizing the Main Ideas

The basic units of DNA and RNA are nucleotides. Each nucleotide consists of a sugar unit connected to a phosphate group and to a nitrogenous base. In DNA the sugar is deoxyribose and the bases are adenine (A), guanine (G), cytosine (C), and thymine (T). In RNA the sugar is ribose and the bases are adenine, guanine, cytosine, and uracil (U). Alternating sugar and phosphate groups form the backbone structure of nucleic acids and the bases extend from the backbone. The three-dimensional structure of DNA is a double helix of two strands held together by hydrogen bonds between base units. The principle of complementarity governs base pairing—A and T (U in RNA), G and C—and is the basis for DNA replication and information transfer.

The linear sequence of bases in DNA carries genetic information in the form of genes. Genes encode the directions for building proteins, which are involved in all life functions.

Information on DNA is transcribed into a single complementary messenger RNA (mRNA) molecule. Triplets of mRNA bases called codons specify an amino acid. Translation occurs with the help of transfer RNA (tRNA) and ribosomal RNA (rRNA) in the ribosomes. Each tRNA molecule carries a triplet of bases called an anticodon at one end and a specific amino acid at the other. The ribosome reads mRNA's message one codon at a time and a complementary tRNA anticodon pairs with each codon, adding an amino acid to a polypeptide chain. Polypeptides form proteins.

Proteins are polyamides produced via condensation reactions between amino acids. Alpha (α) amino acids are organic molecules in which a carboxylic acid group, —COOH, and an amino group, —NH$_2$, are bonded to the same intermediate carbon atom —CHR—. Amino acids linked by peptide bonds form polymeric molecules called polypeptides (if the chain is relatively short) or proteins (if the chain is long). All proteins in humans are composed of polymers formed by the same collection of 20 α-amino acids. Simple proteins contain only amino acids; conjugated proteins possess other groups as well.

The primary structure of proteins corresponds to the sequence of amino acids joined by condensation reactions. A protein's secondary structure is the shape adopted by the polymer's backbone. Common shapes for proteins or parts of proteins are the α helix and the β pleated sheet structures. The tertiary structure of a protein refers to the manner in which it is folded. This structure is determined by the interactions between the side groups of the component amino acids. The quaternary structure of the protein refers to the arrangement of the component polymer chains and/or the non-polypeptide components relative to each other. The destruction of the secondary, tertiary, and quaternary structure of a protein is called denaturation.

Proteins have many different functions in the body. Proteins that catalyze specific biochemical reactions are called enzymes. A reactant molecule, the substrate, fits into the active site of the enzyme. During the reaction, partial bonds form between the substrate and the enzyme, thereby decreasing the amount of energy that is required before the substrate can react.

Mutations are changes in the sequence of bases that code for a particular protein. Mutations can alter the shape and therefore the function of a protein. Silent mutations do not affect overall protein function. Insertions and deletions can affect protein function minimally or render a protein inactive. If an inactive protein is vital to life, the mutation is lethal. Mutations are the basis for many diseases.

Our understanding of the genetic code has made cloning and DNA analysis possible. Cloning uses genetic material to make exact copies of organisms. DNA analysis allows forensic investigators to use DNA from biological samples to identify, convict, or exonerate individuals.

Lipoproteins are a collection of protein molecules surrounding a lipid particle. They provide a mechanism by which hydrophobic molecules such as cholesterol can be transported through the bloodstream, even though they are insoluble in it.

Human sex hormones and cholesterol all contain the steroid structure of four fused hydrocarbon rings. Both oral contraceptives and the RU-486 abortion pill are synthetic variants of natural steroid sex hormones. Cholesterol is formed by the body from saturated fatty acids.

Key Terms

DNA	nitrogen base	double helix	transcription
RNA	guanine	principle of	transfer RNA
nucleic acid	adenine	complementarity	ribosomal RNA
nucleotide	cytosine	chromosome	translation
deoxyribose	thymine	gene	codon
ribose	uracil	messenger RNA	ribosome

anticodon   polypeptide   incomplete protein   mutation
gene expression   α helix structure   active site   mutagen
protein   β pleated sheet   substrate   steroid
α-amino acid   structure   simple protein   anabolic steroid
peptide bond   salt bridge   conjugated protein   androgen
dipeptide   essential amino acid   coenzyme   estrogen
tripeptide   complete protein   lipoprotein   progestin

Web Sites of Interest

To link to Web sites of interest, go to www.whfreeman.com/ciyl2e,
Chapter 9, and select the site you want.

For Further Reading

"Chemistry of Colors and Curls," *Science News,*
160, August 25, 2001, pp. 124–126. This
interesting article examines styling, coloring,
and perming hair at the molecular level.

"Eating the Genes," *Technology Review,*
July/August, 2001, p. 90. This article is an essay
on the impact genetically modified foods can have
in protein-deprived developing countries.

"Grains of Hope," *Time,* July 31, 2000, pp. 50–57.
This article outlines how genetically modified
foods, including rice, can help to alleviate hunger
and health problems in developing countries.

"A New Breed of High-Tech Detectives," *Science,*
289, August 11, 2000, pp. 850–857. A thorough
examination of how forensic scientists use the
latest tools for collecting and analyzing evidence,
including DNA. It also discusses privacy issues.

Technology Review, September/October 2000. This
issue has a large collection of articles devoted to
issues relating to genetic technology.

Review Questions

1. What is *DNA*?

2. What is *RNA*?

3. Why can DNA and RNA be viewed as polymers?

4. How are DNA and RNA similar to condensation polymers?

5. How are the subunits of DNA and RNA polymers different? How are they alike?

6. What nitrogen bases are found in DNA molecules?

7. What nitrogen bases are found in RNA molecules?

8. What is meant by the term *base sequence*?

9. Describe the structure associated with a DNA molecule and explain why DNA adopts this structure.

10. How many letters are in the DNA alphabet?

11. What is a *gene*?

12. What does it mean to say that a gene is *expressed*?

13. Identify the different regions on a gene that are important for its expression.

14. What is *messenger RNA*?

15. What is a *codon*?

16. What is the relationship between DNA and mRNA?

17. What is *base pairing*?

18. What role does transfer RNA play in protein synthesis?

19. What role does ribosomal RNA play in protein synthesis?

20. What is an *α-amino acid*?

21. What is the basic structural unit of proteins?

22. Identify the two amino acids below:

$$H_2N-CH-\overset{\overset{\displaystyle O}{\|}}{C}-OH \qquad H_2N-CH-\overset{\overset{\displaystyle O}{\|}}{C}-OH$$

$$\underset{\underset{\displaystyle OH}{|}}{CH_2} \qquad \qquad \underset{\underset{\displaystyle H}{|}}{H_3C-C-CH_3}$$

(a)　　　　　　　　(b)

23. What is an *endorphin*?

24. What is the *primary structure* of a protein?

25. How does the *secondary structure* of a protein differ from the *tertiary structure*?

26. What attractive forces produce the α *helix* structure and the β *pleated sheet* in a protein?

27. What characteristics of side groups make them hydrophobic? Which category of side groups act as hydrophobic areas?

28. What characteristics of side groups make them hydrophilic? Which category of side groups act as hydrophilic areas?

29. What is a *complete protein*?

30. What is the relationship between an *essential amino acid* and complete proteins?

31. How do dietary pairings help to provide complete protein nutrition?

32. How does a simple protein differ from a *conjugated protein*?

33. What components other than amino acids can be part of a conjugated protein?

34. What roles do vitamins play in protein function?

35. Identify the proteins below as transport, structural, defense, or enzymatic proteins:
a) Amylase catalyzes the breakdown of starch into glucose units.
b) Transferrin carries iron in the body.
c) Actin is involved in the movement of muscle fiber.
d) Hemoglobin carries oxygen in the blood.

36. What is an *enzyme*?

37. What is a *substrate*?

38. What is a *coenzyme*?

39. What is a *lipoprotein*? Describe its structure.

40. What is a *mutation*?

41. Identify three types of mutations.

42. What is a *mutagen*?

43. What is a genetic disease? Provide an example.

44. Why can DNA be used to individualize evidence in criminal cases?

Understanding Concepts

45. Draw the basic subunit of a DNA polymer.

46. Draw the basic subunit of an RNA polymer.

47. Explain how the base sequence on a DNA molecule is like an alphabet.

48. Explain how the structure of DNA carries information. Provide an example.

49. What is the relationship between a codon and an amino acid? What is the relationship between a codon and an anticodon?

50. What is the relationship between a protein and amino acids?

51. Explain how mRNA is constructed from DNA.

52. Determine the mRNA base sequences for the DNA base sequences below:
a) GGC-GAA-TAT
b) AAA-AAG-CAG
c) GGA-GAT-GCG

53. Use Table 9.1 to determine the sequence of amino acids in the mRNA base sequences in question 52.

54. What role does hydrogen bonding play in base pairing?

55. What role does hydrogen bonding play in maintaining the DNA helix structure?

56. Outline the sequence of events that begins with information on a DNA molecule and ends with the production of a protein.

57. How does a gene activator protein switch a gene on or off?

58. How can mutations arise?

59. Join the amino acids in question 22 by making peptide bonds. What are the two possible ways they can join? What are the shorthand designations for the two dipeptides?

60. Draw the structure of the dipeptide formed by combining two molecules of the amino acid cysteine.

61. Draw the structures of the two possible dipeptides formed by combining one molecule of alanine with one cysteine.

62. How many different tripeptides could be formed by combining three different amino acids A, B, and C? Solve this problem by drawing them.

63. What effect does the primary structure of a protein have on its final three-dimensional shape?

64. What effect does the primary structure of a protein have on its function?

65. How does an enzyme work to catalyze a reaction?

66. Enzymes are proteins. Protein structure is affected by heat. Explain why high fevers can be especially dangerous.

67. What property differences might you predict between two proteins if one has a mostly α helix structure and the other has a mostly β pleated sheet structure?

68. Why are some mutations called silent?

69. Explain how mutations are related to protein structure.

70. Identify the mutations in the sequences below:

    Normal sequence: She ran and sat too.
    a) Sae ran and sat too.
    b) She ran dna sat too.
    c) She ran and too sat.

71. Explain why genetic diseases can be passed to offspring.

72. Both DNA and proteins can be used to classify evidence. Explain how they are used to do this and why the two methods are different.

Synthesizing Ideas

73. Some genes need to be on at all times, but other genes are only activated under certain circumstances. Think about this and identify three processes in which each of these two types of genes might be involved.

74. Milk is high in protein and is sometimes administered in cases of metal poisoning when the substance has been eaten. Why do you think milk is used this way?

75. Adding salt to a mixture of egg albumin and water causes the protein to coagulate. Which side-group interactions do you think are affected by the addition of the salt?

76. Explain how the denaturation of a protein differs from the digestion of a protein.

77. Based upon your understanding of protein function in the body and temperature influence on protein activity, what effects would you predict for a person suffering from hypothermia, an abnormally low body temperature?

78. Explain why lack of iron in the body leads to an inability to carry oxygen.

79. Pet stores sell a product that reportedly removes stains and odors caused by animal urine. The product contains an enzyme. Explain how this product might work.

80. Why might the deletion or insertion of three bases in a DNA sequence not be as detrimental as inserting or deleting two bases in a DNA sequence?

81. Why do dentists put a lead apron over patients when doing dental X-rays?

82. Use the resources on the Web site for this book to provide two examples, other than those in the chapter, of how genetic information can be used to benefit humanity.

83. Use the resources at the Web site for this book to answer the following questions: What issues of privacy have arisen over genetic information? How can genetic information be misused? How could you prevent the misuse of genetic information in the examples you provided?

84. How many carbon atoms are contained within the four fused rings (A, B, C, and D) of a steroid? How many hydrogen atoms in total would be bonded to this ring system?

85. Using the structural diagram for testosterone in Figure 9.27 as a guide, draw the molecule that would be obtained if the double bond within the rings were to be hydrogenated.

86. The RDA for iron is 15 milligrams per day for young women. Given that there are about 4 grams of this element present in the body of an average woman, estimate how long the average iron atoms remains in a woman's body once it is absorbed. Assume for simplicity that 10% of the iron intake is absorbed.

■ Group Activity: Extracting the DNA from Wheat Germ
Adapted from an experiment provided by Kay Calvin, University of Western Ontario

As for all plants and animals, the genetic information for wheat plants is contained in its DNA. In this experiment, you will extract strands of DNA from the germ part of wheat plants and be able to see them. The human DNA that is present in your cells differs only slightly—but obviously in some important ways—from the DNA of wheat germ and other plants.

For this activity, your group will need the following supplies:

■ salt (about 3 teaspoonfuls)
■ lemon juice (1/4 of a teaspoon)
■ wheat germ (about 50 mL, or 1/5 of a cup)
■ rubbing alcohol (about 35 mL, or 1/6 of a cup)
■ water
■ dishwashing liquid (a few drops)
■ 2 beakers or transparent glass tumblers
■ 2 cotton dishwashing cloths
■ measuring cup or equivalent
■ teaspoon

One person should be in charge of the supplies and measure them out as needed. A second person should read out the instructions and record the observations.

Procedure

1. Measure about 100 mL (2/5 of a cup) of cold water into one of the clear beakers or tumblers. Then add about 1 teaspoon of salt and 1/4 teaspoon of lemon juice to the water. Stir the mixture with the teaspoon until the salt dissolves.

2. Add 50 mL of wheat germ to the beaker, and stir the mixture for about 10 minutes. (This breaks down the wheat germ.)

3. Strain the watery mixture through the cloth, letting the liquid drain into the second beaker. Gently squeeze the cloth to get all of the water out. Discard the liquid, but keep the cloth and its solid contents.

4. Rinse out a beaker, then add about 50 mL (1/5 cup) of clean water to it. Add 1/2 teaspoon of salt, 5 mL (1 teaspoon) of the alcohol, and a few drops of the dishwashing liquid to the water.

5. Using the spoon, gently scrape the wheat germ pulp off the cloth into the first beaker, and stir gently for about 20 minutes.

6. Add 2 teaspoons of salt and stir again for about 10 minutes, then let the mixture stand for a few minutes.

7. If a solid settles to the bottom of the beaker, carefully pour off the liquid into the rinsed second beaker. Discard the solid.

8. If no solid settled, filter the liquid through a new cloth into the rinsed second beaker. If there is solid in the filtered liquid, repeat the process. Discard the solid.

9. Measure a few spoonfuls (10 mL) of the liquid into a cleaned beaker, and add about 30 mL (1/8 cup) of the alcohol to it.

10. As you gently stir the liquid, you will see white strands forming. This is the DNA of the wheat germ!

11. Clean up the equipment and dispose of the materials as indicated by your instructor.

Results:

1. Strands of DNA were isolated from the aqueous solution when alcohol was added to it. Given the acid structure of DNA and the fact that the alcohol has a substantial hydrocarbon component

to its structure, why do you think the DNA is more soluble in water than in the alcohol?

2. Do you think that each of the white strands you see is a single molecule of DNA? *Hint:* Recall the sizes of atoms and molecules compared to the minimum size you can see.

3. Would the sequence of nitrogen bases in wheat germ be identical to that in other types of plants? To that in your body?

Taking It Further

Altering a protein's shape affects its functionality

In general, a protein will not function properly unless its shape is exactly right. The secondary, tertiary, and quaternary structure of proteins can be destroyed by various processes. When a protein loses its functional shape, it is said to be denatured, and any process that leads to that loss is called **denaturation.** A familiar example occurs when an egg is cooked. Heating denatures the protein albumin in the egg white, converting it from a clear liquid into a white solid. At the molecular level, the spherical albumin protein—which has ionic groups on its surface and consequently repels other identical molecules—uncoils and forms networks with neighboring albumin molecules, producing a solid interconnected network. You can prevent an egg that cracks during boiling from losing too much of its contents by adding a little vinegar or salt to the water. These substances quickly denature the escaping uncooked protein, and the coagulated products seal the crack and protect the egg from further loss.

Another example of denaturation occurs in your stomach, where acid denatures the protein in your food, permitting rapid hydrolysis of the peptide bonds to yield individual amino acids.

Generally, the shapes of proteins in our bodies are sensitive to changes in temperature and acidity. This is one of the reasons that fevers can be dangerous. The higher internal temperatures associated with fever cause molecules, including proteins, to move around more than usual. This motion is enough to disrupt bonds between side groups and alter a protein's structure. Once the structure is changed, the protein can no longer do its job in the body. It is not uncommon for a protein to change its shape and even its constitution upon heating. For example, the albumin proteins in raw egg whites are individual, separated spheres. When heated, they unravel and bond with each other to form an interconnected network solid.

Prions are the infectious agents at work in "mad cow disease" and other forms of TSE (transmissible spongiform encephalopathies). Prions are proteins that are have not folded properly, and as a result, they are misshapen. Specifically, some of the alpha regions in the normal protein are converted to the beta form. This now deformed protein then becomes a template, inducing similar refolding among normal proteins of the same type that they encounter. Thus, once prions gain entry into the body, their spread is rapid because they are able to self-propagate. TSEs result from a type of prion produced in the brain, and whose change in shape causes healthy brain tissue to become spongy. Protein misfolding is also known to play a role in Alzheimer's and Parkinson's diseases, and is the cause of Creutzfeld-Jakob disease.

In this chapter, you will learn:

- about vitamins, their sources and why we need them in our diet;

- about antioxidants, how some are used as food additives to preserve the freshness and flavor, while others help prevent cancer and heart disease;

- about the different types of food additives, including natural and synthetic antioxidants, preservatives, food colors, and flavor enhancers;

- how food additives affect your health;

- about pesticides and their effect on the foods we eat;

- why the pesticide DDT has been banned in most countries;

- about the environmental sources and toxicity of dioxin and PCBs;

- about the sources of environmental estrogens and the potential threats to human health associated with them.

What price do you pay for cheap and plentiful food?

The use of synthetic compounds, such as the pesticide being sprayed in this field, increases agricultural production and helps to keep food prices low. However, it may have hidden costs for your body and the environment. In this chapter we examine how chemicals found in small amounts in food and in the environment can sometimes have far-reaching effects. (Digital Vision)

Chemicals in Our Bodies and Our Environment

Vitamins, Food Additives, Pesticides, and More

Though you may hear a great deal about changing your eating habits to avoid "chemicals," as you've seen throughout this book, chemicals make up everything in our world. The last three chapters have focused on the biochemistry of the human body, including the major biological molecules that make you who you are. In addition to the fats, carbohydrates, proteins, and nucleic acids, you have in your body small concentrations of a large number of other chemicals. Some of these substances are natural chemicals, necessary for our health; vitamins fall into this category. On the other hand, as a consequence of living in modern society, our bodies also contain low levels of many synthetic chemicals, especially food additives and pesticides, that we absorb from our food. In this chapter, we shall explore the nature, origin, benefits, and potential health risks from both the natural and synthetic chemicals we live with.

Vitamins

Vitamins are essential, noncaloric nutrients composed of organic compounds that enable the body to absorb, digest, and build what it needs in association with other nutrients and to preserve its existing components. Although essential, vitamins are needed in only very small amounts—milligrams, or in a few cases just micrograms—compared to our much more massive requirements for proteins, carbohydrates, fats, and even minerals. Like the essential amino acids and essential fats, our body cannot synthesize most vitamins from other materials, and so we must be supplied with them directly in our diet, or through supplements. The word *vitamin* comes from the fact that these substances are <u>vit</u>al to our bodies, and that the earliest known vit<u>amin</u>s were <u>amine</u>s (though we now know that many vitamins are not amines).

Vitamins and minerals are both important to a person's well-being. A healthy diet that includes lots of fresh vegetables and fruits, whole grains, and complete protein sources such as fish, meat, and soy provides the recommended daily amount of vitamins and minerals. Many people, however, have diets that put them at risk for vitamin deficiency conditions. One way to compensate is to take vitamin supplements. There are also many well-researched nutritional plans available from

Whenever you see this icon in this chapter, go to
www.whfreeman.com/ciyl2e

Figure 10.1 Water-soluble vitamins.

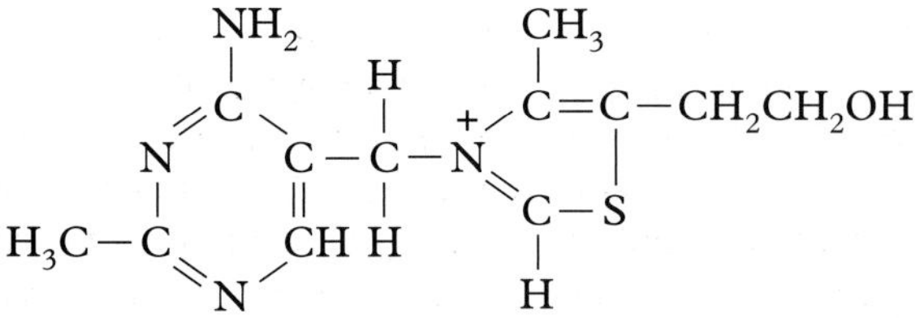

(a) Thiamine (vitamin B$_1$)

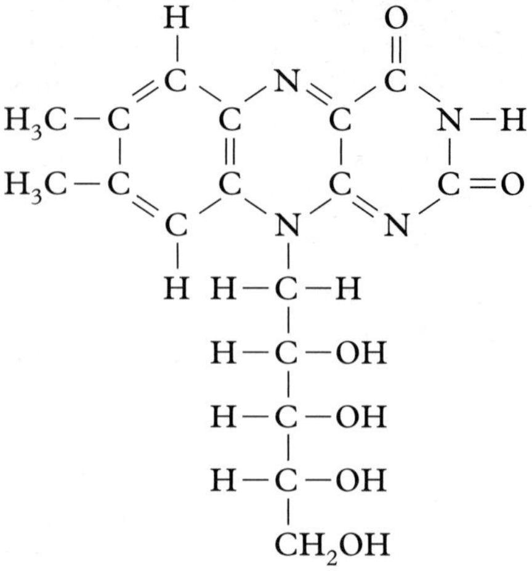

(b) Riboflavin (vitamin B$_2$)

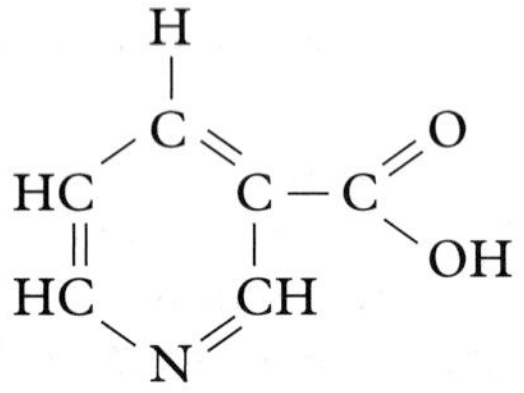

(c) Niacin (vitamin B$_3$)

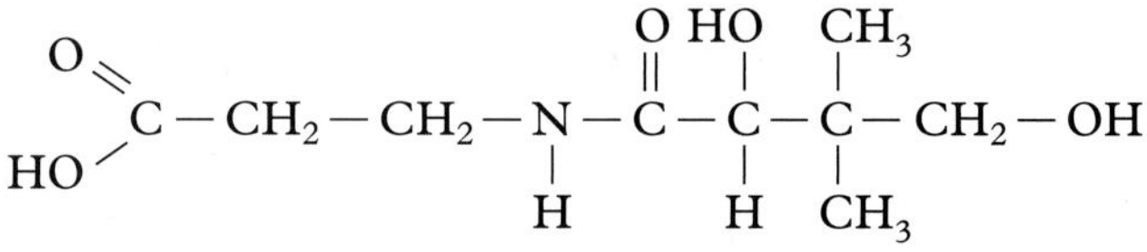

(d) Pantothenic acid (vitamin B$_5$)

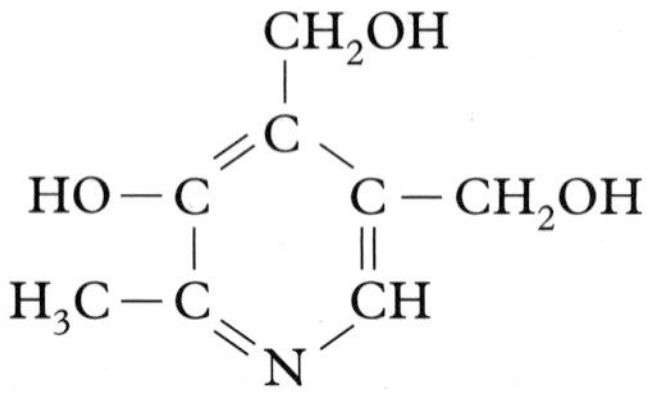

(e) Pyridoxine (vitamin B$_6$)

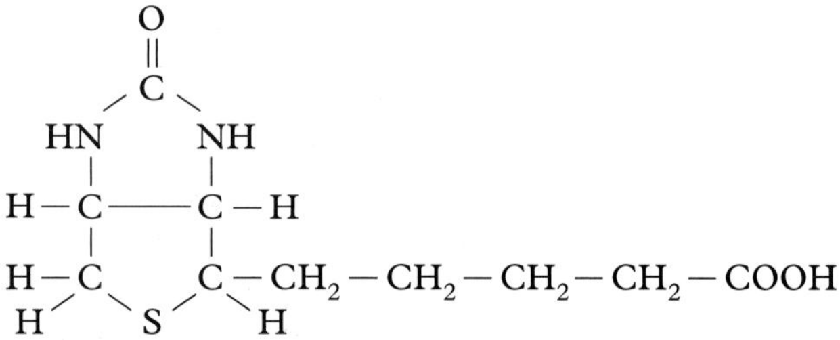

(f) Biotin (vitamin B$_7$)

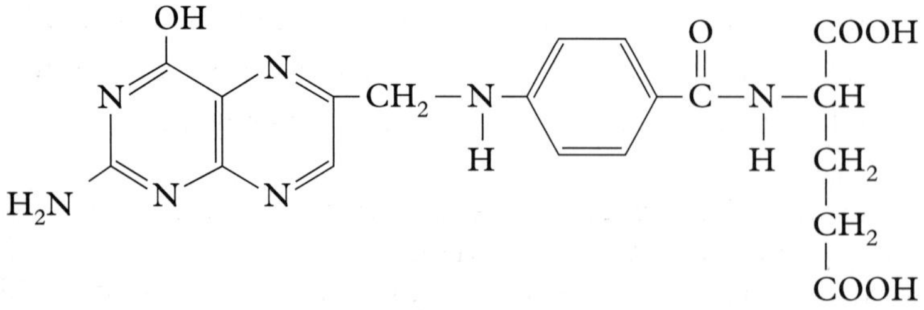

(g) Folic acid (vitamin B$_9$)

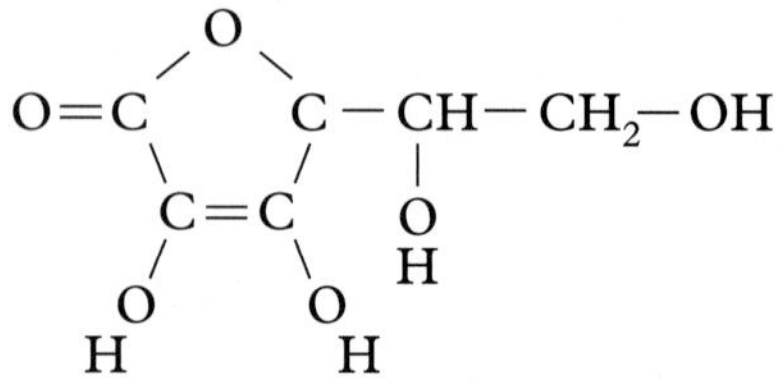

(h) Vitamin C (ascorbic acid)

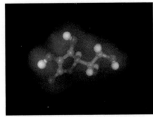

View the structures of the water-soluble molecules vitamin B and vitamin C at Chapter 10: Visualizations: Media Link 1.

Most mammals other than primates (a group that includes humans) can synthesize vitamin C in their bodies.

???????????????? **?** ????????????
Fact or Fiction !

People of British origin are sometimes called by the slang term *limeys* because their sailors ate limes.

True. The British navy solved the problem of sailors on long voyages contracting scurvy by supplying them with lime juice.

Figure 10.2 Citrus fruits.
(Photodisc/Punchstock)

doctors. Some people need nutritional supplements because of chronic diseases, age, pregnancy, or medications that they take.

Vitamins are divided into two groups: *water soluble* and *fat soluble.* Solubility determines how they are absorbed into, and transported around, the bloodstream, whether they can be stored in the body, and how easily they are lost from the body.

10.1 Vitamin C and members of the vitamin B family

B and C vitamins are water soluble, since they are relatively small molecules that contain carboxylic acid groups or other types of oxygen or nitrogen atoms, all of which can hydrogen-bond to water (see Figure 10.1). Vitamin C, for example, is water soluble because it contains many —OH groups. Water-soluble vitamins are generally absorbed directly into the bloodstream, where they travel freely. Toxicity is not usually a concern with water-soluble vitamins because they are not stored in tissue to any great extent, and excess amounts are excreted in urine. Because vitamins are not stored, a constant supply of them is required in the diet. Since these vitamins dissolve in water, their concentration is diminished in foods that are cooked in water.

Vitamin C, also known by the chemical name *ascorbic acid,* is required in the body for the proper formation of collagen, which, as mentioned in Chapter 9, section 9.19, is an important structural protein that is the foundation for skin and connective tissue. If the body synthesizes collagen in the absence of adequate vitamin C, the collagen will contain too few —OH groups to properly hydrogen bond and form strong fibers. The disease called *scurvy* results, with its consequent skin lesions and tooth loss due to weakened blood vessels. People who have no access to citrus fruits and vegetables for extended periods often contract scurvy (see Figure 10.2). The disease was common among sailors and soldiers on extended missions in the old days.

Vitamin C is also involved in the immune system response that protects the body against infection, which is why it is a popular cold remedy. Research about its effectiveness in this area has been inconclusive, and people are advised to refrain from taking massive amounts of vitamin C. Vitamin C is also an antioxidant, as discussed in section 10.2.

The other water-soluble vitamins are all members of the B family, which includes the following:

- Thiamin (vitamin B_1)
- Riboflavin (vitamin B_2)
- Niacin (vitamin B_3)
- Pyridoxine (vitamin B_6)
- Vitamin B_{12}
- Folic acid or folate (vitamin B_9)
- Pantothenic acid (vitamin B_5)
- Biotin (vitamin B_7)

As indicated by their structures (see Figure 10.1), these vitamins are mainly relatively small molecules. The B vitamins act as *coenzymes.* As explained in Chapter 9, section 9.20, coenzymes are substances that cannot be synthesized by the human body, but are required to combine

As indicated in Table 10.1, the dietary source of Vitamin B_{12} is animal products. As a consequence, vegans require supplemental supplies of this vitamin since they do not eat any animal products.

with certain enzymes in order to make those enzymes active. Coenzymes are vital in metabolic activity and are involved in the production of energy, the synthesis of RNA and DNA, and the processing of proteins. The most important sources, roles, and deficiency symptoms of the water-soluble vitamins are listed in Table 10.1.

Adequate levels of *folic acid*, vitamin B_9, are especially important for women who are pregnant because B_1 helps prevent having babies with devastating birth defects such as *spina bifida*. Folic acid forms coenzymes that are vital to the synthesis of amino acids, DNA, and RNA. Sufficient amounts of folic acid prevent a collection of birth defects

Table 10.1 Properties of the water-soluble vitamins

Water-soluble vitamin	Major roles	Deficiency symptoms	Significant sources
(a) Thiamine (B_1)	Nerve function, energy generation, supports normal appetite	Edema, enlarged heart, heart damage, weakened muscles, paralysis	Pork, whole and enriched grains, legumes, sunflower seeds
(b) Riboflavin (B_2)	Energy generation, supports normal vision and skin health	Light sensitivity, skin inflammation and lesions	Liver, milk, yogurt, whole and enriched grains, leafy greens
(c) Niacin (B_3)	Energy generation, supports skin health, supports nervous and digestive systems	Diarrhea; inflammation of mouth, gums, and tongue; dermatitis; fatigue; depression	All protein-containing foods, especially chicken, beef, tuna, milk, eggs; enriched grains, peanuts
(d) Pantothenic acid (B_5)	Energy metabolism	Nausea, headache, fatigue, insomnia	Liver, mushrooms, whole grains, avocados, broccoli
(e) Pyridoxine (B_6)	Fat and amino acid metabolism, helps to make red blood cells	Headache, anemia, nausea, smooth tongue, cracks at corners of mouth, dermatitis, muscle disruption	Meat, fish, poultry, whole grains, legumes, leafy greens, seeds
(f) Biotin (B_7)	Coenzyme in glucose production and fat synthesis	Dermatitis, depression, hair loss, loss of appetite, nausea	Organ meats, fish, egg yolks, whole grains, soybeans
(g) Folic acid (B_9)	Synthesis of RNA and DNA, new cell formation	Anemia (large cell type), impaired growth, diarrhea, frequent infections, smooth tongue, depression, confusion, weakness	Liver, leafy greens, legumes, seeds, fortified grains
Vitamin B_{12}	Coenzyme in folic acid metabolism, nerve function, new cell synthesis	Pernicious anemia, anemia, degeneration of nerve function, smooth tongue, fatigue, hypersensitive skin	Animal products including cheese, milk, eggs
(h) Vitamin C	Collagen synthesis, antioxidant, amino acid metabolism, immune system support, assists in wound healing, iron absorption	Anemia (small cell type), bleeding gums, loosened teeth, muscle degeneration, fragile bones	Citrus fruit, strawberries, greens, broccoli

*Letters refer to the structures shown in Figure 10.1. The structure of vitamin B_{12} is very complex, so we have not illustrated it.

called *neural tube defects,* NTDs. The neural tube forms very early in the life of the embryo, and later becomes the spinal cord, spine, and brain. These components can be exposed to amniotic fluid if the neural tube fails to close properly, an event that is more likely to happen if folic acid levels are low. This exposure can lead to spina bifida, the improper development of the spinal cord and backbone, and to the incomplete development and encapsulation of the brain. These embryonic developments occur between the 3rd and 4th week of pregnancy, which is before most women even know they are pregnant. Because many pregnancies are unplanned and most women are unaware that they are pregnant during this critical phase, the governments of the United States and Canada have now legislated the fortification of grain products such as cereals, flour, and pasta with folic acid to ensure that everyone receives adequate levels of the vitamin. In Canada, for example, the number of NTDs in babies has dropped by half since 1998 when folic acid fortification was begun. There is also evidence that folic acid can help prevent Alzheimer's disease, heart disease, and stroke. These effects may be due to the ability of folic acid to reduce blood levels of the amino acid homocystein, high levels of which can damage coronary walls.

10.2 Vitamins A, D, E, and K

The fat-soluble, water-insoluble vitamins are A, D, E, and K. Fat-soluble vitamins are carried by fat-transporting proteins in the blood. They can be stored with other lipids in fatty tissue, so fat-soluble vitamins can build up to toxic concentrations if you take mega-doses indiscriminately. You can survive for weeks without eating foods that contain the fat-soluble vitamins because of the stored supply. Deficiencies result when the diet consistently lacks these vitamins or when they are dissolved in undigested fat and lost from the digestive tract. Fat-soluble vitamin deficiency occurs, for example, in people who use mineral oil as a laxative or eat lots of certain fat substitutes. Diseases that interfere with fat absorption can also cause deficiencies of these vitamins. The most important sources, roles, deficiency symptoms, and chemical structures of the fat-soluble vitamins are listed in Table 10.2.

Figure 10.3 Eggs are a good source of vitamin E. (Royalty-Free/Corbis)

Vitamin E, an antioxidant *Vitamin E,* known chemically as α-*tocopherol,* is fat soluble but water insoluble, which is not surprising given that it has mainly a hydrocarbon-like structure and possesses only one —OH group (see Table 10.2). Vitamin E is present in foods such as eggs, vegetable oils, nuts, and other fatty plant-based foods (see Figure 10.3). Though grains contain naturally high levels of vitamin E, it is easily destroyed in food-processing activities such as the bleaching of flour.

Table 10.2 Properties of fat-soluble vitamins

Fat-soluble vitamin	Major roles	Deficiency symptoms	Significant sources
Vitamin A 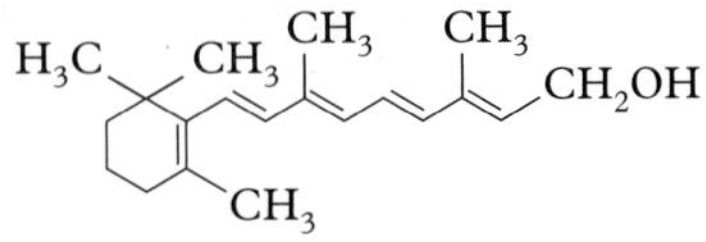	Vision, growth, reproduction, immune function, cell development, bone and tooth growth	Night blindness, impaired growth, anemia, diarrhea, frequent infections	Liver, fortified milk, cheese, eggs, orange-colored fruits and vegetables, leafy greens, broccoli
Vitamin D 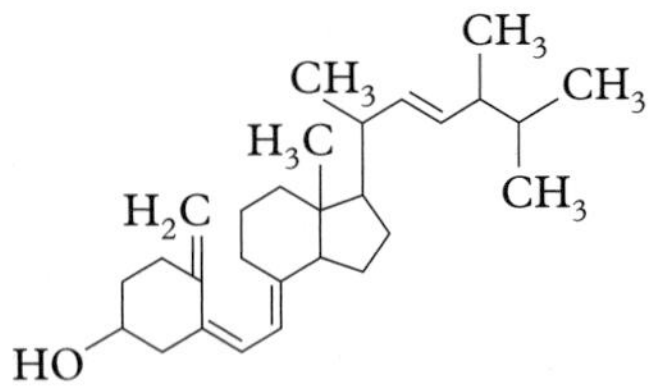	Absorption of calcium and phosphorus, bone maintenance	In children, misshapen bones and retarded growth; in adults, softening of bones, lax muscles	Tuna, salmon, fish oils, fortified milk, egg yolk, butter, margarine, cereals; synthesized in the body with help of sunlight
Vitamin E 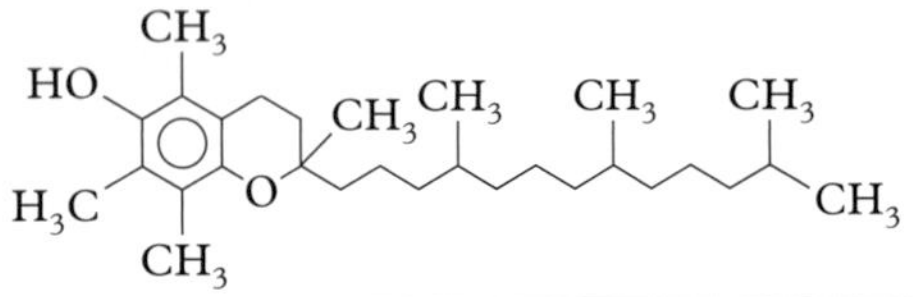	Antioxidant, protects cell membranes	Breakdown of red blood cells, anemia, nerve damage	Egg yolks, leafy greens, whole grain products, wheat germ, nuts, seeds, corn oil, safflower oil, soybean oil
Vitamin K 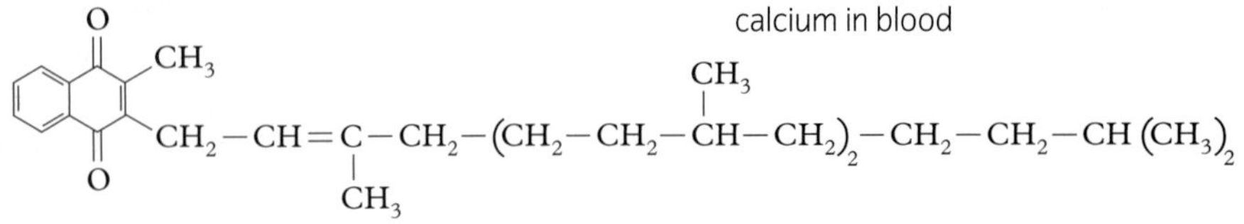	Syntheses of blood-clotting proteins, regulation of calcium in blood	Hemorrhaging	Beef liver, egg yolk, legumes, leafy greens, cabbage family; synthesized by bacteria in digestive tract

The natural function of vitamin E in the plants where it occurs is to prevent oxidation. Similarly, the vitamin acts as an **antioxidant** when you consume it. You may have seen this term in discussions of health advantages and nutritional supplements. What exactly is an antioxidant and how does it work?

As we've seen, the body uses oxygen, O_2, to produce energy by oxidizing glucose. Sometimes, the biochemical processes involving oxygen produce highly unstable, highly reactive molecules called *free radicals*. Other factors such as pollution, tobacco, radioactivity, and some medical conditions also increase free radical formation in the body. The technical definition of a free radical and a discussion of how these substances behave chemically are found in the *Taking It Further* section at the end of this chapter.

Free radicals damage biological tissue by disrupting cell membranes. This disruption affects what and how the cell transports substances across the membrane. Portions of the membrane that consist of polyunsaturated fatty acids are particularly sensitive to free radicals. Free radicals are also destructive to proteins and DNA. The accumulated damage due to free radical activity in the human body is called "oxidative stress" since the free radicals promote oxidation. Antioxidants are substances that prevent this oxidative damage from occurring. Vitamin E, for example, helps to prevent degradation of cell membranes in regions containing $C=C$ bonds by reacting with free radicals and thereby deactivating them. Following this process, the vitamin E molecules are reactivated by vitamin C, which is thereby consumed in the process.

Recall from Chapter 7 that polyunsaturated fatty acids are those containing two or more $C=C$ bonds.

The symptoms of vitamin E deficiency include scaly skin, sterility, and decay of muscle tissue. There is some evidence that supplementing with vitamins C and E together reduces the risk for Alzheimer's disease. Although research in the past indicated that a high intake of vitamin E lowers the risk of coronary heart disease, more recent evidence suggests that dosages beyond the recommended daily allowance may increase overall mortality and should be avoided.

Some antioxidants operate by a different mechanism, reacting with O_2 molecules to keep them from being incorporated into free radicals or by preventing the production of free radicals in the first place. In addition to antioxidant vitamins such as vitamin C and vitamin E, enzymes, selenium, and *phytochemicals*—that is, chemicals that occur in small quantities in plant foods (particularly highly colored fruits and vegetables)—also fight free radical activity in the body. A number of antioxidants have the same two-ring structure with multiple —OH groups as vitamin E and probably operate in the same way it does, but they differ in the structure of the remainder of the molecule. Antioxidants called *flavonoids* are present in plant-derived foods such as tea (especially green tea) and red wine, and are also present in chocolate and to a lesser extent in coffee. Because the protein in milk will interact with phenols and thereby deactivate their antioxidant ability, drinking tea or coffee to which milk has been added, or eating milk chocolate, are not efficient ways to increase antioxidant levels in your body.

One of the most effective antioxidants in foods is *lycopene*, the carotene-like pigment that gives tomatoes their red color. Lycopene occurs in all tomato products, and also in watermelon and pink grapefruit. Apparently the heating involved in producing tomato juice and sauce, or in cooking the tomatoes, changes the lycopene into a form that the body can absorb more easily. There is some evidence that high levels of dietary lycopene can decrease the risk of prostate cancer. It is claimed that the highest concentration of antioxidants in foods occurs in blueberries, due largely to the compound anthrocyanin, which imparts the blue color to the fruit.

■ Chemistry in Your Home: The antioxidant properties of vitamin C

When apples are peeled, the flesh is exposed to the air. Oxygen in the air goes to work immediately, causing browning via oxidation processes. You can examine the antioxidant properties of vitamin C by watching its effects on a peeled apple. Peel an apple and cut it into four parts. Brush the first piece with some water; brush the second piece with some lemon juice; brush the third piece with a vitamin C tablet that has been dissolved in water; the fourth is your control. Leave the apple pieces exposed to the air and record changes in their color over a day. Which pieces brown first? Which brown last? How do the rates of browning differ? What can you say about the relative antioxidant properties of water, lemon juice, and vitamin C?

Vitamins A, D, and K *Vitamin A* is present in milk, eggs, and liver. However, much of our vitamin A is supplied indirectly, in the form of beta-carotene, from vegetable sources such as carrots. Once in our bodies, the beta-carotene is converted by enzymes into vitamin A. One major role of vitamin A is to produce the substance by which our eyes detect light and allow us to have vision. Thus, it *is* true that eating carrots improves our eyesight, particularly at night; indeed, night blindness is a symptom of vitamin A deficiency. Vitamin A is also required for growth in children.

Strictly speaking, *vitamin D* isn't a vitamin since we can form it in our own bodies. The ultraviolet component of sunlight converts a cholesterol derivative in exposed skin into a precursor of vitamin D. The vitamin is subsequently converted into the hormone that regulates the uptake of calcium through the intestinal wall into the bloodstream and thereby controls calcium levels in the construction of bone (see Figure 10.4). A deficiency in vitamin D leads to defective bone formation and the disease called *rickets*. In the past, this condition was common in children who lived in northern climates with so much cloud cover that they received insufficient ultraviolet light to synthesize the vitamin themselves, especially if the children were kept indoors during the winter. Today, vitamin D is routinely added to milk to ensure that its calcium content is properly deployed and that children do not develop rickets.

Vitamin K is needed for making the proteins that are involved in blood clotting. A small percentage of newborn babies are deficient in Vitamin K, and consequently a small injection of it is routinely given to newborns. People with heart disease and some surgical patients may need to take "blood-thinning" drugs that interfere with vitamin K activity in

???????????????? **?** ????????????
Fact or Fiction **?**

Eating too many carrots can turn your skin orange.

True. The beta-carotene your body doesn't need from the over-consumption of carrots can give your skin an orange-colored glow, especially in the palms of your hands. But an adult would need to eat a lot of carrots to achieve this. (In fact, the author's daughter, when she was a baby, liked carrots so much that her fingertips turned orange!)

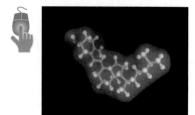

View the structure of the fat-soluble molecule vitamin D at Chapter 10: Visualizations: Media Link 2.

Notice the similarities in the structures of vitamin D (Table 10.2) with the general steroid unit illustrated on page 360.

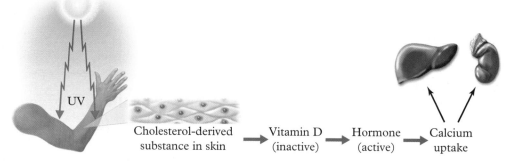

Figure 10.4 Formation and activation of vitamin D in the body.

UV — Cholesterol-derived substance in skin → Vitamin D (inactive) → Hormone (active) → Calcium uptake

In extreme cases of rickets, the bones are so soft that the weight of the body deforms them.

order to prevent the formation of blood clots. *Warfarin* is a synthetic compound that prevents the formation of the clot-producing proteins. Patients taking warfarin are routinely monitored to ensure that their blood-clotting ability is neither too high nor too low. Ironically, warfarin is used to poison mice and rats, who are particularly susceptible to it and die from internal bleeding after they have consumed it.

▶ Discussion Point: Should vitamin supplements be regulated?

Are vitamins biologically active compounds that should be regulated by the Food and Drug Administration? Or are they naturally occurring food supplements that should not be regulated like drugs? Use the resources on the Web site for this book to examine the controversy and develop arguments for and against government regulation of vitamins.

A recent survey found that one-third of Canadians are vitamin D-deficient during winter months, thus risking their bone health.

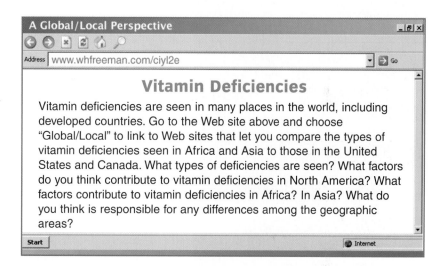

A Global/Local Perspective

Address www.whfreeman.com/ciyl2e

Vitamin Deficiencies

Vitamin deficiencies are seen in many places in the world, including developed countries. Go to the Web site above and choose "Global/Local" to link to Web sites that let you compare the types of vitamin deficiencies seen in Africa and Asia to those in the United States and Canada. What types of deficiencies are seen? What factors do you think contribute to vitamin deficiencies in North America? What factors contribute to vitamin deficiencies in Africa? In Asia? What do you think is responsible for any differences among the geographic areas?

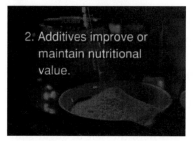

CHEMISTRY IN ACTION

2: Additives improve or maintain nutritional value.

What food additives are also nutrients? Find out at Chemistry in Action 10.1.

Food Additives

10.3 Antioxidants prevent foods from spoiling

Food can spoil as a result of chemical reactions. In particular, unsaturated fat molecules can decompose—become rancid—and produce foul-smelling, foul-tasting, and therefore unwanted materials such as carboxylic acids, aldehydes, and ketones if the fat molecules react with the oxygen in air.

Unsaturated fat molecules + O_2 → → Aldehydes, acids, ketones

The sequence of reactions that fat molecules undergo is complex. Overall the process corresponds to oxidation at the position of the double bonds. Antioxidants prevent the process from occurring by interfering with one of the first stages in the oxidation reaction sequence. As we have seen, nature has devised its own antioxidant—vitamin E—for such purposes; it occurs naturally in all vegetable oils, where it serves to stabilize them.

A number of synthetic mimics of vitamin E, each having a structure similar to it, have been synthesized by chemists and can be produced more cheaply than the vitamin itself:

- *BHA* (<u>b</u>utylated <u>h</u>ydroxy<u>a</u>nisole)
- *BHT* (<u>b</u>utylated <u>h</u>ydroxy<u>t</u>oluene)
- *Propyl gallate*

These antioxidants are used in small amounts in packaged food products that are susceptible to oxidation (see Figure 10.5).

The sequence of oxidation reactions that lead to rancidity is catalyzed by tiny concentrations of metal ions that inadvertently are present in the foodstuffs. These metals find their way into foods from the machinery used in harvesting and processing and from metals in soil. The metal ions can be stopped from catalyzing oxidation by adding to the food a *sequestrant*, which is any substance that strongly ties the ions up by binding to them chemically. This process reduces their reactivity to other components of the food. Citric acid is an example of such a sequestrant, as is the anion of *EDTA* (which stands for the mouthful <u>e</u>thylene<u>d</u>iamine<u>t</u>etraacetic <u>a</u>cid). Many foods such as salad dressings contain the calcium disodium salt of EDTA, which binds to metal ions and thereby prevents spoilage.

10.4 Dehydration and food preservatives

Food generally comes under attack from microorganisms—bacteria, fungi, and molds. If these microorganisms are allowed to multiply, the toxic substances they secrete will eventually spoil the food.

One of the simplest ways to preserve food is to *dehydrate* it because microbes cannot grow in the absence of water. It has been common for eons to dry grains, fruit, and even meat in order to preserve them. Adding common salt, NaCl, whether as the solid or in a very concentrated solution (think of the liquid in which dill pickles are stored) is also effective in dehydration since it draws water to itself and therefore out of the substance. Concentrated sugar solutions will similarly draw water from a substance and are used to preserve certain fruits. The phenomenon at work here is **osmosis,** the tendency of water to travel through a membrane from a less concentrated to a more concentrated solution. Here water travels from the food, where the salt or sugar solution is not concentrated, to the more concentrated salt or sugar solution, diluting it (see Figure 10.6). Freeze-drying preserves food by removing all the water in a process that involves a vacuum.

Figure 10.5 Antioxidants as food preservatives. (George Semple for W. H. Freeman and Company)

Figure 10.6 Osmosis as a means of food preservation. Water moves through the cucumber's cell membrane from a less concentrated to a more concentrated solution.

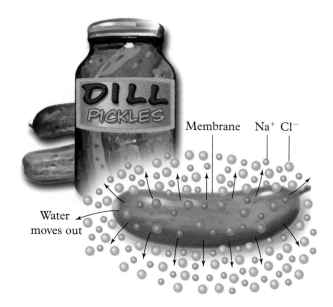

Organic acids Certain acids that are toxic to many microorganisms, but not particularly toxic to humans, can be added to food as preservatives. Probably the most common such additive is *benzoic acid,* molecules of which consist of a benzene ring attached to the acid —COOH group.

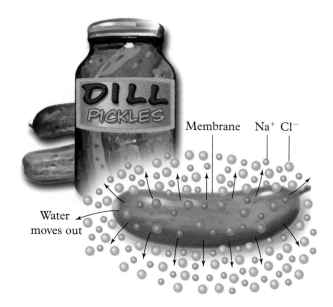

Benzoic acid

It is used in many food products, including jams, pickles, olives, and certain cola drinks. Benzoic acid is a natural component of raspberries, so it is a preservative invented by nature!

In some products, benzoic acid is supplied indirectly, in the form of *sodium benzoate.* This ionic compound is formed industrially from benzoic acid by adding sodium hydroxide. However, in the acidic environment of the food product, it reverts to benzoic acid. Benzoic acid and sodium benzoate are effective as preservatives only under acidic conditions, because it is the molecular form of the acid, not the ionic version, that is toxic to the microbes.

Another acid used as a food preservative is *propionic acid,* or an ionic compound formed from it, such as calcium or sodium *propionate.* Propionic acid occurs as a natural preservative in Swiss cheese and is even present in our bodies naturally. The propionates are added to baked goods, cheeses, and other no-liquid products. *Sorbic acid* and its ionic *sorbate* compounds (involving calcium, sodium, or potassium) are also commonly used in the same environments, but they function effectively only against molds and fungi, not bacteria. Sorbic acid and potassium sorbate occur naturally in fruit. Other weak acids that act as preservatives against microbial growth include *acetic acid,* as found in vinegar, and *citric acid,* as occurs in citrus fruits.

Because all the organic acids and their salts mentioned above occur naturally in some foods, and are metabolized by your body, there is little concern about their use as food additives. They are all classified as substances "*generally regarded as safe*" ("GRAS") by the U.S. Food and Drug Administration. Nevertheless, some people oppose their artificial introduction into foods, perhaps partially because they have such "chemical-sounding" names.

The salts of two *inorganic* acids, sulfurous and nitrous acid, are used in specialized applications to preserve food and can be of health concern to some individuals, as we see below.

Sulfites *Sodium bisulfite,* $NaHSO_3$, contains the acid HSO_3^-, known as the *bisulfite ion.* The acid kills bacteria and is used to protect wine, other grape products, and certain other food products.

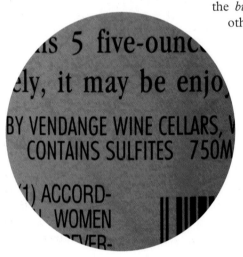

Figure 10.7 Sulfites, preservatives in wine. (George Semple for W. H. Freeman and Company)

The bisulfite ion can be generated in solution by dissolving the solid salt sodium bisulfite or by dissolving *sulfur dioxide* gas, SO_2, in water. Sulfur dioxide produces *sulfurous acid,* H_2SO_3, which almost completely ionizes to give the bisulfite ion under all but very acidic conditions:

$$SO_2(g) + H_2O(aq) \rightarrow H_2SO_3 \rightarrow H^+ + HSO_3^-$$

Another way of producing the bisulfite ion in solution is to dissolve an ionic compound containing the *sulfite* ion, SO_3^{2-}. The ion reacts with water to produce bisulfite:

$$SO_3^{2-} + H_2O \rightarrow HSO_3^- + OH^-$$

Some people are quite allergic to these sulfur-containing substances, so wines and foods that use them as preservatives must be labeled as "containing sulfites" (see Figure 10.7).

Nitrites The other inorganic acid used as a preservative is *nitrous acid,* HNO_2, in the form of its salts, the nitrites, all of which contain the *nitrite ion,* NO_2^-. *Sodium nitrite,* $NaNO_2$, is used to preserve cooked and cured meats such as ham, bacon, hot dogs, sausages, and luncheon meats. It is especially useful since it prevents the growth of the bacterium that causes *botulism.* It also has a cosmetic function in that it turns these meats pink or reddish, which is more attractive to consumers than is their natural brown or gray color.

The nitrite ion oxidizes the iron in the blood and muscle tissue in the meat. This reaction both provides the reddish color and prevents the botulism bacteria from gaining the iron it needs to thrive. Indeed, one danger to humans is that nitrite will oxidize the iron in our blood and prevent it from carrying oxygen. Nitrite is a natural component in our bodies, and normally 1–2% of the hemoglobin in our blood is oxidized by the nitrate to *methemoglobin*, which cannot carry oxygen. In most people, including infants, enzymes prevent the methemoglobin level from increasing beyond this level. However, some newborn infants, and a very few adults, do not possess these enzymes, and so their methemoglobin can build up to dangerous levels. Beyond a 10% conversion of hemoglobin to methemoglobin, the skin turns blue, and at 40%, death can occur. Fortunately, the occurrence of this type of *blue baby syndrome* is now rare in the developed world, since most sources of nitrite have been reduced.

Few people are susceptible to the problem with methemoglobin, and this is not the major health worry about nitrites in meat. Rather, there is concern that nitrite could be converted in the human body into carcinogens called *nitrosamines*. Whether or not this represents a real danger to human health is a controversial issue. Much of the nitrite in our bodies is actually formed from the *nitrate ion*, NO_3^-, which has one more oxygen per ion than nitrite. The nitrate ion is present in many foods, including preserved meats, in much higher concentrations than nitrite.

10.5 Additives are used to enhance food's appearance, texture, and flavor

The list of ingredients on packaged foods tells you its components, in descending order of amounts (see Figure 10.8). In many cases, the main ingredients—such as sugar, flour, water, and oil—are followed by a list of chemical names. Some of the names of these additives will now be familiar to you from our discussion of antioxidants and preservatives, and others refer to ingredients such as sweeteners and emulsifiers that we discussed back in Chapter 7. There are a few other categories we have not previously discussed:

- **Colors and bleaching agents:** These are used to impart to the food whatever color consumers expect and consider appetizing. For example, the natural orange-yellow substance beta-carotene is used to give margarine—which would otherwise be white—its color. About 30 other substances, half of them synthetic and half naturally occurring, are used as food colors. Ones that have been found to be carcinogenic in test animals have been banned. The seven food colorings that have been certified for use in the United States, and that actually are used currently, are listed in Table 10.3.
- **Thickening agents and texture modifiers:** Examples include natural products such as *carrageenan*, a carbohydrate extracted from Irish moss; *xanthan gum; locust bean gum;* and modified natural substances, such as *propylene glycol alginate* and various cellulose derivatives.

FRENCH DRESSING

INGREDIENTS: CANOLA OIL, HIGH FRUCTOSE CORN SYRUP, WATER, VINEGAR, SALT, PAPRIKA, XANTHAN GUM, POLYSORBATE 60, AQUARESIN PAPRIKA, TOMATO PASTE, NATURAL FLAVORS, POTASSIUM SORBATE (PRESERVATIVE), GARLIC POWDER, PROPYLENE GLYCOL ALGINATE, CALCIUM DISODIUM EDTA TO PROTECT FLAVOR, BETA CAROTENE (FOR COLOR).

Figure 10.8 Ingredients in a processed food. Such foods may contain several nonfood additives. (George Semple for W. H. Freeman and Company)

CHEMISTRY IN ACTION

Emulsifiers

What food additives assure product consistency? Find out at Chemistry in Action 10.2.

Food Additives

Table 10.3 Food colorings certified and commonly used in the U.S.

Official FD&C name	Common name	Hue	Common food uses
Blue No. 1	Brilliant Blue FCF	Bright blue	Beverages, dairy product powders, jellies, confections, condiments, icings, syrups, extracts
Blue No. 2	Indigotine	Royal blue	Baked goods, cereals, snack foods, ice cream, confections, cherries
Green No. 3	Fast Green FCF	Sea green	Beverages, puddings, ice cream, sherbet, cherries, confections, baked goods, dairy products
Red No. 40	Allura Red AC	Orange-red	Gelatins, puddings, dairy products, confections, beverages, condiments
Red No. 3	Erythrosine	Cherry-red	Cherries in fruit cocktail and in canned fruits for salads, confections, baked goods, dairy products, snack foods
Yellow No. 5	Tartrazine	Lemon yellow	Custards, beverages, ice cream, confections, preserves, cereals
Yellow No. 6	Sunset Yellow	Orange	Cereals, baked goods, snack foods, ice cream, beverages, dessert powders, confections

Source: U.S. FDA Web site www.cfsan.fda.gov/~lrd/colorfac.html

- **Flavor enhancers:** The famous example here is *MSG*, which stands for *monosodium glutamate*, used in Chinese food and many soups, etc. MSG is the sodium salt of the naturally occurring amino acid glutamic acid (see Figure 9.14). Some people are affected—for example, by headaches—after consuming food containing MSG. An inspection of the packaged foods in your kitchen cupboards would probably reveal the other common flavor enhancers, *disodium guanylate* and *disodium inosinate*. All three exaggerate the flavor of meat and allow manufacturers to use less of the "real thing," or inferior-quality ingredients, in a dish or a product. They are ionic compounds produced from organic acids and occur at small concentrations in some natural products.

Choose two different packaged food items and identify the additives listed on their labels. What purpose does each additive serve? Are the same additives in both products? Are both products subject to the same type of degradation or contamination? Do any products labeled *No preservatives* contain added citric acid, acetic acid, or vitamin E?

Pesticides: An Introduction

Since the mid-20th-century advent of the massive use of synthetic organic compounds in agriculture, people have worried about the possible health effects of residual quantities of them in food. Industrial chemicals and chemical by-products, with names like PCBs and dioxins, have also become a concern since human exposure to them—although tiny—occurs mainly through foods.

In this section, we shall investigate the nature of the various types of pesticides that are used both agriculturally and domestically, and the sources of the industrial chemicals and by-products of most concern.

10.6 Pesticides kill undesirable organisms

A **pesticide** is a substance that kills or otherwise controls an organism that humans find undesirable. All chemical pesticides share the common property of blocking a vital metabolic process of the organisms to which they are toxic. Unfortunately, pesticides are toxic not only to the target organism but also to some extent to other forms of life, including humans.

The various categories of pesticides are listed in Table 10.4. The most important types of pesticides are:

- **Insecticides**—substances, such as DDT and malathion, that kill insects
- **Herbicides**—substances, such as 2,4-D, that kill plants, especially weeds

Most households contain at least one pesticide, whether it be weed killers for the lawn and garden, algae controls for the swimming pool, flea powders for pets, or sprays to kill insects.

Collectively, insecticides and herbicides, together with *fungicides,* less prevalent substances that are used to control the growth of various types of fungus, represent the bulk of the 1 billion kilograms of pesticides that are used annually in North America. Almost half this amount is used in agriculture. The greatest use of insecticides occurs in the growing of cotton, whereas the majority of herbicide use comes in the growing of corn and soybeans. Fungicides are often used for protecting stored seeds before planting. Other fungicides, such as powdered elemental sulfur, are sprayed on plants to kill fungi before they penetrate the stem or leaf.

Table 10.4 Pesticides and their targets

Pesticide type	Target organism
acaricide	mites
algicide	algae
avicide	birds
bactericide	bacteria
disinfectant	microorganisms
fungicide	fungi
herbicide	plants
insecticide	insects
larvicide	insect larvae
molluscicide	snails, slugs
nematicide	nematodes
piscicide	fish
rodenticide	rodents

Note that almost all the terms end in the suffix –*cide*, which means "to kill."

Almost since their introduction, synthetic pesticides—that is, those made by chemical processes—have been a concern because of the potential impact on human health of eating food contaminated with these chemicals. About half the foods eaten in the United States contain measurable levels of at least one synthetic pesticide. For that reason, many such substances have been banned or restricted in their use. Nevertheless, a report by the U.S. National Academy of Science pointed out that pesticide regulation to date has not paid enough attention to the protection of human health, especially that of infants and children, whose growth and development are at stake.

By way of contrast, other scientists have pointed out that living plants themselves manufacture insecticides in order to discourage insects and fungi from consuming them, and that consequently we are exposed in our food supply to much higher concentrations of these natural pesticides than to synthetic ones. Natural pesticides are not necessarily any less toxic than are synthetic ones.

Unfortunately, persistent pesticides migrate from the areas on which they were applied into the rest of the environment. For example, some fraction of any applied pesticide will evaporate and travel in air, eventually to be deposited on other land or in natural waters. Similarly, other pesticide molecules will seep down into the soil and eventually into underground water. By both these routes, pesticides may eventually show up in our water supplies.

Insecticides

Insecticides of one type or another have been used by society for thousands of years. One principal motivation for using them is to control disease: human deaths due to insect-borne diseases through the ages have greatly exceeded those attributable to the effects of warfare. The use of insecticides has greatly reduced the incidence of diseases transmitted by insects: malaria, yellow fever, bubonic plague, sleeping sickness, and recently the West Nile virus scarcely exhaust the list. People also try to control insects such as the mosquito and the common housefly simply because their presence is annoying. The other principal motivation for insecticide usage is to prevent insects from attacking food crops. However, even with extensive use of pesticides, about one-third of the world's total crop yield is destroyed by pests or weeds during growth, harvesting, or storage.

10.7 DDT is an infamous insecticide

The insecticide DDT has had a tumultuous history. It was hailed as "miraculous" in 1945 by Sir Winston Churchill because of its use in the Second World War. It was very effective against the mosquitoes that carry malaria and yellow fever, against body lice that can transmit typhus, and against plague-carrying fleas. Malaria-reduction programs, one component of which was the use of DDT, saved the lives of more than five million people in the postwar period.

The possible role of DDT in affecting human development is discussed later in this chapter.

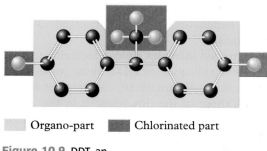

Organo-part   Chlorinated part

Figure 10.9 DDT, an organochlorine insecticide.

Unfortunately, DDT initially was widely overused, particularly in agriculture. As a result, its environmental concentration rose rapidly and it began to affect the reproductive abilities of birds, which indirectly incorporated it into their bodies. By 1962, DDT was being called an "elixir of death" by Rachel Carson in her influential book *Silent Spring,* because of its role in decreasing the populations of certain birds, such as the bald eagle, whose dietary intake of the chemical was very high.

DDT is an example of compounds called *organochlorines*—organic compounds whose molecules contain chlorine covalently bonded to some of the carbon atoms. Structurally, DDT is a substituted ethane: at one carbon, all three hydrogens are replaced by chlorine atoms, while at the other, two of the three hydrogens are replaced by a benzene ring. As you can see in Figure 10.9, each of the rings contains a chlorine atom directly opposite the ring carbon that is joined to the ethane unit.

One reason DDT was such an ideal insecticide was its persistence: one spraying gave protection from insects for weeks to years, depending upon the method of application.

The environmental levels of DDT, and of the related product called DDE into which it is transformed in the environment, have declined in countries where their use has been restricted or banned. However, these substances still enter the environment everywhere as a result of long-range transport by wind from developing countries in which DDT is still in use to control malaria and typhus and for some agricultural purposes.

As a result of the decline in DDE levels, *bald eagles* have made a comeback around Lake Erie and elsewhere. Similarly, the population of *arctic peregrine falcons,* a bird that was driven to near extinction due to the effects of DDE, has now largely recovered.

10.8 Organochlorine molecules accumulate in living matter

Many organochlorine compounds, including DDT and other organochlorine pesticides, are thousand to millions of times more concentrated in the tissues of fish than in the waters in which they swim. Hydrophobic substances like DDT are particularly liable to exhibit this phenomenon. There are several reasons for this **bioaccumulation** of chemicals in biological systems.

In the first place, most organochlorines are inherently much more soluble in hydrocarbon-like media, such as the fatty tissue in fish, than they are in water. Thus, when water passes through a fish's gills, the compounds selectively diffuse from the water into the fish's fatty flesh and become more concentrated there. Such a process is called **bioconcentration.**

Fish also bioaccumulate organic chemicals from the food they eat and from their intake of particles suspended in water and deposited on sediments onto which the chemicals have adsorbed. In many such cases, the chemicals are not metabolized by the fish: the substance simply accumulates in the fatty tissue of the fish, becoming more concentrated over time.

The average concentration of many chemicals also increases dramatically as one proceeds up a **food chain,** which is a sequence of species, each of which feeds upon the one that precedes it in the chain. A **food web** incorporates interlocking food chains. Over a lifetime, a fish eats many times its weight in food from the lower levels of the food chain, retaining rather than eliminating most of the organochlorine chemical content from these meals.

A chemical whose concentration increases along a food chain is said to be **biomagnified.** The biomagnification of DDT along some of the Great Lakes food chains is shown in Figure 10.10: compare the *herring gull's* high level of DDT with those of the fish below it in the chains leading up to it. Fish at the top of the aquatic part of the chain biomagnify DDT considerably, so that even higher concentrations are found in the birds of prey that feed on them.

In general, the bioaccumulation of a substance arises from the combination of the two processes, bioconcentration and biomagnification. The bioaccumulation of organochlorines in fish and other animals is the

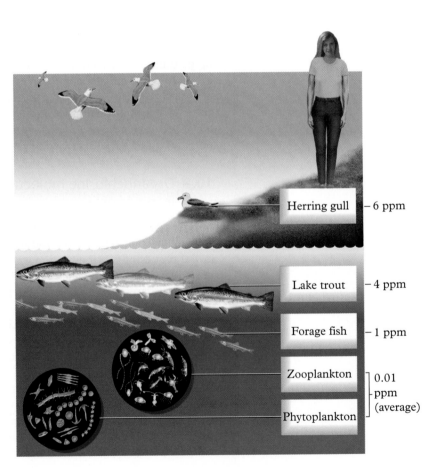

Figure 10.10 Biomagnification of DDT in a food chain. (Source: *The State of Canada's Environment. 1991.* Ottawa: Government of Canada)

Herring gull — 6 ppm

Lake trout — 4 ppm

Forage fish — 1 ppm

Zooplankton 0.01 ppm (average)

Phytoplankton

Chapter 10: Chemicals in Our Bodies and Our Environment

reason that most of the human daily intake of such chemicals enters via our food supply rather than from the water we drink.

Humans are at the top of the food chain (see Figure 10.10). We eat fish, such as salmon, from lakes in which DDT and other organochlorines have biomagnified. We consume meat and dairy products, in the fatty components of which are present these same substances, having been carried up the terrestial food chain. Thus, 99% of North Americans still have detectable levels—averaging 3 ppm—of DDT and its metabolites in their bodies. The good news is that the levels of DDT in humans has declined drastically since DDT's period of greatest use in the 1950s and 1960s.

DDT is one of the "dirty dozen" chemicals designated by the United Nations Environmental Program as **Priority Organic Pollutants,** or POPs, that are being banned or phased out by international agreement. Of the twelve, only DDT will not be totally banned. Its use will be largely restricted to the control of disease. Although many groups pressured the United Nations to include a total ban on DDT, others were strongly opposed to a complete ban, given that it is so effective in small amounts against malaria, a disease which kills 1 million children annually in Africa. Switching to alternatives would probably be beyond the financial means of many poor countries. Groups supporting a total, immediate ban countered that some countries, such as Mexico, have already eliminated malaria without the use of DDT, and that the mosquitoes responsible for carrying most of the disease are already resistant to DDT in some parts of the world, such as India.

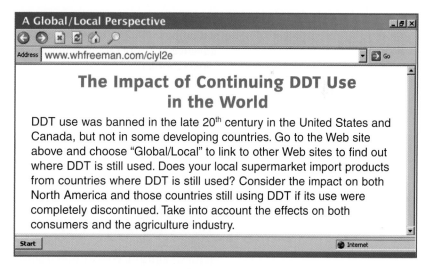

A Global/Local Perspective

Address www.whfreeman.com/ciyl2e

The Impact of Continuing DDT Use in the World

DDT use was banned in the late 20th century in the United States and Canada, but not in some developing countries. Go to the Web site above and choose "Global/Local" to link to other Web sites to find out where DDT is still used? Does your local supermarket import products from countries where DDT is still used? Consider the impact on both North America and those countries still using DDT if its use were completely discontinued. Take into account the effects on both consumers and the agriculture industry.

10.9 Organophosphate and carbamate insecticides are toxic

Organophosphate pesticides are derivatives of *phosphoric acid,* $O=P(OH)_3$, in which two of the hydrogen atoms have been replaced by methyl or ethyl groups and the third by a longer, more complicated organic group which varies from one pesticide to another. The oxygen to which this complicated group is attached, and/or the doubly bonded oxygen, are replaced by sulfur in some variations.

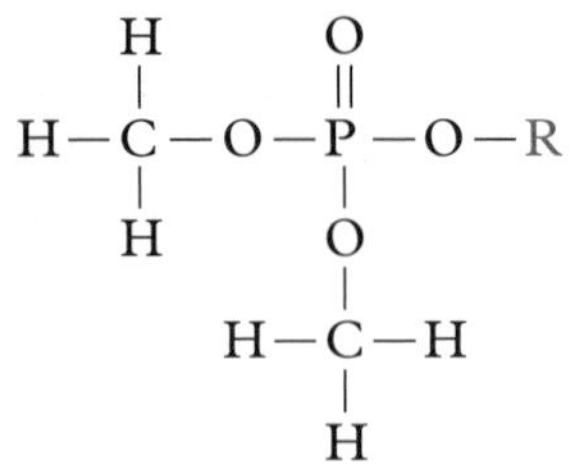

$$
\begin{array}{ccccccccc}
 & & \text{H} & & & \text{O} & & & \\
 & & | & & & || & & & \\
\text{H} & - & \text{C} & - & \text{O} & - & \text{P} & - & \text{O} - \text{R} \\
 & & | & & & & | & & \\
 & & \text{H} & & & & \text{O} & & \\
 & & & & & & | & & \\
 & & & & & & \text{H} - \text{C} - \text{H} & & \\
 & & & & & & | & & \\
 & & & & & & \text{H} & &
\end{array}
$$

An organophosphate structure

The organophosphate pesticides are toxic to insects because they inhibit enzymes in the nervous system, and thus they function as nerve poisons. The mode of action of **carbamate** insecticides is similar to that of the organophosphates except that a carbon atom rather than a phosphorus atom attacks the enzyme:

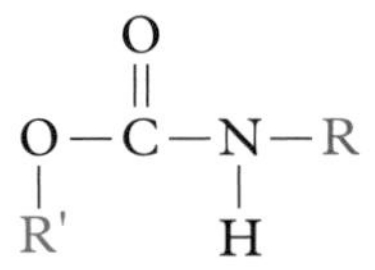

$$
\begin{array}{ccccc}
 & & \text{O} & & \\
 & & || & & \\
\text{O} & - & \text{C} & - & \text{N} - \text{R} \\
| & & & & | \\
\text{R}' & & & & \text{H}
\end{array}
$$

A carbamate structure

Organophosphate and carbamate pesticides are nonpersistent: they decompose within days or weeks. In this respect, they represent an advance over organochlorines since they have a much lower tendency to bioaccumulate.

On the other hand, organophosphates and carbamates are generally much more *acutely* toxic—that is to say, they affect health immediately—to humans and other mammals than are organochlorines. Many organophosphates represent an immediate danger to the health of those who apply them and to others who may come into contact with them. Exposure to these chemicals by inhalation, swallowing, or absorption through the skin can lead to immediate health problems.

U.S. children generally have their greatest exposure to organophosphates through the food they eat, with exposure from drinking water only a small fraction of that from food. A 2003 report found that preschool children (in Seattle) who consumed mainly organic fruits, vegetables, and juices had much lower exposure to organophosphates, as measured in their urine, than did children with conventional diets.

There is evidence that organophosphates have chronic as well as acute health problems associated with them. Several recent studies found links between indoor use of insecticides—organophosphates especially—and the incidence of leukemia and brain cancer in children. The U.S. EPA has placed organophosphates in their highest priority group in their current re-examination of pesticides.

Recall that LD_{50} is a measure of the lethal dose, as we discussed in Chapter 6, section 6.6. The higher its value, the less toxic the substance.

You are probably familiar with one or more of the organophosphate insecticides. The most important organophosphate insecticide having a very low toxicity to mammals is *malathion*. Though malathion is not particularly toxic to mammals (its $LD_{50} = 885$ mg/kg, which is fairly high), it is fatal to many insects since they metabolize it in a different way. It is used in domestic fly sprays and to protect agricultural crops. In combination with a protein bait, low concentrations of malathion have been sprayed from helicopters over California, Florida, and Texas to combat infestations of the Mediterranean fruit fly, a dangerously destructive pest. The spraying is controversial, since many people do not wish to have the areas they inhabit sprayed with any pesticide, no matter how low its human toxicity.

Diazinon is another organophosphate commonly used for insect control in homes, in gardens, in shrubs, and on pets. Although it has a low toxicity rating ($LD_{50} = 300$ mg/kg), it has been found to be toxic to birds, so its use in residential applications is being phased out.

One of the most useful, though dangerous, organophosphate insecticides is *parathion*. It is very toxic ($LD_{50} = 3$ mg/kg, so it takes rather little to be lethal) and is probably responsible for more deaths of agricultural field workers than any other pesticide. Since it is nonspecific to insects, its use can inadvertently kill birds and other nontarget organisms. Bees, which are often economically valuable, are also indiscriminately destroyed by parathion. It is now banned in some Western industrialized countries but is still widely used in developing countries.

Chlorpyrifos ($LD_{50} = 135$ mg/kg) was commonly sprayed by exterminators to control cockroaches, ants, termites, and other insects. However, it has been withdrawn for domestic use in the U.S. by its manufacturer due to health concerns, especially those involving fetal and childhood exposure. The removal of this pesticide from the market was in response to the results of experiments on rats and reported cases of unintentional poisonings involving humans. Detectable traces of chlorpyrifos were estimated to occur in 93% of Americans. Its uses had been restricted as of the late 1990s by the EPA, as had that of *methyl parathion*, which is parathion with methyl rather than ethyl groups. Indeed, most indoor uses of organophosphates have been eliminated by regulatory action in recent years.

An important organophosphate having intermediate toxicity to mammals is *dichlorvos;* its LD_{50} is 25 mg/kg. A relatively volatile insecticide, dichlorvos is used as a domestic fumigant released from impregnated flypaper hung from ceilings and light fixtures. The chemical slowly evaporates, and its vapor kills flies in the room. Plastic is impregnated with dichlorvos for use in flea collars, and the chemical is also used in some aerosol sprays used to control insects.

Important examples of the carbamate pesticides are _Carbaryl_ ($LD_{50} = 307$ mg/kg), and *Aldicarb* ($LD_{50} = 0.9$ mg/kg); Aldicarb is very toxic indeed to humans. Carbaryl, a widely used lawn and garden insecticide, has a low toxicity to mammals but is fatal to honeybees.

In summary, although organophosphates and carbamates solve the problem of environmental persistence and accumulation associated with organochlorine insecticides, this solution sometimes comes at the expense of dramatically increased acute toxicity to the humans and animals who encounter them while the chemicals are still in the active form. These less persistent insecticides—together with the pyrethrin derivatives mentioned below—have largely replaced organochlorines in residential uses. Organophosphates and carbamates are a particular problem in developing countries, where widespread ignorance about their hazards has led to sickness and many deaths among agricultural workers. Because these workers are either unaware of the dangers or because the climate is too hot, many do not wear the appropriate clothing to protect them from direct exposure to these chemicals. The types of pesticides used in developing countries are also more likely to be those that are highly toxic (so much so that many have been banned elsewhere in the world) because they are easier to use and less expensive. Estimates by the United Nations and the World Health Organization put the number of persons who suffer acute illnesses from short-term exposure to pesticides in the millions annually; 10,000–40,000 die each year from the poisoning, about three-quarters of these in developing countries. About 99% of the deaths from pesticide poisonings occur in developing countries. But even in the U.S., about 20,000 people receive emergency medical care annually for actual or suspected poisoning from pesticides; about 30 people die from pesticide poisoning annually.

▶ **Discussion Point:** Should flea collars impregnated with chemicals be used on pets?

Flea collars are popular pet-care products. Use the resources on the Web site for this book to examine the components of flea collars and their effects. Do you think they may pose long-term health risks to pets? Does the presence of flea-collar chemicals in the home create health risks for the inhabitants? Are alternatives available? How effective are the alternatives? Why are fleas a problem for pets and their owners? Develop arguments for and against the use of flea collars.

10.10 Natural and green pesticides

As we noted earlier, for their own self-protection, many plants themselves manufacture certain molecules that either kill or disable insects that prey on them. Chemists have isolated some of these compounds so that they can be used to control insects in other contexts. Examples are *nicotine* (see Chapter 8), *rotenone* (see below*)*, the *pheromones,* and *juvenile hormones.*

One group of natural pesticides that humans have used for centuries is the *pyrethrins*. The original compounds, the general structure for which is illustrated in Figure 10.11, were obtained from the flowers of a species of chrysanthemum. In the form of dried, ground-up flower heads, pyrethrins were used in Napoleonic times to control

Figure 10.11 Pyrethrin, a natural insecticide from chrysanthemums. R_1 and R_2 are organic groups. (Photo: Photodisc Green/Getty Images)

body lice, and they are still used in flea sprays for animals. They are generally considered to be safe to use. Like organophosphates, they paralyze insects, though they usually do not kill them. Unfortunately, these compounds are unstable in sunlight. For this reason, chemists have developed synthetic pyrethrin-like insecticides that are stable outdoors and so can be used in agricultural applications. They usually are given names ending in *-thrin* to denote their nature (for example, *permethrin*).

Green chemistry is the science of designing chemical products and processes that reduce or eliminate the use and generation of hazardous substances. This 1991 initiative of the U.S. EPA has since become a worldwide movement. In the area of pesticides, green chemistry research has developed pesticides and procedures that are safer for humans than those available previously, and has developed safer ways of producing existing pesticides.

An example of a low-toxicity insecticide produced by green chemistry is *hexaflumuron*. This compound is used against termites, and interrupts their molting process. A termite molts in response to its body's growth. When the insect is unable to molt, it dies. The hexaflumuron is spread throughout the termite colony when termites that have ingested the bait return to the nests and share it with the other insects. Hexaflumuron is used in much smaller quantities than the traditional pesticides that were used for termite control, is not harmful to most beneficial insects, and was the first substance to be classified as a reduced-risk pesticide by the EPA.

A second example of an insecticide developed by green chemistry are the *diacylhydrazines*, which are organic derivatives of hydrazine, H_2N-NH_2. These compounds are used to control caterpillars by interrupting the chemical signal that tells them to resume feeding after the molting process has ended. The insect dies of starvation or dehydration. Only insects that have a molting stage are affected by these compounds. Since their use reduces pesticide risks to human health and to nontarget organisms, they are also classified by the EPA as reduced-risk pesticides.

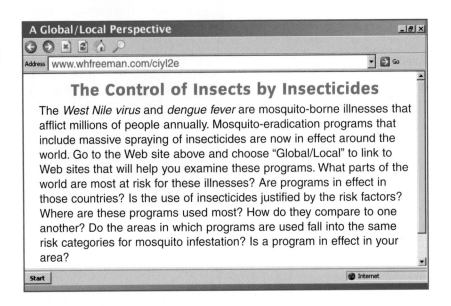

A Global/Local Perspective

Address www.whfreeman.com/ciyl2e

The Control of Insects by Insecticides

The *West Nile virus* and *dengue fever* are mosquito-borne illnesses that afflict millions of people annually. Mosquito-eradication programs that include massive spraying of insecticides are now in effect around the world. Go to the Web site above and choose "Global/Local" to link to Web sites that will help you examine these programs. What parts of the world are most at risk for these illnesses? Are programs in effect in those countries? Is the use of insecticides justified by the risk factors? Where are these programs used most? How do they compare to one another? Do the areas in which programs are used fall into the same risk categories for mosquito infestation? Is a program in effect in your area?

Start — Internet

Herbicides, PCBs, Dioxins, and Furans

Herbicides are chemicals that destroy plants. They are often employed to kill weeds without causing injury to desirable vegetation—for example, to eliminate broad-leaf weeds from lawns without killing the grass. Herbicides are also used to eliminate undesirable plants from roadsides, railway and power-line rights-of-way, and so on, and sometimes to defoliate entire regions. Herbicides are the most widely used type of pesticide in North America, with sales now exceeding $4 billion annually. The agricultural use of herbicides has replaced human and mechanical weeding in developed countries. This practice has resulted in greatly increased productivity per acre and it has sharply reduced the number of people employed in agriculture.

Organic herbicides now dominate the weed-control market. These compounds are *selective;* that is, they are much more toxic to certain types of plants than to others, so they can be used to eradicate weeds while leaving crops unharmed.

10.11 Phenoxy herbicides

Phenoxy herbicides were introduced at the end of the Second World War and are still used on a massive scale. Environmentally, the by-products contained in the commercial products are often of greater concern than the herbicides themselves.

The molecular structure of phenoxy herbicides is based upon a benzene ring connected by an ether oxygen to a short-chain carboxylic acid, usually acetic acid. Either two or three of the five hydrogen atoms of the benzene ring in phenoxyacetic acid are replaced by chlorine atoms in commercial herbicides. The numbering scheme for the benzene ring begins at the carbon attached to the oxygen. The numbers in the names of the chemicals indicate the carbon atoms to which the chlorine atoms are bonded. The *2,4-D* compound, whose full name is 2,4-dichlorophenoxy acetic acid, is used to kill broad-leaf weeds in lawns, golf course fairways and greens, and agricultural fields. The *2,4,5-T* herbicide, whose full name is 2,4,5-trichlorophenoxy acetic acid, is effective in clearing brush, for instance, on roadsides and power-line corridors.

The herbicides known as *MCPA, dichlorprop, silvex,* and *mecoprop* are identical to 2,4-D or 2,4,5-T with minor modifications of the molecular structure.

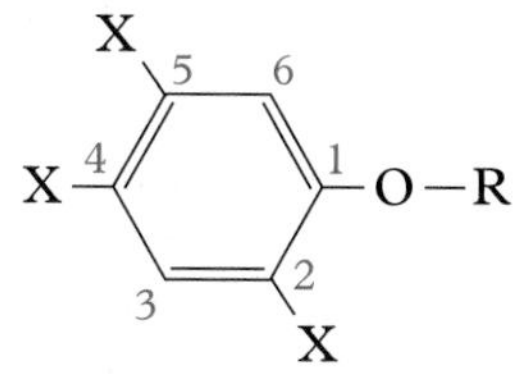

Phenoxy herbicides (general formula)
X=H, Cl, CH$_3$

2,4-D is detectable in 53% of Americans.

Huge quantities of 2,4-D are used in developed countries for the control of weeds in both agricultural and domestic settings. In some communities, its use on lawns has become controversial because of its possible effects on human health.

10.12 Dioxin is highly toxic

In the process used to produce 2,4,5-T commercially, a side reaction produces a highly toxic compound, dioxin, as a trace contaminant. The toxic compound has the structure shown below:

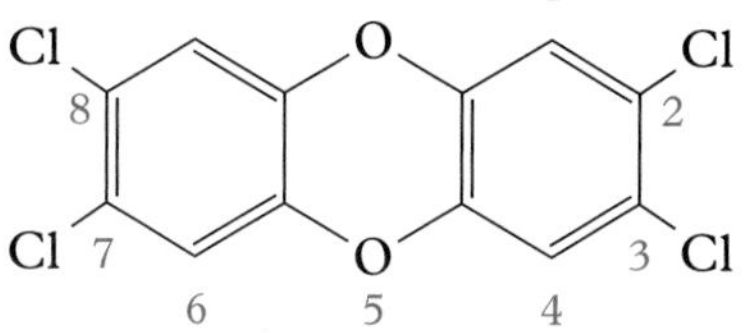

Dioxin

The middle ring is called the *p-dioxin* unit, and two benzene rings are attached to it, so the three-ring structure is called *dibenzo-p-dioxin*. Since four chlorines are present, at the 2-, 3-, 7-, and 8-positions, the compound is named *2,3,7,8-tetrachlorodibenzo-p-dioxin,* or *2,3,7,8-TCDD.* (No wonder it is simply called dioxin!) The same three-ring structure,

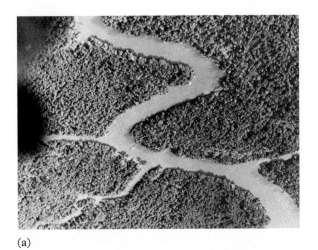

(a)

(b)

Figure 10.12 A forest in South Vietnam (a) before and (b) after spraying with Agent Orange. (AP/Wide World)

but with different numbers of chlorines at the various points of substitution, are also toxic substances and are also found as environmental contaminants.

Agent Orange, a combination of the herbicides 2,4-D and 2,4,5-T, along with a relatively high contamination of the latter by 2,3,7,8-TCDD, was used by the United States for the defoliation of military targets in the Vietnam War (see Figure 10.12). A number of both civilians and soldiers who were present during these applications believe that some of the numerous chronic health problems that they have experienced arose from exposure to dioxin from this source, although this is a very controversial issue.

Because commercial supplies of 2,4,5-T are inevitably contaminated with some 2,3,7,8-TCDD, the environment becomes contaminated with this dioxin as the weed killer is applied to the land. The manufacture and use of 2,4,5-T in North America were phased out in the mid-1980s because of concerns about the dioxin content, however small. We shall see later that there are other sources of dioxins that contaminate the environment, so ending 2,4,5-T production does not completely solve the problem.

10.13 Glyphosate is a herbicide that kills almost all plants

Currently, the most widely used herbicide is *glyphosate.* Its molecules are structurally similar to organophosphate insecticide molecules except that one oxygen of the four that normally surround phosphorus is replaced by an organic group—in this case, the simple amino acid *glycine* (shaded below).

Glyphosate

Glyphosate is widely used as a herbicide (Roundup) and is rather nontoxic: its LD_{50} value is high. Although it kills almost *all* plants, some strains of soybeans have been genetically engineered so that they are resistant to glyphosate; consequently, it can be used as a weed killer where that crop is grown.

10.14 PCBs are multiuse organochlorine compounds

The well-known acronym *PCBs,* or <u>p</u>oly<u>c</u>hlorinated <u>b</u>iphenyls, stands for a group of industrial organochlorine chemicals that became a major environmental concern in the 1980s and 1990s. They are *not* pesticides, but they found many other applications.

PCB mixtures were commercially attractive because they are chemically inert liquids and are difficult to burn, have low vapor pressures, are inexpensive to produce, and are excellent electrical insulators. As a result of these properties, they were used extensively as the coolant fluids in power transformers and capacitors.

Like many other organochlorines, they are very persistent in the environment and they bioaccumulate in living systems. As a result of careless disposal practices, they have become a major environmental contaminant. Due both to their own toxicity and to that of their contaminants, PCBs in the environment have become a cause for concern because of their potential adverse impact on human health, particularly with regard to growth and development.

The main structural component of all PCB molecules is a pair of benzene rings that are connected to each other by a single bond. The substitution of several or all of the ten hydrogen atoms in biphenyl by chlorine atoms gives rise to 209 possible molecules, all known as PCBs; one example is illustrated below:

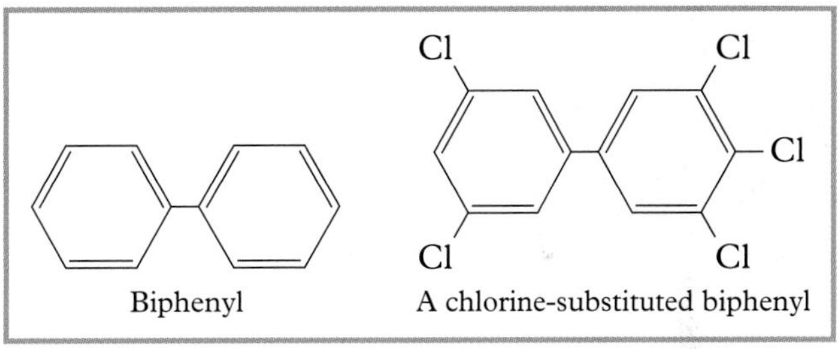

Biphenyl   A chlorine-substituted biphenyl

10.15 PCBs in the environment are recycled for years

PCBs were released into the environment during their production, their use, their storage, and their disposal. Because of their stability and extensive usage, along with careless disposal practices, PCBs became widespread and persistent environmental contaminants. When their accumulation and harmful effects became recognized, uses in which their disposal could not be controlled were terminated. Although North American *production* of PCBs was halted in 1977, old supplies of the substances still remain in use in many electrical transformers.

If they are released into the environment, PCBs persist for years because the molecules are so resistant to breakdown by chemical or biological agents. They are not very soluble in water, but the tiny amounts of PCBs present in surface waters are constantly being volatilized and subsequently redeposited on land or in water after traveling in air for a few days (see Figure 10.13). By such mechanisms, PCB molecules have

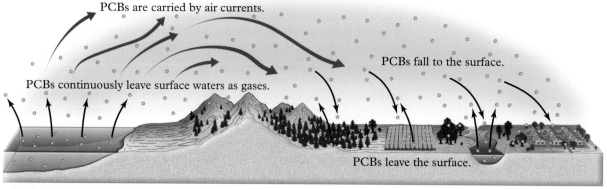

PCBs are carried by air currents.

PCBs fall to the surface.

PCBs continuously leave surface waters as gases.

PCBs leave the surface.

∘ PCBs

Figure 10.13 Recirculation of PCBs already in the environment.

been transported worldwide. Indeed, there are measurable background levels of PCBs even in polar regions and at the bottom of oceans. This environmental load of PCBs will continue to be recycled among air, land, and water, including the biosphere, for decades to come.

10.16 PCBs are contaminated by furans

Strong heating of PCBs in the presence of a source of oxygen can result in the production of small amounts of *dibenzofurans*. These compounds are structurally similar to dioxins; they differ in that they are missing one oxygen in the central ring. The dibenzofurans give rise to health effects similar to those of dioxins and are discussed in a later section. The basic furan ring contains five atoms, one of which is oxygen and the other four of which are carbon atoms that participate in double bonds:

O

Furan ring

Like the dioxins, the dibenzofuran environmental contaminants have chlorine atoms substituted for hydrogen at some or all of the eight possible positions. A typical member of the furan family is illustrated below:

Cl

Cl   O

Cl

Cl   Cl

Cl

A chlorinated dibenzofuran

Almost all commercial PCB samples are contaminated with some furans, though usually only a very tiny amount is found in the originally manufactured liquids. However, if the PCBs are heated and if some oxygen is present, some of the PCB molecules are converted to furans, greatly increasing the level of contamination. The furan concentration in used PCB cooling fluids is found to be greater than in the virgin materials, presumably due to the moderate heating that the fluid undergoes during its normal use. Furan production also occurs if PCBs are burned with anything but an unusually hot flame.

10.17 Environmental sources of dioxins and furans

In addition to the sources mentioned above, furans and dioxins are also produced inadvertently as the by-products of a myriad of processes, including the bleaching of paper pulp, the incineration of garbage, and the recycling of metals. Currently, incinerators are the largest anthropogenic source of dioxins in the environment.

Pulp and paper mills that still use chlorine in the bleaching of the pulp are major dioxin and furan sources. These contaminants result from the reaction of the chlorine with some of the organic molecules produced from the pulp. The most abundant dioxin produced by the bleaching process is the highly toxic 2,3,7,8-TCDD, significant quantities of which were released into the environment, in both paper and pulp mill effluent, in North America and elsewhere. Most pulp and paper mills have now switched their bleaching agent from elemental chlorine, Cl_2, to chlorine dioxide, ClO_2, and their combined furan and dioxin output is now much smaller, perhaps even undetectable.

There is evidence that very small concentrations of dioxins—particularly highly chlorinated ones—were present in the environment in the pre-industrial era, presumably as a result of natural phenomena such as forest fires and volcanoes. However, chemical analysis of soils and the sediments in lakes shows that the greatest anthropogenic input of dioxins and furans to the environment in developed countries began in the 1930s and 1940s and peaked in the 1960s and 1970s. Deposition still continues today, but at a slower rate because of deliberate steps taken by industrialized nations to reduce the production and dispersal of these toxic compounds.

10.18 Dioxins, furans, and PCBs in our diet

Once created and released into the environment, dioxins, furans, and PCB molecules are transported from place to place mainly through the atmosphere. As a consequence of their widespread occurrence in the environment and their tendency to dissolve in fatty matter, they bioaccumulate in the food chain. More than 90% of human exposure to dioxins is attributable to the food we eat, particularly meat, fish, and dairy products. Typically, dioxins and furans are present in fish and meat at levels of a few picograms (pg, or 10^{-12} gram) per gram of the food; in other words, they occur at levels of a few parts per trillion. Though this doesn't sound like very much, we have to consider it in a wider context.

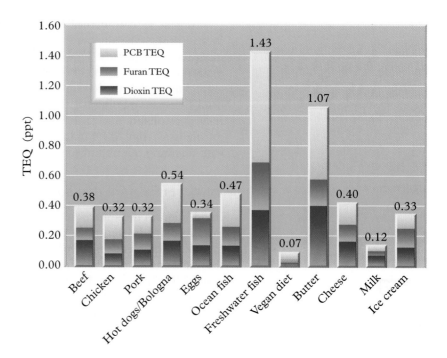

Figure 10.14 TEQ concentration in different foods collected at supermarkets in the United States in 1995. (Source: A. Schecter et al., *Chemosphere*, 34:1437–1447.)

Since most organisms, including humans, have a mixture of many dioxins, furans, and PCBs stored in their body fat, it is important to measure *net* toxicity. Scientists often report the combined concentrations of these organochlorines in terms of the *equivalent* amount of 2,3,7,8-TCDD that, by itself, would produce the same toxic effect. On the **toxicity equivalency,** or TEQ, scale, the toxicity of each dioxin, furan, and PCB molecule is rated relative to that of 2,3,7,8-TCDD, which is arbitrarily assigned a value of 1.0. The toxicity of any sample is reported as the concentration of the equivalent amount of 2,3,7,8-TCDD.

The toxicity values for contamination of various types of foods purchased in U.S. supermarkets in 1995 is shown in Figure 10.14. The values are given as parts per trillion, ppt, of the equivalent concentration of 2,3,7,8-TCDD that matches the actual toxicities of the foods. Contributions from each of the three categories—PCBs, dioxins, and furans—are shown for each food. Notice that freshwater fish contain the highest amounts of toxicity from both PCB and furan sources. The composite *vegan* diet (that is, all vegetable, fruit, and grain, with no animal products at all) has a very low TEQ compared to that with animal-based components.

10.19 The health effects of PCBs, furans, and dioxins

Evidence about toxicity is derived from two sources: experiments on animals that have been deliberately exposed to the chemicals and statistical studies of humans who have been accidentally exposed to them.

Most PCBs are not *acutely* toxic to humans. Though PCBs in high doses cause cancer in test animals, most groups of people who have been exposed to relatively high concentrations of PCBs—as a result of their

Chapter 10: Chemicals in Our Bodies and Our Environment

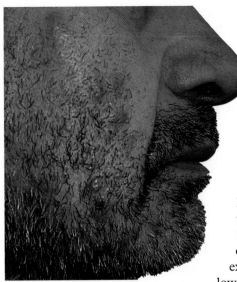

Figure 10.15 A man with chloracne. (DermNetNZ.org, New Zealand Dermatological Society)

employment in electrical capacitor plants, for example—have not experienced a higher overall death rate. The most common reaction to exposure is *chloracne,* a disfiguring acne that is characteristic of exposure to organochlorine compounds (see Figure 10.15).

PCBs are of concern because of their toxic effect on human and animal reproduction. The concern is particularly great for people who have consumed large quantities of fish in which PCBs have accumulated in the fatty tissue. According to one well-documented study, the children born to women who consumed large amounts of fish from Lake Michigan suffered some growth retardation and scored significantly lower on certain memory tests. These effects have been blamed on the transfer of PCBs from mother to child during pregnancy. By 11 years of age, the effects of the prenatal PCBs were still apparent only in the fraction of the group that had been the most highly exposed to PCBs before birth. These children averaged 6 points lower than the others on IQ scores and their most affected mental processes were memory and attention span.

The most dramatic effects yet observed on human health from exposure to PCB mixtures occurred in Japan in 1968 and in Taiwan in 1979. Two groups of people unintentionally consumed PCBs that had accidentally been mixed with cooking oil. Since the PCBs had been heated, the level of dibenzofuran contamination was much greater than in freshly prepared commercial PCBs. The thousands of people who consumed the contaminated oil suffered health effects far worse than did workers at PCB manufacturing and handling plants with comparable PCB levels in their bodies. From this difference, it has been concluded that furans and dioxins were responsible for about two-thirds of the effects, with the PCBs themselves responsible for the remainder.

Like the children in the Lake Michigan study, the cognitive development of children born to the most highly exposed Taiwanese mothers was significantly lower than that of children of unexposed mothers or children born before the accident occurred. This was the case even if birth occurred long after consumption of the contaminated oil. Interestingly, children whose fathers, but not mothers, had consumed the oil showed no detrimental effects. We can conclude that it is exposure of the unfertilized egg or the developing unborn child to the chemicals that causes the effects, not exposure of the sperm.

In 2000, the U.S. Environmental Protection Agency (EPA) issued a 2000-page draft report concerning the health risks of dioxins. The report concluded that 2,3,7,8-TCDD is a human carcinogen, and it is likely that other dioxins are as well, so the mixture of PCDDs to which people are exposed is a "likely human carcinogen." The report also discussed noncancerous effects from such compounds—such as the effects on reproduction, sexual development, and the immune system (some of which are discussed below)—that may also prove a threat to human health. Experiments in the past had indicated that 2,3,7,8-TCDD was the most potent carcinogen known and causes cancer at

Herbicides, PCBs, Dioxins, and Furans

multiple possible sites in test animals. Based upon the EPA's cancer-risk estimate, an upper limit of one thousand of the one million new cancer cases diagnosed each year in the United States could arise from the intake of a mixture of dioxins, furans, and PCBs.

Vigorous debate continues in the scientific, industrial, and medical communities regarding the environmental dangers of PCBs, furans, and dioxins. In one camp are those who feel that the dangers from these chemicals have been wildly overstated in the media and by some special-interest groups. They point to the very low concentrations of these substances that exist in the environment, to the lack of human fatalities that have resulted from them, and to the enormous economic costs associated with instituting effective controls on them. At the other extreme are those who point to the substantial biomagnification and high toxicity per molecule of these substances, and to their presence in almost all environments. They consider the detrimental effects such as cancer and birth deformities caused by these chemicals in wildlife to be "warning canaries" that signal potential ill effects in humans. Discovering where the truth lies between these opposing viewpoints presents a challenge even for well-educated students, to say nothing of the public at large.

Environmental Estrogens

In 1996, the controversial book *Our Stolen Future* by Theo Colburn and her associates stirred public interest in a new threat to the health of wildlife and possibly of humans. The book reported the finding that certain synthetic organic chemicals in the environment contribute to infertility and may also increase the rate of cancer in reproductive organs.

10.20 Certain organic chemicals interfere with natural estrogen

Most of the concern about synthetic organic chemicals in human reproduction centers upon interference with *estrogen,* a female sex hormone. Recall from Chapter 9, section 9.25 that hormones flow through the bloodstream from the point of their production and storage to their target organs, including those involved in sexual reproduction in both females and males. Certain environmental chemicals, even ones that bear little structural resemblance to estrogen itself, can bind to the estrogen receptor and thereby either mimic or block the action of the hormone itself. Other compounds can accelerate the breakdown of the natural hormones. These so-called **environmental estrogens** include the organochloride insecticides such as DDT, as well as some PCBs and dioxins, and a variety of common industrial organic compounds that contain oxygen.

Several non-organochlorine environmental estrogens are of concern. *Bisphenol-A* is a widely used substance that is polymerized industrially into polycarbonate plastics and some epoxy resins. Some of the compound is leached if the food containers made using the resin are heated to high temperatures. Although these same resins are used to line aluminum beer and soda cans, bisphenol-A apparently does not leach from these containers. Bisphenol-A also could potentially leach from dental sealants that are made from resins prepared from it, though the evidence that this occurs now is in doubt.

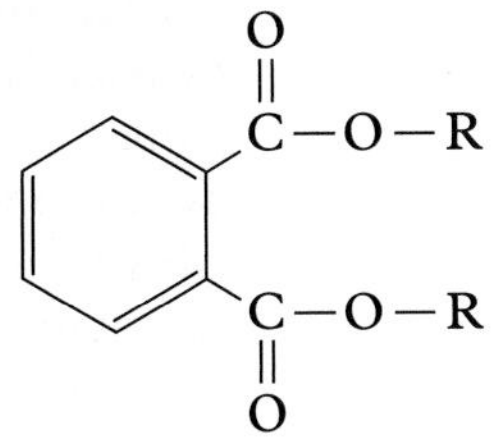

Phthalate esters

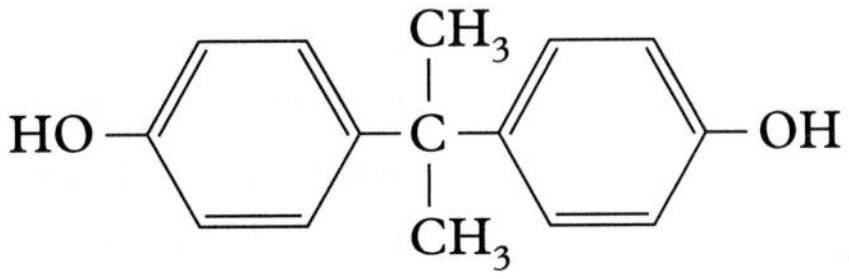

Bisphenol-A

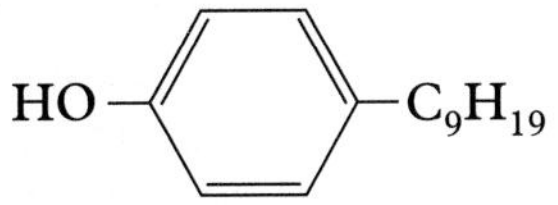

Nonylphenol

Another important environmental estrogen is *nonylphenol,* which displays some structural features in common with bisphenol-A. It, too, occurs in the environment as a result of the breakdown of a larger molecule that is used in detergents, spermicides, and some plastics.

Phthalate esters also have estrogenic action. They are widely used as plasticizers in common plastics (see Chapter 5), from which they can leach into the environment.

10.21 Can environmental estrogens affect health?

The most famous example of the environmental effects of hormone-like chemicals upon wildlife involves alligators in Lake Apopka, Florida. In 1980, massive amounts of DDT and its analogs were spilled into the lake. In the mid-1980s, very few alligator eggs were hatching and few hatchlings survived of those that were born, thereby threatening the future population of the colony. Furthermore, the surviving hatchlings had abnormal reproductive systems and therefore were unlikely themselves to be able to reproduce. The ratio of natural estrogen to the male sex hormone testosterone was greatly elevated in the young alligators. Presumably as a consequence, the penises of male alligators were reduced in size compared to the norm. These effects were apparently

Recall from section 10.7 that DDT is transformed into DDE in the environment.

caused by DDE, which has been found to inhibit binding of male hormones to their receptor. Similarly, reproductive problems, such as embryo mortality and deformities, of birds in the Great Lakes area have been traced to the hormonal activity of pollutants such as PCBs and dioxins.

The most devastating consequences of environmental estrogens do *not* occur in the mammals that originally ingest them. Rather, when the compounds are transferred from the mother to the fetus, they disrupt the hormone balance in the recipient. This disruption causes reproductive system abnormalities or changes that will result in cancer when the offspring grows to adulthood.

There is good evidence that high concentrations of environmental estrogens have caused reproductive problems in wildlife and laboratory animals. However, it is not certain that comparable effects occur in humans at the levels to which we are exposed. Much of the human evidence concerning the possible effects of estrogen mimics was obtained from the experience of women who between 1948 and 1971 took the synthetic estrogen *DES* (diethylstilbestrol) to prevent miscarriage. Many of the daughters of these women are sterile, and a small fraction of them have developed a rare vaginal cancer. The male offspring of the women who took DES have an increased incidence of abnormalities in their sexual organs, have decreased average sperm counts, and may have an increased risk of testicular cancer.

A decline in male sperm counts and an increase in the rate of testicular cancer are often cited as examples of the effects of environmental estrogens. However, both sperm counts and testicular cancer rates vary significantly between geographical regions, and the variations are not closely linked to differences in pollution levels. Some recent research is more convincing that environmental estrogens in humans can affect sexual characteristics. Girls who were exposed prenatally to high levels of DDE reach puberty almost a full year before those with the lowest exposures. Boys are not affected in this way.

Some scientists discount *any* adverse effects of synthetic chemicals acting as environmental estrogens by pointing out that we all ingest much greater quantities of plant-based estrogen mimics called **phytoestrogens.** Common sources of these natural chemicals include all soy products, broccoli, wheat, apples, and cherries. Indeed there is some evidence that phytoestrogens have a *protective* effect against some types of cancers. However, phytoestrogens are quickly metabolized by the body and perhaps do not survive long enough to exert effects on a developing fetus, for example.

● Tying Concepts Together: Tiny concentrations of chemicals in our bodies

Though the subjects of this chapter—vitamins, persistent organochlorine molecules, and both natural and environmental hormones—may seem quite different from one another, they have in common the fact that they are present in

Chapter 10: Chemicals in Our Bodies and Our Environment

our bodies in tiny concentrations. The fact that we are able to even detect their presence is a triumph of modern chemistry. However, the mere detection of a particular environmental chemical in our bodies is of no particular significance in and of itself.

The effect on our health, whether negative or positive, of a substance depends not only upon its composition but also upon its concentration. We are accustomed to this concept in the case of medicinal drugs, where a small amount is beneficial but an overdose can prove deadly. While vitamins in small amounts are certainly beneficial to our health, at least in some cases too high a dose can be harmful. Vitamin A, in particular, is toxic in large amounts, as hunters who have eaten the livers of certain wild animals (in which the vitamin is highly concentrated) have discovered. Thus it is possible to "have too much of a good thing." We also need to be cautious in artificially increasing hormone levels too high, as deleterious effects on health can result.

While it is unlikely that the presence of synthetic organochlorine compounds in your body improves your health, it is not clear that they damage everyone. Some scientists believe that chemicals such as the dioxins have no effect until their concentration reaches a *threshold* value, only beyond which does harm increase in proportion to the amount present. The level of environmental estrogens in our bodies may lie below their threshold for biological action.

Scientists have pointed out that the average level of dioxins found in human flesh is a few times lower than the minimum dose found to cause reproductive problems in rats, so widespread problems in humans are not expected.

Summarizing the Main Ideas

Vitamins are essential substances needed in small quantities for good health. Water-soluble vitamins consist of C and the B group; fat-soluble vitamins are A, D, E, and K. Vitamin E is an antioxidant and helps prevent free radicals from attacking cell membranes.

Manufacturers often add synthetic antioxidants, with structures similar to that of Vitamin E, to packaged foods. Similarly, sequestrants like citric acid and salts of EDTA are added to reduce the activity of metal ions that could catalyze oxidation reactions in foods.

Many food preservatives are found naturally in certain fruits and other plants, and consist of certain organic acids and/or their salts. Dehydration also preserves foods. Sulfites and nitrites are inorganic ions used to prevent the growth of microorganisms as well, though these ions may cause human health problems.

The main categories of pesticides are insecticides and herbicides. Insecticides, such as DDT, are used to control insects that carry disease and prevent crops from being destroyed by them.

Many organochlorine compounds undergo bioaccumulation in the environment and consequently are present in animals at levels far exceeding general environmental concentrations. Bioconcentration occurs as compounds diffuse into living matter; biomagnification occurs when they are accumulated along a food chain.

Organophosphate and carbamate insecticides do not bioaccumulate, but some of them are acutely toxic.

Massive amounts of herbicides are used to kill weeds, particularly on farms in developed countries. Phenoxy herbicides such as 2,4-D are used on lawns. The herbicide 2,4,5-T is invariably contaminated with the highly toxic compound 2,3,7,8-TCDD. The latter is one member of a series of chemicals called dioxins.

PCBs are organochlorine compounds used in a variety of applications. Their inappropriate disposal into the environment has caused concentrations there to build up, since they are so persistent. In addition, they biomagnify in food chains. Most of the toxicity in commercial PCB samples may result from the presence of furans, which are dioxin-like compounds formed when PCBs are heated.

Human exposure to PCBs, furans, and dioxins collectively is measured by the TEQ value, which reports the equivalent amount of 2,3,7,8-TCDD that would give the same level of toxicity. The human health effects of these compounds are not precisely known and are controversial. Most of the serious effects show up not in the adults who ingest the substances but in the children that the women subsequently give birth to.

Environmental estrogens are synthetic compounds that interfere with the natural hormones in the body and can result in reproductive abnormalities. Examples include DDE, some PCBs and dioxins, and certain industrial organic compounds containing oxygen. The evidence that environmental estrogens could be a health problem to humans is based mainly upon studies of wildlife.

Key Terms

vitamin	herbicide	biomagnification	chlorpyrifos
antioxidant	bioaccumulation	priority organic	methyl parathion
osmosis	bioconcentration	pollutants	green chemistry
pesticide	food chain	organophosphate	toxicity equivalency
insecticide	food web	carbamate	environmental estrogen
			phytoestrogen

Web Sites of Interest

To link to Web sites of interest, go to www.whfreeman.com/ciyl2e, Chapter 10, and select the site you want.

For Further Reading

R. Carson, *Silent Spring*. Houghton Mifflin, Boston, 1962. Written over the period 1958–1962, this seminal book examines the effects of pesticides on songbird populations in the United States. The dramatic decline in their numbers yielded the "silence" of the title. Many consider this the book that started the modern environmental movement.

T. Colburn et al., *Our Stolen Future*. Dutton Publishers, 1996. This book identifies ways in which environmental pollutants are disrupting human reproductive patterns and causing such problems as birth defects, sexual abnormalities, and reproductive failure.

J. McClintock, "Silent Summer," *Discover*, July 2000, pp. 76–79. This article discusses the problems associated with mosquito-eradication programs targeting the carriers of West Nile virus in the United States.

J. McLachlan and S. Arnold, "Environmental Estrogens," *American Scientist,* September–October 1996, pp. 452–461. The feminization of wildlife species is traced to pesticides and other chemicals that behave like estrogens. This article examines how the chemicals work and why they don't have to resemble the estrogen structure to be effective.

Review Questions

1. What is a *vitamin*?

2. What roles do vitamins play in human bodies?

3. What are the two basic classifications of vitamins?

4. Which class of vitamins has the potential to be toxic? Why?

5. Identify specific roles in the human body of each of the following vitamins: A, B, C, D, E, and K.

6. What is a *coenzyme*?

7. What is an *antioxidant*?

8. What is a *free radical*?

9. What factors are responsible for free radical formation in the human body?

10. What is *oxidative stress*?

11. What are *phytochemicals*? Where are they found?

12. Why is it true that eating carrots is good for your eyesight?

13. Identify the chemicals that are synthetic versions of vitamin E. What roles do they play?

14. What is the origin of metal ions in food?

15. What is a *sequestrant*? Identify a common example.

16. What is the role of a *food preservative*?

17. What is *osmosis*?

18. Why are some acids used as preservatives?

19. Identify some common acids used as preservatives.

20. Identify two inorganic salts that are used as preservatives.

21. What health concerns are associated with nitrites?

22. What additives, other than preservatives, may be found in processed food? What roles do these additives play?

23. Define each of the following terms: pesticide, insecticide, herbicide, fungicide.

24. Why are pesticides toxic?

25. What damaging effects are associated with DDT use?

26. Define the following terms: food chain, food web, bioaccumulation, bioconcentration, biomagnification.

27. What does it mean to say that humans are at the "top" of the food chain?

28. How do organophosphates and carbamates damage target organisms?

29. Why are organophosphates and carbamates "acutely" toxic to humans?

30. Identify some natural insecticides and their sources.

31. What are the components of *Agent Orange*?

32. What was *Agent Orange* used for? What were some of the consequences of its use?

33. What is *glyphosate*? How is it used?

34. What are *PCBs*? How were they used?

35. What processes produce dioxin and furan by-products?

36. Identify some anthropogenic sources of dioxin.

37. Why do dioxins, furans, and PCBs bioaccumulate in the food chain?

38. What foods are the primary sources of dioxins in humans?

39. What does *toxic equivalency* measure?

40. What is *chloracne*?

41. What are *environmental estrogens*? Why are they a health issue?

42. What chemicals act as environmental estrogens?

43. What is *DES*? Why was it used?

44. What health effects are attributed to the use of DES?

45. How are bisphenol-A, nonylphenol, and phthalates used?

46. What are *phytoestrogens*?

Understanding Concepts

47. Why are vitamins often called "trace" or "micro" nutrients?

48. Use the structures of vitamin C and vitamin A to explain their solubility properties.

49. How do fat- and water-soluble vitamins differ in the way they are transported in the blood?

50. How does cooking foods in water impact their vitamin content?

51. Would it be easier to develop a vitamin A deficiency or a vitamin C deficiency? Explain.

52. Explain how antioxidants work to prevent free radical damage in the body.

53. Explain how free radicals damage biological tissue.

54. How does the body synthesize vitamin D?

55. What is the relationship between vitamin D and bone development?

56. Why is milk a good product to add vitamin D to?

57. Identify the mechanisms by which food spoils.

58. Describe how food becomes rancid.

59. Identify a nonchemical strategy to prevent rancidity. Why does it work?

60. Explain why dehydration is an effective food preservation technique.

61. How does the preservative action of propionates differ from that of sorbates?

62. How does the addition of nitrites create a pink color in meat?

63. Why is pesticide regulation an important issue?

64. What businesses might consider insecticides important to their economic well-being?

65. How does DDT's solubility relate to its toxicity?

66. Why are fat-soluble substances more likely to bioaccumulate than water-soluble substances?

67. Explain how bioaccumulation and bioconcentration occur as one proceeds up the food chain.

68. What is the relationship between a *food chain* and a *food web*?

69. How do organochlorines, organophosphates, and carbamates differ in their toxicities?

70. Why is malathion more toxic to insects than to humans? Do you think it would be toxic to rats? Explain.

71. Compare the environmental impacts of organochlorines, organophosphates, and carbamates.

72. How do the toxicities to mammals of natural pesticides compare to those of synthetic pesticides? Are natural pesticides less toxic to humans?

73. Why have PCBs become an environmental issue?

74. Explain why PCBs are found in all parts of the world.

75. What health effects are associated with exposure to PCBs?

76. Why are environmental estrogens able to affect hormonal activity?

77. What evidence indicates that environmental estrogens can affect health?

Synthesizing Ideas

78. What would happen to vitamin D production in the absence of sunlight? What factors may limit sun exposure? Explain.

79. Explain the relationship between the role of vitamin E in the human body and the symptoms seen when there is a deficiency.

80. EDTA is sometimes used as an antidote to heavy metal poisoning. Explain why it is effective in this role.

81. Honey has been shown to have antibiotic properties. Why might this be so?

82. An emergency medical manual suggests packing an open wound with pure cane sugar until medical help is available. What purpose does the sugar serve? (*Hint:* Osmosis is involved.)

83. Draw the structure of benzoic acid in acidic and in alkaline solutions. Why might its effectiveness as a preservative be affected by the difference in structure?

84. Compare the health issues associated with the use of inorganic preservatives and organic acid preservatives. Explain why there is a difference.

85. Compare the types of products in which sulfites are commonly used to those in which nitrites are commonly used. Are they different? Why?

86. What properties of DDT cause its persistence? What are the advantages and disadvantages of this persistence?

87. Why is DDT *not* excreted in the urine? What are the biological implications of this fact?

88. Compare the impact on a food chain of a persistent insecticide and a nonpersistent insecticide.

89. Use your understanding of the material in this chapter to explain the following observations:
a) Chocolate can be toxic to dogs but not to humans.
b) Arsenic can be toxic to humans but not to owls.

90. Describe the impact that nonspecific insecticides can have on ecosystems.

91. Explain why PCBs and their by-products persist in the environment despite the halting of production in 1977.

92. Compare the chemical exposure from the following diets:
a) average American diet: includes all meats, fish, poultry, dairy products, eggs
b) a vegan diet: no meats, fish, poultry, milk, dairy, eggs, or any animal-derived products; lots of fruits, vegetables, whole grains, beans, soy products
c) a vegetarian diet: no meats, fish, or poultry; lots of fresh fruits, vegetables, dairy products, eggs

93. Freshwater fish have a much higher TEQ than saltwater fish. What factors contribute to this?

Taking It Further

The nature and reactivity of free radicals

A free radical is an atom or a molecule having an odd number of electrons. Since they do not possess an even number of electrons, as do almost all other molecules, one of their valence shell electrons does not become part of either a two-electron chemical bond or a nonbonding electron pair. Such atoms and molecules are highly reactive, because by definition not every one their component atoms has an octet (or duet) of electrons in its valence shell.

Worked Example: Free radicals

Is CH_3 a free radical?

Solution: In order to decide whether or not a given molecule is a free radical, we need to deduce the number of electrons that it contains. We can do this by adding up the atomic numbers of the constituent elements, since the number of electrons an atom possesses equals its atomic number (see Chapter 3, section 3.1). Thus for CH_3, since the atomic number of carbon is 6 and that of hydrogen is 1, the total number of electrons is $6 + 3 \times 1 = 9$. Since 9 is an odd number, CH_3 is a free radical.

Exercise 10.1

Determine which of the following atoms and molecules are free radicals:

a) NH_3 b) NO   c) atomic H

Because they contain an outer shell with fewer than eight electrons, most free radicals are highly reactive. By extracting an electron—with or without an associated atom—from a molecule, free radicals achieve a stable octet. After acquiring the electron, the former free radical is more stable and much less reactive. However, the atom or molecule from which the electron was extracted is now itself a free radical and can continue the process of reaction with another molecule! For example, the free radical OH reacts with a molecule of methane, CH_4, by extracting a hydrogen atom with its electron, thereby converting CH_4 to CH_3, a free radical:

$$OH^{\bullet} \;+\; CH_4 \;\rightarrow\; H_2O \;+\; CH_3^{\bullet}$$
$$\text{free radical} \qquad\qquad\qquad \text{free radical}$$

The dot shown as a superscript after the OH or CH_3 represents the odd electron that is unbonded and thus signifies that the molecule is a free radical. The free radical CH_3 produced in the reaction above is highly reactive and will attack another molecule, thereby continuing the chain of reactions. Such destructive chains often continue until thousands of molecules have taken part in them. If oxygen molecules are available, free radicals will often add O_2 to themselves, producing oxygen-containing free radicals. As a result of this and subsequent reactions, the original nonradical substance is gradually oxidized. By these means, methane in air is gradually oxidized to carbon dioxide, and polyunsaturated oils become rancid.

Antioxidants are molecules that react with free radicals but in the process form free radicals that are unusually unreactive. In this way, antioxidants such as vitamin E prevent the consecutive chain of reactions from proceeding further and oxidizing the material. For convenience, we can represent the complex structure of vitamin E simply as X since the hydroxyl group is the part of it that reacts. When a free radical, $F^{\bullet}$, reacts with a molecule X—O—H of vitamin E, it extracts the hydrogen atom—with its electron—and thereby converts vitamin E into a free radical, $X-O^{\bullet}$:

$$X-O-H \;+\; F^{\bullet} \;\rightarrow\; X-O^{\bullet} \;+\; F-H$$
$$\text{antioxidant} \quad \text{free radical} \quad \text{free radical} \quad \text{nonradical}$$
$$\text{(unreactive)}$$

Since the free radical formed by vitamin E is one of the few free radical types that is unreactive, the reaction chain does not continue beyond this point, and the oxidation process thereby is stopped. The original form, X—O—H, of vitamin E is eventually re-formed by the reaction of the X—O˙ radical with vitamin C.

1. Why are *free radicals* very reactive?
2. Why are free radicals destructive?
3. How do free radicals initiate chain reactions?
4. How do antioxidants prevent chain reactions from continuing?
5. Determine whether any of the following are free radicals:
   a) CH_4 b) NH_4 c) NO_2 d) NO_3^-

In this chapter, you will learn:

- what a bottle of water, an icy driveway, and seawater have in common;
- why fluoride is added to water;
- why there is acid in your stomach;
- what is meant by pH and what it measures;
- how antacids work;
- how acids and bases are used to preserve foods;
- what the difference between hard and soft water is;
- why ammonia works well for cleaning your house;
- how plants can help us determine the presence of acids and bases.

How can something so small contribute to forming something so large?

If you've ever seen a cave full of stalactites and stalagmites, you may find it hard to believe that small drops of water are able to produce such large structures. Inside caves, such as this one in Bisbee, Arizona, processes involving ions dissolved in water produce the columns that we call stalactites and stalagmites. In this chapter, we will explore how salts, acids, and bases behave in water.

(Peter L. Kresan)

Chemistry in Water
Salts, Acids, and Bases

Although ions were introduced in Chapter 3, most of the chemistry we have discussed so far has involved molecules, which by definition are electrically neutral. However, many processes that take place in *aqueous solutions*—that is, ones in water—involve ions. In this chapter, we explore the chemistry of ions dissolved in water, including in our own blood. We will consider the interactions of ions with each other and with water molecules, the formation of ions, and their interconversion when they transfer hydrogen ions, H^+, between them. Along the way you will learn about many things in your everyday world, including how stomach antacids work, how your blood is buffered, why cleaning agents often contain ammonia or sodium hydroxide, why some fluoride is good for your teeth, and what the ppm analysis of ions in mineral water means.

Salts

As you saw in Chapter 3, ions are tightly held within crystals by *ionic bonding*, the attraction of oppositely charged particles for each other. Recall also that most solid compounds consisting of ions are called *salts*. The only ionic solids that are not called salts are those in which the cation is H^+ or the anion is OH^-. The salt you are probably most familiar with is sodium chloride, which consists of sodium ions and chloride ions.

Most ionic compounds do not melt until they reach very high temperatures, when there is enough energy available for the ions to move somewhat from their optimum positions in the crystal lattice. For example, the melting point of NaCl is 801°C.

11.1 Many salts readily dissolve in water

Given that the separation of ions from a crystal lattice requires a great deal of energy, it is somewhat surprising to discover that many ionic compounds readily dissolve in water without any apparent input of energy. Indeed, our common experience is that a large amount of table salt readily dissolves in a cup of water, even cold water. We may well ask what interactions could possibly stabilize the individual ions in water to such an extent that they would compensate for the electrostatic attractions between neighboring opposite charges that exist in the ionic crystal lattice. These attractions must be largely absent in solution since the ions are on average greatly separated from each other.

Whenever you see this icon in this chapter, go to
www.whfreeman.com/ciyl2e

Overall, solid ionic compounds dissolve in water by the successive dissociation of individual ions, both cations and anions, from the ionic lattice as their attraction to the ions of opposite charge is replaced by their attraction to a number of water molecules which approach and surround them (see Figure 11.1). The conceptual model for ionic substances dissolved in water is that of individual ions, both cations and anions, each surrounded by a layer of water molecules. Recall from Chapter 6, section 6.10, that the oxygen atom in a water molecule carries a partial negative charge and the hydrogen atom carries a partial positive charge. The attractions between the charges on the atoms of the water molecules with those of the ions compensate for the loss of ion–ion attractions in the solid.

In the case of dissolved cations (see Figure 11.1a), the water molecules orient themselves with their hydrogen atoms pointing away from the ion. The cation is stabilized by electrostatic attraction between its positive charge and the partial negative charge of the oxygen atom. The arrangement of atoms ensures that the region of space in which the nonbonding electrons of oxygen are most likely to be found, namely on the side of the oxygen atom away from the hydrogens to which it is bonded, lies as close as possible to the positively charged ion. In a sense,

Cations are repelled by the hydrogen ions on the far side of the water molecules, but they lie farther away from the cations than from the oxygen, so the attraction of the cations to the oxygen outweighs the repulsion.

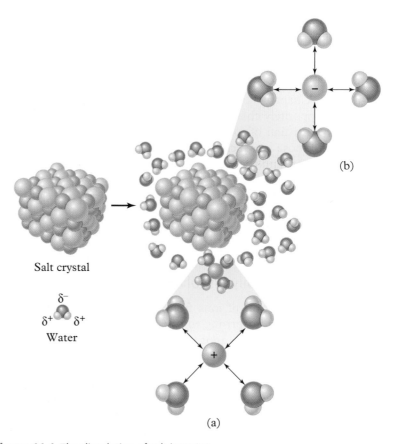

Salt crystal

δ^-
δ^+ δ^+
Water

(b)

(a)

Figure 11.1 The dissolution of salt in water.

Chapter 11: Chemistry in Water

this attraction is similar to that in a hydrogen bond, where the partial positive charge of a hydrogen atom is attracted to the partial negative charge of an oxygen atom.

An anion, on the other hand, is stabilized in an aqueous solution by the electrostatic attraction between its negative charge and the partial positive charges of the hydrogen atoms of H_2O molecules (see Figure 11.1b). Notice that for anions the hydrogen atoms of the H_2O molecules in the surrounding layer are positioned closer to the anions than are the oxygen atoms.

11.2 Salts have varying degrees of solubility in water

A substance that will ultimately dissolve—given enough time and enough stirring—in a liquid is called the **solute.** The dissolving liquid is called the **solvent,** which commonly is water. The maximum amount of solute that will dissolve in a given quantity of solvent varies with the identity of the two substances and with temperature; often it is quite limited. For example, although a large quantity of common table salt will dissolve in water, only a very small amount of calcium sulfate, $CaSO_4$, will do so—even though *other* salts containing calcium or sulfate ions readily dissolve. The *maximum* concentration of a substance that can be achieved in a solution is called the substance's **solubility.** The common unit for solubility is *grams of solute dissolved per liter,* or *per kilogram, in the resulting solution.* For example, the solubility at room temperature of sodium chloride is 265 grams of the salt per kilogram of solution.

Solutions containing a relatively high concentration of solute are said to be *concentrated* solutions, whereas those with very little dissolved solute are said to be *dilute.* There is no precise boundary dividing these two types of solutions.

Solutions that contain the maximum amount of dissolved solute at a given temperature are said to be **saturated.** The solubilities of many salts increase sharply as the temperature increases. Sodium chloride is an exception to this generality, as its solubility increases only by about 10% in going from 0°C to 100°C. The solubilities of a few salts decrease with increasing temperature.

The energy of electrostatic attraction between oppositely charged particles—which tends to hold the ions together in the solid, rather than letting them dissolve—is proportional to the mathematical *product* of their charges. As a general rule of thumb, salts of +1 cations with −1 anions are quite soluble in water. However, salts with greater products of the two ion charges are generally less soluble. Consequently many salts such as calcium carbonate, $CaCO_3$, and calcium phosphate, $Ca_3(PO_4)_2$, are quite insoluble in water, since their component ions have charges of +2 and −2 and +2 and −3, respectively. Ion charges are not the *only* factor determining solubility, however. Properties such as ion size are also important. Thus, many sulfate salts, even when the cation has a charge of +2 or +3 as in *magnesium sulfate,* $MgSO_4$, or *aluminum sulfate,* $Al_2(SO_4)_3$, are quite soluble in water. In contrast, most

Supersaturated solutions are those that temporarily contain more dissolved solute than do saturated ones. They are unstable to the eventual precipitation of the excess solute. Solutions having less than the maximum dissolved solute are called *unsaturated*.

(a)

(b)

Figure 11.2 Formation of a precipitate. (a) Solubility limit is not exceeded. (b) Solubility limit is greatly exceeded. (George Semple for W. H. Freeman and Company)

hydroxides, such as *magnesium hydroxide,* $Mg(OH)_2$, are almost insoluble even though the anion charge is only -1.

When an aqueous solution evaporates, only the water—and *not* the ions—departs from the liquid state and enters the air as a gas. You may have noticed that when water containing dissolved table salt evaporates, a white powder—solid sodium chloride—is left behind. In general, the concentration of ionic solutes *increases* as the evaporation of a solution proceeds. Eventually, often well before full dryness is achieved, the solution will become so concentrated that the solubility limit of the ionic substance is exceeded. In such a situation, the solid form of the substance forms spontaneously, even in the presence of water (see Figure 11.2). If further evaporation occurs, more and more solid is formed, and often it can be seen with the naked eye as a solid lying at the bottom of the solution (where it deposits since it is more dense than the solution). We call a solid that forms because its solubility is exceeded a **precipitate** and say that it has been precipitated from the solution.

As you know, human sweat is fairly salty. When we touch an object, some of the sweat on our fingers is transferred to it. The liquid in the sweat slowly evaporates, leaving a tiny amount of solid sodium chloride on the object. One way that crime investigators develop fingerprints is by spraying the object with dilute *silver nitrate,* $AgNO_3$, which converts salt to *silver chloride,* $AgCl$.

Figure 11.3 Fingerprints made by reacting the salts in sweat with silver nitrate. (Mikael Karlsson)

$$NaCl + AgNO_3 \rightarrow NaNO_3 + AgCl$$

When exposed to sunlight, the silver chloride decomposes to metallic silver, which can be observed as the swirls and lines of the fingerprint (see Figure 11.3).

Crime scene investigators often use *black lights*—which are lamps that emit ultraviolet (UV) rather than visible light. Many materials—including blood, urine, and semen—reemit the UV as visible light, allowing investigators to detect tiny amounts of substances of interest when the scene is darkened. Forensic scientists often dust a crime scene with a dye that emits visible light when a black light is shone upon it, thereby allowing fingerprints to be more easily observed.

11.3 Seawater, drinking water, and bottled water

Aqueous solutions containing ions are everywhere in your world. Even unpolluted fresh water contains ions produced by the dissolving of natural substances from the air and from the soil.

The most salty solution you are likely to encounter is seawater. One kilogram of seawater contains about 35 grams of various dissolved salts. These compounds are composed mainly of sodium Na^+ and chloride Cl^- ions, lesser concentrations of magnesium Mg^{2+} and sulfate SO_4^{2-} ions, smaller concentrations of Ca^{2+} and K^+, and tiny amounts of many other ions including Br^-. *Sea salt* is the solid that results when water is completely evaporated from seawater, and so it differs somewhat from normal table salt, which is almost exclusively Na^+ and Cl^- ions. Although sodium chloride is much more abundant in seawater than other salts, it is the compounds of calcium that are the first to precipitate when seawater is evaporated. These are the salt deposits you see in rock pools by the sea as the deposits gradually dry out. It is not until more than 90% of the water has evaporated that NaCl begins to precipitate.

The concentrations of ions in *fresh* water are, of course, much lower than those in seawater, but they are not zero. For example, the average concentration of the calcium ion, Ca^{2+}, in North American river water and drinking water obtained from municipal sources is about 15 *milli*grams per liter of the liquid, and that of the sodium ion, Na^+, is about 6 milligrams per liter. To get a sense of how small this concentration is, realize that one-thousandth of a teaspoonful of salt dissolved in a liter of water corresponds to a concentration of about 5 milligrams per liter.

Quoting the mass of solute that is dissolved in a standard quantity of solution is a common way to state the **concentration** of aqueous solutions such as drinking water. Thus bottles of mineral water sometimes list salt concentrations as mg/L. More commonly stated, however, are concentrations on the **parts-per-million** (ppm) **scale.** For liquids and solids, ppm concentrations represent the mass of the dissolved substance contained in 1 million grams of the solution.

The relationship between mg/L and ppm concentration units can be deduced as follows. The density of water, and even that of solutions containing small amounts of dissolved substances, is very close to 1.0 gram per milliliter. Therefore 1 liter of water or a dilute solution weighs 1000 grams, and so 1000 liters of it weighs 1 million grams. Consequently, the concentration of a substance on the parts per million scale is identical to the number of grams of it dissolved in 1000 liters of water. If we divide the mass of the substance *and* the volume of water both by the same factor of 1000, the magnitude of the concentration does not change. Thus we conclude that *the numerical concentration of any solute dissolved in water is the same on the milligrams-per-liter scale as it is on the parts-per-million scale.*

The concentrations for the ions commonly dissolved in some bottled waters, along with U.S. government standards for drinking water, are listed in Table 11.1. The concentration of many ions is much higher in **mineral water,** which is obtained from natural sources in which a much higher amount of salts (>500 ppm) has dissolved than in regular water. The drinking water standard for sulfate of 250 ppm is sometimes exceeded in mineral waters. This is of importance to people in whom sulfate causes a laxative effect when the levels exceed about 500 ppm. Indeed, *magnesium sulfate*, $MgSO_4$ known as *Epsom salts,* is an old-fashioned cure for constipation. If the mineral water is obtained from an area having limestone, it contains dissolved CO_2. When the bottle is first opened, the carbon dioxide gas fizzles as it leaves the liquid.

Since 1 mL = 1 cm³, density expressed as g/mL is the same as g/cm³.

Some health aspects of the sodium ion were discussed in Chapter 3.

Exercise 11.1

Labels of bottled drinking water advertising the product to be "low sodium" or "sodium free" typically state that it contains 1.25 milligrams of sodium per 250 mL serving. What is the ppm concentration of Na^+ in such water?

Table 11.1 Drinking water standards and ion concentrations in some bottled waters

Ion	U.S. standard (ppm)	Bottled water, brand A (ppm)	Bottled water, brand B (ppm)	Mineral water, brand C (ppm)
HCO_3^-		357	NL*	1580
Ca^{2+}		78	66	89
Mg^{2+}		24	17	104
SO_4^{2-}	250	10	25	112
N as NO_3^-		1	1.7	8
Na^+		5	4	425
K^+		1	1	25
F^-	0.8–2.4	0.1	0.1	0.7
Cl^-	250	4	6	137
Mineral salts		309	260	1690
$CaCO_3$		NL	200	NL

*NL means *Not listed*. No entry in the U.S. standards column means there is no standard for the ion.

11.4 Fluoride in water can help prevent tooth decay

The level of *fluoride* ion, F^-, in water differs significantly from place to place. Its natural source is the slow weathering of the almost insoluble ionic mineral *fluorapatite*, $Ca_5(PO_4)_3F$, which occurs naturally in many soils. The weathering results in an average freshwater concentration of about 0.1 ppm of the ion, but concentrations of 10 ppm and higher are not unknown.

Decades ago, it was noticed that people living in communities whose water contained higher fluoride levels, greater than about 1 ppm, had on average many fewer dental cavities. And studies in the 1950s showed that there was a dramatic decrease in childhood tooth decay in communities with naturally low flouride that raised it artificially to the 1 ppm level.

Apparently the fluoride ion from drinking water can replace the **hydroxide ion,** OH^-, in *apatite,* $Ca_5(PO_4)_3OH$, the mineral that makes up tooth enamel. The apatite is thereby converted to fluorapatite. Apatite has the same formula as fluorapatite, but has OH^- instead of F^- and is more soluble in acid (see Figure 11.4). This conversion makes the teeth more resistant when the enamel is attacked and dissolved by carboxylic acids produced by bacterial action on food in the mouth. In addition, fluoride ions inhibit the conversion of carbohydrates to the carboxylic acids that attack tooth enamel. For these reasons, many communities in the United States, Canada, Australia, New Zealand, and some other countries add fluoride, sometimes in the form of its sodium salt, NaF, to drinking water supplies in order to artificially raise the fluoride ion's concentration to about 1 ppm. It is now more common in large cities to add instead a source of *hexafluorosilicate ion,* SiF_6^{2-}, which

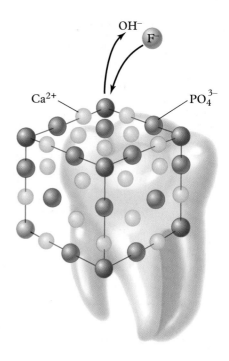

Figure 11.4 Fluoride ions replacing hydroxide ions in apatite. The resulting fluorapatite makes teeth more resistant to acid attack.

Salts

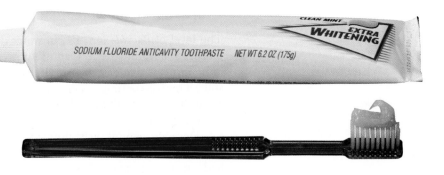

Figure 11.5 Sodium fluoride in toothpaste helps prevent cavities. (Corbis/Punchstock)

The MCL (maximum contaminant level) for fluoride ion in drinking water is 4 ppm.

????????????????? **?** ????????????
Fact or Fiction ❓

Most North Americans brush their teeth only once a day.

False. About half brush twice a day, and 10–25% brush even more often.

reacts with water to produce fluoride ions. In some other countries, such as Germany, Mexico, and France, sodium fluoride is added to table salt rather than to water.

In recent decades, most North Americans have also received fluoride from their toothpaste, usually in the form of sodium fluoride or *sodium monofluorophosphate,* Na_2PO_3F, or *stannous fluoride,* SnF_2 (see Figure 11.5). Many children receive topical fluoride treatments from their dentists. In addition, for many people the fluoride ion ingested from food exceeds that taken in from water. The addition of fluoride ions to public supplies of drinking water continues to be a controversial subject in some communities, because at higher concentrations it causes mottling of teeth. Some discoloration is noted even at 2 ppm levels. At very high concentrations, fluoride ion can be poisonous and even carcinogenic. In addition, some people feel that it is unethical to force everyone to drink water to which a medicinal substance has been added.

▶ **Discussion Point:** Should public water supplies be fluoridated?

Fluoridation of public water supplies has reduced the incidence of dental decay where it has been used, but critics claim there are long-term negative health effects. Use the resources listed on the Web site for this book to develop an argument for or against the mandatory fluoridation of public water supplies.

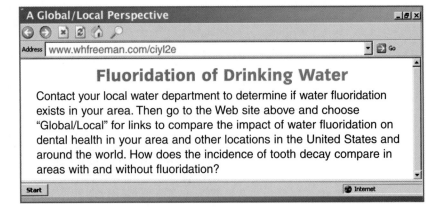

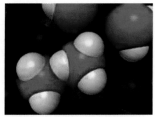

View the molecular activity in an acidic solution at Chapter 11: Visualizations: Media Link 1.

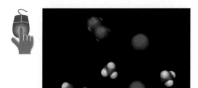

View the formation of acid molecules at Chapter 11: Visualizations: Media Link 2.

Acids

Acid is a category of substances that are of great importance in the chemistry of water and its solutions. You encounter acids every day—in the soft drinks you consume, in the liquids you use to clean windows and toilets, in the citrus fruits you eat, and so on. Acids are important in industry, too. They find extensive use in petroleum refining, in mining, in steel making, and in many other industries. Chemists use various levels of sophistication to describe the reactions of acids. We shall focus here on the simplest way of understanding the behavior of acids.

11.5 Acids increase hydrogen ion concentrations

Acids have several characteristics that are readily observable:

- They have a sour taste; indeed their name is derived from the Latin *acidus,* which means "sour."
- They turn litmus paper red.
- They react with many metals to produce hydrogen gas, H_2, and with calcium carbonate to produce carbon dioxide gas, CO_2.

These properties all arise from the tendency of acids to release hydrogen ions, H^+, in solution, as we discussed in Chapter 6. In fact, *an **acid** is defined as a substance that increases the concentration of H^+ ions in solution,* particularly an aqueous solution.

Consider as an example *hydrogen chloride,* HCl, which in the gas phase consists of covalent molecules, H—Cl. When the gaseous compound is dissolved in water, virtually all the HCl molecules **ionize,** that is to say they produce ions, in this case H^+ and Cl^-. The resulting solution is called *hydrochloric acid* (see Figure 11.6):

$$HCl(g) \xrightarrow{\text{water}} H^+ + Cl^-$$

For simplicity, chemists often do not specifically designate the ions as being in an aqueous solution, that is, we don't follow their formula with (aq). If you see an ion's formula written without reference to its state, assume that it is in aqueous solution.

In this process of ionization of HCl, the hydrogen atom loses its half-share of the pair of electrons that comprised the covalent bond in the molecule. It is as if the covalent bond between the hydrogen and the chlorine were converted into an ionic bond by the transfer of *both* bonding electrons to the chlorine, followed by a separation of the resulting ions in water.

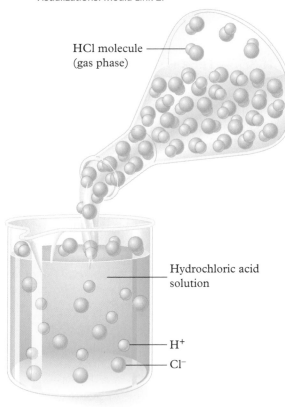

HCl molecule (gas phase)

Hydrochloric acid solution

H$^+$

Cl$^-$

Figure 11.6 Ionization of an acid. HCl in the gas phase ionizes into an aqueous solution of H$^+$ and Cl$^-$.

Acids

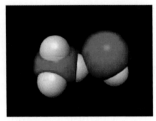

View the molecular-level activity of HCl in water at Chapter 11: Visualizations: Media Link 3.

As with HCl, the ionization of other acids produces H^+ ions and anions. For example, nitric acid, HNO_3, yields H^+ and the nitrate ion, NO_3^-, which we have encountered previously:

$$HNO_3 \rightarrow H^+ + NO_3^-$$

Exercise 11.2

Write the reaction equation for the ionization of hydrobromic acid, HBr.

We should note that the species H^+, which is simply the nucleus of a hydrogen atom, does *not* in fact exist on its own in aqueous solutions, even though we seem to suggest that it does when we write reactions for acids. Like other ions in an aqueous solution, H^+ is not an isolated species but is surrounded by a number of water molecules that have strong attractive interactions with it.

11.6 Polyprotic acids release more than one hydrogen ion

The *protic* part of the term *polyprotic* refers to the hydrogen ion, H^+, which is simply a proton if the nucleus does not also contain any neutrons. Recall that *poly* means many. Thus, a polyprotic acid contains many protons—or at least more than one!

Some acids contain more than one hydrogen atom that can ionize in water to produce H^+ ions; these are called **polyprotic acids.** Sulfuric acid, H_2SO_4, is an important polyprotic acid that contains two ionizable hydrogens. If only one hydrogen were to be ionized, the reaction would be:

$$H_2SO_4 \rightarrow H^+ + HSO_4^-$$

We deduced the formula HSO_4^- by subtracting one hydrogen from the reactant H_2SO_4 and assumed that both electrons of the covalent bond it formed stayed with the HSO_4 unit. The remaining hydrogen atom, that associated with the HSO_4^- ion, is similarly lost in what is called the *second stage of ionization* (see Figure 11.7):

$$HSO_4^- \rightarrow H^+ + SO_4^{2-}$$

It is sometimes more convenient to combine two (or more) sequential reaction stages into a single, net equation. To do this, we combine together the reactants on the left side of the successive reactions and place the result at the left side of the *overall* (or net) *reaction equation.* Then we do the same for the right sides of the equations. For the present example of sulfuric acid ionization, we obtain:

$$H_2SO_4 + HSO_4^- \rightarrow 2\ H^+ + HSO_4^- + SO_4^{2-}$$

Figure 11.7 The ionization of a polyprotic acid. Ionization produces more than one hydrogen ion from each acid molecule.

As a final step, we *cancel any terms appearing equally on both sides,* and so obtain the balanced equation for the overall reaction. In our example, the ion HSO_4^- appears on both sides, so is canceled, giving:

$$H_2SO_4 \rightarrow 2\,H^+ + SO_4^{2-}$$

Exercise 11.3

Write the three ionization steps for phosphoric acid, H_3PO_4, and from them deduce the overall equation for the process.

11.7 Strong acids are completely ionized in solution

There are only a few acids that behave in the simple manner described above in that they ionize completely (or almost so) in water. Such acids are called **strong acids.** The most important members of this small set are the three below:

- Hydrochloric acid, HCl
- Nitric acid, HNO_3
- Sulfuric acid, H_2SO_4

The term *strong* as used by chemists does *not* denote how concentrated an acid solution is. Rather it indicates that all the hydrogens are ionized in aqueous solutions of it. The reverse reaction, to reconstitute the acid, does not occur in dilute solutions of strong acids.

Hydrochloric acid, HCl, finds many practical uses. It is sold as a concentrated solution of HCl in water as *muriatic acid* (see Figure 11.8). It reacts with both metal oxides and calcium carbonate, and the reaction products are soluble in water, so it is used to clean the surfaces of metals, masonry, and cement. It also is the acid that is present in your stomach and that activates digestive enzymes, as discussed later. *Nitric acid,* HNO_3, is used in the manufacture of nitrogen-containing fertilizers and explosives such as ammonium nitrate. *Sulfuric acid,* H_2SO_4, also is used to make fertilizers and explosives and to clean steel, and is the acid in car batteries. Ionic compounds that contain the *bisulfate ion,* HSO_4^-, such as $NaHSO_4$, are useful acids in powdered cleaners, for example, since they occur in the solid form; they are contained in many chemistry sets sold to children. The bisulfate ion generates the hydrogen ion as follows:

$$HSO_4^- \rightarrow H^+ + SO_4^{2-}$$

Bisulfate compounds also are used as toilet bowl cleaners. Liquid versions of such products often contain hydrochloric acid.

Some cleansers for ceramic materials such as toilet bowls contain phosphoric or certain other acids whose cleansing actions are not due to the action of H^+ but to the fact that ions of metals such as iron and

The acidification of rain by sulfuric and nitric acid is discussed in Chapter 14.

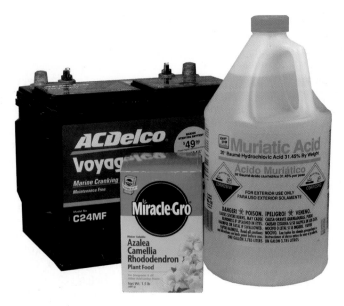

Figure 11.8 Some consumer products that contain acids. (George Semple for W. H. Freeman and Company)

calcium become water soluble when they bind to the anion part of the acid—for example, the phosphate ion, PO_4^{3-}.

11.8 Molarity is a concentration scale

Because the number of atoms or molecules or ions in any finite sample of a material is so huge, chemists usually state them by reference to a specified large number. The unit they use is called the **mole.** A mole always refers to 6.02×10^{23} particles, such as molecules or ions, in the same way that a dozen always means 12 of something, such as eggs, donuts, or anything else (see Figure 11.9). Because particles—like eggs—are of different sizes, a mole of one substance may be greater in mass than a mole of a different substance, but both will contain the same number of particles: 6.02×10^{23}.

We shall have little cause to work with moles, since this text does not emphasize calculations of the *quantities* of chemicals that react with each other. However, the one area where we cannot ignore the mole concept is with respect to hydrogen ion concentration in aqueous solutions, since it gives rise to the important concept of pH, which we discuss in the next section. The **molarity** scale reports *concentrations of dissolved substances in terms of the number of moles of the substance that are present in one liter of a solution*. It happens that one mole of H^+ ions weighs one gram, so *the numerical value for the H^+ molarity concentration is simply the same as the number of grams of this ion that are dissolved in one liter*. Note that this simplification does *not* apply to species other than H^+.

A dozen small eggs have a different mass than a dozen large ones; similarly, one mole of one substance has a different mass than one mole of a different substance.

Taking It Further

For information and problems related to the mole concept, go to Taking It Further at www.whfreeman.com/ciyl2e.

Figure 11.9 The mole is a means of counting. One mole is 6.02×10^{23} particles of anything, just as a dozen is 12 of anything. (a) One mole of water molecules (6.02×10^{23} molecules). (b) One mole of sugar molecules (6.02×10^{23} molecules).
(Part a, Larry Stepanowicz/Visuals Unlimited; part b, George Semple for W. H. Freeman and Company)

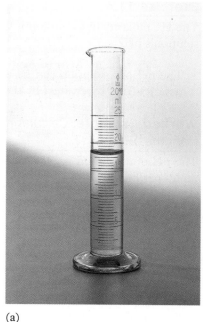

(a)

(b)

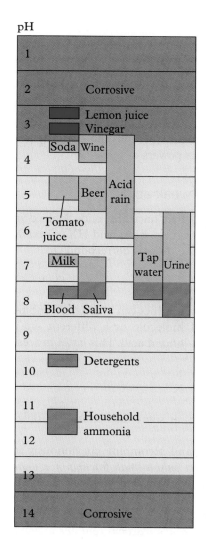

Figure 11.10 The pH scale.

11.9 The pH scale measures H^+ concentrations

The concentration of H^+ ions is an important property of many natural aqueous solutions, including your blood. Because the concentrations of the dissolved acids vary greatly, the concentrations of hydrogen ion in aqueous solutions span many powers of ten. For this reason, chemists have devised a shorthand technique for reporting H^+ concentrations.

Following this technique, chemists convert the value for the molarity of H^+ into a power of ten. The power is almost always negative, since H^+ molarities are usually less than 1.0. *This power, without the negative sign, is the **pH** of the solution.* For example, if the molarity of H^+ ions is 10^{-2} moles per liter, the pH of the solution is 2.

The lower the pH, the higher the concentration of H^+ ions in a solution. Thus the pH for solutions that are highly acidic is low, not high as you might intuitively expect (see Figure 11.10).

Worked Example: Calculating the pH of a solution from its molarity

The H^+ molarity in a solution is 0.00010 mole per liter. What is the solution's pH?

Solution: First we convert 0.00010 into a power of ten. Since the decimal place has to be moved four places to the right to convert the number to 1.0, the value of 0.0001 in powers of ten is 10^{-4}, that is, 1.0×10^{-4}. Since the power of ten is -4, the pH of the solution is 4.

Acids

Review scientific notation in Appendix A if you are confused about the procedure of converting a number to a power of ten, and vice versa.

Exercise 11.4

The H^+ molarity in a solution is 0.0000010 mole per liter. What is the solution's pH?

The pH of a solution can be converted back to the standard form for molarity (moles per liter) by the reverse process.

Worked Example: Calculating the hydrogen ion molarity from the pH

The pH of a solution is 3. What is the solution's H^+ concentration in moles per liter?

Solution: Since the pH is 3, the molarity is 1.0×10^{-3}. In standard form, this is equivalent to 0.001 mole per liter.

Exercise 11.5

The pH of a solution is 5. What is the solution's H^+ concentration in moles per liter?

We will discuss pH further when we consider bases in section 11.15 and buffers in section 11.21. In *Taking It Further* on the Web site for this book, we will explain how to deal with noninteger pH values and hydrogen ion concentrations that are not exact powers of ten.

11.10 Chemical equilibrium and weak acids

In all the examples above, we assumed that once the acids ionized in water, no further processes occurred. For the general acid HX, where X represents an atom or group of atoms, the process corresponds to:

$$HX \rightarrow H^+ + X^-$$

In fact, with *most* acids, the hydrogen ion product can readily recombine with the anion (from the original HX molecule or a different one), thereby re-forming a molecule of the un-ionized acid. This latter process is termed the *reverse reaction:*

$$HX \leftarrow H^+ + X^-$$

Once re-formed, the acid molecule can, at some later moment, ionize again to produce H^+ and the anion.

Weak acids are those that ionize and recombine continuously in water, with the result that *at any instant only a small fraction of a weak acid exists in ionized form.* After a short initial period in which the reactions start and speed up, the two opposing processes thereafter proceed at equal rates, and the concentrations of the various chemical species level off and become constant *even though the reactions continue to occur. When a reaction proceeds in both directions at equal rates, the condition is called* **equilibrium.** We indicate equilibrium in a reaction by the use of *two* arrows in opposite directions, one directly above the other, in the

We discussed equilibrium in a physical process in Chapter 2, section 2.6.

View the dissociation of acetic acid in water at Chapter 11: Visualizations: Media Link 4.

reaction equation for the process. The reaction equation for the general case of a weak acid HX is

$$HX \rightleftharpoons H^+ + X^-$$

It is important to keep in mind that equilibrium describes the rates of two opposing processes. It does *not* mean that half the acid occurs in the un-ionized form HX and half in the ionized form; it indicates only that the *rates* of the forward and reverse reaction are equal. Indeed, for most weak acids, much less than 50%—often only about 1%—of the acid is ionized at any instant.

An example of a weak acid is *hydrofluoric acid,* HF:

$$HF \rightleftharpoons H^+ + F^-$$

Many common weak acids are carboxylic acids, the most important of which is *acetic acid,* CH_3COOH. As with other carboxylic acids, it is only the hydrogen bonded to the oxygen of the $-COOH$ group that can ionize.

Worked Example: Writing reaction equations for ionization of polyprotic acids

Sulfurous acid, H_2SO_3, is a weak acid. Write the reaction equations for both its stages of ionization.

Solution: Ionization of one hydrogen from H_2SO_3 will produce H^+ and the anion that corresponds to the remainder of the acid, namely HSO_3^-. The first stage of ionization therefore is:

$$H_2SO_3 \rightleftharpoons H^+ + HSO_3^-$$

We show the process as an equilibrium, since H_2SO_3 is stated to be a weak acid.

The second stage of ionization corresponds to the loss of H^+ by the anion formed in the first stage, and it also is an equilibrium:

$$HSO_3^- \rightleftharpoons H^+ + SO_3^{2-}$$

Generally, the extent of ionization in the second and any subsequent steps is very tiny indeed for weak polyprotic acids.

Exercise 11.6

Write the reaction equations for all three stages of ionization of phosphorous acid, H_3PO_3, a weak acid.

11.11 Acids give food and beverages their tartness

Food and drinks with low pH taste sour to us because the H^+ ions react temporarily with molecules on the side of the tongue, causing them to change shape somewhat, an action which sends the message "sour" to our

Figure 11.11 Some common foods and beverages containing weak acids. (George Semple for W. H. Freeman and Company)

brain. The sour or tart taste in food and beverages and the tang in sweet juices come from the presence of weak acids (see Figure 11.11). For example, *citr*us fruits—such as oranges, lemons, limes, and grapefruit—all contain *citr*ic acid (Chapter 6, section 6.20). The sour or tangy component of the overall flavor in these fruits is also due partially to *ascorbic acid,* a weak acid also known as vitamin C (see Chapter 10, section 10.1).

The pH of lemon juice and of lime juice is about 2, which is quite acidic, whereas that of oranges and tomatoes is close to 4. Since these pH values of 2 and 4 correspond to H^+ concentrations of 0.01 and 0.0001 moles per liter respectively, the level of hydrogen ions in lemon and lime juice is 100 times that in oranges and tomatoes. Bottled soft drinks are made acidic (pH about 3) by the inclusion of a weak acid as one ingredient—either citric acid in most fruit drinks or phosphoric acid in colas.

Wine contains about 0.7% weak acids, which give rise to a number of versions of a somewhat sour taste. In chardonnay, for example, the buttery taste is due to lactic acid, whereas the citrus flavor is due to citric acid. The substance that gives sour milk its sharp taste is lactic acid.

Vinegar, which has a pH of about 3, is a dilute (5%) solution in water of acetic acid, CH_3COOH, the substance produced by the fermentation of sugars when they are oxidized beyond the alcohol stage. Indeed, the oxidation of alcohol—for example, that in red wine—produces acetic acid, which is the reason that red wine eventually becomes too sour to drink if it is exposed to air for a long period of time.

$$CH_3CH_2OH + O_2 \rightarrow CH_3COOH + H_2O$$
$$\text{alcohol} \qquad \text{in air} \qquad \text{acetic acid}$$

■ Chemistry in Your Home: Effects of acids and bases on hair

You can examine the difference in acid and base effects on human hair by soaking one strand of your hair in lemon juice and water (acid) and another hair in baking soda and water (base) for 15 minutes. Rinse carefully and dry. Examine the hairs to see how they have changed and how they differ. It might help to look at the hairs under a magnifying glass. Acids cause hair cuticles to shrink, whereas bases cause them to swell and soften.

11.12 The reactivity of H^+ makes it very important

The hydrogen ion owes its great importance in chemistry to its ability to react with other substances. Solutions having a low pH, and therefore having a high level of H^+ ions, are quite corrosive to many materials. Acids will attack metals; in the process, they produce hydrogen gas, H_2. For

This reaction is illustrated and discussed in more detail in Chapter 12, section 12.1.

example, acid solutions will react with and thereby dissolve metals such as zinc and iron. In fact, there is a well-known experiment in which a small iron nail is placed in a bottle of a soft drink, where it is seen to slowly disappear by reacting with the acid. The net reaction is:

$$2\,H^+ + Fe\,(s) \rightarrow H_2\,(g) + Fe^{2+}$$

In many of its reactions with other substances, H^+ attaches itself—by forming a covalent chemical bond—to an anion, thereby producing a different ion or a molecule. An important example of this process occurs when a hydrogen ion reacts with substances that contain the *carbonate ion*, CO_3^{2-}. If one H^+ ion from an aqueous acidic solution bonds to such an ion, the *bicarbonate ion*, HCO_3^-, is produced:

$$H^+ + CO_3^{2-} \rightarrow HCO_3^-$$

The charge on the resulting ion is one unit less negative than on the reactant carbonate anion, because one negative charge has been canceled by the positive charge of the hydrogen ion.

The reaction of H^+ with the carbonate ion occurs whether the carbonate ion is dissolved in solution or is present in a solid. For example, if an acidic solution is poured onto crystals of *calcium carbonate*, $CaCO_3$, the H^+ ions in the acid solution will react with the carbonate ion in the portions of the solid they have contact with, and transform the carbonate ions into bicarbonate ions. Because the compound of calcium with the bicarbonate ion, namely $Ca(HCO_3)_2$, is soluble in water, no solid is evident once this transformation is complete. The reaction is:

$$CaCO_3(s) + H^+ \rightarrow Ca^{2+} + HCO_3^-$$

Each bicarbonate ion can react subsequently with an additional hydrogen ion, thereby producing *carbonic acid*, H_2CO_3:

$$H^+ + HCO_3^- \rightarrow H_2CO_3$$

Carbonic acid is not very soluble in water; it decomposes to release water and carbon dioxide gas that can be seen as bubbles escaping from the liquid:

$$H_2CO_3 \rightarrow H_2O + CO_2\,(g)$$

Adding up the above three reactions, we see that the overall process is

$$CaCO_3\,(s) + 2\,H^+ \rightarrow Ca^{2+} + CO_2\,(g) + H_2O$$

Thus, adding acid to calcium carbonate—or any other compound containing the carbonate ion—produces bubbles of CO_2.

The reaction discussed above can be readily observed by pouring an acidic liquid, such as vinegar, onto a substance containing calcium

Taking It Further

For more information and problems related to Hydrogen ion transfer, go to Taking It Further at www.whfreeman.com/ciyl2e.

carbonate. Indeed, this process is the basis of the activity below in which "naked eggs" are created. The acetic acid in the vinegar in which the eggs are placed reacts over a day or two with the eggshells, which are largely composed of calcium carbonate. The bubbles initially seen on the shells consist of carbon dioxide formed by the reaction of the outer-most part of the shell that has already reacted. Eventually the entire shell disappears.

■ Chemistry in Your Home: Making naked eggs

You can remove the shells from raw eggs by placing several of them in a container (so they don't touch each other) and covering them with vinegar. Notice that bubbles form on the shells within a few minutes. Place the container in a refrigerator for a day. Carefully remove the eggs from the container, and replace the liquid with fresh vinegar. Put the eggs back in the container, and store for another day in the refrigerator. Carefully remove the eggs from the container and rinse them with water, discarding any eggs that are oozing their inside material through the membrane. Gently squeeze the membrane of a translucent raw egg to see how flexible it is.

Manufacturers of carbonated beverages use pure CO_2 under high pressure to force excess gas to dissolve in the drinks and then seal the containers before it can escape. When the container is opened, the pressure is released. Consequently bubbles of carbon dioxide form in the liquid, eventually rise to its surface, and enter the air, in accordance with the reaction above.

11.13 Gastric juice has a very low pH

At first thought, it is rather shocking to hear that the pH of the fluid in your stomach is normally about 1–2, which is up to ten times the H^+ concentration of lemon juice. No wonder vomit tastes so sour! The acidity is due to the strong acid HCl, which is secreted by cells in the stomach walls and comprises about 0.5% of gastric juice. The function of this acidity is to suppress the growth of bacteria and to facilitate the digestion of food by enzymes in the stomach. The stomach's lining is not itself rapidly destroyed by the acid because it is protected by a mucous lining that is constantly being replaced. In addition, the lining provides bicarbonate ions, HCO_3^-, which immediately react with the acid in the reaction we encountered previously:

$$H^+ + HCO_3^- \rightarrow H_2CO_3$$

The amount of acid present in the stomach varies considerably, depending upon the amount and nature of the food present in it. Although gastric juice is normally very acidic, overeating or tension can cause the stomach to produce too much acid and make us suffer from the feelings called heartburn and acid indigestion. The conventional remedy for excess stomach acidity is to ingest a substance that will react

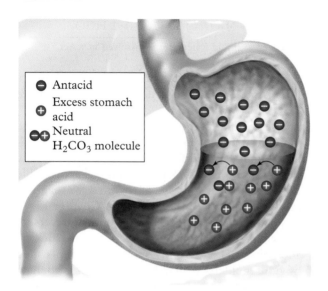

- ⊖ Antacid
- ⊕ Excess stomach acid
- ⊖⊕ Neutral H_2CO_3 molecule

Figure 11.12 The action of an antacid in the stomach.

with hydrogen ions and reduce their concentration in the stomach (see Figure 11.12). Thus it is common to use antacids such as calcium or magnesium carbonate, which, although almost insoluble in water, will "dissolve" in acidic solutions by reacting with H^+ according to the reaction of $CaCO_3$ with H^+ discussed in the previous section. The burp you experience after taking a carbonate-containing antacid is actually the release of the CO_2 gas from your stomach. The chloride ion in HCl originates with sodium chloride, NaCl, in the diet. Some physicians advise patients with chronic excess acidity to reduce their consumption of salt, and consequently of chloride ion, to reduce the production of HCl.

Some antacids are ionic compounds containing the bicarbonate ion. For example, *sodium bicarbonate*, $NaHCO_3$, also known as bicarbonate of soda, baking soda, or bicarb, was a traditional antacid before commercial preparations became available. Other preparations, such as Alka-Seltzer, are powders containing sodium bicarbonate and a solid weak acid. When the preparation is stirred in water, the acid produces H^+, which reacts with some of the bicarbonate ion to produce bubbles of carbon dioxide, and the mixture fizzes. The antacid effectiveness actually originates from the *remaining* bicarbonate ion, that is, the part that does *not* produce the fizzing. The fizzing is simply a device to reassure the consumer that the mixture is working.

Some people who often suffer from heartburn use drugs such as Zantac and Pepcid AC instead of antacids. Such drugs do not directly neutralize HCl; rather they decrease the amount of HCl that is secreted into the stomach and thereby prevent the buildup of excess acidity.

The combination of sodium bicarbonate and a solid weak acid is also present in baking powder. When this powder is mixed with a liquid containing water, bubbles of carbon dioxide are produced and expand as the temperature rises, producing baked goods greater in volume and less dense than the original dough.

11.14 Calcium, bicarbonate, and carbonate ions are common in the environment

The bicarbonate ion is usually the dominant anion in natural waters such as rivers and lakes, including those used to supply bottlers of drinking water. The bicarbonate ion is formed in natural waters by the exposure of the waters to the air and to rocks and soil. First, some of the carbon dioxide from the air dissolves in the water. Then, some calcium carbonate from limestone rocks and soil reacts with the dissolved carbon dioxide to produce calcium ions and bicarbonate ions, according to the following net reaction:

Acids

$$CO_2(g) + CaCO_3(s) + H_2O(aq) \rightarrow \underbrace{Ca^{2+} + 2\ HCO_3^-}$$

from air   from rocks   liquid water   ions dissolved in water

Owing to this reaction, much more calcium is present in natural waters exposed to limestone than there would be if no carbon dioxide were available.

Most calcium ions enter natural waters as a result of their exposure to $CaCO_3$ in the form of limestone or to mineral deposits of *calcium sulfate*, $CaSO_4$. Water that contains appreciable amounts of Ca^{2+} usually also contains the magnesium ion, Mg^{2+}. Water that contains appreciable amounts of calcium and magnesium ions is called **hard water.** The removal of these ions, to form soft water, is discussed in Chapter 13.

When water is heated, the net reaction shown above is reversed. Many of us are familiar with the scale that forms on kettles, showerheads, steam irons, and hot-water taps. This substance, often called "calcium" or "lime" by householders, is mainly the calcium carbonate that precipitates when hard water is heated and its dissolved calcium bicarbonate is converted back to carbon dioxide and calcium carbonate:

$$Ca^{2+} + 2\ HCO_3^- \rightarrow CO_2(g) + CaCO_3(s) + H_2O$$

Most of the scale can be cleaned from these devices by applying vinegar. The acetic acid in the vinegar supplies H^+ to react with the calcium carbonate.

The last reaction above also occurs in cold water when evaporation occurs, and the solubility of calcium carbonate is consequently exceeded. For example, when a stream of underground water containing dissolved calcium bicarbonate flows down through pores in the rock to the roof of an underground cave, the drips of water slowly evaporate and icicle-like *stalactites* of calcium carbonate form from above (see the chapter-opening photograph). Water drops that fall onto the cave floor forms columns called *stalagmites* when the water evaporates.

Bases

11.15 Bases increase hydroxide ion concentrations

The second most important ion in aqueous chemistry, following H^+, is the hydroxide ion, OH^-. When it, rather than H^+, dominates water solutions, they are said to be *alkaline* or *basic*. Notice that the hydroxide ion is the portion of the water molecule that is left behind if H^+ is removed from it.

Bases are the active component of many consumer products, particularly household cleaners (see Figure 11.13). Substances that we call bases have two main observable characteristics when they are dissolved in water:

- They have a bitter taste and a slimy feel.
- They turn litmus paper blue.

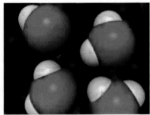

View hydroxide ions in a basic solution at Chapter 11: Visualizations: Media Link 5.

Figure 11.13 Some consumer products containing bases. (George Semple for W. H. Freeman and Company)

At the molecular level, these properties arise because of the presence of the hydroxide ion, OH^-. Indeed, *a **base** is defined as a substance that increases the concentration of the hydroxide ion in solution.*

As solids, many bases consist of ionic lattices in which the anion is the hydroxide ion. When such bases are dissolved in water, OH^- ions are released directly. An example is *sodium hydroxide*, NaOH, also known as *lye*, a white solid that consists of Na^+ and OH^- ions (see Figure 11.14):

$$NaOH(s) \xrightarrow{water} Na^+ + OH^-$$

Sodium hydroxide is the most important base in chemistry and is produced synthetically in massive amounts. Strong bases such as sodium hydroxide are used in drain cleaners and oven cleaners to react with grease buildup and with protein-containing material such as hair, since these substances decompose in the presence of OH^- ions. In particular, the fat that constitutes much of the water-insoluble grease reacts with OH^- to produce glycerol and free fatty acids. The acids then react with additional hydroxide ion to yield sodium salts of the acids. The water-soluble salts and glycerol can then be easily wiped or flushed away. Unlike strong acids—which would also be effective cleaners—they do not react so readily with and corrode the metals in the plumbing, oven walls, etc. They are, however, quite corrosive to living matter, such as your skin and eyes, as indicated on the warning label on consumer products that contain them.

Figure 11.14 Lye (NaOH), a strong base. (Photos by George Semple for W. H. Freeman and Company)

Lattice structure of NaOH

Commercial lye solution

Bases

Other important bases that contain hydroxide ion and operate in a similar manner are *potassium hydroxide*, KOH, and *lithium hydroxide*, LiOH. Together with NaOH, they are **strong bases,** since all the dissolved solid is present in solution as hydroxide ions and ions of the metal.

Exercise 11.7

Write the reaction corresponding to the interaction of KOH with water.

Some common bases have limited solubility in water, and consequently they produce rather little hydroxide ion; examples are *magnesium hydroxide*, $Mg(OH)_2$, and *calcium hydroxide*, $Ca(OH)_2$. The old name for the latter is slaked lime; dilute solutions of it in water are called limewater. Slaked lime and limewater are the cheapest bases to produce in large quantities and were widely used industrially in the past for reasons of economy. These names derive from the fact that calcium hydroxide can be produced from *lime* (also called quicklime)—*calcium oxide*, CaO,—by reaction with water:

$$CaO + H_2O \rightarrow Ca(OH)_2$$

The calcium oxide itself is produced by strongly heating *limestone,* a form of calcium carbonate ($CaCO_3$):

$$CaCO_3 \rightarrow CaO + CO_2$$

The insolubility of magnesium hydroxide provides a way to extract magnesium ion, Mg^{2+}, from seawater and to convert it into magnesium metal; indeed, this is how most magnesium is made. First, limewater is added to a sample of seawater, producing the solid $Mg(OH)_2$, which is then filtered from the mixture. Later, large quantities of magnesium hydroxide isolated in this manner are acidified to release Mg^{2+} ion in concentrated form:

$$Mg^{2+} + Ca(OH)_2 \rightarrow Mg(OH)_2\,(s) + Ca^{2+}$$
$$Mg(OH)_2\,(s) + 2\,H^+ \rightarrow Mg^{2+} + 2\,H_2O$$

The concentrated magnesium ion solution is then made to undergo electrolysis (discussed in Chapter 12) to produce elemental Mg.

11.16 Weak bases also produce hydroxide ion from water

In addition to bases that contain hydroxide ion in their solid, ionic form, there are a number of substances that themselves contain *no* OH^- but that produce it when they are dissolved in water. These substances *react with water molecules* to extract H^+ from the H_2O, leaving OH^-, which enters the solution. Most such reactions are equilibrium processes, since a reaction in both the forward and reverse directions occurs; thus the substances that react are called **weak bases.**

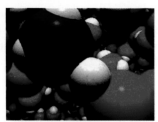

View the dissociation of ammonia in water to form a basic solution at Chapter 11: Visualizations: Media Link 6.

One important weak base is the gas *ammonia,* NH_3. After it dissolves in water, molecules of it can each extract an H^+ ion from a water molecule (see Figure 11.15). As a result of this extraction, the ammonia molecule is converted into the *ammonium ion,* NH_4^+, and a hydroxide ion is released into solution:

$$NH_3(g) \xrightarrow{\text{water}} NH_3(aq)$$
$$NH_3(aq) + H_2O \rightleftharpoons NH_4^+ + OH^-$$

Ammonia dissolved in water is often sold as a household cleaner, especially for windows, since aqueous solutions containing hydroxide ion remove grease readily. Indeed, most detergent preparations contain components that are weak bases.

Most substances that act as bases as a result of their reaction with water are anions. An important example is the carbonate ion, CO_3^{2-}, in a soluble salt such as sodium or potassium carbonate:

$$CO_3^{2-} + H_2O \rightleftharpoons HCO_3^- + OH^-$$

As a result of the above reaction, an aqueous solution prepared by dissolving solid sodium carbonate, $NaCO_3$, or potassium carbonate, K_2CO_3, in water is quite basic. It contains appreciable concentrations of

Figure 11.15 Ammonia. Dissolved in water, NH_3 produces a weak base solution. (Photo by George Semple for W. H. Freeman and Company)

Bases

CO_3^{2-}, HCO_3^-, and OH^- ions as well as of Na^+ or K^+ ion. In the old days, pioneers extracted potassium carbonate from wood ashes by rinsing them with water, thereby obtaining a basic solution that they used to make soap (as discussed in Chapter 13).

Exercise 11.8

Phosphate ion, PO_4^{3-}, acts as a weak base in aqueous solution. Write the equilibrium reaction in which it reacts with water.

The Reactions of Acids and Bases

11.17 Acids and bases react to neutralize each other

Although it is possible to create aqueous solutions having a high concentration of H^+ ions and other aqueous solutions having a high concentration of OH^-, it is *not* possible for a solution to have appreciable concentrations of *both* H^+ and OH^- ions *simultaneously*. This is because *these ions react together rapidly* and almost completely, with the H^+ ion adding to the OH^- to form additional molecules of water:

$$H^+ + OH^- \rightarrow H_2O$$

The ions are said to **neutralize** each other by this process.

The neutralization reaction written above is called a **net ionic reaction** since it summarizes the reactions of the ions whose nature is changed. Neutralization reactions can also be written from the viewpoint of the original acids and bases that gave rise to the H^+ and OH^- ions respectively. For example, the reaction between the strong acid HCl and the strong base NaOH can be considered to be a sequence of three steps, the first two corresponding to the creation of ions in aqueous solution. These are followed by a third reaction in which H^+ and OH^- combine:

$$HCl \rightarrow H^+ + Cl^-$$
$$NaOH \rightarrow Na^+ + OH^-$$
$$H^+ + OH^- \rightarrow H_2O$$

Adding up these three reactions and canceling common terms gives us the net reaction:

$$HCl + NaOH \rightarrow H_2O + Na^+ + Cl^-$$

Of course, a solution containing equal amounts of Na^+ and Cl^- ions is the same as that formed by dissolving table salt in water. In the case of the reaction above, the solid salt can be obtained by evaporating or boiling the water.

Figure 11.16 A neutralization reaction between a strong acid (HCl) and a strong base (NaOH).

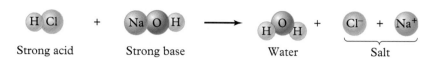

Strong acid   Strong base   Water   Salt

It is really quite amazing that two substances as corrosive and reactive as concentrated hydrochloric acid and concentrated sodium hydroxide will combine to produce the relatively innocuous compounds sodium chloride and water! This is an example of the principle that the chemical compound formed by combining two others usually does *not* have properties that are either intermediate between the two or similar to those of either one.

The reaction of HCl and NaOH is a specific example of *the general reaction in which an acid and a base combine together to form water and a salt* (see Figure 11.16):

$$\text{ACID} + \text{BASE} \rightarrow \text{WATER} + \text{SALT}$$

Thus we can add to our list of characteristic properties that acids are neutralized by bases, and vice versa.

Exercise 11.9

Write the reaction in which the acid HNO_3 is neutralized by the base KOH. What is the formula of the salt formed in this process?

The neutralization reaction occurs even when either the acid and/or the base is weak. For example:

$$HF + KOH \rightarrow H_2O + KF\,(aq)$$

All the carboxylic acids we discussed in Chapter 6 are weak acids; their reaction with sodium hydroxide converts them to the corresponding sodium salts. If we represent the carboxylic acid in general as HX, its neutralization reaction with NaOH is:

$$HX + NaOH \rightarrow H_2O + NaX$$

The salts NaX are ionic, since they are composed of the ions Na^+ and X^-, though the anion here is a complex organic structure. Thus if we were to add sodium hydroxide to acetic acid, which we can conveniently abbreviate as HAc, the product would be the salt *sodium acetate*, NaAc. This salt is used in food products; for example, it is on the surface of "salt-and-vinegar" potato chips. Once it encounters your mouth, the anion Ac^- extracts an H^+ ion from the water molecules in your saliva, producing acetic acid—which in a water solution is vinegar:

$$Ac^- + H_2O \rightleftharpoons HAc + OH^-$$

Thus you actually manufacture the vinegar flavor yourself!

11.18 Antacids are activated by acid

The acid–base neutralization reaction also occurs when the base is a solid that is insoluble in water. Thus magnesium hydroxide, $Mg(OH)_2$, as a colloidal suspension in water is used as a stomach antacid (*milk of magnesia*), since it dissolves when it neutralizes the acid:

$$Mg(OH)_2\,(s) + 2\,H^+ \rightarrow Mg^{2+} + 2\,H_2O$$

If taken in excessive amounts, magnesium compounds act as laxatives, so extensive use of magnesium hydroxide or magnesium carbonates as antacids can have undesirable side effects. Indeed, soluble magnesium salts such as $MgSO_4$ (*Epsom salts*) are sometimes used deliberately as laxatives to relieve constipation.

Aluminum hydroxide, $Al(OH)_3$, another compound insoluble in water, acts as an antacid in the same way as magnesium hydroxide. Bases that are almost insoluble in water are used as antacids since they only become activated when they encounter acid. Swallowing a *strong* base such as sodium hydroxide would be very dangerous—it is highly corrosive to the digestive system and could also increase the stomach's pH to dangerous levels.

Exercise 11.10

Aluminum hydroxide is a component of some antacids, such as Amphojel. By analogy with the action of magnesium hydroxide, write the equation for a chemical reaction that illustrates how it operates to reduce acidity.

View the self-ionization of water at Chapter 11: Visualizations: Media Link 7.

11.19 Water can ionize itself

Aqueous solutions always contain *some* H^+ and OH^- ions, albeit often a relatively small number, because water molecules undergo a **self-ionization reaction** in which an H_2O molecule produces one H^+ ion and one OH^- ion:

$$H_2O \rightarrow H^+ + OH^-$$

Consequently, even pure water contains a (very) tiny concentration of these two important ions.

The self-ionization reaction corresponds to the *reverse* of the neutralization reaction involving the H^+ and OH^- that we discussed earlier. In fact, in *all* aqueous solutions, equilibrium is quickly established between the forward and reverse of this reaction:

$$H_2O \rightleftharpoons H^+ + OH^-$$

There are many, many more water molecules present than there are H^+ and OH^- ions. Indeed, in pure water and in neutralized aqueous solutions formed from strong acids and bases, the H^+ and OH^- concentrations of ions produced by this reaction amount only to 10^{-7} moles per liter; in other words, one hydrogen ion and one hydroxide ion for every 10 million molecules of water. Thus the pH of pure water, and of such neutralized aqueous solutions, is equal to 7.

As a consequence of the operation of this equilibrium reaction, the molarity concentrations of H^+ and OH^- ions in aqueous solutions are interconnected by a simple mathematical formula:

$$\text{concentration of } H^+ \times \text{concentration of } OH^- = 10^{-14}$$

This equation confirms our statement that there cannot be any aqueous solutions that simultaneously have high concentrations of both H^+ and OH^-. For example, if a solution has a pH value of 2, then its H^+ concentration is 10^{-2}, and so its OH^- concentration is only 10^{-12}, since according to the equation above:

$$\text{concentration of } OH^- = 10^{-14} \div \text{concentration of } H^+$$
$$= 10^{-14} \div 10^{-2} = 10^{-12}$$

Solutions in which the H^+ concentration is greater than that of OH^-, that is, solutions with more H^+ than in pure water, are known as **acidic solutions.** Since their H^+ concentration is greater than 10^{-7} moles per liter, *acidic solutions have pH values <u>less</u> than 7.* Solutions which have OH^- concentrations *greater* than 10^{-7} moles per liter are called **alkaline** or **basic solutions.** Because the hydroxide ion concentrations in basic solutions are higher than in pure water, the hydrogen ion concentrations must be lower than the value of 10^{-7} moles per liter. Since their H^+ concentration must be *less* than 10^{-7} moles per liter, *the pH is <u>greater</u> than 7 for basic solutions:*

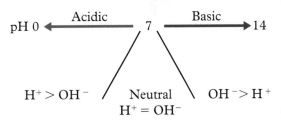

pH < 7 solutions are *acidic*

pH = 7 solutions are *neutral*

pH > 7 solutions are *alkaline* (basic)

11.20 Some natural substances are indicators

Many natural and synthetic materials take on different colors, depending on whether they are in an acidic or a basic environment. These materials are called **indicators** since they can be used to signal whether a given environment is acidic or basic. The most famous of these materials is the one we mentioned in listing characteristics of acids and bases: *litmus paper* is re**d** under aci**d** conditions and **b**lue under **b**asic ones (see Figure 11.17a). Other common materials that are acid–base indicators include the juice of grapes and of red cabbage; the latter is purple at low pH and becomes yellow and then blue as the solution is made more and more basic (see Figure 11.17b). The color of some flowers, including hydrangeas, is quite different depending upon whether the soil in which the plants grow is acidic or alkaline (see Figure 11.18).

■ Chemistry in Your Home: Testing common substances to see if they are acids or bases

Using vinegar and ammonia, you can determine which substances in the list below are acid–base indicators. Place about two teaspoons of

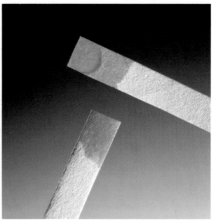

(a)

(b)

Figure 11.17 Acid–base indicators. (a) Litmus turns red in acidic solutions and blue in basic solutions. (b) Red cabbage juice changes from purple to yellow to blue as pH increases. (Part a, Richard Megna/Fundamental Photographs; part b, Ken Karp for W. H. Freeman and Company, from *Molecules, Matter, and Change*, 4th ed., 2000, L. L. Jones and P. W. Atkins.)

Figure 11.18 Hydrangeas. These flowers are blue when grown in acidic soil and pink when grown in neutral or alkaline soil. (Diane Hirsch/Fundamental Photographs)

vinegar in one small cup or disposable paper cup and two teaspoons of ammonia in a second cup. Add a test substance and watch for any color change. Rinse the cups and try again with another substance. Make a table indicating the color of each substance in acidic (vinegar) and in basic (ammonia) solutions. You can try the following substances, but don't limit yourself to them. Many highly colored fruits, vegetables, and even flowers may be acid–base indicators: beet juice, blueberries, tumeric (a spice), cherries, plums. For each fruit or vegetable, cut or shred approximately one-quarter cup of each, cover with a small amount of rubbing alcohol, let it sit for 10 minutes, then strain out the solid material. Use the liquid for the pH test. To test red cabbage juice, shred a few red cabbage leaves, place in a small amount of rubbing alcohol, and test the liquid.

11.21 Buffer solutions resist changes to their pH

As we have seen, the pH of pure water is 7. If a small amount of a strong acid is dissolved in a liter or so of water, the pH decreases significantly—often to about 1 or 2. Similarly, the pH of water increases by many units when a strong base is dissolved in it. However, these very acidic or very basic solutions undergo large changes in their pH when additional acid or base is added to them. For example, adding a small amount of a strong base such as NaOH to a strong acid solution substantially increases its pH, perhaps even beyond 7, since the OH^- ions react to neutralize many of the H^+ ions (see Figure 11.19a).

It comes as something of a surprise, then, to find that certain types of aqueous solutions *resist* changes to their pH values when a strong acid or base is added to them (see Figure 11.19b)! Such solutions are said to be *buffered* against significant changes to their pH and so are called **buffer solutions.** Your blood is a buffer solution; it resists significant changes to its pH level, which is held at about 7.4. Chemists have been

The Reactions of Acids and Bases

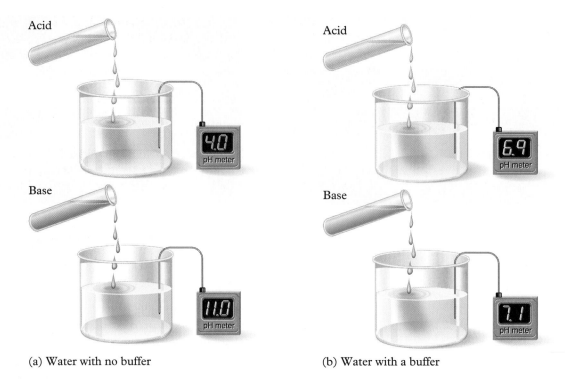

Acid

4.0
pH meter

Base

11.0
pH meter

(a) Water with no buffer

Acid

6.9
pH meter

Base

7.1
pH meter

(b) Water with a buffer

Figure 11.19 The effects of a buffer. (a) The pH of water changes dramatically when a few drops of strong acid (red) or strong base (blue) are added. (b) The pH of a buffered solution, originally 7.0, barely changes when a few drops of strong acid or strong base are added.

able to devise many synthetic buffers now that they understand how such systems operate.

In order for a solution to operate as a buffer against a decrease in pH when some acid is added to it, *the solution must contain a substance that can react with the acid or with the H^+ ions* that the acid generates. The substance that possesses these properties clearly is a *base*, since we know that bases neutralize acids. At the same time, *the solution must also contain an acid that can react with any added base or the OH^- ion it generates.* However, we cannot choose just *any* acid and *any* base as the components for a buffer, since most such combinations react with each other instantly to produce a salt and water.

Buffer solutions consist of a <u>weak</u> acid <u>and</u> a salt of the same weak acid, or of a <u>weak</u> base <u>and</u> its salt. An ion from the salt acts as a weak base in the former case and as a weak acid in the latter case, so in both instances there is both a weak acid and a weak base present. Consider, for example, an aqueous solution that contains some of the weak acid HF *and* of its salt NaF, which in solution is completely ionized to Na^+ and F^- ions. The buffering action of this solution operates as follows:

- If acid is added to the buffer solution, almost all the additional H^+ from the external acid combines with some of the F^- ion in the buffer to form additional HF. This reaction prevents any significant increase in acidity and decrease in pH:

$$\underset{\text{added}}{H^+} \quad + \quad \underset{\text{in buffer}}{F^-} \quad \rightarrow \quad HF$$

- If base is added to the buffer solution, almost all the additional OH^- reacts with the weak acid HF to produce more F^- (and water), thereby preventing a rise in basicity and pH level:

$$\underset{\text{added}}{OH^-} \quad + \quad \underset{\text{in buffer}}{HF} \quad \rightarrow \quad F^- \quad + \quad H_2O$$

In the main buffer in blood, the weak acid is carbonic acid, H_2CO_3, and the salt is $NaHCO_3$, which is dissociated to Na^+ ions and HCO_3^- ions. The pH of human blood must be maintained at close to 7.4 for good health. If it becomes less alkaline and so has a lower pH, the condition known as *acidosis* results. If the blood is too alkaline and so has too high a pH, the result is *alkalosis*. Alkalosis can be produced during *hyperventilation* (breathing too fast), when too much carbon dioxide is thereby released from the blood into the lungs, causing more H^+ to be used up by HCO_3^- ion when it replenishes the H_2CO_3. Symptoms of alkalosis include lightheadedness, dizziness, and numbness of hands and feet.

When blood becomes too acidic due to a buildup of lactic acid, the condition called *lactic acidosis* results. Recall from section 6.21 in Chapter 6 that lactic acid is produced in cells in humans when you exercise, and causes muscle soreness. Lactic acidosis can occur as a side effect of kidney diseases, liver damage, cell damage due to certain medications including those used to treat HIV and AIDS, and from reduced oxygen intake as may occur with lung disease. Symptoms include nausea, vomiting, and lethargy, and if lactic acid amounts in the blood reach extremely high (toxic) levels, it is possible for the condition to be life threatening.

Exercise 11.11

Deduce the two chemical reaction equations that would describe how the H_2CO_3 / HCO_3^- buffer in blood resists changes in pH when additional H^+ or OH^- is added to it.

Summarizing the Main Ideas

When ionic substances dissolve in water, their ions separate and each is surrounded by a layer of water molecules. The attraction between the charges of the ions and the partial charges of the atoms in water

Strong acids such as HCl *cannot* be used to form a buffer since by definition the corresponding anion Cl^- will not react with H^+ ions to form un-ionized HCl. Similarly, strong bases cannot be used.

Do you ever wonder why runners continue to jog or move around after a race? Professional athletes often exercise immediately after a game or race in order to reduce the amount of lactic acid in their bloodstream. Light exercise (such as using an exercise bike) that does not involve muscle strain actually helps remove the lactic acid from their systems and lessens the muscle soreness associated with lactic acid buildup.

molecules compensates for the lost attraction between opposite charges that occur at the close range in the solid.

Solutions that have dissolved the maximum amount of a solute are said to be saturated. When, due to evaporation or a change in temperature, the maximum solubility is exceeded, the excess solute forms a solid called a precipitate. The concentration of salts and ions in water is often expressed in terms of grams of solute per million grams of solution, or ppm. For water solutions, this concentration is equivalent to the milligrams-per-liter scale.

Acids are substances that increase the concentration of H^+ ions in water. Bases increase the concentration of OH^- ions in water. Polyprotic acids have more than one hydrogen atom that can form H^+ ions and detach from the acid. Weak acids and bases are ones in which the reverse of the ionization reaction proceeds at the same rate as the forward reaction, so that in water solution an equilibrium—an equality between the rates of the two processes—is established. For strong acids and bases, only the forward, ionization reaction is important.

Chemists use the term *mole* to denote a particular number of atoms or molecules or ions, namely 6.02×10^{23} of them. Molarity refers to the number of moles of a solute dissolved in a liter of a solution. The pH of a solution is the negative of the exponent when the hydrogen ion molarity is expressed as 10^x. The lower the pH, the greater the concentration of hydrogen ions.

When the ion H^+ adds to the anion OH^-, water is produced; consequently, high concentrations of the two ions cannot coexist simultaneously. This neutralization reaction is reversed to a tiny extent during the self-ionization process for water. The neutralization reaction can also be viewed as one between the original acid and base pair; the reaction of an acid with a base produces a water molecule and a salt.

Buffer solutions resist large changes to their pH by reacting with added H^+ or OH^-. The hydroxide ion reacts with the un-ionized weak acid, and the hydrogen ion reacts with the negative ion of the weak acid that originates from a salt of the acid.

Key Terms

solute	hydroxide ion	weak acid	self-ionization reaction
solvent	acid	equilibrium	acidic solutions
solubility	ionize	hard water	alkaline (basic) solution
saturated solution	polyprotic acid	base	indicator
precipitate	strong acid	strong base	buffer solution
concentration	mole	weak base	
parts-per-million scale	molarity	neutralization reaction	
mineral water	pH	net ionic reaction	

Web Sites of Interest

To link to Web sites of interest, go to www.whfreeman.com/ciyl2e, Chapter 11, and select the site you want.

For Further Reading

A. Mancuso, "Fluoride: A Question in the Water," *Natural Health Magazine,* March–April 1998, pp. 53–63.

Review Questions

1. Describe the general composition of a salt.
2. Describe the process of dissolution of a generic salt AB.
3. What is a *solute*? What is a *solvent*?
4. What is the relationship between a solute, a solvent, and a solution?
5. How does a *concentrated solution* differ from a *dilute solution*?
6. What is a *saturated solution*?
7. What is a *precipitate*?
8. What does *ppm* mean in terms of the amount of a solute dissolved in 1 liter of water?
9. What is *mineral water*? How does mineral water differ from tap water?
10. What is the name and formula for the mineral component of tooth enamel?
11. How does hydroxyapatite differ from fluorapatite?
12. Define *acid* chemically. What are the most observable characteristics of acids?
13. Define *base* chemically. What are the most observable characteristics of bases?
14. Describe how a strong acid such as HCl ionizes in water. Write the corresponding chemical reaction.
15. Explain how a weak acid such as HF ionizes in water. Write the corresponding chemical reaction.
16. How does a *strong acid* differ from a *weak acid*?
17. What is a *polyprotic acid*?
18. What is chemical *equilibrium*?
19. For each substance below, identify a consumer product in which it is found:
    a) a strong acid
    b) a weak acid
    c) a strong base
    d) a weak base
20. What is a *mole*? How is a mole like a dozen?
21. What does the term *molarity* mean? What are the units of molarity?
22. What does the pH of a solution tell you about its acidity or basicity?
23. What types of acids are found in foods?
24. What types of acids are found in the human body?
25. Identify three reactions in which hydrogen ions, H^+, can participate.
26. What are the functions of *stomach (gastric) acid*?
27. The pH of stomach acid is about 1. Why isn't the stomach damaged by such strong acid?
28. Why do some antacids cause you to burp?
29. How does baking powder change the texture of dough?
30. Identify the major anions and cations in bottled and natural water.
31. What ions are present in *hard water*?
32. Outline the general reaction equation for the neutralization of a strong acid by a strong base.
33. What is an *indicator*? Provide two examples of substances that are indicators.
34. What is a *buffer solution*?

35. Explain why water doesn't randomly surround ions in solution.

36. Based upon the "solubility rules" presented in section 11.2, predict which of the following salts is likely to be freely soluble in water:
    a) Al_2O_3
    b) KBr
    c) $CaSO_4$

37. Describe what happens to an aqueous salt solution as the water is evaporated.

38. What is the relationship between 10 ppm of sodium ion and 10 mg/L of sodium ion?

39. If the calcium ion concentration on the analysis on a bottle of spring water is listed as 66 ppm, how many milligrams of calcium are present in a 200-mL glass of it?

40. If the sulfate ion concentration in mineral water is 112 ppm, how much of this water would you have to drink to consume 50 milligrams of sulfate?

41. Explain how fluoride in drinking water works to prevent tooth decay.

42. Why can fluoride be added to drinking water in the form of its sodium salt, NaF?

43. Why should you be suspicious of an ingredient label on a bottle of water that reads "zero" sodium content?

44. Explain how acids and bases differ.

45. Compare the numbers of molecular and ionic components of a strong acid solution and a weak acid solution.

46. Describe the ionization of a *polyprotic acid* and write the corresponding chemical reaction(s).

47. Why is it incorrect to refer to equilibrium when describing a strong acid?

48. Examine the pH scale in Figure 11.10 and examine the type of products that are acidic and the type of products that are basic. Are they similar in use?

49. How does the molarity of an acidic solution relate to the pH of the solution?

50. By what factor does the hydrogen ion concentration of a pH=10 solution differ from that of a pH=5 solution? Which is more acidic? Which has the higher hydrogen ion concentration?

51. Is an aqueous solution of pH of 6 acidic, basic, or neutral? Per liter, does it contain more or fewer hydrogen ions than one having a pH of 4?

52. What is the pH of an aqueous solution that has an H^+ molarity of 0.01? 0.0000001? 1×10^{-9}?

53. A chemist dissolves 0.010 mole of HCl in 100 mL of water. What is the molarity of HCl in the solution? What is the solution's pH?

54. Why is it incorrect to say that calcium carbonate "dissolves" in acid?

55. Describe how an antacid works to neutralize stomach acid.

56. The commercial antacid Maalox contains both $Al(OH)_3$ and $Mg(OH)_2$. By writing out the names for these two bases, show how the product name reflects its ingredients.

57. Predict the products of the reaction between vinegar, CH_3COOH, and baking soda, $NaHCO_3$. Is a fizz produced? Explain.

58. Is human blood normally acidic, basic, or neutral? Explain.

59. Why does the application of vinegar remove gray scale deposits from a teapot?

60. Ammonia, NH_3, does not contain hydroxide ions yet it produces hydroxide ions in water. Explain.

61. Fluoride ion acts as a weak base in water. Write the equilibrium reaction in which it reacts with water.

62. Explain how water produce both H^+ and OH^-. Write the reaction illustrating this.

63. Write the chemical reaction in which NaOH neutralizes the acid HBr. What is the salt that is produced?

64. How are indicators useful in monitoring the progress of an acid–base reaction?

65. Identify the components of a generic buffer system. Explain how buffers work to resist changes in pH.

Synthesizing Ideas

66. Would drinking only pure water with no detectable salt content be healthy? Explain your answer.

67. Why is a unit as small as ppm a reasonable unit of measurement for dissolved salts?

68. The pH of a solution is 4. What is the molarity of hydrogen ions in the solution, expressed both as a decimal number and as a power of ten? Repeat the calculations for a pH of 3.

69. Explain why strong acids cannot be buffers.

70. Ingesting an extraordinary amount of antacid tablets can cause a condition called *alkalosis*, an increased blood pH. Explain why this might happen.

71. Oxalic acid, $H_2C_2O_4$, is a weak acid in which both hydrogens can ionize. Write an equation for each stage of ionization, and combine the equations to find the net reaction if the acid ionizes completely.

72. Outline the processes by which sulfuric acid, H_2SO_4, reacts with sodium hydroxide, NaOH. What is the net ionic reaction for the neutralization?

73. Which of the following pairs of substances would act as a buffer solution if they were to be dissolved in the same sample of water?
    a) HCl and NaCl
    b) HF and NaF
    c) HCl and HF

74. When solutes are dissolved in water to much smaller concentrations than we have discussed, they are often described in terms of the number of *micrograms* (10^{-6} gram) per liter. What would be the corresponding "parts per" scale with equal numerical values for such concentrations?

75. Given your knowledge of acid–base chemistry, can you rationalize why weak acids formed in the mouth would be able to dissolve natural tooth enamel more easily than they dissolve fluorapatite? Write the reaction in which apatite reacts with H^+.

76. A buffer is prepared by dissolving in water some of the weak base ammonia, NH_3, and some of its salt, NH_4Cl, which dissociates to NH_4^+ and Cl^- ions. What in the solution would react with added strong acid to dispose of the new H^+ so formed? What would react with the OH^- ion from added base? Write the chemical equations that operate in these two situations.

77. Mad Dawg chewing gum produces a blue foam that oozes from the mouth after a few chews. The gum contains sodium bicarbonate, citric acid, and malic acid. Explain why the acids and base in the gum don't react with each other until they are in the mouth.

Taking It Further

Deducing pH values

We have already seen (see section 11.9) how to deduce the pH values for aqueous solutions whose H^+ molarities are exact integer powers of ten—such as 10^{-3} or 10^{-7}. What about situations in which the molarity falls somewhere *between* integers—for example, what is the pH if the H^+ molarity is 3.2×10^{-4}? Furthermore, what is the significance of pH values that have decimal places, for example, a pH of 4.6?

To work with molarities in general requires us to work with non-integer powers of ten, and to do so conveniently, we use a scientific calculator that has *log* and *10^x* buttons. *The pH of a solution is defined as the negative of the base 10 logarithm of the hydrogen ion molarity.*

Consequently, to obtain the pH, enter the value of the molarity either in conventional or in exponential mode, and then press the *log* (or log_{10}) button. The result of the calculation is the *negative* of the pH. For example entering 3.2×10^{-4}, or its equivalent 0.00032, and pressing *log* produces the result -3.50, so the pH of the solution is 3.50. (The result -3.50 means that the number 3.2×10^{-4} as a power of ten is equal to $10^{-3.50}$.)

Similarly, to obtain the H^+ molarity *from* the pH value, enter the pH value and press the $+/-$ key (which turns it into a negative value), and then press the *10^x* button (access to which may require first pressing a *SHIFT* or *2nd FUNCTION* button on your calculator). When you follow these steps for a pH of 4.29, you find that the corresponding molarity is 5.1×10^{-5}. (This result means that $10^{-4.29}$ is equal to 5.1×10^{-5}.)

Exercise 11.12

Calculate the pH values for aqueous solutions that have the following H^+ molarities:
a) 1.4×10^{-3} **b)** 6.4×10^{-6} **c)** 9.4×10^{-11}

Exercise 11.13

Calculate the molarity of H^+ for aqueous solutions for which the pH is equal to
a) 3.83   **b)** 7.20   **c)** 10.86

You can also obtain the hydrogen ion concentration and the pH value for a solution from its OH^- concentration by the algebraic equation relating H^+ and OH^- concentrations discussed in section 11.19. The following Worked Example illustrates the procedure.

Worked Example: Calculating hydrogen ion concentrations in basic solutions

If a strong base is dissolved in water so that the resulting OH^- ion concentration is 10^{-3} moles per liter, what is the H^+ concentration and what is the pH of the solution?

Solution: From the information, we have:

$$OH^- \text{ concentration} = 10^{-3}$$

The equation relating H^+ and OH^- concentrations is:

$$\text{concentration of } H^+ \times \text{concentration of } OH^- = 10^{-14}$$

Since the unknown here is the H^+ concentration, we rearrange the equation to solve for it:

$$H^+ \text{ concentration} = 10^{-14} / OH^- \text{ concentration}$$

Substituting the first equation into this last one, we obtain:

$$\text{H}^+ \text{ concentration} = 10^{-14} / 10^{-3}$$
$$= 10^{-11}$$

Since the H^+ concentration is 10^{-11} moles per liter, the pH of the solution is 11.

1. If a strong base is dissolved in water so that the resulting OH^- ion concentration is 10^{-5} moles per liter, what is the H^+ concentration and what is the pH of the solution?
2. Develop a table showing the relationship between pH and hydrogen ion concentrations for integer pH values between 0 and 14.
3. What is the hydrogen ion concentration of a solution with a pH of 0?
4. Develop the algebraic equation that allows you to calculate the hydroxide ion concentration of a solution when its pH is known.

In this chapter, you will learn:

- about electrochemical cells;
- what an electrolyte is;
- how the different types of batteries work, such as those used in flashlights, computers, and cars;
- about fuel cells and their potential in the future;
- what electrolysis is;
- the ways in which hydrogen is used as an energy source.

All three electronic devices being used by this man—cell phone, personal digital assistant, and laptop computer—currently are powered by batteries.
In the near future, lightweight fuel cells will probably replace batteries in such applications. We shall explore the chemistry behind both batteries and fuel cells in this chapter. (Javier Pierini/Getty Images)

Batteries, Fuel Cells, and the Hydrogen Economy

Oxidation and Reduction

When you turn on a flashlight or start your portable music player, there is no immediately obvious source of electricity to make them operate. Batteries, of course, do the work, but where does the electricity in batteries come from? Oxidation processes, which we have seen in relation to the combustion of fossil fuels (see Chapter 4), are the basis of the production of electricity in batteries and fuel cells. Oxidation and reduction reactions are also involved in generating some elements from their naturally occurring compounds. We shall discuss these topics in this chapter, and also see how oxidation reactions involving elemental hydrogen could replace fossil fuel combustion in generating energy for the economy of the future.

In Chapter 11, we found that the chemical reactions of acids with bases involved H^+, the nucleus of a hydrogen atom. In this chapter, we shall discover that oxidation and reduction reactions in water solution involve the transfer between ions and molecules of the other particle that makes up the hydrogen atom, namely the electron, symbolized here as e^- to emphasize its electrical charge.

Oxidation-Reduction Reactions

12.1 The transfer of electrons between atoms can be described by half-reactions

In Chapter 11, we referred to a famous experiment in which an iron nail dissolves in a soft drink. This may sound far-fetched, but you can see it for yourself in the following activity.

■ Chemistry in Your Home: Dissolving an iron nail in a soft drink

Even a dilute acid solution such as phosphoric acid in cola drinks will react with a metal to produce hydrogen gas. Obtain a clean iron nail. (Make sure that all surface dirt, rust, etc. is removed.) Place the nail in a bottle of cola (Coke, Pepsi, etc.), cap the bottle, and let it sit, monitoring the reaction for several days. How do you know that hydrogen gas is being produced? What happens to the nail? Where does the iron go?

As we saw in Chapter 11, one of the characteristics of acids is their ability to react with many metals, producing hydrogen gas. The small bubbles on the surface

Whenever you see this icon in this chapter, go to
www.whfreeman.com/ciyl2e

Figure 12.1 Phosphoric acid in cola reacts with an iron nail to produce hydrogen gas. (Photo by George Semple for W. H. Freeman and Company)

of the nail in the cola signal that the metal is reacting with the acid present in the liquid, releasing H_2 (see Figure 12.1). The iron metal gradually disappears, since each iron atom enters the solution as an iron ion, Fe^{2+}. The net ionic reaction for the process is:

$$2\,H^+\,(aq) + Fe\,(s) \rightarrow H_2\,(g) + Fe^{2+}\,(aq)$$

Let's reconsider this reaction from the viewpoint of the two elements involved, hydrogen and iron.

- In order to form its 2+ ion, each atom of iron that reacts in the metal must *lose* two electrons since it is electrically neutral in the free metal.
- In order to form an H_2 molecule, in which there is a total of two electrons, the two electron-less H^+ ions must *gain* a total of two electrons.

Thus we see that the net reaction involves no interchange of atoms between the reactants but *simply a transfer of electrons between them.* We could conveniently write the net ionic reaction as two steps, each of which involves only one element. In the first step, an iron atom at the surface of the nail loses two electrons and enters the solution as an ion:

$$Fe\,(s) \rightarrow Fe^{2+} + 2\,e^-$$

In the second step, a pair of hydrogen ions in the solution receives the two electrons and forms a molecule of hydrogen gas:

$$2\,H^+ + 2\,e^- \rightarrow H_2\,(g)$$

These two equations represent **half-reactions,** so called since neither alone is a viable process. There must be one species that gives up electrons and another that receives them (see Figure 12.2). The two half-reactions occur simultaneously. When the two half-reactions are summed together, and the common $2e^-$ present on the left and right sides of the resulting equation are canceled, the correct ionic net reaction is obtained:

$$Fe\,(s) + 2\,H^+ + 2\,e^- \rightarrow Fe^{2+} + 2\,e^- + H_2\,(g)$$

or

$$Fe\,(s) + 2\,H^+ \rightarrow Fe^{2+} + H_2\,(g)$$

12.2 Oxidation involves electron loss, reduction involves electron gain

Electron transfer reactions are not limited to acids, nor even to aqueous solutions. For example, if a piece of magnesium metal is sufficiently

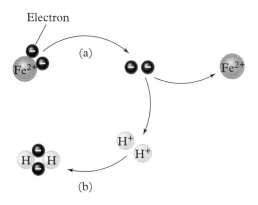

Electron

(a)

(b)

Figure 12.2 The half-reaction concept. Half-reactions describe the transfer of electrons between substances. (a) The oxidation half of the reaction involves the loss of electrons from an atom to form a positive ion. (b) The reduction half of the reaction involves the acceptance of electrons by a positive ion.

Figure 12.3 Magnesium metal burning in air. When magnesium transfers electrons to oxygen, heat and light are produced. (Charles D. Winters/Photo Researchers)

heated in air or oxygen, it bursts into flame and undergoes combustion (see Figure 12.3). Indeed, the bright sparks of some fireworks displays are white-hot grains of magnesium metal that are undergoing this reaction and emitting some of their output energy as visible light. The product of the reaction between magnesium and oxygen is *magnesium oxide*, MgO:

$$2\ Mg\,(s) + O_2\,(g) \rightarrow 2\ MgO\,(s)$$

We can understand this overall process in terms of half-reactions. Magnesium oxide is an ionic compound, composed of equal numbers of Mg^{2+} cations and O^{2-} anions. In forming its ion, each atom of magnesium loses two electrons:

$$Mg\,(s) \rightarrow Mg^{2+} + 2\ e^-$$

At the same time, each of the two oxygen atoms in each oxygen molecule that reacts gains two electrons, so the O_2 unit as a whole gains four electrons:

$$O_2\,(g) + 4\ e^- \rightarrow 2\ O^{2-}$$

The number of electrons lost and electrons gained must be the same since they are simply being transferred in the reaction. This is the reason the reaction requires twice as many magnesium atoms as oxygen molecules.

In terms first developed in Chapter 2, the magnesium metal is undergoing oxidation here, since it is reacting with elemental oxygen. Thus, in this reaction, the oxidation of magnesium is equivalent to a loss of electrons by it. Indeed, at the atomic level, *every* oxidation of a substance by O_2 involves electron loss by the substance that is oxidized. From the viewpoint of the substance undergoing oxidation, it does not matter to its end product (the cation) whether electrons were taken from it by oxygen or by some other substance. For example, magnesium's electrons could be taken by H^+, since placing the metal in an acid solution would produce the same result:

$$2\ H^+ + Mg\,(s) \rightarrow H_2\,(g) + Mg^{2+}$$

What is important to magnesium in both reactions—that with oxygen or that with acid—is that it loses two electrons.

For these reasons, chemists have revised and generalized the concept of **oxidation** so that this term now is defined as *the loss of electrons by a substance.* Thus the magnesium atoms undergo oxidation in their reaction with oxygen, and so do the magnesium atoms in their reaction with hydrogen ions. The other half-reaction that must accompany each oxidation half-reaction is *the gain of electrons by a substance.* This process is the generalized definition of **reduction.** Thus, the atoms of oxygen are said to be reduced in their transition from O_2 to O^{2-}, and the hydrogen ions in H^+ are reduced when they are converted to the electrically neutral atoms of the H_2 molecule. Oxidation and reduction are processes that always go hand in hand, since one substance must give up electrons before another substance can add them. Oxidation-reduction reactions, or **redox reactions** as they are often called, are as important in chemistry as acid-base reactions.

In redox reactions, the substance that pulls electrons from the other material, oxidizing it, is called the **oxidizing agent.** Not only O_2 but also H^+ and several other substances—for example the gases *ozone,* O_3, and *chlorine,* Cl_2—are common oxidizing agents. Substances that readily yield some of their electrons, allowing other chemicals to be reduced, are called **reducing agents.** Fuels such as natural gas that react with oxygen are reducing agents. Many metals are good reducing agents, since they readily form cations and thereby give up some electrons. We have seen that sodium metal reacts readily with chlorine gas to form sodium chloride; we can now understand that this process is a redox reaction. Since sodium atoms each lose one electron in becoming ions, they are oxidized. At the same time, they are the reducing agents in the process since they supply electrons to the atoms of chlorine in the chlorine molecules:

$$Na\,(s) \rightarrow Na^+ + e^- \qquad \text{oxidation half-reaction}$$
$$Cl_2\,(g) + 2\,e^- \rightarrow 2\,Cl^- \qquad \text{reduction half-reaction}$$

Since chlorine takes electrons away from sodium and thereby oxidizes it, chlorine is the oxidizing agent in this process.

Worked Example: Identifying roles and half-reactions for substances in redox reactions

Aluminum metal reacts with molecular oxygen, O_2, in the air to form the compound aluminum oxide, Al_2O_3, which consists of Al^{3+} and O^{2-} ions. Write the half-reactions corresponding to oxidation and reduction in this process. Which of the original reactants is the oxidizing agent and which is the reducing agent? Which substance is oxidized, and which is reduced, during the reaction?

Solution: Let's first consider the aluminum. In the reaction, it is converted from aluminum atoms, which are electrically neutral (since they are part of the free element), into aluminum ions, each having a $+3$ charge. Thus the aluminum atoms are oxidized since each loses three electrons in the process. The oxidation half-reaction must be:

$$Al\,(s) \rightarrow Al^{3+} + 3\,e^-$$

The oxygen atoms in O_2 are electrically neutral, and each gains two electrons in becoming oxide ions since the charge on the latter is -2. Thus oxygen is reduced in the reaction. Since one molecule of O_2 contains two oxygen atoms, the half-reaction must be a four-electron process, and it is a reduction since electrons are gained:

$$O_2\,(g) + 4\,e^- \rightarrow 2\,O^{2-}$$

The substance that forces another to give up electrons is the oxidizing agent, so in this reaction it is oxygen. The substance that supplies electrons is the reducing agent, so here it is aluminum.

Exercise 12.1

Calcium metal reacts with liquid bromine, Br_2, to form the compound $CaBr_2$, which consists of Ca^{2+} and Br^- ions. Write the half-reactions corresponding to oxidation and reduction in this process. Which of the original reactants is the oxidizing agent and which is the reducing agent? Which substance is oxidized, and which is reduced, during the reaction?

12.3 The simple zinc–copper electrochemical cell

In the examples we've given so far, the oxidation-reduction reactions involve the interaction of a metal with a nonmetal. Redox reactions also sometimes occur when both reactants are metals or when both are nonmetals. As we shall see, batteries are usually composed of metals and their compounds that readily undergo redox reactions.

For example, if you place a strip of zinc metal in an aqueous solution that contains copper ions, Cu^{2+}, a redox reaction occurs. Over time, each zinc atom at the metal's surface gives up two electrons to a copper ion with which it comes into contact. In the process, the zinc atom is oxidized to zinc ion Zn^{2+}, which enters the aqueous solution, replacing the copper ion. Having undergone reduction by gaining two electrons, the copper atom becomes attached to the metal surface (see Figure 12.4). The two half-reactions are:

$$Zn\,(s) \rightarrow Zn^{2+} + 2\,e^- \qquad \text{oxidation half-reaction}$$
$$Cu^{2+} + 2\,e^- \rightarrow Cu\,(s) \qquad \text{reduction half-reaction}$$

The net ionic reaction is obtained by adding the two half-reactions and canceling the terms involving free electrons:

$$Zn\,(s) + Cu^{2+} + 2\,e^- \rightarrow Zn^{2+} + 2\,e^- + Cu\,(s)$$
$$Zn\,(s) + Cu^{2+} \rightarrow Zn^{2+} + Cu\,(s)$$

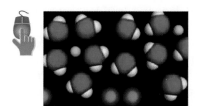

View an aqueous solution of copper nitrate and the redox reaction between copper and zinc at Chapter 12: Visualizations: Media Link 1.

Figure 12.4 The redox reaction between copper and zinc. Copper ions in a solution (blue) form solid copper metal on a strip of zinc. As the copper ions are used up, the solution becomes colorless. (Chip Clark for W. H. Freeman and Company. From *Chemistry: Molecules, Matter, and Change*, 4th ed., 2000, L. L. Jones and P. W. Atkins.)

Oxidation-Reduction Reactions

Chemists have found that this reaction can occur even if the zinc metal does not touch the solution containing the copper ions! All that is required is for an electrical connection to exist between the zinc metal and the copper solution, so that the electrons being given up by the zinc atoms can travel to the copper ions without much resistance.

Thus, it is possible to have the oxidation and reduction half-reactions occur in two separate containers, connected electrically by a metal wire, as illustrated in Figure 12.5. One end of the wire is connected to a strip of metal—almost any metal will do—that dips into an aqueous solution containing copper ions (right side of figure). It is at the surface of this strip of metal that the copper ions receive two electrons each and deposit themselves as copper atoms. Over time the metal strip becomes covered by a layer of pure copper metal. Consequently, the solution into which it dips becomes less and less concentrated in copper ions. The zinc strip is connected to the other end of the wire and also dips into an aqueous solution in a separate container (left side of figure). At the same time that the copper is being deposited on the metal strip in the right container, more and more zinc atoms from the zinc strip lose two electrons each. The zinc atoms enter the solution as zinc ions. Two electrons then travel along the wire to the copper solution. The solution into which the zinc strip dips becomes more and more concentrated in zinc ions and the zinc strip gradually wears away. The reaction stops when all the zinc strip has dissolved or all the copper ions have been converted to copper metal, whichever comes first.

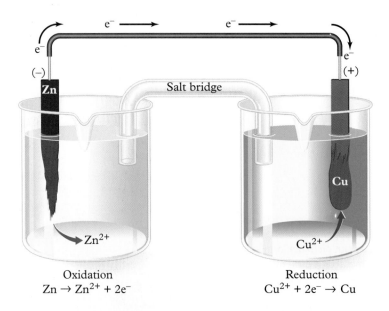

Figure 12.5 An oxidation-reduction reaction occurring in separate compartments linked by an electrical connection.

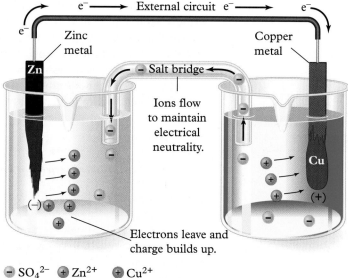

Figure 12.6 The principle of ion flow in an electrochemical cell. Ions flow between solutions via a salt bridge to maintain electric neutrality.

Remember that in cells electrons do not flow through the solution, but along the external wire. Ions flow within the electrolyte solution, but not along the wire.

Of course, an aqueous solution cannot contain exclusively or predominantly ions of one charge; there must be an equal number of positive and negative charges so that it is electrically neutral overall (see Figure 12.6). Initially the copper solution must have contained anions as well as Cu^{2+} cations. For example, if copper sulfate was dissolved in water to create the solution, these anions would be sulfate ions, SO_4^{2-}:

$$CuSO_4(s) \rightarrow Cu^{2+} + SO_4^{2-}$$

Over time, fewer and fewer sulfate ions would be needed in the copper solution to balance the Cu^{2+} ions, and more and more would be needed in the other solution to balance the increasing number of Zn^{2+} ions being created there. Consequently, the two solutions must come into sufficient contact to allow the sulfate anions to *migrate* from the copper area to the zinc one. The solutions need not be mixed for this to occur as long as they can pass anions between them. Often, this can be accomplished by use of a **salt bridge** joining the solutions. The bridge can consist of an aqueous solution or a water-based gel through which ions can pass from one electrode to the other. In other setups, the two electrodes simply dip into a single solution, so no salt bridge is necessary.

The setup shown in Figure 12.6 is a fairly general one used to study redox reactions and to produce electricity from chemical reactions. As a whole, such units are called **electrochemical cells.** The sites at which the oxidation or reduction reactions occur are called the **electrodes.** These are the strips of metal dipping into the solution, or conducting mixture, and connected by a wire. The ion-containing solution into which the electrodes dip is known as the **electrolyte.** The *negative* electrode consists of a component, such as a metal, that is easily oxidized, and a material that collects the electrons; in many cases, these two components are the same material. The *positive* electrode consists of a metal that collects electrons from the wire and passes them to the substance that can be reduced, such as a metal ion. The wire connecting the electrodes and any electrical devices placed along the wire are collectively known as the *external circuit*. During the spontaneous reaction, called the **discharge** of the cell, electrons flow from the negative to the positive electrode.

A battery in which the reactants have been consumed is said to be "discharged."

Batteries

12.4 Batteries generate energy by redox reactions

When the reaction between zinc metal and copper ion occurs, heat is given off. In fact, this is true for most redox processes. In the cell setup of Figure 12.6, the metal wire joining the two electrodes becomes very hot while the reaction proceeds, because the electrons are propelled through the wire and expend some of their energy overcoming the resistance of the wire.

■ Chemistry in Your Home: Feeling the heat of a redox reaction

You can feel the heat generated as electrons move through a wire. Cut a 6-inch (15-cm) piece of aluminum foil and fold it lengthwise several times to make a thick "wire." Tape one end of the foil to the positive end of a size D battery and the other end of the foil to the negative end of the battery. Attaching the foil this way creates a circuit through which electrons can flow. Hold the "wire" for a minute or two. What do you feel?

The energy produced in the zinc–copper reaction is essentially wasted if it just heats up the wire. However, if an electrical motor were placed along the wire, the reaction energy could be used to drive it. In other words, much of the energy given off by the reaction in an electrochemical cell can be obtained as *electrical energy* rather than as heat.

Since electrical energy can be obtained from any electrochemical cell, redox reactions provide a convenient way of generating electricity when connecting an electrical device to power lines is inconvenient. In some instances, several cells are connected together to provide sufficient power. Commercial electrochemical cells, or combinations of cells, are usually called **batteries.** Modern households typically contain several dozen batteries, ranging from the obvious ones in flashlights and laptop computers to those "hidden away" in products such as watches, clocks, stoves, heart pacemakers, and portable music players. (see Figure 12.7). Indeed, the battery in a laptop computer is usually its heaviest component.

Figure 12.7 Batteries come in many shapes and sizes. (George Semple for W. H Freeman and Company)

◼ Chemistry in Your Home: Building a battery from fruit

As unlikely as it sounds, you can build a battery from fruit. Straighten a clean steel paper clip and insert one end through the rind of a lemon so it penetrates into the fruit. Insert a piece of clean, bare copper wire into the lemon about 1 inch (2 cm) from the paper clip. Bend the ends of the wire toward each other and touch them to your tongue several times. Do you feel a tingle? Do you experience a metallic taste? Repeat the experiment using an apple, banana, or any other fruit or vegetable you would like. Which fruit juices work as electrolytes? Why do some fruits facilitate electron transfer between the metal inserts, but not others?

Over 3 billion batteries are sold in North America annually. The worldwide market for batteries reached about $40 billion in the late 1990s.

Although the zinc–copper cell we described in section 12.3 could be used as a battery, it is somewhat inconvenient to use since it involves a liquid solution and a relatively expensive metal, copper. Several popular batteries do use zinc metal as the reducing agent (the substance that gives up electrons and becomes oxidized), but with various oxidizing agents (the substances that receive the electrons). All have the advantage of using an electrolyte that is not liquid and using an oxidizing agent that is less expensive than copper. Several of the common batteries used today are described in Taking It Further I, II, III, and IV at the end of this chapter.

12.5 Batteries can power electric cars

Electric cars are increasingly in the news as we try to find ways to reduce air pollution. Most of the purely electric cars that have already been marketed use the same sort of lead–acid batteries that gasoline-powered vehicles have traditionally employed to operate the starter motor, and quite a number of batteries need to be connected together to produce the required power. In the future, electric cars may well use nickel–cadmium, nickel–metal hydride, and lithium-based batteries.

The nickel- and lithium-based batteries are described in Taking It Further sections II and III respectively.

The practical difficulties that currently discourage widespread adoption of battery-powered electric vehicles are their high cost, the short driving range before the batteries need recharging, the length of the battery recharge period, and the excessive weight of the batteries. The vehicles have the attraction of zero pollution emissions during their operation, little operating noise, and low maintenance costs.

Electric cars are not entirely pollution-free, however. Some pollution is emitted into the environment if the electricity required to charge the batteries is generated from fossil fuel combustion. Some economists and engineers have predicted that lead pollution stemming from the manufacture, handling, disposal, and recycling of lead–acid batteries used to power electrical cars would raise lead emissions into the environment by an amount that would greatly exceed that which used to be associated with leaded gasoline. Critics of this analysis counter that lead emissions during recycling are now low and that not all the lead lost in the processing steps will be emitted into the environment.

Batteries

Several automobile manufacturers now offer "hybrid-electric" vehicles, which have electric motors that run from power stored in batteries such as the nickel–metal hydride battery. In the Toyota Prius, for example, the vehicles have a small gasoline engine that charges the battery and supplements its power at high road speeds. The energy lost by a conventional gasoline engine during idling is eliminated since the electric motor is switched off at these times. In all hybrid-electric vehicles, the battery is recharged without fuel use during braking by using the energy of the motion of the vehicle that is usually lost as heat. Because of these features, the efficiency of the electric motor and its lack of pollutant emissions, these vehicles are much more energy-efficient and "cleaner" than conventional vehicles, albeit more expensive to buy at the present time.

Fuel Cells

12.6 Fuel cells provide continuous electric power

The great disadvantage of batteries as power sources is that they contain a finite, small amount of reactants and eventually become fully discharged and useless. This disadvantage is overcome to some extent by rechargeable batteries, but these devices take considerable time to recharge. A better solution to power vehicles may be the fuel cell.

A **fuel cell** is an electrochemical cell in which the reactants are continuously replenished while the unit delivers electrical power, so it can be continuously operated and never becomes discharged (see Figure 12.8). Thus the fuel cell operates like a battery, but without the battery's limitations. The most convenient reactants for fuel cells are gases or liquids that flow over solid electrodes and react at their surfaces.

The oxidizing agent used in fuel cells is molecular oxygen, O_2, either as the pure gas or as a component of air. The oxygen is reduced in the fuel cell to water. In a fuel cell having an acidic electrolyte, the reduction half-reaction is:

$$O_2(g) + 4\,H^+ + 4\,e^- \rightarrow 2\,H_2O$$

There are several potential choices for the reducing agent in a fuel cell, including liquids such as methanol (CH_3OH), gasoline, and gases such as methane. Currently, the most likely candidate is hydrogen gas, H_2, since it has been the fuel used in most fuel cell technology developed thus far. When it is oxidized in an acidic environment, each molecule of H_2 yields two electrons and two hydrogen ions:

Figure 12.8 Step by Step
A proton-exchange membrane hydrogen/oxygen fuel cell. This cell produces water and never becomes discharged.

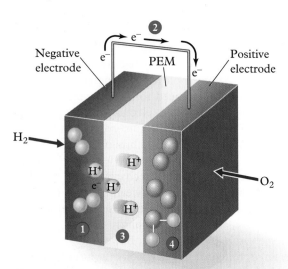

❶ Hydrogen gas is converted to H^+ and e^- by a catalyst on the surface of the proton exchange membrane (PEM).

❷ e^- cannot penetrate the membrane and moves through an external circuit.

❸ H^+ moves through the membrane.

❹ A catalyst on the surface of the positive electrode converts H^+, O_2, and e^- to water.

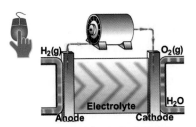

View the components and the action of a hydrogen fuel cell at Chapter 12: Visualizations: Media Link 2.

Prototype buses powered by PEM fuel cells, developed by Canada's Ballard Power Systems, have been running as test vehicles in Chicago and Vancouver.

$$H_2(g) \rightarrow 2\,H^+ + 2\,e^-$$

In the *hydrogen–oxygen fuel cell*, hydrogen gas enters the cell and passes over one electrode, where the oxidation occurs. Air or oxygen passes over the other electrode. The oxygen is converted to water, which is then emitted as a gas from the fuel cell. The electrodes are composed of graphitic carbon, onto the surfaces of which small grains of platinum have been dispersed as catalysts and at which the reactions occur.

12.7 Fuel cells have great potential

Although commercial electric cars are currently powered by batteries, it is widely anticipated that in the future all-electric vehicles will use fuel cells instead. The Daimler-Chrysler company has announced that it will begin commercial production of fuel cell cars in about 2010. It has already unveiled prototype vehicles (see Figure 12.9). Many other auto manufacturers are currently pursuing the development of electric cars that use fuel cells.

▶ **Discussion Point:** Should electric cars replace gasoline-powered vehicles?

California and some other states have mandated that a percentage of new cars sold in their areas must be electric vehicles powered by either batteries or fuel cells. Develop arguments supporting and opposing this regulation. Use economic and environmental factors in your arguments. Would you be in favor of mandating that all new cars sold in North America at some date in the near future be emission-free?

Figure 12.9 A fuel cell vehicle. Prototype vehicles such as these are already being road-tested. (Leslie Eudy/NREL/PIX)

Fuel cells are also attractive for use in small electric power plants, because their pollutant emissions are so much less than those from fossil fuel combustion (about 1% of the nitrogen oxides and sulfur oxides, for example) and also because of their high efficiency in converting fuel to electricity. By-product waste heat from the fuel cell can be recovered and used. The general process by which a fuel is converted to electricity and the waste heat usefully employed is called *cogeneration.*

Although fuel cells currently are quite expensive compared to their alternatives for powering vehicles and power plants, they have several real advantages:

- Their high efficiency in extracting useful energy from fuels, and hence the relatively smaller amount of carbon dioxide produced per unit of energy obtained
- The lack of pollutants generated and emitted when they operate
- Their silent, low-maintenance operation
- Their ability to use hydrogen produced not only from fossil fuels but also from renewable sources

Once manufacturing costs have decreased, fuel cells are likely to be an important way to generate electricity from fuels. However, most initial uses of fuel cells will be to power consumer electronic devices such as laptop computers and cell phones. In these devices, the convenience of fuel cells, which provide power for longer periods than do batteries, is a more important factor than cost.

Electrolysis: Producing Metals from Ores

In discussing the recharging of batteries, we noted that a redox chemical reaction could be made to occur if electrical power was used to force electrons from one material to another. This technique of passing an electrical current through a material to drive a chemical reaction is called **electrolysis** (see Figure 12.10). The reaction does not occur spontaneously, and the electrical energy required to force it to occur can be substantial. Electrolysis is used to prepare and purify some materials, including metals.

Consider aluminum as an example. This element occurs naturally as its ore, called *bauxite,* the formula for which is Al_2O_3. The aluminum in this oxide is present as Al^{3+}, and so it must be reduced to become the free metal. Each aluminum ion must add three electrons, since the atoms in a free metal are electrically neutral:

Figure 12.10 Electrolysis.

$$Al^{3+} + 3e^- \rightarrow Al(s)$$

Unfortunately, no inexpensive chemical reducing agent is available to drive this reduction. For this reason elemental

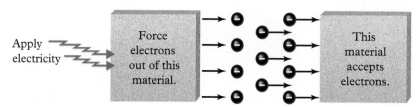

Figure 12.11 Electroplating. This reaction deposits a film of metal, such as chrome, on another object. (Photo by Don W. Fawcett/Photo Researchers)

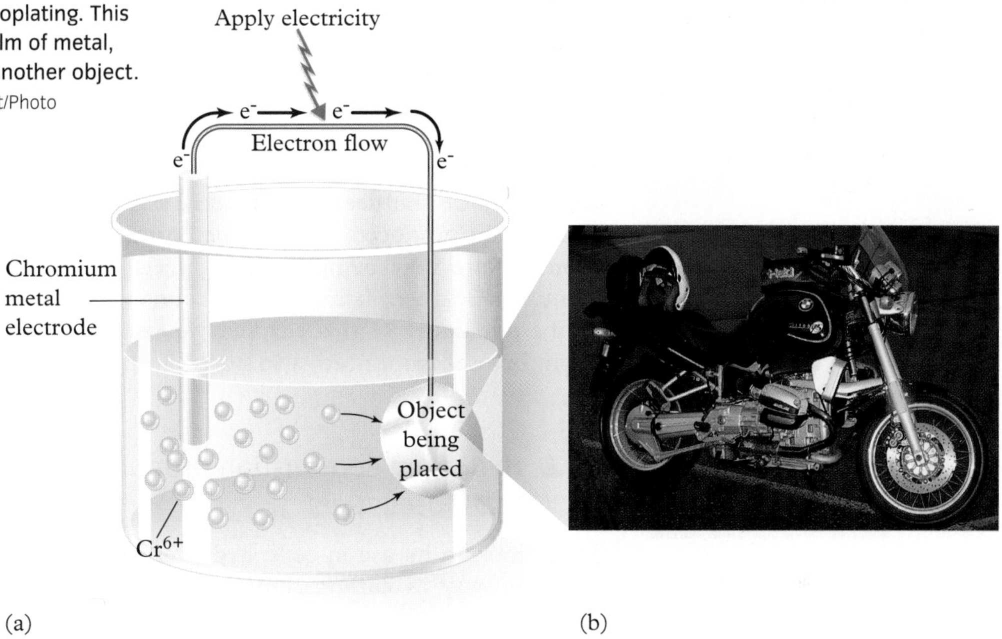

Apply electricity

Electron flow

e⁻ → e⁻ →

Chromium metal electrode

Object being plated

Cr^{6+}

(a)

(b)

aluminum was a very expensive material for many years until a feasible electrolyte was developed and cheap electricity became available. Aluminum now is produced inexpensively in vast quantities by electrolysis of its oxide, dissolved in a molten salt of aluminum.

Other elements in their free state are also produced by electrolysis. Magnesium metal is produced from molten *magnesium chloride,* $MgCl_2$. Similarly, *elemental fluorine* gas, F_2, and elemental *chlorine,* Cl_2, are prepared by electrolysis of their respective salts.

Electrolysis is also used in **electroplating,** the technique by which a thin film of a metal is deposited on another object. This is accomplished by using the object to be plated as an electrode in a cell at which reduction is forced to occur (see Figure 12.11a). For example, chromium metal can be plated onto a metal object by placing the object in a solution of chromium ions, Cr^{6+}. Electricity is forced through the object, which acts as the electrode. The electrons are accepted by the ions, which are deposited as chromium atoms onto the object to be plated (see Figure 12.11b). The other electrode, from which the electrons are obtained, is made from chromium and resupplies chromium ions to solution as the process continues.

Hydrogen: Fuel of the Future?

As we mentioned earlier, hydrogen gas is likely to have a key role in fueling the cars and power plants of the future. Most of the ways in which hydrogen is produced and used are redox processes, involving cells. Some futurists believe that the world will eventually have a hydrogen-based economy.

Figure 12.12 Hydrogen is an important fuel when weight is a factor, as in rockets. (Courtesy of NASA)

The idea that hydrogen would be the ultimate fuel of the future goes back at least as far as 1874, when it was mentioned by a character in the novel *Mysterious Island* by Jules Verne. Indeed, hydrogen has already found use as fuel in applications for which lightness is an important factor, since *per gram* it generates more energy than any other fuel. For example, hydrogen powered the Saturn moon rockets and powers the U.S. space shuttles (see Figure 12.12). The questions of how hydrogen will be produced, transported, stored, and used in the hydrogen economy are now being explored. Iceland has already begun implementing a 30-year plan to run the country on hydrogen-derived energy. In the remainder of this chapter, we explore ways in which hydrogen might be used, produced, and stored in the future.

12.8 Hydrogen combustion produces fewer pollutants than fossil fuels

In addition to its use in cells, hydrogen gas can be burned as a fuel in the same way as carbon-containing compounds. Hydrogen gas combines with oxygen gas to produce water, and in the process it releases a substantial quantity of heat energy:

$$\overset{\text{spark}}{2\,H_2(g) + O_2(g) \rightarrow 2\,H_2O(g) + \text{heat energy}}$$

Hydrogen gas can be combined with oxygen to produce heat by conventional flame combustion or by low-temperature combustion in catalytic heaters. Compared to fossil fuels, hydrogen fuel requires less mass for the amount of energy that is generated, and its combustion produces a smaller amount of polluting gases. Nevertheless, using hydrogen in fuel cells is a more efficient use of its energy than burning it.

Although it is sometimes stated that hydrogen combustion produces only water vapor and no pollutants, this is not actually true. Since combustion involves a high-temperature flame, some of the nitrogen contained in the air that is used as the source of oxygen reacts with the oxygen to form nitrogen oxides. Some *hydrogen peroxide*, H_2O_2, is released as well. Thus, hydrogen-burning vehicles are not zero-emission systems. However, by combusting hydrogen in a gaseous mixture that has a large excess of oxygen, the flame temperature is kept low and consequently more than 90% of nitrogen oxide formation is eliminated. Some analysts believe that vehicles burning hydrogen in an internal combustion engine will enter the market long before those based upon fuel cells.

Of course, even electric vehicles are not really pollution-free if a fossil fuel is burned to generate the electricity to charge the battery, since the fossil fuel's combustion in a power plant yields nitrogen oxides that are released into the atmosphere. Carbon dioxide gas is also released from the fossil fuel, the significance of which we shall discuss in Chapter 16.

The role of nitrogen oxides in air pollution is discussed in Chapter 14.

Chapter 12: Batteries, Fuel Cells, and the Hydrogen Economy

12.9 Hydrogen fuel is produced from water or methane

It is important to realize that hydrogen is not an energy *source*, since it does not occur as a free element at Earth's crust. Hydrogen gas is an energy *vector* or *carrier* only; it must be produced, usually from water and/or a fossil fuel such as methane. In the process, large amounts of energy and/or other fuels are consumed. We will look at a number of possible methods for producing hydrogen for energy.

Electrolysis The most expensive commercial way to produce hydrogen is by electrolysis of water, using electricity generated by some energy source:

$$2\ H_2O\,(l) + \text{electrical energy} \rightarrow 2\ H_2\,(g) + O_2\,(g)$$

Unfortunately, about half the electrical energy is inadvertently converted to heat and therefore wasted in this process. And if the electricity is generated by burning a fossil fuel, carbon dioxide and air pollutants such as nitrogen oxides are emitted into the air during combustion.

A hope for the future is that it will be economical to use solar energy to drive the electrolysis of water. Currently, prototype plants in Saudi Arabia and in Germany use electricity from solar energy to produce hydrogen (see Figure 12.13). Energy is later recovered from the stored hydrogen by reacting it with oxygen. Another possibility is to use excess electricity from hydroelectric, nuclear power, or wind power installations—in other words, power that is generated but not required for immediate use—to produce hydrogen in off-peak demand periods.

Sunlight-induced decomposition of water Even better than using solar electricity to electrolyze water would be using sunlight to directly decompose water into hydrogen and oxygen, but no practical, efficient method to do this has yet been devised. One of the difficulties is that H_2O does not directly absorb sunlight. Consequently some substance

Figure 12.13 A prototype plant in Germany that uses sunlight to generate hydrogen from water.
(The Solar Hydrogen Project in Neunburg vorm Wasd, Germany/Courtesy of E.ON Energie AG)

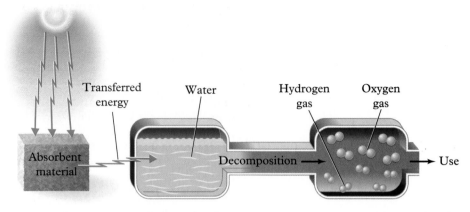

Figure 12.14 Sunlight used to decompose water. Because water doesn't absorb visible or UV light, an intermediate absorbent molecule is necessary.

must be found that can absorb the sunlight and then transfer its energy to the decomposition process (see Figure 12.14). So far, the substances devised for this purpose are very inefficient in converting sunlight. In addition, since the light-absorbing substances and others required in the process are not 100% recoverable at the end of the cycle, they must be continuously resupplied, making the whole process uneconomical.

In principle, the capture of sunlight's heat can produce temperatures hot enough to decompose water into hydrogen and oxygen. Recent research in Israel used a *solar tower* of mirrors to concentrate the energy in sunlight by a factor of 10,000 and thereby produce temperatures of about 2200°C in a reactor. The project succeeded in splitting about one-quarter of water vapor into H_2 and O_2. It is not clear, however, that this process can be made economically viable.

Hydrogen from a fossil fuel Hydrogen gas can be produced by reacting a fossil fuel such as coal, petroleum, or natural gas with water to form hydrogen and carbon dioxide. In the process, the energy value of the fuel is transferred to the hydrogen atoms of water by converting them to hydrogen gas. In effect, the carbon in the original fuel is oxidized, and the hydrogen is reduced. The net reaction, in which carbon is oxidized and hydrogen reduced, is as follows for the specific case of coal (here assumed to be mainly graphite, C):

$$C(s) + 2\,H_2O(g) \rightarrow 2\,H_2(g) + CO_2(g)$$

As much carbon dioxide is produced in this way as would be obtained by combustion of the fossil fuels in oxygen.

12.10 Hydrogen storage presents challenges

Hydrogen gas is superior to electricity as an energy carrier in some ways, since its transmission over long distances consumes less energy and since batteries are not required to store it. However, hydrogen storage presents its own challenges.

Storing hydrogen as a liquid or compressed gas In rocketry applications, hydrogen is stored as a liquid. Since hydrogen's boiling point of 20 K, or $-253°C$ (at 1 atmosphere pressure) is so low, much energy is required both to liquefy it and to keep it cold enough to remain liquid. Consequently, liquid hydrogen is used in only a few specialized situations where its "lightness" (high energy output per gram) makes it vital.

Hydrogen could be stored as a compressed gas, in much the same way as methane is stored in the form of natural gas. However, compared to CH_4, a much greater amount of H_2 gas needs to be stored in order to release the same amount of energy when combustion occurs. Compared with methane, the combustion of one molecule (or one mole) of hydrogen generates only about one-quarter the energy. Thus the bulky nature of hydrogen gas limits its applications even though it is a highly efficient fuel from a mass viewpoint.

Hydrogen gas is considered to be a dangerous fuel because it is highly flammable and explosive; it ignites more easily than do most conventional fuels. A tragic example of the danger of hydrogen fuel occurred in January 1986 when the space shuttle *Challenger* and its crew were lost because the shuttle's gas tanks ruptured and the hydrogen caught fire and exploded. On the positive side, spills of liquid hydrogen evaporate rapidly and rise high into the air.

Storing hydrogen in carbon or a metal A more practical and safer way to store hydrogen for use in small vehicles in the future could be as a substance absorbed by a solid. For example, tiny fibers made of graphite, a light material, can store up to three times their mass in hydrogen as molecules wedged between the graphite layers. Carbon nanotubes (see Chapter 5, section 5.19) may serve this purpose even more efficiently in the future.

Many metals, including alloys, absorb hydrogen gas *reversibly,* that is to say, they release it when a little energy is applied, as a sponge absorbs and releases water. The metals form *hydrides* by incorporating the small atoms of hydrogen in "holes" in the crystalline structure. For example, titanium metal absorbs high-pressure hydrogen to form the hydride TiH_2, a compound in which the density of hydrogen atoms (that is, the mass of hydrogen per volume) is twice that of liquid H_2. Heating the solid hydride gradually releases the hydrogen as a molecular gas, which then can be burned in air or oxygen for power. Research continues in the effort to find a lightweight metal alloy that can efficiently store hydrogen without making the whole system excessively heavy. Even existing metal-hydride systems, however, are lighter than are the pressurized tanks needed to store liquid hydrogen.

Hydrogen storage in fuel-cell–powered vehicles In prototype fuel-cell buses, compressed hydrogen is stored in tanks under the roof of the bus. Transporting hydrogen around in individual cars and trucks is not

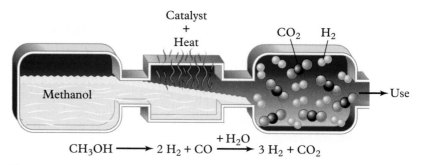

$$CH_3OH \longrightarrow 2\,H_2 + CO \xrightarrow{+\,H_2O} 3\,H_2 + CO_2$$

Figure 12.15 Hydrogen obtained from methanol via a re-forming reaction.

very practical, however. Systems are being developed that allow hydrogen to be extracted as needed from liquid fuels, which are much more convenient to transport and store, as well as being energy-dense, since they provide much more energy per volume than do gases.

For example, the hydrogen required in a fuel cell vehicle may be obtained by onboard decomposition of liquid methanol to produce hydrogen as needed, using the following so-called **re-forming reaction** (see Figure 12.15):

$$CH_3OH \rightarrow 2\,H_2 + CO$$
$$CO + H_2O \rightarrow CO_2 + H_2$$
$$\text{overall} \quad CH_3OH + H_2O \rightarrow CO_2 + 3\,H_2$$

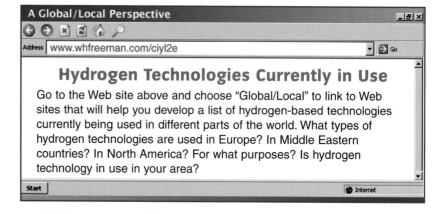

A Global/Local Perspective

Address www.whfreeman.com/ciyl2e

Hydrogen Technologies Currently in Use

Go to the Web site above and choose "Global/Local" to link to Web sites that will help you develop a list of hydrogen-based technologies currently being used in different parts of the world. What types of hydrogen technologies are used in Europe? In Middle Eastern countries? In North America? For what purposes? Is hydrogen technology in use in your area?

▶ **Discussion Point:** What impact will a hydrogen-based economy have?

Use the resources listed on the Web site for this textbook and consider how human activities would be impacted by conversion to a hydrogen economy. What problems do you foresee? What benefits?

● Tying Concepts Together: Redox reactions and electrical energy

We have seen in this chapter that many chemical reactions involve the transfer of electrons between atoms. When the reacting substances are in physical contact, the energy released by the reaction is given off as heat, sometimes accompanied by light. However, people have learned to capture this energy in a more useful form—electricity. Electrical power generated by chemical reactions has been available as batteries, in which a fixed amount of reactants is available. In modern times, the fuel cell—in which the reactants are supplied continuously—is becoming available. The hydrogen economy of the future is based upon the creation and supply of hydrogen gas as a reducing agent that can be oxidized by the oxygen in air at any convenient time to produce energy, especially in the form of electricity.

Each redox reaction has a spontaneous direction, the one in which it occurs if no external energy is applied to resist it. If enough electrical power is applied to a cell, electrons can be forced to travel in the direction that is the reverse of the spontaneous direction. In this way, batteries that are almost or completely discharged can be recharged. Similarly, a metal can be forced to travel as ions through the electrolyte from one electrode to the other, plating the latter. By forcing a non-spontaneous reaction to occur by electrolysis, free elements including hydrogen can be obtained from their compounds.

Summarizing the Main Ideas

Many chemical reactions consist of a transfer of electrons between ions, atoms, or molecules. The electron-loss portion of the process, the oxidation, and the electron-gain portion, the reduction, always occur together. The two portions, called half-reactions, together constitute a redox reaction. Oxidizing agents are substances capable of extracting electrons from other substances, such as metals. In turn, the metals are the reducing agents in the redox process, since they supply the electrons.

In an electrochemical cell, the two half-reactions of a redox process occur at separate sites, called electrodes, which are connected electrically by a wire. Ions travel between the electrodes in a conducting material called the electrolyte.

The energy released by a redox reaction may be captured as electrical energy by placing a motor along the wire joining the electrodes. Batteries consist of one or more electrochemical cells joined together to produce electricity on demand.

Fuel cells are electrochemical cells in which the reactants are continuously supplied and which therefore produce power continuously. The reduction half-reaction in a fuel cell is the conversion of O_2 into water. At the other electrode, hydrogen gas, H_2, or alternatively another fuel is oxidized. Fuel cells hold promise for the future, in vehicles and power plants, mainly because they are energy efficient and relatively non-polluting.

Hydrogen, H_2, may be the main fuel in the future. If it is burned in air, it does produce nitrogen oxides, so combustion is not a nonpolluting way of gaining its energy.

Hydrogen gas can be produced from the electrolysis of water, or perhaps in the future from the sunlight-induced decomposition of water. Currently, most of it is made by reacting fossil fuels, usually methane, with steam.

Hydrogen can be stored as a liquid or as a highly compressed gas, though both methods are relatively expensive in terms of the energy that they consume. A more practical way to temporarily store hydrogen may be in a solid such as graphite fibers or a metal. Alternatively, hydrogen can be stored in chemical combination in an organic compound such as methanol until it is required, at which time it can be produced by the re-forming reaction.

Key Terms

half-reaction	oxidizing agent	electrode	fuel cell
oxidation	reducing agent	electrolyte	electrolysis
reduction	salt bridge	discharge	electroplating
redox reaction	electrochemical cell	battery	re-forming reaction

Web Sites of Interest

To link to Web sites of interest, go to www.whfreeman.com/ciyl2e, Chapter 12, and select the site you want.

For Further Reading

"Batteries Today," *Chemistry in Britain*, March 2000, pp. 34–39. This article has a good list of Web sites and a discussion of the current state of battery technology.

S. Dunn, "The Hydrogen Experiment," *World Watch*, November/December 2000, pp. 14–25. This article describes Iceland's 30-year plan to move to a hydrogen-based economy.

P. Fairley, "Power to the People," *Technology Review*, May 2001, pp. 71–77. This article gives fuel cell basics and discusses the possibility of individual consumer power generation.

L. B. Lave, C. T. Hendrickson, and F. C. McMichael, "Environmental Implications of Electric Cars," *Science, 268*, May 19, 1995, pp. 993–995. This article contains some surprising ideas about the potential environmental impact of electric automobiles.

"Releasing the Potential of Clean Power," *Chemistry and Industry*, October 18, 1999, pp. 796–799. This article discusses recent developments in fuel cells.

Review Questions

1. Define the following terms:
   a) *oxidation*   b) *reduction*
   c) *half-reaction*   d) *redox reaction*

2. Write the half-reactions by which the following ions form from their elements:
   a) Fe^{2+} b) Na^+ c) O^{2-} d) Cl^-

3. What does an *oxidizing agent* do? Identify some common oxidizing agents.

4. What does a *reducing agent* do? Identify some common reducing agents.

5. Identify the oxidizing and reducing agents in the reactions below:
   a) $CS_2 + 3\ O_2 \rightarrow CO_2 + 2\ SO_2$
   b) $Cr_2O_3 + 2\ Al \rightarrow Al_2O_3 + 2\ Cr$
   c) $WO_3 + 3\ H_2 \rightarrow W + 3\ H_2O$

6. What is an *electrochemical cell*?

7. Define the following terms:
   a) *electrode*   b) *electrolyte*   c) *salt bridge*

8. What role does each of the components in the previous question play in an electrochemical cell?

9. What occurs when an electrochemical cell is discharging?

10. What is a *battery*?

11. What is a *fuel cell*?

12. What is *electrolysis*?

13. What is *electroplating*?

14. What does the term *hydrogen economy* mean?

15. Write the reaction for the combustion of hydrogen.

16. Why is hydrogen gas in some ways a better energy carrier than electricity?

17. Describe the ways in which hydrogen can be stored.

Understanding Concepts

18. Write the net ionic reaction for the following:
    a) Zinc metal dissolves in acid to produce Zn^{2+} ions and hydrogen gas
    b) Lithium metal reacts with oxygen gas to produce solid lithium oxide, Li_2O

19. Identify the chemical and physical classes of substances (for example, acids, bases, metals) that commonly participate in redox reactions.

20. In the reaction of metallic iron with acid discussed in the text, namely $2\ H^+(aq) + Fe\,(s) \rightarrow H_2\,(g) + Fe^{2+}\,(aq)$, which substance is the oxidizing agent? The reducing agent? Which substance is oxidized in the reaction? Which is reduced?

21. Why does electron transfer occur between two metal strips in a solution even though they are not touching?

22. What is the role of a salt bridge in an electrochemical cell?

23. Why is it inconvenient to use a liquid electrolyte in a battery?

24. How does a fuel cell differ from a battery?

25. Why is hydrogen gas a desirable choice as a reducing agent in fuel cells?

26. What is a *proton-exchange membrane*? What is its role in a fuel cell?

27. What makes fuel cells attractive for use in small vehicles?

28. How is pure aluminum produced from its ore?

29. Why can hydrogen gas be used as a fuel?

30. Is it true that the combustion of hydrogen gas is a pollution-free process? Explain.

31. How is hydrogen combustion accomplished at low temperatures?

32. How does an *energy source* differ from an *energy carrier*?

33. What role can solar energy play in hydrogen production?

34. What are some problems associated with using sunlight to generate hydrogen gas directly from water?

35. Compare the advantages and disadvantages of storing hydrogen as a gas and as a liquid.

36. Explain how hydrogen gas can be stored in a solid.

37. Describe how hydrogen gas can be produced from methanol. Why is this called a *re-forming reaction*?

38. Why is it impractical to use methanol as the fuel in a fuel cell?

Synthesizing Ideas

39. Which of the following are oxidation-reduction reactions? Explain your choices.
    a) The production of NaCl from Na metal and Cl_2 gas: $2 Na(s) + Cl_2(g) \rightarrow 2 NaCl(s)$
    b) The production of carbon dioxide and water when glucose reacts with oxygen: $C_6H_{12}O_6 + 6 O_2 \rightarrow 6 CO_2 + 6 H_2O$
    c) The neutralization of hydrochloric acid by sodium hydroxide to produce salt and water: $HCl + NaOH \rightarrow NaCl + H_2O$

40. Aluminum atoms at the surface of the elemental metal react with atmospheric oxygen to form a coating of aluminum oxide, Al_2O_3, which adheres to the metal and prevents further reaction. Write this overall reaction in terms of two half-reactions. Which is the oxidizing agent and which is the reducing agent?

41. Explain why an electrochemical reaction between two different metal strips in an electrolytic solution stops after a period of time.

42. Draw a atomic-level view of a solution at the beginning, during, and after the electrochemical reaction between zinc and copper.

43. Could you stop the transfer of electrons between two different metals in a solution? Explain.

44. Compare the advantages and disadvantages of battery-driven electric cars.

45. Do the products of the oxidation of methanol (CH_3OH), methane (CH_4), octane (C_8H_{18}), and hydrogen gas (H_2) differ?

46. Explain how a hydrogen–oxygen fuel cell works.

47. Outline the advantages of using fuel cells to power vehicles and power plants.

48. An alkaline electrolyte can be used in the hydrogen–oxygen fuel cell to replace the acidic environment. Deduce the two balanced half-reactions and the balanced overall reaction for such a fuel cell. Assume that the reaction of O_2 with water and electrons produces hydroxide ions, OH^-, and that these hydroxides travel to the other electrode, where they react with hydrogen to give up electrons and produce more water. (This reaction is the basis for fuel cells that provide electricity in space vehicles.)

49. What would be the most practical storage system for hydrogen use in small vehicles? Explain your answer.

50. Are fuel cells really "pollution free"? Explain.

51. Use the resources listed at the end of the chapter to investigate and describe the processes by which hydrogen can be produced. Compare the advantages, disadvantages, and environmental impact of each process.

52. Explain how the density of hydrogen gas stored in a solid can be greater than the density of liquid hydrogen.

■ Group Activity: Electrochemical reactions

In this chapter, you have learned that a current will flow through an electrical conductor attached to two different metals, provided that ions can flow between the metals. In this activity, your group will discover whether or not electrochemical reactions occur between the different metals present in common coins, provided that a conducting solution (a salt bridge) is present between them.

For this activity, each group will need:

- a roll of paper towels
- salt
- water
- ten coins each of pennies, nickels, dimes, and quarters
- aluminum foil
- flat glass container such as a baking/casserole dish
- voltmeter or volt-ohmmeter that reads low voltages and low currents
- galvanized nail (optional)
- lemon (optional)

Procedure

Assign one member of the group to mix several teaspoons of salt with water in the glass dish, and stir until the salt has dissolved. Another group member should tear off a length of paper towel with a dozen sheets, and fold the sheets back and forth along the perforation so they alternate in direction. One member of the group will be responsible for setting up the coins in the necessary arrangement and one group member should record all observed voltage and current readings.

1. Soak the pile of paper towels in the salt solution until each sheet is thoroughly wet. Discard the excess solution.

2. Place a nickel between the first and second layers of paper towel sheets, and then place a penny on the top of the paper towel layer above the nickel. Touch the penny with one contact wire from the meter and the nickel from the other. Record the readings you obtain from the voltmeter, including current and voltage values.

3. Replace the penny with a dime and repeat the experiment. Record your results.

4. Repeat the experiment with various other combinations of coins and record the results.

5. Repeat the experiment using the same type of coin for both layers. Record your results.

6. Repeat the experiment using a sheet of aluminum foil rather than the nickel between the first and second layers. Record your results.

7. Using the entire length of paper towels, place pennies on the top layer and fill all the layers beneath with nickels. Record the current and voltage between the top and bottom layers of coins. Is it greater, the same, or less than with a single layer of each coin?

Results

1. What conclusions can you draw from the readings you have taken from the voltmeters with the different combinations of coins and metals?

2. Did you observe any trends?

3. Which combination of coins and metals gave the highest readings?

4. Which combination of coins and metals gave the lowest readings?

Optional activity

Cut the lemon in half. Insert the nail and a penny in separated parts of the exposed fruit. Record the current and voltage. Given that lemon juice contains an acid, why do you think that salt is unnecessary in the lemon cell?

Taking It Further I

Flashlight batteries come in several versions

In the common dry-cell flashlight battery, the zinc electrode is actually the metal container in which the other materials are present (see Figure 12.16a). As the battery is used to supply power, the zinc metal becomes oxidized, as discussed previously:

$$Zn\,(s) \rightarrow Zn^{2+} + 2\,e^-$$

Because the zinc container dissolves as the reaction proceeds, the batteries are prone to leaking when they become almost fully discharged.

Flashlight batteries are called *dry cells* because the electrolyte is a moist paste and not a liquid per se, although it does contain some water. The paste also contains the oxidizing agent, *manganese dioxide*, MnO_2, a substance that contains the Mn^{4+} ion. In the operation of the battery, this ion is reduced to Mn^{3+}, in the form of its compound $MnO(OH)$:

$$\underbrace{Mn^{4+}}_{\text{as } MnO_2} + e^- \rightarrow \underbrace{Mn^{3+}}_{\text{as } MnO(OH)}$$

This reduction is actually accomplished by the half-reaction:

$$MnO_2\,(s) + H_2O\,(aq) + e^- \rightarrow MnO(OH)\,(s) + OH^-$$

The electrode at which this half-reaction occurs is a graphite carbon rod, around which the paste containing manganese dioxide is present. The water comes from the electrolyte paste.

The electrolyte in this type of battery is the salt *ammonium chloride*, NH_4Cl. In fact, the ammonium ion, NH_4^+, acts not only as the electrolyte but also as a reactant, combining with the hydroxide ion that is produced in the reduction, generating ammonia, and re-forming the water consumed in the reduction half-reaction:

$$NH_4^+ + OH^- \rightarrow NH_3\,(g) + H_2O\,(aq)$$

The ammonia gas is reabsorbed into the electrolyte by combining with the zinc ion. Adding together the last two equations show us that the overall reduction half-reaction is:

$$MnO_2\,(s) + NH_4^+ + e^- \rightarrow MnO(OH)\,(s) + NH_3\,(g)$$

In the battery described above, the electrolyte is somewhat acidic since the ammonium ion is a weak acid. In the alternative alkaline version of this battery (see Figure 12.16b), the electrolyte is moist, concentrated *potassium hydroxide*, KOH. The hydroxide ion produced during the reduction half-reaction reacts with the zinc ion, Zn^{2+}, produced in the oxidation to form insoluble *zinc hydroxide*, $Zn(OH)_2$.

Alkaline flashlight batteries are more expensive, but they have several advantages. The zinc metal is in the form of fine granules suspended in a gel. The manganese dioxide is mixed with carbon powder and compressed into a cylindrical electrode. Since zinc granules are more reactive than strips of the metal, the alkaline battery delivers power more quickly than the dry-cell flashlight battery. In addition, the container is steel and does not participate in the reaction, so the cell is less prone to leaking. For these reasons, the alkaline version of the flashlight battery has a longer shelf life and provides several

In heavy-duty batteries capable of high current output, $ZnCl_2$ is used as the electrolyte.

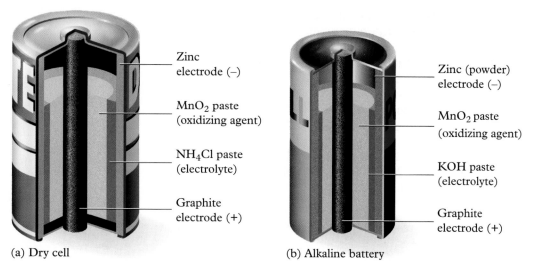

Zinc electrode (−)		Zinc (powder) electrode (−)	
MnO_2 paste (oxidizing agent)		MnO_2 paste (oxidizing agent)	
NH_4Cl paste (electrolyte)		KOH paste (electrolyte)	
Graphite electrode (+)		Graphite electrode (+)	

(a) Dry cell   (b) Alkaline battery

Figure 12.16 Two common types of battery. (a) A dry-cell battery. Dry cells produce ammonia gas as they discharge. (b) An alkaline battery. Alkaline batteries do not produce ammonia gas as they discharge and they have a longer shelf life.

times more electrical current in high-demand applications than the cheaper dry cell.

Currently under development is a "super iron" version of the alkaline battery, which replaces the MnO_2 component with a salt of the ferrate ion, FeO_4^{2-}. During discharge, the Fe^{6+} ion contained within the ferrate is converted to Fe^{3+}. This new battery would have the advantage of being lighter than current batteries, since the manganese oxide component of the current battery is relatively dense.

Worked Example: Deducing net overall redox reactions

What is the net overall reaction for the nonalkaline version of the flashlight battery?

Solution: First write out the two balanced half-reactions individually:

oxidation $Zn(s) \rightarrow Zn^{2+} + 2e^-$

reduction $MnO_2(s) + NH_4^+ + e^- \rightarrow MnO(OH)(s) + NH_3(g)$

Before combining the two half-reactions, we must ensure that the number of electrons lost in the oxidation process equals the number gained in the reduction, since no electrons can be created or destroyed in the process. In the half-reaction equations, two electrons are involved in the oxidation whereas only one is involved in the reduction. Consequently the reduction half-reaction must occur twice as often as the oxidation, so we should multiply all terms in it by 2 before combining it with the oxidation:

reduction $2[MnO_2(s) + NH_4^+ + e^- \rightarrow MnO(OH)(s) + NH_3(g)]$

When we add the left sides of the two half-reactions, and their two right sides, we obtain as a result:

$$Zn(s) + 2\,MnO_2(s) + 2\,NH_4^+ + 2\,e^- \rightarrow$$
$$Zn^{2+} + 2\,e^- + 2\,MnO(OH)(s) + 2\,NH_3(g)$$

After canceling the common $2\,e^-$ term, we obtain the net overall reaction for the cell:

$$Zn(s) + 2\,MnO_2(s) + 2\,NH_4^+ \rightarrow$$
$$Zn^{2+} + 2\,MnO(OH)(s) + 2\,NH_3(g)$$

Exercise 12.2

Deduce the net overall reaction that occurs in the alkaline flashlight battery.

1. What is a *dry cell*? Where are dry cells used?
2. Compare the alkaline and nonalkaline versions of dry-cell batteries.

Taking It Further II

Nicad is a rechargeable battery

The *nicad* battery, used in a variety of electronic equipment, is named after its reactive materials, nickel and cadmium. The oxidation involves the conversion of metallic cadmium, one of the electrodes, into cadmium ion, Cd^{2+}, in the form of insoluble *cadmium hydroxide*, $Cd(OH)_2$:

$$Cd(s) \rightarrow \underbrace{Cd^{2+}}_{\text{as } Cd(OH)_2} + 2\,e^-$$

This oxidation is actually accomplished by the half-reaction:

$$Cd(s) + 2\,OH^- \rightarrow Cd(OH)_2(s) + 2\,e^-$$

> The charge on the cation in hydroxides such as $Cd(OH)_2$ can be deduced from the facts that the charge on each hydroxide ion is -1 and that the metal charge must cancel the charge from all the hydroxides.

Although the reduction electrode is metallic nickel, it does not undergo reaction. In the half-reaction, Ni^{3+} is reduced to Ni^{2+}; these ions are present as their hydroxide compounds $Ni(OH)_3$ and $Ni(OH)_2$, respectively. The half-reaction is:

$$\underbrace{Ni(OH)_3(s)}_{\substack{\text{contains} \\ Ni^{3+}}} + e^- \rightarrow \underbrace{Ni(OH)_2(s)}_{\substack{\text{contains} \\ Ni^{2+}}} + OH^-$$

> Nickel–metal hydride batteries are similar to nicad batteries, except that the cadmium electrode is replaced by a metal alloy (M) that readily forms the metal hydride (MH). The oxidation reaction at the alloy is
> $$MH + OH^- \rightarrow M + H_2O + e^-$$

All three hydroxides involved in the reactions are insoluble and stick to the corresponding electrodes. Thus, if electricity is forced to pass between the electrodes in the direction *opposite* from that which

forced to flow in the direction opposite to that which occurs spontaneously, each of the half-reactions is forcibly reversed and the original reactants are readily re-formed. In other words, the nicad battery can be *recharged* once some or nearly all of the original reactants have been consumed and electricity generated. The discharge/recharge cycle can be repeated about a thousand times in nicad batteries. Just as the battery gives off energy when it is discharging, it requires energy to reverse the reaction.

1. What is a *nicad* battery? Where are nicads used?

2. How can a nicad battery be recharged?

Exercise 12.3

Deduce the overall reaction that occurs when the nicad battery is producing electricity. Write it also in reverse to show the process when the battery is being recharged.

Taking It Further III

The lithium-ion battery

The *lithium-ion battery* is a commercial battery that is useful in consumer electronic devices such as portable computers and cell phones, since lithium has a low density and so the battery is light. In making a rechargeable lithium battery, a challenge arose from the fact that elemental lithium metal is very reactive, especially with water, and thus would react with any moisture that might enter the battery. Therefore, it was necessary to devise a process in which the metal itself was never formed. The solution was to make a battery in which both the oxidizing and the reducing agents are lithium ions, Li^+, but are contained in different electrolytic environments so that they are chemically distinct and current will flow between them. One electrode consists of lithium ions packed between the layers of graphite. The other electrode consists of lithium ions present in a metal oxide such as CoO_2. The electrolyte is a solution of a Li^+ salt in an organic solvent. During discharge of the battery, lithium ions from the graphite electrode migrate to the metal oxide electrode and an electron flows in the same direction in the external circuit. In a sense, the material being oxidized is the negatively charged carbon, and that being reduced is the metal oxide.

Lithium-ion batteries are also the power source used in pacemakers, which maintain a steady heartbeat in people with heart disease. Before the development of lithium-ion batteries, the inability to hermetically seal pacemakers from the warm, salty environment in the body would make them stop working suddenly within 1 to 2 years of being implanted. The introduction of the more reliable and long-lived lithium batteries revolutionized pacemaker technology and extended the lives of many persons afflicted with heart disease.

1. Why isn't solid lithium used in batteries?

2. How does the design of a lithium-ion battery overcome the problem of forming highly reactive lithium metal?

Taking It Further IV

Lead-acid storage batteries

The battery used in cars, known as the *lead-acid storage battery,* is also rechargeable. Indeed, it is partially discharged every time the engine is started and becomes recharged automatically later when the engine is running since electricity is produced by the generator driven by the car's engine.

Both electrodes in this battery are elemental lead, Pb (see Figure 12.17). At the electrode that undergoes oxidation when the battery *produces* electricity, metallic Pb is oxidized to Pb^{2+} in the form of the insoluble salt $PbSO_4$, which clings to the electrode:

$$Pb\,(s) + SO_4^{2-} \rightarrow \underbrace{PbSO_4\,(s)}_{\text{contains } Pb^{2+}} + \; 2\;e^-$$

At the other electrode, Pb^{4+} in the form of PbO_2 is reduced to Pb^{2+}, again in the form of $PbSO_4$:

$$\underbrace{PbO_2\,(s)}_{\substack{\text{contains}\\ Pb^{4+}}} + SO_4^{2-} + 4\;H^+ + 2\;e^- \rightarrow \underbrace{PbSO_4\,(s)}_{\substack{\text{contains}\\ Pb^{2+}}} + \; 2\;H_2O$$

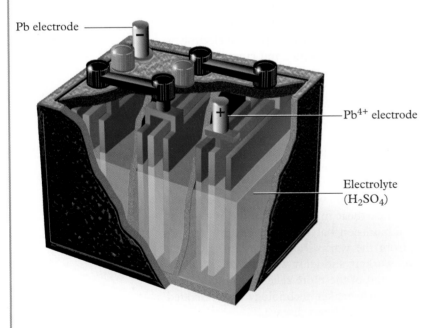

Figure 12.17 A lead-acid storage battery. Such batteries are used in many vehicles and for heavy equipment.

Chapter 12: Batteries, Fuel Cells, and the Hydrogen Economy

The net overall reaction, obtained by summing these two half-reactions, is:

$$Pb\,(s) + PbO_2\,(s) + 4\,H^+ + 2\,SO_4{}^{2-} \rightarrow 2\,PbSO_4\,(s) + 2\,H_2O$$

Sulfuric acid, H_2SO_4, is both the electrolyte *and* a reactant in the lead-acid storage battery. It is consumed in the battery's operation and also becomes diluted by the water that is a product of the reaction. However, the acid is reconstituted in concentrated form when the battery is recharged by reversing the overall reaction.

The lead-acid storage battery is not used just to start gasoline-powered vehicles. Versions of it are used as the power source for golf carts, electric wheelchairs, and forklift trucks. One of the disadvantages of this battery is its weight. The cell electrodes are made of a very dense element, lead, and six individual cells must be packed together to produce one 12-volt battery.

1. What is a *lead-acid storage battery*? Where are these batteries used?

2. What roles does sulfuric acid play in a lead-acid battery?

3. A hydrometer is an instrument that measures the density of a solution relative to pure water. What information could a hydrometer reading provide about the electrolyte fluid in a lead storage battery?

In this chapter, you will learn:

- about sources for drinking water;
- how drinking water is disinfected to make it safe to consume;
- which chemicals cause groundwater contamination;
- how surface water becomes polluted;
- how soaps and detergents work.

When you drink a glass of water or swim in a pool, how can you know that the water is clean and safe?

Pure water is only a small percentage of the total amount of water on Earth; the rest must be purified for our use. In this chapter we study water as an increasingly precious resource that is threatened by pollution and overuse. (Bob Daemmrich/Stock Boston/PictureQuest)

CHAPTER 13

Fit to Drink

Water Sources, Pollution, and Purification

> *"Water, water everywhere, nor any drop to drink."*
> —Samuel Coleridge,
> Rime of the Ancient Mariner (1798)

We all need fresh water to drink at least every few days or we will die. And the water must be clean or it will make us ill, sometimes fatally. Though you use only about seven liters of water for drinking and cooking purposes each day, you also bathe and wash clothes and dishes in it and flush toilets with it, so your total personal daily use of this resource amounts to about 300 liters, or about 100,000 liters per year (see Figure 13.1). Vastly larger amounts are also used on your behalf by industry in producing goods and services and by agriculture for irrigation. For example, thousands of liters of water are required to produce one kilogram of aluminum or cotton or beef or even rice.

Ironically, we seem to be surrounded by water—as streams, rivers, lakes and oceans—but we cannot drink much of it since it is impure or salty (see Table 13.1). The pollution of natural waters by both biological and chemical contaminants is a worldwide problem: few populated areas, whether in developed or developing countries, do not suffer from one form of water pollution or another. In this chapter we explore the sources and purification of water and some of the reasons that many natural waters are polluted.

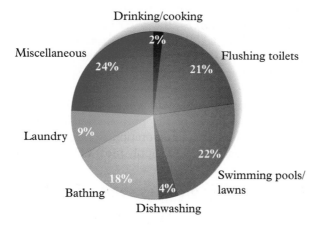

Figure 13.1 The ways in which a U.S. household uses water. Percentages indicate average daily water use per person.

 Whenever you see this icon in this chapter, go to
www.whfreeman.com/ciyl2e

Table 13.1 Earth's water resources

Source	Volume (thousands of cubic kilometers)	Percent of total
Total	1,403,477	100
Ocean	1,370,000	97.6
Ice and snow	29,000	2.07
Groundwater to 1 km	4,000	0.28
Lakes, reservoirs (fresh water)	125	0.009
Atmosphere	113	0.008
Saline lakes	104	0.007
Soil moisture	65	0.005
Plants	65	0.005
Swamps, marshes	3.6	0.003
Rivers, streams	1.7	0.0001

Adapted from W. P. Cunningham, M. A. Cunningham, and B. W. Saigo, "Environmental Science—A Global Concern," McGraw-Hill, New York, New York.

Drinking Water: Sources

13.1 Earth's fresh water

Although water covers more than two-thirds of Earth's surface, 98% of it is salt water and cannot be used directly for drinking, irrigation, or most industrial purposes. Unfortunately, most of the 2% of Earth's water that is fresh is currently inaccessible, since it occurs as glaciers or ice caps, is far underground, or is heavily polluted. Currently, about one-third of the world's population lives in countries that already experience some shortage of fresh water; this fraction will probably rise to two-thirds by 2025 (see Figure 13.2).

Surface fresh waters, the water that resides on Earth's surface in rivers, streams, and lakes, are important as a major source of drinking water. They are used also as habitats for the plant and animal life they contain and as sources of recreation and transportation. Surface water is produced by precipitation and by the melting of glaciers. The water in lakes, streams, and rivers accounts for only 0.01% of the total water on Earth; about two-thirds of the precipitation that falls returns to the atmosphere through direct evaporation or through plants. Nevertheless, surface fresh water is used to supply most of us with drinking water.

13.2 Groundwater provides some of Earth's water supply

The great majority of the available fresh water lies underground, half of it at depths exceeding one kilometer. **Groundwater** is the name given to the fresh water that lies underground; that which lies within one

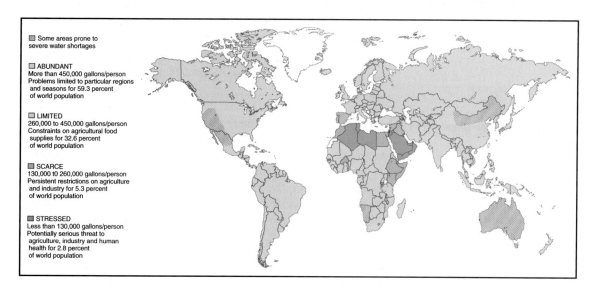

Figure 13.2 Estimated annual water availability in 2025. (From "Making Every Drop Count," by Peter H. Glieck. © 2001 by *Scientific American.* All rights reserved. February 2001, p. 43.)

kilometer of the surface makes up 0.3% of the world's total water supply (see Table 13.1). The ultimate source of groundwater is precipitation that falls onto the surface.

Some groundwater is contained in layers that are composed of porous rocks such as sandstone, or in highly fractured rock such as gravel or sand. If such water is bounded at its lower depths by a layer of clay or impervious rocks, then it constitutes a permanent reservoir—a sort of underground lake—called an **aquifer.**

Groundwater has traditionally been considered a pure form of water. Because of the filtration through soil and its long time isolated underground, it normally contains much less natural organic matter and many fewer disease-causing microorganisms than does water from lakes or rivers. However, some groundwater is naturally too salty or too acidic for either drinking or irrigation purposes, and may contain too many sodium, sulfide, or iron ions to be used.

Groundwater in aquifers can be extracted by wells and is the main supply of drinking water for almost half the population of North America. In the United States, groundwater supplies about 39% of the water used for public supplies. Many U.S. cities, including Miami, Memphis, San Antonio, and Tucson, rely almost entirely on groundwater for their drinking water.

In the United States, particularly in the western states, much of the groundwater is used for irrigation purposes. The massive extraction of water from American aquifers has given rise to fears about future supplies of fresh water (and about the sinking of land above the aquifers), since such aquifers are replenished only very slowly. For example, the Ogallala aquifer of the southern U.S. would need several thousand years to be replenished, and given its current rate of use, will be essentially empty in

Drinking Water: Sources

a few years. In addition, the contamination of groundwater by chemicals is becoming a serious concern in many areas, as we shall see later.

Drinking Water: Removal of Ions, Gases, Solids, and Organic Compounds

The quality of raw (untreated) water, whether surface water or groundwater, that is intended eventually for drinking varies widely, from almost pristine to highly polluted. Because both the type and quantity of pollutants and of natural substances that are considered undesirable in drinking water vary, the processes used in purifying raw water also differ from place to place. Figure 13.3 shows schematically the most commonly used procedures for treating relatively unpolluted raw water, in the order in which they generally occur; refer to this figure as you read the text descriptions that follow. Keep in mind, however, that not all water undergoes all these steps; which ones are used will depend largely on the quality of the starting water. Since the bulk of water purified by these techniques will not be used for drinking but for washing, etc., some of the purification steps address these other needs.

13.3 Aeration removes gases

Aeration is commonly used in the improvement of water quality. Municipalities aerate drinking water, especially that which is drawn from aquifers, in order to remove dissolved gases such as the foul-smelling hydrogen sulfide, H_2S, and organic compounds containing sulfur, plus volatile organic compounds, some of which have a detectable odor (Figure 13.3, step ❶). Aeration also produces reactions that

Figure 13.3 Step by Step The steps by which raw water is made drinkable.

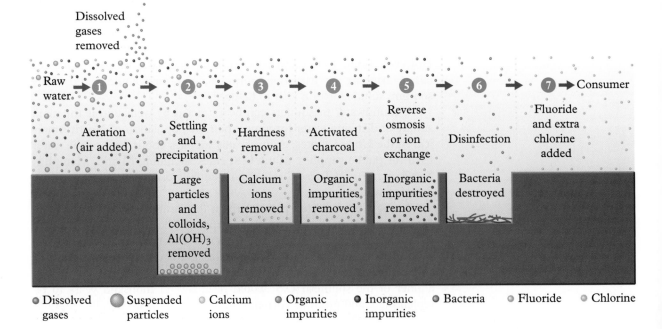

convert the most easily oxidized organic material to CO_2, thus eliminating these organics from the water. As we shall see later, most of the remaining organics can be removed by subsequently passing the water over activated carbon, although this process is relatively expensive.

13.4 Settling and precipitation remove particles

Surface waters inevitably contain particles and colloidal material, the result of its passing over rock beds and of the existence of various life forms within it. The colloids contain inorganic materials such as *silica*, SiO_2, from the weathering of rocks, and organic materials, including bacteria, associated with the decomposition of animal matter.

A micrometer, symbolized μm, is 10^{-6} meters.

Most municipalities allow raw water to settle, which permits large particles (> 1 μm in diameter) to settle out or to be readily filtered (Figure 13.3, step ❷). However, much of the insoluble matter will *not* precipitate spontaneously since it is suspended in water in the form of colloidal particles. Recall from Chapter 1, section 1.8, that colloidal particles have diameters ranging from 1 to 1000 nanometers, that is, 0.001 to 1 μm. In water they consist of *groups* of molecules or ions that are weakly bound together and that exist as a unit instead of breaking up and dissolving as individual ions or molecules. The surfaces of inorganic colloidal particles often carry a negative electric charge arising from ionic groups. The charges on the surface of one particle repel like charges on neighboring particles, preventing aggregation and subsequent precipitation of the particles.

The colloidal material in drinking water is removed for both aesthetic and health reasons. The water is greatly clarified, and therefore more aesthetically pleasing, once this colloidal material has been removed. To capture the charged colloidal particles, *aluminum sulfate*, $Al_2(SO_4)_3$, is added to the water. If the pH is subsequently increased to neutral or alkaline pH values (7 and higher) by the addition of *lime*, $Ca(OH)_2$, the aluminum is converted to its insoluble, gelatinous hydroxide $Al(OH)_3$. For simple aluminum ions, Al^{3+}, the reaction is:

$$Al^{3+} + 3\ OH^- \rightarrow Al(OH)_3\,(s)$$
$$\uparrow$$
$$\text{from } Ca(OH)_2$$

The Al-containing cations and the hydroxide form an insoluble network structure that attracts and entraps the suspended particles in a gel-like material dispersed in the water. Much of the aluminum-containing solid, which has incorporated the charged colloidal particles, slowly precipitates to the bottom of the water tank. The remainder is removed by filtering the water through sand.

13.5 Removing calcium and magnesium ions

As we saw in Chapter 10, raw water from areas with limestone bedrock will contain significant levels of Ca^{2+} and Mg^{2+} ions and is called *hard water.* Calcium can be removed from such water by the addition of phosphate ion, since calcium and phosphate ions combine to form an insoluble salt, which precipitates (Figure 13.3, step ❸). More commonly, calcium ion is removed by precipitation and filtering of the insoluble salt $CaCO_3$. The carbonate ion that is required to combine with the calcium is either added as *sodium carbonate,* Na_2CO_3, or if sufficient HCO_3^- is naturally present in the water, hydroxide ion can be added in order to convert the bicarbonate ion to carbonate:

$$OH^- \quad + \quad \underset{\text{bicarbonate ion}}{HCO_3^-} \quad \rightarrow \quad \underset{\text{carbonate ion}}{CO_3^{2-}} \quad + \quad H_2O$$

followed by precipitation:

$$Ca^{2+} + CO_3^{2-} \rightarrow CaCO_3(s)$$

Magnesium ion precipitates as *magnesium hydroxide,* $Mg(OH)_2$, when the water is made sufficiently alkaline. Water that contains very little calcium or magnesium, or that has had these materials removed from it, is said to be *soft.*

13.6 Activated carbon removes organic compounds

Many municipalities, and even some individual households, pass their drinking water through granulated charcoal filters in order to remove small concentrations of pollutants that remain after water has been treated by conventional means (Figure 13.3, step ❹ and Figure 13.4). The material usually used for this purpose is **activated carbon,** sometimes called **activated charcoal** (discussed in Chapter 5, section 5.20). This material is granulated charcoal—with grains about 1 millimeter in diameter—that has been treated with steam. Charcoal has a low density due to its highly porous nature: its structure is like that of a sponge, though the holes are too small to be seen with the naked eye. Because of this porosity, it has a very large surface area, and so it can adsorb considerable amounts of other substances (see Figure 13.5).

The pollutants are adsorbed from the water onto the surface of the carbon particles, and so are removed from water as it passes over them, thereby improving its taste, odor, and purity. Activated carbon is particularly useful in removing dissolved organic molecules from water, including pesticides and chlorinated solvents. Given sufficient contact time, it does remove the chloroform that is produced during the chlorination stage of water purification (Figure 13.3, step ❻), though the removal is probably incomplete if only small tap filters are used.

Although the total surface area of the activated carbon particles is very large because of its many holes and internal channels (see Figure 13.5), eventually the surfaces become covered with adsorbed molecules. The charcoal unit is then discarded, in the case of small domestic filters. In larger commercial units, the charcoal can be incinerated, or regenerated by heating it to a high temperature in air and steam, a process that oxidizes the adsorbed molecules.

Figure 13.4 Activated charcoal. The activated charcoal column purifies water just like a home filtration system. (Joel Gordon Photography)

Recall from Chapter 5 that an adsorbed substance is one that is weakly attached to a surface.

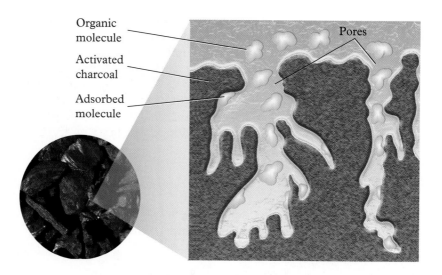

Figure 13.5 Activated charcoal. Contaminants in water are adsorbed onto the surface of the charcoal particles. (Photo by George Semple for W. H. Freeman and Company)

Organic molecule

Activated charcoal

Adsorbed molecule

Pores

▪ Chemistry in Your Home: Charcoal purification

You can examine the purification properties of charcoal using a charcoal barbecue briquette or charcoal used in fish tanks. Grind a briquette or place some fish tank charcoal in a jar. Mix several drops of red, yellow, and blue food coloring with a small amount of water. Keep some of the food coloring solution apart for comparison, and add some of it to the jar with the charcoal. Cover the jar and shake it vigorously to mix the solution and charcoal pieces. Allow the jar to sit so that settling occurs. Compare the color of the solution with the charcoal to that of the solution without the charcoal. Is there a difference? Were certain colors removed better than others? Which colors were not filtered? Which colors were effectively filtered?

13.7 Reverse osmosis removes ions

Water can be purified of most contaminant ions, molecules, and small particles, including viruses and bacteria, by passing it through a membrane in which the individual holes, called *pores*, are of uniform and microscopic size. To be effective in providing a barrier, the pore size of the membrane must be smaller than the contaminant size.

The ultimate in membrane filtration occurs in the widely used technique called **reverse osmosis** (Figure 13.3, step ⑤). Here, water is forced under high pressure to pass through a *semipermeable membrane,* a very thin sheet of an organic plastic, such as cellulose acetate. Since only water can pass through the pores, and ions or most other contaminants cannot, the liquid on the other side of the membrane is pure water (Figure 13.6). The solution on the contaminant-trapping side of the membrane becomes more and more concentrated in contaminants as time goes on, and is discarded.

All particles, molecules (including even small organic molecules) and ions down to less than 0.001 μm in size are removed by reverse

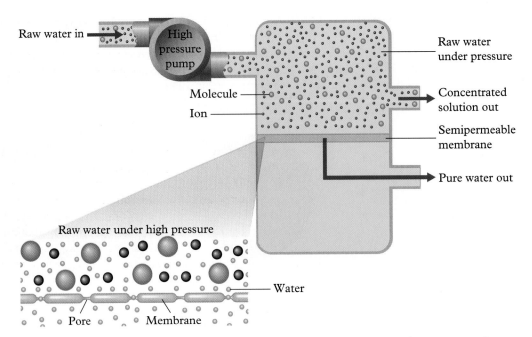

Figure 13.6 Reverse osmosis. Applying pressure to a membrane removes ions from water.

Labels in figure:
Raw water in — High pressure pump — Raw water under pressure — Molecule — Ion — Concentrated solution out — Semipermeable membrane — Pure water out — Raw water under high pressure — Water — Pore — Membrane

osmosis. It is particularly useful for removing alkali and alkaline earth metal ions, as well as the salts of heavy metals. Thus it is useful in hospitals and kidney dialysis units for producing water that is particularly free of ions.

The procedure is called *reverse* osmosis because, by use of pressure, the natural phenomenon of osmosis—by which pure water would migrate through the membrane *into* the solution, thereby diluting it—is reversed. Because energy is used to produce the pressure that reverses this natural tendency, reverse osmosis is an expensive procedure. Water destined for drinking purposes is commonly pretreated by passing it over activated carbon to remove most of the impurities before the semi-purified liquid is treated by reverse osmosis.

Some domestic consumers of drinking water have installed small under-the-sink reverse osmosis units to further purify their drinking water, since the technique is particularly efficient at removing heavy metal ions (such as those of lead), anions (such as nitrate and fluoride), brackish water ions (sodium and chloride), and hard water ions (calcium and magnesium). However, the amount of water discarded in reverse osmosis is greater than the amount that passes through the membrane, which can make its cost prohibitive. Reverse osmosis does not necessarily purify water of all organic and biological contaminants, since some are molecules small enough to pass through the membrane.

Reverse osmosis is also used for **desalination,** the production of fresh water from salty seawater by the removal of ions. The process is used in large- and medium-scale installations in Florida, Israel, Malta,

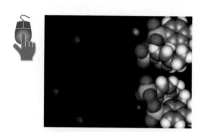

View the action of calcium and magnesium ions on an ion exchange column at Chapter 13: Visualizations: Media Link 1.

Saudi Arabia, and elsewhere, but it is too expensive for widespread use in developing countries. Small reverse osmosis units are used on boats to desalinate seawater and in medical facilities for producing water that is particularly free of ions. Desalination currently supplies only about 0.1% of current fresh water.

13.8 Ion exchange changes the ions in water

The concentration of ions in hard water—Ca^{2+} and Mg^{2+}, and also often ions of iron—can be high enough to cause difficulties when the water is used for washing, especially clothes, for reasons that are explained in section 13.24. The difficulties are eliminated if these ions are *replaced* by ions such as sodium. **Ion exchange** is a process by which water is passed through an apparatus containing a supply of sodium or other ions. The calcium, magnesium, and iron ions are exchanged for sodium ions, so that the water exiting the apparatus is no longer hard but rather is soft. Ion exchange can also remove metal ions that are pollutants, such as lead. In combination with activated carbon, it is one of the two water-purifying processes used in most filter systems on drinking water taps.

The *resin*, the material in which the process takes place, is a natural or synthetic solid that contains many sites that hold ions relatively weakly, so one type of ion can be exchanged for another of the same charge sign that happens to pass by it (see Figure 13.7). Most commonly, the resin initially contains sodium ions, Na^+, that are held by ionic bonding to anions (A) that are components of polymeric structures. When a cation such as calcium passes by, it can displace two nearby sodium ions. The $+2$ calcium ion, being more highly charged, is attracted more strongly to the resin anions and remains with the resin. The sodium ions are released into the water solution traveling along the solid resin:

$$Na_2A\,(s) \,+\, Ca^{2+}\,(aq) \rightarrow CaA\,(s) \,+\, 2\,Na^+\,(aq)$$

In this way, hard water entering the ion exchange apparatus is converted to soft water exiting it.

Of course, the ability of the ion exchange apparatus to soften water depends on most of its cation sites being occupied by sodium ions, not calcium, magnesium, or iron ions. For domestic water purifiers with disposable filters, this is not a problem, provided that the unit is not used for too long a time before the filter is replaced. However, if a resin is to be used continuously, it must be regenerated from time to time with sodium ions. This is usually accomplished by passing a very concentrated solution of sodium chloride through the apparatus. Due to the continuous stream of Na^+ ions passing each cation site, hard water ions are eventually replaced by two sodium ions. The resin is then ready to be reused. The salty solution that exits the apparatus when recharge is occurring contains all the calcium, magnesium, and iron ions that had previously been removed from the water that was filtered and must be discarded.

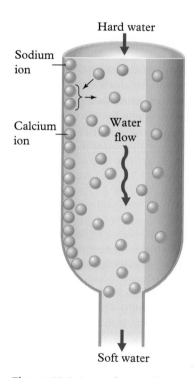

Hard water

Sodium ion

Calcium ion

Water flow

Soft water

Figure 13.7 Ion exchange. The resin in an ion exchange column attracts calcium ions and releases sodium ions.

Figure 13.8 Solar stills in Mexico. Solar stills can be used for desalination. (Gerald & Buff Corsi/Visuals Unlimited)

13.9 Water can be deionized by distillation

A technique used to deionize water is **distillation,** in which the raw water is boiled and the steam collected and condensed to what is called *distilled water.* The ions do not evaporate with the water that becomes steam but stay in the remaining liquid. Consequently, the water obtained by condensing the steam in a separate container is ion-free. The remaining liquid becomes more and more concentrated in ions and is eventually discarded. Because so much energy is required to boil the water, this method of producing deionized water is quite expensive.

In areas of the world with plentiful sunshine and land, but only salt water, a **solar still** can be used to desalinate water by distillation (see Figure 13.8). Here the water is not boiled but heated sufficiently by the Sun's rays that it evaporates rapidly. The humid air containing the evaporated water rises, and the water recondenses on the sloping glass surfaces and runs down to troughs, where it is collected and then transported or stored for later use.

● Tying Concepts Together: Arsenic in drinking water

The presence of significant levels of arsenic, As, in drinking water supplies is a significant and controversial environmental issue. Although arsenic has been used for millenia as a poison, the major health problem stemming from its presence at low levels in drinking water is cancer: arsenic is known to cause lung, skin, and liver cancer.

Drinking water, especially that derived from groundwater, is a major source of arsenic for many people. Another source is food, especially meat and seafood. In some cases, high arsenic levels in water occur naturally, whereas in other cases they stem from the use of arsenic in pesticides; from its unintended release during the mining and smelting of gold, lead, copper, and nickel; and from the combustion of coal, of which it is a contaminant.

The World Health Organization has set 10 ppb as the acceptable limit for arsenic in drinking water. About 1% of Americans currently consume drinking water that has arsenic levels of 25 ppb or more. In the last days of the Clinton administration in 2000, the U.S. maximum allowable limit for arsenic in drinking water was lowered from 50 ppb to 10 ppb. Although the Bush administration at first withdrew this regulation, they later concluded that it was warranted and it will become law, effective 2006.

Like calcium and magnesium, arsenic can be removed from drinking water by precipitating it in the form of one of its insoluble salts. The arsenic in water normally exists as the arsenate ion, AsO_4^{3-}. Since the salt formed between the ferric ion, Fe^{3+}, and arsenate is insoluble, the soluble salt *ferric chloride*, $FeCl_3$, is dissolved in the water, and the precipitated ferric arsenate is filtered from the resulting mixture:

$$Fe^{3+} + AsO_4^{3-} \rightarrow FeAsO_4(s)$$

Arsenic cannot be removed from water by cation exchange, since it occurs as an anion, not a cation.

> Recall from Chapter 10 that *ppb* means *parts per billion,* the number of grams of the pollutant present in one billion, 10^9, grams of water.

Drinking Water: Disinfection by Chlorination

Most of the treatments we have described so far do not rid water of harmful bacteria and viruses, especially those arising from both human and animal fecal matter. Therefore, the **disinfection** of water intended for drinking and other human uses is usually necessary (Figure 13.3, step ⑥). A variety of alternative techniques, ranging from chemical ones such as *chlorination* and *ozonation* to physical ones such as ultraviolet irradiation, are discussed in this section and the next. On a small scale, an alternative when none of these techniques can be used or is reliable is to boil the water for several minutes, since this kills the pathogens, though this energy-intensive procedure is not feasible on a mass scale. We begin with the most common method of disinfection, chlorination.

> Because of the lack of disinfected drinking water, water-borne diseases are the major cause of sickness and death in many developing countries, accounting for tens of thousands of death per *day*.

13.10 The disinfecting agent in chlorination is hypochlorous acid

The most common water purification process used in North America is **chlorination.** The active purification agent in chlorination is *hypochlorous acid*, HOCl. About half of the U.S. population uses surface water, and one-quarter of the population uses groundwater, that is disinfected by HOCl. This neutral, covalent compound kills microorganisms, readily passing through their cell membranes. In addition to being effective, disinfection by chlorination is relatively inexpensive. Chlorination is more common than ozonation in North America than elsewhere because generally the water is less polluted to begin with.

Hypochlorous acid is not stable in concentrated form, so it cannot be stored. For large-scale installations—municipal water treatment plants—it is generated by dissolving *molecular chlorine* gas, Cl_2, in water. At moderate pH values, the equilibrium in the reaction of chlorine with water is achieved in a few seconds:

$$Cl_2(g) + H_2O(aq) \rightleftharpoons \underset{\text{hypochlorous acid}}{HOCl(aq)} + H^+ + Cl^-$$

A dilute aqueous solution prepared by dissolving chlorine gas in water contains very little aqueous Cl_2, most of the chlorine having been converted to HOCl and Cl^-.

13.11 Chlorination in swimming pools

Water in swimming pools must be continuously disinfected, or microorganisms quickly build up to dangerous levels. In small-scale applications of chlorination such as domestic swimming pools, the handling of cylinders of Cl_2 is inconvenient and dangerous. The chlorine can be produced as needed on the spot by the electrolysis of salty water, which converts Cl^- to Cl_2. More commonly, hypochlorous acid is generated from the salt *calcium hypochlorite*, $Ca(OCl)_2$ (see Figure 13.9), or is supplied as an aqueous solution of *sodium hypochlorite*, NaOCl. In water, an acid–base reaction occurs to convert most hypochlorite ion, OCl^-, to the corresponding weak acid HOCl:

$$\underset{\text{hypochlorite ion}}{OCl^-} + H_2O \rightleftharpoons \underset{\text{hypochlorous acid}}{HOCl} + OH^-$$

Because of its electrical charge, hypochlorite ion is less able than is hypochlorous acid to penetrate bacteria and kill them. As a consequence, close control of the ratio of hypochlorous acid to hypochlorite is necessary so that a high level of HOCl is maintained. To do this, the pH must be closely controlled, since the equilibrium between HOCl and OCl^- shifts rapidly in favor of the ion between pH values of 7 and 9. Thus, to maximize the amount of hypochlorous acid, it is beneficial for water to be somewhat acidic. On the other hand, acidic water can corrode the pool construction materials. As a compromise between these two opposing requirements, the pH in swimming pools is usually kept just above 7. Swimming pool acidity can be adjusted by the addition of *sodium bisulfate*, $NaHSO_4$, which contains the acidic anion HSO_4^-, or a base such as sodium carbonate, Na_2CO_3. A solution of $NaHCO_3$, which contains the anion HCO_3^- that neutralizes either excess acid or excess base, may also be used.

Maintenance of an alkaline pH in swimming pool water also prevents the conversion of dissolved ammonia, NH_3, to the *chloramine* compounds NH_2Cl, $NHCl_2$, and especially NCl_3, which is a powerful eye irritant:

Aqueous solutions of calcium hypochlorite are sold in stores as chlorine bleach.

Figure 13.9 Chemicals that chlorinate this home swimming pool are delivered by a floating dispenser. (George Semple for W. H. Freeman and Company)

$$\underset{\text{ammonia}}{NH_3} \quad + \quad \underset{\text{hypochlorous acid}}{3 \; HOCl} \quad \rightarrow \quad \underset{\text{nitrogen trichloride}}{NCl_3} \quad + \quad 3 \; H_2O$$

As we shall see later, smaller concentrations of chloramines in drinking water are desirable to provide residual protection during storage and transmission.

Chlorine must be constantly replenished in outdoor pools since the ultraviolet light in sunshine is absorbed by and decomposes the hypochlorite ion:

$$\underset{\text{hypochlorite ion}}{2 \; OCl^-} \quad \xrightarrow{\text{ultraviolet light}} \quad \underset{\text{chloride ion}}{2 \; Cl^-} \quad + \quad O_2$$

If hypochlorite is destroyed, more will form from HOCl, decreasing the concentration of the active form of chlorine in water.

13.12 Chlorination has some drawbacks

Since HOCl is a chlorinating agent, a significant drawback to the use of chlorination in disinfecting water is the inevitable simultaneous production of chlorinated organic substances, some of which are toxic.

If the raw water contains phenol or a phenol derivative, chlorine atoms readily substitute for the hydrogen atoms on the ring, giving rise to chlorinated phenols. Not only do these compounds have an offensive odor and taste, but they are toxic as well. Some communities switch from chlorine to chlorine dioxide when their supply of raw water is temporarily contaminated with phenols.

A more general problem with chlorination of water is the production of _trihalomethanes_ (THMs), whose general formula is CHX_3, where the three X atoms are chlorine, bromine, or a combination of the two. The compound of most concern is _chloroform,_ $CHCl_3$, which is produced when hypochlorous acid reacts with organic matter that is dissolved in the water:

$$HOCl + \text{organic matter} \rightarrow CHCl_3$$

The presence of chloroform, even at very low levels of approximately 30 ppb, raises the fear that chlorinated drinking water may pose a health hazard. Currently, the limit of THMs in drinking water in the United States is set at 80 ppb by the Environmental Protection Agency (EPA), compared to that of 100 ppb in Canada and that specified by the World Health Organization. The 80–100 ppb limit is set to regulate not only the THM chemicals themselves but also indirectly the production of _other_ chlorinated organic by-products. As of the early 1990s, about 1% of the larger U.S. drinking water utilities that used surface waters had average THM levels exceeding 100 ppb.

The THM and chlorinated organic content of chlorinated water could be decreased by using activated carbon to remove dissolved organic com-

Phenol molecules contain a benzene ring bonded to an —OH group.

pounds before the water is chlorinated. Activated carbon can also be used to remove THMs and other chlorinated organics after chlorination, though THMs are not very efficiently adsorbed by the carbon. Because using activated carbon is an expensive process, it is often restricted to large cities.

Chloroform is a suspected liver carcinogen in humans, and it may also have negative reproductive and developmental effects. A recent analysis of epidemiological studies concluded the risk of bladder cancer in humans increased by 21%, and that of rectal cancer by 38%, for Americans who had drunk chlorinated surface water in the past. A similar study in Ontario found even higher risk factors for bladder cancer among people who drank water for 35 years or more with THM levels greater than 50 ppb, and for colon cancer when the concentration exceeded 75 ppb.

In addition to chloroform, several other chlorinated organic compounds formed during chlorination that cause DNA mutations and could initiate cancer have been detected in water. It is not clear whether the main carcinogen in the chlorinated drinking water is THM itself or some other chlorination by-product present in trace amounts. If the amount of this other substance produced is always proportional to the amount of THMs produced, cancer rates would still be expected to correlate with THM levels even though it was not THM itself that produced the effect. A recent study showed that exposure to chloroform by skin contact and inhalation of gases that evaporate from the hot water during showers and baths contributes about as much to one's intake of THMs as does drinking the water itself.

The same health risks do *not* usually apply to chlorinated well water, since its chlorinated organic content is much less—only 0.8 ppb on average, versus 51 ppb for surface water. The concentration of organic by-products is much lower because well water contains much smaller amounts of organic matter in the first place.

Because of the risks of cancer, some communities have switched to water disinfection by ozone or chlorine dioxide, since these agents produce little or no chloroform. The extent of chlorination has already been reduced in most American communities.

13.13 Residual chlorine protects water in storage

An advantage chlorination has over disinfection by other methods is that some chlorine remains dissolved in water after it has left the purification plant. Thus the water is protected from subsequent bacterial contamination before it is consumed. Indeed, some chlorine is usually added to water purified by the other methods to provide this protection (Figure 13.3, step ⑦). There is very little danger of significant chloroform production in the purified water since its organic content has been virtually eliminated.

Residual chlorine in water is usually converted to the chloramines NH_2Cl, $NHCl_2$, and NCl_3, which, as previously discussed, are produced from the reaction of hypochlorous acid with dissolved ammonia gas. Although not as fast as HOCl in disinfecting water, such **combined chlorine** is longer lived and thus provides longer residual protection. Ammonia is added, in a 3:1 ratio, to purified drinking water in order to convert the residual chlorine to the combined form, in accordance with the reaction equation on page 495.

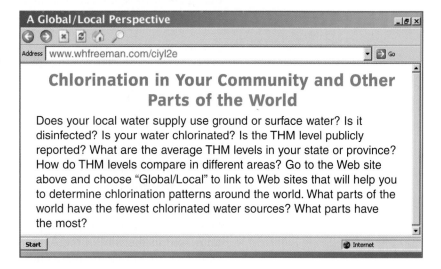

A Global/Local Perspective

Address www.whfreeman.com/ciyl2e

Chlorination in Your Community and Other Parts of the World

Does your local water supply use ground or surface water? Is it disinfected? Is your water chlorinated? Is the THM level publicly reported? What are the average THM levels in your state or province? How do THM levels compare in different areas? Go to the Web site above and choose "Global/Local" to link to Web sites that will help you to determine chlorination patterns around the world. What parts of the world have the fewest chlorinated water sources? What parts have the most?

Start | Internet

Drinking Water: Disinfection by Methods Other Than Chlorination

13.14 Ozone is used to disinfect some water supplies

In some localities, particularly in France and other parts of Western Europe but also in some North American cities—Montreal and Los Angeles are examples—ozone is used to disinfect drinking water. *Ozone,* O_3, has a very short lifetime, so it cannot be stored or shipped. It must be generated on-site by a relatively expensive process involving electrical discharge (20,000 volts) in dry air. The resulting ozone-laden air is bubbled through the water; about 10 minutes of contact is sufficient. Ozone is more effective than hypochlorous acid at killing waterborne viruses. However, since the lifetime of ozone molecules is short, there is no residual protection in the purified water against future contamination during storage and transmission.

Ozone has other drawbacks as well. If any organic matter is present in water, ozone may react with it and form tiny amounts of partially oxidized organic compounds, such as formaldehyde and other small aldehydes, some of which are toxic.

13.15 Chlorine dioxide can disinfect water

Chlorine dioxide gas, ClO_2, is used in more than 300 North American and several thousand European communities to disinfect water. Since chlorine dioxide is *not* a chlorinating agent—that is, it does not generally introduce chlorine atoms into the substances with which it reacts—and since it oxidizes the dissolved organic matter, much smaller amounts of toxic organic chemical by-products are formed than if chlorine is used. As is the case with ozone, ClO_2 cannot be stored. Chlorine dioxide is explosive in the high concentrations that its practical use calls for, and so it must be generated on-site. This is accomplished by using an oxidizing agent to remove the extra electron in the corresponding chlorite ion, ClO_2^-, supplied, for example, as the salt *sodium chlorite:*

$$ClO_2^- \quad \rightarrow \quad ClO_2 \quad + \quad e^-$$

$$\text{chlorite ion} \qquad \text{chlorine dioxide}$$

Some of the chlorine dioxide in these processes is converted to ClO_2^- and ClO_3^- (chlorate) ions; the presence of these species in the final water has raised health concerns.

13.16 Irradiation with ultraviolet light eliminates toxic organisms

Ultraviolet light can be used to disinfect and purify water. Powerful lamps containing mercury vapor, which emit ultraviolet light, are immersed in the water flow. About ten seconds of irradiation is usually sufficient to eliminate toxic microorganisms, including the parasite *Cryptosporidium* that survives chlorination. The germicidal action of the light arises from the disruption of the DNA in the microorganism such that its subsequent replication is impossible and the cell is inactivated.

An advantage of ultraviolet disinfection technology is that it is not restricted to large installations. Small units can be employed to serve small population bases or even individual wells, both in developed and developing countries.

> ▶ **Discussion Point:** Should chlorination be phased out?

Develop arguments both for and against the phasing out of chlorination and its replacement by chlorine dioxide, ozonation, or ultraviolet treatment of drinking water. For example, are all forms of disinfection equally effective? What are their health advantages and disadvantages? How do their costs compare? Would you still allow some chlorine to be added to the disinfected water to provide continuing protection during water storage and transmission?

13.17 The advantages of disinfecting drinking water

The disinfection of water by chemical methods has been one of the most beneficial applications of chemistry to public life. Disinfection is extremely important in protecting public health and saves countless more lives—by a very large factor—than are affected negatively by it, even when chlorination by-products are considered. For example, both typhoid and cholera were widespread in Europe and North America a century ago, but they have been almost completely eradicated in the developed world thanks largely to chlorination and the other methods for disinfecting drinking water. (Improved sanitation in general has also helped.) The same is not true in many developing countries. For example, in the early 1990s there were more than half a million cases of cholera in Peru, where the water

Figure 13.10 A community water source in Côte d'Ivoire, West Africa. Water supplies in developing countries are often unpurified.
(Charles O. Cecil/Visuals Unlimited)

was not chlorinated, partially because of fears of chlorination by-products. Overall, about 20 million people, most of them infants, die each year from waterborne diseases in underdeveloped countries, where water purification is often erratic or even nonexistent (see Figure 13.10). Under no circumstances should effective disinfection of water be abandoned because of concern for the by-products of chlorination!

Water contamination problems are not limited to developing countries, however. The pathogen Cryptosporidium was responsible for the death of 50 people and for hundreds of thousands of cases of diarrhea in Milwaukee in 1993. There were also outbreaks in Oxford, England, in 1989 and North Battleford, Saskatchewan, in 2001. This deadly parasite is largely resistant to standard methods of disinfection, such as chlorination and chlorine dioxide, and is so small—less than 5 μm in diameter—that it easily passes through standard filters. The most effective disinfectants against Cryptosporidium, and against microorganisms in general, are ozonation and ultraviolet treatment.

In addition, disinfection equipment can break down and, if not monitored properly, can result in nondisinfected water being distributed to the public. This happened in 2000 in the town of Walkerton, Ontario, when chlorination equipment failed. A deadly bacteria was then spread by the municipal water supply, resulting in sickness for hundreds of people and seven deaths.

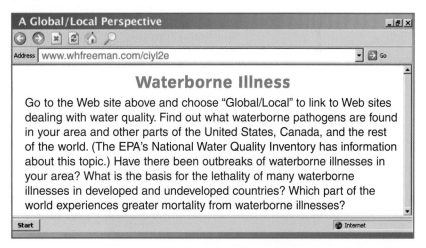

A Global/Local Perspective

Address www.whfreeman.com/ciyl2e

Waterborne Illness

Go to the Web site above and choose "Global/Local" to link to Web sites dealing with water quality. Find out what waterborne pathogens are found in your area and other parts of the United States, Canada, and the rest of the world. (The EPA's National Water Quality Inventory has information about this topic.) Have there been outbreaks of waterborne illnesses in your area? What is the basis for the lethality of many waterborne illnesses in developed and undeveloped countries? Which part of the world experiences greater mortality from waterborne illnesses?

Groundwater Pollution by Organic Compounds

Chemical contamination of groundwater was not recognized as a serious environmental problem until the 1980s. Groundwater contamination was neglected because it was not immediately visible—it was "out-of-sight, out-of-mind"—even though groundwater is a major source of drinking water. We ignored the long-range consequences of our waste-disposal practices, which had been contaminating groundwater for half a century. Unfortunately, though surface water can be cleaned up relatively easily and quickly, groundwater pollution is a much harder, much more expensive long-range problem to solve.

13.18 Organic compounds can leach into the soil

The contamination of groundwater by organic chemicals is now a major concern. Liquid containing dissolved matter that drains (leaches) from a terrestrial source is called a **leachate.** Municipal landfills (formerly called garbage dumps) and industrial waste-disposal sites are often the source of the contaminants. In many rural areas, shallow aquifers have been contaminated by organic pesticides, such as weedkillers leached from the surface. However, the typical organic contaminants in most major groundwater supplies are (1) trihalomethanes; (2) chlorinated solvents, especially *trichloroethene*, C_2HCl_3, and *perchloroethene*, C_2Cl_4; and (3) BTX hydrocarbons benzene, toluene, and xylene from gasoline.

These chemicals are commonly found in groundwater at sites where manufacturing and/or waste disposal occurred, especially from 1940 to 1980. In that period, little attention was paid to the ultimate fate of the organic chemicals that had entered the ground from leaking chemical waste dumps, leaking underground gasoline storage tanks, leaking municipal landfills, and accidental chemical spills. In fact, many substances do decay rapidly or are immobilized in the soil. Consequently, the number of compounds that are sufficiently persistent and mobile to eventually contaminate groundwater is relatively small.

Gasoline enters the soil via surface spills, leakage from underground storage tanks, and pipeline ruptures. Before 1980, underground gasoline storage tanks were made from steel; almost half of them were sufficiently corroded to leak by the time they were 15 years old (see Figure 13.11). Once they descend to groundwater, the water-soluble components of gasoline are leached into the water and can migrate rapidly in the dissolved state. The BTX component often occurs at 1–50 ppb concentrations in groundwater. However, molecules such as toluene and xylene that have methyl or ethyl side chains on the benzene rings are rapidly degraded by aerobic bacteria and consequently are not long-lasting.

The MTBE component of gasoline (discussed in Chapter 6) is more soluble than are the hydrocarbons but is not readily biodegraded. However, it is not highly toxic, its main problem being the odor and taste that it gives to the water. MTBE contamination of well water, albeit at low levels, has become a concern in the United States.

The BTX fraction of gasoline was discussed in Chapter 4, section 4.20. Chlorinated solvents were discussed in Chapter 5, section 5.3.

Figure 13.11 Gasoline tanks which had corroded while underground.
(Courtesy of Office of Underground Storage Tanks, EPA)

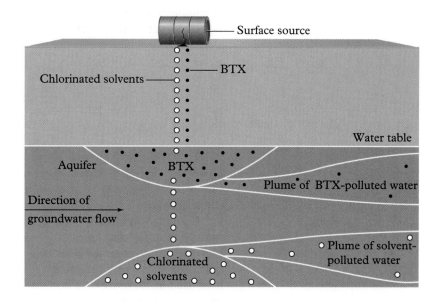

Figure 13.12 Pollution from organic compounds. Organic contaminants pollute aquifers through movement of toxic plumes. (Source: C. Baird, *Environmental Chemistry*, 2nd ed., W. H. Freeman and Company, 1999.)

The subsequent behavior of the organic compounds that do migrate to the water table depends largely on their density relative to the density of water, which is 1.0 g/mL. Liquids that are *less* dense ("lighter") than water form a mass that floats on the top of the water table. All small and medium-sized hydrocarbons belong to this group, including the BTX fraction of gasolines and other petroleum products. In contrast, polychlorinated solvents are *more* dense ("heavier") than water, and so they tend to sink deeply into aquifers; important examples are methylene chloride, chloroform, carbon tetrachloride, trichloroethene, and perchloroethene. These compounds also share the properties of being persistent in soil and being slightly soluble in water. The nonchlorinated traditional organic mixtures called *creosote* and *coal tar* also belong to the heavier-than-water group.

Although the oily liquid blobs that water-insoluble organic compounds form are generally found in an aquifer either directly below their original point of entry into the soil or close to it, they are capable of some horizontal motion. In a process that often takes decades or centuries to complete, these compounds very gradually dissolve in the water that passes over the blob, and so provide a continuous supply of contaminants to the groundwater. Thus plumes of polluted water grow, in the direction of the water's flow, contaminating the bulk of the aquifer (see Figure 13.12). Many wells have had to be closed because of such contamination.

13.19 Cleaning up groundwater contamination

In the last two decades, considerable energy and money have been spent in the United States on attempts to control aquifer pollution by organic compounds. Unfortunately, no easy cure to the problem of contamination of groundwater has yet been found. Control usually consists of

pump-and-treat systems which pump contaminated water from the aquifer, treat it to remove its organic contaminants, and return the cleaned water to the aquifer or to some other water body. The volume of water that must be pumped and treated in a given aquifer is huge. Recontamination of water returned to the aquifer by additional dissolution from the blob will occur. Consequently, these treatment systems must operate in perpetuity, and thousands of them are already spread across the United States.

Bioremediation is the term applied to the decontamination of water or soil using biological rather than strictly chemical or physical processes. Recently, there has been interesting progress reported in using bioremediation to cleanse water of chlorinated ethene solvent contamination.

Owing to the high cost and limited effectiveness of many groundwater cleanup technologies, the inexpensive process of natural attenuation—allowing natural biological, chemical, and physical processes to treat groundwater contaminants—has become popular. Indeed, it is now used at more than 25% of the Superfund program's contaminated sites in the United States and is the leading method to remedy the contamination of groundwater from leaking underground storage sites.

Because we are now aware of the consequences—including high remediation costs—of the uncontrolled disposal of organic wastes, most large corporations in developed countries have become much more responsible in their disposal of chemicals. Unfortunately, the collective discharges from small sources—including many municipalities as well as small industries and farms—have not yet been controlled in like manner. Similarly, the huge number of existing septic tanks are collectively a major source of nitrate, bacteria, viruses, detergents, and household cleaners to groundwater.

Water Pollution by Nitrogen Compounds

13.20 Nitrogen fertilizers cause serious pollution

The three most important elements that plants receive from soil are nitrogen, phosphorus, and potassium; all three are taken in through the roots as ions.

Potassium occurs in soil as the simple K^+ ion, and phosphorus as phosphate ion, PO_4^{3-}. For nitrogen, the situation is more complicated. If nitrate ion, NO_3^-, is available in the soil, plants will absorb it as their source of nitrogen, though some plants can also absorb ammonium ion, NH_4^+. However, not all soil contains enough NO_3^-, so farmers use commercial fertilizers to supply the nitrogen necessary for crop growth. The nitrate ion that occurs naturally as *sodium nitrate*, $NaNO_3$, is mined from deposits in Chile and can be used as fertilizer added to soil. However, nitrogen now is often supplied in fertilizers for farms as liquid ammonia, NH_3, or for home garden use as a salt of the corresponding ammonium ion. One important fertilizer is *ammonium nitrate*, NH_4NO_3, which contains nitrogen in two forms, both ammonium ion and nitrate ion.

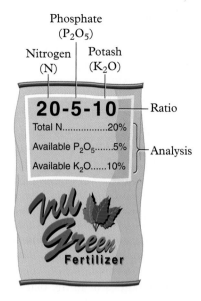

Phosphate
(P$_2$O$_5$)

Nitrogen
(N)

Potash
(K$_2$O)

20-5-10 ——— Ratio

Total N.................20%

Available P$_2$O$_5$.......5%   }——— Analysis

Available K$_2$O......10%

Mil Green
Fertilizer

Figure 13.13 A bag of commercial fertilizer.

"Slow release" nitrogen supplies the element as an organic compound such as *urea*, OC(NH$_2$)$_2$, that slowly decomposes in the soil and thereby slowly releases nitrogen in a form that can be absorbed by plants. Homeowners and golf course operators use high-nitrogen fertilizers to obtain vigorous growth of grass on lawns.

The packages of commercial fertilizers available to home gardeners and farmers list the concentrations of the three elements N, P, and K, usually in that order. Regardless of the chemical form in which the elements are supplied, the latter two are calculated as if they were composed of P$_2$O$_5$ and K$_2$O respectively, even though they are not (see Figure 13.13). Thus *10-20-15* fertilizer contains 10% nitrogen, as much phosphorus *as if* 20% of the contents were P$_2$O$_5$, and as much potassium *as if* the contents contained 15% K$_2$O. *Bone meal* fertilizer is powdered animal bones, which, like teeth, are mainly composed of Ca$_5$(PO$_4$)$_3$OH. In starting or transplanting plants, a fertilizer that is high in phosphorus content usually is applied since vigorous root growth, with its heavy phosphorus demand, is desired at that time. Later, when green leaf growth is more important, fertilizers that are highest in nitrogen are applied.

13.21 Nitrate is a contaminant of well water

One *inorganic* groundwater contaminant of potential human health concern, especially to babies, is the nitrate ion, which commonly is found in both rural and suburban aquifers. Concern has been expressed about the increasing nitrate ion levels in drinking water, particularly in rural well water. About 9% of shallow aquifers—from which well water is often extracted—in the United States now have nitrate levels that exceed the EPA's 10 ppm nitrogen limit for drinking water. U.S. public water supplies rarely exceed this limit, partially because they are drawn from deeper aquifers. Deep aquifers are generally less contaminated because of their depth, because they are remote from major sources of contamination, and because their water can undergo natural denitrification in the low-oxygen conditions that occur there. Water contaminated with high levels of nitrate normally is not used for human consumption, at least not in public supplies, because nitrate removal is expensive.

Nitrate in groundwater originates mainly from four sources:

- Runoff of nitrogen fertilizers, both inorganic ones and animal manure, applied to cropland
- Atmospheric deposition
- Human sewage deposited in septic systems
- Cultivation of the soil

The main source of nitrate is runoff from agricultural lands into rivers and streams. In most cases, the original forms of nitrogen become oxidized in the soil to nitrate, which then migrates down to the groundwater where it dissolves in water and is diluted. Years ago, oxidized animal wastes (manure) and unabsorbed ammonium nitrate and other nitrogen fertilizers were thought to be the main culprits. These materials can provide the bulk of the nitrogen. However, it is now known that the intensive cultivation of land, even *without* the application of fertilizer or manure, provides aeration and moisture, which facilitate the oxidation of nitrogen compounds into nitrate in decomposed organic matter in the soil.

Areas having not only a high nitrogen input but also well-drained soils and little woodland that could intercept the fertilizers draining from cropland have the greatest risk for nitrate contamination of groundwater. In urban areas, the use of nitrogen fertilizers on lawns and golf courses is a significant source of the ion.

13.22 Nitrate has potential health effects

Excess nitrate ion in drinking water is a potential health hazard. It can result in **methemoglobinemia,** or *blue baby syndrome,* in newborn infants, as well as in adults with a particular enzyme deficiency. The pathological process, in brief, runs as follows.

Bacteria—for example, in unsterilized baby bottles or in the baby's stomach—convert some of the nitrate ion to nitrite ion, NO_2^-:

$$\overset{\text{bacteria}}{NO_3^- \;\; \rightarrow \;\; NO_2^-}$$
$$\text{nitrate ion} \qquad \text{nitrite ion}$$

The nitrite combines with the hemoglobin in blood and prevents the proper absorption and transfer of oxygen to the body's cells. The baby turns blue and suffers respiratory failure. In almost all adults, the oxidized hemoglobin is readily converted back to its oxygen-carrying form and the nitrite is readily converted back to nitrate. In addition, nitrate is mainly absorbed in the digestive tract of adults before reduction to nitrite can occur. The occurrence of methemoglobinemia is now relatively rare in industrialized countries, but it still occurs in some developing countries.

Excess nitrate ion in drinking water is also of concern because of its potential link with stomach cancer. Although recent investigations have failed to establish a solid relationship between nitrate levels in drinking water and the incidence of stomach cancer, a study reported in 2001 found that the risk of bladder cancer among women in Iowa who drank water from municipal supplies (having elevated nitrate levels) were almost three times as likely to be diagnosed with bladder cancer than those who drank from supplies having the lowest nitrate levels.

Surface Water Pollution by Phosphates: Soaps and Detergents

Beyond its use for drinking, for cooking, and in toilets, one of our principal uses of water domestically is for washing—of our bodies, our clothes, our dishes, our cars, and so on. We inevitably use a soap or detergent with water to accomplish the washing. As we'll see below, the type of water that is used has a great effect on how the cleansing agents

Soap molecules tend to form clusters, with the hydrophobic parts pointing inward and the ionic portions pointing outward, even in water that contains no grease.

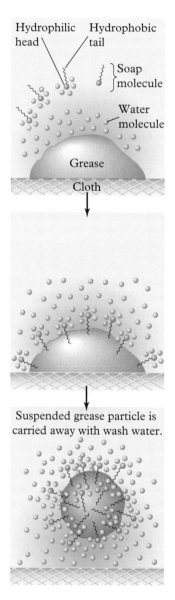

Hydrophilic head   Hydrophobic tail

} Soap molecule

Water molecule

Grease

Cloth

Suspended grease particle is carried away with wash water.

Figure 13.14 Soap molecules interacting with water molecules and grease molecules to remove dirt.

work. In turn, the phosphate ion in these materials often causes water pollution. To understand all these complications, we first need to investigate the chemical nature of soaps and detergents.

13.23 The hydrophobic portion of soap attracts grease

Soap is a substance containing the sodium (or potassium) salts of long-chain fatty acids.

$$H-\overset{\overset{\displaystyle H}{|}}{\underset{\underset{\displaystyle H}{|}}{C}}-\overset{\overset{\displaystyle H}{|}}{\underset{\underset{\displaystyle H}{|}}{C}}-\overset{\overset{\displaystyle H}{|}}{\underset{\underset{\displaystyle H}{|}}{C}}-\overset{\overset{\displaystyle H}{|}}{\underset{\underset{\displaystyle H}{|}}{C}}-\overset{\overset{\displaystyle H}{|}}{\underset{\underset{\displaystyle H}{|}}{C}}-\overset{\overset{\displaystyle H}{|}}{\underset{\underset{\displaystyle H}{|}}{C}}-\overset{\overset{\displaystyle H}{|}}{\underset{\underset{\displaystyle H}{|}}{C}}-\overset{\overset{\displaystyle H}{|}}{\underset{\underset{\displaystyle H}{|}}{C}}-\overset{\overset{\displaystyle H}{|}}{\underset{\underset{\displaystyle H}{|}}{C}}-\overset{\overset{\displaystyle H}{|}}{\underset{\underset{\displaystyle H}{|}}{C}}-\overset{\overset{\displaystyle H}{|}}{\underset{\underset{\displaystyle H}{|}}{C}}-\overset{\overset{\displaystyle O}{||}}{C}-O^- \, Na^+$$

Thus its structure consists of a hydrocarbon component, which terminates not in a —COOH group as it does in an acid, but rather in a COO⁻ ionic group that in the solid forms an ionic bond with the Na⁺ or K⁺ ion. A fatty acid molecule itself would not be very soluble in water since its behavior is dominated by its long hydrophobic chain of carbon atoms.

The mechanism by which soap in water removes grease and grease-containing dirt is based on this dual structure. The long carbon chain of a soap molecule attracts other hydrophobic substances, such as droplets of oil and grease, that might be suspended in the water or present at the surface of cloth. The carbon chains attract and break up the grease particles and then cluster together in the water, trapping the droplets of grease in the interior of the cluster (see Figure 13.14). The hydrophilic, ionic portions of the soap molecules are located at the outer portion of the cluster, where they are stabilized by the water molecules. The soap-and-grease clusters form an emulsion with the water and can be rinsed away, leaving clean clothes, dishes, or skin.

13.24 Soap is manufactured from fats

The preparation of soap from natural materials was discovered at least as long ago as Roman times—indeed, soap was a commercial product in the Middle Ages. The source of the fatty acid in soap was animal fat, which, as we saw in Chapter 7, section 7.17, consists of molecules that are a triester of glycerol with three fatty acids. The fat can be decomposed back into glycerol and free fatty acids by heating. Nowadays, high-temperature, high-pressure water is used to split the fat into fatty acids and glycerol. Fat can also broken down more quickly by reacting it with a basic substance in addition to heating it. The naturally available base is a mixture of sodium and potassium carbonate, a dilute solution of which can be obtained by mixing water with wood ashes. The presence of the base also converts the resulting freed fatty acid into its sodium or potassium salt, which corresponds to the soap:

$$\text{fat} + \text{base} \rightarrow \text{glycerol} + \text{fatty acid}$$
$$\text{fatty acid} + \text{base} \rightarrow \text{soap}$$

The ancient Romans initially used soap only for medicinal purposes rather than as a cleaning agent.

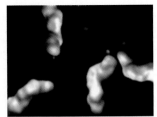

View the structure of a soap molecule and the formation of soap scum at Chapter 13: Visualizations: Media Link 2.

Salt is then added, and the soap and glycerol separate from the rest of the alkaline solution. The soap separates from the more dense glycerol, which sinks to the bottom of the watery reaction mixture. The aqueous mixture can be distilled to recover the glycerol, a valuable by-product of the soap-making process. Soaps made in the old days contained some of the unreacted base and consequently were corrosive to the skin.

The same overall process of producing soap from fat, *saponification,* is still used today, although the base used is a pure one. Most—though not all—of the glycerol is removed from the product before it is sold. Most modern soaps are sodium salts of specific fatty acid mixtures. The *soft soaps* (semisolids) used in shaving lather and liquid soaps are mainly potassium salts. Hard soaps that are denser than water are made to float by blowing air into them before the liquid mass is allowed to solidify. Most modern soaps have other substances added to them during manufacture, such as perfumes, oils, dyes, creams, and even abrasives in some cases. Some soaps have free fatty acids added to them to make them especially mild.

Unfortunately, calcium and magnesium ions that are dissolved in water form insoluble salts with the anions of soap molecules, giving rise to the production of a slimy, gray, gummy scum or curd in washwater. Scum is objectionable not only because it results in a deposit on the object being cleaned. This conversion by calcium and magnesium of the soap into a substance that does not clean also prevents the removal of the dirt from materials being washed.

Recall that water is said to be hard if it contains substantial concentrations of calcium and/or magnesium ions; consequently, hard water deactivates soap and increases soap scum production. Many people prefer soft water, which contains little Ca^{2+} and Mg^{2+}, for washing their hair since no residues from the soap will remain. People used to collect rainwater and use it specifically for washing hair and clothes because it is soft water. Rainwater is formed by the recondensation of water that has evaporated from Earth's surface; any ions present in the surface water do not evaporate along with the H_2O molecules and so do not become part of the rainwater. We have already seen in section 13.8 how hard water can be artificially softened—for example, by ion exchange.

13.25 Synthetic detergents

To solve the problem of cleaning in hard water, in the 1950s chemists developed **synthetic detergents** that clean like soap but do *not* form insoluble precipitates (scum) in the presence of Ca^{2+} and Mg^{2+} ions. Instead, the detergent molecules bind to the ions and form *soluble* substances. Most synthetic detergents are chemically similar to soap in that they consist of sodium ions ionically bonded to anions that have a large organic hydrophobic component to which grease and oil are attracted. Synthetic detergents differ from soap in that they contain the sulfonate group $-SO_3^-$, rather than the carboxylate group $-COO^-$, bonded to the organic portion. In today's detergents, the hydrocarbon portion that forms the hydrophobic tail consists of a benzene ring to which is also bonded a long, unbranched chain of 12 carbon atoms. These synthetic cleaners are called *linear alkylsulfonate detergents* and have the structure shown below:

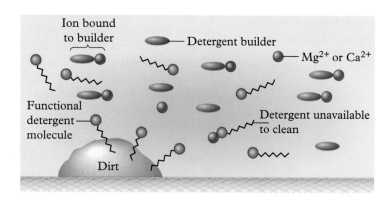

In the original synthetic detergents, the chains attached to the benzene ring were branched rather than unbranched ones. Such detergents are very stable and do not *biodegrade*—that is, break down into simpler substances with the aid of microorganisms present in the environment—as readily as those having unbranched chains do. In response to environmental problems such as foaming water resulting from buildup of synthetic detergents in rivers and lakes, detergents containing branched chains were replaced in the 1960s by those with unbranched, "linear" ones.

13.26 Detergent builders enhance cleaning power

In reacting with calcium and magnesium ions, some of the detergent is used up. For this reason, detergent manufacturers add to their formulations inexpensive substances called **builders.** The main function of a builder is to strongly bind the calcium and magnesium ion by forming soluble substances in the water, and thereby allow more of the detergent to operate as a cleansing agent (see Figure 13.15). The most effective builders are salts containing polyphosphate ions, which are anions containing several phosphate units linked by shared oxygens. Another role of the builder is to make the washwater somewhat alkaline. Thus they themselves have some cleansing action since the hydroxide ion they produce helps to solubilize grease and so lift it from the surface to which it is attached.

Another common additive to detergents is a **brightener.** Brighteners are chemicals that absorb ultraviolet rays from sunlight and artificial light, and re-emit them as visible blue light. This emission makes clothes look brighter, and white areas less yellowed.

Figure 13.15 The action of detergent builders.

Originally, *sodium tripolyphosphate* (STP), $Na_5P_3O_{10}$, was used as the builder in most synthetic detergent formulations:

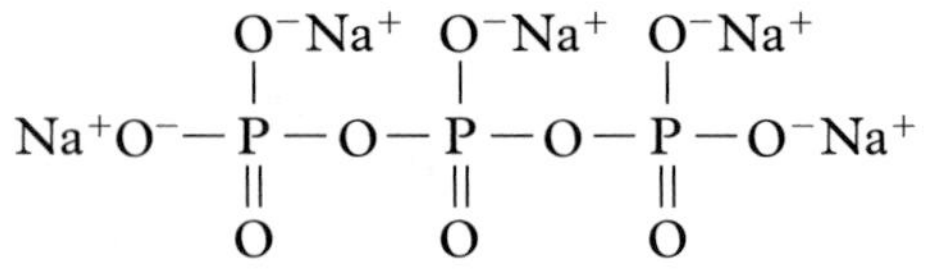

Sodium tripolyphosphate, STP

As shown above, STP contains a chain of alternating phosphorus and oxygen atoms, with one or two additional oxygens attached to each phosphorus. In solution, one tripolyphosphate ion can bond to one calcium ion by forming interactions (shown below as dashed lines) between three of its oxygen atoms and the metal ion:

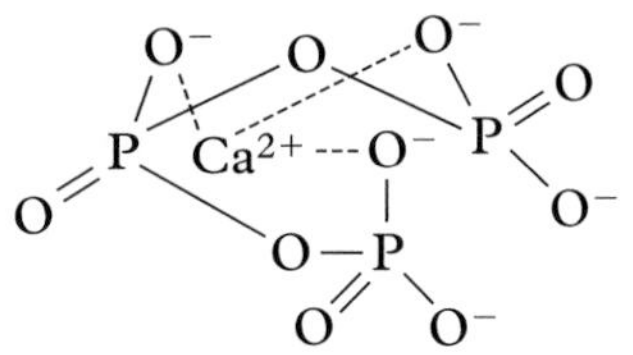

STP binding a calcium ion

Chelate is derived from the Greek word meaning "claw," which is appropriate considering the action of these substances.

Substances such as STP that have more than one site of attachment to the metal ion, thereby producing ring structures that each incorporate the metal, are called **chelating agents.** Because several bonds are formed, chelates are very stable and do not normally release their metal ions back into a free form.

When washwater containing STP is discarded, the excess polyphosphate enters waterways, where it slowly reacts with water and is transformed into phosphate ion, PO_4^{3-}:

$$P_3O_{10}^{5-} \quad + \quad 2\,H_2O \quad \rightarrow \quad 3\,PO_4^{3-} \quad + \quad 4\,H^+$$

tripolyphosphate ion   phosphate ion

Most detergent formulations sold for use in automatic dishwashers still contain polyphosphate.

Because of environmental concerns we discuss below, phosphates are now used only sparingly as builders in detergents in many areas of the world. Builders that now are used include sodium citrate, sodium carbonate (washing soda), and sodium silicate.

Substances called *zeolites* are also employed as detergent builders. Zeolites are minerals consisting of sodium, aluminum, silicon, and oxygen. In the presence of calcium ion, they exchange their sodium ions for Ca^{2+}. Detergents containing zeolites list *aluminosilicate* as an ingredient. One disadvantage to zeolites is that they are insoluble, so their use increases the amount of wastewater sludge that must be removed at wastewater treatment plants.

You can determine if your water is relatively hard or soft by examining the suds produced when detergent or soap is added to the water. Put about two teaspoons of your water into two different jars. Add one drop of detergent to one jar and one drop or a small amount of solid soap to the other jar. Cover the jars and shake vigorously. Hard water produces very few suds when detergent is added and produces a noticeable scum in the presence of soap. What do your jars look like? How high are the suds levels? Do you think your water is relatively hard or relatively soft? If you want to compare your results to true hard and soft water, repeat the above steps with distilled water (soft) and water to which 1 teaspoon of Epsom salts has been added (hard).

13.27 Phosphates from point and nonpoint sources

One of the world's most famous cases of water pollution involves Lake Erie, which in the 1960s was said to be dying. Indeed, the author of this book can recall visiting a once-popular beach on Erie's north shore in the early 1970s and being repulsed by the sight and smell of dead, rotting fish on the shoreline. Lake Erie's problems stemmed primarily from an excess input of phosphate ion in the waters of its tributaries.

The phosphate sources that affected Lake Erie, and in general are still the main sources of pollution to waterways in many parts of the world, are the polyphosphates in detergents, in raw sewage, and in the runoff from farms that use phosphate fertilizers. Phosphate ion enters waterways from both point and nonpoint sources (see Figure 13.16). **Point sources** are specific sites such as towns, cities, and factories that individually discharge a large quantity of a pollutant. For example, some municipalities are point sources of phosphates if they discharge untreated sewage into a waterway, since phosphates are a component of human wastes. **Nonpoint sources**

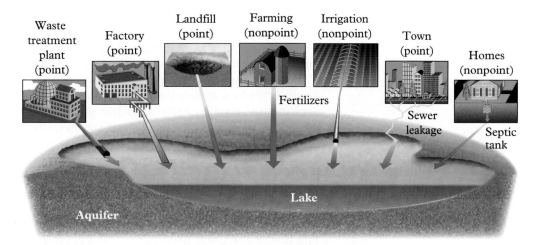

Figure 13.16 Point and nonpoint sources of pollution.

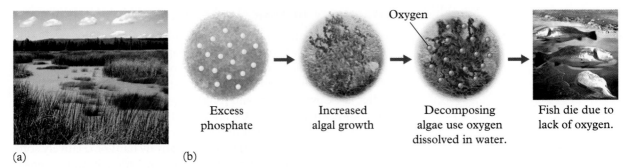

(a)

(b)

Oxygen

Excess phosphate

Increased algal growth

Decomposing algae use oxygen dissolved in water.

Fish die due to lack of oxygen.

Figure 13.17 Phosphate pollution. (a) An algal bloom resulting from excessive phosphate in the water. (b) The process by which excess phosphate destroys organisms in a lake. (Part a, Sylvester Allred/Fundamental Photographs)

are the numerous entities such as farms, home septic tanks, and housing developments and golf courses (both of which contribute fertilizers and pesticides). A nonpoint source provides a much smaller amount of pollution than a point source. In the case of farms, phosphates come from runoff water that contains animal wastes, fertilizers, and milkhouse washwater (produced when phosphate-rich detergents are used to clean milking equipment). However, because of the large number of locations involved, nonpoint sources can generate greater total amounts of pollution than point sources. Indeed, now that sewage treatment plants and detergent controls have been instituted around the Great Lakes, for example, much of the remaining phosphate pollution comes from nonpoint agricultural sources.

The oversupply of phosphates can be an environmental problem for freshwater lakes and rivers. Lakes commonly contain an excess of other dissolved nutrients, so phosphate ion usually functions as the limiting (or controlling) nutrient for algal growth (see Figure 13.17a). The *limiting nutrient* is the substance which runs out first and which therefore limits the growth of the plant. The larger the supply of the phosphate ion, the more abundant the growth of algae, and its growth can be quite abundant indeed. When such a vast amount of excess algae eventually dies and starts to decompose by consuming oxygen gas dissolved in the water, the water becomes depleted of dissolved oxygen, and fish and other aquatic organisms die (see Figure 13.17b). The lake water also becomes foul-tasting, green, and slimy, and masses of dead fish and aquatic weeds rot on the beaches (see Figure 13.18).

To correct the existing phosphate problem in the Great Lakes area, more than $8 billion has been spent in building sewage treatment plants to remove phosphates from wastewater before it reaches the tributaries and the lakes themselves. In addition, the levels of phosphates in laundry detergents were restricted in Ontario and in many of the states that border the Great Lakes. The total amount of phosphorus entering Lake Erie has now decreased by more than two-thirds. As a result, Lake Erie has sprung back to life: its once fouled beaches are regaining popularity with tourists and its commercial fisheries have been revived.

(a)

(b)

Figure 13.18 (a) Lake Erie during its polluted phase in the 1970s. (b) Lake Erie after cleanup. (Part a, Jane Forsyth Stone Laboratory, courtesy of Ohio Sea Grant; part b, courtesy of Ohio Sea Grant)

▶ **Discussion Point:** Should phosphates be put back into detergents?

Some policymakers believe that the optimum solution to pollution by detergents is to reintroduce the use of polyphosphates, thereby replacing the other builders, and then to remove phosphate at wastewater treatment plants. In this way, excessive amounts of other builders would not enter the environment. This solution would work in urban areas and other municipalities in which wastewater is collected and processed. Phosphate ion would be removed from wastewater by the addition of sufficient calcium ion so that insoluble calcium phosphates would be formed as precipitates that could then be readily removed. Do you agree with this plan to reintroduce phosphates into detergents or not? List two arguments in its favor, and two against the idea.

Summarizing the Main Ideas

Our sources of fresh drinking water are primarily drawn from surface waters and from groundwater. Aquifers, supplies of groundwater contained in porous rock and bounded beneath by impervious material, are one of the main sources of useful water.

Most drinking water is treated before it can be safely consumed. Gases and volatile organic compounds can be removed from raw water by aeration. Colloidal particles do not spontaneously gather together and precipitate since they are electrically charged and repel each other. Substances such as aluminum hydroxide that form voluminous precipitates are used to capture the colloidal particles and remove them, clarifying the water.

Hard water contains substantial amounts of Ca^{2+} and Mg^{2+}. These ions can be removed from raw water by adding anions to form insoluble compounds, such as calcium phosphate, calcium carbonate, and magnesium hydroxide, which then precipitate.

Once the water has been chemically treated, small residual quantities of dissolved organic compounds can be removed by passing the raw water over activated carbon. The many grooves and channels in its porous structure give this material a huge surface area, which enables it to adsorb many organic molecules.

The removal of ions from raw water can be accomplished by the process of reverse osmosis, in which H_2O molecules are forced under pressure to pass through a membrane containing very small holes. Ions can also be removed by distillation. Hard water and pollutant ions can be replaced in water by sodium and/or chloride ions by ion exchange.

The disinfection of water involves the destruction of toxic microorganisms dissolved or suspended in it. The most common technique for disinfection is chlorination, in which the active agent is HOCl. This substance is produced in water by bubbling Cl_2 gas into it or by dissolving a salt of the acid in it. In either case, the pH of the water must be controlled to optimize the relative concentration of HOCl over other chlorine-containing substances. The residual chlorine in the water is combined with added ammonia to form chloramines, which impart some residual protection of the water during transmission and storage against reinfection. A drawback to chlorination is the formation of the by-products chloroform and other trihalomethanes, and possibly other carcinogenic chlorine-containing organic compounds, in the water by the reaction of HOCl with dissolved organic matter.

Disinfection can also be also accomplished by passing ozone or chlorine dioxide through the water or by irradiating it with ultraviolet light.

Groundwater can also be seriously polluted by organic compounds that enter the soil at the surface and that are sufficiently stable to survive migration downwards. Organic compounds less dense than water float near the top of aquifers in blobs, which slowly and continuously contaminate the passing water. Organics that are more dense than water form a blob near the bottom of the aquifer and similarly contaminate the passing water.

In soil, nitrogen fertilizers and other nitrogen compounds are converted into nitrate ion, which then can pass through the soil into groundwater supplies and contaminate them.

Soaps are the sodium or potassium salts of fatty acids. They contain a long hydrocarbon-like chain to which a COO^- group is attached. Calcium and magnesium salts of the fatty acid anions are insoluble in water and form gummy deposits when soap is dissolved in hard water. Synthetic detergents combine a hydrocarbon portion, including a benzene ring, with a sulfonate group. They form soluble salts with calcium and magnesium ion, and have replaced soaps for many cleaning operations. Builders such as polyphosphate ion salts or zeolites are usually added to detergents to form soluble salts with the calcium and magnesium ions, leaving detergent molecules available for cleaning purposes.

Excessive use of phosphates in detergents, and other sources of this ion, led in the 1970s to serious water pollution problems. Phosphates have both been largely eliminated from detergents and removed in wastewater treatment plants to overcome this problem.

Key Terms

surface fresh waters	desalination	leachate	builder
groundwater	ion exchange	pump-and-treat	chelating agent
aquifer	distillation	bioremediation	point source
aeration	solar still	methemoglobinemia	nonpoint source
activated carbon	disinfection	soap	
activated charcoal	chlorination	saponification	
reverse osmosis	combined chlorine	synthetic detergent	

Web Sites of Interest

To link to Web sites of interest, go to www.whfreeman.com/ciyl2e, Chapter 13, and select the site you want.

For Further Reading

P. H. Gleick, "Making Every Drop Count," *Scientific American,* February 2001, pp. 40–45.

J. W. M. La Riviere, "Threats to the World's Water," *Scientific American,* September 1989, pp. 80–90.

P. Sampat, "The Polluting of the World's Major Freshwater Sources," *World Watch,* January/ February 2000, pp. 10–22.

Review Questions

1. Which activities contribute most to domestic water use? What types of use constitute the largest percentage? The smallest percentage?

2. What percentage of drinking water is supplied by surface fresh water? By groundwater?

3. What percentage of Earth's water can be used directly for drinking?

4. Where is most of Earth's fresh water found?

5. What is *surface fresh water*?

6. How is surface fresh water produced?

7. What is *groundwater*?

8. Where is groundwater found?

9. What is the source of groundwater?

10. Why was groundwater traditionally considered to be pure?

11. What is an *aquifer*?

12. Which components of groundwater impart detectable odors?

13. What are the sources of colloidal particles in groundwater?

14. Which ions are identified with hard water?

15. What is *activated charcoal*?

16. What is *reverse osmosis*?

17. What is *desalination*?

18. Which ions are commonly removed by ion exchange?

19. What is a *solar still*?

20. What is the objective of disinfecting drinking water? Identify three methods of disinfection.

21. What is *chlorination*? Identify one way that water can be chlorinated.

22. What are *THMs*? Why is the presence of THMs in chlorinated water a concern?

23. What is a *leachate*?

24. Identify four pollutants that contaminate surface and groundwater.

25. Identify the main sources of the following pollutants in surface and groundwater:
    a) phosphates   b) nitrates
    c) organic compounds

26. What are some of the health effects associated with excess nitrates in drinking water?

27. What is a *soap*? Describe the molecular structure of soap.

28. What is *saponification*?

29. What is a synthetic detergent? Describe the general structure of a synthetic detergent.

30. What properties are associated with detergent builders? Provide an example of a detergent builder.

31. What does the term *biodegradable* mean?

32. What is a *chelating agent*? Provide an example.

33. What is a *zeolite*?

Understanding Concepts

36. Outline the process by which untreated water is made fit for drinking.

37. Explain the effect that aeration has on water quality.

38. Why is colloidal material removed from drinking water?

39. What properties of colloids are used to facilitate their removal from drinking water?

40. Outline the process by which ions are removed from hard water.

41. How is activated charcoal used to treat water?

42. Describe the process of reverse osmosis.

43. Will reverse osmosis remove biological contaminants from water? Explain.

44. Explain the process by which hard water is softened.

45. What impact might softened water have on one's daily sodium intake?

46. Explain how deionized water is produced.

47. How does the ion content of distilled water compare to that of softened water?

48. Describe the process of distillation.

49. Under what conditions would the use of solar stills make sense?

50. Identify some of the health risks associated with commonly used disinfection methods.

51. What is *combined chlorine*?

52. Why is ammonia often added to drinking water that has already been purified?

53. Why is water disinfection an important public health issue?

34. Provide an example of both a point and nonpoint pollution source. What category describes your household?

35. What environmental problems were associated with phosphates?

54. How do the health risks associated with chlorinated water compare to the health benefits?

55. What are some of the difficulties with the pump-and-treat methods of purifying groundwater?

56. How do excess phosphates impact freshwater sources?

57. How does excess nitrate impact the seawater into which it flows?

58. What do the N,P,K percentages on a bag of 5-10-30 fertilizer represent?

59. What role does nitrogen-fixing bacteria play in nature?

60. What is *blue baby syndrome*? Explain its relationship to nitrates.

61. How does soap clean?

62. How does hard water affect the cleansing ability of soaps?

63. Why is rainwater soft water?

64. How do synthetic detergents overcome the problem of scum formation in hard water?

65. What are the roles of *detergent builders*? How can they act as chelating agents?

66. Identify some of the environmental issues associated with the use of detergent builders.

67. Which components of gasoline contribute to aquifer pollution?

68. Explain how point sources of pollution differ from nonpoint sources.

Synthesizing Ideas

69. Use the Web sites listed at the textbook's Web site to determine where the drinking water in your area comes from. What is the source of the water? Is it near or far from where the water is actually used? Is the water supply under any stress? What levels of contaminants and pollutants are found in your water?

70. Why is activated charcoal especially effective for adsorption of particles?

71. Do you think there may be health impacts when all ions are removed from drinking water? Should people on sodium-restricted diets use water softeners? What information would you need to make an informed decision?

72. Water polluted by inorganic ions could be purified by freezing it, since the ice that first forms does not contain any ions. Why do you think this technique is not used on a mass scale to purify salty water?

73. What chemical species are formed in chlorinated pool water when the pH is acidic? When the pH is basic?

74. Why is pH control of water important in chlorinated swimming pools?

75. How does the production of chlorinated organic by-products in water disinfection by ClO_2 compare to that using $HOCl$?

76. Why are the health risks associated with chlorinated surface water different from those associated with chlorinated groundwater?

77. Identify four methods by which water can be disinfected and describe how each works. Compare the advantages and disadvantages of each disinfection method.

78. Draw the molecular-level view of soap removing a grease stain.

79. Describe, in words, the chemical reactions involved in the preparation of soap.

80. Compare the structures, chemical and physical properties, and cleansing abilities of soaps and synthetic detergents.

81. How do the structures of branched and unbranched linear alkylsulfonate detergents affect their environmental impact?

82. Think about the role of chelating agents and discuss how they might be useful in food and medical products.

83. Describe the methods currently in place to deal with aquifer pollution. Are they adequate? Explain.

In this chapter, you will learn:

- how smog is produced and how it affects our lives and our agriculture;

- about adapting cars and power plants to reduce toxic emissions;

- about acid rain and its ecological and aquatic effects;

- about the hazardous effects to your health posed by air pollution, including indoor toxins.

When was the last time you paid attention to car exhaust?

Your atmosphere is filled with chemicals that you never see but that have an impact on you and your environment. In this chapter we'll learn about pollution in the lowest layer of the atmosphere—the one in which life on Earth resides. (Nello Giambi/Getty Images)

Dirty Air, Dirty Lungs
Air Pollution

In Shakespeare's *Hamlet* (Act II, Scene 2), Hamlet laments:

> [This] most excellent canopy, the air, look you, this brave o'erhanging firmament, this majestical roof fretted with golden fire—why, it appears no other thing to me than a foul and pestilent congregation of vapours.

Though Hamlet's perception of the air was determined by his state of mind, his description of it could apply literally to much of the air around us today. The best-known example of air pollution is the smog that occurs in many cities throughout the world. The most obvious manifestation of smog is a yellowish-brownish-gray haze, which is due to the presence in air of gases and small water droplets containing products of chemical reactions that occur between pollutants in air. The substances that react in air to produce smog are mainly emissions from cars, diesel trucks, and power plants. Indeed, the operation of motor vehicles produces more air pollution than any other single human activity. The haze, familiar to most of us who live in urban areas, also now extends periodically to once pristine areas such as the Grand Canyon in Arizona. Smog often has an unpleasant odor due to some of its gaseous components. More seriously, the substances in smog can affect human health and can cause damage to plants, animals, and some materials.

Our atmosphere is made up of several different layers, demarcated by their distance above Earth's surface (see Figure 14.1). In this chapter, we examine the chemistry of ground-level air pollution—that is, in the air just

View representations of gas molecules in the atmosphere at 5, 20, and 40 km above Earth's surface at Chapter 14: Visualizations: Media Link 1.

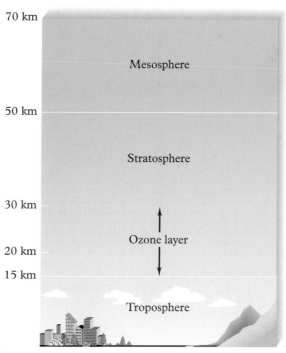

Figure 14.1 The structure of the atmosphere from ground level to 70 kilometers above ground level.

Whenever you see this icon in this chapter, go to www.whfreeman.com/ciyl2e

517

above Earth's surface, the region called the troposphere—including that which we encounter indoors. We will consider the detrimental effects of polluted air on humans, plants, and materials. Air in the stratosphere suffers from different types of contamination; we consider it in Chapter 15.

Urban Ozone: The Photochemical Smog Process

14.1 Photochemical smog production

Many urban centers undergo episodes of air pollution involving relatively high levels of ground-level *ozone*, O_3, a gas that is an undesirable constituent of the air we breathe. The ozone is not emitted as such, but rather is produced as a result of chemical reactions of pollutants. The result of the process that transforms the emissions into ozone is **photochemical smog,** so called because the chemical reaction requires the absorption of sunlight to drive it. The process of photochemical smog formation involves hundreds of different reactions occurring simultaneously.

The *original* pollutant reactants in photochemical smog are *nitric oxide*, NO, and gaseous hydrocarbons. The other vital ingredients in photochemical smog are sunshine, oxygen, and water vapor. Both nitric oxide and hydrocarbons are emitted into the air from internal combustion engines. Hydrocarbons are also present in urban air as a

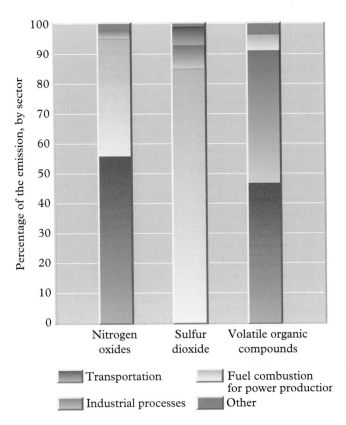

Figure 14.2 North American emissions of primary air pollutants from various sectors. (Source: Environmental Protection Agency, *1999 National Air Quality Trends Report*.)

result of the evaporation of solvents, liquid fuels, and other organic compounds. Collectively, carbon-containing substances that readily vaporize into air are called **volatile organic compounds,** or VOCs.

The final products of the photochemical smog process are ozone, nitric acid, and partially oxidized organic compounds. The overall reaction can be approximately expressed in an equation as:

$$\text{VOCs} + \text{NO} + \text{O}_2 + \text{sunlight} \rightarrow$$
$$\text{mixture of O}_3, \text{HNO}_3, \text{and partially oxidized organics}$$

Some of the products are gases, and some are dissolved in tiny particles that are suspended in air. Among the substances in the aqueous droplets formed by condensation of water vapor into a mist are carboxylic acids, hydrogen peroxide, and a variety of other partially oxidized and in some cases partially nitrated organic compounds.

Substances such as NO, hydrocarbons, and other VOCs that are initially emitted directly into air are called **primary pollutants;** the substances into which they are transformed, such as O_3 and HNO_3, are called **secondary pollutants.** A summary of the emissions of the primary pollutants sulfur dioxide, nitrogen oxides, and VOCs from various sources in the United States and Canada is given in Figure 14.2. The process by which the primary pollutants are converted in air into secondary ones is illustrated in Figure 14.3.

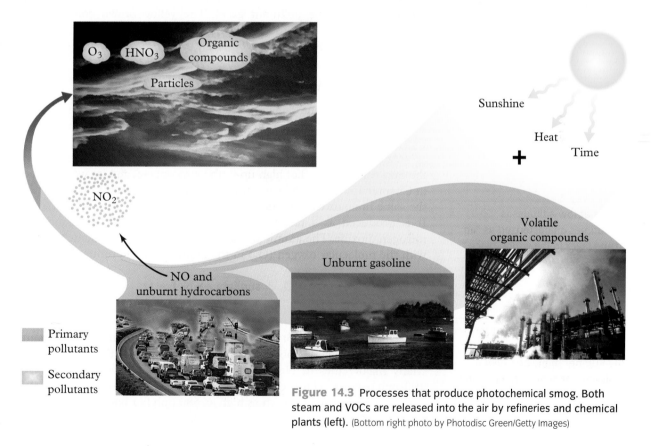

Figure 14.3 Processes that produce photochemical smog. Both steam and VOCs are released into the air by refineries and chemical plants (left). (Bottom right photo by Photodisc Green/Getty Images)

Urban Ozone: The Photochemical Smog Process

14.2 Burning fuel produces nitrogen oxides

We have seen above that nitrogen oxides play a major role in creating smog. Where do these gases come from? Nitrogen oxide gases are produced whenever a fuel such as gasoline, coal, or natural gas is burnt in air. Some nitrogen oxides are produced from the small amount of nitrogen compounds in the fuel. However, a much greater amount comes from the nitrogen already in the air. The fuel combustions occur at such high temperatures that some of the nitrogen and oxygen gases in the air passing through the flame combine to form nitric oxide, NO:

$$N_2 + O_2 \xrightarrow{\text{very hot flame}} 2\,NO$$

This reaction between atmospheric N_2 and O_2 is negligibly slow except at the very high temperatures such as those that occur in the modern combustion engines in vehicles and in power plants. The higher the flame temperature, the more NO that is produced. At high temperatures, NO molecules could react with each other to reverse this reaction and revert back to molecular nitrogen and oxygen if given sufficient time. However, the exhaust gases from combustion cool too quickly for this reverse reaction to occur completely.

The nitric oxide emitted into air is gradually converted to *nitrogen dioxide*, NO_2, over a period of minutes to hours depending upon the concentration of the pollutant gases. The yellow-brown color in the atmosphere of a smog-ridden city is due in part to the nitrogen dioxide present. Collectively, NO and NO_2 in air are referred to as **NO_x.**

Nitric oxide is also produced naturally in air by this reaction during thunderstorms. Lightning provides the energy for the process.

14.3 Ground-level ozone is a form of pollution

We are warned of the danger of depletion of Earth's ozone layer, but we're also told that ozone is an air pollutant. These seem like contradictory concerns: Do we want ozone in our air or don't we?

Ozone, O_3, is a necessary protection against dangerous radiation from the Sun when the gas lies high up in the atmosphere, and, as we shall see in Chapter 15, this is where we must be concerned about its depletion. At the same time, ozone is an undesirable constituent of air at low altitudes. It is produced as a result of the sunlight-induced decomposition of nitrogen dioxide. The atomic oxygen (O) released in this reaction reacts with the molecular oxygen (O_2) in the atmosphere to produce ozone:

$$NO_2 + \text{sunlight} \rightarrow NO + O$$
$$O + O_2 \rightarrow O_3$$

Ozone at low altitudes is sometimes characterized as "an ozone layer in the wrong place."

The concentration of ozone in polluted air often amounts to about 100 parts per billion (ppb), which is about three times that in clean air. Many major cities in North America, Europe, and Japan exceed the 100-ppb ozone level for several days each summer. The levels of ozone in Los Angeles air used to reach as high as 680 ppb but have now declined.

Recall from Chapter 2 that for gases *parts per billion* refers to the number of molecules of a pollutant present in every billion molecules of air.

Table 14.1	Health-based U.S. national air quality standards	
Pollutant	**U.S. EPA standard**	
Sulfur dioxide	140 ppb (daily average)	
	30 ppb (annual average)	
Nitrogen oxides	53 ppb (annual average)	
Carbon monoxide	9 ppm (8-hour average)	
	35 ppm (1-hour average)	
Ozone	120 ppb (1-hour maximum)	
	80 ppb (proposed 8-hour average)	
Lead	1.5 micrograms per cubic meter (quarter-year average)	
Particulates less than 10 micrometers in diameter	150 micrograms per cubic meter (daily average)	
	50 micrograms per cubic meter (annual average)	
Particles less than 2.5 micrometers in diameter	65 micrograms per cubic meter (proposed daily average)	
	15 micrograms per cubic meter (proposed annual average)	

Source: U.S. Environmental Protection Agency Office of Air Quality.

The background ozone level in clean air is about 30 ppb.

Many countries individually have established goals for maximum allowable ozone concentrations in air, averaged over a 1-hour period (see Table 14.1). The standard in the United States is 120 ppb, whereas that in Canada is 82 ppb. In 1997, the U.S. government attempted to adopt a standard in which the average ozone level over an *eight*-hour period, rather than that over one hour, would be the quantity that is regulated. The average eight-hour limit was set at 80 ppb. Generally speaking, the longer the period over which the concentration is averaged in a regulation, the lower is the stated limit, since it is presumed that exposure to a high level of an air pollutant is acceptable if it occurs only for a short time. However, the new U.S. standard was challenged in court and has not yet been implemented.

14.4 Photochemical smog occurs worldwide

In order that a city be subject to photochemical smog, several conditions must be fulfilled. First, there must be substantial vehicular traffic in order to emit sufficient NO and other VOCs into the air. Second, there must be warmth and ample sunlight in order for the crucial reactions, some of which need light to be initiated, to proceed. Finally, there must be relatively little movement of the air mass so that the reactants are not dispersed but can react together. Vertical air movement is limited if a *temperature inversion* occurs. In this condition, warm air lies

Figure 14.4 Smog in Mexico City. The combination of pollutants, geography, and climate gives Mexico City the worst air pollution in the world. (Photodisc Green/Getty Images)

above colder air. The cold air does not rise because its density is greater than that of the warm air that lies above it. In other words, only warm air rises. For reasons of geography (for example, the presence of mountains, which prevent horizontal air movement) and dense population, cities such as Los Angeles, Denver, Mexico City, Tokyo, Athens, Sao Paulo, and Rome all fit the bill splendidly and are subject to frequent smog episodes. Indeed, the photochemical smog phenomenon was first observed in Los Angeles in the 1940s and has generally been associated with that city ever since.

Although the quality of air is improving with time in most developed countries, it continues to worsen in the larger cities of developing countries. Mexico City is now generally considered to have the worst urban air pollution in the world (see Figure 14.4). Indeed, in the center of the city residents can purchase pure oxygen from booths to help them breathe more easily! The air in Mexico City is so polluted by ozone and other components of smog, and by other substances, that it causes thousands of premature deaths annually.

Because gaseous pollutants are transported over long distances by air currents, many geographic areas which themselves generate few emissions are subject to regular episodes of high ground-level ozone and other smog components. When hot summertime weather conditions produce large amounts of ozone in urban areas but do not allow much vertical mixing of air masses as they travel to rural sites, elevated ozone levels are often observed in eastern North America and Western Europe in zones that extend for 1000 km (600 miles) or more. In the United States, considerable ozone is transported from its origin in the Midwest to surrounding states and Canadian provinces, especially around the Great Lakes. Thus, ozone control is a *regional* rather than a local air quality problem.

14.5 Ozone affects crops

Ground-level ozone affects some agricultural crops because it attacks plants. The ozone reacts with the ethene gas that the plants emit, gener-

Figure 14.5 Corn plants affected by smog. The visible tissue damage is caused by high concentrations of ground-level ozone. (Courtesy of Sarnia-Lambton Environmental Association and The University of Guelph)

ating substances that then damage plant tissue (see Figure 14.5). The rate of photosynthesis is slowed, and consequently the total amount of plant material is reduced. The collective damage to North American crops—for example, alfalfa in the United States and white beans in Canada—is estimated to be $3 billion a year.

Elevated levels of ozone also affects materials: it hardens rubber, reducing the useful life span of consumer products such as automobile tires, and bleaches color from materials such as fabrics. We will discuss the effects on human health of the ozone from photochemical smog later in this chapter.

14.6 Limiting VOC emissions can reduce smog

In order to improve the air quality in areas that are subject to photochemical smog, the quantity of reactants that are emitted into the air must be reduced. The control strategies that have been used on vehicles in the United States have resulted in some reduction in ozone levels, notwithstanding the huge increase in total vehicle-miles driven—up 100% in the last 25 years—that has occurred.

It is the nitrogen oxides, rather than the hydrocarbons, that usually determine the overall rate of the smog-formation reaction. This situation exists because in the absence of pollution controls, there is an *over-abundance* of hydrocarbons relative to the amount of nitrogen oxides required. Consequently, a small reduction in hydrocarbon emissions simply reduces the large excess of this material without significantly slowing down the reactions. NO_x reduction, rather than VOC reduction, would be much more effective in reducing ozone in almost all of the eastern United States, though it is more expensive to achieve.

Although the common perception is that trees fight pollution by giving off oxygen, in fact deciduous trees and shrubs also emit reactive hydrocarbons that contain C=C bonds. In urban atmospheres, the concentration of these compounds normally is much less than that of the **anthropogenic** hydrocarbons—that is, those that result from human activities—so it is not until the latter are reduced substantially that the influence of the natural substances becomes noticeable. Some urban areas such as Atlanta, Georgia, and others located in the southern United States incorporate or border upon heavily wooded areas whose trees emit enough reactive hydrocarbons to sustain smog and ozone production. In such areas, only the reduction of emissions of nitrogen oxides will reduce photochemical smog production substantially.

Hydrocarbons with C=C bonds are the most reactive type of VOC in photochemical smog processes. However, other VOCs play a significant role after the initial stages of a smog episode since they are brought into the reaction by substances generated in the smog. For this reason, emissions of *all* reactive VOCs must be controlled in areas with serious

(b)

Figure 14.6 Los Angeles (a) 1972 and (b) 2004. (Part a, Gene Daniels, EPA/NARA; part b, courtesy of Hannah Thonet for W. H. Freeman & Company.)

(a)

photochemical smog problems if the reduction of emissions is to make a significant difference. Gasoline is now formulated in order to reduce its evaporation, since gasoline vapor has been found to contribute significantly to atmospheric concentrations of hydrocarbons.

Two-cycle engines, such as those that power many motorboats, also emit significant quantities of unburned gasoline. New regulations in California, with Los Angeles especially in mind, limit the use of hydrocarbon-containing products such as barbecue starter fluid, household aerosol sprays, and oil-based paints that contain a hydrocarbon solvent that evaporates into the air as the paint dries. The air quality in California has improved because of current emission controls. However, the increase in vehicle-miles driven and the hydrocarbon emissions from nontransportation sources such as solvents have thus far prevented a more complete solution (see Figure 14.6).

14.7 Catalytic converters help reduce vehicle emissions

Control of NO_x emissions from gasoline-powered cars and trucks has been achieved using **catalytic converters** placed just ahead of the mufflers in the exhaust system (see Chapter 2, Figure 2.17). The original converters controlled only carbon-containing gases in the emissions, including carbon monoxide. However, the modern converter uses a surface impregnated with a platinum–rhodium catalyst to also change nitrogen oxides back to elemental nitrogen and oxygen:

$$2\,NO \xrightarrow{\text{catalyst}} N_2 + O_2$$

The same catalyst combination oxidizes the carbon-containing gases to CO_2 and water:

$$2\ CO + O_2 \rightarrow 2\ CO_2$$
$$\text{hydrocarbons} + O_2 \rightarrow CO_2 + H_2O$$

An oxygen sensor in the vehicle's exhaust system is monitored by a computer chip that controls the intake air/fuel ratio of the engine to ensure that the correct amount of oxygen remains in the exhaust gases. In this way, a high percentage of conversion of the subsequent pollutants occurs in the converter. If the air/fuel ratio is too high or too low, then one or the other of the reactions will not occur, and pollutants will pass unchanged through the converter and be emitted into the atmosphere.

Once an engine has warmed up, properly working catalytic converters eliminate 80–90% of the hydrocarbons, CO, and NO_x before the exhaust gases are released into the atmosphere. However, before the engine has warmed up, and also during sudden acceleration or deceleration, the converters cannot operate effectively. Under these conditions, there are bursts of emissions from the tailpipe. Indeed, about 80% of all the emissions from catalyst-equipped cars are produced in the first few minutes after the car has been started. At these times, sulfur compounds—such as hydrogen sulfide, H_2S, which often gives vehicle emissions their characteristic odor of rotten eggs—are produced from the sulfur-containing components in the fuel, and also emitted.

The small amounts of sulfur that remain in gasoline and diesel fuel after refining can partially deactivate catalytic converters. This occurs because the particles produced from the sulfur-containing molecules during the fuel's combustion can become attached to the catalyst metal sites, thereby deactivating them. The maximum sulfur levels in gasoline, amounting to several hundred parts per million in the past, were reduced to 30 ppm by the end of 2004 in both the U.S. and Canada, and in the European Union to 50 ppm by 2005. The maximum level in diesel fuel for new on-road vehicles dropped from 500 ppm to 15 ppm in the U.S. More active, and hence more effective, metal catalysts will be available for diesel engines as a result of this decrease in sulfur levels.

Possible ways that start-up emissions from cars could be decreased include devising a converter that works at lower temperatures or is preheated, or recirculating the exhaust until the converter is hot.

14.8 Nitric oxide is also emitted from power plants

In North America, approximately equal amounts of NO_x are emitted from vehicles and from the electric power plants that burn coal, oil, or natural gas in order to produce electricity (see Figure 14.2). Taken together, these two sources generate the majority of the anthropogenic sources of these gases.

To reduce their NO_x production, some power plants use special burners designed to lower the temperature of the flame. Nitric oxide formation in power plants can also be greatly reduced by having the combustion occur in two stages. In the first, high-temperature stage, no excess oxygen beyond that required for fossil fuel combustion is allowed to be present. This greatly minimizes oxygen's tendency to react with N_2

and produce NO. In the second stage, additional oxygen is supplied to complete the fuel combustion but under lower temperature conditions, so that again little NO is produced since its rate of formation declines rapidly with temperature. Other power plants use large-scale catalytic converters in order to change NO_x back to N_2 before the release of emission gases into air.

Acid Rain

14.9 Acid rain has a low pH

A serious environmental problem facing many regions of the world today is **acid rain.** This generic term covers a variety of phenomena, including *acid fog* and *acid snow,* all of which correspond to atmospheric precipitation with a pH less than 5. As will be discussed later in this chapter, acid rain has a variety of ecologically damaging consequences. The presence of acidic particles in air probably also has direct effects on human health.

The phenomenon of acid rain refers to precipitation that is significantly *more* acidic than natural (unpolluted) rain. Natural rain itself is usually mildly acidic due to the presence in it of dissolved atmospheric carbon dioxide, which forms *carbonic acid,* H_2CO_3:

$$CO_2(g) + H_2O(aq) \rightleftharpoons H_2CO_3(aq)$$

The weak acid H_2CO_3 partially ionizes to release a hydrogen ion, with a resultant reduction in the pH of the system:

$$H_2CO_3(aq) \rightleftharpoons H^+ + HCO_3^-$$

Due to this source of acidity, the pH of unpolluted, natural rain is about 5.6. Only rain that is appreciably more acidic than this—that is, with a pH of less than 5—is considered to be truly acid rain. Strong acids such as HCl released by volcanic eruptions can temporarily produce natural acid rain in regions such as Alaska and New Zealand.

The *primary* pollutants NO and *sulfur dioxide,* SO_2, *themselves* do not make rainwater particularly acidic. However, they are converted over hours or days into the secondary pollutants *sulfuric acid,* H_2SO_4, and *nitric acid,* HNO_3, both of which are very soluble in water and are strong acids. Indeed, virtually all the acidity in acid rain is due to the presence of sulfuric and nitric acids (see Figure 14.7). In eastern North America, sulfuric acid predominates over nitric acid in acid rain because many electrical power generating plants use high-sulfur coal. During the combustion of the coal, the sulfur is converted to sulfur dioxide, which then is emitted into the air and eventually becomes converted to sulfuric acid. In western North America, nitric acid produced from NO_x vehicle emissions is the predominant acid, since the coal mined and burned there is low in sulfur.

Generally speaking, acid rain falls far downwind from the source of the primary pollutants, as the acids are created over a period of days, during the transport of the air mass that contains the primary pollu-

Figure 14.7 Formation of acid rain. The primary pollutants NO_2, SO_2, and SO_3 are converted into secondary pollutants (HNO_3 and H_2SO_4) which dissolve in rainwater. (Bottom right photo by Photodisc Green/ Getty Images)

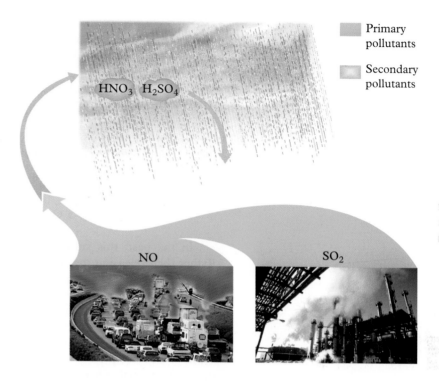

tants. Thus acid rain is a pollution problem that does not respect state or national boundaries.

14.10 Sulfur dioxide pollution and control

On a global scale, SO_2 is produced by volcanoes and by the oxidation of sulfur gases produced during the decomposition of plants. Because this natural sulfur dioxide is mainly emitted high into the atmosphere or far from populated centers, the background concentration of the gas in clean urban air is quite small.

However, a sizable amount of sulfur dioxide is presently emitted as a pollutant into ground-level air, particularly over land masses in the northern hemisphere. The largest anthropogenic source of SO_2 is the combustion of coal (see Figure 14.2), a solid which, depending upon the geographic area from which it is mined, contains 1 to 6% sulfur, as we discussed in Chapter 4. In many countries, including the United States, coal is used primarily to generate electricity.

Recently, **clean coal technologies** have been developed whereby coal can be utilized in ways that are cleaner—with respect to the air pollution it causes—and often more energy efficient than those used in the past. In the various technologies, the cleaning can occur precombustion, during combustion, postcombustion, or by conversion of the coal to another fuel. In precombustion cleaning, the mineral component of the coal—which contains some but not all of its sulfur component—is removed by grinding the raw coal into small particles and then separating and removing the heavier, sulfur-containing, mineral-rich ones.

Alternatively, the sulfur dioxide gas produced in coal combustion can be removed as it forms in combustion, or later as it exits with the other combustion gases, by reacting it with *calcium carbonate*, $CaCO_3$, to form solid calcium salts.

Individual sites that emit large amounts of SO_2 are also associated with **nonferrous smelting,** the conversion of ores—other than those of iron—to free metals. Many valuable and useful metals, such as copper and nickel, occur in nature as *sulfide ores*, that is, ores that contain the sulfide ion, S^{2-}. In the first stage of their conversion to the free metals, they are usually heated in air (roasted) to remove the sulfur, which is converted to SO_2. For example,

$$2\,NiS\,(s)\,+\,3\,O_2\,(g)\,\rightarrow\,2\,NiO\,(s)\,+\,2\,SO_2\,(g)$$

In the past, this sulfur dioxide was usually emitted into the air. Rather than releasing the sulfur dioxide into the air as a pollutant, ores such as copper sulfide can be smelted using pure oxygen instead of air, and the very concentrated sulfur dioxide that is obtained from the reaction can be readily extracted, liquefied, and sold as a by-product. In another nonpolluting alternative process, the SO_2 from conventional roasting processes using air can be passed over a catalyst that converts much of it to sulfur trioxide. Water can then be added to the SO_3 to produce commercial concentrated sulfuric acid:

$$2\,SO_2\,(g)\,+\,O_2\,(g)\,\rightarrow\,2\,SO_3\,(g)$$
$$SO_3\,(g)\,+\,H_2O\,(aq)\,\rightarrow\,H_2SO_4\,(aq)$$

If no sulfur dioxide controls are implemented, and the pollutant gases from smelting are simply emitted into the air, the SO_2 can cause devastation to the plant life in the surrounding area unless extremely high smokestacks are used. The tallest such stacks in the world, reaching 400 meters high, are located at Sudbury, Ontario (see Figure 14.8). However,

Figure 14.8 A sulfur dioxide–emitting smelting plant in Sudbury, Ontario. Tall smokestacks reduce local pollution but can create problems far from the source.
(William J. Weber/Visuals Unlimited)

Chapter 14: Dirty Air, Dirty Lungs

using tall stacks simply solves a local SO_2 problem at the expense of creating a problem downwind. For example, emissions from mainland North America, including Sudbury, can be detected in Greenland.

Because of federal regulations, the amount of sulfur dioxide emitted into the air in North America has fallen substantially—by about 20% from 1980 levels by 1994 in the United States and 45% by 2000 in Canada. The 1991 Air Quality Accord between the United States and Canada required both countries to reduce substantially their sulfur dioxide emissions. Such emissions in the United States are now restricted in accordance with the Clean Air Act (see Tables 14.1 and 14.2). Indeed, the emissions of SO_2 fell by 35% in the United States in the period from 1970 to 1997.

The reductions in sulfur dioxide emissions by power plants in the U.S. Midwest have been achieved at lower-than-expected cost, due in part to the availability of cheap, low-sulfur coal and inexpensive post-combustion scrubbers, and in part to the implementation of a system of tradable emission permits. This system delegates a certain amount of "allowable emissions" for each industry. If an industry needs to exceed their allowed level, this permit system allows them to buy emission allowances from other industries. Conversely, if they have extra "allowable emissions," industries are able to sell them off on the open market through the Chicago Board of Trade.

Global SO_2 emissions are predicted to keep rising until about 2020, due mainly to increased releases from Asia. China has become a world leader in emitting sulfur dioxide, and that country now has a serious acid rain problem. Some of the acid rain from China is carried by wind to Japan and on occasion all the way to North America. In contrast to China, Japan initiated tight controls on SO_2 emissions in the 1970s, and by 1980 its power plants had almost eliminated such emissions by the widespread installation of postcombustion scrubbers.

Table 14.2 Anthropogenic sources of air pollution

Pollutant	Sources
Sulfur dioxide	Electric utility plants burning fossil fuel (mainly coal and oil) produces 70% of emissions; metal smelting
Nitrogen oxides	Fuel combustion and transportation produces 95% of emissions
Carbon monoxide	Vehicle exhaust produces 60% of emissions nationally (in urban areas may be responsible for up to 95% of emissions); industrial processes; fuel combustion in boilers and incinerators; natural sources such as wildfires
Ozone	Produced by reaction of NO_x and volatile organic compounds in the presence of sunlight; refineries, factories, consumer/commercial products, motor vehicles, chemical plants, and solvents are sources of volatile organic compounds
Particulates (small)	Fuel combustion, industrial processes, residential fireplaces and woodstoves

Figure 14.9 Average pH values of acid precipitation in different regions of central North America, 1985. (Source: Redrawn from E. G. Nisbet. 1991. *Leaving Eden: To Protect and Manage the Earth.* Cambridge, U.K.: Cambridge University Press.)

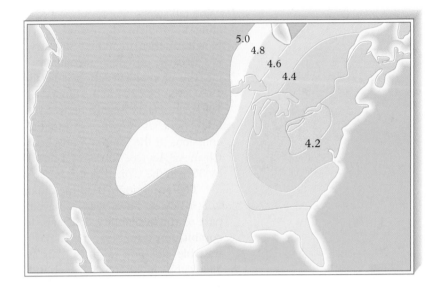

14.11 The acidity of rain varies

Figure 14.9 shows a contour map of the average pH in precipitation for different regions of North America. At each point along any one of the solid lines, the annual average pH has the same value shown; thus the contours connect contiguous regions having the same acidity in rain. Inside each contour, the rainfall's pH falls even further, until the value of the next contour is reached. In North America, the greatest acidity occurs in the eastern United States and in southern Ontario, since both regions lie in the path of air originally polluted by emissions from power plants in the Ohio Valley. Currently the average pH of rainfall in the eastern United States lies between 3.9 and 4.5.

Exercise 14.1

Calculate the H^+ ion concentration, in moles per liter, that would be present in acid rain that had a pH of 3. Repeat the calculation for rain that is barely acidic, having a pH of 5. What is the *ratio* of the H^+ ion concentrations in the two samples? *Hint:* Consult Chapter 11, section 11.9, if you do not recall how to interconvert pH and H^+ concentration values.

Although sulfur dioxide emission levels have fallen significantly in recent decades in North America, there has been little corresponding change in the pH of the precipitation. Acidity has stayed relatively constant over this period because of the simultaneous reduction of ash and other solid particles emitted from smokestacks; these basic substances had previously neutralized a fraction of the sulfur dioxide and sulfuric acid. In addition, the total nitrogen oxide emissions—much of them eventually converted into nitric acid—have not decreased significantly in the last few decades.

14.12 The ecological effects of acid rain

Whether or not acid precipitation affects biological life in a given area depends strongly upon the composition of the soil and bedrock in that region. Most strongly affected are areas having granite or quartz bedrock, since the soil there has little capacity to neutralize the acid. The largest areas susceptible to acid rain are the Precambrian shield regions of Canada and Scandinavia.

In contrast, if the bedrock is limestone or chalk, the acid can be efficiently neutralized since these rocks are composed of calcium carbonate, $CaCO_3$, which acts as a base and reacts with acid and neutralizes it; the overall net reaction is:

$$CaCO_3(s) + 2 H^+(aq) \rightarrow Ca^{2+}(aq) + CO_2(g) + H_2O(aq)$$

Thus the calcium carbonate rock dissolves, producing carbon dioxide. The calcium ions replace the hydrogen ions in the water, and thereby reduce the acidity. This same reaction is responsible for the deterioration of limestone and marble statues; fine detail, such as ears, noses and other facial features, are gradually lost as a result of reaction with acid and with sulfur dioxide itself (see Figure 14.10).

Because of acid rainwater falling and draining into them, tens of thousands of lakes in the shield regions of both Canada and Sweden have become strongly acidified, as have lakes in the United States, Great Britain, and Finland. Lakes in Ontario are particularly hard hit, since

Figure 14.10 Corrosion by acid rain. A statue (a) before and (b) after acid rain degradation. (Part a, NYC Parks Photo Archive/Fundamental Photographs; part b, Kristen Brochmann/ Fundamental Photographs)

(a)

(b)

Figure 14.11 Neutralization of an acidified lake by addition of limestone. (Tero Niemi/Naturbild)

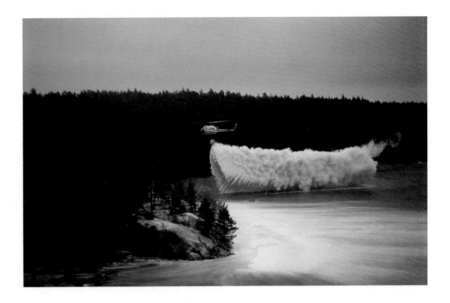

they lie directly in the path of polluted air and because the soil there contains little limestone. In a few cases, attempts have been made to neutralize the acidity by artificially adding limestone or calcium hydroxide to the lakes (see Figure 14.11); however, this process must be repeated every few years to sustain an acceptable pH.

14.13 Aluminum is made soluble by acid rain

Acid precipitation can affect aquatic life both directly and indirectly. For example, acidity from precipitation leads to the deterioration of soil. When the pH of soil is lowered, plant nutrients such as the cations potassium, calcium, and magnesium are leached from it—in other words, they are transferred into the water that passes through the soil.

Acidified lakes characteristically have elevated concentrations of dissolved ionic aluminum, Al^{3+}. Under normal conditions, when the water has a pH closer to 7, the aluminum is immobilized in the rock by its insolubility. However, this ion is leached from rocks by reaction with the hydrogen ions, H^+, in acidic water that passes over them. Scientists believe that both the acidity itself *and* the high concentrations of aluminum are together responsible for the devastating decreases in fish populations that have been observed in many acidified water systems (see Figure 14.12).

Different types of fish and aquatic plants vary in their tolerances for aluminum and acid, so the biological composition of a lake changes as it gradually becomes increasingly acidic. Generally speaking, fish reproduction is severely diminished even at low levels of acidity, which, however, can be tolerated by adult fish. Very young fish, hatched in early spring, also are subject to the shock of very acidic water when the acidic winter snow melts in a short time and enters the water systems all at once.

Figure 14.12 Acidity and
aluminum ions contribute to fish
kills.

Few species survive and reproduce when the pH drops much below 5; healthy lakes have a pH of about 7 or a little higher. As a result, many lakes and rivers in affected areas are now devoid of their valuable fish. For example, some rivers in Nova Scotia are too acidic for Atlantic salmon. The water in many acidified lakes is crystal clear due to the death of most of the flora and fauna.

14.14 Acid rain and air pollution affect forests

Air pollution can have a severe effect on trees. Forest decline occurs mainly at high altitudes and was first observed on a large scale in western Germany. However, the cause-and-effect relationship behind this phenomenon has been difficult to untangle. As we have seen above, acidification of the soil can leach nutrients from it and, as occurs in lakes, solubilize aluminum. This element may interfere with the uptake of nutrients by trees and other plants. Apparently both the acidity of the rain falling on affected forests *and* the ground-level ozone in the air to which they are exposed pose significant stresses to the trees. These stresses alone will not kill them, but when they are combined with drought, temperature extremes, disease, or insect attack, the trees become much more vulnerable.

Forests at high altitudes are most affected by acid precipitation, possibly because they are exposed to the base of low-level clouds, where the acidity is most concentrated (see Figure 14.13). Fogs and mists are even more acidic than precipitation, since there is much less total water to dilute the acid. For example, white birch trees along the shores of

Figure 14.13 Forest dieback resulting from acid rain. (Rob and Ann Simpson/Visuals Unlimited)

The width of a human hair is about 100 micrometers.

Lake Superior are experiencing dieback in regions where acid fog occurs frequently. Deciduous trees (those that lose their leaves annually) affected by acid rain gradually die from their tops downward; the outermost leaves dry and fall prematurely and are not replenished the following spring. The trees become weakened as a result of these changes and become more susceptible to other stressors.

Particulates in Air Pollution

14.15 Particulates are tiny particles suspended in air

The black smoke released into the air by a diesel truck or bus is the most obvious form of air pollution that most of us routinely encounter. The smoke is composed of *particulate matter.* **Particulates** are the tiny solid or liquid particles that are suspended in air. Most of them are so small that they are individually invisible to the naked eye. Collectively, however, particulates form a haze that restricts visibility. Indeed, on many summer days the sky over North American and European cities is milky white rather than blue, owing to the scattering of sunlight by particulates suspended in the air.

The particles that are suspended in a given mass of polluted air are neither all of the same size or shape nor do they all have the same chemical composition. The *diameter* of particulates—their side-to-side length—is perhaps their most important property. The diameter values are usually quoted in *micrometers,* or *microns,* symbolized as μm (a micrometer is one-millionth of a meter, 10^{-6} m). Each micrometer is 1000 times as large as the nanometer (nm) unit of length we used in Chapter 1. The smallest suspended particles are about 0.002 μm in diameter and are only a little larger than the biggest gaseous *molecules*. The upper limit for suspended particles corresponds to diameters of about 100 μm (that is, 0.1 mm). When atmospheric water droplets coalesce to particles bigger than this limit, they are raindrops and fall out of the air quickly. Figure 14.14 shows the ranges of particle size for the common types of suspended particulates.

Qualitatively, individual particles are classified as *coarse* or as *fine* depending upon whether their diameters are greater or less than 2.5 μm, respectively. About 100 million particles of diameter 2.5 μm would be required to cover the surface of a small coin.

There are many common names for atmospheric particles: *dust* and *soot* refer to solids, whereas *mist* and *fog* refer to liquids, fog denoting a high concentration of water droplets. An **aerosol** is a collection of particulates, whether solid particles or liquid droplets, dispersed in air. A true

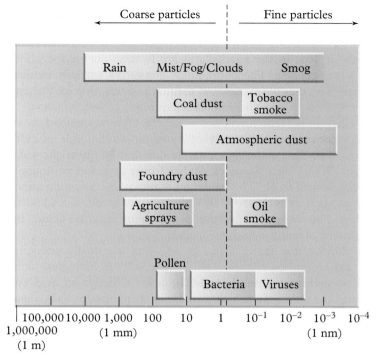

Coarse particles | Fine particles

Rain Mist/Fog/Clouds Smog

Coal dust | Tobacco smoke

Atmospheric dust

Foundry dust

Agriculture sprays | Oil smoke

Pollen

Bacteria Viruses

100,000 10,000 1,000 100 10 1 10^{-1} 10^{-2} 10^{-3} 10^{-4}
1,000,000 (1 mm) (1 nm)
(1 m)

Particle diameter in micrometers (μm)

Figure 14.14 Sizes of common airborne fine and coarse particulates. (Adapted from J. G. Henry and G. W. Heinke. 1989. *Environmental Science and Engineering*. Prentice-Hall.)

aerosol (as opposed, say, to the output of fairly large droplets from a hair-spray dispenser) has small particles: their diameters are less than 100 μm.

Intuitively, you might think that all particles should settle out rapidly under the influence of gravity and be deposited onto Earth's surface, but this is *not* true for the smaller ones. The small ones fall so slowly that they are suspended almost indefinitely in air (unless they stick to some object they encounter). As we shall see later, the very small ones aggregate to form larger ones, usually still in the fine size category. Thus fine particulates usually remain airborne for days or weeks, whereas coarse particulates settle out fairly rapidly. Particles are also removed from air by their absorption into falling raindrops.

14.16 Sources of coarse particulates

Larger coarse particles can come from natural sources such as volcanic eruptions or human activities such as stone crushing in quarries and land cultivation, which result in particles of rock and topsoil being picked up by the wind. Most coarse particulates originate from the disintegration of larger pieces of matter. Pollen released from plants also consists of coarse particles in the 10–100 μm range (see Figure 14.14). Volcanic ash particles are mostly of coarse size.

Mineral pollutants constitute one source of the coarse particulates in air. Because many of the large particles in atmospheric dust, particularly in the air of rural areas, originate as soil or rock, their elemental composition is similar to that of Earth's crust: high concentrations of Al, Ca, Si, and O in the form of aluminum silicates. Coarse particles in many areas are basic, reflecting the calcium carbonate and other such salts in soils. Near and above oceans, the concentration of solid NaCl is very high, since sea spray leaves sodium chloride particles airborne when the water in the spray droplets evaporates.

14.17 Sources of fine particles

Whereas coarse particles often result from the mechanical breakup of larger ones, fine particles are formed mainly by chemical reactions and by the coagulation of even smaller particles, including molecules in the vapor state. The incomplete combustion of carbon-based fuels such as coal, oil, gasoline, and diesel fuel produces many small soot particles, which are miniature crystals of carbon. Consequently, one of the main sources of carbon-based atmospheric particulates is the exhaust from vehicles, especially those having diesel engines.

In smoggy urban areas, such as Los Angeles, up to half the organic compounds present in the particulates are formed during the photochemical smog process. The compounds consist of partially oxidized hydrocarbons, such as aldehydes and carboxylic acids, that have chemically incorporated oxygen and sometimes nitrogen as well.

The other important fine particulates that are suspended in the atmosphere consist predominantly of inorganic compounds of sulfur and of nitrogen. The sulfur compounds are derived from the sulfur dioxide gas, SO_2, which becomes oxidized over a period of hours or days to sulfuric acid in air. The oxidation of nitrogen-containing atmospheric gases such as NH_3, NO, and NO_2 results in nitric acid as the end-product. The suspended fine particles in many areas are acidic, due to their content of sulfuric and nitric acids.

Both sulfuric and nitric acids in air often eventually encounter *ammonia* gas, NH_3, that is released as a result of biological decay processes occurring at ground level. The acids undergo an acid–base reaction with the ammonia, which transforms them into the salts *ammonium sulfate* and *ammonium nitrate*. For example,

$$H_2SO_4(aq) + 2\ NH_3(g) \rightarrow \underset{\text{ammonium sulfate}}{(NH_4)_2SO_4(aq)}$$

Recall from Chapter 3, section 3.30, that the sulfate ion is SO_4^{2-} and the nitrate ion is NO_3^-.

Aerosols dominated by such oxidized sulfur compounds are called **sulfate aerosols.**

14.18 Particulate concentration affects air quality

Particles whose diameters are somewhat less than that of the wavelength of visible light can interfere with the transmission of light in air. A high concentration in air of particles of diameters between 0.1 μm and 1 μm produces a *haze*. The enhanced haziness in summertime over much of North America is due mainly to sulfate aerosols arising from industrialized areas in the United States and Canada. Fine particles also are largely responsible for the haze in locations subject to episodes of photochemical smog. The smog aerosols contain nitric acid that has been neutralized to ammonium nitrate, plus carbon-containing products that are generated in the photochemical smog reactions.

When air quality is monitored, the most common measure of the concentration of suspended particles is a **PM index,** which is the mass of particulate matter that is present in a given volume. The units for PM are *micrograms of particulate matter per cubic meter of air,* that is, μg/m^3. Because smaller particles have a greater detrimental effect on human health than do larger ones, usually only those having a specified diameter *or less* are collected and reported; this cutoff diameter, in μm, is shown as a subscript to the PM index.

In recent years, government agencies in many countries including the United States have monitored PM_{10}, that is, the total concentration

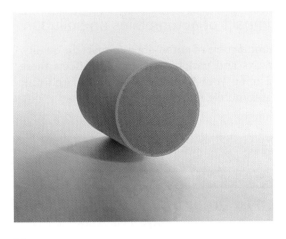

(a)

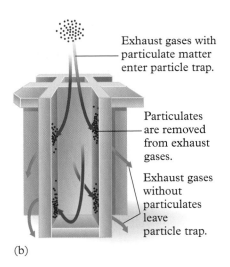

Exhaust gases with particulate matter enter particle trap.

Particulates are removed from exhaust gases.

Exhaust gases without particulates leave particle trap.

(b)

Figure 14.15 (a) A clean particle trap for a diesel-powered vehicle, manufactured by Corning. (b) Particulate matter is removed from exhaust by the filter. (Part a, courtesy of Corning Incorporated, Environmental Technologies; part b adapted from a drawing supplied by Corning Incorporated)

of all particles having diameters less than 10 μm. Increasingly, regulators now are using the $PM_{2.5}$ index, which includes all and only fine particles.

Since most fine particle air pollution results ultimately from chemical reactions among gaseous molecules, the main remedy is to control the emissions of the primary *gaseous* pollutants, namely sulfur dioxide and the reactants NO and VOCs that produce photochemical smog.

Diesel engines directly emit particles in both the fine and coarse ranges as soot. Recently, particulate emissions from diesel trucks have begun to be controlled, by a combination of engine modifications and the use of **particle traps** or *filters* (see Figure 14.15). In light-duty diesel vehicles, the traps operate by preventing organic particulate emissions from escaping from the exhaust system.

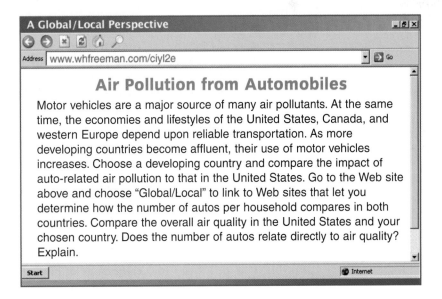

A Global/Local Perspective

Address www.whfreeman.com/ciyl2e

Air Pollution from Automobiles

Motor vehicles are a major source of many air pollutants. At the same time, the economies and lifestyles of the United States, Canada, and western Europe depend upon reliable transportation. As more developing countries become affluent, their use of motor vehicles increases. Choose a developing country and compare the impact of auto-related air pollution to that in the United States. Go to the Web site above and choose "Global/Local" to link to Web sites that let you determine how the number of autos per household compares in both countries. Compare the overall air quality in the United States and your chosen country. Does the number of autos relate directly to air quality? Explain.

Start | Internet

Particulates in Air Pollution

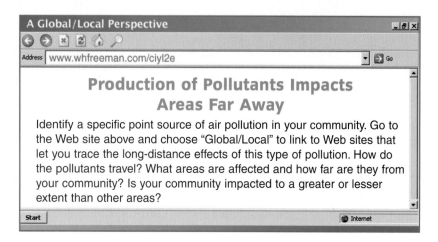

A Global/Local Perspective

Address www.whfreeman.com/ciyl2e

Production of Pollutants Impacts Areas Far Away

Identify a specific point source of air pollution in your community. Go to the Web site above and choose "Global/Local" to link to Web sites that let you trace the long-distance effects of this type of pollution. How do the pollutants travel? What areas are affected and how far are they from your community? Is your community impacted to a greater or lesser extent than other areas?

The Health Effects of Outdoor Air Pollutants

We started this chapter by mentioning that air pollution can affect human health. But how can we determine what the health effects of air pollution are? Generally speaking, we are most concerned with the effects of *long-term* (chronic) exposure to *low* levels of pollution. It is difficult to extrapolate from short-term clinical studies of high-level pollution to the long-term exposures at low levels for a number of reasons. In particular, for some pollutants there may exist a **threshold concentration,** or exposure below which a particular health effect does not occur. In these cases predictions obtained by assuming direct proportionality between exposure and effect would be unwarranted. In addition, chronic exposure could produce deleterious effects that do not come into play when exposure, even intense exposure, to the pollutants occurs only for brief periods of time.

As would be expected, the major effects of air pollution on human health occur in the lungs. For example, although air pollution may or may not *cause* asthma, asthmatics suffer worse episodes of their disease when the sulfur dioxide, ozone, or particulate concentration rises in the air that they breathe. As we shall discuss, the most serious health problems are those that arise from the combination of high concentrations of particulate matter and sulfur dioxide or its oxidation products. Table 14.3 summarizes some of the major health effects of common air pollutants.

Table 14.3 Health effects of pollutant gases and particulates

Pollutant	Health effects
Sulfur dioxide	Respiratory illness; aggravates existing cardiovascular disease; constricts air passages; can trigger asthma attacks
Nitrogen oxides	Irritate lungs; lower resistance to respiratory infections
Carbon monoxide	Reduces oxygen delivery to organs and tissues
Ozone	Induces inflammation in lung tissue; reduces lung function; affects impaired respiratory systems; chest pain; coughing; nausea; pulmonary congestion
Particulates (small)	Affect breathing and respiratory system; damage lung tissue; wheezing; premature death

Figure 14.16 Attempts to reduce the effects of particulate pollution in Beijing, China. (Kevin Lee/Getty Images)

14.19 Soot-and-sulfur smog is a problem

The word *smog* is a combination of *smoke* and *fog*, which were the constituents of the original, nonphotochemical type of smog. In the middle decades of the 20th century, several Western industrialized cities experienced wintertime episodes of such smog so severe that the death rate increased noticeably. For example, in December 1952 about 4,000 people in London, England, died within a few days as a result of the high concentrations of these pollutants that had built up in a stagnant, foggy air mass trapped close to the ground. Those most at risk were elderly persons already suffering from bronchial problems and young children. A ban on household coal burning, from which most of the pollutants originated, has now largely eliminated such problems in London.

Today, due to pollution controls, *soot-and-sulfur smog* is no longer a major problem in Western countries. However, in some areas of Eastern Europe, such as southern Poland, the Czech Republic, and eastern Germany, the quality of winter air was very poor until recently due to the burning of large amounts of high-sulfur (up to 15% S) brown coal. Indeed, even though the average SO_2 level in Prague had decreased by about 50% from the early 1980s to the early 1990s, the pollutant levels were still so high that four out of five children admitted to the hospital in the early 1990s were there for treatment of respiratory problems.

European cities are not the only ones affected by air pollution; levels of both sulfur dioxide and particulate matter regularly exceed World Health Organization guidelines in Beijing, Seoul, and Mexico City (see Figure 14.16). Particulate level guidelines are also exceeded regularly in Bangkok, Bombay, Cairo, Calcutta, Delhi, Jakarta, Karachi, Manila, and Shanghai. In many of these megacities, coal is still the predominant fuel; in some cases diesel-powered vehicles substantially worsen the problem.

14.20 Health effects of ozone and carbon monoxide

Photochemical smog, arising from nitrogen oxides and VOCs, is now more important than sulfur-based smog in most cities, particularly those of high population and vehicle density. As we have seen, it consists of gases, such as ozone, and an aqueous phase, containing water-soluble organic and inorganic compounds, in the form of suspended particles.

Ozone itself is a harmful air pollutant. In contrast to sulfur-based chemicals, its effect on the robust and healthy is as serious as on those with preexisting respiratory problems. Experiments with human volunteers

The Health Effects of Outdoor Air Pollutants

have shown that ozone produces transient irritation in the respiratory system, giving rise to coughing, nose and throat irritation, shortness of breath, and chest pains upon deep breathing. Even healthy, young people often experience such symptoms while cycling or jogging outdoors during smog episodes. It is not yet clear what, if any, long-term lung dysfunction results from exposure to ozone, and indeed this is a controversial subject among scientists.

One anticipated effect of ozone exposure is a decreased resistance to disease from infection because of the destruction of lung tissue. Many scientists believe that chronic exposure to high levels of urban ozone leads to the premature aging of lung tissue. At the molecular level, ozone readily attacks substances containing components with C=C bonds, such as occur in tissues of the lung.

Most industrialized nations have enacted standards (see Table 14.1) which regulate the maximum concentrations in air of sulfur dioxide, nitrogen dioxide, and carbon monoxide as well as ozone since all these gases cause adverse health effects at sufficiently high concentrations. For example, several North American studies have statistically linked the rate of hospitalization for congestive heart failure among elderly people to the daily carbon monoxide concentration in outside air. People such as traffic police who work outdoors in areas of high vehicular traffic can be exposed to elevated CO levels for long periods. The introduction of *oxygenated* substances, which are hydrocarbons in which some of the atoms have been substituted with oxygen, into American gasoline reduces CO emissions from vehicles. Largely as a result of the presence of oxygenated substances in gasoline and the use of catalytic converters, the average outdoor CO levels in the U.S. fell by 37% between 1986 and 1995.

14.21 High particulate concentrations are a health risk

Particulate matter in the form of smoke from burning coal has been an air pollution problem for many hundreds of years, especially in the United Kingdom. Indeed, unsuccessful attempts to control coal burning and to punish offenders had begun there in the 13th century.

As we noted earlier, serious episodes of soot-and-sulfur smog have been largely eliminated in Western industrialized countries. Nevertheless, the air pollution parameter that correlates most strongly with increases in the rate of disease or mortality in most such regions is $PM_{2.5}$. Particle-based air pollution appears to have a greater effect on human health than that produced directly by pollutant gases.

One reason that particles have this effect could be that they carry toxic substances on their surfaces into the lungs. Substances that temporarily stick to the surface of the particle are said to be **adsorbed,** whereas those that dissolve into the body of a particle are said to be **absorbed** by it (see Figure 14.17). In addition, many insoluble airborne particles are surrounded by a film of water, which can itself dissolve other substances. If toxic substances were present in air and inhaled simply as gases, they would be filtered out in the upper respiratory system.

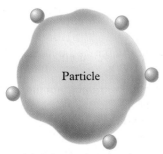

(a) Adsorbed molecules

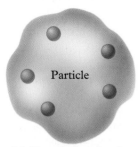

(b) Absorbed molecules

Figure 14.17 The difference between (a) adsorbed and (b) absorbed molecules.

Large particles are of less concern to human health than are small ones for several general reasons:

- Since coarse particles settle out of the air quickly, human exposure to them via inhalation is reduced.

- When inhaled, coarse particles are efficiently filtered by the nose, including its hairs, and the throat; they generally do not travel as far as the lungs. In contrast, inhaled fine particles usually travel to the lungs. Toxic substances adsorbed onto the surfaces of the particles can also affect the lung tissue.

- The amount of surface area per gram of large particles is *smaller* than that of small particles. Thus, gram for gram, the ability of large particles to transport adsorbed gas molecules to any part of the respiratory system, and there to catalyze chemical and biochemical reactions, is correspondingly smaller.

- Devices that are used to remove particulates from air are efficient only for coarse particles. Thus although a device may remove 95% of the total particulate mass, the fractional reduction of fine particles is much smaller.

Some scientists do not believe the causal link between particulate air pollution and human mortality has been proven. They point out that most people spend most of their time indoors, and that consequently their personal exposure to particulates is not tightly linked to outdoor pollution levels. In addition, no biological mechanism has as yet been established to account for the apparent effect of the particles upon health.

The main evidence linking the deterioration of human health to airborne particulates comes from statistical studies correlating death rates with their particulate air pollution levels in different cities. In such studies, the rates of death—either total rates or rates from specific diseases such as lung cancer—are plotted against the average concentration of particulates in order to determine whether they are related to each other.

One important study related particulate air pollution levels to mortality across 50 metropolitan areas in the United States for the 1982–1989 period. The study group consisted of half a million adults. The overall mortality rates were significantly correlated with the $PM_{2.5}$ (fine particle) levels (see Figure 14.18). Several studies in North America have established that *daily* mortality rates, including that of very young children who succumb to sudden infant death syndrome, correlate with PM_{10} values. There is no threshold below which fine particles do not adversely affect mortality. Asthmatic individuals appear to be adversely affected by acidic sulfate aerosols, even at very low concentrations. However, to put the effects of air pollution into perspective, the health risk from smoking greatly exceeds that from air pollution in all categories.

Nevertheless, the U.S. EPA in 1997 decided to limit $PM_{2.5}$ levels to an average of no more than 15 $\mu g/m^3$ annually and 65 $\mu g/m^3$ daily. The EPA estimated that the new particulate standards could prevent 15,000

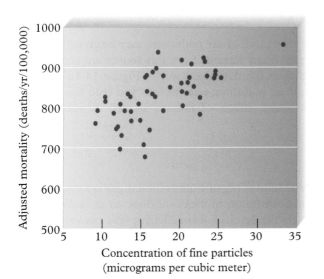

Figure 14.18 Mortality rates correlated with fine particle levels for 50 U.S. cities. (Source: C. A. Pope et al. 1995. Review of epidemiological evidence of health effects of particulate air pollution. *Inhalation Toxicology* 7, pp. 1–18.)

premature deaths, as well as 250,000 person-days of aggravated asthma annually. Like the 1997 ozone standards, however, these regulations have been challenged in court and are not yet implemented.

> ▷ **Discussion Point:** The costs of cleaning the air
>
> The U.S. Clean Air Act is designed to set standards for air pollution levels and to present guidelines for vehicle manufacturing and industrial activities. Use the resources listed on this textbook's Web site to review the Clean Air Act and find information about emission standards for both vehicles and industrial activities. Discuss the potential economic impact of the Clean Air Act. Explore the potential costs to comply with the standards set by the Clean Air Act. Can a business be closed down because of the costs associated with compliance? How is this cost passed along to the consumer?

Indoor Air Pollution

The levels of some common air pollutants often are *greater* indoors than outdoors. Since most people spend more time indoors, exposure to indoor air pollutants is an important environmental problem. Indeed, the inadequate ventilation practices encountered in developing countries that burn coal, wood, crop residues, and other unprocessed biomass fuels create smoke and carbon monoxide pollution that leads to widespread respiratory problems and ill health, particularly in women and young children.

Though high levels of indoor air pollutants may be more frequent in developing countries, many hazards from this source are common in homes and apartments in the developed world. Figure 14.19 summarizes sources of indoor air pollution; refer to it as you read this section. Table 14.3 includes health effects of indoor, as well as outdoor, air pollutants.

14.22 Formaldehyde is found indoors

The most controversial indoor organic air pollutant gas is *formaldehyde,* $H_2C{=}O$. While its concentration in clean outdoor air is normally too small to be important, the level of formaldehyde gas *indoors* is often much greater.

The chief sources of indoor exposure to this gas are emissions from cigarette smoke and from synthetic materials used in *urea-formaldehyde* foam insulation and as adhesives in plywood and particleboard (chipboard). Formaldehyde itself is also used in the dyeing and gluing of carpets, carpet pads, and fabrics. In the first few years after their creation, such materials release small amounts of free formaldehyde gas into the surrounding air. Consequently, new prefabricated structures such as mobile homes that contain chipboard generally have

Figure 14.19 Common sources of pollution in the home.

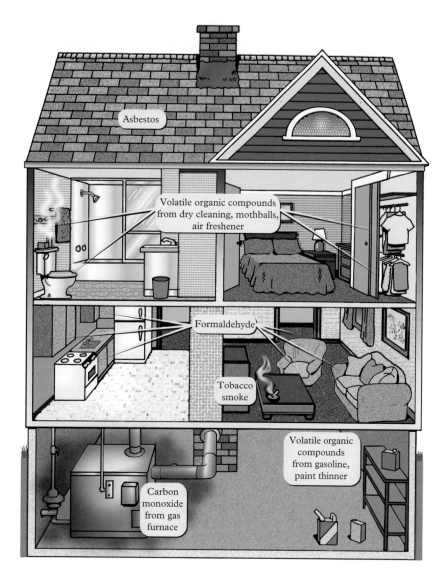

much higher levels of formaldehyde in their air than do older, conventional homes. Many manufacturers of pressed-wood products have now modified their production processes in order to reduce the rate at which formaldehyde is released.

Formaldehyde has a pungent odor that can often be detected in stores that sell carpets and synthetic fabrics. At levels slightly higher than the odor detection threshold, many people report problems of irritation to their eyes, especially if they wear contact lenses, and to their noses, throats, and skin. The formaldehyde in cigarette smoke can cause eye irritation. Formaldehyde in air may cause children to develop more respiratory infections and allergies as well as asthma, although evidence for these effects is very controversial.

Indoor Air Pollution

Formaldehyde is established as a carcinogen in test animals and was classified as a *probable human carcinogen* by the U.S. EPA in 1987. The expected cancer sites are in the respiratory system, including the nose. However, studies of human populations exposed to formaldehyde have led to no clear-cut conclusions concerning an increase in cancer frequency due to nonoccupational exposure to formaldehyde.

14.23 Carbon monoxide can be deadly

We discussed the formation and environmental hazards of carbon monoxide in some detail in Chapter 2, so here we will just summarize its main indoor effects. Carbon monoxide is a colorless, odorless gas whose concentration indoors can be greatly increased by the incomplete combustion of carbon-containing fuels such as wood, gasoline, kerosene, or gas. High indoor concentrations usually are the result of a malfunctioning combustion appliance, such as a kerosene or gas heater or stove or furnace. In developing countries, carbon monoxide poisoning is a serious hazard when biomass fuels are used to heat poorly ventilated rooms in which people sleep.

Low-priced, easily installed carbon monoxide detectors suitable to warn residents in homes and offices when high CO levels occur are now on the market and have been installed in many homes that use natural gas for heating. Average indoor and outdoor CO concentrations usually amount to a few parts per million, though elevated values are common in parking garages due to the carbon monoxide emitted by motor vehicles.

14.24 Nitrogen oxides are indoor air pollutants

The gases NO and NO_2 are not only outdoor air pollutants; they are indoor pollutants as well. This is particularly true in homes and apartments that are heated by natural gas furnaces, and especially where gas appliances are used for cooking food and as space and water heaters. The temperature of the flame that is produced when gas burns in air is high enough to produce some nitric oxide, which is partially converted to nitrogen dioxide. Homes that use natural gas for cooking or that have a kerosene stove have average NO_2 levels that are similar to those in outdoor air for urban areas and about three times those for homes without these appliances.

Nitrogen dioxide can dissolve in living tissue and can react with it, and so it is expected to have a negative effect on human health, especially in the respiratory system since it is present in the air. One study found that an increase of 15 ppb in the average NO_2 concentration in a home leads to about a 40% increase in symptoms of illness in the lower respiratory system among children aged 7 to 11 years. However, the conclusions of many studies of the effects on respiratory illness in children from exposure to NO_2 produced from gas appliances are not consistent, so it has not yet been definitively determined whether or not indoor exposure to this gas causes illness.

Nitrogen oxide produced inside the body plays several vital roles. NO is a chemical messenger, or "signaling molecule," and helps regulate blood pressure by relaxing blood vessel walls.

During sexual stimulation in males, nerves in the penis release nitric oxide, which relaxes muscle tissue at the base of the penis, allowing blood to flow into the shaft. The drug Viagra works by preventing the breakdown of nitric oxide and thus keeps blood vessels relaxed and maintains the blood flow necessary for an erection.

Figure 14.20 Smokers intake cigarette smoke into their mouths and noses. (Photodisc Red/Getty Images)

14.25 Health effects of secondhand smoke

It is well established that smoking tobacco is the leading cause of lung cancer and is one of the main contributors to heart disease. Nonsmokers are often exposed to cigarette smoke, although in lower concentrations than smokers since it is diluted by air. This **environmental tobacco smoke (ETS)**—also known as *secondhand smoke*—has been the subject of many investigations in order to determine whether or not it is harmful to people who are exposed to it (see Figure 14.20).

ETS consists of both gases and particles. The concentration of some toxic products of partial combustion is actually *higher* in sidestream smoke—that is, the smoke that leaves the smoldering cigarette between puffs—than in mainstream smoke, the substance the smoker inhales directly. This occurs because combustion occurs at a lower temperature—and consequently is less complete, producing more pollutants—in the smoldering cigarette than in one through which air is being inhaled. However, the sidestream smoke is usually diluted by ambient air before being inhaled, so the concentrations of pollutants reaching the lungs of nonsmokers are much lower than those reaching the lungs of smokers themselves.

The chemical constitution of tobacco smoke is complex: it contains thousands of components, several dozen of which are carcinogens. The gases include carbon monoxide, nitrogen dioxide, formaldehyde, cadmium, polycyclic aromatic hydrocarbons (mentioned in Chapter 5), and other VOCs. The particulates, called **tar,** contain nicotine and the less volatile hydrocarbons, much of these in the fine particle range.

The gaseous components of ETS, especially formaldehyde, hydrogen cyanide, acetone, toluene, and ammonia, cause most of the odor and cause irritation to the eyes and respiratory system of many people. Exposure to ETS aggravates the symptoms of many who suffer from asthma or from *angina pectoris,* chest pains brought on by insufficient oxygen reaching the heart. Some recent studies have established a correlation between the rate of acute respiratory illness and the level of indoor $PM_{2.5}$, the total concentration of fine particulates from all indoor sources, including tobacco smoke. **Passive smoking**—inhalation of sidestream as well as already exhaled smoke—is believed by scientists to cause bronchitis, pneumonia, and other infections in up to 300,000 infants, and several thousand instances of sudden infant death, in the United States each year.

In 1993, the U.S. EPA classified ETS as a *known human carcinogen* and estimated that it causes about 3000 lung cancer deaths annually. ETS is also considered to be responsible for killing as many as 60,000 Americans annually from heart disease. An analysis of all recent studies on passive smoking led to the conclusion that the risks of acquiring lung cancer and heart disease each are increased by about one-quarter for nonsmoking spouses of smokers.

14.26 Asbestos fibers damage the lungs

The term *asbestos* refers to a family of six naturally occurring silicon-based minerals that are fibrous. Chemically, they are composed of long double-stranded networks of silicon atoms connected through intervening oxygen atoms; the net negative charge of this silicate structure is neutralized by the presence of cations such as magnesium.

The most commonly used form of asbestos, *chrysotile*, $Mg_3Si_2O_5(OH)_4$, is a white solid whose individual fibers are curly (see Figure 14.21a). Chrysotile is mined mainly in Quebec and is the principal type of asbestos used in North America. In the past, it was employed in huge quantities because of its resistance to heat, its strength, and its relatively low cost. Common applications of asbestos included its use as insulation and spray-on fireproofing material in public buildings, in automobile brake-pad lining, as an additive to strengthen cement used for roofing and pipes, and as a woven fiber in fireproof cloth.

The use of asbestos of all types has been sharply reduced because studies on the health of asbestos miners and other asbestos workers have shown it to be a human carcinogen. It causes *mesothelioma*, an incurable cancer of the lung, abdomen, and heart. Airborne asbestos fibers and cigarette smoke act **synergistically**—that is, their combined effect is greater than the sum of their individual effects—in causing lung cancer. There is much controversy concerning whether chrysotile should be banned outright from further use and whether or not existing asbestos in buildings should be removed. Many experts feel that asbestos should be left in place unless it becomes damaged enough that there is a chance that its fibers will become airborne. Indeed, its removal can increase dramatically the levels of airborne asbestos in a building unless extraordinary precautions are taken not to allow detached fibers from the asbestos being removed to escape into the air. One scientist stated: "Removing asbestos is like waking up a pit bull terrier by poking a stick in its ear. We should let sleeping dogs lie." However, some environmentalists feel that asbestos is a ticking time bomb—that it should be removed as soon as possible, as one can never predict when building insulation will be damaged.

Most of the initial concern about asbestos was related to *crocidolite*, blue asbestos (see Figure 14.21b). Evidence implicating this material as

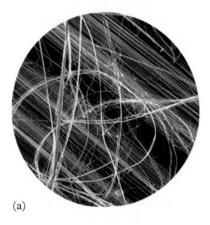 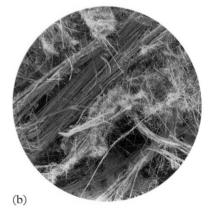

Figure 14.21 The structure of asbestos fibers. (a) Chrysotile and (b) crocidolite asbestos. (Part a, Dee Breger/Photo Researchers; part b, Dr. Jeremy Burgess/Photo Researchers)

(a)　　　　　　　(b)

a cause of cancer in humans was already well established several decades ago. A material with thin, straight, and relatively short fibers, it more readily penetrates lung passages and is a more potent carcinogen than the white form. Crocidolite, which is mined in South Africa and Australia, has not been used much in North America.

▶ Discussion Point: Asbestos removal

Asbestos removal from older buildings has become a priority based upon evidence that asbestos is a health hazard. Removal, however, can pose its own health risk. Use the resources on the Web site for this textbook to identify the health hazards associated with both keeping and removing asbestos. Could removing asbestos pose more of a health hazard than keeping it in place? Are there alternatives to asbestos removal? Develop an argument for and an argument against the removal of asbestos from older buildings.

■ Chemistry in Your Home: The air pollutants in your home

Indoor pollution arises from many different sources. Using the information in this chapter, make a list of the sources and types of pollution in your home or dorm. What types of pollutants are most prevalent in your environment? How can you decrease your exposure to indoor pollutants?

Summarizing the Main Ideas

Photochemical smog is a phenomenon in which nitric oxide and VOC pollutants are converted in air by sunlight and warmth over a period of hours and days. It consists of a complex mixture of ozone, nitric acid, and partially oxidized organic compounds. The nitrogen oxides are produced by the reaction of the nitrogen and oxygen in air at high temperatures. This reaction can be reversed in catalytic converters on vehicles and by comparable processes in power plants, re-forming N_2 so that the oxides themselves are not emitted into the air.

Substances emitted directly into the environment are called primary pollutants; substances chemically produced from them are secondary pollutants.

Acid rain is the acidification of atmospheric precipitation by strong acids, principally sulfuric and nitric acids, that are produced from the primary pollutants SO_2 and NO. To be considered acid rain, the pH of precipitation must be significantly less than that (of about 5) produced by carbonic acid derived from atmospheric carbon dioxide.

Particulates are tiny solid or liquid particles that are suspended temporarily in air. Fine particulates have diameters less than 2.5 μm, are relatively long-lived, and result mainly from the chemical reaction of gaseous pollutants. Coarse particles have diameters greater than 2.5 μm, are relatively short-lived, and result mainly from the mechanical breakup of larger ones, such as those in soil.

The concentration in air of particulate matter is usually reported by PM_x values, where x represents the maximum diameter in micrometers of the particulate matter considered. PM_x values correspond to the mass in micrograms of this component of the particulate matter that is present in one cubic meter of air. The concentration of particulate matter in air can be lowered by reducing the emissions of the primary pollutants which would form it or by trapping primary particles, such as the soot produced in diesel engines, before it is emitted into the air.

Numerous statistical surveys show that human health is degraded significantly by long-term exposure to atmospheric pollution, both gases and particulates. Gases can also be carried to the lungs if their molecules are temporarily adsorbed onto the surface of particles small enough to reach the lungs. Fine particles are generally more of a long-term health concern than are coarse ones. The rate of mortality from lung cancer and pulmonary disease increases in proportion to the average outdoor levels of small particulates. Asthma attacks also increase when the particulate levels are high. Ozone gas causes short-term breathing problems at least, and perhaps long-term effects as well.

Important indoor air pollutants include formaldehyde gas, the main source of which is emissions from synthetic materials; carbon monoxide from incomplete combustion; environmental tobacco smoke; and asbestos.

Key Terms

photochemical smog	anthropogenic	aerosol	absorbed
volatile organic compound	catalytic converters	sulfate aerosol	environmental tobacco smoke
primary pollutant	acid rain	PM index	tar
secondary pollutant	clean coal technology	particle trap	passive smoking
NO_x	nonferrous smelting	threshold concentration	synergistic effect
	particulate	adsorbed	

Web Sites of Interest

To link to Web sites of interest, go to www.whfreeman.com/ciyl2e, Chapter 14, and select the site you want.

For Further Reading

D. W. Dockery et al., "An Association Between Air Pollution and Mortality in Six U.S. Cities," *New England Journal of Medicine, 329* (24), 1993, pp. 1753–1759. The paper presents findings of a study relating particulate air pollution to mortality. The study gathered particulate pollution data and mortality statistics from 8111 living adults who were followed through a 12-year period.

W. R. Ott, "Everyday Exposure to Toxic Pollutants," *Scientific American,* February 1998, pp. 86–91. This article discusses the emerging issues related to exposure to indoor pollutants, especially exposure of children. If you access the article through the Scientific American Web site (www.sciam.com) you can link to other sites that deal with indoor pollution.

J. Raloff, "Lemon-Scented Products Spawn Pollutants," *Science News, 158,* December 9, 2000, p. 375. This short article summarizes an experiment that examines the formation of particulates indoors as a result of the reaction between volatile components of lemon-scented products and indoor ozone. The reactants are similar to those emitted by evergreens, which react with ozone in a similar fashion to produce particulate haze around forests.

Review Questions

1. Where in the atmosphere is air pollution, such as smog, found?

2. Identify the major components of outdoor air pollution.

3. Explain what is meant by the term *photochemical smog.*

4. What is a *volatile organic compound*? Provide an example.

5. How are volatile organic compounds related to smog?

6. How does a *primary pollutant* differ from a *secondary pollutant*? Provide an example of each.

7. How are nitrogen oxide gases formed?

8. Which two sources produce most of the urban NO?

9. To what does the term NO_x refer?

10. What conditions work to create photochemical smog in a city?

11. Describe some of the detrimental effects of ozone.

12. What are *anthropogenic hydrocarbons*? Provide an example.

13. Identify some natural sources of hydrocarbons.

14. Explain what a *catalytic converter* is and describe how it works.

15. Describe, in words, how a two-stage combustion process in power plants helps to minimize the amount of NO_x that power plants produce and emit.

16. What is *acid rain*?

17. Is the natural pH of rain water relatively acidic or relatively basic? Explain.

18. What natural phenomena can produce acid rain?

19. What makes acid rain acidic?

20. What natural phenomena produce SO_2?

21. Identify the primary anthropogenic sources of SO_2.

22. Why do fogs and mists tend to be more acidic than rain?

23. What is a *particulate*?

24. What are the size categories that describe particulates?

25. How do *coarse particles* differ from *fine particles*?

26. What is an *aerosol*?

27. How does *soot* differ from *mist*?

28. Identify some sources of coarse and fine particulates.

29. How do the sources of coarse and fine particulates differ?

30. How does the general composition of a coarse particle differ from that of a fine particle?

31. What is a *sulfate aerosol*?

32. How are sulfate aerosols formed?

33. What is the *PM index*?

34. What does the subscript on PM tell you?

35. Identify the major components of indoor air pollution.

36. What is meant by the term *threshold pollutant concentration*?

37. Explain how *adsorption* differs from *absorption*.

38. Identify the main sources of indoor exposure for each of the following pollutants:
    a) formaldehyde
    b) Carbon monoxide
    c) ETS
    d) asbestos
    e) nitrogen oxides

39. Describe the physical effects of exposure to each of the following pollutants:
    a) formaldehyde
    b) Carbon monoxide
    c) ETS
    d) asbestos
    e) nitrogen oxides

40. Explain what the term *passive smoking* means.

41. Identify some of the components of cigarette smoke.

Understanding Concepts

42. What are the initial reactants in the formation of photochemical smog?

43. What weather and geographical features favor the formation of photochemical smog?

44. Why would a reduction in NO_x emissions instead of VOC emissions be an effective way to decrease ground-level ozone?

45. What roles do deciduous trees play in the production of ground-level ozone?

46. Describe some of the strategies that have been developed to reduce urban ozone levels. Which, if any, have been effective?

47. Explain why ozone is both a pollutant and a necessary component of the atmosphere. What is the difference between ozone that pollutes and ozone that protects?

48. How does a catalytic converter control NO_x emissions?

49. Why are catalytic converters in diesel engines less effective than those in gasoline engines?

50. Describe the strategies that have been developed to address the problem of acid rain.

51. Use chemical reactions to show how acid rain is neutralized by limestone.

52. Describe the effects of acid rain on:
    a) dissolved aluminum
    b) fish populations
    c) trees

53. Why are coarse particles deposited on the ground more quickly than fine particles?

54. What is a *particle trap*? Explain how a particle trap reduces particulate emissions.

55. What is the relationship between the temperature of a catalytic converter and its efficiency?

56. What would a subscript of 40 on the PM index indicate?

57. How is outdoor air quality determined?

58. How is indoor air quality determined?

59. Compare the major components of indoor and outdoor air pollution in terms of their chemical composition and state (that is solid, liquid, or gas). Do they differ?

60. Describe the major health effects of indoor and outdoor air pollution. Where in the body do the major effects occur for each type of pollution?

61. What is the relationship between the threshold pollutant concentration and negative health effects?

62. Which populations are most at risk of being affected by air pollution? Why are these particular populations most vulnerable?

Synthesizing Ideas

63. Why is sunlight required in the chemical process that produces photochemical smog?

64. Why do you think that ozone affects the properties of rubber?

65. Explain why the dominant acid component of acid rain is different in eastern and in western North America.

66. Explain why the acidification of a lake can lead to a decline in plant, aquatic, and animal life.

67. How do the biological effects of smaller particles differ from those of larger particles? Why do you think the biological effects are different?

68. Explain the relationship between $PM_{2.5}$ levels and mortality rates as suggested by research results. (Refer to the *New England Journal of Medicine* article in *For Further Reading*.)

69. Many homes are now built to be as airtight as possible in order to achieve maximum energy efficiency and insulation. Based on the material in this chapter, write a brief paragraph comparing the potential indoor pollution problems associated with older and newer homes. (For more information refer to the *Scientific American* article in *For Further Reading*.)

70. Use the following information to calculate the mass of fine particles inhaled by an adult each year. A person inhales 350 liters of air every hour. The average $PM_{2.5}$ index of the air is 10 micrograms per cubic meter (1000 liters = 1 cubic meter).

In this chapter, you will learn:

- about the ozone layer, what it is and how it protects us by filtering UV light from the sun;

- how UV light causes skin cancer and cataracts, and how it affects plants and animals;

- how ozone is continually created and destroyed, and why a hole forms in the ozone layer over Antarctica for several months each year;

- which chemical compounds increase the destruction of ozone, which are safer replacement compounds, and what the international community is doing to save the ozone layer.

It sometimes seems like a nuisance to use sunscreen when you spend time in the Sun; why is it so important?

Chemicals in sunscreen protect your skin from the harmful effects of the Sun in the same way the ozone layer protects Earth from harmful solar rays. In this chapter we examine the chemicals in the ozone layer—what they are, how they react, and more importantly, how human activities can affect them. (Richard Rickel/Black Star Publishing/PictureQuest)

A Thin Veil of Protection

Stratopheric Chemistry and the Ozone Layer

When we talked in Chapter 14 about ground-level air pollution from ozone, we contrasted it to the "good" ozone that exists high in Earth's atmosphere. This so-called **ozone layer** has been called "Earth's natural sunscreen" since it filters out harmful rays from sunlight before they can reach the surface of our planet and cause damage to humans and other life forms. The ozone layer is located at an altitude of approximately 15 to 30 kilometers, in the lower part of the **stratosphere** (see Figure 15.1). Even there, only a very small fraction of the air molecules are those of *ozone*, O_3, rather than O_2 or N_2. If all the ozone were to be brought down to ground level, this overhead layer of pure ozone would be only 3.5 mm thick.

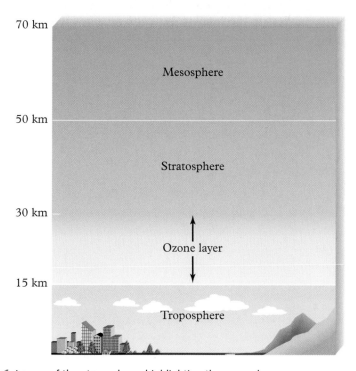

Figure 15.1 Layers of the atmosphere, highlighting the ozone layer.

Whenever you see this icon in this chapter, go to
www.whfreeman.com/ciyl2e

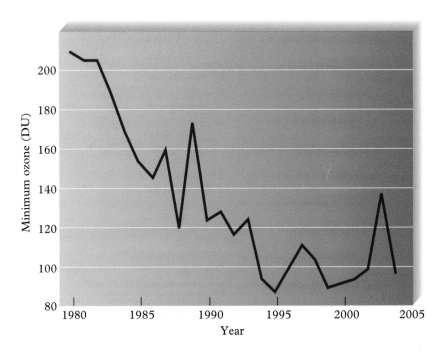

Because of stratospheric winds, ozone is transported on air currents from tropical regions, where most of it is produced, toward polar regions. Thus, ironically, the closer to the equator (where the sunlight is strongest) you live, the *less* the total amount of ozone that protects you from ultraviolet light. A major reduction in the quantity of stratospheric ozone would threaten life as we know it. Thus the appearance in the mid-1980s of a large hole in the ozone layer over Antarctica represented a major environmental crisis.

In this chapter, we shall investigate the chemical processes that occur in the stratosphere—those involved in the production of the ozone layer, as well as those underlying the phenomenon of ozone depletion. We shall then investigate the sources of the chemicals that cause the depletion and what has been done to stop the problem.

The Ozone Layer

15.1 There is an ozone hole over Antarctica

The ozone level above any location varies naturally with season. Ozone normally peaks in the spring and reaches a minimum in the fall.

The total amount of ozone in the atmosphere that lies over a given point is measured in Dobson units. On average, this total ozone amounts to about 350 Dobson units for those who live in temperate regions. Scientists studying the atmosphere have found that since the late 1970s, the total amount of stratospheric ozone above Antarctica in the springtime (September to November) has been falling each year, though almost normal levels are recovered by each midsummer. By the mid-1980s, the springtime loss in the total overhead ozone amounted to over 50% (see Figure 15.2). It is therefore appropriate to speak of a hole in the ozone

Figure 15.2 Minimum concentration of total overhead ozone above Antarctica each spring in recent years. (NASA Web site http://toms.gsfc.nasa.gov/multi/min_ozone.jpg)

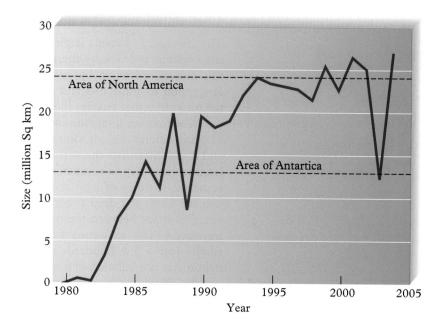

Figure 15.3 The variation with year of the size of the Antarctic ozone hole. (NASA Web site http://toms.gsfc.nasa.gov/multi/oz_hole_area.jpg)

layer, which now appears each spring over the Antarctic and lasts for several months. The geographic area covered by the ozone hole has increased substantially since it began, and is now about as large as the North American continent (see Figure 15.3).

Initially it was not clear whether the hole was due to a natural phenomenon involving meteorological forces or to a chemical mechanism involving air pollutants. In the latter possibility, the suspect chemical was chlorine, produced mainly from synthetic chlorofluorocarbon (CFC) gases. Large quantities of these gases, which were used in various appliances such as air conditioners, had been released into the air as a consequence of their use and dispersed by winds around the world. In the 1970s, in fact, scientists had predicted that the chlorine from this source would destroy ozone—but only to a small extent and only after several decades. The discovery of the Antarctic ozone hole came as a complete surprise to everyone. Subsequent research revealed that the hole is indeed a result of chlorine pollution.

As a consequence of these discoveries, governments worldwide moved quickly to legislate a phaseout in production of the responsible chemicals so that the situation did not become much worse, with the development of even more severe ozone depletion over populated areas and the corresponding threat to the health of humans and other organisms. Even so, the Antarctic hole will probably continue to reappear each spring until the middle of this century, and a hole of corresponding size may one day appear above the Arctic region.

Ozone is being depleted not just in the air above Antarctica, but to some extent worldwide. The greatest depletion in midlatitude regions occurs in the winter–spring season. There is some depletion in the summertime, when most people's exposure to sunlight is at a maximum.

15.2 Molecules selectively absorb light energy

The chemistry of ozone depletion, and of many other processes in the stratosphere, is driven by energy associated with light from the Sun. For this reason, we begin by investigating the relationship between light absorption by molecules and the resulting activation, or energizing, of the molecules that enables them to react chemically.

When we see a rainbow, we become aware that sunlight is made up of many different colors. One end of the rainbow is violet, and the other end is red. Scientists characterize different colors of light by their wavelength, since they have established that light is a wave phenomenon. In some ways, waves of light are similar to the successive rings of waves you see when you drop a stone into the water of a still pond. The distance in space between the peaks of successive waves is known as the *wavelength*. The analogous wavelengths of visible light are incredibly small—just a few hundred nanometers, where, as we saw earlier, a nanometer is a billionth of a meter (10^{-9} m).

An object that we perceive as black in color absorbs all wavelengths of visible light, which is that from about 400 nm (violet light) to about 750 nm (red light). Substances differ enormously in their propensity to absorb light of a given wavelength. *Diatomic molecular oxygen, O_2,* does not absorb visible light significantly, but it does absorb some types of **ultraviolet (UV) light,** which has wavelengths that lie between about 50 and 400 nm (see Figure 15.4) and which also is a component of sunlight. Notice that the UV region begins at the violet edge of the visible region, hence the name ultraviolet. (We will discuss the individual components—UV-A, UV-B, and UV-C—in section 15.3.)

When a molecule absorbs visible or UV light, the bonding pattern of some of its electrons changes, and consequently its properties change somewhat. Indeed, as we shall see below, the bond between two atoms might break as a result of the extra energy associated with the light, and the molecule will come apart at that junction.

An *absorption spectrum,* such as that illustrated in Figure 15.5, graphs the fraction of light that is absorbed by molecules of a substance as a function of wavelength. O_2 molecules efficiently absorb light between 70 and 250 nm. A minuscule amount of absorption continues beyond 250 nm, but in an ever-decreasing fashion. The fraction of light absorbed by O_2 varies quite dramatically with wavelength. This sort of selective absorption behavior occurs for *all* molecules, although the specific regions of strong absorption and of zero absorption vary widely.

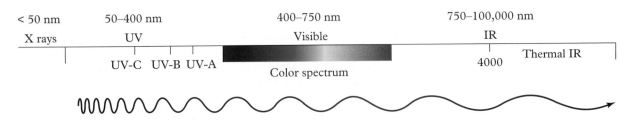

Figure 15.4 Light is classified according to its wavelength.

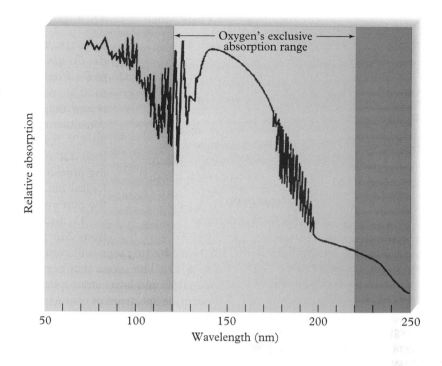

Figure 15.5 Absorption spectrum of diatomic oxygen. Oxygen alone filters UV light from sunlight in the 120- to 220-nm range. (Adapted from T. E. Graedel and P. J. Crutzen, *Atmospheric Change: An Earth System Perspective,* New York: W. H. Freeman and Company, 1993.)

■ Chemistry in Your Home: Glows in your mouth or in your hand *Based on an experiment provided by Professor Maria Dean, Coe College*

Energy is often visible to the naked eye in the form of light. One example of this is triboluminescence, which is the emission of light when certain materials are crushed, scratched, or rubbed. Other types of luminescence you may be familiar with are bioluminescence (when the ocean's surface seems to glow in the moonlight due to the presence of microorganisms) and fluorescence (often used in overhead office lighting).

Using WintOGreen LifeSavers, you can observe the phenomenon of triboluminescence in your own room. If you are alone, you'll need a mirror to observe the light. Turn off all the lights—the room must be completely dark. (A bathroom without windows may be an even better location.) Allow your eyes to acclimate to the darkness for about 10 seconds. Then, unwrap the candy and bite into it. (You can also hit the candy with a hammer, as an alternative to biting into it.) What do you see?

Another option for observing the phenomenon of triboluminescence is by use of Curad bandages (the Band-Aid brand will not work in this experiment). Once the room is dark and you have given your eyes time to acclimate, quickly pull apart the paper wrapper by pulling apart the tabs at the one end of the bandage. (Note: This only works when you use the *outside* wrapper of the bandage.) What did you observe? Was it the same or different from what you observed with the WintOGreen LifeSavers? If the light was a different color in the two cases, what does this say about the wavelengths involved? From the colors of the light, can you find approximately where in the visible spectrum (see Figure 15.4) the emissions occurred?

The Ozone Layer

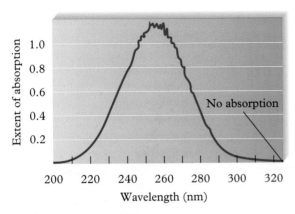

Figure 15.6 **Absorption spectrum of ozone. Ozone filters UV light in the 220- to 320-nm range.** (Adapted from M. J. McEwan and L. F. Phillips, *Chemistry of the Atmosphere*, Edward Arnold Publishers, 1975.)

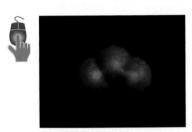

View representations of oxygen and ozone molecules at Chapter 15: Visualizations: Media Link 1.

15.3 O_2 and O_3 filter sunlight

As a result of its light absorption characteristics, the O_2 gas that lies above and in the stratosphere filters from sunlight all of the UV light from 120 to 220 nm. Ultraviolet light with wavelengths *shorter* than 120 nm is filtered by O_2 and other constituents of air such as N_2. Consequently, no UV light having wavelengths that are shorter than 220 nm reaches Earth's surface. This screening protects our skin and eyes, and in fact all biological life, from extensive damage by this part of the Sun's output.

O_2 also filters some, but not all, of the UV light from 220 to 240 nm. The ultraviolet light in the 220- to 320 nm range is filtered from sunlight mainly by the ozone molecules, O_3. The ozone that performs this screening is located through the middle and lower stratosphere—in other words, the ozone layer. The absorption spectrum of ozone in this wavelength region is shown in Figure 15.6. Since its molecular constitution and bonding are different from those of diatomic oxygen, ozone's light absorption characteristics are quite different.

Ozone, aided to some extent by O_2 at the shorter wavelengths, filters out all of the Sun's ultraviolet light in the 220- to 290-nm range, which overlaps the 200- to 280-nm region known as **UV-C** (see Figure 15.4). However, ozone absorbs only a fraction of the Sun's UV light in the 290- to 320-nm range, since, as you can infer from Figure 15.6, its inherent ability to absorb light of such wavelengths is quite limited. The remaining amount of the sunlight of such wavelengths, 10–30% depending upon latitude, penetrates the atmosphere through to Earth's surface. Thus ozone is never *completely* effective in shielding us from light in the **UV-B** region, defined as that which lies from 280 to 320 nm. Since the absorption by ozone falls off quickly with wavelength in this region, the fraction of UV-B from the Sun that reaches the troposphere increases with wavelength.

Because neither ozone nor any other constituent of the clean atmosphere absorbs significantly in the **UV-A** range, 320–400 nm, most of this light does penetrate to Earth's surface. UV-A is the least biologically harmful type of ultraviolet light.

The Biological Consequences of Ozone Depletion

The reduction in stratospheric ozone concentration allows more UV-B light to penetrate to Earth's surface. A 1% decrease in overhead ozone is predicted to result in a 2% increase in UV-B intensity at ground level. This increase in UV-B is the principal environmental concern about ozone. Exposure to UV-B causes human skin to sunburn and suntan. Overexposure can lead to skin cancer, the most prevalent form of cancer. Increasing amounts of UV-B may also adversely affect the human immune system and the growth of some plants and animals.

15.4 Skin cancer and cataracts are related to UV exposure

Most biological effects of sunlight arise because UV-B can be absorbed by DNA molecules, which may then undergo damaging reactions. The absorption of UV radiation can cause two thymine bases to become covalently bonded. This linkage disrupts the local DNA structure (see Figure 15.7) and confuses DNA polymerase, which then makes mistakes when synthesizing a new DNA strand (see Chapter 9, section 9.3).

Most skin cancers in humans are due to overexposure to UV-B in sunlight, and so any decrease in ozone is eventually expected to yield an increase in the incidence of this disease. Fortunately, the great majority of skin cancer cases are of a slowly spreading type that can be treated. These cancers currently affect about one in four Americans at some point in their lives. Indeed, the incidence of skin cancer is approximately equal to the combined incidences of all other cancers.

The incidence of the often fatal **malignant melanoma** form of skin cancer, which affects about 1 in 100 North Americans, is thought to be related to short periods of very high UV exposure, particularly early in life. Most susceptible are fair-skinned, fair-haired, freckled people who burn easily and who have moles with irregular shapes or colors. The lag period between first exposure and melanoma is 15–25 years. If malignant melanoma is not treated early, it can spread via the bloodstream to body organs such as the brain and the liver.

The incidence of malignant melanoma is related to latitude. White males living in sunny climates such as Florida or Texas are twice as likely to die from this disease as those in the more northerly states. However, part of this difference is probably due as much to different patterns of personal behavior, such as clothing that leaves much of the body exposed, as it is to increased UV-B content in southern sunlight. Curiously, indoor workers—who are only intermittently exposed to the Sun—are more susceptible than are tanned, outdoor workers!

If you look in your pharmacy or supermarket, you'll see an enormous array of sunscreens offering to protect you against the Sun's rays. However, the effectiveness of these products varies greatly. In fact, the use of sunscreens that block UV-B but not also UV-A may actually lead to an *increase* in skin cancer, since sunscreens allow people to expose

Nonmalignant skin cancer is thought to be initiated by sunlight-induced mutation at DNA sites having adjacent pyrimidine bases (cytosine and/or thymine). The mutation prevents the normal self-destruction of the cell that occurs if it is subsequently badly injured by additional UV exposure.

The phrase "full spectrum" is sometimes used to denote sunscreens that block UV-A as well as UV-B light.

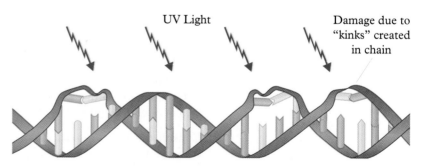

UV Light

Damage due to "kinks" created in chain

Figure 15.7 UV damage in DNA molecules. New chemical bonds form, creating kinks in the DNA chain.

their skin to sunlight for prolonged periods without burning. The substances used in sunscreen lotions and lip balms either reflect sunlight or absorb its UV component before it can reach the skin. The chemicals chosen for sunscreens are ones that do not undergo chemical reactions when they absorb sunlight, because this would quickly reduce the effectiveness of the application and because the reaction products could be toxic to the skin.

Sunscreens were one of the first consumer products to use *nanoparticles,* that is, tiny particles only a few dozen or a few hundred nanometers in size. Since such particles are so tiny and do not absorb or reflect visible light, the sunscreens appear transparent. The SPF (Sun Protection Factor) of a sunscreen measures the multiplying factor by which you can stay exposed to the Sun without burning. Thus an SPF of 15 means you can stay in the Sun 15 times longer than you could have without the sunscreen. To receive that protection, however, you must reapply the sunscreen every few hours, or even more frequently if any of the following conditions apply: it is windy, you are at a high altitude, you are in the tropics, you have been swimming, or you have become sweaty.

It is predicted that there will eventually be a 1–2% increase in malignant skin cancer incidence for each 1% decrease in overhead ozone. Because there is a long time lag between exposure to UV light and the subsequent manifestation of nonmalignant skin cancers, it is unlikely that effects from ozone destruction are observable as yet. The rise in the skin cancer rate that has occurred in many areas of the world is probably due instead to greater amounts of time spent outdoors in the sun over the past few decades. For example, the incidence of skin cancer among residents of Queensland, Australia, most of whom are light-skinned, rose to about 75% of the population as lifestyle changes increased their exposure to sunlight years before ozone depletion began. As a consequence of its experience with skin cancer, Australia has led the world in public health awareness of the need for protection from ultraviolet exposure.

■ Chemistry in Your Home: How much UV are you exposed to?

Assess your exposure to ultraviolet radiation by considering the following:

a) Amount of time you spend outdoors (including time of day, season, latitude, altitude, and whether snow, sand, or water are part of your environment)
b) Use of sunlamps or tanning salons
c) Use of sunscreens
d) Use of sunglasses that filter 99–100% of UV
e) Use of clothing that covers the body
f) Use of wide-brim hats

Consult the Web sites listed under UV/sunscreens/skin cancer at the Web site for this textbook to determine the extent of your exposure. Using the Web resources, assess your risk of skin cancer based upon your total exposure. How does your risk compare to the average risk?

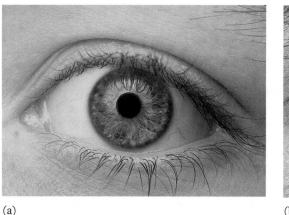
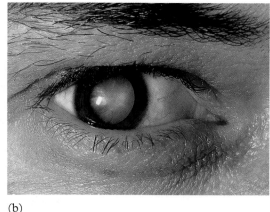

(a)   (b)

Figure 15.8 (a) A normal human eye and (b) a human eye with cataract. (Part a, Martin Dohrn/Photo Researchers; part b, Sue Ford/Photo Researchers)

Beyond skin cancer, UV exposure has been linked to a number of other human conditions. The front of the eye is the one part of the anatomy where ultraviolet light can penetrate the human body. However, the cornea and lens filter out about 99% of UV from light before it reaches the retina. Over time, the UV-B absorbed by the cornea and lens produces substances that attack the structural molecules and can produce cataracts, a condition in which the lens becomes opaque, causing partial or complete blindness (see Figure 15.8). Indeed, there is evidence that increased UV-B levels give rise to an increased incidence of cataracts, particularly among the nonelderly. Increased UV-B exposure also leads to a suppression of the human immune system, probably with a resulting increase in the incidence of infectious diseases.

15.5 Increased UV exposure affects plants and animals

Humans are not the only organisms affected by ultraviolet light. Increases in UV-B exposure can interfere with the efficiency of photosynthesis, and plants may respond by producing less leaf, seed, and fruit. All organisms that live in the first 5 meters or so below the surface in bodies of clear water will also experience increased UV-B exposure as a consequence of ozone depletion and may be at risk. If the production of the microscopic plants called *phytoplankton* near the surface of seawater is affected by increased UV-B, the marine food chain for which they form the base would be affected in turn. The interrelationship between plant production and UV-B intensity is particularly complex since UV also affects the survival of insects that feed off the plants.

The recent worldwide drop in the population of frogs and other amphibians and the increase in deformities in frogs that do survive have also been linked to increasing levels of UV. In particular, the UV-induced mortality of amphibian eggs laid in the springtime in shallow waters may be combining with habitat destruction to cause this decline.

The Interplay of Light with Chemistry in the Stratosphere

15.6 Light's energy varies with wavelength

Ultraviolet light is dangerous to living matter, whereas visible light generally is not. The difference comes about because the amount of energy imparted to a molecule when it absorbs light varies dramatically with the wavelength of the light. Although a beam of light appears to be a continuous source of energy, Albert Einstein and other scientists discovered a century ago that light's energy can only be absorbed by matter in finite packets, now called *photons*. A given molecule in a gas can absorb one photon from a light beam and must dispose of it in some manner before another photon can be absorbed.

The quantity of energy, E, that is associated with a photon depends on the light's wavelength in an inverse way: the longer the wavelength, the smaller the amount of energy absorbed. Consequently, infrared light, which has wavelengths greater than those in visible light, imparts relatively little energy to a molecule when a photon of it is absorbed. In contrast, ultraviolet light imparts more energy than visible light, since its wavelengths are shorter. Furthermore, UV-C photons are more energetic than UV-B ones, which in turn are more energetic than UV-A, since the UV-C wavelength is shorter than UV-B, etc. Overall, the order of photon energies is as follows:

$$UV\text{-}C > UV\text{-}B > UV\text{-}A > visible > infrared$$

In Chapter 14 we saw that light energy can initiate certain chemical reactions, such as some of those involved in smog formation, and that these processes are called photochemical reactions. Of course, in order for a sufficiently energetic photon to supply the energy for a reaction, it must first be absorbed by the molecule. There are many wavelength regions in which molecules do not absorb significant amounts of light. Thus, for example, because ozone molecules do not absorb visible light near 400 nm, shining light of this wavelength on them does not cause them to decompose, even though 400-nm photons carry sufficient energy to dissociate them.

15.7 Ozone is created in the stratosphere

In this section and the next, we will discuss the formation and destruction of ozone in the stratosphere. As we shall see, the formation reaction generates sufficient heat to warm this region of the atmosphere.

Above the stratosphere and in its higher parts, there are sufficient UV-C photons in sunlight to dissociate appreciable numbers of O_2 molecules:

$$O_2 \xrightarrow{\text{UV-C light}} O + O$$

At any instant the concentration of O_2 molecules is relatively large and the concentration of atomic oxygen is small. Therefore the stratospheric oxygen atoms that are created by the photochemical decomposition of O_2 are likely to collide and react with undissociated, intact diatomic oxygen molecules. In this process, ozone molecules are created:

$$O + O_2 \rightarrow O_3 + \text{heat}$$

Indeed, this reaction is the source of *all* the ozone in the stratosphere. During daylight hours, ozone is constantly being formed by this process. Above the stratosphere, oxygen exists mainly as oxygen atoms rather than as O_2 or O_3 molecules.

The release of heat by the ozone-formation reaction results in the temperature of the stratosphere as a whole being higher than that of the air below or above it. Within the stratosphere, the air at any given altitude is cooler than that which lies above it, so there is a kind of temperature inversion (Chapter 14, section 14.4) in the stratosphere. Consequently, vertical mixing of air in the stratosphere is a very slow process compared to mixing in the troposphere. The air in this region therefore is *stratified*—and hence the name *strat*osphere.

> The temperature in much of the stratosphere exceeds that near the top of the troposphere because of the heat generated by the ozone formation reaction.

15.8 Stratospheric ozone is constantly destroyed and re-formed

As we have seen (Figure 15.6), ozone absorbs UV light with wavelengths shorter than 320 nm—in other words, UV-B and most UV-C light. The energized molecules produced by absorption of these sunlight photons undergo dissociation. Thus, the absorption of a UV-C or UV-B photon by an ozone molecule in the stratosphere results in the decomposition of that molecule:

$$O_3 \xrightarrow{\text{UV-B or UV-C light}} O_2 + O$$

This process accounts for some of the ozone destruction in the stratosphere. Notice that only one oxygen atom is produced: the ozone molecule has *not* received enough energy to split into three separate oxygen atoms.

Most oxygen atoms produced in the stratosphere by photochemical decomposition of ozone or of O_2 subsequently react with intact O_2 molecules to re-form ozone, in accordance with the reaction at the top of the page. However, some of the oxygen atoms instead react with intact ozone molecules and in the process destroy them, since they are converted to O_2:

$$O_3 + O \rightarrow 2\,O_2$$

In effect, the unbonded oxygen atom takes one of the two oxygen atoms at the ends of the molecule from the ozone molecule. However, few

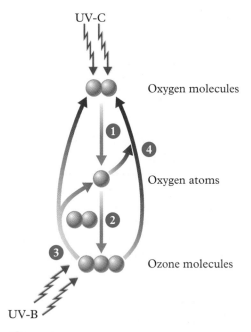

1. UV-C causes decomposition of oxygen molecules into oxygen atoms.

2. Oxygen atoms combine with oxygen molecules to produce ozone.

3. UV-B causes the decomposition of some ozone molecules into oxygen atoms and oxygen molecules.

4. Some ozone molecules combine with oxygen atoms to produce oxygen molecules.

Figure 15.9 Step by Step The Chapman cycle. UV light causes the noncatalytic production and destruction of ozone in the stratosphere.

collisions of an ozone molecule with an oxygen atom occur with sufficient energy to allow this reaction to occur. Consequently, most encounters of atomic oxygen with ozone molecules do not result in reaction. However, there are a number of molecules, such as those of *nitric oxide,* NO, that are present in tiny concentrations in the stratosphere that can catalyze this reaction. As a consequence, ozone destruction by the above reaction is an important process in the stratosphere.

To summarize the processes we have been describing, ozone in the stratosphere is constantly being formed, decomposed, and re-formed during daylight hours by a series of reactions that proceed simultaneously. Ozone is produced in the stratosphere because there is adequate UV-C from sunlight to dissociate some O_2 molecules and thereby produce oxygen atoms, most of which collide with other O_2 molecules and form ozone. The ozone gas filters UV-B and UV-C from sunlight but is destroyed temporarily by this process or by reaction with oxygen atoms. (The average lifetime of an ozone molecule at an altitude of 30 kilometers is about half an hour.) The ozone production and destruction processes, known as the **Chapman cycle,** are summarized in Figure 15.9.

Even in the "ozone layer" portion of the stratosphere, O_3 is not the gas of greatest abundance or even the dominant oxygen-containing species. Its concentration never exceeds 10 ppm, which is much, much lower than the concentration of either N_2 or O_2. Thus the term *ozone layer* is something of a misnomer. Nevertheless, this tiny concentration of ozone is sufficient to filter all the UV-C and most of the UV-B from the sunlight that reaches the stratosphere, before it reaches the lower atmosphere, where life resides.

The Ozone Hole and Other Sites of Ozone Depletion

As we discussed at the beginning of this chapter, stratospheric ozone over Antarctica is reduced by about 50% for several months each year due mainly to the action of chlorine. An episode of this sort, during which there is said to be a hole in the ozone layer, can occur from September to early November, corresponding to spring at the South Pole. Extensive research in the late 1980s led to an understanding of the chemistry of this phenomenon.

15.9 The ozone-destruction cycle over the South Pole

Most of the ozone destruction over the South Pole during an ozone hole episode occurs via a sequence that starts with the reaction of a chlorine *atom* with an ozone molecule:

$$Cl + O_3 \rightarrow ClO + O_2$$

Next in the sequence, two *chlorine monoxide*, ClO, molecules combine temporarily to form a *dichloroperoxide* molecule, ClOOCl :

$$2\ ClO \rightarrow ClOOCl$$

This reaction step occurs half as often as the previous one, since it requires two ClO molecules as reactants. The dichloroperoxide molecules absorb UV light and decompose over the course of several sequential reactions. The net decomposition reaction is:

$$ClOOCl \xrightarrow{\text{UV light}} \rightarrow \rightarrow 2\ Cl + O_2$$

Thus, ClO returns to the ozone-destroying form of chlorine, Cl.

A sequence of arrows indicates a reaction that occurs by several sequential steps.

Exercise 15.1

Add the last two reactions together (ignoring the fact that the latter one is itself the sum of several reactions). Then, multiply the ozone destruction step by two (since two ClO molecules are involved in the formation of ClOOCl). Add this to the other two reactions that you have already summed. Now, cancel the common terms from both sides of the equation to establish the overall (or net) reaction.

The overall reaction for the complete catalytic cycle is

$$2\ O_3 \rightarrow 3\ O_2$$

Thus, ozone is converted back to diatomic molecular oxygen in the overall process. No chlorine-containing species appear in the overall equation; chlorine is simply a catalyst in this process and is regenerated after each ozone molecule is destroyed by the reaction.

By the catalytic cycle we have just described, each chlorine atom destroys about 50 ozone molecules per day during the spring. About three-quarters of the ozone destruction in the Antarctic ozone hole occurs by this mechanism, in which chlorine is the only catalyst. A minor route for ozone destruction in the ozone hole involves the destruction of 1 ozone molecule by a chlorine atom and 1 by a bromine atom. In a process analogous to that involving ClO molecules, ClO and BrO molecules are produced and then collide with each other and rearrange their atoms, eventually yielding O_2 and atomic chlorine and bromine.

15.10 Inactive chlorine can be temporarily activated

The chlorine-catalyzed process for ozone destruction can operate only when chlorine exists in the form of the atom Cl or the molecule ClO.

Total Ozone for September 25, 2005

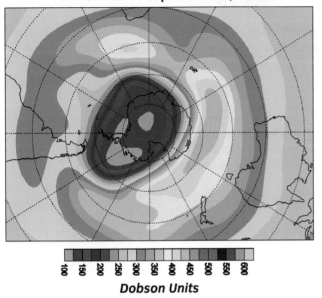

Dobson Units

Figure 15.10 The Antarctic ozone hole on September 25, 2005. (© The Meteorological Service of Canada. Reproduced with the permission of Environment Canada, 2005.)

However, in most of the stratosphere most of the time, the great majority of the chlorine exists in a form that is inactive as a catalyst for ozone destruction. The two main *catalytically inactive* molecules containing chlorine in the stratosphere are *hydrogen chloride* gas, HCl, and *chlorine nitrate* gas, $ClONO_2$.

Special polar weather conditions in the lower stratosphere permit the conversion of all the chlorine that is stored in the catalytically inactive forms, HCl and $ClONO_2$, into the active forms, Cl and ClO. This high concentration of active chlorine causes the large, though temporary, annual depletion of ozone that is the Antarctic hole. A color contour map of ozone levels above Antarctica on September 25, 2005 is shown in Figure 15.10. Note the low levels of ozone right over the South Pole. The various continents are outlined in black.

In most parts of the world, even in winter, the stratosphere is cloudless. The condensation of water vapor into liquid droplets or solid crystals that would constitute clouds doesn't normally occur in the stratosphere since the concentration of water in that region is exceedingly small. However, the temperature in the lower stratosphere drops so low ($-80°C$) over the South Pole in the sunless winter months that condensation does occur.

During the winter months, there is total darkness over the Antarctic. As we saw earlier, production of atomic oxygen from O_2 and O_3 depends on light energy. Therefore during this time, the dark stratosphere is very cold because the usual stratospheric warming mechanism—the release of heat by the $O_2 + O$ reaction—is absent. In turn, because the polar stratosphere becomes so cold during the total darkness at midwinter, the air pressure drops. This pressure phenomenon, in combination with Earth's rotation, produces a **vortex,** a whirling mass of air in which wind speeds can exceed 300 km (180 miles) per hour. Since matter cannot readily penetrate the vortex, the air inside it is isolated and remains very cold for many months. At the South Pole, the vortex is sustained well into the springtime (October). The particles produced by condensation of the gases within the cold vortex form *polar stratospheric clouds,* or PSCs, during the winter and spring months. PSCs can sometimes also be observed over nonpolar areas (see Figure 15.11).

Chemical reactions that lead ultimately to ozone destruction occur in a thin aqueous layer that is present at the surface of PSC ice crystals. Both hydrogen chloride and chlorine nitrate gases dissolve in the aqueous layer, then react together in a series of steps. The overall result is the production of *diatomic molecular chlorine gas,* Cl_2, which escapes from the particle into the surrounding air:

Chapter 15: A Thin Veil of Protection

Figure 15.11 Polar stratospheric clouds, photographed over Swedish Lapland.
(David Hay Jones/Photo Researchers, Inc.)

$$\text{HCl (g)} + \text{ClONO}_2\text{(g)} \xrightarrow{\text{particle surface}} \rightarrow \rightarrow \text{Cl}_2\text{(g)} + \text{HNO}_3\text{(aq)}$$

During the dark winter months, the molecular chlorine accumulates and eventually becomes the predominant chlorine-containing gas in the Antarctic stratosphere. Once a little sunlight reappears in the very early Antarctic spring, or the air mass moves to the edge of the vortex where there is some sunlight, the Cl_2 is decomposed by the light into atomic chlorine, Cl:

$$\text{Cl}_2 \xrightarrow{\text{sunlight}} 2\ \text{Cl}$$

The chlorine atoms then participate in a catalytic cycle of ozone destruction, as we described in section 15.9. In the lower stratosphere above Antarctica, an overall ozone destruction rate of about 2% per day occurs each September due to the combined effects of the various catalytic reaction sequences. As a result, by early October almost all the ozone is wiped out between altitudes of 15 and 20 km, the region in which its concentration normally is highest over the South Pole. This result is illustrated in Figure 15.12.

The destruction cycles largely cease, and the ozone concentration builds back up toward its normal level, a few weeks after the PSCs have disappeared and the vortex has ceased. The ozone hole closes for another year, though nowadays the ozone never quite returns to its natural levels, even in the fall. Furthermore, before the ozone level builds back up in the spring, some of the ozone-poor air mass can move away

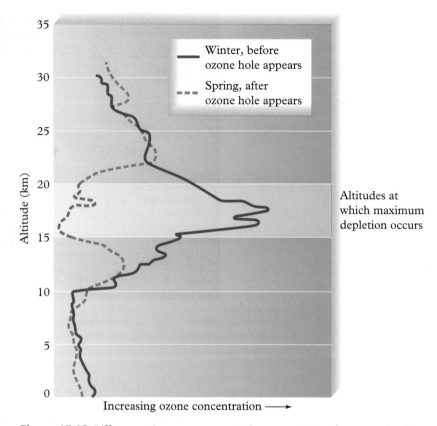

Figure 15.12 Differences in ozone concentrations over McMurdo, Antarctica, in August (late winter) and November (midspring) of 1987. (Adapted from B. J. Johnson et al., *Geophysical Research Letters, 19: 1105–1108,* 1992. Copyright by the American Geophysical Union.)

from the Antarctic and mix with surrounding air, temporarily lowering the stratospheric ozone concentration over adjoining geographic regions, such as Australia, New Zealand, and the southern portions of South America.

15.11 Measuring the size of the Antarctic ozone hole

Because the stratospheric concentration of chlorine continued to increase almost until the end of the 20th century, the extent of Antarctic ozone depletion increased from the early 1980s until at least the late 1990s. There are several relevant measures of the extent of the depletion. One is the surface (ground) *area* covered by low overhead ozone, as shown in Figure 15.3. This area grew rapidly and approximately linearly from 1981 through 1994, and to a lesser extent in more recent years.

The amount by which the total overhead ozone above the Antarctic decreases each spring has stopped becoming greater and may have leveled off (see Figure 15.2). However, the *length of time* that ozone depletion occurs each year has increased; the ozone hole in 2000 lasted longer than any previous one. Some reduction in ozone levels is now seen both in midwinter and in the summer as well as the spring, and indeed there is now some persistence of the depletion from one year to the next. There has also been an increase in the vertical region over which significant ozone depletion occurs in the stratosphere.

15.12 Stratospheric Arctic ozone depletion is increasing

Given the similarity in climate, it is perhaps surprising that an ozone hole did not start to form above the Arctic region in the 1980s, as occurred in the Antarctic. Until recently, very little springtime loss of stratospheric ozone has occurred over the Arctic region. The reason for this difference is that Arctic stratospheric temperatures do not fall as low or for as long in the winter. Consequently, polar stratospheric clouds form less frequently over the Arctic and do not last as long. In the past, the vortex containing the cold air mass above the Arctic usually broke up by late winter, before much sunlight returned to the polar region in the spring. Thus, although chlorine becomes activated during the polar winter above the Arctic as it does above the Antarctic, it has mostly been transformed back to inactive forms before it could destroy much ozone.

Unfortunately, there are ominous signs that springtime temperatures above the Arctic are now decreasing, with the result that ozone depletion there is accelerating in the lower stratosphere. Several springtime episodes of partial ozone depletion over the Arctic region have occurred recently. The Arctic vortex in the winter and spring of 1995–1996 was exceptionally cold and persistent, resulting in significant chlorine-catalyzed losses of ozone as late as mid-April. This amount of ozone depletion was about the same as that over the South Pole in the early 1980s, when the Antarctic hole first began to form. Consequently, some scientists have stated that an Arctic ozone hole formed in March of 1996. However, since overhead ozone is never 100% depleted, even over Antarctica, the definition of what conditions constitute a hole is somewhat arbitrary.

Even though stratospheric chlorine concentrations have begun to fall, changing weather conditions above the Arctic mean that springtime ozone holes may form there over the next decade or two. Since both the depletion of ozone and the increase in atmospheric carbon dioxide levels themselves *cool* the stratosphere, the lifetime of Arctic PSCs could be extended, and more ozone depletion would then occur. In the future, falling stratospheric chlorine concentrations will eventually reduce the ozone depletion over the Arctic. However, in the next decade or two, changing weather conditions above the Arctic will probably be a more important factor, and will probably increase depletion. Since 1997 (with the exceptions of 2000 and 2003), there have been only small springtime ozone losses above the Arctic. Although low ozone levels are likely to occur again in the coming winters over the next few decades, the

Indeed, increased stratospheric cloud lifetimes led to substantial ozone depletion in the winter of 2004–5.

current theoretical models predict that the Arctic hole will never rival that of the Antarctic in terms of the magnitude of ozone depletion. Students interested in following year-to-year developments of this phenomenon can consult the National Oceanic and Atmospheric Administration Web site at www.cpc.ncep.noaa.gov/products/stratosphere/tovsto.

15.13 Stratospheric ozone has decreased in nonpolar areas

As noted earlier, since the 1980s the stratospheric ozone concentration over nonpolar areas has decreased by several percent worldwide. The greatest amount of this depletion occurred over northern midlatitude regions, including North America. The greatest depletion occurs in the March–April period, and the least in the early fall.

Scientists have had a much harder time tracking down the source of this midlatitude ozone depletion than the source of depletion over polar regions. As in Antarctica, almost all the ozone loss over nonpolar regions occurs in the lower stratosphere. Some scientists speculate that reactions leading to ozone destruction could occur not only on PSC ice crystals but also on the surfaces of other particles that are present in the lower stratosphere. In particular, the reactions could occur on cold liquid droplets consisting mainly of *sulfuric acid*, H_2SO_4, that occur naturally in the lower stratosphere at all latitudes. The liquid droplets would have to be cold enough for significant uptake of gaseous HCl to occur, or no net reaction would take place.

The dominant, though erratic, source of H_2SO_4 at these altitudes is by the direct injection of *sulfur dioxide*, SO_2, gas emitted from volcanoes, followed by its oxidation to the acid. Indeed, substantial though temporary declines in stratospheric ozone occurred worldwide in the years following two large volcanic eruptions in recent years: the 1991 massive eruption of the Philippine volcano Mt. Pinatubo and the 1982 eruption of the Mexican volcano El Chichon. These eruptions both temporarily increased the concentration of sulfuric acid droplets in the lower stratosphere, which in turn led to increased amounts of chlorine activation and ozone destruction.

Although large volcanic eruptions can account for ozone depletion in the high-chlorine lower stratosphere, this explanation does not account entirely for the overall trend of decreasing ozone in the last two decades. Much of the gradual decline over midlatitudes is believed to be due to other factors, such as springtime dilution of ozone-depleted polar air and its transport out of the polar regions, changes in the solar cycle, and both natural and anthropogenic changes in the pattern of atmospheric winds and temperatures.

The Chemicals That Cause Ozone Destruction

15.14 Natural and synthetic chlorine and bromine compounds

Some chlorine has always existed in the stratosphere as a result of the slow upward migration of *methyl chloride* gas, CH_3Cl, produced naturally at Earth's surface. The gas forms mainly in the oceans, as a result of the interaction of chloride ion in seawater with decaying vegetation. Only a portion of the methyl chloride gas emitted from the oceans is destroyed in the troposphere. When intact molecules of it reach the stratosphere, they are decomposed, and atomic chlorine, Cl, is produced.

In recent decades, stratospheric chlorine resulting from natural CH_3Cl has been completely overshadowed by much larger amounts of chlorine produced from synthetic chlorine-containing gaseous compounds that are released into air during their production or use. The particular compounds at fault are those that do not have a *sink*—a process by which the substance is removed—in the troposphere. After a few years of traveling in the troposphere, they begin to diffuse into the stratosphere, where eventually they undergo decomposition and release their halogen atoms.

Though our discussion has focused on chlorine so far, bromine also is released into the atmosphere in synthetic organic compounds and also has ozone-destruction power. The way in which the net stratospheric concentration of these two elements has changed over the course of the last quarter century, and is projected to vary in this century, is shown by the yellow curve in Figure 15.13. (The concentration of the two elements is expressed as the equivalent of chlorine alone in terms of their joint ozone-destruction power.) The peak chlorine concentration in the stratosphere occurred about 1999 and was almost four times as great as the natural level arising from natural sources of chlorine and bromine.

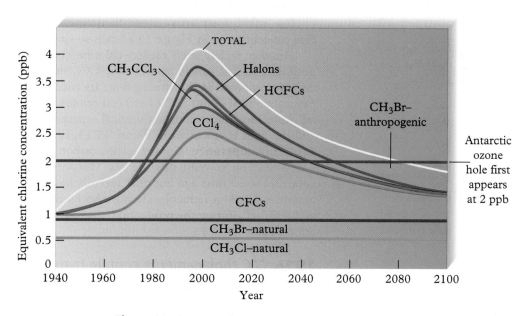

Figure 15.13 Stratospheric concentrations of chlorine equivalents from various sources measured from 1940 and projected through the next 100 years. (Source: Dr. Donald Wuebbles, University of Illinois, Urbana, IL)

The Chemicals That Cause Ozone Destruction

15.15 CFCs and carbon tetrachloride

The recent increase in stratospheric chlorine is due primarily to the use and release of *chlorofluorocarbons*, CFCs, compounds that we discussed in Chapter 5, section 5.2. The growth and decline of CFCs in the stratosphere can be seen in Figure 15.13. The CFC concentration in any year corresponds to the vertical distance within the area marked "CFCs" on the graph—that is, between the horizontal green line and the light blue curve that lies above it.

CFCs have no tropospheric sink because they do not undergo any of the normal removal processes: they are not soluble in water and thus they are not rained out from air; they are not attacked by any other tropospheric gases and so do not decompose; and they are not photochemically dissociated by either visible or UV-A light. Therefore, all CFC molecules rise to the stratosphere. Eventually they migrate to the middle and upper parts of the stratosphere, where there is sufficient unfiltered UV-C from sunlight to photochemically decompose them, thereby releasing chlorine atoms. For example, for molecules of the compound known as CFC-12, the reaction is

$$\underset{\text{(CFC-12)}}{CF_2Cl_2} \xrightarrow{\text{UV-C}} CF_2Cl + Cl$$

Eventually another Cl is released when the CF_2Cl molecules decompose. The CFCs must rise to the mid-stratosphere before decomposing, since UV-C does not penetrate to lower altitudes. Because vertical motion in the stratosphere is slow, their atmospheric lifetimes are long: 60 years on average for CFC-11 molecules, 105 years for CFC-12. It is because of the long stratospheric lifetimes of such gases that the chlorine concentration in Figure 15.13 falls so slowly.

Another widely used carbon–chlorine compound that lacks a tropospheric sink is *carbon tetrachloride*, CCl_4, which also is photochemically decomposed in the stratosphere. Its concentration since the 1970s extending until about 2020 has been remarkably constant, as you can tell from the almost constant vertical separation between the dark red and the light blue curves in Figure 15.13. Like CFCs, it is an **ozone-depleting substance.** Commercially, carbon tetrachloride was used as a dry-cleaning solvent; this use was discontinued in most developed countries some time ago but until recently has continued in many other countries. Carbon tetrachloride was also used to manufacture CFC-11 and CFC-12, during the production of which some CCl_4 was lost to the atmosphere.

15.16 CFC replacements contain hydrogen

The compounds being used as the direct replacements for CFCs all contain hydrogen atoms bonded to carbon. A majority of such molecules will be removed from the troposphere because they are decomposed by the sequence of reactions that begins when they are attacked

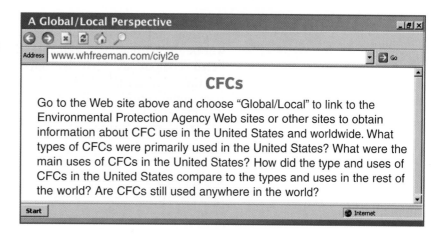

A Global/Local Perspective

Address www.whfreeman.com/ciyl2e

CFCs

Go to the Web site above and choose "Global/Local" to link to the Environmental Protection Agency Web sites or other sites to obtain information about CFC use in the United States and worldwide. What types of CFCs were primarily used in the United States? What were the main uses of CFCs in the United States? How did the type and uses of CFCs in the United States compare to the types and uses in the rest of the world? Are CFCs still used anywhere in the world?

Start   Internet

by an OH molecule, the hydroxyl free radical. For example, because methyl chloride, methyl bromide, and *methyl chloroform,* CH_3CCl_3, each contain hydrogen atoms, a fraction of such molecules are removed in the troposphere before they have a chance to rise to the stratosphere.

The temporary replacements for CFCs employed in the 1990s and the early years of the 21st century contain hydrogen, chlorine, fluorine, and carbon; they are called *HCFCs,* **hydrofluorochlorocarbons.** One HCFC in major use is CHF_2Cl, the gas called *HCFC-22* (or just CFC-22). It is employed in most domestic air conditioners and in some refrigerators and freezers. Since it contains a hydrogen atom and thus is mainly removed from air before it can rise to the stratosphere, its long-term ozone-reducing potential is small—only 5% of that of CFC-11. Notice the contribution of HCFCs to the curve in Figure 15.13; they are significant only from the late 1990s until about 2030.

Since the volume of HCFC consumption would presumably rise with increasing world population and affluence, reliance exclusively on HCFCs as CFC replacements would eventually lead to a renewed

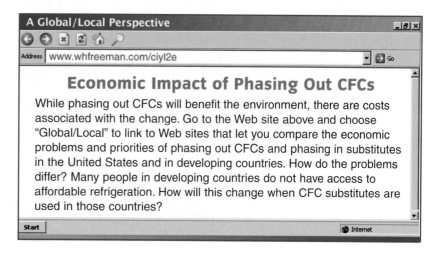

A Global/Local Perspective

Address www.whfreeman.com/ciyl2e

Economic Impact of Phasing Out CFCs

While phasing out CFCs will benefit the environment, there are costs associated with the change. Go to the Web site above and choose "Global/Local" to link to Web sites that let you compare the economic problems and priorities of phasing out CFCs and phasing in substitutes in the United States and in developing countries. How do the problems differ? Many people in developing countries do not have access to affordable refrigeration. How will this change when CFC substitutes are used in those countries?

Start   Internet

The Chemicals That Cause Ozone Destruction

buildup of stratospheric chlorine. Products that are entirely free of chlorine, and that therefore pose no hazard to stratospheric ozone, will be the ultimate replacements for CFCs and HCFCs. *Fully* fluorinated compounds—that is, ones consisting only of carbon and fluorine—are unsuitable since they have no tropospheric sinks, and they would contribute to global warming for very long periods of time (see Chapter 16).

Hydrofluorocarbons, *HFCs*—substances that contain hydrogen, fluorine, and carbon—are the main long-term replacements for CFCs and HCFCs. The compound CH_2F—CF_3, called *HFC-134a,* is already being used rather than CFC-12 as the working fluid in refrigerators and in some types of air conditioners, including those in automobiles.

15.17 Are CFCs really to blame?

The theory that CFCs cause most of the stratospheric ozone depletion was questioned in the past by some commentators because natural sources, especially seawater and volcanoes, spew much more chlorine into the atmosphere than do CFCs. These observers concluded that ozone depletion must therefore be due to natural causes.

This argument is invalidated by a number of related factors. Natural sources emit almost all their chlorine into the troposphere rather than into the stratosphere. The sodium chloride emitted into ground-level air over oceans and the HCl emitted into the high troposphere and the very low stratosphere by volcanoes are both water soluble; hence they are rained out before they can rise to the stratosphere, where they could destroy ozone. Thus the total stratospheric chlorine introduced by natural processes is a small fraction of that which originates from CFCs.

15.18 Halons and methyl bromide contain bromine

Halon chemicals are bromine-containing, hydrogen-free substances such as CF_3Br and CF_2BrCl. Because they have no tropospheric sinks, they eventually rise to the stratosphere. There they are photochemically decomposed, with the release of atomic bromine (and chlorine). As we have mentioned, bromine can participate in catalytic ozone-destruction cycles, so halons also are ozone-depleting substances. Bromine from halons will continue to account for a significant fraction of the ozone-destroying potential of stratospheric halogen catalysts for decades to come (see Figure 15.13).

As with methyl chloride, large quantities of *methyl bromide,* CH_3Br, are also produced naturally. A sizable fraction of atmospheric CH_3Br is due to release during fires involving grass, crops, and trees. The average atmospheric lifetime of a methyl bromide molecule, CH_3Br, is about 1 year. Consequently, some methyl bromide molecules released at ground level eventually make their way to the stratosphere, where each one eventually decomposes to release atomic bromine. Methyl bromide is produced synthetically for use as a soil fumigant, as discussed in Chapter 5, and on that account its release into the troposphere has been increasing.

15.19 International agreements protect the ozone layer

The use of CFCs in most aerosol products was banned in the late 1970s in North America and some Scandinavian countries. This decision was based on predictions made by Sherwood Rowland and Mario Molina, chemists at the University of California, concerning the effect of chlorine on the thickness of the ozone layer. There was no experimental indication of any depletion at the time of their predictions. Subsequent research by Molina's group revealed the ClOOCl mechanism that operates in the polar ozone hole. Rowland and Molina, together with the German chemist Paul Crutzen, were jointly awarded the Nobel Prize in Chemistry in 1995 to honor their work in researching the science underlying ozone depletion.

In contrast to almost all other environmental problems, such as global warming (Chapter 16), international agreement on remedies to stratospheric ozone depletion has been obtained and successfully implemented in a fairly short period of time. The growing awareness during the 1980s of the seriousness of chlorine buildup in the atmosphere led to international agreements to phase out CFC production in the world. The breakthrough came at a conference in Montreal, Canada, in 1987 that gave rise to the *Montreal Protocol*. This agreement was strengthened at several follow-up conferences. As a result of this international agreement, all ozone-depleting chemicals are now destined for phaseout in all nations. All legal CFC production in developed countries ended in 1995. Developing countries have been allowed until 2010 to reach the same goal. Unfortunately, some of the CFC produced in developing countries and Russia is smuggled into developed countries.

Developed countries have agreed to end production of HCFCs by 2030, and developing countries by 2040, with no increases allowed after 2015. Halon production was halted in 1994 by the terms of the Montreal Protocol. However use of existing stocks continues, as do releases from fire-fighting equipment, so the atmospheric concentration of this chemical has continued to rise.

Methyl bromide has been added to the list of ozone-depleting substances that will be banned. Developed countries were scheduled to phase out its use by 2005. However, led by the United States, more than a dozen such countries applied for and received exemptions for 2005 and 2006 for limited "critical use exemptions" by the Ozone Secretariat of the United Nations from the scheduled 2005 phaseout of methyl bromide by such nations. The exemptions were intended to give farmers and other users of the gas more time to develop and adopt cost-effective substitutes. Developing countries are scheduled to phase out their use of methyl bromide by 2015, with uses from 2002 onwards not to exceed their individual mid-1990s levels.

Methyl bromide is widely used to sterilize soil before planting crops such as tomatoes, strawberries, grapes, tobacco, and flowers and to fumigate some crops such as dried fruit and nuts after their harvest. Currently practicable alternatives to the use of methyl bromide in these applications, and to control termites, are apparently less effective and more costly, although there has been some recent success in substituting methyl iodide for it.

▷ Discussion Point: Should agricultural methyl bromide be phased out?

Methyl bromide was scheduled to be phased out of use in the United States by 2005. While many people believe the environmental impact of methyl bromide is well-documented, there is much debate about the extent of that impact and the economic price that will be paid by U.S. farmers. Two of the arguments are that U.S. farmers will be less competitive and that they will be prevented from exporting to certain markets. Do you believe that the economic impact is more important than the environmental impact? Visit the relevant Web sites at www.whfreeman.com/ciyl2e and discuss the pros and cons of banning methyl bromide use in the United States.

Now that new releases are restricted, the stratosphere will eventually be cleaned of its excess chlorine and bromine. The gases HCl and HBr eventually diffuse from the stratosphere back into the upper troposphere, dissolve in water droplets, and are subsequently carried to lower altitudes and transported to the ground by rain. Thus although the lifetime of chlorine and bromine in the stratosphere is decades long, it is not infinite: they are eventually removed.

As a direct result of the reduced production and use of ozone-depleting substances, the tropospheric concentration of chlorine peaked in 1994, and the stratospheric chlorine equivalent level peaked at about 4 ppb at the turn of the 21st century (see Figure 15.13). The slowness in the subsequent decline in chlorine levels is due to the following:

- The long time it takes molecules to rise from the troposphere to the middle or upper stratosphere, where they encounter UV-C photons and are dissociated to atomic chlorine
- The slowness of the removal of chlorine and bromine from the stratosphere
- The continued input of some chlorine and bromine into the atmosphere

Because ozone is formed (and destroyed) in rapid natural processes, its level responds very quickly to a change in stratospheric chlorine concentration. Thus the Antarctic ozone hole probably will not continue to appear after the middle of the 21st century, that is, once the chlorine equivalent concentration is reduced back to the 2 ppb level it had in the years before the hole began to form (dark blue line in Figure 15.13). Without the international agreements, catastrophic increases in chlorine, to many times the present level, would have occurred, particularly since CFC usage and atmospheric release in developing countries would have increased dramatically. A further doubling of stratospheric chlorine levels would probably have led to the formation of an ozone hole each spring over the Arctic region and possibly to an enlargement of the Antarctic hole. Thus, the phaseout of CFCs and other ozone-destroying substances and the consequent eventual disappearance of the ozone holes represent a major environmental success story.

Summarizing the Main Ideas

The ozone layer, which filters out the harmful rays of sunlight, is located in the middle and lower parts of the stratosphere. The stratosphere is the region of the atmosphere that lies above the troposphere, approximately 15–50 kilometers above ground level. For the last quarter century, stratospheric ozone over Antarctica has been largely destroyed each spring, producing an ozone hole. To a lesser extent, ozone depletion occurs in nonpolar regions as well.

Ultraviolet light has shorter wavelengths than does visible light. Molecules differ in the wavelengths of light they absorb in the UV and visible regions of the spectrum. O_2 and O_3 filter UV-C and some UV-B from sunlight by absorbing it. Excessive exposure to UV-B can produce skin cancer and perhaps affect our immune systems. The increased exposure to UV-B that would result is the main environmental concern about ozone depletion.

Stratospheric ozone is created when O_2 molecules absorb UV-C from sunlight and decompose into oxygen atoms, which then combine with intact O_2 molecules to produce ozone, O_3. This gas accumulates in the lower and middle stratosphere, though it is by no means the dominant gas there. The release of heat by the ozone-creating reaction produces the temperature increase in this region.

Molecules of ozone in the stratosphere are destroyed when they absorb ultraviolet light or when they collide with free oxygen atoms. They are also destroyed in reaction cycles that involve catalysts such as atomic chlorine. In the first step of the most important cycle, a chlorine atom abstracts an oxygen atom from ozone, producing ClO and O_2. During the subsequent steps, which operate during ozone hole conditions, two ClO molecules combine, and the product ClOOCl subsequently is decomposed photochemically, yielding O_2 and two Cl atoms.

Most stratospheric chlorine normally is tied up as the catalytically inactive forms HCl and $ClONO_2$. In the Antarctic ozone hole, chlorine is converted to the active forms Cl and ClO by means of the reaction of the inactive forms on the liquid surfaces of frozen particles. The latter form only when the stratospheric air temperature becomes very cold, due to its isolation in a vortex above the South Pole.

In recent years, the chlorine level in the stratosphere has become so large that when chlorine above the South Pole is temporarily activated, much of the ozone is destroyed by the catalytic reaction cycle. Once the vortex is destroyed in the spring, the chlorine reverts to its inactive forms, and the ozone hole heals.

A full ozone hole has not formed in the Arctic stratosphere because the vortex usually breaks up before sunlight has had a chance to initiate reactions that destroy much ozone. However, partial depletion of ozone has occurred there in recent early springs.

Chlorine and bromine atoms occur naturally in the stratosphere owing to the release of methyl chloride and bromide from the oceans and the eventual rise of some of these emissions to the stratosphere, where they are decomposed into atoms.

The anthropogenic chemicals that supply chlorine or bromine to the stratosphere and thereby cause ozone destruction are called ozone-depleting substances. These include CFCs, carbon tetrachloride, methyl chloroform, HCFCs, halons, and methyl bromide. The Montreal Protocol agreements have resulted in a phaseout in the production and use of most of these compounds. The chlorine content of the stratosphere has recently peaked, but it will take about half a century to fall enough so that the Antarctic ozone hole is healed.

Key Terms

ozone layer	UV-B light	vortex	hydrofluorochlorocarbon
stratosphere	UV-A light	ozone-depleting	hydrofluorocarbon
ultraviolet (UV) light	malignant melanoma	substance	
UV-C light	Chapman cycle		

Web Sites of Interest

To link to Web sites of interest, go to www.whfreeman.com/ciyl2e, Chapter 15, and select the site you want.

For Further Reading

"Ozone Depletion: 20 Years After the Alarm," *Chemical and Engineering News*, August 15, 1994, pp. 8–13. This article is an overview of the developments since the first concerns about ozone depletion and CFCs arose in 1974.

B. Rensberger, "A Reader's Guide to the Ozone Controversy," *Skeptical Inquirer, 18,* Fall 1994, pp. 488–497. Accessibly written article that covers most of the aspects of the debate over ozone.

R. S. Stolarski, "The Antarctic Ozone Hole," *Scientific American,* January 1988, pp. 30–36. Good article to use for comparative discussions about scientific predictions, computer models, accuracy of data, instrumentation, research methods, policy responses, etc., then and now.

O. B. Toon and R. P. Turco, "Polar Stratospheric Clouds and Ozone Depletion," *Scientific American,* June 1991, pp. 68–74. Use this article with the Cambridge multimedia tour of the ozone hole (see this textbook's Web site, Ozone Hole) to look at the role that polar stratospheric clouds play in ozone depletion.

Review Questions

1. What is the *ozone layer*?

2. Sketch the regions of the atmosphere from ground level to 50 km (30 miles). Where is the ozone layer located?

3. Where is ozone produced in the atmosphere?

4. What is the *ozone hole*?

5. Has ozone depletion occurred only above Antarctica?

6. How does visible light differ in wavelength from ultraviolet light?

7. What is the source of both visible and ultraviolet light?

8. Which range of wavelengths is associated with ultraviolet light?

9. How can UV light affect molecules?

10. How does the absorption spectrum of oxygen, O_2, differ from ozone, O_3?

11. Explain how UV-A, UV-B, and UV-C differ.

12. Describe some of the biological damage that occurs upon exposure to ultraviolet radiation.

13. What is *malignant melanoma*? What is the relationship between malignant melanoma and ultraviolet radiation?

14. What impact is a 1% decrease in ozone levels predicted to have on cases of skin cancer?

15. Why does it take so long to see the impact of decreased ozone on skin cancer cases?

16. What negative health effects, besides skin cancer, are associated with decreased levels of ozone?

17. What negative environmental effects are associated with decreased levels of ozone?

18. What is a *photon*?

19. What is the relationship between the energy of a photon and its wavelength?

20. Arrange the following in order of decreasing energy: UV-C, infrared light, UV-A, visible light, UV-B.

21. What is the source of all ozone in the stratosphere?

22. What is the *Chapman cycle*?

23. What are the catalytically inactive forms of chlorine in the stratosphere? Why are they catalytically inactive?

24. What is a *vortex*?

25. How have ozone concentrations in nonpolar regions changed since the 1980s?

26. Identify the primary natural and anthropogenic sources of chlorine and bromine in the atmosphere.

27. Has the chlorine level in the stratosphere peaked yet? If so, when did it peak?

28. What is a *CFC*? What types of chemicals are proposed as long-term CFC replacements?

29. When light interacts with a CFC molecule, which chemical bond is broken?

30. What is a *tropospheric sink*?

31. CFCs have no tropospheric sink. What are the consequences of this?

32. Where in the atmosphere are CFCs produced? How do they get to the stratosphere?

33. What are the primary sources of CFCs in the atmosphere?

34. Why are CFC lifetimes in the stratosphere so long?

35. Provide five examples of *ozone-depleting substances*.

36. What environmental problems may be posed by HFCs?

37. What is a *halon*?

38. What is the *Montreal Protocol*?

39. How is methyl bromide used?

Understanding Concepts

40. Where is the ozone layer thickest? Where is it thinnest?

41. Explain how oxygen, O_2, and ozone, O_3, molecules protect Earth from ultraviolet (UV) exposure.

42. Of UV-A, UV-B, and UV-C, which type penetrates the atmosphere to the greatest extent? Why?

43. Compare the relative biological effects of UV-A, UV-B, and UV-C. Which is most dangerous? Why?

44. Does the absorption of a photon mean that a molecule will necessarily undergo a chemical reaction? Explain.

45. How is ozone created in the stratosphere? Explain and write the important chemical reactions involved in the process.

46. Sketch the atmospheric regions from ground level to 50 km (30 miles) and describe:
    a) the relative amounts of ozone in each section
    b) the range of ultraviolet light that penetrates each section

47. Is it possible to destroy all the ozone in the atmosphere permanently? Explain.

48. What impact does increased UV penetration have on ozone production in lower altitudes?

49. What polar atmospheric conditions produce the ozone hole in the winter? How are polar stratospheric clouds formed?

50. What role does sunlight play in the formation and disappearance of the Antarctic ozone hole?

51. Compare the ozone holes over Antarctica and the Arctic. Which is more severe? Why?

52. Why does a polar vortex eventually affect ozone levels in other parts of the world? What effect can it have?

53. Describe the trend in global stratospheric ozone levels since the 1980s.

54. What gases are being phased out according to the Montreal Protocol and subsequent amendments? Why are the phaseout timelines different for different countries?

55. Why does it take so long for chlorine levels to decline in the atmosphere?

Synthesizing Ideas

56. Compare the environmental impacts of infrared and ultraviolet light.

57. Use the absorption spectra of diatomic oxygen and ozone in Figures 15.5 and 15.6 to answer the following:

a) Which wavelengths of light are most effectively absorbed by ozone?

b) Which wavelengths of light are most effectively absorbed by O_2?

c) Does ozone effectively absorb ultraviolet light? If so, which types?

d) Does diatomic oxygen effectively absorb ultraviolet light? If so, which types?

e) How do the absorption characteristics of diatomic oxygen and ozone compare?

58. How do ground-level activities influence the catalytic destruction of ozone?

59. What mechanism accounts for the majority of ozone destruction in the Antarctic ozone hole? Write the chemical reactions for ozone destruction there.

60. How do scientists determine the extent of ozone depletion in the ozone hole?

61. How do HCFCs, HFCs, and CFCs differ? Are HCFCs and HFCs considered suitable long-term replacements for CFCs?

62. No controls on the release of CH_3Cl, CH_2Cl_2, or $CHCl_3$ have been proposed. What does that imply about their atmospheric lifetimes and therefore their tendency to migrate to the stratosphere, compared to CFCs, CCl_4, and methyl chloroform?

63. Hydrocarbons such as propane and butane have been used as coolants in the past. Unfortunately, they tend to form explosive mixtures with air, which can pose a danger if the cooling system develops a leak. Identify some pros and cons of using such a coolant in the air conditioning systems of automobiles in the absence of CFCs. Consider both the larger environmental impact and the more personal risk.

In this chapter, you will learn:

- about global climate changes in the 20th and early 21st centuries, and some of the associated consequences;
- about the causes of global warming, including greenhouse gases, emissions from fossil fuels, deforestation and anaerobic decomposition of living matter;
- how the effects of global warming create environmental factors that further amplify global warming;
- how clouds and aerosols counter some global warming;
- about the predictions of future global warming, including rising temperatures, rising sea levels, changing agriculture patterns, and changes to human health;
- about methods of controlling future CO_2 levels.

Is the flooding from Hurricane Isabel in North Carolina a consequence of global warming?

Many scientists attribute such dramatic weather events to increasing temperatures in the Earth's lower atmosphere. In this chapter we will see how the greenhouse effect warms the air closest to the Earth and discuss the possible consequences of global warming. (Mike Appleton/Corbis)

Global Warming and the Greenhouse Effect

The climate is changing. Within your lifetime, the air has warmed by about 0.5°C and shows every sign of continuing to become hotter. Although half a degree may seem small, the climate of the world changed more during the 20th century than in any comparable period in recorded history. We are aware that the global change that has occurred so far includes not only increases in global air temperatures, but also increased annual precipitation in most locations worldwide, increased sea levels, and a variety of other effects. Some of this rapid change in Earth's climate, called **global warming,** was due to natural climate cycles, but much of it in the last few decades was probably caused by the actions of humans. After almost 5 billion years of it being controlled almost entirely by nature, we have become a dominant force in changing Earth's weather. Since it is difficult to predict the consequences of our actions, we are engaged in a gigantic experiment on our environment.

In this chapter, we shall discuss the factors that determine world climate, the changes in climate that have occurred recently, and the nature and origin of the chemicals whose emission into the air are believed by many scientists to be driving the warming of the atmosphere. We will also consider what climate changes await us in the future if emissions continue, how the changes might affect us, and what actions we could take to reduce these emissions.

Global Changes in Weather in the 20th Century

16.1 Global warming today exceeds that in the past

In the 20th century, Earth's surface air temperature—averaged over night and day, over all seasons, and over all places on the planet—increased. However, the change was not uniform, and some regions even became cooler! Although most of the United States and Canada, and indeed most of the world, became warmer—as indicated by the red dots in Figure 16.1—the Arctic region warmed most of all. Portions of the southeast and south-central United States experienced a slight cooling (blue dots). Most of the change in air temperatures corresponded to an increase in the daily *minimum*—the low temperature experienced just before dawn, rather than in the daily high temperatures, usually achieved in the afternoon.

Air temperature did not increase *continuously* throughout the 20th century. The trends of change were different in four periods, as illustrated in Figure 16.2. The greatest rate of heating occurred during the last quarter of the century. As you can see from the figure, the average air temperature *declined* slightly in the preceding

Whenever you see this icon in this chapter, go to
www.whfreeman.com/ciyl2e

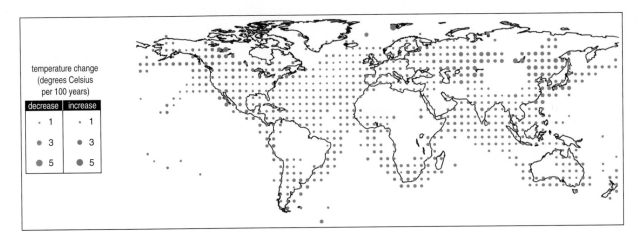

Figure 16.1 Trends in average annual global temperature from 1901 to 1998. (Source: *American Scientist*, Vol. 87, 1999, p. 536.)

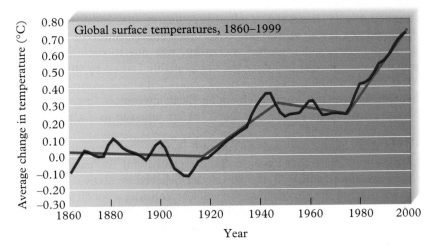

Figure 16.2 Trends in average global surface temperatures (blue line), 1860–1999. The red line shows average trends in four characteristic periods. (Adapted from World Wildlife Fund report.)

period (1940–1975), as it also did at the beginning of the century. There was also a warming period from 1915 to 1940. Nine of the hottest ten years on record have occurred between the years 1994 and 2004. The hottest years were 1998 and 2002.

16.2 Global precipitation has changed

One aspect of climate is the amount of precipitation—rain and snow—that falls at various locations on Earth. In the 20th century, the total annual precipitation around the world increased, as you can tell from Figure 16.3, since the green dot areas (increased precipitation) greatly outnumber the tan dot areas (decreased precipitation). Many areas just north and just south of the equator, especially in Africa, became much

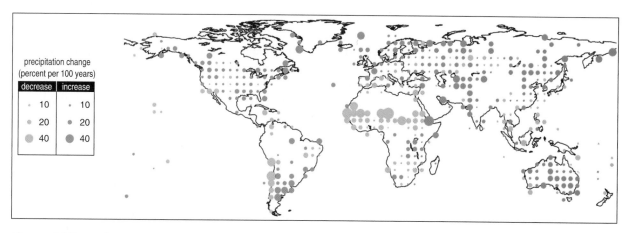

Figure 16.3 Trends in average annual global precipitation from 1901 to 1998. (Source: *American Scientist,* Vol. 87, 1999, p. 536.)

drier, with disastrous consequences for food production. By contrast, most temperate regions of North America and Europe became somewhat wetter. The overall increase in precipitation is expected, since warming of the air warms the surface waters of lakes and oceans, and warmer water evaporates faster and thereby increases the water content of the atmosphere.

16.3 What are other signs of global warming?

Throughout the world there are many indications that global warming is occurring:

- *Winters have become shorter by about 11 days.* In the northern hemisphere, spring has been arriving sooner and autumn has been starting later. Over the last three decades, the advent of spring—as observed by the appearance of buds, the unfolding of leaves, and the flowering of plants—has advanced by an average of 6 days in Europe, while the start of autumn—as defined by the date at which leaves turn color and begin to fall—has become delayed by about 5 days. Consequently, there are fewer "frost days" now than there used to be. The change in behavior of the plants has been driven by the increase in average daily air temperatures.

- *Earth's ice cover is shrinking fast.* Glaciers, polar ice caps, and polar sea ice are melting and disappearing at unprecedented rates, as a consequence of global warming. For example, the remaining glaciers in Glacier National Park in the U.S. Rocky Mountains could disappear in 30 years if current melting rates continue. Sea ice in the Arctic summer has decreased by 20% in the 1979–2003 period alone. Warmer weather has also delayed the seasonal formation of sea ice. All these changes have caused a sharp decline in some populations of Antarctic penguins and Arctic caribou.

- *Warming water is killing much of the coral in ocean reefs and threatening sea life.* Coral reefs in tropical waters nurture and protect fish and attract scuba divers. As water warms, corals bleach themselves by expelling the algae that give them color and provide nutrition. Over 95% of the coral is already dead in some parts of the Seychelles islands. Thus far, reefs in the central Pacific Ocean have escaped bleaching, and it is just beginning in the Caribbean. The death of the coral reefs not only affects the tourist trade but also threatens fishing for species that depend on the reefs for food. Beaches will erode if the reefs break up and no longer provide protection.

- *Mosquito-borne diseases have reached higher altitudes.* Because of warmer temperatures, mosquitoes are now able to survive in regions where they formerly were not viable. As a result, mosquitoes have carried malaria to higher mountain regions in parts of Africa and *dengue fever* to new regions in Central America. Outbreaks of malaria have occurred in Texas, Florida, Michigan, New York, New Jersey, and even southern Ontario in the last decade. Warmer weather and changing precipitation patterns allowed the West Nile virus, another mosquito-borne disease, to become established in the New York City area in the late 1990s and now throughout most of the U.S. and southern Canada.

- *Rising sea levels are threatening to engulf Pacific islands.* As we shall see later in the chapter, warming the air eventually leads to a rise in sea levels. The average level of the sea has risen by about 10 centimeters since 1940, a sharp increase in rate for this natural process.

- *Extreme weather is becoming more common.* The frequency of extreme and violent weather events has increased in many areas of the world. Such events include blizzards and storms with heavy snow and freezing rain in northern areas, but record heat waves, hurricanes, and drought in others. For example, the number of heat waves lasting 3 days or longer almost doubled in the United States between 1949 and 1995. Moreover, the frequency of storms with heavy or extreme precipitation has increased in the United States and in many other countries. The economic damage caused by storms in the 1990s greatly exceeded that in previous decades.

The devastation of New Orleans by Hurricane Katrina in 2005 provided a terrifying example of the power of extreme weather.

The Causes of Global Warming

Now that we have established that Earth's climate *is* changing, the logical question to ask is what is causing the change. To answer it, we shall look into the science behind the factors that control global temperatures. We shall consider the greenhouse effect, the natural mechanism that keeps the oceans from freezing over and keeps us at a comfortable temperature, but one that is threatened with disturbance by our emissions of pollutant gases into the air.

Chapter 16: Global Warming and the Greenhouse Effect

16.4 Earth balances energy received from the Sun

Earth's surface and atmosphere are kept warm primarily by energy received from the Sun in the form of light. You are already familiar with the fact that light is a form of energy—your body becomes hot when you sunbathe on a beach, and a car's metal surface becomes hot when the vehicle sits in bright sunlight.

Light having wavelengths *greater* than 750 nanometers (red light), but less than about a million nanometers, is called **infrared light;** we experience it as radiant heat. This is the type of light that is transmitted by the remote-control devices used to communicate between you and your TV or your automobile. Sunlight contains infrared light in the 750–4000-nm region. For convenience, we shall discuss infrared wavelengths in units of micrometers; as we have seen previously, 1 μm equals 10^{-6} (a millionth) of a meter, equivalent to 1000 nanometers. Thus the infrared (IR) light in sunlight ranges from 0.75 to 4.0 μm (see Figure 16.4).

Like any warm body, Earth *emits* energy in the form of light. The energy emitted by Earth is not visible light; rather it is infrared light having wavelengths from 4 to 50 μm. This range is called the **thermal infrared** region, since this energy is a form of heat—the same kind of heat energy a heated metal pot radiates. As you can see in Figure 16.4, the energy that Earth emits is rather longer in wavelength than is the IR light that we receive *from* the Sun.

Since infrared light is the only way in which the planet can release energy, the amount of energy that our planet absorbs from sunlight and the amount that it releases by emitting infrared light must be equal if Earth's temperature is to remain constant. The infrared light is emitted both at Earth's surface and by its atmosphere, though in different amounts. This is because the release rate of the infrared light is temperature sensitive: the release rate increases sharply with increased temperature.

Refer to Chapter 15, Figure 15.4, to see how IR light's wavelengths compare to those of visible and UV light.

Figure 16.4 Wavelength distribution for light emitted by the Sun (dashed curve) and by Earth's surface and troposphere (solid curve). (Redrawn from J. Gribben, "Inside Science: The Greenhouse Effect," *New Scientist,* supplement to issue of Oct. 22, 1988.)

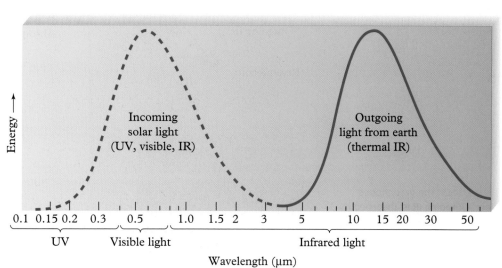

16.5 The greenhouse effect traps heat

Not *all* the IR light emitted from Earth's surface and lower atmosphere escapes into space. The molecules of certain gases that are naturally present in the air can absorb thermal infrared light. Much of this absorbed IR energy is immediately converted into heat. Consequently, the temperature of the air surrounding the absorbing molecules increases. This phenomenon, in which the temperature of Earth's surface and the nearby air is increased by trapped outgoing energy, is called the **greenhouse effect**. You could also call it the "blanket effect," since the primary way that a blanket keeps you warm is also by trapping some of your body heat rather than allowing it to escape into the air.

You might think that the temperature would increase without limit as the air traps more and more of the outgoing infrared light. However, as they warm, the air and the surface emit more and more energy as infrared light. This additional infrared light is re-emitted in all directions, randomly. Some of this additional thermal IR is redirected back toward Earth's surface and reabsorbed (see Figure 16.5), further heating both the surface and the air at the surface. The remainder of the IR light that is absorbed by molecules—that is, the fraction not re-emitted as IR light—simply heats the air at the altitude at which the light is absorbed. The equilibrium between the incoming solar energy and the outgoing infrared energy is reestablished by the additional IR that happens to be emitted upward into space. We therefore refine the definition for the greenhouse effect; it is *the absorption and conversion to heat and the subsequent re-emission of some of this trapped energy back toward Earth, with the consequent warming of the surface and lower atmosphere.* The greenhouse effect is responsible for the average temperature at Earth's surface being +15°C rather than −15°C, the temperature it would be if there were no IR-absorbing gases in the atmosphere. The very fact that our planet is

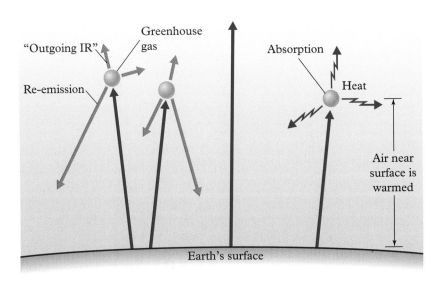

Figure 16.5 The greenhouse effect. Outgoing IR is absorbed by greenhouse gases and converted to heat, thereby raising the temperature of surrounding air.

not entirely covered by a thick sheet of ice is due to the natural operation of the greenhouse effect. The surface is warmed as much by this mechanism as it is by the solar energy it absorbs directly.

The natural greenhouse effect is beneficial, not detrimental, to life on Earth. However, the concentrations of all the natural greenhouse gases in air have increased appreciably in the last 2 centuries, as a direct result of human activity. In addition, several new, persistent, synthetic compounds that act as greenhouse gases have been introduced into the air. Environmental scientists worry that further increases in the concentration of greenhouse gases will result in the conversion of even more of the outgoing thermal infrared energy into heat and that this will increase the average air temperature significantly beyond today's 15°C. This phenomenon is called the **enhanced greenhouse effect,** to distinguish it from the one that has been operating naturally for millennia. Let us look next at the various gases that enhance the greenhouse effect.

Greenhouse Gases

16.6 Greenhouse gases absorb thermal infrared light

A **greenhouse gas** is a component of our atmosphere that efficiently absorbs thermal infrared light emitted by Earth. Neither single atoms nor diatomic molecules containing two identical atoms can absorb infrared light. Consequently, the main components of dry air—namely, N_2, O_2, and argon—*cannot* absorb IR and consequently are *not* greenhouse gases.

The most important greenhouse gas in Earth's atmosphere is *water vapor*, $H_2O(g)$. Water vapor in air arises primarily from the evaporation of liquid or solid water on Earth's surface or from the evaporation of water droplets in the air. Its concentration varies considerably at given times and places.

The second most important greenhouse gas is *carbon dioxide*, CO_2. Most of the CO_2 in air arises from natural biological sources, namely the decay or combustion of plant matter, but a growing fraction—currently about one-third—arises from the combustion of fossil fuels, as we will see.

Greenhouse gases do not absorb all or even most wavelengths of thermal infrared light. The absorption spectrum in the infrared region for carbon dioxide is shown in Figure 16.6. The fraction of light transmitted all the way through the gas, and therefore *not* absorbed, is very high *except* near 4 μm and near 15 μm, where CO_2 molecules do absorb the light.

The effect of the collective absorption by *all* greenhouse gases currently in our atmosphere of the infrared light leaving Earth is illustrated in Figure 16.7. The red curve at the top is the amount of light emitted from the surface and lower atmosphere at each wavelength in the thermal IR region, as predicted from Earth's temperature. The jagged blue line represents the amount that actually leaves the upper atmosphere. Thus, the vertical separation between the blue and the red lines is the amount of light that is absorbed by greenhouse gases at each specific wavelength. For

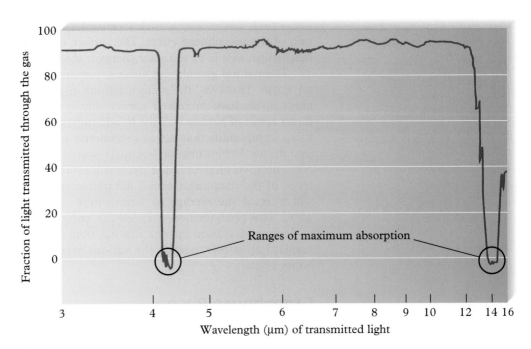

Figure 16.6 The infrared absorption spectrum for CO_2. The smaller the fraction of light transmitted, the greater the absorption by CO_2 gas. (Source: A. T. Schwartz et al., 1994. *Chemistry in Context: Applying Chemistry to Society.* American Chemical Society. Dubuque, IA: Wm. C. Brown Communications, Inc.)

example, there is a large gap between the curves in the 14–16-μm region. This means that much less IR light of these wavelengths escapes the atmosphere than is originally emitted by the surface and lower atmosphere. This absorption is due to carbon dioxide molecules, since the absorption occurs at just the wavelengths that we expect this gas to absorb based upon its spectrum (see Figure 16.6). The large atmospheric loss of IR light near 7 μm and beyond 18 μm is due to its absorption by the water vapor in the air.

The third most important greenhouse gas is *methane*, CH_4, which strongly absorbs IR light having wavelengths near 8 μm and is responsible for some of the "bite" taken out of Earth's emission at that wavelength (see Figure 16.7). Like carbon dioxide, it is a natural constituent of Earth's atmosphere, but its concentration has increased substantially because of human activities.

There are two other natural greenhouse gases:

- *Nitrous oxide,* N_2O, is a minor by-product of biological processes involving nitrogen compounds in soil.
- *Ozone,* O_3, is a minor constituent of air that is produced in small concentrations by natural atmospheric processes and by air pollution. It absorbs IR light near 9 μm.

Other than ozone, no other natural greenhouse gas absorbs infrared light in the 8–13-μm region. Since the atmosphere in that wavelength region acts like a glass window, transmitting most of the light that shines upon it (see Figure 16.7), the 8–13-μm region is called the **atmospheric window.** Introducing into the atmosphere pollution gases that absorb strongly in the window region is particularly dangerous since

Chapter 16: Global Warming and the Greenhouse Effect

Figure 16.7 Greenhouse gases selectively absorb some of the infrared light leaving Earth's surface and atmosphere. (Source: E. S. Nesbit, 1991. *Leaving Eden*. Cambridge, U.K.: Cambridge University Press.)

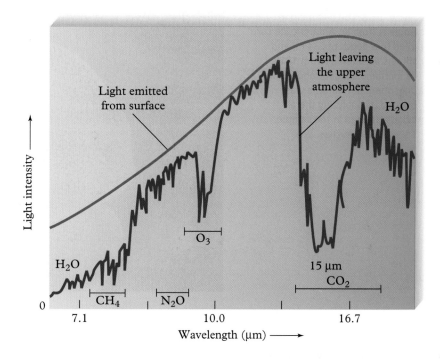

there is such an abundance of IR energy (emitted by Earth's surface) available in that region. In the absence of such gases, the energy in this region wouldn't be absorbed by anything, and consequently the IR energy would simply exit the atmosphere. The IR energy becomes trapped in the atmosphere when gases are present that will absorb it.

In later sections, we discuss the reasons that emissions of specific greenhouse gases have risen since the industrial age began and the extent of the increases.

16.7 Carbon dioxide varies in concentration

The concentrations of atmospheric gases are now monitored by sampling clean air at various locations around the globe and using sophisticated equipment to analyze the samples. Scientists determine the composition of the atmosphere in past centuries by analyzing air bubbles that are trapped in the ice of glaciers, since these bubbles were formed when a new layer of ice formed annually on top of an existing glacier (see Figure 16.8). The lower the ice sample in the glacier, the older the ice and thus the older the air in its bubbles.

Analysis of ice core samples from Antarctica and Greenland indicate that the atmospheric concentration of carbon dioxide in pre-industrial times (approximately before 1750) was about 280 ppm. It then began to increase sharply, since the burning of coal was central to the Industrial Revolution in Europe, which began at about this time. Since then, the concentration of atmospheric CO_2 has increased about 30%, to 376 ppm (in 2003), and is growing at an annual rate of about 1.6 ppm. In the last

Figure 16.8 The handheld ice sample is from an Antarctic ice core drilled to a depth of 234 m, and contains air bubbles from 1819. The methane concentration in the atmosphere at that time was obtained by analyzing the air pockets trapped in the sample. (CSIRO/Photo Researchers, Inc.)

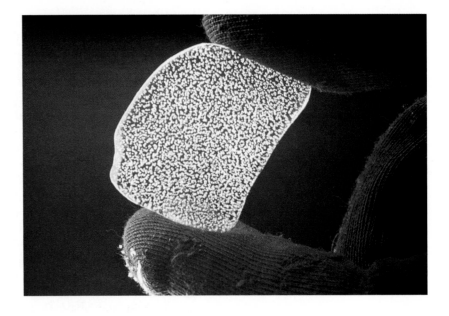

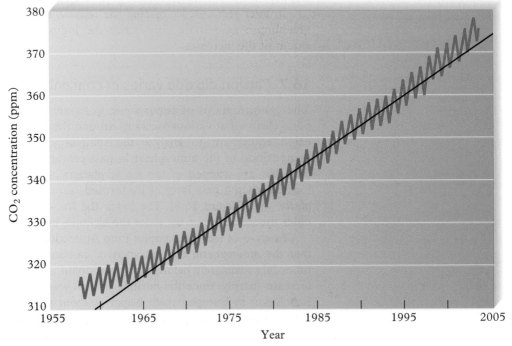

Figure 16.9 Annual atmospheric CO_2 concentration trends (red line) for the last half century. The black line represents the linear trend from **1970–2000.** (Modified from D. Keeling and T. Whorf [Scripps Institute of Oceanography], http:cdiac.esd.ornl.gov/trends/CO2/slo-mlo.htm.)

quarter of the 20th century, the average annual concentration of carbon dioxide increased more or less linearly with time (see full black line in Figure 16.9). This followed a period of at least 15 years in which the growth had been sharper than linear. Sharper-than-linear growth seems to have restarted since 2001. The actual carbon dioxide concentrations are shown by the red line in Figure 16.9. The concentration oscillates somewhat during each year. The levels are lowest in the northern

How was the relationship between carbon dioxide and global temperature detected? Learn about the experiment at Chemistry in Action 16.1.

hemispheric summer, when extraction of the gas from the air by photo-synthesizing plants is greatest, and highest in the winter, when decay of vegetation, which releases the gas, is greatest.

16.8 Anthropogenic sources of carbon dioxide

Much of the considerable increase in anthropogenic contributions to the carbon dioxide concentration in air is due to the combustion of fossil fuels—chiefly coal, oil, and natural gas. On average, each person in the industrial countries accounts for an annual release into the air of about 11 tonnes (metric tons) of CO_2 from the combustion of these carbon-containing fuels! Some of this *per capita* output comes about directly, for example from the exhaust gases released when vehicles are driven and homes are warmed by burning a fossil fuel. The remainder of the emissions are indirect, arising when energy is used to produce and transport goods, to heat and cool factories, classrooms, and offices, to produce and refine oil—in fact, to accomplish virtually any constructive economic purpose in an industrialized society.

Some carbon dioxide—a few percent of the anthropogenic emissions—is released into the atmosphere when rock (limestone) containing *calcium carbonate*, $CaCO_3$, is heated to produce the quicklime, *calcium oxide*, CaO, that is used in the manufacture of cement:

$$CaCO_3(s) \xrightarrow{\text{heat}} CaO(s) + CO_2(g)$$

A significant amount of carbon dioxide is also added to the atmosphere when forests are cleared and the trees burned, often in order to provide land for agricultural use. This activity occurred on a massive scale in temperate climate zones in past centuries. For example, immense deforestation accompanied the settlement of the United States and southern Canada. Deforestation has now shifted largely to the tropics. The greatest single amount of current deforestation occurs in Brazil. Overall, deforestation accounts for about one-quarter of the annual anthropogenic release of CO_2, most of the other three-quarters originating mainly with the combustion of fossil fuels.

Global annual *emissions* of carbon dioxide from fossil fuel combustion and cement production grew almost linearly with time from the late 19th century until about the 1940s. As shown in Figure 16.10, the CO_2 emission rates after 1950 grew rapidly, due to economic expansion in the developed world. In the last decade or two of the 20th century, the CO_2 emission rate showed slower growth, due to the balance between increasing use of fuels in developed and many developing countries, and the great decline in fuel use in the countries of the former Soviet bloc that accompanied the rapid decay of their economies.

16.9 Carbon dioxide becomes dissolved in oceans

The annual inputs and outputs of carbon dioxide to and from our atmosphere, averaged over the 1990s, are summarized in Figure 16.11.

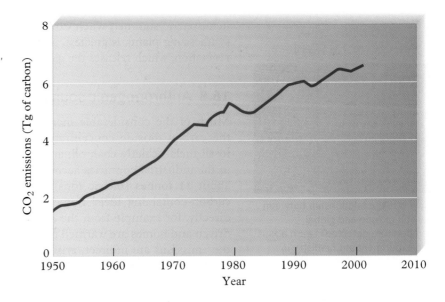

Figure 16.10 World carbon dioxide emissions from fossil fuel burning, 1950–1999. (L. R. Brown et al., *Vital Signs 2000,* W. W. Norton and Co., New York, 2000.)

Figure 16.11 Annual net movement of CO_2 to and from the atmosphere over land, sea, and air due to anthropogenic emissions in the 1990s. Units are gigatonnes. (Adapted from J. Houghton et al., *Climate Change 2001: The Scientific Basis,* Cambridge University Press, 2001.)

Collectively, fossil fuel combustion and cement production released 23 gigatonnes (Gt, or billions of tonnes) of carbon dioxide per year into the air. On average, 12 Gt, or about 50%, remained in the air and increased the atmospheric CO_2 concentration because it did not find an immediate sink—that is, it was not removed from the atmosphere by some process.

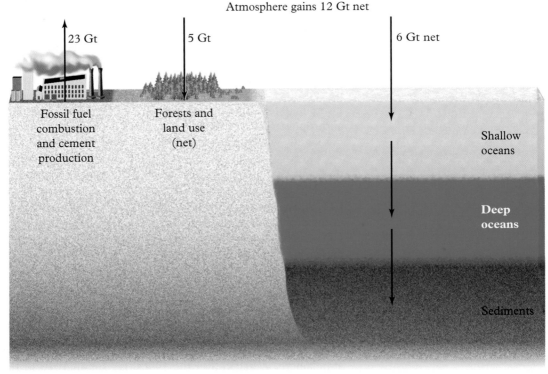

Chapter 16: Global Warming and the Greenhouse Effect

Though tropical deforestation also released considerable carbon dioxide into the air, this contribution was more than matched by the net withdrawal of CO_2 by growth in temperate zone forests and storage of carbon in their soils. Scientists believe that there has been faster growth in temperate zone forests and elsewhere due to the "fertilization effect" of carbon dioxide: because of the higher atmospheric level of CO_2, the rate of photosynthesis has increased, withdrawing the gas faster from air. The net effect of deforestation and increased growth was a withdrawal of about 5 Gt of carbon dioxide annually from the air.

About half the anthropogenic CO_2 emissions of 23 Gt are removed within a few years from the atmosphere, going into the surface waters of the ocean (6 Gt) and into new plant growth (5 Gt), leaving about 12 Gt to add to the existing atmospheric load. The shallow ocean and increased plant growth are only *temporary* sinks for the gas, however, since over a period of decades this gas naturally re-enters the atmosphere. Carbon dioxide molecules find a *permanent* sink only over a period of many decades and centuries, as the gas that is present in the surface ocean layer slowly makes its way to the intermediate and lower ocean depths and then is deposited as calcium carbonate in the sediments.

Because of the slowness in the permanent sink mechanism, the half of CO_2 emissions that don't find a temporary sink remain in the atmosphere for about a century on average. Consequently, over a timescale of decades and centuries, carbon dioxide accumulates in air. For example, the emissions that occurred in the 1990s increased the carbon dioxide level in air by 15 ppm.

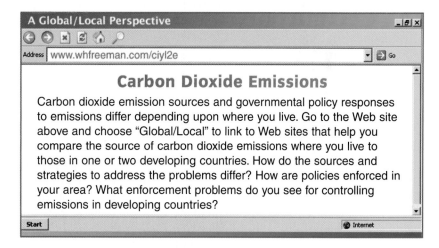

Carbon Dioxide Emissions

Carbon dioxide emission sources and governmental policy responses to emissions differ depending upon where you live. Go to the Web site above and choose "Global/Local" to link to Web sites that help you compare the source of carbon dioxide emissions where you live to those in one or two developing countries. How do the sources and strategies to address the problems differ? How are policies enforced in your area? What enforcement problems do you see for controlling emissions in developing countries?

16.10 Methane contributes to global warming

The atmospheric concentration of methane has more than doubled since pre-industrial times, and almost all of this increase occurred in the 20th century. The rise in the atmospheric CH_4 level is presumed to be the consequence of a number of human activities, as discussed below. By the 1990s, however, the rate of increase in the concentration had fallen to almost zero. This decline is not yet fully understood.

Although the concentration of methane is much less than that of carbon dioxide, it is an important greenhouse gas. In general, we cannot judge from concentration alone the relative importance of substances in causing global warming because different greenhouse gases vary widely in the ability to absorb thermal IR light. Per molecule, increasing the amount of methane in air causes 21 times the warming effect as does adding more carbon dioxide, since CH_4 molecules on average absorb a greater fraction of the thermal IR light that passes through them. Consequently, although the CO_2 concentration has increased 80 times as much as the CH_4 concentration, methane has been much less important in producing global warming.

In contrast to the century-long lifetime of carbon dioxide emissions, molecules of methane in air have an average lifetime of only about a decade. The dominant sink for atmospheric methane, accounting for about 90% of its loss from air, is its oxidation to carbon dioxide.

16.11 Methane comes from natural and anthropogenic processes

As was the case for carbon dioxide, post–World War II methane emission rates increased much faster than had previously been the case.

Most of the methane produced from plant decay results from the process of **anaerobic decomposition,** which is the decomposition of formerly living matter in the absence of air, that is, under oxygen-starved conditions. This process converts cellulose (whose composition we'll simplify as CH_2O) into methane and carbon dioxide:

$$2 \ CH_2O \rightarrow CH_4 + CO_2$$

The anaerobic decomposition of plant material occurs on a huge scale where plants decay under water-logged conditions—for example, in natural wetlands, such as swamps and bogs, and in rice paddies. Indeed, the original names for methane were *swamp gas* and *marsh gas*. Wetlands are the largest *natural* source of methane emissions.

The expansion of wetlands that occurs by the deliberate flooding of land to produce more hydroelectric power adds to the total natural emissions of the gas. Deep, small reservoirs produce and emit much less methane than do shallow ones that contain large volumes of flooded biomass, such as those in the Brazilian Amazon, especially if the trees are not first removed. Indeed, the global warming effect of the methane and carbon dioxide produced by a large, shallow reservoir created to generate hydroelectric power can, for many years, exceed the carbon dioxide that would be emitted if a coal-fired power plant were used instead to generate the same amount of electrical power! Thus, hydroelectric power is not a zero-emission form of energy production if land is flooded to create it.

Ruminant animals—including cattle, sheep, and certain wild animals—produce huge amounts of methane as a by-product in their stomachs when they digest the cellulose in their food. The animals subsequently emit the methane into the air by belching or flatulence. The decrease in the population of some methane-emitting wild animals (for example, buffalo) in recent centuries has been far exceeded by the huge increase in the population of cattle and sheep. The net result has been a large increase in emissions of methane from animal sources.

The anaerobic decomposition of the organic matter in landfills is another important source of methane in air. Food waste produces the greatest amount of methane. In some communities, methane from landfills is collected and burned to generate heat, rather than being allowed to escape into the air. Combustion of the methane produces an equal number of molecules of carbon dioxide, but because the *per molecule* effect of CO_2 molecules is so much lower than the effect of CH_4 molecules, the net greenhouse enhancement from the emission is greatly reduced.

Some methane is released into air when natural gas pipelines leak, when coal is mined and the CH_4 trapped within it is released into the air, and when the gases dissolved in crude oil are released or incompletely flared into the air when the oil is collected or refined.

16.12 The nitrous oxide concentration in air has increased

Per molecule, N_2O is 206 times as effective as CO_2 in increasing global warming. As with methane and carbon dioxide, the atmospheric concentration of nitrous oxide was constant until the Industrial Revolution, at which time it began to increase, though it has increased by only 13% since that time. Presently, it is increasing at an annual rate of about 0.25%.

Most *natural* nitrous oxide gas comes from its release from the oceans and from the soils of tropical regions. The gas is a by-product of the biological denitrification process in aerobic (oxygen-rich) environments and the biological nitrification process in anaerobic (oxygen-poor) environments (see Figure 16.12). In **nitrification,** nitrogen in the form of ammonia or the ammonium ion is converted mostly to nitrite (NO_2^-) and nitrate (NO_3^-) ions. In **denitrification,** nitrogen in the form of the nitrate ion is converted mostly to molecular nitrogen, N_2. Nitrous oxide is a minor by-product of both processes.

Overall, the increased use of nitrogen-based fertilizers for agricultural purposes probably accounts for the majority of the 40% of nitrous oxide emissions that are anthropogenic. A few years ago it was discovered that

Figure 16.12 Nitrification and denitrification.

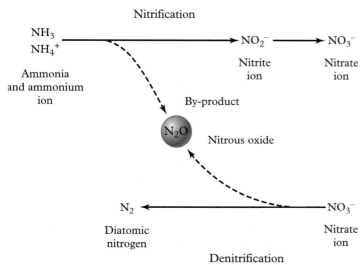

nylon production leads to emissions of nitrous oxide and that these emissions were a significant source of the gas to the atmosphere. Nitrous oxide from nylon-producing sources has now been greatly reduced by passing the gaseous emissions over a catalytic converter that converts them to N_2 and O_2.

Nitrous oxide is *not* formed as a by-product of the chemical combination of the N_2 and O_2 in air that occurs during combustion. However, some of the NO produced from atmospheric N_2 during gasoline combustion in automobiles is converted to N_2O rather than to N_2 in the catalytic converters that are currently in use, and the gas is subsequently released into air. Some newer catalysts developed for use in automobiles do not suffer from this flaw of producing and releasing nitrous oxide during their operation.

16.13 CFCs and their replacements

The chlorofluorocarbons (CFCs) discussed in Chapter 15 have a great potential to induce global warming, since they are both very persistent in air and they absorb strongly in the 8–13-μm window region of thermal IR light. Each CFC molecule has the potential to cause the same amount of global warming as do tens of thousands of CO_2 molecules. However, the *net* effect of CFCs on global temperature is reduced, by about half, by the cooling that they induce in the stratosphere by their destruction of ozone as discussed in Chapter 15.

The influence of CFCs on climate in the future will be reduced as a result of the *Montreal Protocol,* the international agreement that banned further production of these chemicals in developed countries after 1995, as discussed in Chapter 15. Most replacements for CFCs have shorter atmospheric lifetimes and absorb thermal infrared light less efficiently, and thus on a molecule-for-molecule basis they pose less of a greenhouse threat. As well, modern refrigeration equipment is now designed to leak much less gas into the air as compared to the older equipment.

The industrial gas sulfur hexafluoride, SF_6, is a minor greenhouse gas.

16.14 Water vapor contributes to global warming

Although human activities, such as the burning of fossil fuels, produce water as a by-product, the concentration of water vapor in air is determined primarily by temperature and by other aspects of the weather. Virtually all the H_2O in the troposphere arises from the evaporation of liquid and solid water on Earth's surface and in clouds. As we saw in Chapter 2, the rate at which this water evaporates and the maximum amount of water vapor that an air mass can hold both increase sharply with increasing temperature. Thus, the rise in air temperature that is caused by increases in the concentration of the other greenhouse gases, and by other global warming factors, heats the surface water and ice and thereby causes more evaporation to occur.

The increase in water vapor concentration from global warming produces an *additional* amount of global warming, comparable in magnitude to the initial amount, because water vapor is a greenhouse gas. This behavior of water is an example of the phenomenon called feedback, which is found to occur in many contexts. **Feedback** is the response of a system (be it a house, a lake, your body, or the atmosphere) to a change that has occurred; with **positive feedback,** the system's response accelerates the pace of the change within that system. This occurs because the interaction between the agent of change and the system produces a result that further amplifies the change in the system. In the case of the water vapor, the increase of the amount of water vapor in the atmosphere that results from global warming itself increases the amount of global warming. A more familiar example of positive feedback is the screeching noise that a public-address system makes when some of the loudspeaker output is fed into the microphone and is thereby amplified over and over.

16.15 Aerosols contribute to our climate

Aerosols, the tiny particles suspended in the air that we discussed in Chapter 14, play an important role in modifying our climate. Because they had not been seriously considered in the past, scientists overestimated the amount of global warming to be expected, since overall aerosols exert a *cooling* effect on the overall climate.

All solids and liquids—including atmospheric particles—have some ability to reflect light (see Figure 16.13). Many atmospheric particles reflect incoming sunlight. Some of the light is directed back into space and so is unavailable later for absorption—and heat production—at the surface, so particles that reflect sunlight can be said to cool the atmosphere.

Some types of aerosol particles absorb certain wavelengths of light (see Figure 16.13). Once absorbed, the energy that was associated with the light is rapidly converted into heat, which is then shared with the surrounding air molecules when they collide with the hot particle. Thus, the

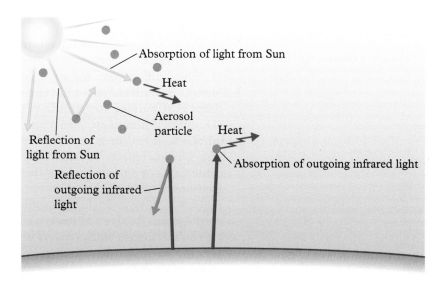

Figure 16.13 Reflection and absorption of light by suspended aerosols.

absorption of light by a particle leads to the *warming* of the air immediately surrounding it. The absorption of sunlight, with consequent warming, is significant only for dark-colored particles such as those composed primarily of (black) soot and of ash particles from volcanoes. Sulfate aerosols do not absorb sunlight since none of their constituents—water, sulfuric acid, and the ammonium salts thereof—absorb it. Only if tropospheric sulfate aerosols incorporate some soot will absorption of sunlight by these particles be significant.

A short-term, dramatic example of the effects of atmospheric aerosols upon climate occurred as a consequence of the massive eruption of substances into the atmosphere by the Mount Pinatubo volcano in the Philippines in 1991. The massive amount of SO_2 gas that the volcano blasted directly into the lower parts of this region produced a sulfuric acid aerosol, which remained in the stratosphere for several years, during which time it efficiently reflected sunlight back into space. This effect from the Pinatubo eruption lowered average surface air temperatures globally by about 0.2°C in 1992 and 1993 since less sunlight reached the surface. Many regions, including North America, experienced several cool summers in this period.

The sulfate-rich tropospheric aerosols, discussed in Chapter 14, that daily are produced from air pollutants, especially over urban areas in the Northern Hemisphere, *reflect* sunlight back into space more effectively than they absorb it. Consequently, less sunlight is available to be absorbed by the surface and in the lower troposphere and converted to heat. Thus the net effect of sulfate aerosols is to cool the air near ground level, and thereby to offset some of the effects of global warming induced by greenhouse gases.

Since the sulfate aerosol droplets are removed efficiently during rainstorms, their lifetime in the lower troposphere is only a few days or a week; in contrast to greenhouse gases, the sulfate aerosol does *not* accumulate. Although the sulfate aerosol has a short lifetime, new supplies of it are constantly being formed from the sulfur dioxide pollution that pours into the atmosphere from coal burning and other activities on a daily basis. Thus a steady amount of the aerosol exists in the troposphere, postponing the full effects of global warming induced by the rise in greenhouse gas concentrations.

The bulk of the anthropogenically produced aerosol in North America is centered above the Ohio Valley as a result of extensive high-sulfur coal burning for electric power production in that region. The SO_2 emitted from these plants and other sources is converted in air to sulfuric acid and then to sulfate aerosols, as discussed in Chapter 14. The aerosol reflects sunlight mostly above that area and cools the immediate region, outweighing there the global warming produced by greenhouse gases. Other regions in which the cooling effect from aerosols locally outweighs the heating effect due to greenhouse gases are south-central Europe and eastern China, both of which are also areas of high sulfur dioxide emissions. As we shall see later, however, *on a global basis* the warming effects of the greenhouse gases outweigh the cooling effects of aerosols.

16.16 Additional cloud cover

The largest remaining uncertainty in the prediction of future climate is the effects of clouds. Since they are composed of water droplets and crystals, clouds reflect sunlight back into space, and this cools Earth's surface. However, they also reflect outgoing infrared light back to the surface and lower atmosphere, and the redirected IR light heats the surface (see Figure 16.13).

In clouds that are low in the atmosphere, not far from the surface, the cooling mechanism prevails since they reflect incoming sunlight more efficiently than they reflect outgoing IR light. Indeed, we are all aware that when clouds pass overhead during a sunny day, Earth's surface cools rapidly. However, the opposite is true for high clouds. They reflect and re-emit IR light more efficiently than they reflect sunlight, and their presence warms the air.

Because the atmosphere will contain more and more water as the surface heats due to increased evaporation, scientists know that the total amount of cloud cover will increase. However, it is not yet known whether the *additional* cloud cover at low altitudes will exceed the additional cloud cover higher up, thereby reducing the amount of global warming caused by greenhouse gases, or whether the new high cloud cover will exceed that at low altitudes, thereby adding to greenhouse warming. Most of the scientists who remain skeptical about global warming (see next section) believe that the new cloud cover will occur mainly at low levels and will counter almost all the global warming produced by greenhouse gases.

16.17 The impact of humans on global warming

There is no *definitive* proof that human activities are responsible for any of the global warming that has occurred, nor is it likely that definitive proof can ever be obtained for this sort of complex scientific problem. Indeed, a small group of scientists remain highly skeptical of the evidence gathered thus far and of the predictions about future warming. You can explore their arguments by consulting the appropriate Web sites and articles cited at textbook's Web site. However, the great majority of atmospheric researchers agree with the United Nations–sponsored Intergovernmental Panel on Climate Change (IPCC), which has studied this problem intensely. This international group of scientists concluded in their 2001 report that "most of the observed warming over the last 50 years is likely to have been due to the increase in greenhouse gas concentrations."

Figure 16.14 presents the estimated portion of global warming or cooling between 1750 and 2000 caused by each of the factors we have discussed in this chapter. The greatest heating from a greenhouse gas came from the increased carbon dioxide concentrations in air. The contribution from methane was one-third that from CO_2, and even less came from the other greenhouse gases. Figure 16.14 does not show water vapor. As we have seen, the contribution from water vapor arises

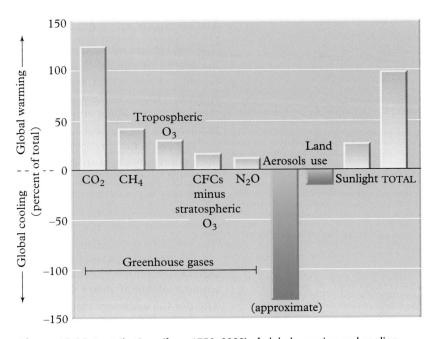

Figure 16.14 Contributions (from 1750–2000) of global warming and cooling produced by the various factors discussed in this chapter. Note that some factors individually exceed 100% but are partially canceled by others. (Adapted from J. T. Houghton et al., *Climate Change 2001: The Scientific Basis*, Cambridge University Press, 2001.)

solely from the heating induced by the other factors, and so its effect has been apportioned among the factors listed.

Variations in the intensity of sunlight received on Earth have also produced a small amount of the total global warming from 1750 to 2000. However, as we shall see, this effect dominated global warming until recently.

The amount of cooling that results from aerosols and the cloud changes they produced is highly uncertain. The best estimate is that it canceled about half the warming produced by increases in greenhouse gas emissions.

Although the global warming of the last few decades is attributable mainly to increasing concentrations of greenhouse gases, this is not believed to be the case for the 100 years before that. Some of the cooling in the late 19th century probably arose from changes in land use, particularly in North America and other midlatitude regions that increased the amount of sunlight reflected from the surface. However, the trends—whether cooling or heating—in air temperature until the mid-1960s were dominated by changes in sunlight intensity. The changes in solar energy probably gave rise to most of the two periods of cooling and the first period of heating of air temperatures in the 1860–1975 period (see Figure 16.2). Anthropogenic effects (greenhouse gases and atmospheric aerosols) provided only a slow but steady warming contribution, though one that accelerated around 1965 and has dominated global warming since then.

Chapter 16: Global Warming and the Greenhouse Effect

Predictions of Future Global Warming

16.18 Even moderate increases in CO_2 emissions sharply increase its concentration

Because carbon dioxide has such a long lifetime in the atmosphere, a century or more on average, the gas accumulates in air. Thus, almost all the mass of CO_2 emissions from the 1990s, for example, that did not find a temporary sink will remain in the air for decades to come, adding to the bulk of the emissions from the 1980s, the 1970s, and previous years.

The growth pattern of the CO_2 concentration in air is determined mainly by the pattern of CO_2 emissions. Suppose, for example, that the same amount of carbon dioxide emissions was added to the air each year and did not find a temporary sink. The total amount of CO_2 in air—and hence its concentration—would then annually increase by a constant amount. The carbon dioxide concentration increases linearly with time in this case, as shown in Figure 16.15a. For example, if the world were able to hold its carbon dioxide emissions constant at their value in the year 2000, then the CO_2 concentration would increase linearly and would become slightly greater than 500 ppm in 2100.

Another scenario, which for some time periods has been more realistic than the situation just described, is that the CO_2 *emissions* were not the same each year, but themselves increased linearly, that is, by a constant amount k each year. Thus, if the emissions one year amounted to *A*, the next year they were *A* + *k*, and the following year *A* + *2 k*, etc. In this case, the growth in CO_2 *concentration* is much sharper than linear: the resulting plot of CO_2 concentration curves upward, as illustrated in Figure 16.15b.

Experts predict that if no deliberate actions are taken to reduce them, carbon dioxide emissions will continue to increase, by approximately constant annual amounts through the 21st century, and by 2100 will be almost three times those of today. The predicted curve for the carbon dioxide concentration rises sharply upward because of these linearly increasing emissions, in agreement with our analysis for Figures 16.15b, and achieves a value of about 700 ppm by 2100, almost double the current value and equal to 2.5 times the pre-industrial value. In other scenarios developed by the IPCC, the CO_2 concentration by 2100 is predicted to reach 560–820 ppm, double or triple the pre-industrial levels, depending upon whether or not any steps are taken to reduce emissions.

16.19 Global air temperature is likely to increase

According to computer simulations of Earth's climate, the average global air temperature will be 1.9–2.9°C higher by 2100 than it was in 1990, if the IPCC predictions for growth in carbon dioxide levels are valid. An increase of 2 or 3 degrees may seem small, but our current average air temperature is less than 6°C warmer than that in the coldest periods of the ice ages! These estimates of air warming depend greatly

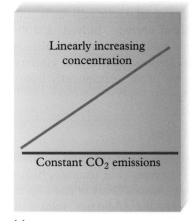

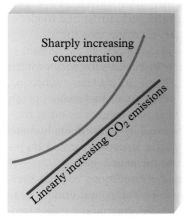

(a)

(b)

Figure 16.15 CO_2 concentration related to emissions. (a) Constant emissions of CO_2 produce a linearly increasing concentration of the gas. (b) Linearly increasing emissions produce sharply increasing concentrations.

The warming over some areas, including the United States and Canada, should be noticeably faster than the average rate for the globe.

on whether emissions are controlled or not: the temperature rise could be as small as 1.4°C or as high as a staggering 5.8°C. At a minimum, however, the world is predicted to warm more than twice as fast in this century as it did in the last. It should be noted, however, that the modeling of world climate—even using the fastest computers—is a difficult exercise, and scientists are uncertain as to the accuracy of these predictions.

More extreme weather may result from global warming, and it is the phenomenon that will affect many of us the most. The frequency of days with very high temperatures is predicted to increase, as are the number of days without frost. There may well be enough melting of ice in the Arctic region for the Northwest Passage to be used for commercial transport, since the warming in winter of all Arctic regions is projected to be much greater than the global average. Indeed, a 2004 international report confirmed that this polar warming has already begun in earnest, and has resulted in widespread melting of glaciers, thinning of sea ice, and rising temperatures of the permanently frozen land in the far north, called the permafrost.

16.20 New methane emission sources

Some scientists speculate that the rate of release of methane into air could greatly increase as a *consequence* of global warming. For instance, higher temperatures would accelerate the anaerobic biomass decay of plant-based matter, including that in landfills. In turn, the additional release of methane due to a warmed atmosphere would cause a *further* rise in temperature. This is another example of the phenomenon of positive feedback.

In addition, much methane is currently immobilized in the permafrost of far northern regions. Melting of the permafrost due to global warming could release large amounts of this methane into the air.

Clathrate formulas have a midline dot connecting the separate formulas for the two types of molecules that form the weak combination.

Monumental amounts of methane are also trapped at the bottom of the oceans, on continental shelves, as *methane hydrate,* a substance with the formula $CH_4 \cdot 6H_2O$. This is an example of a *clathrate compound,* a rather remarkable structure that forms when small molecules occupy vacant spaces ("holes") in a cagelike solid structure formed by other molecules (see Chapter 4, section 4.14, and Figure 4.4). In methane hydrate, CH_4 molecules are "caged" in a 3-D ice lattice structure formed by the water molecules, giving a 6:1 ratio of water to methane.

Clathrates form and remain stable under conditions of high pressure and low temperature, as are found in cold waters and under ocean sediments. If seawater warmed by the enhanced greenhouse effect penetrates to the bottom of the oceans, the clathrate compounds could decompose and release their own methane, as well as reservoirs of pure methane currently trapped below them, into the air above.

16.21 Global warming could be accelerated

In the long term, a dramatic—though unlikely—effect of substantial global warming would be a substantial change in the circulation pat-

The 2004 film *The Day After Tomorrow* explored the unrealistic effects on the climate of the U.S. from a global-warming induced worldwide rapid cooling followed by an equally rapid change in North Atlantic water circulation. In reality, such dramatic changes in climate and water circulation patterns would take decades to slowly develop.

terns of water in the Atlantic Ocean. Currently, warm surface waters flow northward from the tropics into the North Atlantic, bringing heat to Europe and to a lesser extent to eastern North America. A rapid rise in temperature and rainfall levels could drastically change this circulation pattern, as geological records indicate has happened in the past. In this case, European air temperatures would cool significantly.

A few scientists believe that several positive feedback mechanisms *could* combine to trigger an unstoppable warming of the globe: this worst-case scenario is called the *runaway greenhouse effect*. Such climate change would threaten all life on Earth, as the temperature would rise markedly, ocean currents probably would shift, and rainfall patterns would be very different from those we know.

Predicted Effects of Future Global Warming

Those of us who currently suffer through severe winters each year may look forward to the warmer climate associated with the enhanced greenhouse effect. After all, in the 11th and 12th centuries, an increase of a few tenths of a degree was sufficient for farming to take place on the coast of Greenland, for vineyards to flourish extensively in England, and for the Vikings to travel the North Atlantic and settle in Newfoundland.

However, the climate changes predicted for the 21st century and beyond do not present a uniformly pleasant prospect. The *rate* of change in our climate, which to date has been modest, will be dramatic by the middle of the century. Indeed, the rapid rate of global change will probably be the greatest problem with which humanity will have to contend. A more gradual transition, even to the same end result, would be much easier to handle, not only for humans but for all living organisms on the planet.

It is very difficult for scientists who use computer programs to model the climate—even with the assistance of the fastest computers in the world—to make definitive statements about what changes will occur in particular regions in the future. We know that there will be substantial changes in the climate, but we are unable to specify exactly what they will be. In this section, we examine some of the side effects of increasing air temperatures that have been predicted consistently.

16.22 Sea levels are predicted to rise

Without controls on CO_2 emissions, global sea levels are predicted to rise by an average of about 50 centimeters by 2100—in addition to the 10–25-cm rise experienced over the last 100 years.

Recall that density was discussed in Chapter 2, Section 2.1

Although some of the predicted rise in sea level is due to the melting of glaciers, most arises from the **thermal expansion** of seawater. The expansion occurs because the density of water *decreases* gradually as water warms beyond 4°C, the temperature at which it reaches its maximum density. Since the density of a sample of a substance is its mass divided by its volume, and since the mass of a given sample of water cannot change, the volume it occupies must increase with rising temperature as its density decreases. Thus, as seawater warms, the volume occupied by a gram or kilogram of it increases; the only way that this can occur is if the top of the water—the sea level—increases.

Although air and land surfaces are quick to warm with an increase in average global temperature, the same is not true of seawater. It takes many centuries for an increase in air temperature to gradually make its way down lower and lower into the ocean. For this reason, the rise in sea levels resulting from any particular amount of global warming is largely delayed for many years. Consequently, even if atmospheric carbon dioxide levels did not increase at all beyond today's values and no further global warming were to occur, sea levels would *continue* to rise for centuries yet to come, as lower and lower layers of the oceans became heated—and expanded—by the global warming of air that has already occurred.

Although an increase of about half a meter in sea level does not seem gigantic, remember that half the world's population lives in coastal areas, some of which are rather low-lying. For example, in Bangladesh, some people currently live on land that would be flooded by a rise in sea level of this size. Damage to such countries from tropical storms would increase because the higher sea levels make the land more vulnerable to damage. The intrusion of salt water into fresh groundwater resources is another problem caused by rising sea levels. Small island states in the Indian and Pacific oceans and in the Caribbean are likely to be particularly vulnerable. Other areas that could be seriously affected include the Netherlands, the Nile delta in Egypt, the eastern coastline of China, and river deltas in southern and southeast Asia, eastern Africa, portions of the Mediterranean, and even the Mississippi delta in the United States.

In contrast to some fears expressed in the 1980s, it now seems unlikely that air temperature increases in the 21st century will be enough to cause rapid melting of the Antarctic and Greenland ice sheets. Though scientists are uncertain whether global warming led to the 2002 dramatic breakup of the Larsen Ice Shelf off Antarctica (see Figure 16.16), they see the event as evidence of a continuing threat. Parts of such ice sheets sit on land above sea level, and their transfer into the oceans would cause a major increase in sea levels.

16.23 Changes in precipitation would affect agriculture

Global rainfall should increase as a consequence of global warming, since more water will evaporate from rivers, lakes, and the soil at the higher surface temperatures. The global average precipitation increases by about 2% for every degree (C) rise in temperature. Annually, the number of days having intense rain showers is predicted to increase, as is the amount of rain delivered by tropical storms. An increase in the average atmospheric temperature means that more energy is contained in the air and water at Earth's surface, so more violent weather disturbances could result.

Although the world *overall* will become more humid and rainier, some areas will likely become drier. To make matters worse, most areas of the

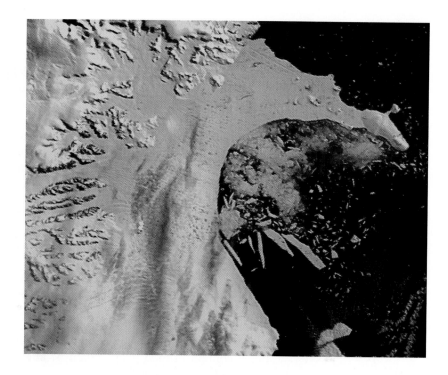

Figure 16.16 Breakup of the Larsen Ice Shelf in Antarctica, 2002. (NASA)

world that currently suffer from drought are predicted to become even drier. Many climate simulations predict that the southeastern regions of the United States and Mexico will become drier. Ironically, areas just north of these regions, from the northeastern United States and southern Ontario to the northern Great Plains states, are predicted to have significant increases in winter precipitation by 2100. Winters in California and other West Coast states are predicted to become much wetter in some simulation models. Northern Europe should also become somewhat wetter, at least in winter.

The number of countries classified as *water-scarce* or *water-stressed* is expected to double within a few decades. Significant decreases in runoff of precipitation into rivers are predicted for Australia, India, southern Africa, and most of Southern Europe and South America. In some parts of the world, such as China and parts of the United States, *increased* river runoff from increased precipitation will actually increase freshwater availability, though perhaps at the cost of more flooding. Thus, some areas will be subject to more frequent and more intense floods or droughts than in the past, and some will experience fewer.

Although most regions of the world will experience an increase in precipitation during the 21st century, this does not necessarily mean that the soil will contain more moisture. Indeed, the increase in the rate of evaporation of moisture from soil, due to higher air and surface temperatures, will more than offset the increased rainfall in most areas. However, increased atmospheric levels of CO_2 will lead to a higher rate of photosynthesis and improved water-use efficiency by plants. This positive fertilization effect on plants will cancel out some of the negative effects of decreased soil moisture.

Predicted Effects of Future Global Warming

There will be longer frost-free growing seasons at northern latitudes but increased chances that heat stress will affect crops grown there. Food production in temperate areas will probably also be affected by the attack of insects that in the past have been killed off in large measure during the winters but that could survive and flourish under warmer conditions.

Because temperature and moisture changes will occur more than in the past, some ecosystems will be destabilized. The species composition of forests is likely to change, especially in regions far removed from the equator. For example, the hardwood forest in eastern North America may be at risk of extinction if climate zones shift more quickly than replacement species of trees can move in. Coastal ecosystems such as coral reefs are also particularly at risk.

> ## ▶ Discussion Point: Might global warming be good for agriculture?
>
> Some organizations disagree that the consequences of global warming will only be negative. They argue that increased carbon dioxide levels will stimulate plant growth and food supplies will increase. Discuss the potential agricultural benefits and disadvantages of global warming. Do you think that the nutritional content of food plants might be affected by changes in carbon dioxide levels and growth rates?

16.24 Changes in climate may affect human health

Some scientists believe that human health will be adversely affected by global warming. The expected doubling in the annual number of very hot days in temperate zones will affect people who are especially vulnerable to extreme heat, such as the very young, the very old, and those having chronic respiratory diseases, heart disease, or high blood pressure. The problem is particularly acute for poor people, who have less access to air conditioning. As a consequence of the warmer climate, episodes of photochemical smog will increase in most locations. Domestic violence and civil disturbances might also increase, as they tend to occur more frequently in hot weather. On the other hand, there should be a decrease in cold-related illnesses because of the milder winters. Indeed, the rise in winter temperatures is predicted to reduce the winter mortality rate in large, temperate zone cities more than it will increase the summer mortality rate.

As we noted earlier in the chapter, global warming may extend the range of insects carrying diseases, such as malaria, and may intensify transmission in regions where such diseases already are prevalent (see Figure 16.17). A warming climate not only extends the range of the mosquitoes, but also allows them to proliferate faster, bite more frequently, and incubate more quickly the pathogens inside them that actually cause the disease. Some scientists have predicted that malaria could claim an additional million victims annually if the temperature

More than ten thousand elderly people died from a heat wave in France during the very hot summer of 2003.

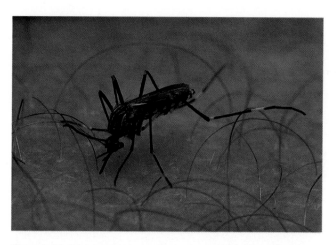

Figure 16.17 A mosquito of the type that carries Malaria.
(Digital Vision/Getty Images)

rise is sufficient to allow parasite-bearing, cold-blooded mosquitoes to spread into areas not now affected. In North America, suitable habitat for the mosquitoes is being lost, so fortunately it should not be as much of a problem for Americans and Canadians.

A different type of mosquito carries the dengue and yellow fever viruses, and, as we have previously mentioned, its range has already increased with warming and will probably continue to do so. Warm winters followed by hot, dry summers also facilitate the transmission of mosquito-borne encephalitis. In addition, there is evidence that cholera rates increase with warming of ocean surface waters because higher temperature increases coastal blooms of algae, which are breeding grounds for the disease.

Some experts in disease control discount these predictions regarding the spreading of tropical diseases, arguing that other factors such as increased rainfall could well negate or even reverse the effects of increases in air temperature. Indeed, malaria was widespread in North America and Europe in the past, but changes in public health and lifestyle—leading to less contact with mosquitoes—resulted in its near eradication. These experts do not believe that increasing the air temperature will mean that malaria and dengue fever will stage a comeback overall. While malaria may become more of a problem in some areas, such as the southern United States, changing weather patterns will mean that it becomes less of a problem in others.

Ways to Control CO_2 Emissions

16.25 Energy usage and CO_2 emissions vary widely

As we have seen, carbon dioxide emissions are closely tied to energy use. Commercial energy use depends upon many factors, including not only a country's population, geography, and climate, but also the cost of energy and the strength of the country's economy. Although increases in energy use are sometimes thought to be directly tied to population, globally they are more strongly related to economic development. The most important factor determining total energy use by a country is its *gross national product* (GNP), not its population. Thus, the sharp rise in global energy usage and carbon dioxide emissions in the second half of the 20th century was due mainly to industrial expansion and to increases in the standard of living in developed countries. Figure 16.18a lists in order the top 20 countries that emit carbon dioxide from fossil fuel sources in 1999 (dark bars), compared to their emissions in 1950 (light bars).

In *per capita* terms, the emissions of carbon dioxide into the air currently amount to about 4 tonnes per person per year when averaged

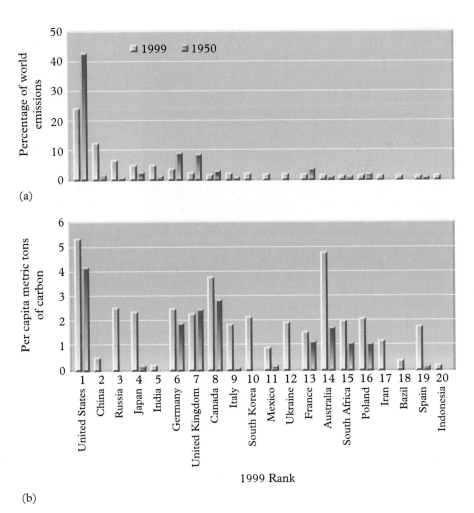

(a)

(b)

Figure 16.18 (a) Total annual emissions of CO_2, expressed as a percentage of the world total, and (b) per capita annual CO_2 emission for the top 20 total emitter countries in 1999, with comparative data for 1950. (Source: Carbon Dioxide International Analysis Center, part of Oak Ridge National Laboratory)

over the global population. Of course people in developed counties have much higher annual averages (11 tonnes of CO_2 per person) due to much heavier energy use, than do those in developing countries (2 tonnes). The length of the bar for each country in Figure 16.18b indicates its emissions per person in 1999 (dark bars) and 1950 (light bars). The United States leads in both total and *per capita* CO_2 emissions. The United States, Canada, and Australia currently have high *per capita* CO_2 emission rates, due in part to the high transportation requirements of these vast lands and to the fact that fossil fuel energy is much cheaper than in European countries.

Because populations of different countries differ so much in size, their greenhouse gas emissions *per capita*, or per dollar of GNP, are no guide to their *total* emissions. Thus, for example, China and India make substantial contributions in total emissions since their populations are so large, even though their *per capita* emission rates are still quite modest.

The increase, by one-third, in the atmospheric CO_2 level experienced to date and the accompanying temperature increase and climate modification has resulted largely from the industrialization and the increase in standard of living in developed countries. Without a significant change in the methods by which energy is produced and stored, these same nations will continue to emit high rates of CO_2 in the future in order to maintain their economic growth. Thus, the energy usage in developed countries may well continue to expand, though only slowly, by about 1.1% per year on average. However, if the Kyoto agreement to limit carbon dioxide emissions—discussed later in this chapter—is successfully implemented, the growth in energy usage by developed countries will decline and may even become slightly negative.

Economic growth in *developing* countries—which contain three-quarters of the world's population—is rising more quickly than in developed ones. And, as we have seen, with economic growth comes expanded energy consumption. Although developing countries collectively used only 30% of the world's commercial energy in 1993, they are expected to consume more than half the total starting sometime in the next two decades. The contributions of developing nations to greenhouse gas emissions will rise accordingly. Between 1994 and 2010, energy use, and thus carbon dioxide emissions from developing countries, collectively are expected to increase 4% annually. Consequently, developing countries will probably contribute half the annual emissions by 2008, and if this rate of growth is sustained, perhaps two-thirds of them by 2030 or earlier.

16.26 Setting a target level for atmospheric CO_2

Even with constant carbon dioxide emissions at current levels or a few percent lower, the carbon dioxide concentration in the atmosphere will continue to grow, as we discussed in our analysis of Figure 16.15. Some policymakers have promoted the idea that, through international agreements or allocation schemes, the world should control future CO_2 emissions so that the atmospheric level of the gas never exceeds some specific *concentration*. Although there is no consensus as to what is the most appropriate target, for our discussion we shall use 550 ppm. This value is twice the pre-industrial value—in other words, a situation in which human actions have doubled the natural atmospheric carbon dioxide concentration.

One way in which global CO_2 *emissions* could rise and fall with time in order to eventually achieve the 550 ppm concentration target is shown in Figure 16.19a. Figure 16.19b shows how the corresponding atmospheric CO_2 *concentration* would change with time for this emission scenario. This emission scenario, proposed by the IPCC, assumes that international agreement on CO_2 emissions can be achieved in the near future. Consequently, the scenario assumes only modest growth in CO_2

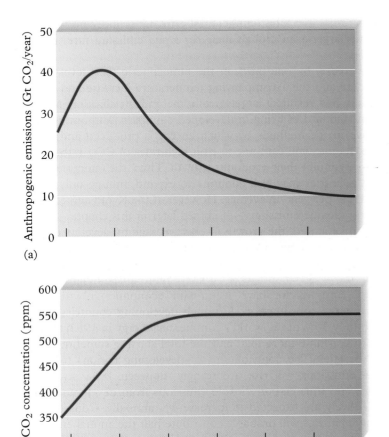

(a)

(b)

Figure 16.19 IPCC projections (a) for carbon dioxide emission rates to meet (b) a concentration limit of 550 ppm. (Adapted from IPCC, *Climate Change 1994*, Cambridge University Press, 1995.)

releases until about 2030, at which point a slow decline would set in. Alternative scenarios, in which CO_2 controls are not implemented until several additional decades later, have been proposed. Under these scenarios, a sharper decline in emissions than those shown in Figure 16.19a would occur later in this century. However, it is *not* possible to defer emission reductions indefinitely if the concentration target is to be achieved. In the 550 ppm scenario the temperature increase relative to that for the year 2000 is just under 2°C. The rise in sea levels would be reduced by about one-third if we embark soon on the scenario in which the 550 ppm concentration of carbon dioxide is never exceeded.

16.27 Some proposals allocate CO_2 emissions

Some national governments and organizations have debated how future CO_2 emissions can be minimized while still allowing economic growth. An international agreement on emissions was reached in 1997 negotiations held in Kyoto, Japan, where 39 industrialized nations agreed to decrease their collective greenhouse gas emissions by 5.2%, compared to 1990 levels, by 2008–2012. As a result of a combination of this agreement and growth in emissions in developing countries, the *per capita* CO_2 emissions in 2010 would have decreased from 11.2 tonnes in 1997 to 10.4 tonnes, while due to economic development, emissions in developing countries would probably have risen from 1.9 to 2.7 tonnes. However, the United States and Australia decided in 2001 that they would not in fact ratify and participate in the Kyoto agreement. The remaining countries that had agreed to reduce their emissions have decided to proceed, though it is not clear how many of them will ultimately meet their obligations. The implementation of the treaty began in 2005 when Russia, the last country to join in the treaty, signed on. As you read this book, an international system of trading greenhouse gas emission credits between countries that will exceed their allowed emissions with ones that have emission allowances to spare is underway.

Developing countries did not agree at Kyoto to limit their future emissions. One of their arguments in defending this decision was that most of the growth to date in greenhouse gas concentrations, and the global change it has wrought, has been due to developed countries.

However, it is widely believed that in the future more drastic cuts to global emissions than those proposed at Kyoto will be necessary and that developing countries will ultimately participate in limiting emissions.

16.28 CO_2 emissions could be minimized

Fuel switching Fossil fuels differ in the amount of carbon dioxide they emit during combustion. Per unit of heat energy produced, natural gas generates less carbon dioxide than does oil by a ratio of about 3:4, and oil in turn gives off less CO_2 than coal, by a ratio of about 2:3. Unfortunately, natural gas pipelines leak, so the greenhouse-enhancing effect of the methane that escapes into the air cancels some of the advantage associated with using methane rather than oil or coal.

The potential of a fossil fuel to generate CO_2, and so to cause atmospheric warming, depends on its carbon content. Consequently, some policymakers believe that **carbon taxes,** that is, *taxes based on the amount of carbon contained in a fuel rather than upon its total mass,* should be instituted as a disincentive to use the less desirable fuels and to minimize overall fossil fuel usage. The imposition of a tax on carbon also allow the market price for fuels to reflect their social as well as their economic costs. The tax rate could rise with time in order to encourage increasing abatement. Switching to fuels that emit no carbon dioxide upon their formulation or combustion, such as hydrogen gas produced by solar or nuclear energy (see Chapter 17), would then become more economically attractive.

Burying carbon dioxide In the future, CO_2 may be removed chemically from the exhaust gas of power plants that burn fossil fuels instead of being released to the atmosphere. The carbon dioxide gas so recovered would then be concentrated and *sequestered,* that is, deposited in a location that would prevent its release into the air. In a sense, the CO_2 would be locked away from the atmosphere. For example, the CO_2 could be sequestered by burial in the deep oceans, where it would dissolve, or in very deep aquifers under land or the seas, or in empty oil and natural gas wells or coal seams.

In very deep oceans (>3500 meters), carbon dioxide's density exceeds that of seawater and consequently it would form a stable "lake" of nearly pure CO_2. In shallower seas, carbon dioxide delivered to the seafloor would react with the solid calcium carbonate formed from seashells, producing soluble *calcium bicarbonate,* $Ca(HCO_3)_2$:

$$CO_2(g) + H_2O(aq) + CaCO_3(s) \rightarrow Ca(HCO_3)_2(aq)$$

For practical purposes, the CO_2, now chemically trapped in the bicarbonate form, would remain indefinitely in the dissolved state.

The energy input required for the CO_2-concentrating phase of these so-called **carbon sequesterization** schemes—some of which are currently being field-tested—would represent a substantial fraction, from one-third to one-half, of the total energy generated, so they would be quite

Figure 16.20 Iron fertilization. Iron sulfate is added to tanks for pumping into the Pacific Ocean.
(Courtesy of Dr. C. S. Law, National Institute of Water and Atmospheric Research [NIWA], New Zealand)

expensive. In addition, only a fraction of CO_2 comes from point sources such as power plants. It seems unlikely that carbon dioxide could be extracted from automobile exhaust and other less centralized, but collectively important, sources. Instead, the fossil fuel would have to be converted at a central location to a non-CO_2-emitting fuel such as hydrogen and the resultant CO_2 sequestered.

Removing CO_2 from the air One possible way to extract some of the carbon dioxide that is already dispersed into the atmosphere and to deposit it in ocean depths is the *iron fertilization* proposal. Large portions of the seas, especially the tropical Pacific and Southern oceans, lack plankton because they are very iron deficient. Artificially adding iron to these areas (see Figure 16.20) would result in massive blooms of plankton, which in the Southern Ocean at least would quickly descend into the deep oceans. The carbon dioxide used in photosynthesis to grow the plankton would then be locked away. Experiments are underway to test the feasibility of this approach.

Carbon dioxide can also be removed from the atmosphere by growing plants deliberately for this purpose. Some utility companies and some countries want CO_2 emission credit for planting forests, which will temporarily sequester carbon dioxide as they grow. Unfortunately, young trees absorb very little carbon dioxide. In addition, the sequestered carbon would have to be stored somehow so that it was not subject to fires or other processes that would convert it to carbon dioxide.

16.29 A final perspective on global warming

We conclude by commenting upon the paradox that faces humanity today concerning the enhancement of the greenhouse effect. On the one hand, there exists an outside chance that doubling or tripling the CO_2 concentration will have no measurable effect on climate. In that case, efforts to prevent such an increase would represent an economic burden for both the developed and the developing worlds, with no real advantages accruing. On the other hand, if the predictions of scientists who model Earth's climate turn out to be realistic, but we do nothing to prevent further buildup of the gases, both present and future generations will collectively suffer from rapid and perhaps cataclysmic changes to Earth's climate. As many observers have pointed out, people are accustomed to purchasing insurance to protect themselves against risks that are much less certain than are those predicted to arise from global warming. There is no *proof*—and there never can be—that additional global warming will occur and that it will have many adverse effects, but the probability is high that both premises will come to pass. It rests in the hands of today's adults, young and old, to decide whether to undertake measures to prevent rapid climate change or to leave to future generations the problem of adapting to whatever results.

A Global/Local Perspective

Address www.whfreeman.com/ciyl2e

Should the U.S. Now Sign the Kyoto Treaty?

Instead of signing the Kyoto Treaty, the U.S. government proposed initiatives to reduce the rate of increase of carbon dioxide emissions. Using the Internet, explore the environmental and economic arguments put forward by various American groups, both governmental and private, that argue for and against the U.S. signing the Kyoto accord and implementing the alternative initiatives. Also explore the initiatives taken by the U.S. states that have decided to implement Kyoto-like plans on their own. Which arguments do you find more persuasive? Do you think the U.S. should participate in upcoming international discussions on emission restrictions in the post-Kyoto period or continue to go it alone?

Start   Internet

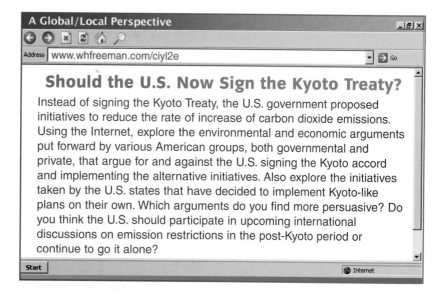

Summarizing the Main Ideas

Global warming is a phenomenon that causes increases in average air temperature and precipitation as well as other observed effects, such as the shortening of the winter season and shrinking of Earth's ice cover.

Earth achieves energy balance by emitting as much energy, in the form of thermal infrared light, as it absorbs from sunlight. The greenhouse effect is the temporary trapping of some of the outgoing IR light and its conversion to heat, causing an increase in the temperature at the surface and in the lower atmosphere. The main greenhouse gases are water vapor, carbon dioxide, methane, nitrous oxide, ozone, and CFCs. An increase in the concentration of any of these gases will enhance the greenhouse effect and cause additional global warming.

Anthropogenic sources of carbon dioxide include the burning of fossil fuels, the manufacturing of cement, and the destruction of forests. About half the new anthropogenic emissions of carbon dioxide remain in the atmosphere for extended periods of time, until they find a permanent sink in the deep ocean. The other half of the gas is present in temporary sinks—dissolved in the upper levels of the ocean, and as biomass.

Most atmospheric methane arises from the anaerobic decomposition of biological matter. Important anthropogenic sources include landfill emissions, rice paddies, livestock, biomass burning, coal mining, and the use of natural gas. The only important sink for methane is its atmospheric oxidation.

The concentration of water vapor in the atmosphere increases as the surface temperature goes up, thereby producing additional global warming; this is an example of positive feedback.

The presence of atmospheric sulfate aerosols cools Earth, since they form small, long-lasting water droplets that reflect sunlight. Their lifetime, however, is short, and consequently, unlike carbon dioxide, they do not accumulate in air.

Carbon dioxide has produced more global warming than all other greenhouse gases combined. A large part of the warming and cooling trends in the first two-thirds of the 20th century was due to fluctuations in the amount of solar energy received by Earth. However, almost all the warming in recent decades is due to increases in greenhouse gas concentrations.

By 2100, the CO_2 levels in air are predicted to be two to three times what they were in the pre-industrial era, a scenario that is predicted to raise air temperatures an additional 2° or 3°C. Additional greenhouse warming may be produced by positive feedback from methane emissions.

Sea levels are expected to rise appreciably as the air temperature increases. Most of the rise will be due to the thermal expansion of water. Global rainfall should increase as the air warms, though some areas will become drier. The potential effects of global warming on human health are uncertain and controversial.

The *per capita* emissions of carbon dioxide into the atmosphere vary greatly between countries and are tied most strongly to gross national product. Energy use—and hence carbon dioxide emissions—in developing countries collectively is increasing by about 4% annually, much faster than in developed countries.

If a target is chosen for the ultimate CO_2 atmospheric concentration, then ultimately the emissions of this gas will have to be restricted. The trading of CO_2 allocation rights between countries may occur in the future.

Carbon dioxide emissions could be reduced in the future by switching much fossil fuel consumption from coal to natural gas. Carbon dioxide produced in large facilities could be sequestered by locking it away in underground locations or in ocean depths.

Key Terms

global warming	enhanced greenhouse	anaerobic decomposition	clathrate compound
infrared light	effect	nitrification	thermal expansion
thermal infrared light	greenhouse gas	denitrification	carbon tax
greenhouse effect	atmospheric window	positive feedback	carbon sequesterization

Web Sites of Interest

To link to Web sites of interest, go to www.whfreeman.com/ciyl2e, Chapter 16, and select the site you want.

For Further Reading

C. Baird and M. Cann, *Environmental Chemistry*, 3rd edition, W.H. Freeman, New York, 2005. See Chapters 4 and 5 of this textbook for further information concerning the topics discussed in this chapter.

E. Claussen, "An Effective Approach to Climate Change," *Science*, October 29, 2004, p. 816. A summary of the current situation, with extensive references to Internet resources on the issues. The author is president of the Pew Center on Global Climate Change.

J. D'Agnese, "Why Is Our Weather So Wild?," *Discover,* June 2000, pp. 73–81. If global warming is causing current weather extremes, then future weather might look a lot worse.

P. Epstein, "Is Global Warming Harmful to Health?," *Scientific American,* August 2000, pp. 50–57. Computer models predict disease increases as Earth's temperature increases. Mosquito-borne diseases such as malaria and dengue fever are projected to spread dramatically.

M. Holloway, "Core Questions," *Scientific American,* December 1993, pp. 34–36. Information from cores drilled in the Greenland ice sheet is used to develop a profile of climate changes over a 250,000-year period.

J. T. Houghton et al., *Climate Change 2001: The Scientific Basis,* Cambridge University Press, 2001. A report of the Working Group of the Intergovernmental Panel on Climate Change. Current scientific consensus about the causes and future of climate change.

R. Kunzig and C. Zimmer, "Carbon Cuts and Techno Fries," *Discover,* June 1998, pp. 60–71. There are two ways to prevent global warming due to carbon dioxide: stop producing so much carbon dioxide or fix it so it won't warm up. The article examines ten ideas to avoid the consequences of carbon loading.

F. Pearce, "Climate Change: Menace or Myth?," *New Scientist,* February 12, 2005, pp. 38–43. A summary of the views of skeptics about greenhouse warming and their viewpoints and sources of financing.

T.M.L. Wigley, "The Climate Change Commitment," *Science,* March 18, 2005, pp. 1766–1769. Discusses the future rises in global mean temperature and sea level that will occur even if the atmospheric composition were to be fixed today.

Review Questions

1. Explain what is meant by the term *global warming.*

2. Identify the phenomena that indicate that global warming is occurring.

3. Identify the causes of global warming.

4. Which range of wavelengths is associated with infrared light? Which is associated with thermal infrared light?

5. What is the *greenhouse effect*? How is the greenhouse effect related to global warming?

6. What is a *greenhouse gas*? Identify five examples.

7. Why is N_2, nitrogen as it is found in the atmosphere, *not* a greenhouse gas?

8. To what does the term *atmospheric window* refer?

9. How is the composition of today's atmosphere different from that of the pre-industrial world?

10. What are the primary anthropogenic and natural sources of the following in the atmosphere?
    a) carbon dioxide
    b) methane
    c) nitrous oxide, N_2O
    d) CFCs

11. What relationship exists between *per capita* energy use and economic development?

12. What are the main sinks for the following?
    a) carbon dioxide
    b) methane
    c) nitrous oxide, N_2O
    d) CFCs

13. Explain how each of the following contributes to or cancels out global warming:
    a) carbon dioxide
    b) methane
    c) nitrous oxide, N_2O
    d) CFCs
    e) water
    f) aerosols

14. What is a *clathrate compound*?

15. Compare the relative amounts of carbon dioxide produced when a unit of heat is produced by combustion of natural gas; of oil; and of coal.

16. What does the term *positive feedback* mean?

17. What effects do volcanic eruptions have on average global temperatures? Are the effects relatively long or short term?

18. What effects do clouds have on global warming?

19. Why does carbon dioxide accumulate in the air over time?

20. What is the predicted increase in air temperature by 2100? What is that prediction based upon?

21. What global warming phenomenon will likely affect us most? Why?

22. How do total emissions differ from *per capita* emissions?

23. What is the *Kyoto agreement*?

24. Identify four ways to control carbon dioxide emissions.

Understanding Concepts

25. What is *greenhouse effect enhancement* and how can it occur?

26. Compare the concentrations of carbon dioxide in the atmosphere before and after the Industrial Revolution. How did it change? What factors contributed to the change?

27. Explain how the combustion of fossil fuel and the production of calcium oxide for cement increase atmospheric concentrations of carbon dioxide.

28. Compare the greenhouse enhancement effects per molecule of carbon dioxide, methane, and nitrous oxide.

29. Compare the processes of *nitrification* and *denitrification*.

30. How does the use of catalytic converters in vehicles contribute to increased levels of nitrous oxide?

31. Compare the relative impact of carbon dioxide and CFCs on global warming.

32. What role does water vapor play in global warming?

33. Explain the role that aerosols play in modifying the climate.

34. Compare the direct and indirect effects of sulfate aerosols in reflecting sunlight.

35. Describe the effects of increased atmospheric carbon dioxide concentrations on air temperature.

36. What impact will global warming have on methane hydrates? How might this affect global temperatures?

37. Explain how global warming could affect the following:
    a) ocean circulation patterns
    b) sea levels
    c) precipitation
    d) agriculture
    e) insect-borne diseases

38. How does the density of seawater vary with temperature?

39. Identify some of the potential consequences of increased sea levels.

40. Why are *per capita* emissions data not a guide for estimating total emissions in different countries?

41. What arguments did developing countries present for not participating in the Kyoto Treaty?

42. The *greenhouse factor* for a gas is a measure of how much of a contribution the gas makes to global warming. The factor depends upon characteristics of the molecule including how effectively it absorbs infrared light. Carbon dioxide is assigned a greenhouse factor value of 1, nitrous oxide a value of 160, and a CFC molecule a value of 25,000. Which molecule contributes most to global warming? How many carbon dioxide molecules have the same impact as one CFC molecule?

Synthesizing Ideas

43. Are the replacements proposed for CFCs themselves greenhouse gases? Why is their emission considered to be less of a problem in enhancing the greenhouse effect?

44. Compare the two mechanisms by which light interacts with atmospheric particles.

45. Describe how the absorption of light by aerosol particles warms the air around them.

46. Explain how a natural event such as a volcanic eruption can cause short-term climate changes. Describe the process by which the changes occur.

47. Explain how sulfate aerosols in the troposphere affect the air temperature at Earth's surface.

48. Based upon the data in the chapter, describe the trends in global temperature changes from the pre-industrial era to the present.

49. Explain how atmospheric substances can have both warming and cooling effects. Currently, what is the net effect of these competing processes on global temperature?

50. Explain the relationship between carbon dioxide emissions and carbon dioxide concentration in the atmosphere.

51. Describe the potential impact of global warming on weather patterns. Use Web sites at www.whfreeman.com/ciy/ae to compare different climate models.

52. Which poses a greater threat to living organisms: a rapid rise in global temperature or a gradual rise in global temperature? Explain.

53. What is the relationship between the temperature of seawater and the volume it occupies? How does this affect sea levels as the temperature of seawater increases?

54. In the presence of a sustained increase on global temperatures, how does the response of seawater differ from the response of air and land surfaces?

55. Increased global temperatures may result in both an increase in the rate of evaporation of moisture and an increase in annual rainfall. How do these affect each other? Which effect is predicted to dominate?

56. Discuss both the negative and positive effects of global warming with relation to precipitation, agriculture, and human health. Use information from Web sites at the end of the chapter to support both arguments.

■ Group Activity: Identifying Ways of Saving Energy and Its Pitfalls

Request one member to record ideas and conclusions generated by the group.

1. In a brainstorming session—that is, one in which ideas are generated but not analyzed—think of at least ten practical ways students like you can reduce their energy use and thereby reduce accompanying carbon dioxide emissions.

2. As a group, discuss the ideas and try to achieve consensus (unanimity) on which five suggestions are the most practical and would have the greatest effect.

3. For each of the five consensus suggestions, decide as a group if any of them would be subject to the Rebound Effect—the tendency to engage in an activity more after a more energy-efficient way has been devised to do it, with the result that total energy usage may actually be increased, not decreased. Use this discussion to create a list of pros and cons for each of your five suggestions.

In this chapter, you will learn:

- about radioactivity: what it is, and why alpha and beta particles and gamma rays are useful medicinal tools but can also harm living organisms;

- about the formation of radon, how it enters the home, and its potential health hazards;

- about fission, fusion, nuclear waste, and the potential to harness nuclear energy;

- about indirect solar energy, such as hydroelectric power, wind power, biomass energy, and wave and tidal power;

- about direct solar energy, such as thermal conversion and photovoltaic power;

- about the advantages and disadvantages of using solar energy.

Is nuclear energy good or bad?

Outside of Harrisburg, Pennsylvania, the reactors at Three Mile Island, the site of the worst nuclear disaster in U.S. history, continue to supply electricity to millions of people. In this chapter we will explore the perils and possibilities of nuclear energy and other alternative energy sources. (Larry LeFever/Grant Heilman Photography)

The Core of Matter

Radioactivity, Nuclear Energy,
and Solar Energy

In discussing energy production in earlier chapters, we concentrated on the use of chemical reactions to produce heat from fossil fuels. A negative aspect of all these processes is that they release carbon dioxide into the air, enhancing the greenhouse effect, as we discussed in Chapter 16. What alternatives to fossil fuel combustion are open to us if we decide that continuing to burn coal, oil, and natural gas is not acceptable? Basically there are two, nuclear energy and solar energy, both of which rely upon physical rather than chemical processes. In this chapter we explore these techniques, including the chemistry that underlies the production and dispersal of the materials required. In both cases we shall concentrate on the ability of these energy sources to produce electricity. We will begin by discussing the phenomenon and dangers of radioactivity.

Radioactivity

17.1 Most elements occur as a mixture of isotopes

When we introduced atomic structure in Chapter 3, we explained that the nucleus of an atom contains two types of particles: *protons* and *neutrons*. Each proton has a $+1$ charge. Neutrons have about the same mass as protons but are electrically neutral and therefore do not affect the chemical properties of the atom. Recall also from Chapter 3, section 3.4 that the sum of the number of protons and neutrons in a nucleus is the *mass number* of that nucleus.

We pointed out in Chapter 3 that nuclei of different atoms associated with the same element can have different numbers of neutrons, and therefore different mass numbers. Nuclei of a given element with different mass numbers are called *isotopes*. Natural samples of most elements consist of a homogeneous mixture of several isotopes. Recall that the mass number is sometimes displayed as a leading superscript to the elemental symbol, thereby generating the *isotopic symbol*.

Consider, as an example, the element carbon. All carbon atoms have 6 protons in their nuclei, as their atomic number is 6, and all have 6 electrons that travel about the nucleus. The great majority of carbon atoms also have 6 neutrons in their nuclei; consequently their mass number is $6 + 6 = 12$, and they are symbolized as ^{12}C. The number of protons in the nucleus is sometimes shown as a leading subscript to the isotopic symbol; thus these carbon atoms would be symbolized either as $^{12}_6C$ or ^{12}C. About 1% of carbon atoms contain 7 rather than 6 neutrons in their nucleus. Their mass number is $6 + 7 = 13$, rather than 12, and therefore their

Whenever you see this icon in this chapter, go to
www.whfreeman.com/ciyl2e

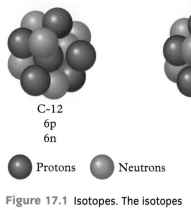

C-12   C-13   C-14
6p   6p   6p
6n   7n   8n

● Protons   ● Neutrons

Figure 17.1 Isotopes. The isotopes of an element vary only in the number of neutrons in the nucleus.

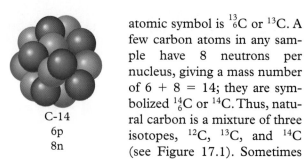

atomic symbol is $^{13}_{6}C$ or ^{13}C. A few carbon atoms in any sample have 8 neutrons per nucleus, giving a mass number of $6 + 8 = 14$; they are symbolized $^{14}_{6}C$ or ^{14}C. Thus, natural carbon is a mixture of three isotopes, ^{12}C, ^{13}C, and ^{14}C (see Figure 17.1). Sometimes for simplicity the mass numbers of isotopes are just displayed informally after the elemental name or symbol, for example carbon-13 or C-13.

The **atomic weight** of each element is the average mass of the natural mixture of its isotopes. For example, the atomic weight of carbon is 12.01, reflecting the fact that 99% of carbon on Earth is ^{12}C and 1% is ^{13}C. (The amount of ^{14}C is negligible in calculating the atomic weight.) Atomic weights for elements are often shown on periodic tables, or on lists of the elements such as that shown facing the inside back cover of this book.

Exercise 17.1

Write out the complete isotopic symbols for atoms of oxygen (atomic number of 8) that have **a)** 8 or **b)** 10 neutrons in their nucleus.

17.2 Radioactive nuclei emit small particles

Although most types of atomic nuclei stay the same indefinitely, some do not. The unstable, or **radioactive,** nuclei spontaneously decompose by emitting a small particle that is very fast moving and consequently carries with it a great deal of energy. In some types of nuclear decomposition processes, the atomic number changes as a result of the emission, and the atoms are converted from those of one element to those of another. Very heavy elements are particularly prone to this type of decomposition. The nuclei that are produced by emission of a particle may or may not themselves be radioactive and undergo decay at a later time.

The most common particles to be emitted by nuclei, and the only ones to be considered in this book, are of three types: alpha, beta, and gamma.

Alpha emission An **alpha (α) particle** is a radioactively emitted particle that has a charge of $+2$ and a mass number of 4—it has two neutrons in addition to the two protons—and is identical to the nucleus of the most common isotope of helium. Thus, an alpha particle is written as $^{4}_{2}He$. The nucleus that remains behind after an atom has lost an alpha particle consequently has an atomic number that is 2 units *smaller* than the original, and it is 4 units *lighter* (see Figure 17.2). For example, when a $^{232}_{90}Th$ (thorium-232) nucleus emits an α particle, the resulting

Recall that the charge of any atomic nucleus is equal to the atomic number of the atom.

Chapter 17: The Core of Matter

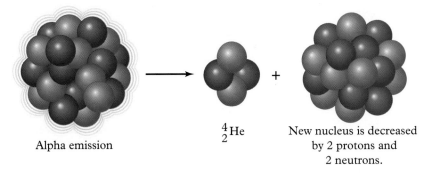

Figure 17.2 Radioactive decay. Alpha emission creates a lighter nucleus.

$\frac{4}{2}$He

Alpha emission

New nucleus is decreased by 2 protons and 2 neutrons.

nucleus has a mass of $232 - 4 = 228$ units and an atomic number of $90 - 2 = 88$. The atom no longer is an atom of the element thorium, atomic number of 90, but one having an atomic number 88, which corresponds to the element radium, Ra. The process can be written as a **nuclear reaction,** which is any process in which one or more nuclei change:

$$^{232}_{90}\text{Th} \rightarrow {}^{228}_{88}\text{Ra} + {}^{4}_{2}\text{He}$$

Notice that both the *total* mass number and the *total* nuclear charge of the substances on each side of the equation individually balance in such equations.

Worked Example: Determining the products of alpha particle decay

What are the element and the isotopic symbol for the product of the process in which the nucleus of an atom of ${}^{235}_{92}\text{U}$ emits an α particle? Write the nuclear reaction for the process.

Solution: First consider the change in mass number. Since an α particle, being a ${}^{4}_{2}\text{He}$ nucleus, has a mass number of 4, we subtract 4 from the mass number, 235, of the original isotope to determine the mass number of the product:

$$235 - 4 = 231 = \text{mass number of the product nucleus}$$

Now consider the number of protons in the product nucleus. This must be equal to 92 (the original nucleus) minus 2 (the number carried off by the α particle). Thus the number of protons in the product nucleus, and its atomic number, is

$$92 - 2 = 90 = \text{atomic number of the product nucleus}$$

Consulting the periodic table, we see that the symbol for element 90 is Th (thorium). The full symbol for the product nucleus is ${}^{231}_{90}\text{Th}$. Having identified the product, we can write the full reaction:

$$^{235}_{92}\text{U} \rightarrow {}^{231}_{90}\text{Th} + {}^{4}_{2}\text{He}$$

Radioactivity

Exercise 17.2

What are the element and the isotopic symbol for the nucleus that is formed when nuclei of each of the following atoms lose an α particle? Write the nuclear reaction for each process.

a) $^{239}_{94}\text{Pu}$ **b)** $^{210}_{83}\text{Bi}$ **c)** $^{179}_{79}\text{Au}$

Beta emission A **beta (β) particle** is an electron. It is formed within the nucleus in the unusual event when a neutron splits into a proton and an electron:

$$\text{neutron} \rightarrow \text{proton} + \text{electron}$$

The proton remains behind in the nucleus but the electron leaves it. Consequently, the atomic number *increases* by one unit; you may imagine this effect as "subtracting a negative particle" from the nucleus (see Figure 17.3). Since the total number of neutrons + protons remains the same, there is *no* change in mass number for the nucleus.

For example, when an atom of the isotope $^{214}_{82}\text{Pb}$ (lead-214) decays radioactively by the emission of a β particle, the nuclear charge of the product is $82 + 1 = 83$, corresponding to the element bismuth. The mass number remains 214:

$$^{214}_{82}\text{Pb} \rightarrow {}^{214}_{83}\text{Bi} + {}^{0}_{-1}\text{e}$$

The symbol ${}^{0}_{-1}\text{e}$ used for the electron here shows its mass number (0) and its charge. You can see from the subscripts and superscripts that the equation is balanced.

Exercise 17.3

What is the element and what is the isotopic symbol for the nucleus that is formed when nuclei of each of the following atoms lose a β particle? Write the nuclear reaction for the processes.

a) $^{14}_{6}\text{C}$ **b)** $^{61}_{27}\text{Co}$ **c)** $^{99}_{42}\text{Mo}$

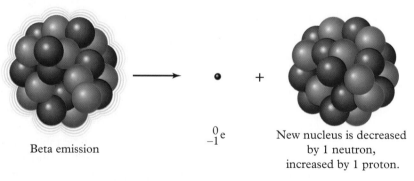

Beta emission ${}^{0}_{-1}\text{e}$ New nucleus is decreased by 1 neutron, increased by 1 proton.

Figure 17.3 Beta emission causes an atom to lose an electron, but the nuclear mass number doesn't change.

Gamma emission A third type of radioactivity occurs when a nucleus emits a gamma "particle," which like an X-ray is actually a huge amount of energy concentrated in one photon. Gamma particles possess no mass and are usually referred to as **gamma rays.** Neither the nuclear mass number nor the nuclear charge changes when a gamma ray is emitted. The emission of a gamma ray often occurs in conjunction with the emission of an alpha or beta particle.

17.3 Radioactivity can be dangerous to living organisms

The α and β particles that are produced in the radioactive decay of a nucleus are not in themselves harmful chemicals, since they are simply the nucleus of a helium atom and an electron. However, they are ejected from the nucleus with an incredible amount of energy of motion, and this energy—as well as that of a gamma ray—can cause harm when it collides with molecules in living organisms and transfers the energy of the molecules to the organisms. In particular, the absorbed energy can break chemical bonds or ionize electrons from atoms or molecules.

The different particles vary in their ability to penetrate living matter (see Figure 17.4). Alpha particles are relatively slow-moving compared to beta particles, since their mass is so much greater. As a consequence, α particles cannot penetrate matter, even a thin sheet of paper. However, if we ingest or inhale material, such as suspended airborne particles containing radioactive elements that emit alpha particles, damage can occur, especially if our DNA is thereby disrupted. Alpha particles that are emitted adjacent to cells inflict particularly severe damage because they concentrate all their energy in a small area of absorption, within about 0.05 mm of their point of emission.

In general, beta particles travel much faster than alpha particles, since they are so much lighter (by a factor of several thousand). Beta particles can travel about 1 meter in air, or several millimeters in water or human tissue, before losing their excess energy. Even a 1-cm sheet of plastic affords protection from them. However, if they are emitted from radioactive material that has been inhaled or ingested, they too can cause considerable damage.

Gamma rays—the high-energy photons emitted by certain types of nuclei—are generally the most dangerous type of radioactivity, since they are able to penetrate matter efficiently and thus do not have to be inhaled or ingested. Gamma rays easily pass through concrete walls—and our skin. Although they pass all the way through us, gamma rays lose some of their energy in the process, and we can be damaged by this transferred energy. However, only a few centimeters of lead, a material particularly good at preventing the transmission of gamma rays, are required to shield us from gamma rays.

Examine the penetrating abilities of alpha, beta, and gamma particles at Chapter 17: Visualizations: Media Link 1.

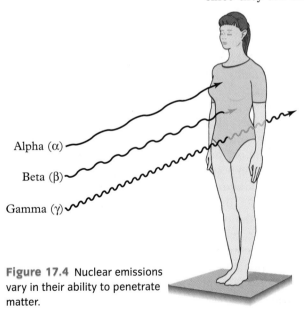

Alpha (α)

Beta (β)

Gamma (γ)

Figure 17.4 Nuclear emissions vary in their ability to penetrate matter.

The emissions from radioactive substances are often called *ionizing radiation,* because in their passage through matter some of their energy ionizes atoms and molecules. The ions so produced are free radicals and hence highly reactive (see Chapter 10, section 10.2). For example, a water molecule can be ionized by an alpha particle, beta particle, gamma ray, or X-ray. The resulting H_2O^+ ion dissociates into a hydrogen ion and the hydroxyl free radical, OH.

$$H_2O + \text{radiation} \rightarrow e^- + H_2O^+ \rightarrow H^+ + OH$$

If the affected water molecule is contained in a cell, the hydroxyl radical can engage in harmful reactions with biological molecules in the cell, such as DNA and proteins.

In some cases, radiation damage is sufficient to kill cells of living organisms. This is the basis of food irradiation, which kills microorganisms and thereby prevents them from spoiling the food. For decades, the U.S. government has allowed certain foods to be irradiated, such as wheat and flour (to control insects) and white potatoes (to inhibit sprouting). In 1985, irradiation of spices, fruits, vegetables, pork, and poultry was approved, as was the irradiation of beef (1997), and fresh eggs (2000). However, irradiation is not widely used on foods in the U.S., in part because of concerns that consumers would resist buying irradiated products. U.S. regulations require irradiated foods sold in stores to be specially labeled, although spices are exempt from such regulations.

Radiation has proved both helpful and harmful to human health as well. Gamma ray scans have proved useful in detecting bone diseases and abnormalities, such as cancer, osteoarthritis, infections, and fractures. Trace amounts of radioactive isotopes are injected into the body, and emit radiation that can be detected and converted to an image. In the example shown Figure 17.5, an isotope that concentrated in bones was used. However, if human beings are exposed to substantial, though sublethal, amounts of ionizing radiation, they can develop *radiation sickness.* The earliest effects to be observed occur in tissues containing cells that divide rapidly, because damage to the cell's DNA or protein can affect cell division. Such rapidly dividing cells are found in bone marrow, where white blood cells are produced, and in the lining of the stomach. It is therefore not surprising to find that early symptoms of radiation sickness include nausea and a drop in white blood cell count. Children are more susceptible to radiation than adults because their tissues are undergoing more cell reproduction. On the other hand, radiation can be effectively used to kill cancer cells since they divide rapidly. Radiation therapy cannot be completely selective in terms of the cells it affects, so it has side effects such as nausea.

Long-term effects from radiation may show up in genetic damage, because chromosomes may have undergone damage

Figure 17.5 Gamma ray scans of a healthy body from the front (left) and back. A radioactive isotope that accumulates in bone had been injected into the body. (Simon Fraser/RVI RMPD/Photo Researchers, Inc.)

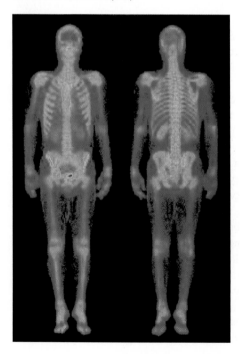

or had their DNA mutated. Such damage may lead to cancer in the radiated person or to effects on his or her offspring if the changes occurred in the testes or ovaries.

17.4 A radioactive isotope decays over time

The radioactive decomposition of the atoms in a sample of an isotope does not occur all at once, any more than the molecules in a chemical reaction react all at once. In fact, for some isotopes, radioactive decay can take a very long time. For example, in a sample of uranium-238 just large enough to be visible, there are about 10^{24} atoms. Only about 10^7 of the nuclei in such a sample decompose in any given second, so it requires billions of years for most of the sample to decompose.

The measure often used to express such decomposition rates is the time period required for half the nuclei in a sample to disintegrate, or its *half-life time period*, $t_{1/2}$. For example, the half-life of uranium-238 is about 4.5 billion years. This means that about half of the U-238 that existed when Earth was formed, by coincidence also 4.5 billion years ago, has now disintegrated. In the next 4.5 billion years, half the *remaining* uranium-238, amounting to one-quarter of the original, will disintegrate, leaving one-quarter of the original amount still intact. After three half-lives (13.5 billion years for U-238) have passed, only one-eighth of the original uranium will remain, and only one-sixteenth will still be there after four half-lives (18 billion years). Most radioactive isotopes have half-lives that are much shorter than that of U-238. For example, radon-222, which we discuss below, has a half-life of only a few days. Figure 17.6 illustrates the general concept of half-life.

Recall that we introduced the term *half-life* in Chapter 8, section 8.5, for chemical and biochemical reactions.

Archeologists often date ancient objects by determining the percent of a characteristic isotope, such as C-14, that is still present in them. From this data, they calculate the number of half-life periods that have expired since the object was in equilibrium with its environment, and hence its age.

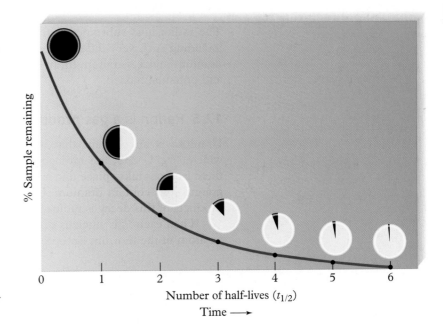

Figure 17.6 Radioactive decay. Nuclear decay follows a predictable curve whether the half-life is measured in fractions of a second or in years.

The concept of half-life can be modeled using several packages (the more the better) of M&M's (the standard milk chocolate type) or Skittles. Before you start the next part, be sure that you have a count of how many total candies you are starting with. Pour the M&M's (or Skittles) into a bowl and mix them by shaking the bowl. Next, spill the contents out onto a (clean) surface and count those with the printed side up ("M" or "S") (these represent the ones that did not "decay" in the first half-life period). Record this value, and place the candies that landed printed side up back into the bowl. Place those showing no label into a separate pile but do *not* eat them yet. Repeat the process, including the mixing, with the candies in the bowl, and record how many candies were printed side up after this second half-life. Repeat the process until you come to a point where no candies land printed side up. Be sure that you are recording how many are landing printed side up each time you spill the candies. Now put all the candies you had at the beginning of the experiment back into the bowl and repeat the whole process one or two more times. (Once you have collected your data you can eat the candy.) Now, calculate the average number of printed-side-up candies that remained after each half-life for all repetitions of the experiment. Next, calculate the ratio of each average to the total number of candies. Compare your ratios to the values expected from probability (0.50, 0.25, 0.125, 0.063 ...). How do your results compare with the expected values? What do you think would happen to your values if you were able to repeat this experiment 100 times with a larger number of candies? Would your ratios get closer to the expected values?

Radon Gas

The radioactive substance that is of greatest concern in many areas, including such very different locations as the basements of homes and uranium mines, is radon. In the next three sections we discuss its origin and potential hazard.

17.5 Radon is a gas produced from uranium decay

Uranium is an element that is composed of several isotopes, all of which, like U-238, are radioactive. The radioactive decomposition of these isotopes takes place under our feet each day, since many rocks and granite soils contain uranium. One of the unwelcome, though natural, products of this decay is radioactive radon gas.

When each $^{238}_{92}U$ nucleus decays, it emits an α particle, and thereby an atom of the thorium isotope $^{234}_{90}Th$ (Th-234) is formed:

$$^{238}_{92}U \rightarrow ^{234}_{90}Th + ^{4}_{2}He$$

This process is the first of 14 sequential radioactive decay processes, shown in Figure 17.7, that this nucleus undergoes. The last of these

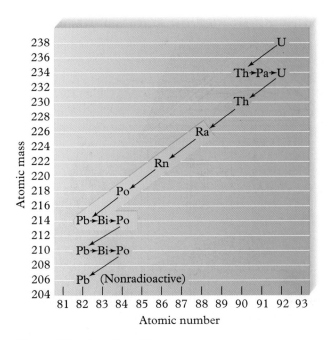

Figure 17.7 A portion of the radioactive decay series of uranium.

The average person receives about 40% of their exposure to radioactivity from radon, about 30% from cosmic rays from outer space, 20% from the decay of natural radioactive isotopes in their bodies, and 10% from medical X-rays.

reactions produces $^{206}_{82}Pb$, which is a nonradioactive (stable) isotope of lead that terminates the sequence.

Of particular interest is that portion, highlighted in Figure 17.7, of the sequence of ^{238}U radioactive decay which involves *radon,* Rn. This element is the only one, other than the helium from the alpha particles, that is gaseous and therefore mobile. The immediate precursor of the radon-222 is radium-226, which decays by emission of an α particle:

$$^{226}_{88}Ra \rightarrow \, ^{222}_{86}Rn + \, ^4_2He$$

The radon-222 isotope has a half-life of 3.8 days, which is time enough to diffuse through the solid rock or soil in which it is formed. Most radon escapes directly into outdoor air since the surface of Earth at which it appears is not covered, for example, by a building. The very small background concentration of radon in air nevertheless yields about half our total exposure to radioactivity since, although radon decays in a few days, it is constantly replenished by the decay of more radium. However, some of this natural radon from the ground can seep into homes and can become a potential health hazard if it collects there. Radon gas can also accumulate to unhealthy levels in caves, including some that are often used for recreational purposes.

17.6 Radon can enter buildings

Most radon that seeps into homes comes from the top meter of soil below and around the house's foundation (see Figure 17.8). Radon that is produced much deeper than this will likely decay to the nongaseous, and therefore immobile, element polonium before it reaches the surface. Radon enters basements of homes through holes and cracks in their concrete foundations. The intake is increased significantly if the air pressure in the basement is low.

Construction materials and water from artesian wells are other potential sources of radon in homes. For example, when well water is heated and exposed to air, as occurs when it exits from a showerhead, dissolved radon is released into the air.

The heaviest member of the noble gas group of elements, radon is chemically inert under ambient conditions and so it remains a monatomic gas. As such, it becomes part of the air that we breathe once it enters our homes. Because of its inertness, physical state, and low solubility in body fluids, radon *itself* does not pose much of a danger to us. The chance that a radon atom will disintegrate during the short time it is present in our lungs is small, and the range of radon's α particles in air before they lose most of their energy is less than 10 cm.

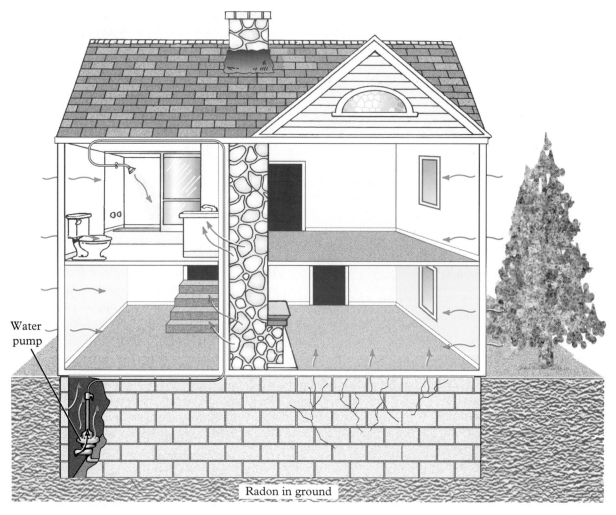

Figure 17.8 Sources of radon in the home. Radon gas can enter homes in many ways.

Water pump

Radon in ground

The danger arises from the radioactivity produced by the three elements formed in sequence (see Figure 17.7) by the disintegration of radon—namely polonium (Po), lead (Pb), and bismuth (Bi). Such "descendants" are termed **daughters** of radon, which is known as the **parent.** In macroscopic amounts, these daughter elements are solids, not gases, and so when they are formed in air from radon, they all quickly adhere to dust particles. Some dust particles in turn adhere to lung surfaces when we inhale them, and it is in this condition that the radon daughters pose a health threat. In particular, both the polonium that is formed directly from ^{222}Rn, namely ^{218}Po, and the ^{214}Po that is formed further along the decay sequence, emit energetic α particles that can cause damage to the bronchial cells near the dust particles. This damage can eventually lead to lung cancer; indeed, radon (or rather its daughters) is the second-leading cause of such cancer, though it follows smoking by a wide margin.

Some of the radon daughters in the uranium radioactive decay sequence in Figure 17.7 disintegrate by β particle emission. However,

Chapter 17: The Core of Matter

the deleterious health effects of such particles are negligible because the α particles carry much more energy than do the β particles, and it is the disruption of cell molecules by the burst of high energy that breaks bonds within these molecules that initiates cancer.

17.7 Is radon dangerous to our health?

Three statistical public health studies published in 1994, one from Sweden, one from Canada, and one from the United States, reached contradictory conclusions about the risk of radon to householders. In the Swedish report, the rate of lung cancer in nonsmokers and especially in smokers was found to increase with increasing levels of radon in their homes. The Canadian study, focusing on residents of Winnipeg, Manitoba, which has the highest average radon levels in Canada, found no linkage between radon levels and lung cancer incidence. The U.S. study, which was conducted among nonsmoking women in Missouri, also found little evidence for a trend of increasing lung cancer rates with increasing indoor radon concentration.

The "radon problem" has received greatest attention in the United States, where currently there are programs in place to test the air in home basements, the area of houses in which the concentration is likely to be greatest, for significantly elevated levels of radon. Once radon is identified, the owners can alter the air circulation patterns in these dwellings to reduce future radon levels in living areas and thereby reduce the lung cancer risk. It has been pointed out, however, that because of the mobility of the U.S. population, on average an individual who happens to live in a high-radon house will be there for only a few years and likely will spend most of his or her life in low-radon houses (since only about 7% of houses have high levels).

An analysis by the U.S. National Research Council concluded that in the mid-1990s, about 22,000 lung cancer deaths per year in the United States, corresponding to about 14% of the total number of deaths from lung cancer, were associated with exposure to radon in indoor air. Almost 90% of these radon victims were smokers, because when combined, radon and cigarette smoke are synergistic in causing the disease. Nevertheless, radon exposure accounted for about one-quarter of the lung cancer deaths for nonsmokers.

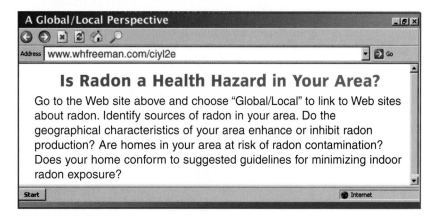

A Global/Local Perspective

Address www.whfreeman.com/ciyl2e

Is Radon a Health Hazard in Your Area?

Go to the Web site above and choose "Global/Local" to link to Web sites about radon. Identify sources of radon in your area. Do the geographical characteristics of your area enhance or inhibit radon production? Are homes in your area at risk of radon contamination? Does your home conform to suggested guidelines for minimizing indoor radon exposure?

Nuclear Energy

As we have seen, most of the energy used by humans originates as the heat generated by the combustion of carbon-containing fuels. However, heat in commercial quantities can also be produced indirectly from certain processes involving atomic nuclei. This power source, called **nuclear energy,** results from the conversion of a small amount of the nuclear mass into energy during the reaction. Since the forces operating between the particles in the nucleus are much greater than chemical bond forces, the energy released per atom in nuclear reactions is immense compared to that obtained in combustion reactions.

There are two processes by which energy is obtained from atomic nuclei, fission and fusion:

- In **fission,** a neutron collides with a type of heavy nucleus that has many neutrons and protons, splitting the nucleus into two similarly sized fragments. Collectively the fragments are more stable energetically than the original heavy nucleus. Owing to this difference in stability of the fragments and the heavy nucleus, energy is released by the process. Fission is the process by which nuclear power is currently produced. The fraction of electricity produced by fission in various countries is listed in Table 17.1.

- **Fusion** is the combination of two very light nuclei to form one heavier one (and another particle). Fusion also results in the release of huge amounts of energy, again since the combined nucleus is more stable than the original, lighter ones. Fusion may become a practical energy source in the future, as we discuss in section 17.14.

17.8 Fission reactors use chain reactions

The most economically useful example of fission is that induced by the collision of a neutron with a ^{235}U nucleus (see Figure 17.9). The typical

Table 17.1 Proportion of electrical energy generated by nuclear power

Country	Proportion
France	76%
South Korea	41%
Sweden	39%
Japan	34%
United Kingdom	22%
United States	20%
Canada	14%
Russia	12%
Netherlands	4%

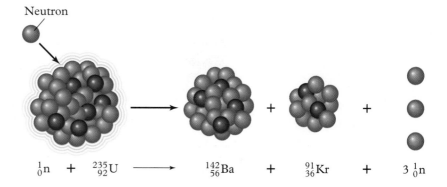

Figure 17.9 A fission reaction of uranium-235.

Neutron

$$\text{}^1_0\text{n} + \text{}^{235}_{92}\text{U} \longrightarrow \text{}^{142}_{56}\text{Ba} + \text{}^{91}_{36}\text{Kr} + 3\,\text{}^1_0\text{n}$$

products of the decomposition of the unstable combination of these two particles are a nucleus of barium, ^{142}Ba, one of krypton, ^{91}Kr, and three neutrons. The symbol for a neutron is $^1_0$n, since its mass number is 1.

$$\text{}^1_0\text{n} + \text{}^{235}_{92}\text{U} \rightarrow \text{}^{142}_{56}\text{Ba} + \text{}^{91}_{36}\text{Kr} + 3\,\text{}^1_0\text{n}$$

Not all the uranium nuclei that absorb a neutron form exactly the same products, but the process always produces two nuclei of about these sizes, along with several neutrons.

Exercise 17.4

For the U-235 fission reaction above, show that the sum of the mass numbers of the products is equal to that of the reactants. Similarly, show that the sum of the atomic numbers of the products is equal to that of the reactants, including neutrons in both cases.

The two new nuclei produced in the fission reaction are very fast moving, as are the neutrons. It is the thermal, or heat, energy due to this excess energy of motion that is used to produce electrical power. In fact, the generation of electricity by both nuclear energy and the burning of fossil fuels involves using the energy source to produce high-temperature steam. The energy possessed by the steam is used to turn large turbines, which in turn generate the electricity (see Figure 17.10). In this manner, heat energy—regardless of its source—is converted into electrical energy. We shall see in section 17.18 that there is a natural upper limit to the efficiency of this conversion.

An average of about three neutrons are produced per ^{235}U nucleus that reacts. Each of these neutrons then could produce the fission of another ^{235}U nucleus, and so on, yielding a **chain reaction** (see Figure 17.11). In atomic bombs, the extra neutrons accelerate the process and thereby induce a very rapid fission of all the uranium in a small volume, so energy is released explosively, rather than gradually as it is in a nuclear power reactor.

The only naturally occurring uranium isotope that can undergo fission is ^{235}U, which constitutes only 0.7% of the native element. The uranium in nuclear power reactors is present as *uranium oxide*, UO_2,

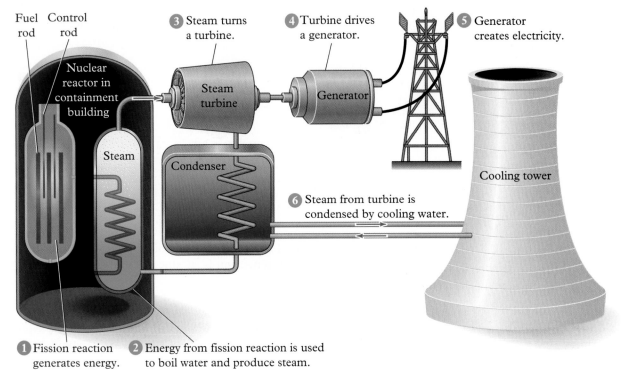

① Fission reaction generates energy.   ② Energy from fission reaction is used to boil water and produce steam.

Figure 17.10 Step by Step Schematic diagram of a nuclear fission reactor.

The *depleted uranium* used for making armor-piercing weapons, especially projectiles used against tanks, is what remains of natural uranium once most of the isotopes U-235 and U-234 have been extracted from it.

contained in a series of enclosed bars called *fuel rods* (see Figure 17.10). Each rod contains several hundred fuel pellets, each one about the size of a pencil eraser. When the uranium fuel in a rod is spent, that is, when it has reacted so much that its ^{235}U content becomes too low for it to continue to be useful as a fuel, it is removed from the reactor.

In most but not all nuclear power reactors (the Canadian *CANDU* system being the main exception), the uranium fuel must be enriched in the fissionable ^{235}U isotope; that is, the abundance of ^{235}U must be increased to 3.0%, compared to 0.7% in the naturally occurring element. The extent of enrichment required for use of uranium in atomic bombs is much greater still; uranium sufficiently enriched for this purpose is called *weapons-grade* material and is 90% or more ^{235}U. Enrichment of specific isotopes in a sample is a very expensive, energy-intensive, lengthy process. The isotopes must be separated from each other on the basis of the small difference in their masses. A chemical process cannot be used for the separation, since all isotopes behave identically chemically.

17.9 Fission produces radioactive by-products

The great majority of uranium is not U-235, but U-238 (99.3% natural abundance; the other uranium isotopes have negligible abundance). Although uranium-238 does not undergo fission, a neutron produced

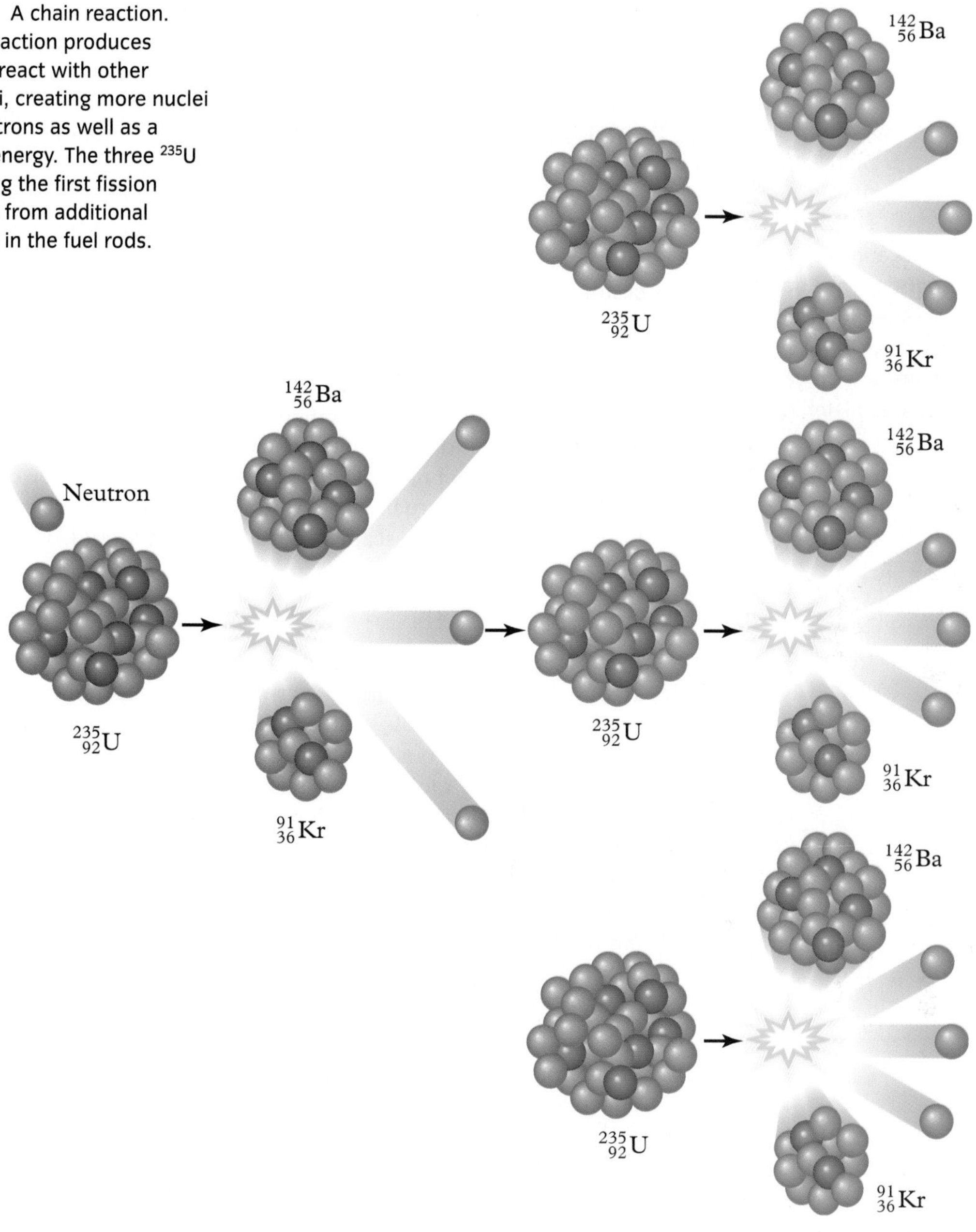

Figure 17.11 A chain reaction. Each fission reaction produces neutrons that react with other uranium nuclei, creating more nuclei and more neutrons as well as a great deal of energy. The three ^{235}U nuclei following the first fission reaction come from additional uranium oxide in the fuel rods.

by the fission of a ^{235}U nucleus that collides with it can be *absorbed* by it. The resulting ^{239}U nucleus is radioactive and emits a β particle, producing ^{239}Np. The ^{239}Np in turn emits another β particle. As a result of this sequence of two nuclear reactions, a nucleus of *plutonium*, ^{239}Pu, is produced, since it is the element having an atomic number 2 greater than that of uranium:

$$\,_0^1n + \,_{92}^{238}U \rightarrow \,_{92}^{239}U \rightarrow \beta + \,_{93}^{239}Np \rightarrow \,_{94}^{239}Pu + \beta$$

Thus, plutonium-239 is produced as a by-product of the operation of nuclear power reactors. Unfortunately, both the nuclei that are the products of U-235 fission and the ^{239}Pu by-product of U-238 neutron absorption are highly radioactive substances. In fact, the spent fuel mixture is much *more* radioactive than the original uranium.

Although many of the radioactive products of U-235 fission decay rapidly, by β emission, others have half-lives of many years. After 10 years, most of the radioactivity remaining in spent fuel rods is due to β emission from strontium-90, ^{90}Sr (half-life of 28 years), and from cesium-137, ^{137}Cs (half-life of 30 years). The dispersal of radioactive strontium and cesium into the environment would constitute a serious environmental problem. Ions of both metals could be readily incorporated into living organisms because strontium and cesium readily replace chemically similar elements that are normally part of the bodies of animals, including humans. For this reason, the radioactive waste from the spent fuel rods of nuclear power plants must be carefully monitored and will eventually have to be deposited in a secure environment from which it cannot escape, as we discuss in section 17.12.

The strontium and cesium are fission products. The Chernobyl nuclear disaster, in which products such as these were dispersed over a wide geographic area, is discussed later in the chapter.

17.10 Plutonium is a radioactive element

Plutonium is not an element that occurs naturally in significant amounts; it is only produced during the fission of other elements. The isotope, plutonium-239, that is produced during uranium fission is an α particle emitter with a long half-life: 24,000 years. After 1000 years, the main sources of radioactivity from spent fuel rods will be plutonium and other such very heavy elements, since most of the medium-sized nuclei produced in fission have much shorter half-lives and will have largely decayed by that time. Consequently, the long-term radioactivity of spent fuel rods can be greatly reduced by chemically removing plutonium and the other very heavy elements from them.

Exercise 17.5
Given that the half-life of ^{239}Pu is 24,000 years, how many years will it take for the level of radioactivity from plutonium in a sample to decrease to 1/128 (about 1%) of its original value?

Since the plutonium-239 that forms as a by-product in fuel rods is itself fissionable, once its concentration in the rods becomes high enough, it too undergoes fission and contributes to the power output of the reactor. The plutonium that accumulates over time in fuel rods can later be chemically removed from the spent fuel by **reprocessing.** In this procedure, plutonium is separated from other elements by exploiting the differences in the solubility of their salts. Since reprocessing uses chemical procedures, it is much less energy-intensive than is the isotope separation process used for uranium isotopes. Fuel from spent fuel rods is reprocessed in France, England, and Russia—though not in the United States or Canada—with a resulting buildup in the supply of plutonium. Comparable quantities of the element have come

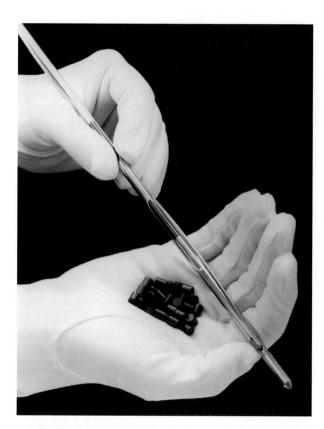

Figure 17.12 Nuclear reactor fuel pellets with their injector rod. (© U.S. Department of Energy/Photo Researchers, Inc.)

from the dismantling of nuclear weapons by the United States and Russia.

Breeder reactors are nuclear power reactors that are designed specifically to *maximize* the production of by-product plutonium. Such reactors actually produce more fissionable material than they consume! Breeder reactors use fuel pellets (see Figure 17.12), which each contain a core of fissionable enriched uranium surrounded by a layer of U-238, which is partially converted by neutrons into plutonium during the operation of the reactor. By separating the plutonium from the spent uranium fuel rods, plutonium-based fuel rods could be made and used subsequently in power reactors. In this way, Earth's rather limited uranium supply could be greatly extended. The breeder reactors are very technologically sophisticated and have been used in France and Japan, but operational problems have placed their future in some doubt.

Aside from questions of health and safety, there are major security problems associated with the handling of plutonium derived from both civilian and military sources. Only a few kilograms of weapons-grade plutonium, which consists of 93% or more of the Pu-239 isotope, is required to make an atomic bomb. A somewhat larger quantity of *reactor-grade* plutonium, which contains more of the other isotopes of plutonium and similar elements, is needed for a bomb. The world's current stockpile of plutonium exceeds 1000 tonnes and continues to grow. Governments are working on security measures to prevent the material from slipping into the hands of terrorists and rogue governments who wish to fashion their own bombs.

Although plutonium-239 is radioactive, it does not pose a threat outside your body since its α particle does not carry enough energy to penetrate your skin. However, when plutonium metal is exposed to air, it forms the oxide PuO_2, a powdery solid. If the plutonium dioxide dust is inhaled, it can lodge in the lungs, where even a very small amount of it can cause lung cancer.

17.11 Uranium mining contaminates the environment

Since naturally occurring uranium slowly decays into other substances that also are radioactive, uranium ore contains a mixture of several radioactive elements. Thus, the large volume of waste material that remains after the uranium is chemically extracted from the ore is itself radioactive. Consequently, environmental contamination during the mining of uranium (see Figure 17.13) by radioactive substances is common.

Figure 17.13 Uranium mining. Mining uranium ore causes radioactive contamination of the surrounding environment. (Dewitt Jones/Corbis)

The radioactive waste released from the original rock ore occurs as a heterogeneous mixture containing both a liquid and a powder, both called *tailings*. The semiliquid tailing mixture is normally held in special outdoor ponds until solids spontaneously separate out by gravity. Pollution of the local groundwater can occur if these ponds leak or overflow. In addition, when solid tailings are exposed to the weather and are partially dissolved by rainfall, they can contaminate local water supplies. When the solid tailings are used as landfill on which buildings are constructed, the radon produced by the radioactive decay of the radium in the tailings can create the problems we discussed in section 17.6.

17.12 The nuclear waste disposal problem is unsolved

Nuclear energy has been used to generate electricity now for many decades and constitutes an appreciable energy source in some countries (see Table 17.1). However, there is still no consensus among scientists and policymakers as to the best procedure for the long-term storage of the radioactive wastes generated by these plants. Initially, spent fuel rods are simply stored above ground—often under cooled water, in what look like swimming pools—for several years or decades, until the level of radioactivity is reduced and the rods have consequently cooled somewhat. At this stage, the rods can be transferred to dry storage—for example, in concrete canisters. If the fuel rods are reprocessed to remove the plutonium, the remaining highly radioactive waste is subsequently resolidified.

Two methods have been proposed to dispose of excess plutonium:

- To mix it with other highly radioactive waste and then to **vitrify** the mixture, that is, to make it into glass, which would be buried far underground in metal canisters.

- To convert it to *plutonium dioxide*, PuO_2, and mix it with uranium oxide to produce a *mixed oxide fuel* that could be used in existing nuclear power plants. Indeed, some mixed oxide fuel is now used in reactors in France, Germany, and Switzerland, and in the future it may be used in the United States, Canada, Belgium, and Japan. The issue of using mixed oxide fuel is very controversial in Great Britain, where one-third of the world's nonmilitary plutonium is presently stockpiled.

Figure 17.14 Underground nuclear waste disposal. Burial of nuclear wastes at depths of 500–1000 meters would isolate them until their radioactivity diminishes to safer levels.

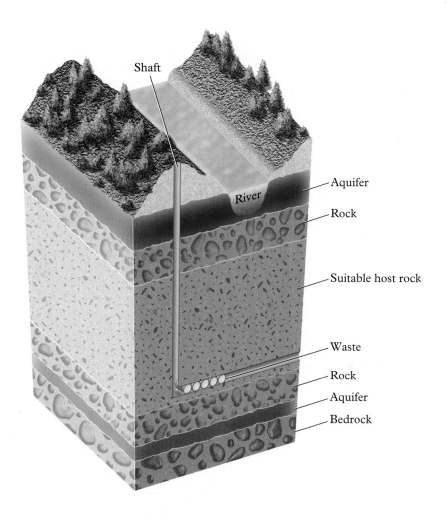

Shaft

River

Aquifer

Rock

Suitable host rock

Waste

Rock

Aquifer

Bedrock

The United States has decided to locate a nuclear waste depository below the surface of Yucca Mountain in Nevada. Planning for this site has gone on for several decades.

Whether or not the plutonium is removed, most nuclear waste disposal plans assume that the solid waste material would be encapsulated and immobilized in a glass or ceramic form and then be buried far below Earth's surface (see Figure 17.14). The container for this ultimate disposal would likely be made of a metal, such as copper or titanium, that is highly resistant to corrosion. The canisters are designed to last for at least several hundreds of years before leakage could occur; by that time, the level of radioactivity from many of the fission products would have declined substantially, though almost all the plutonium would remain. In Sweden, canisters are being designed to last for 100,000 years, at which time the level of radioactivity inside would be no greater than that of uranium ore.

Canisters will ultimately be buried in vaults 300–1000 meters below the surface. The geological features of the burial sites should include high stability (that is, little likelihood of disruption by earthquakes or volcanic activity) and low permeability (that is, low ability of fluids to flow into or out of them) to assure minimal interactions with groundwater and with the biosphere. Deep geological disposal is the only method by which safety requirements can be met without burdening future generations with monitoring and management responsibilities. Some governments do not accept the idea of *permanent* disposal of nuclear wastes. They want to be able to recover the plutonium from spent fuel rods if it turns out to be needed in the future for nuclear power.

17.13 The future of fission-based nuclear energy

Many people are suspicious of nuclear power, thinking that power plants could run out of control and blow up like atomic bombs. However, this process cannot occur with uranium unless it is enriched to 95% U-235, compared to the 3% maximum level used in nuclear fuel.

A more realistic fear is that the highly radioactive fission products contained within operating fuel rods could be spread into the surrounding countryside if a nonnuclear explosion occurs in a power plant. This in fact did occur at one of the nuclear power plants in Chernobyl, Ukraine, in 1985. Engineers at the plant lost control of the reactor during a routine test, resulting in a fire in the reactor that subsequently produced a huge explosion. The force of the blast blew off the heavy plate covering the building, and radioactivity from Cs-137, I-131, and other isotopes was spread over a wide area. Several hundred people became ill from radiation sickness, and a number of them died. Subsequently, many of the children of the area have been treated for thyroid cancer, a disease whose rate jumped sharply because of exposure in the environment to radioactive isotopes, especially ^{131}I. The rate of thyroid cancer among adolescents in the Chernobyl area was still rising in the late 1990s, but had begun to decrease in younger children—who were born after the accident. A rise in the incidence of leukaemia in residents of the Chernobyl area, attributed to exposure to Cs-137, has also been observed.

A less serious accident occurred at the Three Mile Island nuclear power plant in Harrisburg, Pennsylvania, in 1979. This accident also occurred as a result of the loss of the coolant that keeps the nuclear reaction under control. Although a small amount of radioactive isotopes escaped the plant during this incident, it was nowhere near as serious as that at Chernobyl.

The public attitude toward nuclear power produced from fission in North America and many European countries shifted from positive to negative in the last few decades of the 20th century, partially as a consequence of the incidents at Three Mile Island and Chernobyl. No new reactors have been ordered in the United States or Canada since the Three Mile Island incident, and several operating power plants in both countries have been shut down. Some observers believe that the nuclear power industry might revive early in the present century as oil and gas supplies dwindle and as carbon dioxide emission restrictions become more stringent. Indeed, there has been some discussion by the U.S. government of increasing nuclear power production to fulfill the growing energy needs of the country.

17.14 Fusion reactors are another source of power

Fusion—by which two light nuclei combine to form a heavier one—is the source of the energy in stars, including our own Sun, and in hydrogen bombs. Some people believe that fusion ultimately will be a feasible

?????????????????? **?** **????????????**
Fact or Fiction?

A "dirty bomb" is a small nuclear warhead.

False. The explosion associated with a "dirty bomb" is a conventional one arising from a chemical reaction. However, radioactive materials are deliberately mixed together with the explosives in order to disperse them and thereby contaminate a wide area with radioactivity.

source of energy and will solve all the world's energy problems because it will be so plentiful.

The fusion reactions that have the greatest potential as producers of useful commercial energy, since they occur at lower temperatures, involve the nuclei of the heavier isotopes of hydrogen, namely *deuterium*, ^{2}H, and *tritium*, ^{3}H. An abundant supply of deuterium is available, since it is a nonradioactive, naturally occurring isotope (constituting 0.015% of hydrogen). Thus deuterium is a natural component of all water. Tritium, however, is a radioactive element (a β particle emitter) with a short half-life (12 years) that does not occur naturally in useful amounts. To be used in a fusion reaction, tritium would have to be synthesized by the fission of the relatively scarce element lithium.

Examples of fusion reactions that could be used to produce power are given below:

$$\underset{\text{deuterium}}{^2_1\text{H}} \quad + \quad \underset{\text{deuterium}}{^2_1\text{H}} \quad \rightarrow \quad ^3_2\text{He} \quad + \quad ^1_0\text{n}$$

$$\underset{\text{deuterium}}{^2_1\text{H}} \quad + \quad \underset{\text{tritium}}{^3_1\text{H}} \quad \rightarrow \quad ^4_2\text{He} \quad + \quad ^1_0\text{n}$$

Proton   Neutron

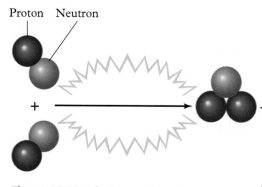

Figure 17.15 A fusion reaction of deuterium, $^2_1$H. Fusion reactions produce tremendous amounts of energy.

The energy that is released when a fusion reaction occurs is about one million times the energy produced per atom in a typical chemical reaction (see Figure 17.15). As a consequence, a huge amount of energy could be obtained from a small mass of deuterium. As in fission, the heat energy released in the process would be used to produce high-temperature steam, and from it electrical power. Since much energy must be supplied to overcome the huge electrostatic repulsion that exists between the positively charged nuclei when they are brought very close together, it has proven difficult to initiate and sustain a controlled fusion reaction that ultimately provides more energy than it consumes. Thus, no feasible fusion power reactors have been constructed, nor are they expected to be in the near future.

The environmental consequences of generating electrical power from fusion reactors should be less serious than those associated with fission systems. The only radioactive waste produced directly in quantity would be tritium, although the neutrons emitted in the process (see equations above) could produce radioactive substances when they are absorbed by atoms in the reactor material. The β particle that tritium emits is not sufficiently energetic to penetrate the outer layer of human skin. Tritium is dangerous nevertheless. Biological systems incorporate it as readily as they do normal hydrogen (^{1}H or ^{2}H) during inhalation, absorption through the skin, or ingestion of water or food. Currently, tritium in drinking water—some of which results from artificial sources and some of which is natural—constitutes the source of about 3% of our average total exposure to radioactivity.

Finally, fusion power reactors will cost very large sums of money to construct, even in comparison to fission power plants.

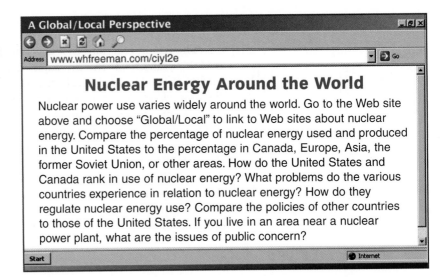

Nuclear Energy Around the World

Nuclear power use varies widely around the world. Go to the Web site above and choose "Global/Local" to link to Web sites about nuclear energy. Compare the percentage of nuclear energy used and produced in the United States to the percentage in Canada, Europe, Asia, the former Soviet Union, or other areas. How do the United States and Canada rank in use of nuclear energy? What problems do the various countries experience in relation to nuclear energy? How do they regulate nuclear energy use? Compare the policies of other countries to those of the United States. If you live in an area near a nuclear power plant, what are the issues of public concern?

▶ Discussion Point: What impact could nuclear power have on your life?

Use the resources on the Web site for this textbook to develop arguments for and against the construction of new nuclear power plants. What would you be willing to do to minimize your use of electrical energy so that additional plants might not be needed?

● Tying Concepts Together: The sources of fission, fusion, and solar energy

It may seem odd that combining two very light nuclei, as occurs in fusion, and splitting a heavy one into two fragments, as occurs in fission, both produce energy. You might have expected one phenomenon or the other, but not opposing trends. This seeming paradox is resolved when we realize that the maximum stability per nuclear particle (protons plus neutrons) occurs for nuclei that have an intermediate size, neither very light nor very heavy. Thus, per nuclear particle, nuclei of atoms such as iron are more stable than very small nuclei, such as hydrogen, or very large nuclei, such as uranium. For that reason, the fission of a heavy nucleus into two fragments of intermediate size, comparable to that of iron, releases energy. Similarly, the fusion of two very light nuclei to produce a heavier one also releases substantial quantities of energy.

As a result of fusion reactions within it that convert isotopes of hydrogen into helium, the Sun becomes extremely hot (see Figure 17.16). The Sun's outer layer of gas achieves a temperature of about 6000°C and emits light, mainly in the visible and infrared regions (see Figure 16.4). The enormous amount of energy that Earth receives as sunlight keeps our planet warm and drives photosynthesis. As we'll see in the next section, the deliberate trapping of sunlight could solve part or all of our energy supply problems in the future without the

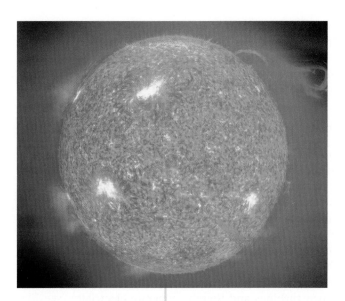

Figure 17.16 Nuclear fusion powers the Sun. (NASA)

environmental problems associated with fission, fusion, or fossil fuel combustion.

Although our Sun will continue to supply us with energy for billions of years to come, it will not last indefinitely. Eventually, it will have consumed all of its hydrogen and will then begin to fuse its helium into carbon and oxygen:

$$7\,{}^{4}_{2}\text{He} \rightarrow \rightarrow \rightarrow {}^{12}_{6}\text{C} + {}^{16}_{8}\text{O}$$

Indeed, most of the carbon atoms in your body originated in such fusion reactions inside stars the size of our Sun, which expelled matter into space as they began to die. In more massive stars, fusion reactions of carbon and oxygen produce elements up to about the size of iron. Elements heavier than iron, and therefore less stable than iron, are produced during the death throes of massive stars when they become *supernovas,* and their own energy is consumed in driving the formation of heavy elements such as gold, lead, and uranium from nuclei of intermediate size. The heavy elements on Earth were all components long ago of large stars that became supernovas and eventually expelled their contents into space.

We have seen how energy is generated by fission and fusion reactions, as well as the consequences of those processes. Next we turn to power that comes from the Sun itself.

Solar Energy

The Sun sends enough energy to Earth in the form of sunlight to supply all of our conceivable energy requirements, now and in the future, if only we could trap it efficiently. It is useful to discuss such **solar energy** in terms of the large energy unit EJ, which is equivalent to 10^{18} joules. Earth intercepts about 3 million EJ annually from sunlight. We would need to absorb and convert only about 0.01% of this amount to satisfy our annual global needs of about 400 EJ, with the United States consuming about 100 EJ of that total. Currently, fossil fuels supply about 75% of global energy. Hydroelectric and nuclear power each supply about 6%, and biomass about 14%.

In addition to being plentiful and reliable, solar energy is **renewable energy,** which is defined as energy that will not run out in the foreseeable future *and* whose capture and use does not result in the direct emission of greenhouse gases. In the long run, solar energy or nuclear energy are the only viable means of generating energy once fossil fuels run out or we decide that carbon dioxide emissions must be greatly reduced.

The E in EJ stands for *exa,* the prefix for 10^{18}.

There are many ways to trap solar energy. Some are direct techniques, in which we collect the Sun's rays using some sort of device such as a solar cell. In indirect techniques, some facet of the natural world, such as the atmosphere, does the initial trapping. We will begin our discussion with the indirect forms.

17.15 Solar energy is used to generate power

Hydroelectric power The Sun's energy evaporates water from oceans, lakes, rivers, and the soil, and transports the H_2O molecules upward in the atmosphere via winds. After the water molecules condense to rain, they still possess considerable *potential energy*, which is defined as energy that is waiting to be released. In this case, the energy arises due to the elevation of the water, since it is attracted to Earth below by gravity. Only a fraction of this potential energy is dissipated when the raindrops fall onto land, or into a body of water, that lies above sea level. We can harness some of the remaining potential energy by forcing the flowing, falling water to turn turbines and thereby generate electricity (see Figure 17.17). This process is known as **hydroelectric power.** We already harness considerable amounts of solar energy in this form.

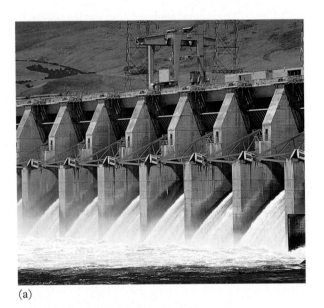

(a)

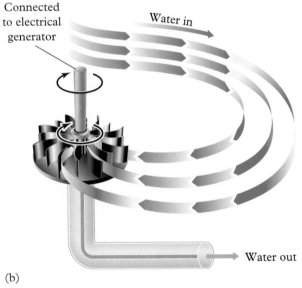

Connected to electrical generator

Water *in*

Water out

(b)

Figure 17.17 Hydroelectric power. (a) A hydroelectric power station in Germany. (b) Water flowing past turbine blades turns the shaft that connects to an electrical generator.

(Part a, Jim Steinberg/Photo Researchers)

Although some small hydroelectric installations use the flow of a river to turn the turbines, most large-scale facilities use dams and waterfalls where the water pressure—and hence the power yield—is much greater. The current annual amount of energy obtained from hydroelectric sites is about 24 EJ; if all potential sites were exploited, about four times this amount is potentially available.

Hydroelectric power is not as pollution-free as it may first appear. Recall from Chapter 16, section 16.11, that the flooding of vegetation on land that is used to provide water storage for new dams can produce enough greenhouse gas emissions—mainly methane—to cancel the CO_2

emission savings that arise from not burning fossil fuel instead to generate the energy. In addition, mercury in the form of ions is freed from the flooded vegetation and soil and dissolves in the water; it can then poison fish in the dam.

Wind power An even larger quantity of solar energy, about 300 EJ, is potentially available as **wind power,** although only 0.05% of that is currently being tapped. Wind is an indirect form of solar energy because it results when air masses that have undergone different amounts of heating by sunlight, and therefore have different pressures, attempt to equalize pressure. As a consequence of the local terrain, some geographical regions experience almost constant windy conditions.

The force of the wind can be exploited to generate power, including electrical energy, in the same way that the force of flowing water is used. For example, the strong, sustained winds in central North America were exploited via windmills to pump water and later to generate small amounts of electricity on farms until the middle of the 20th century. Windmills have been in use in Europe—especially in Holland—for centuries.

In recent decades, the large-scale generation of electricity by means of arrays of huge, high-tech windmills gathered in wind farms has become feasible. The windmills are purposely made large, since the power they generate increases with the propeller length. The energy produced rises sharply with wind speed, so only locations with high-speed winds are useful. Wind power has shown the greatest increase in generating capacity of all energy forms in recent decades; the total capacity in 2002 was about ten times that of the early 1990s. Wind power could be expanded to provide up to one-fifth of the world's electricity. Although it currently generates less than 1% of global electricity, its contribution is predicted to rise to 4% by 2030.

The most elaborate wind farm installations are in Denmark and California (see Figure 17.18). The greatest recent growth in wind power

Figure 17.18 A wind farm in California. (Photodisc Red/Getty Images)

Solar Energy

installations occurred in Germany and India, though there is the potential for this technology to be useful in many parts of the world. Indeed, the cost of generating electricity using modern windmill technology, and feeding it into existing power grids, is already competitive with conventional energy sources in some localities. About 90% of the potential for wind power in the U.S. lies in the Midwest, ranging from North Dakota to Texas. The world's largest wind farm is located in Oregon and Washington. Canada, Russia, and the eastern Africa regions have enough usable wind to meet all their electricity requirements, though there are as yet few installations in these regions.

Disadvantages to wind power include bird kills, the need for large amounts of land, and what some people consider to be the ugly look of the monster-sized windmills. Early problems of excessive noise and of interference with TV signals have been largely overcome.

Biomass energy The world's **biomass**—its plant or animal matter—also constitutes a form of solar energy because it is produced directly or indirectly by photosynthesis. The solar energy is converted to chemical energy, and then to heat energy when the biomass is dried and used for fuel (see Figure 17.19). The annual amount of energy currently produced from this source is 55 EJ; a much larger amount is potentially available. Wood, crop residues, and dung (dried excrement from plant-

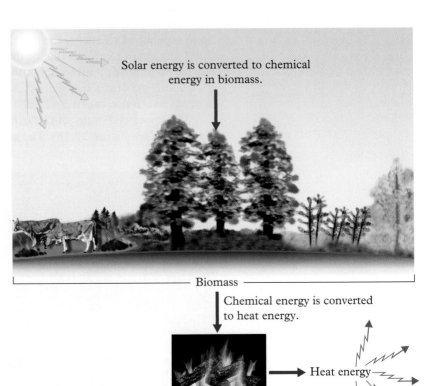

Figure 17.19 Biomass energy. Solar energy is converted to chemical energy in biomass. Burning biomass converts the chemical energy into heat energy.

consuming animals) have been traditional sources of energy in undeveloped countries. However, their domestic and small-scale combustion is very polluting to the air—especially with respect to emissions of fine particulate matter—and is quite inefficient. Generally biomass energy is phased out in favor of commercial energy such as fossil fuels and electricity as a country's economy develops. In developed countries, some households, especially in rural regions, burn wood to produce heat, although this practice has had to be restricted in some U.S. states because it produces so much particulate air pollution. Recently, technology has been developed to use plant biomass in large-scale installations that do not pollute the air. For example, wood chip waste can be burned to produce the heat required to generate steam.

As an alternative to direct burning, wood can be gasified, or digested by bacteria, and then converted into alcohol fuels. Fast-growing trees in plantations on land not suitable or needed for agriculture could be used for this purpose. As we discussed in Chapter 6, section 6.8, crops such as corn and sugarcane are currently grown to produce ethanol for fuel. However, these facilities often consume so much fossil fuel in their operation that relatively little is saved overall in CO_2 emissions.

Tidal and wave power Tides, caused by gravitational effects of the Moon and the Sun on ocean waters, cause large masses of water to be lifted and then lowered twice a day. If the tides in a coastal basin are generally high, a gate that can be opened or closed can be built across the basin. When the tide is coming in, the gate is left open so the water behind it rises. At high tide, the gate is closed. The dammed water leaving the basin turns a turbine, generating electricity (see Figure 17.20).

Three tidal power plants are currently in operation, located in France, Nova Scotia, and Russia. These installations had high capital

Geothermal power, though not solar-based, is another useful form of renewable energy. This heat energy emanates from beneath Earth's surface, and occurs in the form of steam and hot water. Per capita, Iceland leads the world in the production of geothermal energy.

Figure 17.20 A tidal power plant in Nova Scotia. (Stephen J. Krasemann/Photo Researchers)

Solar Energy

costs and can operate only twice daily. Although the energy produced is renewable and pollution-free, sedimentation occurs behind the dam gates, and tidal mudflats are often destroyed as a result of the operation of the facility.

Wave power is generated by using the up-and-down motion of water that results from waves, which are caused by winds and thus are an indirect form of solar energy. The rising wave compresses air in a chamber. The high-pressure air is then released through a valve, turning a turbine to produce electricity. As the wave recedes, air rushes back in through another valve, also spinning the turbine. Currently there are thousands of oceanic navigation buoys whose 60-watt lightbulbs are powered by this mechanism. Large-scale wave power facilities are still in the future.

It is estimated that about 20 EJ of power is potentially recoverable from waves and tides.

17.16 Solar energy can be used directly in two ways

The absorption of energy from sunlight and its conversion to useful forms of energy such as electricity can occur in two general ways:

- **Thermal conversion:** Sunlight, especially its infrared component, which accounts for half its energy content, is captured as heat energy by some absorbing material. Solar energy is an excellent source for heating water and living space, which accounts for up to half of our total energy consumption. An example of simple thermal conversion technology is the solar box cookers used in developing countries (see Figure 17.21). Buildings in temperate climates designed to absorb and retain the maximum fraction of the solar energy that falls on them are also passive thermal conversion systems.

- **Photo-conversion:** The absorption of the ultraviolet, visible, and infrared photons of sunlight excites electrons in the absorbing material to higher energy levels. The excitation subsequently causes a physical or chemical change, rather than a simple degradation to heat.

In the sections that follow, we shall investigate some practical energy sources of each of these two types.

17.17 Thermal conversion can be used to heat water

Solar water heaters are used extensively in Australia, Israel, the southern United States, Japan, and other hot areas that receive lots of sunshine. Solar collectors located on the rooftops of private homes and apartment buildings, as well as some

An everyday example of thermal conversion is when a shiny metal surface, such as the body of a car, becomes very hot when left in sunlight.

Figure 17.21 A solar cooker.
(Courtesy of SUN OVENS International, Inc.)

Chapter 17: The Core of Matter

commercial establishments such as car washes, contain water that is circulated around a closed system by an electrically driven pump. Sunlight is absorbed and converted to heat by a black flat-plate collector, which transfers the absorbed heat to water flowing over it in transparent tubing. Alternatively, the sunlight can be absorbed by thin black plastic tubes through which water is circulated. The warmed water is pumped to an insulated storage tank until it is required for bathing or laundry purposes, or needed to supplement swimming pool water to heat it up.

In more elaborate installations, the hot water is passed through a *heat exchanger*, a system of pipes over which air is passed. The air is warmed by transfer of the heat from the water and can be used to heat the rooms of the building. If not needed immediately, the heat can be stored in other media such as rocks. Usually an electric or fossil fuel backup system is incorporated into these systems in order to provide heat on cloudy days or in high-demand situations.

17.18 Thermal conversion can be used for electricity

Sunlight that is reflected by mirrors and focused onto a receiver that contains water or some other liquid can reach temperatures high enough to produce steam. The steam can be used to generate electricity by turning turbines, in the same way that coal-fired or nuclear plants produce it. Unfortunately, power plants require large amounts of cooling water in order to condense the steam back into the liquid state as part of the energy conversion cycle (similar to that for a nuclear power plant), and many areas, such as deserts, that have abundant land and sunlight have little water available for this purpose.

The fraction of thermal energy that can be extracted from a mass of hot fluid and converted to electricity, rather than degraded to waste heat, is limited by the *second law of thermodynamics:* the hotter the fluid, the greater the fraction. Consequently, it is advantageous to use steam superheated to the highest possible temperature to maximize the fraction of the energy that is transformed to electricity. Steam has been heated as high as 1500°C by focusing sunlight. To obtain such high temperatures, the sunlight receiver must be as small as possible since the rate of its heat loss to the environment by infrared light emission is proportional to its surface area. Several small-scale plants, in which the steam temperature is comparable to the 400–500°C attained in coal-fired facilities, have been constructed in Sicily, the United States, Spain, and Japan.

The **solar thermal electricity** that results from power plants of this type may soon be competitive in price with electricity from conventional sources. This is particularly true if the waste heat, for example, steam whose temperature lies close to the boiling point of water, can also be used for some purpose. This technique of using the waste heat from a heat-to-electricity conversion for a constructive purpose is called the **cogeneration** of energy. (It is a common feature of new power plants fueled by natural gas.)

Solar Energy

17.19 Solar cells produce electricity

Electricity can be produced directly from solar energy using the photo-conversion mechanism we mentioned previously. This application exploits the **photovoltaic effect,** which is the creation of separated positive and negative charges in a solid material as a result of absorption of a photon of light. The photon energy excites an electron of the solid from its normal shell to a higher, empty one. This electron is then free to move within the solid. When the electron leaves, the remainder of the atom is positively charged and called a *hole* (see Figure 17.22, step ❶). The location of the positive hole "moves" by transfer of an electron from an atom adjacent to the initial hole to the atom on which the hole is located, thereby switching the position of the positive charge (step ❷). Successive electron transfers of this type allow the location of the hole to move further. Since both the excited electron and hole are free to move, an electrical current could be made to flow in the solid material (step ❸).

The material used for photovoltaic or solar cells is a **semiconductor,** a solid that conducts electricity in a manner intermediate between a metal (freely conducting) and an insulator (nonconducting). In semiconductors, the energy required to excite an electron from the least stable of the filled energy levels to the most stable of the empty levels is small but not zero. The most common semiconductor used in solar cells is elemental silicon, for which the energy separating the two levels is 2.1×10^{-19} J/atom, an amount possessed by some photons of infrared light.

The light absorption ability of silicon extends to the higher energies associated with visible light, so silicon absorbs most of the photons

Figure 17.22 Step by Step The photovoltaic effect.

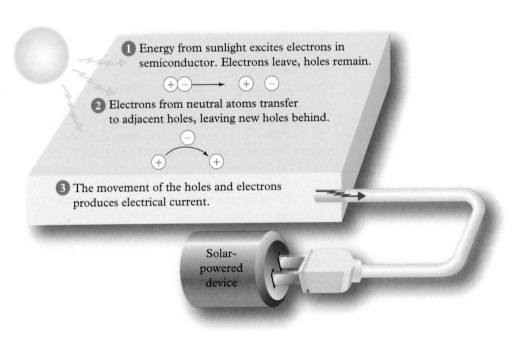

❶ Energy from sunlight excites electrons in semiconductor. Electrons leave, holes remain.

❷ Electrons from neutral atoms transfer to adjacent holes, leaving new holes behind.

❸ The movement of the holes and electrons produces electrical current.

Solar-powered device

Chapter 17: The Core of Matter

Figure 17.23 An array of solar cells in a parking meter in Cincinnati, Ohio. (Visuals Unlimited)

An area of about 100 miles by 100 miles (160 km × 160 km) covered with solar cells would be sufficient to supply all current U.S. electricity needs. However, the cost of constructing these cells would be in the trillions of dollars.

of sunlight. However, all photon energy absorbed in *excess* of 2.1×10^{-19} J/atom is wasted by being converted into heat. A maximum of only 28% of sunlight's energy can be converted to electricity by cells made of crystalline silicon (Chapter 1, Figure 1.4). Amorphous silicon has an even lower efficiency—only slightly more than half that of crystalline silicon—but it is now used extensively because it is so much less expensive to manufacture in thin films. Recently three solar cells, each containing silicon alloyed with another element (such as germanium) so as to absorb at a different wavelength, have been developed to utilize a greater overall fraction of sunlight's energy. If it can be adapted to a large scale, this development should significantly reduce the cost of solar cell electricity.

Solar cells each provide only a tiny electrical current, so to generate electricity in useful quantities, many of them must be joined together in a *solar array* (see Figure 17.23). One problem with the electricity that is generated using solar cells is that it is <u>d</u>irect <u>c</u>urrent, DC, rather than the <u>a</u>lternating <u>c</u>urrent, AC, that is used in power grids and by most equipment and appliances. The DC electricity can be converted to AC, but with the loss of some power (as waste heat). The direct current electricity from solar cells can be used without conversion to produce hydrogen by the electrolysis of water, as mentioned in Chapter 12, section 12.9.

The cost of producing and installing solar cells—and the problem of storing the electricity for use at night and on cloudy days—is the greatest barrier to expanding their use. As with other applications of solar energy, the capital cost in creating the infrastructure required to capture and use the "free" energy of the Sun by solar cells is substantial. The cost of manufacturing the solar cells has continued to fall with time, and photovoltaic solar electricity may become competitive with conventional power generation methods in the future. One important recent advance is the invention of plastic solar cells, in which the light is absorbed and converted to electricity by polymeric organic molecules. Although not as efficient in capturing light as those made with crystalline silicon, they are much cheaper to produce.

Photovoltaic power may become attractive in hot, sunny locations (see Figure 17.24) such as the southwestern United States where the peak power demand, driven by the need for air conditioning, coincides in time with the peak solar energy availability (midday in the summer). Already, solar cell power (plus storage) is cheaper than extending power grid lines a kilometer or more away from an existing network into remote regions and is competitive in cost with the use of diesel generators. Portions of this textbook were written at a seaside location that is within sight of an offshore lighthouse powered by solar cells.

The use of solar cells in developing countries, most of which have sunshine in abundance, could obviate the need for creating power grids that carry electricity over long distances from source to user. This

Solar Energy

represents the greatest potential market for expansion of photovoltaic power. Solar cell electricity is already used to power water pumps, lights, refrigerators, and TVs in some developing countries. More than half of new solar cells are now connected to electricity grid systems. The Million Solar Roofs Initiative of the U.S. Department of Energy is a public–private partnership that aims to overcome barriers to the use of solar technologies. Its goal is the installation of one million solar panels on roofs by 2010. In addition to solar cells—including solar roofing shingles—eligible technologies include those for water, space, and pool heating.

17.20 Solar energy has advantages and disadvantages

In our discussion of solar energy, we have touched on some general comparisons with fossil fuel and nuclear energy. Let us now extend and summarize the main advantages and disadvantages of solar energy.

The general advantages of solar energy:

- It is free and extremely abundant.
- It has low environmental impact.
- Its operating costs are low.
- It does not require large, centralized suppliers and expensive distribution networks.
- It has high public acceptance as a "natural" form of energy.

The general disadvantages of solar energy:

- It is intermittent in its availability and thus requires efficient storage or backup systems so that power can be supplied continuously.
- It is diffuse; sunlight provides a small amount of energy per unit of surface collection area, so large areas of solar collectors are required to harvest the energy.

- It requires high capital costs to construct the energy collection and storage systems; this offsets the "free" aspect of the energy itself for many years until the investment is paid off.

- It generally receives no economic (tax) or regulatory credit from governments in recognition of its low air pollution and greenhouse gas emissions.

Summarizing the Main Ideas

A radioactive nucleus decomposes spontaneously by emitting a small, fast-moving particle. If the emission is an alpha particle, which is the same as a helium nucleus, the nucleus decreases its atomic number by 2 and its mass number by 4. If a beta particle is emitted, the nucleus increases its atomic number by 1, with no change in its mass number. The mass number of an isotope is shown as a leading superscript to the shorthand name of the element; collectively this information is its isotopic symbol. The mass number is also sometimes given following the element name or symbol.

Radioactivity is dangerous to living matter because the emitted particles carry a great deal of energy, which is transferred to matter that they encounter. This energy can result in the breakage of bonds or in ionization.

The nuclei of a radioactive element do not all decay at once, but over time. The half-life characteristic of each isotope is the length of time it takes for half the nuclei to decay.

Radon gas is one of the products of the sequence of radioactive processes initiated when a U-238 nucleus decays. The gas can escape from soil that contained the uranium, and enter the air or buildings constructed on that soil. Radon daughters—elements that result from radon decay—are themselves radioactive. They are more likely than radon to adhere to lung tissue and cause damage when they decay.

Nuclear energy is a power source derived from either the neutron-induced fission of a heavy nucleus into two parts or the fusion of two light nuclei to produce a heavier one. The products of these processes carry much energy, which could be captured as heat. Some of the products of fission are radioactive, and the tailings produced when uranium is mined also contaminate the environment with radioactivity. The plutonium produced by fission of U-238 is fissionable and can be extracted from the spent fuel rods by reprocessing. Breeder reactors can be devised to maximize the production of plutonium.

Solar energy directly or indirectly comes from the Sun. It is renewable since it will not run out and since it emits no greenhouse gases when it is used. Indirect forms of solar energy include wind power, biomass energy, and wave power.

The direct absorption of solar energy can occur by either thermal or photo-conversion mechanisms. Thermal conversion is usually used to obtain hot water but can also be employed to produce electricity from extremely hot steam. Solar cells use the photovoltaic effect, in which a semiconductor absorbs photons, to produce electrical power.

Key Terms

atomic weight	nuclear energy	solar energy	solar thermal electricity
radioactive	fission	renewable energy	cogeneration
alpha particle	fusion	hydroelectric power	photovoltaic effect
beta particle	chain reaction	wind power	semiconductor
gamma ray	reprocessing	biomass	
daughter nucleus	breeder reactor	thermal conversion	
parent nucleus	vitrify	photo-conversion	

Web Sites of Interest

To link to Web sites of interest, go to www.whfreeman.com/ciyl2e, Chapter 17, and select the site you want.

For Further Reading

C. Baird and M. Cann, *Environmental Chemistry*, 3rd edition, W. H. Freeman and Company, New York. Chapter 6 covers renewable energy technologies, and Chapter 13 covers nuclear energy and radioactivity, in greater detail.

W. Hafele, "Energy from Nuclear Power," *Scientific American*, September 1990, pp. 136–144.

P. Weiss, "Oceans of Electricity," *Science News, 159*, April 14, 2001, pp. 234–236. This article describes the technologies being used and developed to convert the motion of waves into electrical power.

Review Questions

1. What particles comprise the nucleus of an atom? Describe the properties associated with each particle.

2. Explain each of the following terms:
   (a) *mass number*
   (b) *isotopic symbol*
   (c) *atomic weight*

3. What do each of the following designations tell you about the identity and mass number of that isotope? Using the information provided and the atomic number of each atom, can you determine the number of neutrons for each isotope listed?
   a) C-12   b) Na-24   c) I-131   d) Co-60

4. Write out the complete isotopic symbols for atoms of chlorine (atomic number of 17) that have a) 20 or b) 22 neutrons.

5. List the properties associated with each of the following:
   a) alpha particle   b) beta particle
   c) gamma radiation

6. Write the nuclear decay reactions and identify the products for the following:
   a) alpha particle emission by U-238
   b) the emission of a beta particle by I-131

7. Describe two biological effects associated with exposure to alpha and beta radiation.

8. What is *ionizing radiation*?

9. What does *half-life* mean when talking about a radioactive isotope?

10. Identify two natural sources of radon gas.

11. Why is there concern about exposure to radon?

12. Explain the relationship between a parent and daughter isotope in a radioactive decay series.

13. What is *nuclear energy*?

14. How does the process of fission differ from the process of fusion?

15. Explain how a *chain reaction* occurs.

16. Identify two commercial processes that produce radioactive contaminants.

17. What is a *breeder reactor*?

18. What role does plutonium play in breeder reactors?

19. How does uranium used in a power plant differ from *weapons-grade* uranium?

20. How is excess plutonium currently stored? What alternatives have been suggested?

21. What are *tailings*? Why are they an environmental concern?

22. How does deuterium differ from tritium?

23. What environmental problems could occur with fusion reactors?

24. What is *solar energy*?

25. What does the energy unit "EJ" mean?

26. What is *renewable energy*?

27. How is hydroelectric power currently produced?

28. What is a *wind farm*?

29. What is *biomass*?

30. What is the *photovoltaic effect*?

31. What is a *semiconductor*?

32. What element is useful as a semiconducting material?

Understanding Concepts

33. How are the following isotopes the same and how are they different?
    a) C-12 and C-14   b) Pb-214 and Pb-208
    c) H-2 and H-3

34. What are the element and isotopic symbols for the nucleus that is formed when each of the following atoms loses an α particle?
    a) $^{221}_{87}$Fr   b) ^{212}Rn

35. What are the element and isotopic symbols for the nucleus that is formed when each of the following atoms loses a β particle?
    a) $^{59}_{26}$Fe   b) ^{63}Ni

36. Explain how radioactive nuclei differ from nonradioactive nuclei.

37. How does a nucleus change after each of the following processes?
    a) emission of an alpha particle
    b) emission of a beta particle
    c) emission of gamma radiation

38. Write the symbol for the particle or atom indicated by a blank for each of the following nuclear reactions:
    a) $^{222}_{86}$Rn $\rightarrow$ $^{4}_{2}$He + _____
    b) $^{214}_{83}$Bi $\rightarrow$ β + _____
    c) $^{236}_{94}$Pu $\rightarrow$ α + _____
    d) $^{9}_{3}$Li $\rightarrow$ _____ + $^{0}_{-1}$e

39. Explain why a thick pair of gloves will protect against alpha radiation but not against gamma radiation.

40. Describe how free radical production occurs in the presence of ionizing radiation.

41. Why does irradiated food have a longer shelf life than nonirradiated food?

42. I-131 has a half-life of 8 days. About how much of a 1.000-gram sample will be left after approximately 1 month?

43. Tc-99 (technetium-99) is used medically for diagnostic procedures. It has a half-life of 6.0 hours. If a patient receives 50 mg of Tc-99, how much will be left after 1 day? After 2 days?

44. What is the relationship between uranium and radon?

45. How do soil properties affect the amount of radon diffused into surrounding air?

46. Why is lung tissue particularly vulnerable to the damaging effects of radon?

47. How does the energy released in a nuclear reaction compare to that released in a chemical reaction?

48. Suppose that a nucleus of $^{235}_{92}$U were to capture a neutron and then undergo fission, so that it produces two neutrons and two identical nuclei. Determine the correct isotopic symbol for the nuclei.

49. Compare the amount of energy released from the same mass of matter undergoing fission versus undergoing a chemical reaction.

50. Why is fission, but not fusion, currently used to produce energy?

51. How could fusion reactions be used to generate power?

52. Why do radioactive strontium and radioactive cesium present health problems if incorporated into the body?

53. Describe some of the problems associated with:
a) the widespread use of nuclear energy
b) nuclear waste disposal

54. Why has it been difficult to initiate and sustain a controlled fusion reaction?

55. Does fusion require radioactive nuclei? Explain.

56. Identify some of the environmental problems associated with generating electrical energy from fusion reactions.

57. Is solar energy really free? Explain.

58. How does renewable energy differ from nonrenewable energy?

59. How is hydroelectric power related to solar energy?

60. Explain how each of the following can be used to generate electricity:
a) the flow of water
b) wind
c) biomass
d) tidal movements
e) solar power

61. Identify some of the advantages and disadvantages associated with:
a) solar power
b) hydroelectric power
c) wind power
d) biomass power
e) solar conversion systems

62. How can biomass be converted to energy in an environmentally friendly way?

63. Describe two mechanisms by which energy from sunlight is converted to more useful forms.

64. How does *thermal conversion* differ from *photo-conversion*?

65. Explain why it is best to use steam as hot as possible in generating electricity by thermal conversion.

66. Why is silicon well suited as a semiconductor material?

67. How do the structural differences between crystalline and amorphous silicon affect its efficiency in converting sunlight to energy?

Synthesizing Ideas

68. Why do alpha and beta particles cause more damage if they are inside the body than if they are outside the body?

69. Write the symbol for the particle or atom indicated by a blank for each of the following nuclear reactions:

(a) $^{214}Pb \rightarrow {}^{214}Po + $ _____

(b) _____ $\rightarrow {}^{234}_{90}Th + \alpha$

70. Outline the process by which radon gas is produced from the radioactive decay of uranium-238.

71. How does the half-life of radon, 3.8 days, affect its ability to diffuse to the surface of soil in which it is produced?

72. Is radon itself a health hazard? What evidence points to radon being dangerous to human health? What evidence suggests that radon may *not* increase the risk of lung cancer? Explain. Use the Web site at the end of the chapter to find information to support your answers.

73. What geological features need to be associated with nuclear waste burial sites? Why are these features necessary?

74. Is the "permanent disposal" of nuclear wastes possible? Explain.

75. Identify the power source that could produce "renewable" energy and explain how that could be done.

76. Describe how a solar cell works. Draw a diagram if that will help in your explanation.

Appendix A
Scientific Notation

In reporting and using numbers that are very large or very small, scientists usually use **scientific notation.** It is easiest to explain this mathematical shorthand with an example.

The number 1000 is equal to $10 \times 10 \times 10$. Since 10 occurs three times in the chain of multiplications, the standard form 1000 is written in scientific notation as 10^3. Similarly 1,000,000 could be written by repeating the number 10 six times in a multiplication chain: $10 \times 10 \times 10 \times 10 \times 10 \times 10$; thus, in scientific notation the number is abbreviated 10^6. In general, the number 1000 . . . 000 is equal to 10^n, where n is the number of zeros in the number and which becomes the exponent of 10, sometimes called the *power* of 10. The approximate size or distance of a wide range of objects in the universe is given in scientific notation in the diagram on the right.

> **Worked Example:** Interconverting standard and exponential forms of powers of 10
>
> **a)** How is the number 10,000 written in scientific notation? **b)** What is 10^7 equal to as a number written out in full?
>
> **Solution: a)** Since 10,000 contains four zeros after the 1, it is equal to 10 raised to the power 4, so it is written 10^4.
>
> **b)** 10^7 is equal to the number 1 followed by seven zeros, that is, 10,000,000.

> **Exercise A.1:**
>
> **a)** How is the number 100,000 written in scientific notation?
> **b)** What is 10^9 when written out in full?

When the number involved is not as simple as just 10 multiplied by itself a number of times, the conversion to scientific notation is a little more complicated, but the principle is the same. In general, to convert a number larger that 1.0 into scientific notation, first write the number so as to include a decimal place if one is not already specified. For example, 3429 would 3429.0. Then *move the decimal point to the left enough times so that there is only one digit before the decimal, keeping track of the number of times you have moved it;* the number in our example is changed to 3.4290. *The number of places that*

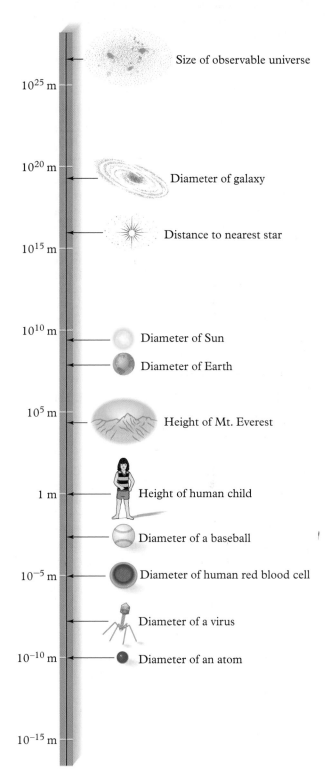

10^{25} m — Size of observable universe

10^{20} m — Diameter of galaxy

— Distance to nearest star

10^{15} m

10^{10} m — Diameter of Sun
— Diameter of Earth

10^5 m — Height of Mt. Everest

1 m — Height of human child

— Diameter of a baseball

10^{-5} m — Diameter of human red blood cell

— Diameter of a virus

10^{-10} m — Diameter of an atom

10^{-15} m

the decimal place has been moved is equal to the power of 10 for the number. In this case, the decimal point was moved three places to the left, so the power of 10 is 3. Since we've moved the decimal point, we now multiply the new figure by 10 raised to this power to get the final answer. Thus, $3429 = 3.429 \times 10^3$.

Worked Example: Interconverting standard and exponential forms of numbers

a) Convert 35,980 into scientific notation. **b)** Convert the number 6.3×10^2 into standard form.

Solution: a) First we write the number with a decimal point: 35980.0. We next move the decimal point so there is only one digit to the left of it; it is now positioned after the 3 and before the 5. To do this requires a shift of 4 places, so the result is 3.5980×10^4.

b) To convert a number in scientific notation back into standard form, first write some zeros (say five of them) after the last digit, so 6.3×10^2 becomes 6.300000×10^2. Then move the decimal point to the right the number of times that is equal to the exponent. Thus in this case we move the decimal point two places to the right, giving 630.0000, or simply 630.

Exercise A.2:

a) Convert the number 7650450 to scientific notation.
b) Convert 4.56×10^5 to standard form.
c) Convert 762.44 to scientific notation.
d) Convert 3.01×10^3 to standard form.

Scientific notation is useful for numbers that are much *less* than 1, as well as for numbers that are much larger than it. Numbers less than 1 are equivalent to 10 raised to a *negative* exponent. (This follows from the rule of algebra that the ratio $10^n / 10^m$ equals 10^{n-m}. For example, 0.1 is equal to the fraction 1/10. Since 1 is equal to 10^0, and 10 is equal to 10^1, the fraction is equivalent to $10^0 / 10^1$. According to the algebra rule, this is equal to 10^{-1}. Similarly, because 0.01 is 1/100 = $1/10^2$, we can express it as 10^{-2}.)

When the number involved is not equal to just 1 divided by a power of 10, the conversion to scientific notation is more complicated. In general, *to convert a number smaller than 1.0 into scientific notation, move the decimal point enough places to the right so that there is now only one nonzero digit before the decimal.* For example, 0.0034 becomes 0003.4. *The number of places that the decimal point moved to the right is equal to the negative exponent of 10 for the number.* Therefore, 0.0034 is 3.4×10^{-3} in scientific notation.

Worked Example: Interconverting standard and exponential forms of numbers less than 1

a) Convert 0.00447 into scientific notation. **b)** Convert the number 2.66×10^{-4} into regular form.

Solution: a) We move the decimal point three places to the right here so it comes just after the first nonzero number, namely the first 4. Since that movement required a shift three places to the right, the appropriate power of 10 is −3. Thus the number in scientific notation is 4.47×10^{-3}.

b) We place some zeros (six of them, say) in front of the digit that precedes the decimal, so our number now reads 0000002.66. Then we move the decimal point four places to the left since the power of 10 is −4. Our answer is 00.000266, or simply 0.000266.

Exercise A.3:

a) Convert the number 0.00000634 to scientific notation.
b) Convert 7.29×10^{-2} back to standard form.
c) Convert 0.0042 to scientific notation.
d) Convert 3.14×10^{-4} back to standard form.

Appendix B

Answers to In-Text Exercises

1.1 a) heterogeneous b) heterogeneous
c) homogeneous

1.2 a) S_8 b) O_3

1.3 a) 2 of nitrogen, 1 of oxygen b) NH_3

1.4 a) 12,8 b) 0.42, 0.92

1.5 1.0×10^{24}

1.6 1.0×10^{27}; 1.0×10^{-6} m^3

1.7 39

1.8 393

2.1 1.7, no

2.2 $2 C_2H_4 + 13 O_2 \rightarrow 8 CO_2 + 10 H_2O$

2.3 a) $CH_4 + 2 O_2 \rightarrow CO_2 + 2 H_2O$

b) $2 C_2H_6 + 7 O_2 \rightarrow 4 CO_2 + 6 H_2O$

2.4 Honey, yeast, water, and flour are reactants.
Bread is the product. The recipe produces one
loaf.

2 tablespoons honey + 1 package yeast + 1 cup
water + 3/4 pound flour $\rightarrow$ 1 loaf bread.

The numbers 2, 1, 1, 3/4, and 1 are analogous to
coefficients. Double the amount of each reactant
to produce 2 loaves.

2.5 Yes, it is balanced.

2.6 $2 C_4H_{10} + 9 O_2 \rightarrow 8 CO + 10 H_2O$

2.7 25 molecules of gas in 1 million molecules of air,
25,000 in 1 billion molecules of air.

2.8 a) 8.85 cm^3 b) 11,300 g

2.9 272 cm^3

2.10 O_2 2×16.00 g = 32.00 g

CO_2 12.01 g + (2 × 16.00 g) = 44.01 g

H_2O (2 × 1.01 g) + 16.00 g = 18.02 g

2.11 2.79 g butane, 4.33 g H_2O

2.12 224.1 g CO, 1.93 g CO

2.13 34.2 g C_2H_6, 61.4 g H_2O

2.14 0.430 g O_2; 1.430 g Fe_2O_3; yes, conservation of
mass

3.1 Boron: a) 5 b) +5 Sulfur: a) 16 b) +16

3.2 31, ^{31}P, 15

3.3 a) 2,8,7 b) 7; 2,5

3.4 a) 2 b) 6

3.5

a) CH_4
$$H-\overset{\overset{\displaystyle H}{|}}{\underset{\underset{\displaystyle H}{|}}{C}}-H$$

b) H_2S $H-\ddot{S}-H$

c) PH_3 $H-\overset{..}{\underset{\underset{\displaystyle H}{|}}{P}}-H$

3.6 a) $:\ddot{F}-\ddot{F}:$ b) $:\ddot{F}-\ddot{C}l:$

3.7 $H-\ddot{F}:$ $:\ddot{F}-H$ Not stable

3.8 a) magnesium oxide, lithium bromide
b) potassium fluoride, calcium sulfide

3.9 a) 10; 2,8 b) 10; 2,8,0 c) 18; 2,8,8

3.10 a) K_2S, potassium sulfide b) Al_2O_3, aluminum
oxide c) AlN, aluminum nitride

3.11 $:\ddot{S}-H$

4.1
$$H-\overset{\overset{\displaystyle H}{|}}{\underset{\underset{\displaystyle H}{|}}{C}}-\overset{\overset{\displaystyle H}{|}}{\underset{\underset{\displaystyle H}{|}}{C}}-\overset{\overset{\displaystyle H}{|}}{\underset{\underset{\displaystyle H}{|}}{C}}-H$$

4.3

Cyclohexane 12

4.4 a)

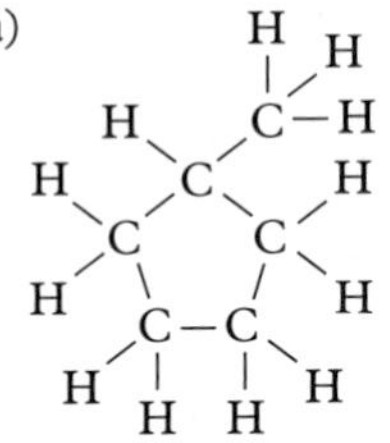

b)

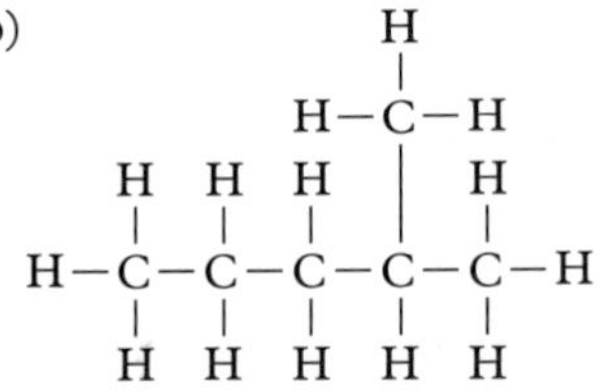

c)

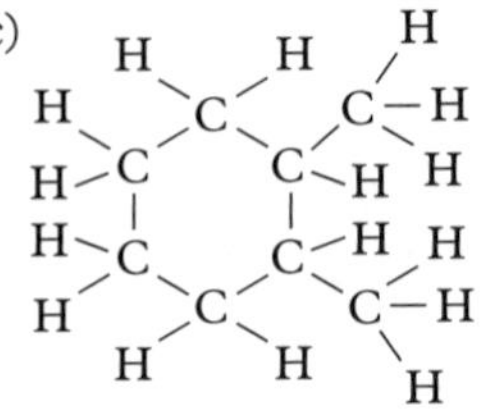

4.5

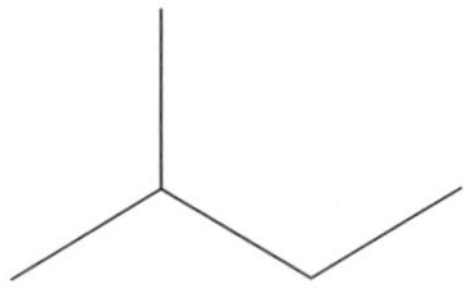

4.6 $H-\overset{..}{N}=\overset{..}{N}-H$

4.7 $H-C\equiv N:$

5.1 $-CH_2-CHBr-$ and $CH_2=CHBr$

5.2

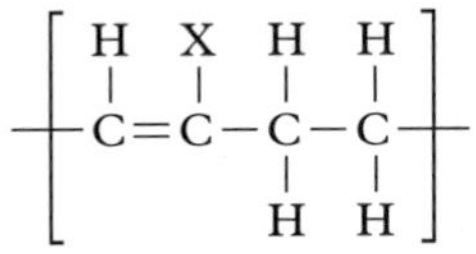

5.3

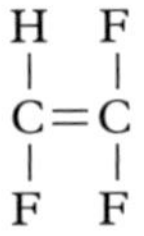

5.4 Reprocessing is physical; the other three are chemical processes.

5.5

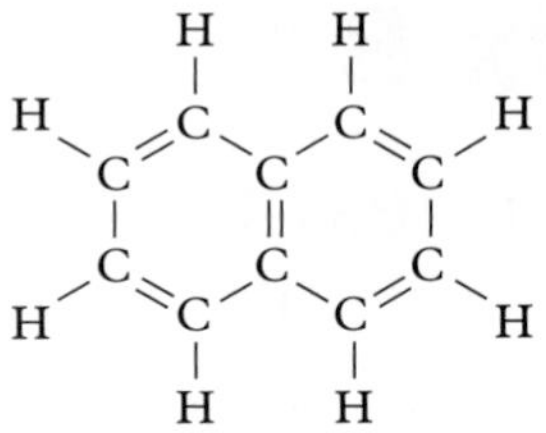

6.1

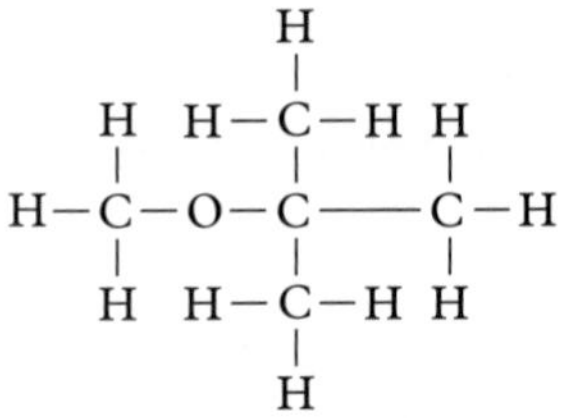

6.2

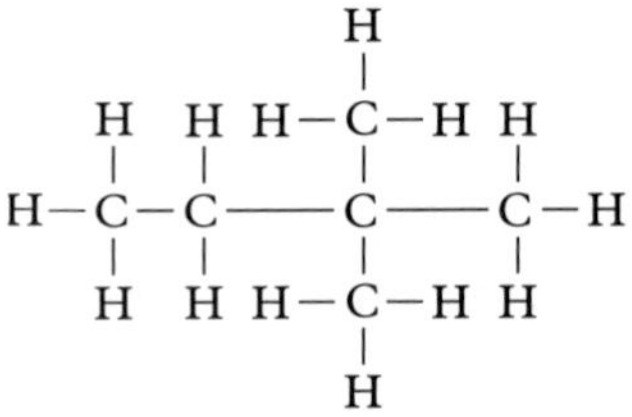

6.3 a) 10 b) 40%

6.4 1.9 times your mass in kilograms; for example, 114 grams if you weigh 50 kg

6.5 E_{10}, E_{85}

6.6

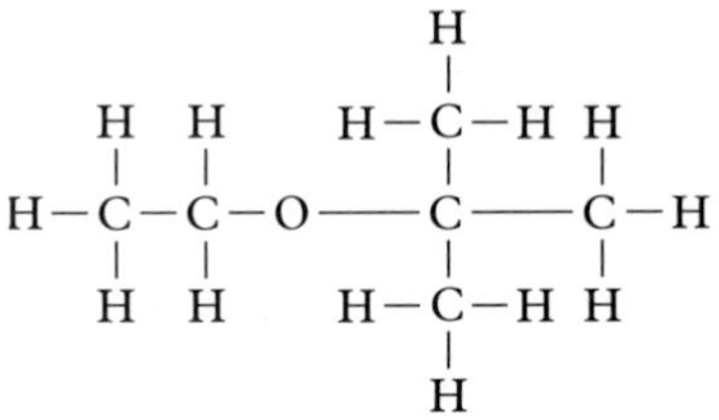

ethyl *t*-butyl ether (ETBE)

6.7 Generic diagram with $R = CH_3CH_2$

6.8

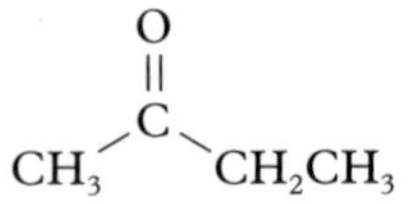

6.9 $CH_2 = CH - CH_3$, hydrogen bonding

6.10 Butyl propionate

6.11 CH_3CH_2COOH, $143°$

6.12 CH_3COOH and $CH_3CH_2CH_2OH$

6.13 a) 1821 kJ / mole   b) 1892 kJ / mole, the ether

6.14 20 kJ, endothermic

7.1 In honey, glucose and fructose are physically mixed, but in sucrose they are chemically combined.

7.2 a)

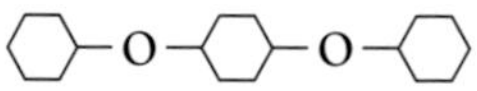

b) Beano contains an enzyme that hydrolyzes raffinose.

7.3

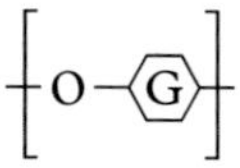

7.4 ii > iii > i

7.5 $\omega - 6$

8.1 180, 360, 540, 720

8.2 The -*ine* ending. Less soluble. So it is soluble in water-based body fluids.

8.3 Tylenol is an alcohol. Midol is not.

8.4 Six-membered N-containing ring, N-CH$_3$ group, COO ester two carbons from N

8.5 Caffeine and acetaminophen are amides, amphetamine is not.

8.6 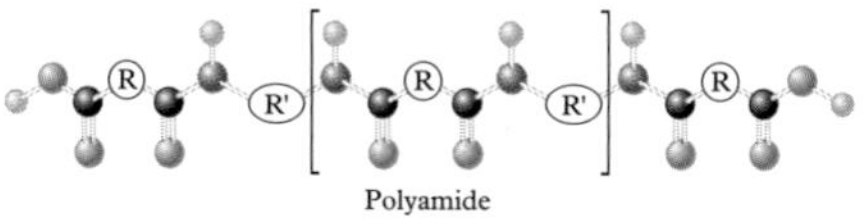
Polyamide

8.7

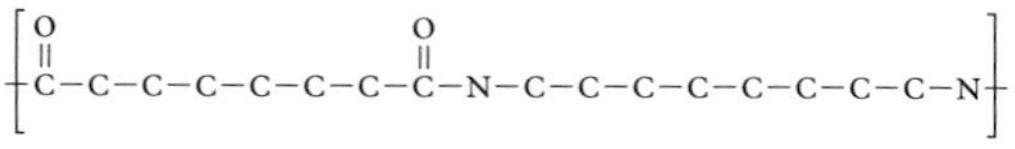

8.8

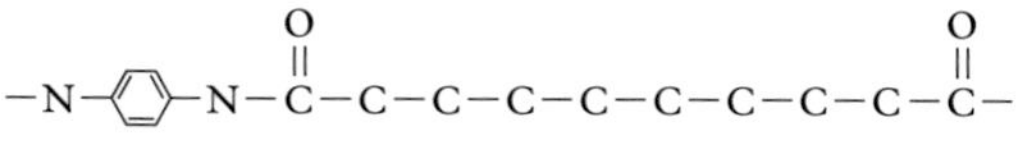

9.1 UGAAGUCCC

9.2 AUG/GAU/GUC/GGG/UAU, Start or Met / Asp / Val / Gly / Tyr

9.3

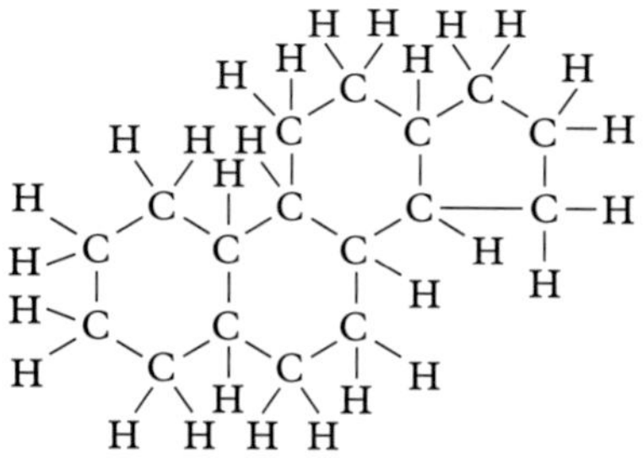

10.1 NO and atomic H are free radicals.

11.1 5.0 ppm

11.2 $HBr \rightarrow H^+ + Br^-$

11.3 $H_3PO_4 \rightarrow H^+ + H_2PO_4^-$

$H_2PO_4^- \rightarrow H^+ + HPO_4^{2-}$

$HPO_4^{2-} \rightarrow H^+ + PO_4^{3-}$

$H_3PO_4 \rightarrow 3\ H^+ + PO_4^{3-}$ (overall)

11.4 The pH is 6.

11.5 0.00001 mole / L

11.6 $H_3PO_3 \rightleftharpoons H^+ + H_2PO^{3-}$

$H_2PO_3^- \rightleftharpoons H^+ + HPO_3^{2-}$

$HPO_3^{2-} \rightleftharpoons H^+ + PO_3^{2-}$

11.7 $KOH \rightarrow K^+ + OH^-$

11.8 $PO_4^{3-} + H_2O \rightleftharpoons HPO_4^{2-} + OH^-$

11.9 $HNO_3 + KOH \rightarrow KNO_3 + H_2O$

KNO_3 is the salt.

11.10 $Al(OH)_3 + 3\ H^+ \rightarrow Al^{3+} + 3\ H_2O$

11.11 $H^+ + HCO_3^- \rightarrow H_2CO_3 + H_2O$

$OH^- + H_2CO_3 \rightarrow HCO_3^- + H_2O$

11.12 a) 2.85   b) 5.19   c) 10.03

11.13 a) 0.00015   b) 6.3×10^{-8} c) 1.4×10^{-11}

12.1 $Ca\ (s) \rightarrow Ca^{2+} + 2\ e^-$ (oxidation)

$Br_2\ (l) + 2\ e^- \rightarrow 2\ Br^-$ (reduction)

Bromine is the oxidizing agent, calcium is the reducing agent.

12.2 $Zn\ (s) + 2\ OH^- + 2\ MnO_2\ (s) + 2\ NH_4^+ \rightarrow Zn(OH)_2\ (s) + 2\ MnO(OH)\ (s) + 2\ NH_3\ (g)$

12.3 Cd (s) + 2 Ni(OH)$_3$ (s) → Cd(OH)$_2$ (s) + 2 Ni(OH)$_2$ (s)
Reverse the arrow to obtain the equation for recharging.

14.1 0.001 mole / L for acid rain with a pH of 3, 0.00001 mole / L for rain with a pH of 5. The ratio is 100 to 1.

15.1 The overall reaction is 2 O$_3$ → 3 O$_2$.

17.1 a) $^{16}_{8}$O   b) $^{18}_{8}$O

17.2 a) Uranium, U $^{235}_{92}$U $^{239}_{94}$Pu → $^{235}_{92}$U + $^{4}_{2}$He

b) Thallium, Tl $^{206}_{81}$Tl $^{210}_{83}$Bi → $^{206}_{81}$Tl + $^{4}_{2}$He

c) Iridium, Ir $^{175}_{77}$Ir $^{179}_{79}$Au → $^{175}_{77}$Ir + $^{4}_{2}$He

17.3 a) Nitrogen, N $^{14}_{6}$C → $^{14}_{7}$N + $^{0}_{-1}$e

b) Nickel, Ni $^{61}_{27}$Co → $^{61}_{28}$Ni + $^{0}_{-1}$e

c) Technetium, Tc $^{99}_{42}$Mo → $^{99}_{43}$Tc + $^{0}_{-1}$e

17.5 168,000 years

Appendix C
Answers to End-of-Chapter Exercises

Chapter 1
The "Elemental" Foundation of Chemistry
Atoms, Molecules, Elements, Compounds, Mixtures, and States of Matter

Review Questions

1. matter and its transformations

3. luster, opaqueness, and malleability

5. developing very small devices able to manipulate matter one atom at a time

7. no

9. A homogeneous mixture has a uniform composition throughout.

11. Emulsions are colloids composed of two liquids. Gels are colloids composed of a liquid dispersed in a solid. Milk and mayonnaise are emulsions; jelly and shaving preparations are gels.

13. solid (ice), liquid (water), and gas (steam)

15. The particles of a gas are a great distance from each other.

17. Gases are composed of independent tiny particles that rapidly travel through empty space in a straight line. A gas quickly expands to fill the allowable volume.

19. Atoms are minute, spherical particles that are indivisible. Molecules are collections of atoms that are bound to each other.

21. molecules that consist of more than two atoms (e.g., P_4, H_2O)

23. a) 2 H atoms b) 2 O atoms c) 2 H atoms and 1 O atom

25. waste, sewage sludge, and fuels

27. a mixtures of metals

29. A substitutional alloy is formed when atoms of the minority metal replace some of the atoms in the regular structure of the majority metal. Examples are brass (zinc, copper) and bronze (tin, lead, copper).

31. a) element b) alloy c) element d) alloy

33. a) table salt b) coffee with completely dissolved sugar c) brass

35. silver, copper, and nickel

Understanding Concepts

37. a) Elements are the fundamental types of matter. Compounds consist of two or more elements in a fixed ratio.

    b) Mixtures, unlike compounds, contain substances in no fixed proportions.

39. solution

41. The substances a, c, d, f, and g are heterogeneous mixtures; b and e are elements.

43. $C_{12}H_{22}O_{11}$

45. When particles are regularly arranged, a crystalline structure is formed. When particles are much less ordered, the material is called amorphous.

47. When a molecular substance is in a liquid crystal state, it has physical properties that are intermediate between those of a solid and a liquid.

49. a) Hg^0, Hg^{2+}, Hg(particulate), $(CH_3)Hg^+$, and $(CH_3)_2Hg$ b) Hg^0, Hg^{2+}, $(CH_3)Hg^+$, and $(CH_3)_2Hg$ c) Hg^0, Hg^{2+}, $(CH_3)Hg^+$, and $(CH_3)_2Hg$

51. a) 100% b) 83% c) 50%

Synthesizing Ideas

53. mechanically strong, not brittle, should not tarnish, and have the appearance and texture of real teeth

Chapter 2
New Identities?
Physical and Chemical Change

Review Questions

1. Physical properties define characteristics of a material that are aspects of the material itself and do

not involve its transformation into another substance. Processes in which the composition or identity of a pure substance is not altered are called physical changes.

3. the ratio of an object's mass to its volume

5. Dry ice is solid carbon dioxide. Cold molecules of carbon dioxide rapidly cool the air surrounding them, resulting in the condensation of water vapor into "clouds" above the dry ice.

7. When a molecule melts, its component molecules move past each other in a medium that contains only molecules of the same type. When a molecule dissolves, its molecules move among not only molecules of the same type but also the molecules of a liquid, such as water, dissolving it.

9. Equilibrium signifies the equality in rate between any two opposing processes. *Volatile* is a term used to describe liquids that evaporate readily.

11. Evaporation is defined as the escape of molecules from the surface of a liquid. Boiling is the complete and rapid conversion, at a specific temperature, of a substance from the liquid to the gaseous state.

13. The chemical properties of a substance describe how it can change into other substances. A chemical change is defined as a process that results in the rearrangement of atoms and a change in the identity of a pure substance.

15. Reactants are substances that are present before a chemical change occurs. Products are substances that are present after the chemical change occurs.

17. the reaction of oxygen with a fuel

19. As a result of the law of conservation of mass, the number of atoms of each type in the products of a reaction must equal the number of atoms of each type in the reactants. When chemical reactions are written to reflect this law, they are said to be balanced.

21. a) 6 oxygen atoms b) 12 carbon atoms, 24 hydrogen atoms, and 12 oxygen atoms c) 2 hydrogen atoms and 1 oxygen atom d) 1 argon atom

23. Soot, an impure form of the element carbon, is another common product of the incomplete combustion of carbon-containing compounds.

25. In every one million molecules of air, there is one molecule of CO.

27. Carbon monoxide interferes with the efficient transport of oxygen molecules in the body. This results in the heart having to work harder to supply the cells with the oxygen necessary for normal function.

29. A catalyst is a substance that speeds up a reaction without itself being consumed by the reaction.

31. Rusting is a process that occurs when atoms on the surface of iron, or its alloys, react with atmospheric oxygen dissolved in water.

33. Yes. It boils at $-33^{\circ}C$.

35. H_2O and CO_2; particles of elemental carbon (soot); wax vapor

Understanding Concepts

37. The forces between the atoms keep them from entering the vapor phase.

39. rubbing alcohol, due to the weaker forces of attraction

41. Yes. It has a noticeable odor and evaporates readily at room temperature.

43. Yes. Oxidation reactions refer to the consumption of oxygen gas by its interaction with other substances.

45. a) $2\ SO_2 + O_2 \rightarrow SO_3$ b) $2\ C_2H_6 + 5\ O_2 \rightarrow 4\ CO + 6\ H_2O$ c) $C_2H_4 + 3\ O_2 \rightarrow 2\ CO_2 + 2\ H_2O$ d) $12\ CO_2 + 11\ H_2O \rightarrow 12\ O_2 + C_{12}H_{22}O_{11}$

47. The combustion of glucose. $C_6H_{12}O_6 + 6\ O_2 \rightarrow 6\ CO_2 + 6\ H_2O$. Both a combustion and an oxidation reaction. The fuel for this reaction is glucose.

49. $2\ Ag + H_2S \rightarrow Ag_2S + H_2$; $2\ Al + 3\ Ag_2S \rightarrow 6\ Ag + Al_2S_3$

51. Both burning and rusting are oxidation processes. Breathing is the process by which oxygen gas is introduced into the lungs so that it can subsequently react with glucose to produce CO_2, H_2O, and energy.

Synthesizing Ideas

53. Determine the mass and volume of each ring. Divide the mass by its volume.

55. At higher altitudes, atmospheric pressure is lower and a lower temperature is required to produce a vapor pressure that is equal to atmospheric pressure.

57. 50,000 ppm

59. Any activity that follows the steps of the scientific method—the gathering of observations, the formation of a hypothesis, the testing of the hypothesis, and the revision and extension of the hypothesis—would use the processes described by the scientific method.

61. When one sweats, water is released to the surface of the skin. Over time, the sweat evaporates and the body is cooled.

Chapter 3
An Insider's Perspective
The Internal Workings of Atoms and Molecules

Review Questions

1. A plasma is a gaseous mixture of electrically charged particles. Electrons and the nucleus of the argon atom are the components.

3. The nucleus is a component of the atom that carries a positive electrical charge equal and opposite to the total charge of the electrons in the atom.

5. Rutherford's experiments indicated that the atom is a sphere containing mostly empty space with the nucleus in the center and the electrons traveling at high speeds around the nucleus. Rutherford bombarded a thin sheet of gold foil with positively charged particles and found that while most of the particles passed straight through the foil, a few were deflected at a huge angle.

7. A proton is a particle contained in the nucleus of an atom that carries an electrical charge equal, but opposite, to that of an electron. A neutron is a particle also contained in the nucleus of an atom. Neutrons have approximately the same mass as a proton but do not have an electrical charge.

9. Isotopes are atoms of a given element with different mass numbers due to varying numbers of neutrons in the atom's nucleus.

11. a model in which electrons are thought of as occupying one of several concentric shells centered on the nucleus

13. 2, 8, 18

15. the outermost occupied shell of the atom

17. Argon extends the lifetime of the filament in a bulb. Helium is used to fill balloons. Neon is used to emit light in commercial signs.

19. All have the same number of electrons in their valence shell. The only exception to this is the element helium.

21. the repeating, or periodic, properties of the elements

23. the number of electrons, the number of protons, the number of neutrons, the number of electrons in the valence shell, the number of shells in the atom, and the relative size of the atom

25. They have the same number of valence shell electrons.

27. The sharing of an electron pair between two adjacent atoms is called a covalent chemical bond. In the hydrogen molecule, both of the single electrons associated with the two hydrogen atoms are shared by the two atoms.

29. fuel

31. Lewis structures are diagrams that show the bonds between atoms and the nonbonding electrons in molecules. Only valence electrons are depicted in Lewis structures. Nonbonding electrons are represented as distinct, separate pairs around the element symbol.

33. H_2O_2

35. as a disinfectant and a bleaching agent

37. A cation is positively charged, while an anion is negatively charged. Ions are represented with either a positive or negative superscript after the symbol for the element indicating the charge on the ion.

39. Metals form positive ions and nonmetals form negative ions.

41. a three-dimensional network composed of layers of ions held together by electrostatic forces

43. maintaining the ion balance in bodily fluids and the normal rhythmic beating of the heart

45. Recommended Dietary Allowance

47. Calcium serves a regulatory function within the cell and provides strength to bones and teeth.

49. Trace minerals are inorganic substances needed in amounts of only a few milligrams per day. Examples of trace minerals include iron, zinc, and selenium.

51. They differ in the quantities required for good health.

53. a) 2 b) 1 c) 3 d) 4

55. The total positive charges of the cation must equal the total negative charges of the anions.

57. lowers

59. the ability of an atom to attract electrons in a chemical bond to itself

Understanding Concepts

61. 12, ^{12}C; 13, ^{13}C; 14, ^{14}C

63. a) $\underline{2}$ b) 2, 8, $\underline{2}$ c) 2, 8, $\underline{7}$ d) 2, 8, $\underline{8}$

65. a) 8 b) 6 c) 2 d) 1

67. The energy level, or shell, is completely filled.

69. The element nitrogen has 7 protons and 7 electrons. It is a member of Group V of the Periodic Table and has 5 valence shell electrons.

71. H_3 is H · H : H and H_4 is H · H : H · H. No, neither follows the rules.

73. 8 minus the number of valence shell electrons

75. a cation with a +2 charge; an anion with a −3 charge

77. a) 0, 1, +1 b) 10, 7, +7 c) 18, 20, +20 d) 10, 9, +9 e) 54, 54, +54

79. Neutral atoms react in order to attain a stable octet configuration.

81. a) NaBr, sodium bromide b) Na_3N, sodium nitride c) MgS, magnesium sulfide d) $MgCl_2$, magnesium chloride e) SrO, strontium oxide f) Sr_3P_2, strontium phosphide

83. a) 17 protons and 17 electrons b) 17 protons, 18 electrons, −1 net charge c) 2 chlorine atoms each sharing 1 electron d) 17 protons, 17 electrons, and 18 neutrons

85.

87. X, X, Y

Synthesizing Ideas

89. a) Single molecule

$$
\begin{array}{c}
\text{H} \qquad\qquad \text{H} \\
\diagdown \qquad\quad \diagup \\
\text{N} - \text{N} \\
\diagup \qquad\quad \diagdown \\
\text{H} \qquad\qquad \text{H}
\end{array}
$$

b) Two separate molecules:

$$
\begin{array}{cc}
\text{H} & \text{H} \\
\diagdown & \diagdown \\
\overset{\times\times}{\text{N}}-\text{H} \quad & \overset{\times\times}{\text{N}}-\text{H} \\
\diagup & \diagup \\
\text{H} & \text{H}
\end{array}
$$

91. Insufficient hemoglobin reduces oxygen-carrying capacity.

Chapter 4
Powering the Planet
Hydrocarbons and Fossil Fuels

Review Questions

1. compounds made up exclusively of carbon and hydrogen

3. 1

5. gas; no strong attractive forces

7. Branched hydrocarbons have carbon chains attached to the main continuous chain of carbon atoms; straight chain molecules do not.

9. There are 3 possible isomers.

$$
\begin{array}{c}
\qquad\quad \text{CH}_2 \qquad\qquad \text{CH}_2 \\
\qquad \diagup \quad \diagdown \qquad \diagup \quad \diagdown \\
\text{H}_3\text{C} \qquad\quad \text{CH}_2 \qquad\qquad \text{CH}_3
\end{array}
$$

$$
\begin{array}{cc}
\text{H}_3\text{C}-\text{CH}_2 & \quad \text{H}_3\text{C} \qquad \text{CH}_3 \\
\qquad\quad \diagdown & \qquad\quad \diagdown \quad \diagup \\
\qquad\quad \text{CH}-\text{CH}_3 & \qquad\qquad \text{C} \\
\qquad \diagup & \qquad\quad \diagup \quad \diagdown \\
\text{H}_3\text{C} & \quad \text{H}_3\text{C} \qquad \text{CH}_3
\end{array}
$$

11. Cycloalkanes are alkanes in which the carbon atoms form ring structures. Cyclohexane.

13. a) decomposition of living matter and termites b) natural gas and oil c) natural gas and oil d) natural gas and oil

15. Alkenes are hydrocarbons containing carbon–carbon double bonds.

17. cigarette smoke, automobile exhaust, and the evaporation of gasoline

19. ring structures with alternating carbon–carbon single and double bonds

21. because fossil fuels are residual by-products of organisms that existed millions of years ago

23. oil and natural gas: 45 years; coal: several centuries

25. methane, ethane, propane, n-butane, and isobutane

27. wells in which the concentration of hydrogen sulfide, H_2S, is very high

29. It must be fractionated into components.

31. a measure of the ability of a gasoline to generate power without engine knocking

33. transportation mishaps

35. short-term: seabirds and shellfish; long-term: mangrove trees, coral reefs, and marshes; bottom-dwelling fish

37. Peat, the partially decomposed remains of land-based woody plants, is geochemically changed to coal.

39. The burning of coal produces substantial amounts of air pollutants.

Understanding Concepts

41. a) $CH_3CH_2CH_2CH_2CH_2CH_2CH_2CH_3$

b) $CH_3CH_2CH_2CH_2CH_2CH_2CH_3$

c) C_5H_{10}

d) $CH_2CHCHCHCH_3$ or
$CH_2=CH-CH=CH-CH_3$

43. a)

b)

45. a) unique isomers b) duplicates c) duplicates

47. because they are nonpolar in character

49. Oil is more easily extracted from Earth than natural gas.

51. They do not sufficiently evaporate in the engine to properly burn.

53. fewer than 5 carbon atoms; fewer than 5 carbon atoms; 5 or more carbons

Synthesizing Ideas

55.
$$H-\ddot{N}\cdots H$$
$$\downarrow$$
$$H$$

57. No. Carbon can only form four bonds.

59. 202

61. Two possible conformations for butadiene are given below:

$$H_2C \diagdown^{CH} \diagdown_{CH} \diagup^{CH_2} \qquad HC \diagup^{CH_2}$$

63. i) $2\,H_2S + 3\,O_2 \rightarrow 2\,SO_2 + 2\,H_2O$; $2\,H_2S + SO_2 \rightarrow 3\,S + 2\,H_2O$ ii) $4\,H_2S + 2\,SO_2 \rightarrow 6\,S + 4\,H_2O$ iii) $2\,H_2S + O_2 \rightarrow 2\,S + 2\,H_2O$

65. Young coal or peat has the highest oxygen content. Hard coal or anthracite has the lowest oxygen content and the highest fuel value.

Chapter 5
From Diamonds to Plastics
Carbon's Elemental Forms, Addition Polymers, and Substituted Hydrocarbons

Review Questions

1. a) 6 b) 5 c) 4 d) 3 e) 2 f) 1

3. CFCs are chlorofluorocarbons. CFCs destroy the ozone layer.

5. Plastics are easily molded and don't rot, rust, corrode, or break easily.

7. Most plastics used today are synthetic polymers.

9. polyethylene, ethylene

11. small molecules that serve as repeating units and combine to form long polymer chains

13. a) $-C-C-C-C-X-X-$ b) $-C=C-C-$ c) $-C=C-$

15. Plasticizers gradually leak out of plastics as they age.

17. Styrofoam consists of a gas trapped within beads of polystyrene.

19. Branched polymers have carbon chain branches; straight-chain polymers do not.

21. Latex consists of a mixture of small latex particles and water.

23. Vulcanized rubber is harder, stronger, resists flowing, and is elastic.

25. Cross-linked polymers have short chains of varying lengths that link or bind adjacent chains, or adjacent regions of the same chain, together. Branched polymers do not.

27. reprocess, depolymerize, transform, burn

29. A diamond is composed of a network of carbon atoms "welded" together.

31. hard, insoluble, poor electrical conductor, good heat conductor

33. gasoline and diesel engine exhaust, tar from cigarettes, and burning wood and coal

35. a coal tar derivative that has been used as a wood preservative

37. solid mixtures of nonmetallic materials

39. a form of carbon consisting of small crystals containing about 10% hydrogen

41. Carbon nanotubes are elemental forms of carbon having a hollow cylindrical structure with closed ends.

43. Nanotechnology is the field of science that deals with devices having the dimensions of nanometers.

Understanding Concepts

45. All are identical because the four bonds form a tetrahedral arrangement.

47. CH_2FCl, CHF_2Cl, and $CHFCl_2$

49. $-CH_2-CCl_2-CH_2-CCl_2-CH_2-$
    $CCl_2-CH_2-CCl_2-CH_2-CCl_2-CH_2-CCl_2-$

51. $-CCl_2-CHCl-$, $CCl_2=CHCl$

53. because the rubber state results from the presence of liquid-like random regions that provide little resistance to distortion

55. All plastics are polymers, but all polymers are not plastics.

57. High-density polymers are composed of unbranched chains and pack tightly together. Low-density polymers consist of branched chains that do not pack as tightly.

59. These polymers are extended networks of carbon atoms, each of which forms a double bond, and which behave like metals. They can be converted into conductors in which the current passes from one end to the other.

61. Both consist of an extended network of carbon atoms covalently bonded to each other.

Synthesizing Ideas

63.

Polymer	Monomer R	Properties
HDPE	Ethene	Hard, waxy, opaque, structural strength, toughness, and rigidity
PVC	Vinyl chloride	Rigid and strong. Can be made flexible by the addition of plasticizers.

Polymer	Monomer R	Properties
POLY-PROPYLENE	Propene	Hard, opaque, structural strength; stronger, more rigid, and can withstand higher temperatures than HDPE. Can be drawn into fibers.
POLY-STYRENE	Styrene	Transparent, hard, brittle, thermoplastic, can be molded into different shapes. Lack of frictional resistance to motion, repels water, and "breathes."
PTFE	Tetrafluoro-ethylene	Very high resistance to chemical attack and thermal decomposition
SARAN WRAP	Vinylidine chloride and vinyl chloride	Flexible and impermeable to oxygen

65. Carbon nanotubes are elemental forms of carbon having a hollow cylindrical structure. Carbon buckyballs are spherical forms of carbon. Possible uses include hydrogen storage, muscle fibers, transistors, and superconductors.

67. a) PVC b) clarity, transparency, flexibility, sterilizability, ease of processing, low cost, compatibility with other medical products, and resistance to chemical stress cracking c) no d) no

Chapter 6
The Flavor of Our World
The Oxygen-Containing Organic Compounds We Drink, Smell, and Taste

Review Questions

1. a solvent for organic compounds; an anesthetic

3. MBTE is a replacement for some of the hydrocarbon content and it introduces oxygen into the fuel. It also improves the octane rating of the gasoline.

5. a) ether b) aldehyde c) carboxylic acid

7. the minimum dose of a substance that is lethal to 50% of the population

9. Alcohol significantly increases an individual's risk of death from cancer and liver disease.

11. chemical reactions that occur in the body that ultimately alter the chemical nature of substances present in the body

13. Denatured alcohol is ethanol mixed with a small amount of organic compounds to make it undrinkable.

15. Ethanol undergoes combustion in oxygen to produce heat.

17. The supply can be continually renewed and there is no net carbon dioxide release during its production–combustion cycle.

19. Ethanol and water are miscible because of hydrogen-bonding interactions between the molecules.

21. Concentrated solutions denature the proteins and disrupt the lipid structures of infection-causing bacteria.

23. They contain free hydroxyl, or —OH, groups.

25. It causes eye, nose, throat, and skin irritation; is toxic in high concentrations; and has been classified as a "probable human carcinogen."

27. Ethanol is first converted into acetaldehyde, which produces the effects associated with a hangover.

29. a) 2   b) 5   c) 1   d) 4   e) 3

31. a reaction that removes a water molecule

Understanding Concepts

33. MTBE introduces oxygen into the fuel mixture and is able to replace some of the aromatic hydrocarbons in gasoline.

35.

37. a)

   b)

39. M_8, M_{90}

41. 18 proof, 7%

43. Since yeast is not active at alcoholic concentrations above about 12%, products with higher alcoholic content are not produced directly by the process of fermentation. Rather, the products of fermentation are distilled to produce solutions with higher concentrations of ethanol.

45. The higher the temperature at which a liquid boils, the greater the attractive forces between the molecules.

47.
   Propionaldehyde

49. butyric acid and butanol

51. Hard contact lenses are made from a polymer known as poly(methyl methacrylate), or PMMA. Soft contact lenses are composed of a variation of PMMA in which one hydrogen of the ester group is replaced with an —OH group, allowing the lens to absorb water.

Synthesizing Ideas

53. Place equal volumes of the liquid at the same temperature and determine which evaporates first or determine the boiling points.

55. A renewable fuel is defined as a fuel having both a continually renewable supply and no net carbon dioxide release during its production–combustion cycle. Renewable fuels include ethanol, methanol, hydrogen, and oils from food products such as soy.

57. 100 mg/kg body weight

59.

61. $CH_3 CH_2 C(H)O$, a ketone

Chapter 7
Health and Energy
Carbohydrates, Fats, and Oils

Review Questions

1. carbon, hydrogen, and oxygen in roughly a 1:2:1 atomic ratio

3. Dextrose circulates in your blood in a concentration of about 0.1%.

5. Sugar molecules containing one ring are called monosaccharides, and those containing two rings are called disaccharides.

7. Both have the formula $C_6H_{12}O_6$.

9. Hydrolysis reaction refers to the reaction of a substance with water.

11. Sucralose, xylitol, and sorbitol

13. a condition that results from the lack of an enzyme called lactase

15. glucose

17. to store and provide glucose

19. When the oxygen atom of the C—O—C linkage lies below one of the rings and in the plane of the other, the rings are said to be in the alpha form. When the oxygen atom of the linkage lies in the plane of both rings, a beta linkage is formed.

21. Insoluble fiber absorbs, but is insoluble in, water. Soluble fiber is very good at binding to water.

23. Paraffin wax consists of long-chain hydrocarbons obtained from oil refining.

25. Saturated and unsaturated fatty acids differ in the nature of their carbon–carbon bonds.

27. An omega−3 fatty acid is a polyunsaturated fatty acid in which the last double bond in the hydrocarbon chain occurs between the third and fourth carbons from the end of the chain.

Understanding Concepts

29. has many available OH groups to hydrogen bond to water molecules

31. same reaction overall, giving the same products

33. The —OH group on the molecule must form a hydrogen bond with either an oxygen or nitrogen atom on the receptor, a second hydrogen bond between an oxygen atom of the molecule and a hydrogen bonded to either an oxygen or nitrogen atom on the receptor is required, the hydrocarbon region behind the two hydrogen-bonded oxygen atoms of the molecule must be hydrophobic, and the molecule must be small and fit into a specific location on the receptor.

35. Carbohydrate loading occurs when the glycogen supply is depleted by a regime of exercise and/or a diet low in carbohydrates over the course of several days. A few days before a competitive event, a high-carbohydrate diet is initiated. The excess glucose is converted to high levels of glycogen.

37. Carbohydrates are partially oxidized.

Synthesizing Ideas

39. converts some or all double bonds to single bonds by H_2 addition

41. Complex carbohydrates are absorbed more slowly than simple carbohydrates.

43. High fat content is associated with meat products.

45. a) lactose (glucose + galactose)

b) a trisaccharide composed of three glucose molecules

47. a) Starch is composed of two glucose polymers, amylose and amylopectin. Glycogen consists only of glucose monomers in a highly branched arrangement.

b) Amylose consists of about 100 glucose monomers linked in unbranched chains. Amylopectin is composed of thousands of glucose monomers in a highly branched structural arrangement.

49. polyunsaturated, monounsaturated, and saturated

51. $C_6H_{12}O_6 + 6\ O_2 \rightarrow 6\ CO_2 + 6\ H_2O$;
$C_{13}H_{28} + 20\ O_2 \rightarrow 13\ CO_2 + 14\ H_2O$; 3.3

Chapter 8

The Chemistry of Medication and Clothing

Condensation Polymers—Especially Those Containing Nitrogen

Review Questions

1. a molecule that contains a nitrogen bonded to one or more hydrocarbon groups

3. Amines correspond to molecules of NH_3 in which one, two, or all three of the hydrogen atoms are replaced by hydrocarbon groups.

5. molecules responsible for transmitting impulses between nerve cells

7. Neurotransmitters travel across a gap to a receptor on an adjacent nerve cell and signal the impulse to travel along that nerve cell.

9. In pyridine, a nitrogen atom replaces one of the carbon atoms.

11. Nicotine is a mild stimulant. It is addictive because a decrease in its blood levels produces withdrawal symptoms.

13. a stimulant and an amine

15. A, because it has a shorter half-life.

17. They increase the concentration of the OH^- when they are dissolved in water.

19. Analgesics relieve pain; an antipyretic relieves fever.

21. An amide is a molecule containing an amino group bonded to a carbonyl carbon. A polyamide is a polymer formed by the condensation of a dicarboxylic acid and a diamine, with the loss of water.

23. A diacid contains two carboxylic acid groups. When dicarboxylic acids combine with diamines, a polyamide is produced.

25. a molecule that contains both an amino group and a carboxylic acid group joined by a carbon chain

27. The polymer chains in polyamides have a high degree of hydrogen bonding between the chains. In contrast, polyesters do not.

29. Polycarbonates, like polyesters, are condensation products of a diacid and a diol. Unlike polyesters, however, polycarbonates use a diacid.

Understanding Concepts

31. b

33. Nicotine is eliminated from the body faster.

35. 180 minutes, 360 minutes

37. 1.5 hours

39. The lower the number of carbon atoms per nitrogen atom, the greater the number of hydrogen bonds per molecule and the greater its solubility in water.

41. Ions are more water soluble.

43. Adrenaline and amphetamine act in a similar way to increase heart rate, wakefulness, energy level, and drive. The effects are similar because the structures of both amines are similar.

45. All three molecules contain a carboxylic acid group and a benzene ring. They have a planar (or flat) region associated with the benzene ring and have similar shapes, sizes, and polarities.

47. Reaction of a diacid and a diol:

Reaction of a monoacid and a diol:

49. Amino acids contain both an amino group and a carboxylic acid group.

Synthesizing Ideas

51. no, significantly lower

53. The polymer chains are not volatile.

55. Neurotransmitters are small molecules with hydroxyl and/or amino functional groups. Binding sites are polar in character and small in size.

Chapter 9
The Molecules That Make You What You Are
Nucleic Acids and Proteins

Review Questions

1. DNA contains the genetic code for protein molecules.

3. DNA and RNA are polymers of nucleotides linked by ester linkages.

5. Both DNA and RNA are composed of nucleotides. In DNA, the sugar of the nucleotide is deoxyribose, while in RNA the sugar is ribose.

7. adenine, guanine, cytosine, and uracil

9. DNA normally exists as a double helix held together by hydrogen bonding between nitrogen bases.

11. linear sequences of bases that direct the protein-producing activities of the cell

13. Base sequences that occur in front of each gene serve a regulatory function. Other regions separate the genes but have an undetermined function.

15. a sequence of three consecutive nitrogen bases on an mRNA molecule

17. a process in which a DNA strand acts as a template to produce a complementary strand of mRNA

19. Ribosomal RNA provides a framework to which the proteins are bound.

21. α-amino acids are the basic structural units of proteins.

23. molecules that relieve pain and produce pleasant sensations

25. The secondary structure is the shape adopted by the protein chain. A protein's tertiary structure results from the further folding and coiling of the secondary structure.

27. Ones dominated by their carbon content are hydrophobic, nonpolar, saturated R groups, or side chains, and aromatic R groups.

29. A complete protein contains all of the essential amino acids in the correct ratio for conversion into human protein.

31. by obtaining the correct balance of essential amino acids

33. metal ions, phosphate groups, sugars, lipids, or coenzymes

35. a) enzymatic b) transport c) structural / contractile d) transport

37. reactant molecules in enzyme-catalyzed reactions

39. a combination of lipids and protein

41. Three types of DNA mutations are substitutions, insertions and deletions, and inversions.

43. Diseases that result from the mutations of DNA in germ cells. Examples include diabetes, sickle-cell anemia, and some forms of cancer.

45.

47. The four bases of DNA—adenine, guanine, cytosine, and thymine—can be arranged into 64 unique 3-letter combinations.

49. A codon is specific for a given amino acid. The codon of mRNA is complementary to an anticodon of transfer RNA.

51. The DNA strand acts as a template to produce a complementary strand of mRNA.

53. a) Pro-Leu-Ile b) Phe-Phe-Val c) Pro-Leu-Arg

55. Hydrogen bonding keeps the double-helical structure of DNA intact.

57. Gene activator proteins recognize and bind to specific DNA sequences and then assist in the binding of RNA polymerase to the DNA.

59.

Ser-Val

Val-Ser

61.

Ala-Cys

Cys-Ala

63. The primary structure defines the final three-dimensional shape.

65. Enzymes lower the activation energy of the reaction.

67. Proteins consisting of an α-helical arrangement are fibrous. Proteins consisting of β-sheets cannot be appreciably stretched.

69. Mutations in a DNA sequence alter the sequence of nitrogenous bases. Mutated codons produce a protein with an altered sequence of amino acids, or primary structure.

71. Genetic diseases result from mutations in germ cells.

Synthesizing Ideas

73. Housekeeping genes maintain the basic functions of the cell. Genes whose expression is regulated include those that encode DNA repair enzymes, enzymes involved in cholesterol synthesis, or heat-shock proteins.

75. The adding of salt to a mixture of egg albumin and water affects both hydrophilic and hydrophobic interactions. As the ionic strength, or polarity, of the water is increased, the position of polar or charged R groups will be favored on the surface of the protein while internal positions will be favored by hydrophobic nonpolar R groups.

77. You would expect bodily processes to slow down.

79. Cleaning products often contain enzymes, such as proteases, amylases, and lipases, that break down starch and fatty stains.

81. to help reduce the exposure of the patient to radiation

83. the diagnosis of the presence of genes responsible for some forms of cancer and the possibility of repairing defective genes

85.

Chapter 10

Chemicals in Our Bodies and Our Environment
Vitamins, Food Additives, Pesticides, and More

Review Questions

1. noncaloric nutrients essential for normal body function

3. fat-soluble or water-soluble

5. A: vision and growth, B: coenzymes, C: formation of collagen, immune system, antioxidant, D: calcium uptake, E: antioxidant, K: blood clotting

7. compounds that prevent oxidative damage

9. biochemical processes that involve oxygen, exposure to pollution, tobacco, radioactivity, sunlight, and the existence of some medical conditions

11. antioxidant chemicals that occur in plants in small quantities

13. BHA, BHT, and propyl gallate are used in packaged foods that are susceptible to oxidation.

15. Sequesterants, like EDTA, form strong chemical bonds to metal ions.

17. the tendency for water to travel through a semipermeable membrane from a less concentrated solution to a more concentrated solution

19. benzoic acid (sodium benzoate), propionic acid (sodium propionate), sorbic acid (sodium sorbate), acetic acid, and citric acid

21. its ability to bind to iron in our blood and the potential conversion to nitrosamines

23. A pesticide is a compound that kills undesirable organisms. An insecticide kills insects. Herbicides kill plants. Fungicides control the growth of fungus.

25. the reproductive abilities of birds, fish, and humans

27. Humans are able to feed on all of the species preceding us in the chain.

29. They are known to have an immediate effect on health.

31. 2,4-D, 2,4,5-T, and 2,3,7,8-TCDD

33. a herbicide that kills almost all plants

35. the bleaching of paper pulp, the incineration of garbage, and the recycling of metals

37. They are hydrophobic.

39. It is a measure of the concentration of various organochlorines in terms of the amount of 2,3,7,8-TCDD that would produce the same toxic effect.

41. Environmental estrogens are environmental chemicals that either mimic or block the action of estrogen or accelerate the breakdown of the natural hormone; they have been linked to infertility and an increased risk of reproductive organ cancer.

43. Diethylstilbestrol is a synthetic estrogen once taken by women to prevent miscarriages.

45. Bisphenol-A is used as a raw material for polycarbonate plastics and epoxy resins. Nonylphenol results from the breakdown of materials used in detergents, spermicides, and some plastics. Phthalates are commonly used as plasticizers.

Understanding Concepts

47. They are needed only in very small amounts.

49. Fat-soluble vitamins are carried by fat-transporting proteins in the blood. Water-soluble vitamins travel freely in the bloodstream.

51. vitamin C, because vitamin A is fat soluble

53. disrupting cell membranes or affecting the operation of cellular processes

55. Vitamin D controls the uptake of calcium and phosphorus.

57. The decomposition of unsaturated fatty acids by oxygen produces rancid foods. Additionally, microorganisms can multiply in foods and secrete toxic substances.

59. vitamin E

61. Propionates are effective against bacteria, molds, and fungi, while sorbates are effective against only molds and fungi.

63. The use of pesticides is a concern because of its potential impact on human health.

65. Because of its low water solubility, DDT accumulates in fat.

67. A substance is bioaccumulated by a combination of the processes of bioconcentration and biomagnification.

69. Organochlorines have been shown to affect reproduction. Organophosphate and carbamate compounds function as nerve toxins.

71. Organophosphates and carbamates are not as environmentally persistent as are the organochlorine compounds.

73. their once widespread use, their persistence in the environment, and their bioaccumulation in living systems

75. PCBs are believed to be detrimental to human and animal reproduction as well as to cognitive development.

77. Environmental estrogens can cause reproductive problems, embryo mortality and deformities, reproductive system abnormalities, and cancer.

Synthesizing Ideas

79. Symptoms of vitamin E deficiency include scaly skin, sterility, and the decay of muscle tissue.

81. the presence of naturally occurring weak organic acids

83. benzoic acid (acidic conditions):

benzoate anion (basic conditions):

Only the molecular (nonionic) form of the acid is toxic to the microbe.

85. Sulfites are ideal preservatives for aqueous solutions. Nitrites are ideal preservatives for solid foods.

87. DDT is hydrophobic. DDT accumulates in the fat, where it can reach toxic concentrations.

89. a) Dogs metabolize theobromine more slowly than humans.

    b) There are different biochemical processes that occur in each.

91. their stability, their extensive usage, and their careless disposal

93. Hydrophobic materials are much less soluble in salt water.

Chapter 11
Chemistry in Water
Salts, Acids, and Bases

Review Questions

1. Salts are solid compounds consisting of a cation and an anion.

3. A solute dissolves in a liquid. A solvent is the dissolving liquid.

5. Solutions that contain a high concentration of solute are concentrated. Those that contain a low concentration of solute are dilute.

7. the solid that forms in a solution when the solid's solubility is exceeded

9. Mineral water is obtained from natural sources and has a higher concentration of dissolved salts than regular water.

11. A fluoride ion replaces the hydroxide ion.

13. A base increases the concentration of the hydroxide ion in solution. Characteristics include its bitter taste, slimy feel, and ability to turn litmus paper blue.

15. $HF \rightleftharpoons H^+ + F^-$

17. acids that contain more than one hydrogen atom

19. a) muriatic acid b) vinegar c) drain cleaners d) ammonia

21. Molarity is the number of moles of a substance that are present in 1 liter of solution, moles / liter.

23. Weak acids are typically found in foods and provide a sour or tart flavor.

25. $2H^+ + Fe \rightarrow H_2 + Fe^{2+}$; $H^+ + CO_3^{2-} \rightarrow HCO_3^-$; $H^+ + HCO_3^- \rightarrow H_2CO_3$

27. The lining provides a source of bicarbonate anions to react with the strong acid to produce a weak acid, carbonic acid.

29. The release and expansion of carbon dioxide produces baked goods with greater volume and less density than the original dough.

31. Ca^{2+} and Mg^{2+}

33. Indicators are materials that take on different colors depending upon whether they are in an acidic or a basic environment. Examples of indicators include litmus paper and the juice from grapes or red cabbage.

Understanding Concepts

35. Water consists of a partially positive region that associates with anions and a partially negative region that associates with cations.

37. The salt precipitates out of the solution.

39. 13.2 mg

41. Fluorapatite is more resistant than apatite to the carboxylic acids produced by bacteria in the mouth, thereby preventing tooth decay. Also, fluoride ions inhibit the conversion of carbohydrates to the carboxylic acids that attack tooth enamel.

43. Sodium is a very common cation in natural water.

45. A strong acid is composed exclusively of ionic components. A weak acid contains mostly molecular components with very few ions present.

47. With a strong acid, the reaction proceeds only in the forward direction.

49. The greater the molarity of the acid, the lower the pH.

51. acidic, less

53. 0.10 M; 1

55. Antacids contain bases that neutralize the acid in your stomach.

57. $HC_2H_3O_2 + NaHCO_3 \rightarrow NaC_2H_3O_2 + CO_2 + H_2O$; yes

59. $HC_2H_3O_2 + CaCO_3 \rightarrow Ca(C_2H_3O_2)_2 + CO_2 + H_2O$

61. $F^- + H_2O \rightleftharpoons HF + OH^-$

63. NaOH + HBr → NaBr + H_2O, sodium bromide

65. Buffer solutions consist of a weak acid and a salt of the weak acid or a weak base and a salt of the weak base. If a hydrogen ion is added to the buffer solution, the large excess of anion reacts with the hydrogen ion. When a base is added to the buffer, it reacts with the molecular form of the weak acid.

Synthesizing Ideas

67. The concentrations of ions in aqueous solutions are generally small and in the milligram-per-liter range.

69. Strong acids ionize completely.

71. $H_2C_2O_4 \rightleftharpoons H^+ + HC_2O_4^-$

$HC_2O_4^- \rightleftharpoons H^+ + C_2O_4^{2-}$

$H_2C_2O_4 \rightleftharpoons 2 H^+ + C_2O_4^{2-}$

73. a) no b) yes c) no

75. A weak acid can react with the hydroxy ion of apatite.

$Ca_5(PO_4)_3OH + H^+ \rightarrow Ca_5(PO_4)_3^+ + H_2O$

77. When they exist as solids, acids and bases do not react. When they ionize in solution, however, they readily react to form a salt and water.

Chapter 12
Batteries, Fuel Cells, and the Hydrogen Economy
Oxidation and Reduction

Review Questions

1. a) the loss of electrons by a substance b) the gain of electrons by a substance c) processes that describe the transfer of electrons between atoms d) oxidation-reduction reactions

3. An oxidizing agent promotes oxidation. H^+, Cl_2, O_2, and O_3 are common oxidizing agents.

5. a) oxidizing agent: oxygen, reducing agent: sulfur b) oxidizing agent: chromium, reducing agent: aluminum c) oxidizing agent: tungsten, reducing agent: hydrogen

7. a) metal surfaces at which oxidation and reduction reactions occur b) ion-containing solutions into which the electrodes in an electrochemical cell are placed c) a device that maintains the anion balance in an electrochemical cell

9. During the discharge of an electrochemical cell, the metal surface of one electrode is oxidized to the corresponding ion, which enters the aqueous solution. The electrons from this oxidation travel to the other electrode, where they are transferred to another ion in solution, resulting in the plating of the metal on the electrode. The salt bridge serves to redistribute the anions in order to maintain positive and negative ion balance.

11. an electrochemical cell that can be continuously operated and does not become discharged

13. a technique in which a thin film of metal is deposited on an object

15. $2 H_2 (g) + O_2 (g) \rightarrow 2 H_2O (g)$ + heat energy

17. a liquid, compressed gas, as a gas between graphite layers or reversibly absorbed by metals or metal alloys, and as an elemental component of liquid fuels

Understanding Concepts

19. Solid metals are oxidized, aqueous acids are reduced, oxygen gas is reduced, and hydroxide is often formed in reduction reactions.

21. There must be an electrical connection between the two metals, and the two metals must spontaneously undergo an oxidation-reduction reaction.

23. evaporation and leakage

25. Hydrogen can be produced from both fossil fuels and from renewable sources, and it has been the fuel used in most fuel cell technology development efforts.

27. Fuel cells eliminate the need to carry and recharge heavy conventional battery systems.

29. Hydrogen gas combines with oxygen to produce water and a substantial amount of energy.

31. catalytic heaters

33. the electrolysis of water to H_2 and O_2

35. The advantage of storing hydrogen as a liquid is the relatively small holding vessels required. The condensation of hydrogen gas, however, requires substantial energy and cost. The advantage of storing hydrogen as a gas is the lack of cooling required to condense the gas to a liquid. The major disadvantage is the bulky nature of the holding vessels required.

37. Hydrogen gas can be obtained from methanol by a re-forming reaction. In a re-forming reaction, liquid hydrocarbons are converted into a hydrogen-containing gas mixture.

$$CH_3OH \rightarrow 2\,H_2 + CO$$

$$CO + H_2O \rightarrow CO_2 + H_2$$

$$\text{Overall: } CH_3OH + H_2O \rightarrow CO_2 + 3\,H_2$$

Synthesizing Ideas

39. a) oxidation-reduction reaction: Na is oxidized and Cl is reduced b) oxidation-reduction reaction: C is oxidized and O is reduced c) *not* an oxidation-reduction reaction

41. An electrochemical reaction stops after a period of time when either the electrode where oxidation occurs is dissolved or the ions undergoing reduction are consumed.

43. In general, it is extremely difficult to stop a spontaneous oxidation-reduction reaction.

45. Unlike hydrogen gas, fuels containing carbon produce carbon dioxide when combusted.

47. elimination of the need to frequently recharge heavy batteries, the high efficiency in extracting useful energy from fuels, the lack of pollutants generated and emitted, the silent and low-maintenance operation, and the ability to use hydrogen from fossil fuels and renewable sources

49. absorbed by a solid material

51. The first method is the conversion of fossil fuels to hydrogen. Major drawbacks to this method include the limited rate of production of electric current and the extensive greenhouse gases created. The second method involves the electrolysis of water. A major drawback to this method is the amount of energy required.

Chapter 13
Fit to Drink
Water Sources, Pollution, and Purification

Review Questions

1. drinking, bathing, cooking, washing, producing goods and services, and agricultural irrigation; industrial and agricultural applications (largest), personal use (smallest)

3. 2%

5. water that resides on the surface of Earth in lakes, rivers, and streams

7. fresh water that lies underground

9. precipitation that falls onto Earth's surface

11. a permanent reservoir, or underground lake

13. inorganic materials from rock and organic materials from the decomposition of animal matter

15. granulated charcoal that has been treated with steam

17. the production of fresh water from salt water by the removal of ions

19. A solar still uses solar energy to desalinate water by distillation.

21. Chlorination is a method of water disinfection that utilizes hypochlorous acid. One method to chlorinate water uses chlorine gas dissolved in water.

23. a liquid that contains dissolved matter that drains from a terrestrial source

25. a) soaps, detergents, and other cleaning products b) nitrogen fertilizers, atmospheric deposition, human sewage, and soil cultivation c) terrestrial waste-disposal sites, underground storage tanks, and surface spillage

27. *Soap* is a term given to the class of compounds that consist of a sodium or potassium salt of a long-chain fatty acid. The general structure of a soap is:

$$R-CH_2-CH_2-CH_2-CH_2-CH_2-C(O)-O^-\ Na^+$$
$$(\text{or } K^+)$$

29. A synthetic detergent is a material that cleans like soap but does not form insoluble precipitates with Ca^{2+} and Mg^{2+} ions. Synthetic detergents contain the sulfonate group rather than the carboxylate group.

31. Biodegradable materials can be broken down into simpler substances by bacteria and other microorganisms present in the environment.

33. Zeolites are materials consisting of sodium, aluminum, silicon, and oxygen that exchange their sodium ions for calcium ions in solution.

35. excessive algae growth resulting in the lack of dissolved oxygen

Understanding Concepts

37. removes dissolved gases and other volatile organic compounds

39. Aluminum hydroxide forms an insoluble network structure that attracts and entraps the suspended colloidal particles.

41. removes small concentrations of organic pollutants from water

43. No, some contaminants are small enough to pass through the membrane.

45. Daily sodium intake is increased.

47. Softened water contains ions; distilled water does not.

49. Solar stills are practical only in those areas of the world with plentiful sunshine and land.

51. compounds in which a chlorine atom is combined with other elements

53. The disinfection of water is essential for the protection of public health and the saving of lives that would otherwise be lost to waterborne diseases.

55. the large volumes of water requiring treatment and the associated expense, and recontamination of the water when it is returned to the aquifer

57. Coral reefs are adversely impacted.

59. Nitrogen-fixing bacteria reduce nitrogen in the air to ammonia (or ammonium salts).

61. Grease particles are incorporated into soap's hydrophobic portion.

63. It is formed from the condensation of evaporated water.

65. to form soluble substances with calcium and magnesium ions present in hard water

67. BTX hydrocarbons and the antiknock additive MTBE

Synthesizing Ideas

69. lakes and rivers, aquifers, man-made reservoirs, and seawater

71. Yes. People on sodium-restricted diets should be careful. Testing this water is highly recommended.

73. acidic: hypochlorous acid and chloramine compounds; basic: hypochlorite

75. HOCl produces chlorinated organic substances. Chlorine dioxide produces fewer toxic organic by-products.

77. Method 1: HOCl. HOCl is a disinfectant with residual power. Its use results in the production of potentially toxic chlorinated organic compounds.

Method 2: Ozone. Ozone is a disinfectant with a very short lifetime and no residual protection. It cannot be stored or shipped, and it is associated with the formation of partially oxidized organic compounds.

Method 3: ClO_2. Chlorine dioxide produces fewer toxic chlorinated hydrocarbon by-products. ClO_2 cannot be stored, is explosive, and results in the formation of ClO_2^- and ClO_3^-.

Method 4: Ultraviolet irradiation. Irradiation eliminates toxic microorganisms, is convenient, and can be used in both large and small installations, but it provides no residual protection.

79. Fat is decomposed into its component fatty acids and glycerol by its treatment with a base and heat. The addition of a sodium or potassium salt to this basic solution converts the free fatty acids into their corresponding sodium or potassium ions.

81. Branched alkylsulfonates are very stable and do not biodegrade, whereas unbranched alkylsulfonates do.

83. While we have stopped the production of many hazardous materials, their constant leaching from landfills, industrial and agricultural sites, and storage tanks continually contaminates both the water and the soil.

Chapter 14
Dirty Air, Dirty Lungs
Air Pollution

Review Questions

1. troposphere

3. Photochemical smog is air pollution caused by chemical reactions requiring sunlight.

5. Volatile organic compounds are the chemical reactants of smog.

7. when fuels are burnt in air

9. NO and NO_2

11. Ozone damages plant tissue, hardens rubber, bleaches colors, and is harmful to humans.

13. deciduous trees and shrubs

15. First stage uses minimum oxygen; second stage is carried out at lower temperatures. In both cases, NO_x formation is minimized.

17. Due to the presence of dissolved atmospheric carbon dioxide, the natural pH of rainwater is a relatively acidic 5.6.

19. Virtually all of the acidity in acid rain results from sulfuric acid and nitric acid dissolved in the water.

21. the combustion of coal and nonferrous smelting

23. tiny solid and liquid particles suspended in air

25. diameters greater than 2.5 μm are coarse; smaller diameters are fine

27. Soot refers to solid particles in the atmosphere, whereas mist refers to liquid atmospheric particles.

29. Coarse particles originate from the mechanical breakup of larger pieces of matter. Fine particles are formed primarily from chemical reactions.

31. Sulfate aerosols contain large amounts of oxidized sulfur compounds.

33. the mass of particulate matter in a given volume of air

35. asbestos, VOCs, formaldehyde, tobacco smoke, and carbon monoxide

37. Adsorption is the attachment of a material to the surface of another. Absorption is the dissolution of a material into another.

39. a) eye, nose, throat, and skin irritation; respiratory infections; allergies; and asthma b) drowsiness, fatigue, and death c) eye and respiratory system irritation, aggravates asthma and angina pectoris, known human carcinogen d) causes mesothelioma e) dissolve living tissue

41. carbon monoxide, nitrogen dioxide, formaldehyde, cadmium, polycyclic aromatic hydrocarbons, and tar

Understanding Concepts

43. warmth, ample sunlight, and very little air movement

45. They release reactive hydrocarbon compounds containing C=C bonds.

47. its location in Earth's atmosphere

49. the use of a less-active catalyst formation and the need for excess oxygen

51. $CaCO_3$ (s) + 2 H$^+$ (aq) → Ca^{+2} (aq) + CO$_2$ (g) + H$_2$O (aq)

53. Large coarse particles settle out rapidly due to gravity.

55. A catalytic converter is most efficient when warm.

57. the amount of photochemical smog, nitrogen oxides, sulfur dioxide, carbon monoxide, lead, ozone, and particulate matter in the air

59. Major components of outdoor air pollution include nitrogen oxides, carbon monoxide, ozone, volatile organic compounds, particulate matter, and sulfur dioxide. Major components of indoor air pollution include formaldehyde, carbon monoxide, volatile organic compounds, nitrogen oxides, tobacco smoke, and asbestos and other particulate matter.

61. the concentration of a particular pollutant below which exposure does not produce a particular health effect

Synthesizing Ideas

63. The formation of photochemical smog involves several hundreds of reactions, many of which require sunlight.

65. Electric power generating plants produce emissions that result in higher concentrations of sulfuric acid in the acid rain that falls in eastern North America.

67. Inhalation exposure to large particles is limited. Conversely, small particles have substantial surface areas per gram and are associated with acute and/or chronic health hazards.

69. Indoor air pollution is a much greater problem in newer, more airtight houses built for energy efficiency.

Chapter 15
A Thin Veil of Protection
Stratospheric Chemistry and the Ozone Layer

Review Questions

1. an area of ozone molecules in the stratosphere that filters out harmful rays from sunlight

3. tropical regions of the stratosphere

5. no

7. sunlight

9. It may break or alter bond properties.

11. UV-C region (200 to 280 nm), UV-B region (280 nm to 320 nm), UV-A region (320 to 400 nm)

13. Malignant melanoma is an often-fatal form of skin cancer believed to be related to short periods of very high UV exposure occurring early in life.

15. the direct relationship between the frequency and duration of exposure events and the degree of nucleic acid base alteration, and the biological redundancy inherent in the DNA molecule

17. a decrease in photosynthesis in plants and a negative impact on phytoplankton and small marine life

19. inverse

21. Ozone is created by the action of UV-C light on oxygen gas.

23. HCl and ClONO$_2$. The chlorine is inactive as a catalyst for ozone destruction.

25. decreased by several percent

27. yes; in about 1999

29. carbon-chlorine

31. CFC molecules rise through the troposphere and travel into the stratosphere.

33. CFCs used in refrigeration or as propellants and those used in dry cleaning

35. CCl$_4$, CFC-11, CFC-12, CH$_3$Cl, and CFC-113

37. a class of chemicals that are bromine-containing, hydrogen-free substances

39. a soil and crop fumigant

Understanding Concepts

41. O$_2$ and O$_3$ filter ultraviolet light having wavelengths from 120 to 320 nm.

43. UV-A radiation is the least biologically harmful form. Overexposure to UV-B radiation can lead to skin cancer. Due to its very short wavelength, UV-C light is the most dangerous.

45. O$_2$ + UV-C light → O + O; O + O$_2$ → O$_3$ + heat

47. No, not as long as UV-C light and free oxygen gas exist.

49. The very cold polar stratosphere produces stratospheric clouds. Chemical reactions occur in these clouds to produce chlorine atoms.

51. Due to a lower stratospheric temperature, the hole over the Antarctic is more severe.

53. Ozone concentrations over nonpolar areas have decreased by several percent.

55. slow vertical motion in the stratosphere

Synthesizing Ideas

57. a) 200 nm and 320 nm   b) 100 nm and 220 nm
c) UV-B and UV-C   d) UV-C   e) Diatomic oxygen absorbs at slightly shorter wavelengths.

59. the reaction of a chlorine atom with a molecule of ozone; Cl + O$_3$ → ClO + O$_2$, 2 ClO → ClOOCl, ClOOCl + UV light → → → 2 Cl + O$_2$

61. CFCs contain chlorine, fluorine, and carbon. HCFCs contain hydrogen, chlorine, fluorine, and carbon. HFCs contain hydrogen, fluorine, and carbon. HCFCs are considered temporary replacements, while HFCs are considered long-term replacements.

63. This would entail a great deal of personal risk and exacerbate the problems associated with fossil fuels.

Chapter 16
Global Warming and the Greenhouse Effect

Review Questions

1. the rapid change in Earth's climate

3. the greenhouse effect

5. the absorption of infrared radiation by some gases, its conversion to heat, and the subsequent reemission of some of this trapped energy back toward Earth

7. Diatomic molecules containing two identical atoms cannot absorb infrared light.

9. The concentration of atmospheric CO$_2$ is higher today.

11. direct relationship

13. Parts a) to e) absorb thermal infrared light and increase global warming, and part f) exerts a cooling effect on the environment.

15. natural gas < oil < coal

17. decrease atmospheric temperatures; short-term

19. its long lifetime in the atmosphere

21. feedback; by producing a dramatic rise in global temperature

23. an agreement to reduce greenhouse gas emissions

Understanding Concepts

25. the human "enhancement" of natural greenhouse gases

27. Carbon dioxide is produced by the combustion of hydrocarbons and when calcium carbonate is heated to produce calcium oxide.

29. In the process of nitrification, nitrogen is converted to nitrate and nitrite. In denitrification, nitrogen is converted to molecular nitrogen.

31. On a per molecule basis, CFCs have the greatest impact, but overall, CO_2 emissions account for a greater portion of the global warming problem.

33. If these particles reflect incoming sunlight, this energy is unavailable for absorption and heat production. If they absorb the light, the air immediately surrounding it is warmed.

35. Increased carbon dioxide concentrations account for the greatest increase in greenhouse gas–related heating.

37. a) A shift in ocean currents could result in significant climate changes for several regions of the Earth. b) rising sea levels c) increase d) higher rate of photosynthesis, a longer frost-free growing season, decreased soil moisture, and increased insect infestations e) increase

39. floods, submerged small island countries, increase tropical storm damage, and contamination of fresh groundwater sources

41. Increased greenhouse gas concentrations to date are due primarily to emissions from developed countries.

Synthesizing Ideas

43. yes; shorter atmospheric lifetimes

45. The energy associated with absorbed light is converted into heat, which is subsequently imparted to nearby air molecules.

47. They reflect sunlight back into space more efficiently than they absorb it, thereby having a cooling effect.

49. Atmospheric substances can either reflect or absorb light. The net effect is an increase in atmospheric heating.

51. increased global atmospheric temperature; a rise in global sea level; a reduction in northern hemisphere snow cover and sea-ice content; a global increase in average water vapor and precipitation

53. As seawater warms, the volume it occupies increases and sea level increases.

55. The increased rate of evaporation of soil moisture, due to the elevated temperature, will be greater than the rate at which the moisture is replaced by increased rainfall.

Chapter 17
The Core of Matter
Radioactivity, Nuclear Energy, and Solar Energy

Review Questions

1. protons and neutrons; same mass; proton charge = +1; neutron charge = 0

3. a) carbon, mass number = 12, neutrons = 6
   b) sodium, mass number = 24, neutrons = 13
   c) iodine, mass number = 131, neutrons = 78
   d) cobalt, mass number = 60, neutrons = 33

5. a) charge = +2, mass number = 4, neutrons = 2
   b) an electron c) a large amount of energy concentrated in one photon

7. cancer and developmental abnormalities

9. the time required for half of the nuclei to disintegrate

11. The alpha particles emitted are believed to produce lung cancer.

13. the conversion of a small amount of nuclear mass into energy

15. A reaction product induces another reactant molecule to undergo the same reaction.

17. a nuclear power reactor specifically designed to maximize the production of plutonium

19. Fuel grade contains 3% ^{235}U; weapons grade contains 90% or more ^{235}U.

21. heterogeneous radioactive mixture of liquid and water wastes that are released from uranium rock ore; the potential contamination of ground and surface water supplies, and the production of radon from radium

23. the incorporation of tritium by biological systems

25. 10^{18} joules

27. The potential energy of water molecules is harnessed by forcing falling water to turn turbines that generate electricity.

29. the world's plant and animal matter

31. a solid material that has electrical conductivity properties between those of a freely conducting metal and a nonconducting insulator

Understanding Concepts

33. a) carbon: 6 electrons, 6 protons; C-12: 6 neutrons, mass number = 12; C-14: 8 neutrons, mass number = 14 b) lead: 82 electrons, 82 protons; Pb-214: 132 neutrons; Pb-208: 126 neutrons c) hydrogen: 1 proton, 1 electron; H-2: 1 neutron; H-3: 2 neutrons

35. a) $^{59}_{27}\text{Co}$ b) $^{63}_{29}\text{Cu}$

37. a) an atomic number two units smaller and four mass units lighter b) an atomic number one unit larger and with the same mass number c) no change

39. Alpha particles have a relatively large mass and are relatively slow moving. Gamma rays have no appreciable mass and are able to penetrate matter efficiently.

41. Irradiation kills microorganisms.

43. 0.20 mg

45. Loose, sandy soil allows the maximum diffusion.

47. much greater

49. about a million times greater for fusion

51. Heat energy would be used to produce high-temperature steam.

53. a) health and security concerns b) no solution to the world's long-term nuclear waste–disposal problem

55. no; one possible fusion reaction involves two nonradioactive deuterium atoms

57. no; high capital investment costs

59. indirect solar energy

61. a) Advantages: abundant, renewable, "natural," low environmental impact, low operating costs, does not require large, centralized generating facilities. Disadvantages: intermittent availability, requirement for backup and storage facilities, large surface area required, high capital costs, the lack of tax and regulatory incentives.

b) Advantages: relatively inexpensive, plentiful, environmentally friendly. Disadvantages: production of the greenhouse gas methane; the release of heavy metals, such as mercury, into the water.

c) Advantages: relatively environmentally friendly, generally low cost. Disadvantages: useful only in areas with high-speed winds, requires large amounts of land, considered unsightly, and potentially harmful to birds.

d) Advantages: plentiful and inexpensive. Disadvantages: very polluting, relatively inefficient.

e) Advantages: abundant, renewable, low environmental impact, low operating costs. Disadvantages: high capital cost.

63. thermal (either direct or indirect) and photo-conversion of the Sun's energy

65. to maximize the fraction of energy that can be converted into electricity

67. Amorphous silicon is less expensive but is about half as efficient as crystalline silicon in converting sunlight to electricity.

Synthesizing Ideas

69. a) $2\ _{-1}^{\ 0}e$ b) $^{238}_{92}\text{U}$

71. It is long enough for this gas to diffuse through the solid rock and/or soil in which it is formed and escape into the atmosphere.

73. high stability (necessary to prevent disruption of the waste by earthquakes and/or volcanic activity) and low permeability (for minimal interactions with groundwater and the biosphere)

75. the Sun, through both direct and indirect ways

Glossary

α-amino acid one in which the amino and acid groups are both bonded to the same intermediate carbon atom

α helix structure a secondary protein structure in which the backbone chain of the polypeptide winds around an (imaginary) axis and thereby forms a helix

absorbed substance a material which enters the structure of another

absorption the process by which molecules are held inside a solid

acid a substance that increases the concentration of H^+ ions in solution

acid rain atmospheric precipitation (rain, fog, snow) with a pH less than 5

acidic solution aqueous solution in which the hydrogen ion concentration is greater than that of hydroxide ion

activated carbon (charcoal) granulated charcoal, with grains about 1 millimeter in diameter, that has been treated with steam and that has a highly porous structure and thus a large surface area

active site a component of a particular shape in the tertiary and often quaternary structure of a pocketlike region on the enzyme's surface where reactions take place

addition polymer a polymer that consists of intact molecules that have combined together

adenine a fused-ring nitrogen base denoted as A

adsorbed substance a material which is weakly attached on the surface of another

adsorption the process by which molecules are weakly held on the surface of a solid

aeration the process of using air to improve water quality by removing dissolved gases and volatile organic compounds

aerosol a collection of particulates, whether solid particles or liquid droplets, dispersed in air

alcohol compounds whose molecules contain the three-atom unit C—O—H

aldehyde an organic molecule in which the carbon of the carbonyl group is bonded to at least one hydrogen atom

alkaline (basic) solution a solution in which hydroxide ions, OH^-, dominate the solution rather than H^+ ions; a solution that has an OH^- concentration greater than 10^{-7} moles per liter

allotropes different forms in which an element exists

alloy a mixture of metals formed by melting them together

alpha (α) particle a radioactively emitted particle that has a charge of +2 and a mass number of 4 and is identical to the nucleus of the most common isotope of helium

amide an organic compound in which an amino group is bonded to the carbon of a carbonyl (C=O) group

amines organic molecules that correspond to ammonia, NH_3, in which one, two, or all three of the hydrogen atoms have been replaced by R groups consisting of chains or rings of carbon atoms (with their associated hydrogen atoms)

amino acid a compound containing both an amino group and a carboxylic acid group

amorphous describes the solid produced by a random arrangement of particles

anabolic steroid a steroid that promotes the growth of muscles

anaerobic decomposition decomposition of formerly living matter in the absence of air, that is, under oxygen-starved conditions

androgen the male sex hormones responsible for the development of male sex characteristics

anion a negatively charged ion

antacid a substance that will react with hydrogen ions and reduce their concentration in the stomach

anthropogenic arising from human activities

anticodon each sequence of three consecutive nitrogen bases on a tRNA molecule that specifies one of the amino acids in a protein

antioxidant a substance that prevents the occurrence of oxidative damage from free radicals

aquifer a permanent reservoir of groundwater consisting of porous or highly fractured rocks, above a clay or impervious rock layer, which contain the water

atmospheric window the infrared light region of 8–13 μm in which no natural greenhouse gases except ozone absorb

atomic number the number of protons in the nucleus

atomic weight average mass of the natural mixture of the isotopes of an element

atoms   from the Greek word meaning "indivisible," the smallest identical particles into which matter can be divided

β pleated sheet structure   a secondary protein structure in which there are N—H—O hydrogen bonds between adjacent polypeptide chains, rather than internal to the chain

balanced equation   one in which the number of each type of atom in an equation is the same in front of and following the arrow

base   a substance that increases the concentration of the OH⁻ ion in solution

battery   a commercial electrochemical cell, or combination of cells, in which the electrochemical reaction is used to generate electricity

beta (β) particle   an electron that is emitted from a nucleus

bioaccumulation   the result of bioconcentration and/or biomagnification that leads to a large increase in concentration of an organic compound in animals, including humans

bioconcentration   the increased concentration of an organic compound in a plant or animal as compared to the aqueous solution from which it came

biomagnification   the increasing concentration of a compound as it travels up a food chain

biomass   plant or animal matter

bioremediation   the decontamination of water or soil using biological processes

boiling   complete and rapid conversion, at a specific temperature, of a liquid into the gas state

bond angle   the angles that the bonds joining two atoms to a common third atom make with each other

bond energy   the amount of energy required to break a bond

bonding pair of electrons   the two electrons which are shared by a pair of atoms and which travel around both the nuclei

branched hydrocarbon   a hydrocarbon with one or more carbon chains attached to the main continuous network of carbon atoms

breeder reactor   a nuclear power reactor that is designed specifically to *maximize* the production of by-product plutonium

buffer solution   a solution which resists changes in pH when a strong acid or strong base is added to it; a solution of a weak acid and a salt of the same weak acid or of a weak base and its salt

builder   an inexpensive material added to a detergent to bind with Ca^{2+} and Mg^{2+} ions to form water-soluble materials

by-product(s)   small amount(s) of alternative substance(s) produced when a main reaction is not the exclusive process that the reactants undergo

carat   gold-alloy classification system; carats of gold = 24 × fraction of mass that is pure gold

carbamate   a pesticide that is an ester of carbamic acid, $HO(C=O)NH_2$

carbohydrate   molecules containing only carbon, hydrogen, and oxygen, in a roughly 1:2:1 atomic ratio; polyhydroxyl organic compounds

carbon sequesterization   the process(es) by which CO_2 formed from the burning of fossil fuel would be concentrated and buried and thus not released into the atmosphere

carbon taxes   taxes based on the amount of carbon contained in a fuel rather than upon its total mass

carbonyl group   a carbon–oxygen double-bonded group, $C=O$

carboxylic acid   an organic molecule in which the carbon of the carbonyl group is bonded to one hydroxyl group and either a hydrogen atom or a carbon-containing group

carcinogenic   capable of causing cancer

catalyst   a substance that can speed up a reaction without itself being consumed

catalytic converter   an apparatus in the exhaust system of vehicles which uses a catalyst such as platinum to speed conversion of hydrocarbons, CO, and NO_x in the exhaust gas into less harmful substances

cation   a positively charged ion

chain reaction   a process by which nuclear fission of an isotope produces not only the daughter nuclei but also neutrons that can continue the fission reaction with more parent nuclei

changes of state   changes of the phase—whether gas, liquid, or solid—in which a substance exists at a given time

Chapman cycle   natural ozone production and noncatalytic destruction processes in the ozone layer

chelating agent   a substance that has more than one site of attachment to a metal ion and thus produces ring structures that incorporate the metal

chemical change   a process in which the identity of a pure substance is altered as a result of the rearrangement of atoms

chemical property a reaction that changes the identity of a pure substance

chemical reaction a chemical change involving molecules in which atoms are interchanged between substances, so that the identity of the substances after the change is different from before it occurred

chemical reaction equation a shorthand way of indicating the nature of a chemical change, i.e., a summary of the interchange of atoms that occurs between pure substances; the reactant(s) are shown on the left, followed by an arrow facing right to the products that are shown on the right

chlorination (of water) the use of hypochlorous acid (HOCl) to disinfect water

chlorofluorocarbons (CFCs) compounds that contain chlorine, fluorine, and carbon, originally used as propellants and refrigerants, that decompose in the upper stratosphere to form species which react with and destroy ozone

chromatin a constituent of the cell nucleus in which histones, long strands of double-helix DNA associated with small proteins, are stored

chromosome a structure into which chromatin is packed and that directs the protein-producing activities of each cell

clathrate compound a compound with a structure that consists of small molecules occupying vacant spaces ("holes") in a cagelike solid structure formed by other molecules

clean coal technology methods by which coal can be used that are less polluting and more energy-efficient than used in the past

codon each sequence of three consecutive nitrogen bases on an mRNA molecule that specifies one of the amino acids in a protein or the beginning or end of the protein sequence

coefficient the number placed in front of a chemical formula in a chemical reaction equation

coenzyme complex organic or organometallic structural components that are required by an enzyme and that cannot be synthesized by the organism

cogeneration the technique of using the waste heat from a heat-to-electricity conversion for a constructive purpose

colloid a heterogeneous mixture consisting of two or more types of materials, with one finely divided material dispersed in the other

combined chlorine chlorine in the form of chloramines, which is more stable than hypochlorous acid

combustion the rapid, self-sustaining reaction of oxygen with a fuel

complete combustion when a substance containing carbon and hydrogen burns to carbon dioxide and water, and no other carbon-containing substance is also produced

complete protein a dietary protein that contains all the essential amino acids in the right ratio for conversion into human protein

complex carbohydrate condensation polymers containing glucose and other sugar units

composite material a solid mixture of nonmetallic materials

compound a material that consists of two or more types of atoms in a fixed ratio, with uniform composition throughout, and that cannot easily be separated into its pure component elements

concentration the amount of solute that is dissolved in a standard quantity of solution

concentration scale a parameter that gives the fraction of a mixture that corresponds to one of its components

condensation polymer a polymer formed by the consecutive linkages of molecules with the accompanying elimination of water or some other small molecule as each addition to the chain is formed

condensation reaction a process in which two molecules react to eliminate water or some other small molecule and combine to form a new compound

condensed formula a way of depicting the structure of organic molecules, carbon atom by carbon atom, with the number of hydrogen atoms attached to each shown after the C symbol

condensed states of matter liquids or solids

conjugated protein a protein that consists of amino acids and some other groups

covalent chemical bond the sharing of two electrons between a pair of adjacent atoms

crystal the solid formed when large numbers of particles are arranged in a regular pattern with the same distances between nearest neighbors throughout the whole

cyclic hydrocarbons molecular structures in which the carbon atoms and their C—C bonds define a closed ring

cytosine a one-ring nitrogen base, denoted as C

daughter (nuclei) the product atom formed by radioactive decay of a nucleus

dehydrating agent a substance that can remove water

denitrification the process(es) by which nitrogen in the form of the nitrate ion, NO_3^-, is converted mostly to molecular nitrogen, N_2

density the ratio of the mass to the volume for a material

deoxyribose a five-carbon, five-membered ring sugar with three hydroxyl groups

deoxyribunucleic acid (DNA) a polymer of nucleotide subunits in which the sugar is deoxyribose

depolymerization the process by which heat or chemical reactions are used to convert a polymer back into monomer components

desalination the production of fresh water from salty seawater by the removal of ions

diamines molecules that contain two amino nitrogen atoms

diatomic molecule a substance that consists of only two strongly bound atoms; the atoms may be the same or different

dicarboxylic acid a compound with two carboxylic acid groups

dipeptide; tripeptide; polypeptide dipeptides contain two amino acids; tripeptides contain three amino acids; polypeptides contain more than three amino acids

discharge (of a cell) the spontaneous reaction in an electrochemical cell in which the electrons flow from the negative to the positive electrode

disinfection the removal of harmful bacteria and viruses from water

distillation a separation process that involves vaporization by boiling of a liquid mixture, followed by the cooling of the vapor in order to cause its condensation back to the liquid state; a process in which raw water is boiled and the steam collected and condensed

double bond the bond formed when a pair of atoms share two pairs of electrons, normally with two electrons being contributed from each atom; it is indicated by two parallel lines between the two atoms (e.g., C=C)

double helix the secondary structure of DNA in which two strands of nucleic acid are wrapped around each other in a helix

elastomer a polymer that can be deformed by stress and that can return to its original shape after the stress is removed

electrochemical cell an apparatus consisting of two electrodes, one positive and one negative, connected by a conducting wire, while being immersed in an ionic solution

electrodes strips of metal, immersed in an ionic solution, at which oxidation and reduction reactions occur

electrolysis the technique of passing an electrical current through a material to drive a chemical reaction

electrolyte the ion-containing solution into which the electrodes are dipped

electron a particle in an atom which has a very small mass and carries a negative electrical charge

electron configuration of an atom the allocation of electrons to the various shells in an atom

electronegativity the relative ability of an atom to attract electrons in bonds to itself

electroplating the technique, using an electrical current, by which a thin film of a metal is deposited on another object

electrostatic force of attraction the attraction between opposite electrical charges

electrostatic force of repulsion the repulsion between particles with the same electrical charge

elements fundamental types of matter that cannot be split into other stable entities

emulsifying agent a substance that is soluble in two substances which are insoluble in each other and that enables them to form an emulsion

emulsion a colloid in which both phases are liquids

endothermic reaction a reaction in which heat is absorbed

enhanced greenhouse effect the enhanced effect that is caused by a significant increase in gases which trap thermal infrared light emitted by Earth

environmental estrogen a compound in the environment that can bind to the estrogen receptor, and thereby either mimic or block the action of the hormone itself

environmental tobacco smoke carcinogenic smoke produced by people smoking that affects nonsmokers exposed to it

equilibrium the state when there is equality in rate between two opposing processes

essential amino acid an amino acid that is required for human life and well-being but cannot be synthesized by the body from other compounds

essential fatty acid a fatty acid that is required by the human body but cannot be synthesized by it from other components

ester a molecule in which the carbon of the carbonyl group is bonded to one oxygen which is bonded to another carbon, and to either a hydrogen atom or a carbon-containing group

estrogens the female sex hormones that are responsible for the development of female sex characteristics

ether compounds whose molecules contain the three-atom unit C—O—C but no other types of carbon-oxygen bonds

exothermic reaction a reaction in which heat is released

extended network a repeating pattern of particles that extends indefinitely in at least two dimensions and usually in all three

fat a triester of glycerol and fatty acids that is solid at room temperature

fatty acid a carboxylic acid in which the carboxyl group is attached to an unbranched carbon chain of medium length

fiber a threadlike strand of solid material

fission a process in which a neutron collides with a type of heavy nucleus and splits the nucleus into two similarly sized fragments and releases energy

flammable describes substances that will burn

food chain a sequence of species, each of which feeds upon the one that precedes it in the chain

food web a series of interlocking food chains

fossil fuel a carbon-containing substance that burns in air to release energy in the form of heat and that is the residual by-product of organisms that lived hundreds of millions of years ago

fuel cell an electrochemical cell in which the reactants are continuously replenished while the unit delivers electrical power; unlike a battery, it can be continuously operated and never becomes discharged

functional group an atom or a bonded collection of atoms that occurs in organic molecules

fungicide a substance that kills fungi

fusion a process in which two very light nuclei combine to form one heavier one with the accompanying release of huge amounts of energy

gamma (γ) particle a huge amount of energy concentrated in one photon

gene linear sequences of bases that form a major part of the structure of chromosomes; these are the subunits of genetic information that can be inherited

gene expression the production of functioning proteins

glass state a solid state that lacks uniform order throughout the material

glass transition temperature the temperature at which the change from the glass state to the rubber state occurs in a given polymer; it is specific to each polymer

global warming global change in Earth's climate that increases the global air temperatures and also increases annual precipitation in most locations, increases sea levels, and causes a variety of secondary effects

green chemistry the science of designing products and processes that reduce or eliminate the use and generation of hazardous substances

greenhouse effect the phenomenon in which the temperature of Earth's surface and the nearby air is increased by trapped outgoing infrared energy

greenhouse gas a component of our atmosphere that efficiently absorbs thermal infrared light emitted by Earth

groundwater the fresh water that lies underground within one kilometer of the surface

group (of periodic table) a vertically arranged set of elements which behave similarly because they have the same number of valence-shell electrons

guanine a fused-ring nitrogen base, denoted as G

half-life period the length of time required for half the total amount of a substance to decompose

half-reaction a way of describing the transfer of electrons between atoms and ions

hard water water that contains appreciable amounts of calcium and magnesium ions

heat of combustion the heat released when a substance is burned

herbicide a substance that kills plants

heterogeneous mixture a mixture of substances that does not have uniform properties throughout

homogeneous mixture a mixture with a uniform composition throughout; also known as a solution

hormone "chemical messenger" molecules that are carried in the bloodstream from one part of the body to another and that regulate the rates of biochemical reactions at a target site

hydrocarbon a compound made up exclusively of carbon and hydrogen

hydroelectric power power produced by harnessing some of the remaining potential energy from the Sun by forcing flowing, falling water to turn turbines and thereby generate electricity

hydrofluorocarbons (HCFs) long-term replacements for CFCs and HCFCs that contain only hydrogen, fluorine, and carbon

hydrofluorochlorocarbons (HCFCs) temporary replacements for CFCs in the 1990s and early 2000s; they contain hydrogen, chlorine, fluorine, and carbon

hydrogen bonding a process by which a covalently bonded hydrogen atom, having a partial positive charge, on one molecule is attracted to an electronegative atom on another molecule; this leads to a weak interaction between the molecules

hydrogenation the catalyzed addition of a molecule of hydrogen gas, H_2, to a $C=C$ double bond in a molecule

hydrolysis reaction reaction of a substance with water

hydrophilic water loving, as in the polar portion of a molecule

hydrophobic water hating, as in the nonpolar portion of a molecule

hypothesis an attempt to explain observations in terms of materials and processes that we already understand

incomplete combustion when a substance containing carbon and hydrogen burns to produce products such as carbon monoxide and soot rather than exclusively carbon dioxide

incomplete protein a dietary protein that does not contain all the essential amino acids in the right ratio for conversion into human protein

indicator a material which is a different color depending on whether it is in an acidic or a basic solution

infrared light light, which we experience as radiant heat, with wavelengths between 750 and 1,000,000 nm (0.75 and 1000 μm)

inner shells of electrons all the shells other than the valence shell that contain electrons

insecticide a substance that kills insects

insoluble a substance that does not dissolve at all, or hardly at all, in another is said to be insoluble in that substance

intermediate a substance that is produced during the initial stages of a sequence of reactions, survives for a short time, and is consumed in a later step; none of it normally remains when the overall reaction sequence is completed

interstitial alloy an alloy in which the small atoms of one material fit in the spaces between the rows of the main metal

ion an electrically charged atom or group of atoms

ion exchange a process by which water is passed through an apparatus containing a supply of sodium or other small ions on a resin and in which calcium, magnesium, and iron ions in the hard water are exchanged for the sodium (or other) ions

ionic bond the attractive interaction between a pair of oppositely charged ions

ionic lattice the 3-D arrangement of layers of ions in a solid

ionization (reaction) the separation of materials in water to give positive and negative ions, all surrounded by a layer of water molecules

isotopes nuclei of a given element with different mass numbers arising from different numbers of neutrons

ketone an organic molecule in which the carbon of the carbonyl group is bonded to two carbon-containing groups

LD$_{50}$ the minimum dose of a substance that is lethal to 50% of a population

leachate liquid material emanating from a source such as a landfill

lethal dose the amount of a substance that is required to kill an organism

Lewis structure diagrams that show bonds and non-bonding electrons in molecules

lipoprotein a protein conjugated with a lipid

major mineral minerals required in quantities of 0.1 gram (100 milligrams) or more per day

malignant melanoma an often-fatal form of skin cancer that is thought to be related to exposure to ultraviolet light

mass number the sum of the number of protons and neutrons in a nucleus

matter anything that has mass and occupies space

messenger RNA (mRNA) a form of RNA that is created within the nucleus and conveys genetic information from DNA to the translation center of the cell

metabolic process a chemical reaction that occurs in the body and that alters the nature of chemicals present there

metalloid a solid which has properties of both a metal and a nonmetal

metals shiny, opaque, malleable, ductile materials which conduct heat and electricity

methemoglobinemia a health problem, in which oxygen cannot combine properly with hemoglobin, caused by excess nitrite ion in ingested water

mineral   inorganic compounds that occur as ions and that are required for good health

mineral water   water obtained from natural sources in which the amount of salts dissolved is much higher than usual

miscible   capable of dissolving in another substance in any proportions to form a (homogeneous) solution

mixed-oxide fuel   a mixture of plutonium dioxide and uranium oxide which can be used in existing nuclear power plants

mixture   a combination of substances in no fixed proportion whose formation involves no fundamental change at the atomic level

molarity   the concentration of dissolved substances in terms of the number of moles of the substance that are present in one liter of a solution

mole 6.02×10^{23} molecules or atoms

molecular formula   the way that chemists indicate the composition of a given molecule—the symbol for each atom followed by a numerical subscript to indicate the number of that type of atom present in one molecule

molecule   a collection of a relatively small number of atoms that are strongly bound to each other and that remain as intact units even when the material is melted or boiled

monomer   a small molecule that successively combines to form part of a very long chain

monosaccharide; disaccharide; polysaccharide   monosaccharides contain one saccharide ring; disaccharides contain two rings; polysaccharides contain more than two rings

mutagen   an agent that causes a mutation in DNA

mutation   a change in the sequence of nitrogen bases on a section of DNA

nanotechnology   the construction of machinery so small that its components have dimensions of nanometers

nanotubes   tubular structures of atoms or molecules which are only a few nanometers in diameter

nearest neighbors   atoms which touch another atom are known as its nearest neighbors

net ionic reaction   a written reaction which shows all the ions that have changed in the process depicted

neurotransmitter   a molecule that is released by one nerve cell and then travels for less than a millisecond across a gap, known as the synaptic area, of a few hundred nanometers, toward receptors on the adjacent cell

neutralization (reaction)   the reaction of H^+ ions with OH^- ions to form water molecules

neutron   a particle in the nucleus of an atom that has about the same mass as that of a proton, but no electrical charge

nitrification   the process(es) by which nitrogen in the form of ammonia or the ammonium ion is converted mostly to nitrite (NO_2^-) and nitrate (NO_3^-) ions

nitrogen base   a molecule consisting of one ring or two fused rings and two nitrogen atoms in one ring; one of the molecules that comprise a nucleic acid

NO_x   collectively, NO and NO_2

nonbonding electrons   valence-shell electrons that are not used for bonding purposes

nonferrous smelting   the conversion of ores, other than those of iron, to the free metals

nonflammable   describes substances which will not burn

nonmetals   elements which do not have the characteristics of metals

nonpoint sources   the numerous entities, such as farms, each of which provides a much smaller amount of pollution than a point source; the combined effect of the many locations can contribute large amounts of pollutants

nuclear energy   the energy that results from the conversion of a small amount of the nuclear mass into energy during processes involving atomic nuclei

nucleic acid   a polymer of nucleotide subunits; either deoxyribonucleic acid or ribonucleic acid

nucleotide   a complex sequence formed by combining three simpler units: a phosphate group, a sugar, and a nitrogenous base

nucleus   a particle in an atom which has a relatively large mass and which carries a positive electrical charge equal to the total negative charge of all the electrons in the atom

observation   what is actually seen and/or measured

octet of electrons   eight electrons in the valence shell; an arrangement that gives rise to special chemical stability

oil   a triester of glycerol and fatty acids which is liquid at room temperature

organic chemistry   the area of chemistry that is concerned with the compounds of carbon

organophosphate   a pesticide that contains a pentavalent phosphorus atom with bonds to a sulfur or oxygen, two small alkoxy groups, and a large R group

osmosis   the tendency of water to travel through a membrane from a less concentrated to a more concentrated solution

oxidation chemical changes that involve the consumption of O_2 by its interaction with another substance, whether or not they involve combustion; an electron transfer reaction involving the gain of electrons by a substance

oxidizing agent in redox reactions, the substance that pulls electrons from the other material and thus oxidizes it

ozone layer a portion of the stratosphere that contains ozone, O_3, which filters out most UV-B and all UV-C from sunlight before it reaches Earth's surface

ozone-depleting substance a material that will react with and decompose stratospheric ozone

parent (nuclei) the original atoms which undergo radioactive decay

particle traps a mechanism (filter) that prevents organic particulate emissions from escaping from the exhaust system

particulates; coarse, fine particulates tiny particles suspended in air; coarse particulates have diameters larger than 2.5 μm, fine have diameters less than 2.5 μm

parts-per-million scale—gases number of molecules of a substance that are present in 1 million molecules of another substance

parts-per-million scale—liquids and solids the mass of the dissolved substance contained in 1 million grams of the solution

passive smoking inhalation of sidestream as well as already exhaled smoke

peptide bond the bond formed when the amino nitrogen of one α-amino acid reacts with the carbonyl carbon of the carboxyl group of another to produce an amide group

periodic table of the elements an arrangement of all the known elements in order of their atomic numbers

pesticide a substance that kills or otherwise controls an organism that humans find undesirable

petroleum a sticky, viscous mixture mainly of alkane hydrocarbons mainly 5 to 20 carbons in size

pH a shorthand way of indicating H^+ concentrations; the power of 10, without the negative sign, for the molarity of an H^+-containing solution

photochemical smog the resulting mixture of substances, including particles and ozone, produced by the action of sunlight on air pollutants

photoconversion a process by which the absorption of the ultraviolet, visible, and infrared photons of sunlight excites electrons in the absorbing material to higher energy levels and causes a physical or chemical change, rather than a simple degradation to heat

photosynthesis a process by which carbon dioxide in the air and water in a plant are combined with an input of energy provided by sunlight to produce oxygen and carbohydrate

photovoltaic effect the creation of separated positive and negative charges in a solid material as a result of absorption of a photon of light

physical changes processes in which the composition (identity) of a pure substance is not altered

physical property a characteristic of a material that is an aspect of the substance itself and that can be observed and specified without involving it in a transformation to another substance

phytoestrogen a plant-based estrogen mimic

plasma a gaseous mixture, consisting of electrically charged particles, formed when atoms are subjected to the application of large amounts of energy

plastic a material that can be molded

plasticizer a liquid that blends easily with a polymer and acts as a lubricant between individual polymer chains; the plastic is thereby softened and the material becomes more flexible

PM index a measure of the concentration of suspended particles, which is the mass of particulate matter that is present in a given volume; the units are micrograms of particulate matter per cubic meter of air, i.e., μg / m^3

point source a specific site, such as a town, city, or factory, that individually discharges a large quantity of a pollutant

polar covalent bond a bond where the electron pair is closer to one atom than the other

polar stratospheric clouds (PSCs) clouds formed by the particles produced by condensation of the gases within the cold stratosphene in the winter and spring

polyamide a polymer with multiple amide linkages

polyatomic ion an ion that consists of several covalently bonded atoms; the extra positive or negative charge is shared among them

polyatomic molecule a substance that consists of more than two strongly bound atoms

polycarbonate a polyester in which the dicarboxylic acid is carbonic acid, $(HO)_2C{=}O$

polycyclic aromatic hydrocarbon a planar molecule which has several six-membered benzene rings fused to each other

polyester a polymer with multiple ester linkages

polymer a substance that contains very long chains with the same structural unit repeating over and over again

polyolefin an addition polymer formed from monomers containing the C=C unit

polyprotic acid an acid that contains more than one hydrogen that can be ionized when it is dissolved in water

polysaccharide a polymer made from several saccharide monomers joined together

polyunsaturated fatty acids fatty acids that contain more than one double C=C bond

positive feedback the operation of a phenomenon that produces a result that further amplifies the result

precipitate a solid that forms from a solution because its solubility has been exceeded

pressure force exerted on a surface

primary pollutants substances such as NO, hydrocarbons, and other VOCs that are initially emitted directly into air

primary structure (of protein) the sequence of amino acids along the protein chain

principle of complementarity the principle that the geometry of the double helix and the nature of the hydrogen bonds formed between the bases in the two strands result in guanine and cytosine always hydrogen-bonding together and adenine always bonding with either thymine or uracil; this is the basis for DNA replication and genetic information transfer

Priority Organic Pollutants (POPs) a dozen chemicals designated to be phased out of use by international agreement

products the substances that are present after a chemical reaction occurs

progestins the female sex hormones that promote and maintain gestation

proof value in North America, two times the percent by volume of ethanol in an ethanol–water solution

proteins polyamides formed from α-amino acids, which are the most important constituent of living things

proton a particle in the nucleus of an atom that carries an electrical charge equal to that of the electron but that is positively rather than negatively charged

pump-and-treat a method of purifying groundwater that pumps contaminated water from the aquifer, treats it to remove its organic contaminants, and returns the cleaned water to the aquifer or to some other water body

pure substance matter that is composed exclusively of one particular element or compound

pyrolysis the process of thermal degradation of a material in the absence of oxygen

quaternary structure (of protein) an arrangement of multiple polypeptide chains and nonpolypeptide units, relative to each other

radioactive the condition in which some isotopes exist in which their nuclei spontaneously decompose by emitting a small particle that is very fast moving and consequently carries with it a great deal of energy

reactants the substances which are present before a chemical reaction occurs

Recommended Dietary Allowance the level of intake of essential nutrients judged to be adequate to meet the known nutritional needs of most healthy persons

redox reaction a complete reaction in which oxidation and reduction occur

reducing agent in redox reactions, the substance that readily yields some of its electrons and thus allows other chemicals to be reduced

reduction an electron transfer reaction involving the loss of electrons by a substance

re-forming reaction a catalyzed reaction that provides, *in situ*, the hydrogen essential for fuel cells from more safely handled chemicals

renewable energy energy that will not run out in the foreseeable future and whose capture and use do not result in the direct emission of greenhouse gases

renewable fuel a material whose supply can be continuously renewed with no net carbon dioxide release during its production/combustion cycle

repeating unit the small structural unit that forms the basis of the polymer chain

repressor protein a protein that recognizes and binds to specific DNA sequences, but instead of assisting RNA polymerase, it blocks its action and turns genes off

reprocessing separating the elements in used fuel rods by chemical methods

reverse osmosis a process in which water molecules are forced under high pressure through a semi-permeable membrane through which ions and most other molecules are too large to pass

ribonucleic acid (RNA) a polymer of nucleotide subunits in which the sugar is ribose

ribose a five-carbon, five-membered ring sugar with four hydroxyl groups

ribosomal RNA (rRNA) a form of RNA that is part of a protein molecule called a ribosome, the actual protein-synthesizing site in a cell

ribosome a structure made up of proteins and rRNA, within which there are specific binding sites for both mRNA and tRNA

RNA polymerase a protein molecule that directs the synthesis of RNA

row or period (of periodic table) the elements that are placed on the same horizontal level of the periodic table

rubber state a solid state of a polymer that is essentially a liquid in the disordered regions but is still a solid in its ordered regions

saccharide from the Latin *saccharum* (meaning "sugar"), a member of the group comprising sugars

salt an ionic compound which contains ions other than H^+ or OH^-

salt bridge the interaction formed between a positive nitrogen atom and a negative carboxylate group in a polypeptide; an aqueous solution or a water-based gel through which ions can pass from one electrode to the other

saponification the production of soap by treating fats (triglycerides) with heat and a base to form soap and glycerol

saturated fatty acids fatty acids that contain only single C—C bonds

saturated solution a solution that contains the maximum amount of dissolved solute at a given temperature

secondary pollutants substances into which primary pollutants are transformed, such as O_3 and HNO_3

secondary structure (of protein) the shape adopted by the backbone of the protein which arises from hydrogen bonding between various amino acid atoms; the two most important are the alpha (α) helix and beta (β) pleated sheet structures

selective catalytic reduction the process by which NO_x is converted to N_2 before the emissions are released into the atmosphere; ammonia is added to the cooled gas stream in the presence of oxygen and the catalyst

self-ionization reaction the reverse of the neutralization reaction, in which water molecules produce H^+ and OH^- ions

semiconductor a solid that conducts electricity in a manner intermediate between a metal (freely conducting) and an insulator (nonconducting); the bonds linking the atoms in the solid are relatively weak

sequestrant a substance that strongly ties ions up by binding to them chemically

shell structure for electrons a way of visualizing the electrons in an atom as occupying concentric regions of space (shells) around the nucleus, with the most tightly held electrons being closest to the nucleus and the most weakly held ones farthest away from the nucleus

simple protein a protein that consists entirely of amino acids

single bond the linking of two atoms by two electrons

soap the sodium or potassium salt of a long-chain fatty acid

solar energy energy sent to Earth from the Sun in the form of sunlight

solar still an apparatus in which salt water is heated sufficiently by the Sun's rays so that it evaporates rapidly; the humid air containing the evaporated water rises, and the pure water recondenses on sloping glass surfaces and runs down to troughs, where it is collected

solar thermal electricity electricity produced by using solar energy to produce superheated steam

solubility the maximum concentration of a substance that can be achieved in a solution; it is usually expressed in terms of grams of solute dissolved per liter, or per kilogram, in the resulting solution

soluble a substance that dissolves in another is said to be soluble in that substance

solute a substance that will ultimately dissolve—given enough time and enough stirring—in a liquid

solution a mixture with a uniform composition throughout, also known as a homogeneous mixture

solvent a substance that readily dissolves many other substances

steroid a molecule that contains an alcohol or ketone group and that contains four hydrocarbon rings, three six-membered and one five-membered, that are consecutively fused to each other in a characteristic geometry

stimulant a substance that stimulates the body's central nervous system

stratosphere an upper portion of the atmosphere that starts about 15 km above Earth and ends at about 50 km

strong acid an acid that completely ionizes when it is dissolved in water

strong base one in which all the dissolved solid is present in solution as hydroxide ions and ions of the metal

structural isomers molecules having the same formula but different internal bonding structures

sublimation the conversion of a solid directly to a gas

substitutional alloy an alloy in which atoms of the minority metal simply replace a few of those in the regular structure of the majority metal when the substance solidifies

substrate the molecule which fits into the active site of an enzyme

sugar molecules that contain one or two rings, each of which contains four or five carbon atoms and one oxygen atom connected by single bonds; most ring carbons are also bonded to a hydroxyl group; all are sweet to the taste

sulfate aerosol an aerosol dominated by oxidized sulfur compounds

surface fresh water the water that resides on Earth's surface in rivers, streams, and lakes; it is produced by precipitation and the melting of glaciers

synergistic effect an effect that occurs when the combined effect of two or more substances is greater than the sum of the effect of each operating alone

synthetic detergent materials that clean like soap but do *not* form insoluble precipitates in the presence of Ca^{2+} and Mg^{2+} ions; they have the sulfonate group, SO_3^-, as the anionic group (hydrophilic portion) attached to a large organic (hydrophobic) portion

tar the particulates in tobacco smoke that contain nicotine and the less volatile hydrocarbons, much of it in the fine-particle range

tertiary structure (of protein) a way in which a chain of amino acids folds; the folding is determined by the nature of interactions between side groups on the amino acid residues with each other and with the surrounding environment

tetrahedron characteristic geometry about all carbon atoms that form bonds to four atoms in organic molecules; all X—C—X bond angles are about 109.5°

thermal conversion a process by which the heat energy of sunlight is captured and used to heat various materials

thermal expansion the increase in volume of a substance as it is heated

thermal IR infrared light with wavelengths between 4 and 50 μm

thermoplastic a material that softens but does not decompose when heated and that becomes hard again when cooled

threshold concentration a level of a chemical, exposure below which a particular health effect does not occur

thymine a one-ring nitrogen base, denoted as T

tidal power power produced by damming water at high tide and then releasing it to turn a turbine and thereby generate electricity

toxic substance a substance that produces harmful effects when taken in a small dose

toxicity equivalency factor (TEQ) the rating of the toxicity of any PCB, dioxin, or furan relative to 2,3,7, 8-TCDD, which is arbitrarily assigned a value of 1.0

trace mineral minerals needed in amounts of only a few *milli*grams per day

trans fatty acid a fatty acid in which some of the double bonds have a trans orientation of carbon chains

trans **orientation** one in which two R groups attached to the different carbons of a C=C double bond lie on opposite sides of the bond

transcription the process by which genetic information is transferred from DNA to mRNA

transfer RNA (tRNA) a form of RNA that interprets the base sequence on mRNA into amino acids

translation the transfer of genetic information from mRNA to a working protein

triester molecules which each have three ester functional groups

triglyceride a fatty acid triester with glycerol

triple bond the bond formed when a pair of atoms share three pairs of electrons; it is indicated by three parallel lines between the two atoms (e.g., C≡C)

ultraviolet light light having wavelengths that lie between about 50 and 400 nm

unbranched (straight-chain) hydrocarbon a hydrocarbon in which all the carbon atoms are along one continuous chain

unsaturated fatty acids fatty acids that contain one or more double C=C bonds

uracil a one-ring nitrogen base, denoted as U

UV-A, UV-B, UV-C regions regions of ultraviolet light: 320–400 nm, 280–320 nm, and 200–280 nm in wavelength, respectively

valence the number of covalent bonds formed by an atom

valence shell of electrons the highest shell, or outermost occupied shell, that has any electrons in a particular atom

vapor pressure the pressure exerted on the walls of a flask by the evaporated molecules of a liquid or solid

vitamins essential, noncaloric nutrients that enable the body to digest, absorb, metabolize, or build what it needs in association with other nutrients and to preserve its existing components

vitrify   to make a material into a glass

volatile   a liquid that evaporates readily

volatile organic compounds (VOCs)   carbon-containing substances that readily vaporize into air

vortex   a whirling mass of air in which wind speeds can exceed 300 km per hour

vulcanization   the process by which natural rubber is heated with elemental sulfur to produce a product that is harder, stronger, more durable, and more elastic than natural rubber

wave power   presently, small amounts of power created when a rising wave compresses air in a chamber that is later released through a valve to turn a turbine

weak acid   an acid that ionizes and whose components recombine continuously in water, with the result that at any instant only a small fraction of it exists in ionized form

weak base   a substance that reacts reversibly with water molecules to extract H^+ from the H_2O, leaving OH^-, which enters the solution

wind power   power produced by the force of wind; used to generate power from windmills, particularly on "wind farms"

Index

Note: Page numbers followed by f indicate figures; those followed by t indicate tables. Page numbers preceded by A indicate appendices.

carboxylic acids,
237–243
esters, 243–248
ethers, 211–214
functional groups in,
212, 248
ketones, 236–237
Ozonation, in water
treatment, 493
Ozone, 590
allowable concentrations
of, 521, 521t
creation/destruction of,
562–564, 564f
effect of on crops,
522–523, 523f
filtering of, 558, 558f
ground-level, 522–523
health effects of,
539–540
health effects of,
558–562
light absoprtion by, 558,
558f
photochemical smog
and, 518, 520–523,
521t
production of, 518
protective function of,
520
Ozone-depleting
substances, 574
phaseout of, 575–577
Ozone depletion, 564–570
biological consequences
of, 558–562
bromine in, 574–575
carbon tetrachloride in,
574
causes of, 570–577
CFCs in, 555, 572–574
chlorine in, 564–568,
570–577, 571f
halons in, 574
increase in, 568–570
international accords
on, 575–577
methyl bromide in,
574–575
over Antarctica,
554–555, 554f,
555f, 564–569,
566f, 577

over Arctic, 569–570
over nonpolar areas,
570
process of, 564–565
seasonal variation in,
554–555, 554f,
555f, 566–568,
568f
vortex effects in, 566
Ozone layer, 554–577
atmospheric location of,
553f
hole in, 554–555, 554f,
555f, 564–570. See
also Ozone
depletion
seasonal variation in,
554–555, 554f,
555f, 566–568,
568f

Painkillers, amines in,
307–310, 308f, 309f
Pantothenic acid, 374f,
375–376, 376t
Paraffin wax, 151
Parathion, 393
Parent nucleus, 630
Parent radon, 630
Particle traps, 537, 537f
Particulates, 534–542
adsorption/absorption
by, 540, 540f
in aerosols, 534–535
climate effects of,
599–600, 599f
in air pollution, 521t,
529t
air quality and, 536–538
definition of, 534
diameters of, 534
filters for, 537, 537f
PM index for, 536–537
size of, 534, 535f
sources of, 535–536
Parts per million (ppm)
scale, 57, 419–420
Passive smoking, 545
Paxil, 296–297
PCBs, 399–404
in food, 401–402
furan contamination of,
400–401

health effects of,
402–404
persistence of, 399–400
recirculation of,
399–400, 400f
sources of, 399
structure of, 399
PCE (perchloroethene),
175–176
in groundwater,
500–502, 501f
Pentane, 132t
octane number for, 153t
Peptide bonds, 341–342
-per, 176
Perchloroethene (PCE),
175–176
in groundwater,
500–502, 501f
Perfumes
aldehydes in, 235–236
esters in, 245, 245t
Periodic table, 81–82, 81f,
92–95
groups in, 92–93, 92f
rows in, 92f, 93–94
Peroxides, 103
Perspiration odor, 240
Pesticides, 387–404
bioaccumulation of,
389–391
bioconcentration of, 389
biomagnification of, 390
carbamate, 391–394
in food chain, 390–391,
390f
fungicides, 387
herbicides, 387,
396–398
insecticides, 387,
388–396
natural, 394–395
organophosphate,
391–394
risks and benefits of,
387–388
PET [poly(ethylene
terephthalate)], 315,
316f
in fibers, 318
Petroleum, 150–152. See
also Fossil fuels;
Gasoline

alkanes in, 150–151,
152–153
hydrocarbons in,
150–153, 152t
refining of, 152
distillation in, 152,
166–168
water pollution from,
155–157
pH, 427–428, 427f
of acid rain, 526, 530
buffers and, 443–445,
444f
calculation of, 449–450
of gastric juice, 432–433
Phenoxy herbicides,
396–397
Phenylalanine, 340f
Phenylethylamine, 307,
307f
Phenylketonuria (PKU),
342–343
Pheromones, 43
Phosphates, water
pollution from,
504–511
Phosphoric acid, 391,
425–426
Phosphorus, dietary
requirement for, 111,
111t
Photochemical smog,
518–526, 521–522.
See also Air pollution
distribution of, 522
formation of, 521–522
hydrocarbons in,
518–519, 518f
nitric oxide in, 518,
518f
ozone and, 518,
520–523
particulates in, 536. See
also Particulates
production of, 518–526
products of, 519
temperature inversion
and, 521–522
threshold concentrations
for, 538
volatile organic
compounds in, 519,
519f